National Aeronautics and Space Administration

Wings In Orbit

Scientific and Engineering Legacies of the Space Shuttle

Foreword: John Young
Robert Crippen

Executive Editor: Wayne Hale

Editor in Chief: Helen Lane

Coeditors: Gail Chapline
Kamlesh Lulla

COVER PHOTOS

Front: *View of Space Shuttle Endeavour (STS-118) docked to the International Space Station in August 2007.*

Back: *Launch of Space Shuttle Endeavour (STS-130) during the early morning hours en route to the International Space Station in February 2010.*

Spine: *A rear view of the Orbiter Discovery showing the drag chute deployed during the landing of STS-96 at Kennedy Space Center in May 1999.*

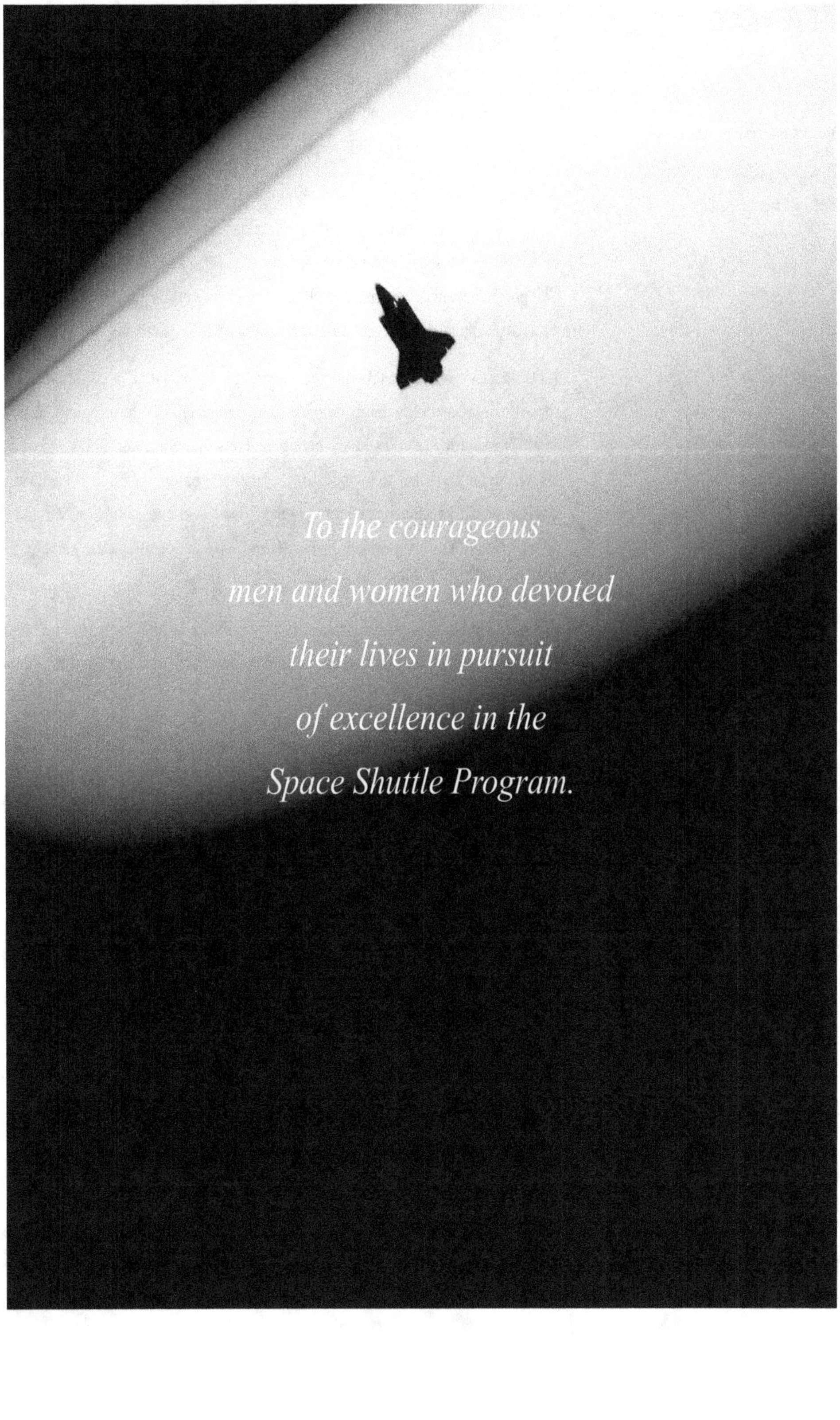

To the courageous
men and women who devoted
their lives in pursuit
of excellence in the
Space Shuttle Program.

Foreword

John Young
STS-1 Commander

Robert Crippen
STS-1 Pilot

We were honored and privileged to fly the shuttle's first orbital flight into space aboard Columbia on April 12, 1981. It was the first time anyone had crewed a space launch vehicle that hadn't been launched unmanned. It also was the first vehicle to use large solid rockets and the first with wings to reenter the Earth's atmosphere and land on a runway. All that made it a great mission for a couple of test pilots.

That first mission proved the vehicle could do the basics for which it had been designed: to launch, operate on orbit, and reenter the Earth's atmosphere and land on a runway. Subsequent flights proved the overall capability of the Space Shuttle. The program went on to deploy satellites, rendezvous and repair satellites, operate as a microgravity laboratory, and ultimately build the International Space Station.

It is a fantastic vehicle that combines human operations with a large cargo capability—a capability that is unlikely to be duplicated in future vehicles anytime soon. The shuttle has allowed expanding the crew to include non-pilots and women. It has provided a means to include our international partners with the Canada arm, the European Spacelab, and eventually the Russians in operation with Mir and the building of the International Space Station. The station allowed expanding that international cooperation even further.

The Space Shuttle Program has also served as an inspiration for young people to study science, technology, engineering, and math, which is so important to the future of our nation.

The Space Shuttle is an engineering marvel perhaps only exceeded by the station itself. The shuttle was based on the technology of the 1960s and early 1970s. It had to overcome significant challenges to make it reusable. Perhaps the greatest challenges were the main engines and the Thermal Protection System.

The program has seen terrible tragedy in its 3 decades of operation, yet it has also seen marvelous success. One of the most notable successes is the Hubble Space Telescope, a program that would have been a failure without the shuttle's capability to rendezvous, capture, repair, as well as upgrade. Now Hubble is a shining example of success admired by people around the world.

As the program comes to a close, it is important to capture the legacy of the shuttle for future generations. That is what "Wings In Orbit" does for space fans, students, engineers, and scientists. This book, written by the men and women who made the program possible, will serve as an excellent reference for building future space vehicles. We are proud to have played a small part in making it happen.

Preface and Acknowledgments

"... because I know also life is a shuttle.
I am in haste; go along with me..."

– **Shakespeare**, *The Merry Wives of Windsor,* Act V Scene 1

We, the editors of this book, can relate to this portion of a quote by the English bard, for our lives have been entwined with the Space Shuttle Program for over 3 decades. It is often said that all grand journeys begin with a small first step. Our journey to document the scientific and engineering accomplishments of this magnificent winged vehicle began with an audacious proposal: to capture the passion of those who devoted their energies to its success while answering the question "What are the most significant accomplishments?" of the longest-operating human spaceflight program in our nation's history. This is intended to be an honest, accurate, and easily understandable account of the research and innovation accomplished during the era. We hope you will enjoy this book and take pride in the nation's investment in NASA's Space Shuttle Program.

We are fortunate to be a part of an outstanding team that enabled us to tell this story. Our gratitude to all members of the Editorial Board who guided us patiently and willingly through various stages of this undertaking.

Acknowledgments: We are grateful to all the institutions and people that worked on the book. (See appendix for complete list.) Each NASA field center and Headquarters contributed to it, along with many NASA retirees and industry/academic experts. There are a few who made exceptional contributions.

The following generously provided insights about the Space Shuttle Program: James Abrahamson, Arnold Aldrich, Stephen Altemus, Kenneth Baldwin, Baruch Blumberg, Aaron Cohen, Ellen Conners, Robert Crippen, Jeanie Engle, Jack Fischer, William Gerstenmaier, Milton Heflin, Thomas Holloway, Jack Kaye, Christopher Kraft, David Leckrone, Robert Lindstrom, William Lucas, Glynn Lunney, Hans Mark, John Mather, Leonard Nicholson, William Parsons, Brewster Shaw, Robert Sieck, Bob Thompson, J.R. Thompson, Thomas Utsman, Edward Weiler, John Young, and Laurence Young.

We also gratefully acknowledge the support of Susan Breeden for technical editing, Cindy Bush for illustrations, and Perry Jackson for graphic design.

Table of Contents

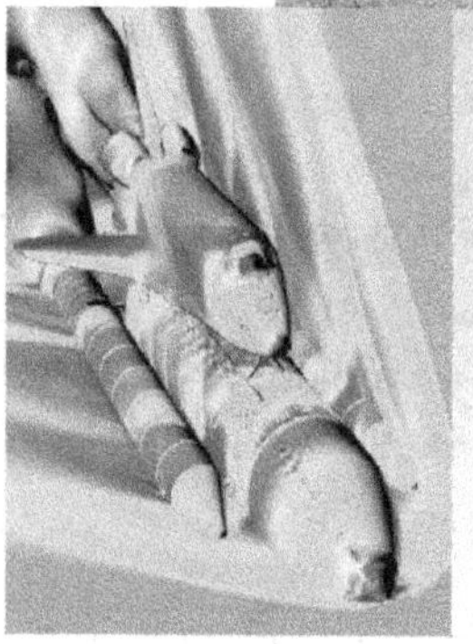

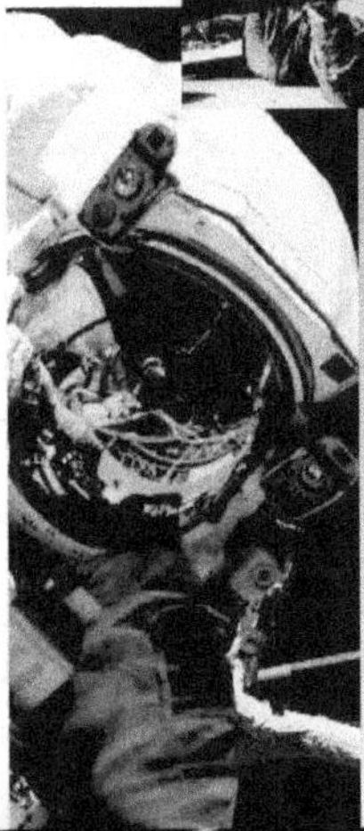

Editorial Board

Witnessing the Launch of the Shuttle Atlantis

Howard Nemerov
Poet Laureate of the United States
1963-1964 and 1988-1990

So much of life in the world is waiting, that
This day was no exception, so we waited
All morning long and into the afternoon.
I spent some of the time remembering
Dante, who did the voyage in the mind
Alone, with no more nor heavier machinery
Than the ghost of a girl giving him guidance;

And wondered if much was lost to gain all this
New world of engine and energy, where dream
Translates into deed. But when the thing went up
It was indeed impressive, as if hell
Itself opened to send its emissary
In search of heaven or "the unpeopled world"
(thus Dante of doomed Ulysses) "behind the sun."

So much of life in the world is memory
That the moment of the happening itself—
So much with noise and smoke and rising clear
To vanish at the limit of our vision
Into the light blue light of afternoon—
Appeared no more, against the void in aim,
Than the flare of a match in sunlight, quickly snuffed.

What yet may come of this? We cannot know.
Great things are promised, as the promised land
Promised to Moses that he would not see
But a distant sight of, though the children would.
The world is made of pictures of the world,
And the pictures change the world into another world
We cannot know, as we knew not this one.

Introduction

Charles Bolden

It is an honor to be invited to write the introduction for this tribute to the Space Shuttle, yet the invitation presents quite an emotional challenge. In many ways, I lament the coming of the end of a great era in human spaceflight. The shuttle has been a crown jewel in NASA's human spaceflight program for over 3 decades. This spectacular flying machine has served as a symbol of our nation's prowess in science and technology as well as a demonstration of our "can-do" attitude. As we face the fleet's retirement, it is appropriate to reflect on its accomplishments and celebrate its contributions. The Space Shuttle Program was a major leap forward in our quest for space exploration. It prepared us for our next steps with a fully operational International Space Station and has set the stage for journeys to deep-space destinations such as asteroids and, eventually, Mars. Our desire to explore more of our solar system is ambitious and risky, but its rewards for all humanity are worth the risks. We, as a nation and a global community, are on the threshold of taking an even greater leap toward that goal.

All the dedicated professionals who worked in the Space Shuttle team—NASA civil servants and contractors alike—deserve to be proud of their accomplishments in spite of the constant presence of skeptics and critics and the demoralizing losses of Challenger (1986) and Columbia (2003) and their dedicated crews. Some of these scientists and engineers contributed to a large portion of this book. Their passion and enthusiasm is evident throughout the pages, and their words will take you on a journey filled with challenges and triumphs. In my view, this is a truly authentic account by people who were part of the teams that worked tirelessly to make the program successful. They have been the heart, mind, spirit, and very soul that brought these amazing flying machines to life.

Unlike any engineering challenge before, the Space Shuttle launched as a rocket, served as an orbital workstation and space habitat, and landed as a glider. The American engineering that produced the shuttle was innovative for its time, providing capabilities beyond our expectations in all disciplines related to the process of launching, working in space, and returning to Earth. We learned with every succeeding flight how to operate more efficiently and effectively in space, and this knowledge will translate to all future space vehicles and the ability of their crews to live and work in space.

The Space Shuttle was a workhorse for space operations. Satellite launching, repair, and retrieval provided the satellite industry with important capabilities. The Department of Defense, national security organizations, and commercial companies used the shuttle to support their ambitious missions and the resultant accomplishments. Without the shuttle and its servicing mission crews, the magnificent Hubble Space Telescope astronomical science discoveries would not have been possible. Laboratories carried in the payload bay of the shuttles provided opportunities to use microgravity's attributes for understanding human health, physical and material sciences, and biology. Shuttle research advanced our understanding of planet Earth, our own star—the sun—and our atmosphere and oceans. From orbit aboard the shuttle, astronaut crews collected hundreds of thousands of Earth observation images and mapped 90% of Earth's land surface.

During this 30-year program, we changed dramatically as a nation. We witnessed increased participation of women and minorities, the international community, and the aerospace industry in science and technology—changes that have greatly benefitted NASA, our nation, and the world. Thousands of students, from elementary school through college and graduate programs, participated in shuttle programs. These students expanded their own horizons—from direct interactions with crew members on orbit, to student-led payloads, to activities at launch and at their schools—and were inspired to seek careers that benefit our nation.

International collaboration increased considerably during this era. Canada provided the robotic arm that helped with satellite repair and served as a mobile crew platform for performing extravehicular activities during construction of the International Space Station and upgrades and repairs to Hubble. The European Space Agency provided a working laboratory to be housed in the payload bay during the period in which the series of space laboratory missions was flown. Both contributions were technical and engineering marvels. Japan, along with member nations of the European Space Agency and Canada, had many successful science and engineering payloads. This international collaboration thus provided the basis for necessary interactions and cooperation.

My personal change and growth as a Space Shuttle crew member are emblematic of the valuable contribution to strengthening the global community that operating the shuttle encouraged and facilitated. I was honored and privileged to close out my astronaut career as commander of the first Russian-American shuttle mission, STS-60 (1994). From space, Earth has no geographic boundaries between nations, and the common dreams of the people of these myriad nations are realizable when we work toward the common mission of exploring our world from space. The International Space Station, the completion of which was only possible with the shuttle, further emphasizes the importance of international cooperation as nations including Russia, Japan, Canada, and the member nations of the European Space Agency join the United States to ensure that our quest for ever-increasing knowledge of our universe continues to move forward.

We have all been incredibly blessed to have been a part of the Space Shuttle Program. The "Remarkable Flying Machine" has been an unqualified success and will remain forever a testament to the ingenuity, inventiveness, and dedication of the NASA-contractor team. Enjoy this book. Learn more about the shuttle through the eyes of those who helped make it happen, and be proud of the human ingenuity that made this complex space vehicle a timeless icon and an enduring legacy.

Magnificent Flying Machine— A Cathedral to Technology

Magnificent Flying Machine—A Cathedral to Technology

Wayne Hale

Certain physical objects become icons of their time. Popular sentiment transmutes shape, form, and outline into a mythic embodiment of the era so that abstracted symbols evoke even the hopes and aspirations of the day. These icons are instantly recognizable even by the merest suggestion of their shape: a certain wasp-waisted soft drink bottle epitomizes America of the 1950s; the outline of a gothic cathedral evokes the Middle Ages of Europe; the outline of a steam locomotive memorializes the American expansion westward in the late 19th century; a clipper ship under full sail idealizes global trade in an earlier part of that century. America's Space Shuttle has become such an icon, symbolizing American ingenuity and leadership at the turn of the 21st century. The outline of the delta-winged Orbiter has permeated the public consciousness. This stylized element has been used in myriad illustrations, advertisements, reports, and video snippets—in short, everywhere. It is a fair question to ask why the Space Shuttle has achieved such status.

The first great age of space exploration culminated with the historic lunar landing in July 1969. Following that achievement, the space policymakers looked back to the history of aviation as a model for the future of space travel. The Space Shuttle was conceived as a way to exploit the resources of the new frontier. Using an aviation analogy, the shuttle would be the Douglas DC-3 of space. That aircraft is generally considered to be the first commercially successful air transport. The shuttle was to be the first commercially successful *space* transport. This impossible leap was not realized, an unrealistic goal that appears patently obvious in retrospect, yet it haunts the history of the shuttle to this day. Much of the criticism of the shuttle originates from this overhyped initial concept.

In fact, the perceived relationship between the history of aviation and the promise of space travel continues to motivate space policymakers. In some ways, the analogy that compares space with aviation can be very illustrative. So, if an unrealistic comparison for the shuttle is the leap from the 1903 Wright Flyer to the DC-3 transport of 1935 in a single technological bound, what is a more accurate comparison?

If the first crewed spacecraft of 1961—either Alan Shepard's Mercury or Yuri Gagarin's Vostok—are accurately the analog of the Wright brothers' first aircraft, the Apollo spacecraft of 1968 should properly be compared with the Wright brothers' 1909 "Model B"—their first commercial sale. The "B" was the product of 6 years of tinkering, experimentation, and adjustments, but were only two major iterations of aircraft design. In much the same way, Apollo was the technological inheritor of two iterations of spacecraft design in 7 years. The Space Shuttle of 1981—coming 20 years after the first spaceflights—could be compared with the aircraft of the mid 1920s. In fact, there is a good analogy in the history of aviation: the Ford Tri-Motor of 1928.

The Ford Tri-Motor was the leap from experimental to operational and had the potential to be economically effective as well. It was a huge improvement in aviation—it was revolutionary, flexible, and capable. The vehicle carried passengers and the US mail.

Top: 1928 Ford Tri-Motor; above: 1909 Wright "Model B." Smithsonian National Air and Space Museum, Washington, DC. (photos by Wayne Hale)

Admiral Richard Evelyn Byrd used the Ford Tri-Motor on his historic flyover of the North Pole. But the Ford Tri-Motor was not quite reliable enough, economical enough, or safe enough to fire off a successful and vibrant commercial airline business; just like the Space Shuttle.

Lower left: 1903 Wright Flyer; right: Douglas aircraft DC-3 of 1935. Smithsonian National Air and Space Museum, Washington, DC. (photos by Wayne Hale)

But here the aviation analogy breaks down. In aviation history, advances are made not just because of the passage of calendar time but because there are hundreds of different aircraft designs with thousands of incremental technology advances tested in flight between the "B" and the Tri-Motor.

Even so, the aviation equivalent compression of decades of technological advance does not do justice to the huge technological leap from expendable rockets and capsules to a reusable, winged, hypersonic, cargo-carrying spacecraft. This was accomplished with no intermediate steps. Viewed from that perspective, the Space Shuttle is truly a wonder. No doubt the shuttle is but one step of many on the road to the stars, but it was a giant leap indeed.

That is what this book is about: not what might have been or what was impossibly promised, but what was actually achieved and what was actually delivered. Viewed against this background, the Space Shuttle was a tremendous engineering achievement—a vehicle that enabled nearly routine and regular access to space for hundreds of people, and a profoundly vital link in scientific advancement. The vision of this book is to take a clear-eyed look at what the shuttle accomplished and the shuttle's legacy to the world.

Superlative Achievements of the Space Shuttle

For almost half a century, academic research, study, calculations, and myriad papers have been written about the problems and promises of controlled, winged hypersonic flight through the atmosphere. The Space Shuttle was the largest, fastest, winged hypersonic aircraft in history. Literally everything else had been a computer model, a wind tunnel experiment, or some subscale vehicle launched on a rocket platform. The shuttle flew at 25 times the speed of sound; regularly. The next fastest crewed vehicle—the venerable X-15—flew at its peak at seven times the speed of sound. Following the X-15, the next fastest crewed vehicle was the military SR-71, which could achieve three times the speed of sound. Both the X-15 and the SR-71 were retired years ago. Flight above about Mach 2 is not practiced today. If the promise of regular, commercial hypersonic flight is ever to come to fruition, the lessons learned from the shuttle will be an important foundation. For example, the specifics of aerodynamic control change significantly with these extreme speeds. Prior to the first flight, computations

The second X-15 rocket plane (56-6671) is shown with two external fuel tanks, which were added during its conversion to the X-15A-2 configuration in the mid 1960s.

for the shuttle were found to be seriously in error when actual postflight data were reviewed. Variability in the atmosphere at extreme altitudes would have gone undiscovered except for the regular passage of the shuttle through regions unnavigable any other way. Serious engineering obstacles with formidable names—hypersonic boundary layer transition, for example—must be understood and overcome, and cannot be studied in wind tunnels or computer simulations. Only by flight tests will real data help us understand and tame these dragons of the unknown ocean of hypersonic flight.

Most authorities agree that getting back safely from Earth orbit is a more difficult task than achieving Earth orbit in the first place. All the tremendous energy that went into putting the spacecraft into orbit must be cancelled out. For any vehicle's re-entry into Earth's atmosphere, this is principally accomplished by air friction—turning kinetic energy into heat. Objects entering the Earth's atmosphere are almost always rapidly vaporized by the friction generated by the enormous velocity of space travel. Early spacecraft carried huge and bulky ablative heat shields, which were good for one use only. The Space Shuttle Orbiter was completely reusable, and was covered with Thermal Protection Systems from nose to tail. The thermal shock standing 9 mm (0.3 in.) off the front of the wing leading edge exceeded the temperature of the visible surface of the sun: 8,000°C (14,000°F). At such an extreme temperature, metals don't melt—they boil. Intense heating went on for almost half an hour during a normal deceleration from 8 km (5 miles) per second to full stop. Don't forget that weight was at a premium. A special carbon fiber cloth impregnated with carbon resin was molded to an aerodynamic shape. This was the

This view of the suspended Orbiter Discovery shows the underside covered with Thermal Protection System tiles.

so-called reinforced carbon-carbon on the wing leading edge and nose cone. This amazing composite was only 5 mm (0.2 in.) thick, but the aluminum structure of the Orbiter was completely reliant on the reinforced carbon-carbon for protection. In areas of the shuttle where slightly lower peak temperatures were experienced, the airframe was covered with silica-based tiles. These tiles were mostly empty space but provided protection from temperatures to 1,000°C (2,000°F). Extraordinarily lightweight but structurally robust, easily formed to whatever shape needed, over 24,000 tiles coated the bottom and sides of the Orbiter. In demonstrations of the tile's effectiveness, a technician held one side of a shuttle tile in a bare hand while pointing a blowtorch at the opposite side. These amazing Thermal Protection Systems—all invented for the shuttle—brought 110 metric tons (120 tons) of vehicle, crew, and payload back to Earth through the inferno that is re-entry.

Nor is the shuttle's imaginative navigation system comparable to any other system flying. The navigation system kept track of not only the shuttle's position during re-entry, but also the total energy available to the huge glider. The system managed energy, distance, altitude, speed, and even variations in the winds and weather to deliver the shuttle precisely to the runway threshold. The logic contained in the re-entry guidance software was the hard-won knowledge from successful landings.

So much for re-entry. All real rocket scientists know that propulsion is problem number one for space travel. The shuttle excelled in both solid- and liquid-fueled propulsion elements.

The reusable Solid Rocket Booster (SRB) motors were the largest and most powerful solid rocket motors ever flown. Solid rockets are notable for their high thrust-to-weight ratio and the SRB motors epitomized that. Each one developed a thrust of almost 12 meganewtons (3 million pounds) but weighed only 600,000 kg (1.3 million pounds) at ignition (with weight decreasing rapidly after that). This was the equivalent motive power of 36,000 diesel locomotives that together would weigh 26 billion kg (57 billion pounds). The shuttle's designers were grounded in aviation in the 1950s and thought of the SRB motors as extreme JATO bottles—those small solid rockets strapped to the side of overloaded military transports taking off from short airfields. (JATO is short for jet-assisted takeoff, where "jet" is a generic term covering even rocket engines.) Those small, strap-on solid rocket motors paled in comparison with the SRB motors—some JATO bottles indeed. Within milliseconds of ignition, the finely tuned combustion processes inside the SRB motor generated internal pressure of over 7 million pascals (1,000 pounds per square inch [psi]). The thrust was "throttled" by the shape in which the solid propellant was cast inside the case. This was critical because thrust had to be reduced as the shuttle accelerated through the speed of maximum aerodynamic pressure. For the first 50 years of spaceflight, these reuseable boosters were the largest solid rockets ever flown.

The Solid Rocket Boosters operated in parallel with the main engines for the first 2 minutes of flight to provide the additional thrust needed for the Orbiter to escape the gravitational pull of the Earth. At an altitude of approximately 45 km (24 nautical miles), the boosters separated from the Orbiter/External Tank, descended on parachutes, and landed in the Atlantic Ocean. They were recovered by ships, returned to land, and refurbished for reuse. The boosters also assisted in guiding the entire vehicle during initial ascent. Thrust of both boosters was equal to over 2 million kg (over 5 million pounds).

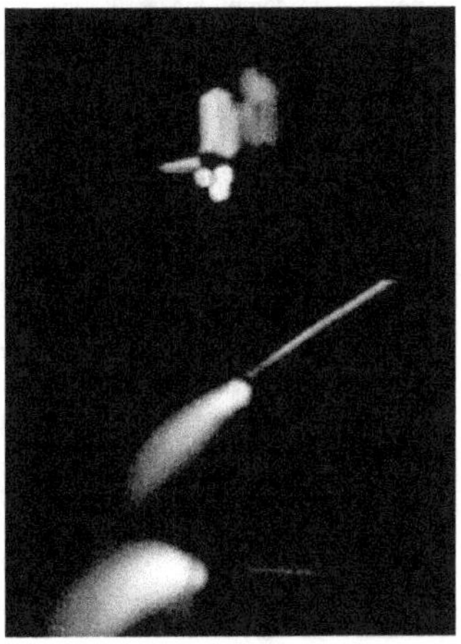

Development of the liquid-fueled Space Shuttle Main Engine was considered an impossible task in the mid 1970s. Larger liquid-fueled rockets had been developed—most notably the Saturn V first-stage engines, the famous F-1 engine that developed three times the thrust of the shuttle main engines. But the F-1 engines burned kerosene rather than hydrogen and their "gas mileage" was much lower than the shuttle main engines. In fact, no more efficient, liquid-fueled rocket engines have ever been built. Getting to orbit requires enormous amounts of energy. The "mpg" rating of these main engines was unparalleled in the history of rocket manufacture. The laws of thermodynamics define the maximum efficiency of any "heat engine," whether it is the gasoline engine that powers an automobile, or a big power plant that generates electricity, or a rocket engine. Different thermodynamic "cycles" have different possible efficiencies. Automobile engines operating on the Otto cycle typically are 15% of the maximum theoretical efficiency. The shuttle main engines operating on the rocket cycle achieved 99.5% of the maximum theoretical efficiency.

To put the power of the main engines in everyday terms: if your car engine developed the same power per pound as these engines, your automobile would be powered by something about the size and weight of a loaf of bread. And it would cost less than $100.00. More efficient engines have never been made, no matter what measure is used: horsepower to weight, horsepower to cost. Nor is the efficiency standard likely to ever be exceeded by any other chemical rocket.

So far, this has been about the basic problem in any journey—getting there and getting back. But the shuttle was a space truck, a heavy-lift launch vehicle in the same class as the Saturn V moon rocket. In fact, over half of all the mass put in Earth orbit—and that includes all rockets from all the nations of the world from 1957 until 2010—was put there by the shuttle. Think of that. The shuttle lofted more mass to Earth orbit than all the Saturn Vs, Saturn Is, Atlases, Deltas, Protons, Zenits, and Long Marches, etc., combined. And what about all the mass brought safely home from space? Ninety-seven percent came home with the shuttle. The Space Shuttle deployed some of the heaviest-weight upper stages for interplanetary probes. The largest geosynchronous satellites were launched by the shuttle. What a truck. What a transportation system.

And Science?

How much science was accomplished by the Space Shuttle? Start with the study of the stars. What has the shuttle done for astronomy? It brought us closer to the heavens. Shuttle had mounted telescopes operated directly by the crew to study the heavens. Not only did the shuttle launch the Compton Gamma Ray Observatory, the crew saved it by fixing its main antenna. Astronauts deployed the orbiting Chandra X-ray Observatory and the international polar star probe Ulysses. A series of astronomy experiments, under the moniker SPARTAN, studied comets, the sun, and galactic objects. The Solar Maximum Satellite enabled the study of our sun. And the granddaddy of them all, the Hubble Space Telescope, often called the most productive scientific instrument of all time, made discoveries that have rewritten the textbooks on astronomy, astrophysics, and cosmology—all because of shuttle.

Don't forget planetary science. Not only has Hubble looked deeply at most of the planets, but the shuttle also launched the Magellan radar mapper

Backdropped by a cloud-covered part of Earth, Space Shuttle Discovery approaches the International Space Station during STS-124 (2008) rendezvous and docking operations. The second component of the Japan Aerospace Exploration Agency's Kibo laboratory, the Japanese Pressurized Module, is visible in Discovery's cargo bay.

to Venus and the Galileo mission to Jupiter and its moons.

In Earth science, two Spacelab Atmospheric Laboratory for Applications and Science missions studied our own atmosphere, the Laser Geodynamic Satellite sphere monitors the upper reaches of the atmosphere and aids in mapping, and three Space Radar Laboratory missions mapped virtually the entire land mass of the Earth to a precision previously unachievable. The Upper Atmosphere Research satellite was also launched from the shuttle, as was the Earth Radiation Budget Satellite and a host of smaller nanosatellites that pursued a variety of Earth-oriented topics. Most of all, the pictures and observations made by the shuttle crews using cameras and other handheld instruments provided long-term observation of the Earth, its surface, and its climate.

Satellite launches and repairs were a highlight of shuttle missions, starting with the Tracking and Data Relay Satellites that are the backbone for communications with all NASA satellites—Earth resources,

Laser Geodynamic Satellite dedicated to high-precision laser ranging. It was launched on STS-52 (1992).

astronomical, and many more. Communications satellites were launched early in the shuttle's career but were reassigned to expendable launches for a variety of reasons. Space repair and recovery of satellites started with the capture and repair of the Solar Maximum Satellite in 1984 and continued with satellite recovery and repair of two HS-376 communications satellites in 1985 and the repair of Syncom-IV that same year. The most productive satellite repair involved five repetitive shuttle missions to the Hubble Space Telescope to upgrade its systems and instruments on a regular basis.

Biomedical research also was a hallmark of many shuttle missions. Not only were there six dedicated Spacelab missions studying life sciences, but there were also countless smaller experiments on the effects of microgravity (not quite zero gravity) on various life forms: from microbes and viruses, through invertebrates and insects, to mammals, primates, and finally humans. This research yielded valuable insight in the workings of the human body, with ramifications for general medical care and disease cure and prevention. The production of pharmaceuticals in space has been investigated with mixed success, but practical production requires lower cost transportation than the shuttle provided.

Finally, note that nine shuttle flights specifically looked at materials science questions, including how to grow crystals in microgravity, materials processing of all kinds, lubrication, fluid mechanics, and combustion dynamics—all without the presence of gravity.

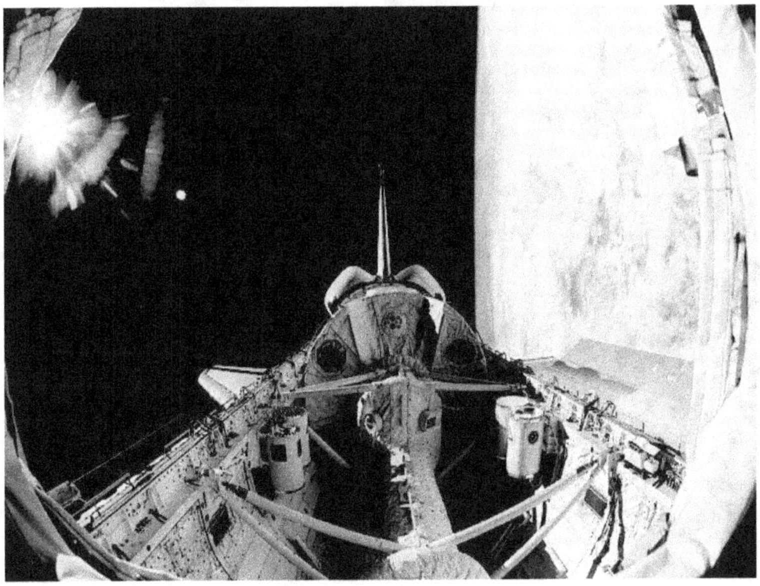

View from the Space Shuttle Columbia's cabin of the Spacelab science module, hosting 16 days of Neurolab research. (STS-90 [1998] is in the center.) This picture clearly depicts the configuration of the tunnel that leads from the cabin to the module in the center of the cargo bay.

Spacewalks

Of all the spacewalks (known as extravehicular activities) conducted in all the spaceflights of the world, more than three-quarters of them were based from the Space Shuttle or with shuttle-carried crew members at the International Space Station (ISS) with the shuttle vehicle attached and supporting. The only "untethered" spacewalks were executed from the shuttle. Those crew members were buoyed by the knowledge that, should their backpacks fail, the shuttle could swiftly come to their rescue.

Space Shuttle Discovery docked to the International Space Station is featured in this image photographed by one of the STS-119 (2009) crew members during the mission's first scheduled extravehicular activity.

The final and crowning achievement of the shuttle was to build the ISS. The shuttle was always considered only part of the future of space infrastructure. The construction and servicing of space stations was one of the design goals for the shuttle. The ISS—deserving of a book in its own right—is the largest space international engineering project in the history of the world. The ISS and the Space Shuttle are two sides of the same coin: the ISS could not be constructed without the shuttle, and the shuttle would have lost a major reason for its existence without the ISS. In addition to the scientific accomplishments of the ISS and the engineering marvel of its construction, the ISS is important as one of the shining examples of the power of international cooperation for the good of all humanity. The shuttle team was always international due to the Canadian contributions of the robot arm, the international payloads, and the international spacefarers. But participation in the construction of the ISS brought international cooperation to a new level, and the entire shuttle team was transformed by that experience.

Anchored to a foot restraint on Space Shuttle Atlantis' remote manipulator system robotic arm, Astronaut John Olivas, STS-117 (2007), moves toward Atlantis' port orbital maneuvering system pod that was damaged during the shuttle's climb to orbit. During the repair, Olivas pushed the turned-up portion of the thermal blanket back into position, used a medical stapler to secure the layers of the blanket, and pinned it in place against adjacent thermal tile.

The Astronauts

In the final analysis, space travel is all about people. In 133 flights, the Space Shuttle provided nearly 850 seats to orbit. Many people have been to orbit more than once, so the total number of different people who have flown to space on all spacecraft (Vostok, Mercury, Voskhod, Gemini, Soyuz, Apollo, Shenzhou, and the shuttle) in the last 50 years is just under 500. Of that number, over 400 have flown on the Space Shuttle. Almost three times as many people flew to space on the

Astronaut Joseph Acaba, STS-119 (2009), works the controls of Space Shuttle Discovery's Shuttle Robotic Arm on the aft flight deck during Flight Day 1 activities.

shuttle than on all other vehicles from all countries of the world combined. If the intent was to transform space and the opening of the frontier to more people, the shuttle accomplished this. Fliers included politicians, officials from other agencies, scientists of all types, and teachers. Probably most telling, these spacefarers represented a multiplicity of ethnicities, genders, and citizenships. The shuttle truly became the people's spaceship.

Fourteen people died flying on the shuttle in two accidents. They too represented the broadest spectrum of humanity. In 11 flights, Apollo lost no astronauts in space—although Apollo 13 was a very close call—and only three astronauts in a ground accident. Soyuz, like shuttle, had two fatal in-flight accidents but lost only four souls due to the smaller carrying capacity. The early days of aviation were far bloodier, even though the altitudes and energies were a fraction of those of orbital flight.

How Do We Rate the Space Shuttle?

Did shuttle have the power of thousands of diesel locomotives? Was it the most efficient rocket system ever built? Certainly it was the only winged space vehicle that flew from orbit as a hypersonic glider. And it was the only reusable space vehicle ever built except for the Soviet Buran ("Snowflake"), which was built to be reusable but only flew once. Imitation is the sincerest form of flattery; the Buran was the greatest compliment the shuttle ever had.

In the 1940s and early 1950s, the world's experimental aircraft flew sequentially faster and higher. The X-15 even allowed six people to earn their astronaut wings for flying above 116,000 m (380,000 ft) in a parabolic suborbital trajectory. If the exigencies of the Cold War—the state of conflict, tension, and competition that existed between the United States and the Soviet Union and their respective allies from the mid 1940s to the early 1990s—had not forced a rapid entry into space on the top of intercontinental ballistic missiles, a far different approach to spaceflight would most likely have occurred with air-breathing winged vehicles flying to the top of the atmosphere and then smaller rocket stages to orbit. But that buildup approach didn't happen. Some historians think such an approach would have provided a more sustainable approach to space than expendable intercontinental ballistic missile-based launch systems. Hypersonic flight continues to be the subject of major research by the aviation community. Plans to build winged vehicles that can take off horizontally and fly all the way to Earth orbit are still advanced as the "proper" way to travel into space. Time will tell if these dreams become reality.

No matter the next steps in space exploration, the legacy of the Space Shuttle will be to inspire designers, planners, and astronauts. Because building a Space Shuttle was thought to be impossible, and yet it flew, the shuttle remains the most remarkable achievement of its time—a cathedral of technology and achievement for future generations to regard with wonder.

The sun radiates on Space Shuttle Atlantis as it is positioned to head for space on mission STS-115 (2006).

The Historical Legacy

Major Milestones

Jennifer Ross-Nazzal
Dennis Webb

Astronauts John Young and Robert Crippen woke early on the morning of April 12, 1981, for the second attempted launch of the Space Shuttle Columbia—the first mission of the Space Shuttle Program. Two days earlier, the launch had been scrubbed due to a computer software error. Those working in the Shuttle Avionics Integration Laboratory at Johnson Space Center (JSC) in Houston, Texas, quickly resolved the issue and, with the problem fixed, the agency scheduled a second try soon after. Neither crew member expected to launch, however, because so much had to come together for liftoff to occur.

That morning, they did encounter a serious problem. With fewer than 2 hours until launch, the crew of Space Transportation System (STS)-1 locked the faceplates onto their helmets, only to find that they could not breathe. To avoid scrubbing the mission, the crew members looked at the issue and asked Loren Shriver, the astronaut support pilot, to help them. Finding a problem with the oxygen hose quick disconnect, Shriver tightened the line with a pair of pliers, and the countdown continued.

At 27 seconds before launch, Crippen realized that this time they were actually going to fly. His heart raced to 130 beats per minute while Young's heart, that of a veteran commander, stayed at a calm 85 beats. Young later joked, "I was excited too. I just couldn't get my heart to beat any faster." At 7:00 a.m., Columbia launched, making its maiden voyage into Earth orbit on the 20th anniversary of Yuri Gagarin's historic first human flight into space (1961).

The thousands who had traveled to the beaches of Florida's coastline to watch the launch were excited to see the United States return to flying in space. The last American flight was the Apollo-Soyuz Test Project, which flew in July 1975 and featured three American astronauts and two cosmonauts who rendezvoused and docked their spacecraft in orbit. Millions of others who watched the launch of STS-1 from their television sets were just as elated. America was back in space.

Like their predecessors, Young and Crippen became heroes for flying this mission—the boldest test flight in history. The shuttle was like no other vehicle that had flown; it was reusable. Unlike the space capsules of the previous generation, the shuttle had not been tested in space. This was the first test flight of the Columbia and the only time astronauts had actually flown a spacecraft on its first flight. The primary objective was to prove that the shuttle could safely launch a crew and then return safely to Earth. Two days later, the mission ended and the goal was accomplished when Young landed the shuttle at Dryden Flight Research Center on the Edwards Air Force Base runway in California. The spacecraft had worked like a "champ" in orbit—even with the loss of several tiles during launch. After landing, Christopher Kraft, director of JSC, said, "We just became infinitely smarter."

Design and Development

It would be a mistake to say that the first flight of Columbia was the start of the Space Shuttle Program. The idea of launching a reusable winged vehicle was not a new concept. Throughout the 1960s, NASA and the Department of Defense (DoD) studied such concepts. Advanced Space Shuttle studies began in 1968 when the Manned Spacecraft Center—which later became JSC—and Marshall Space Flight Center in Huntsville, Alabama, issued a joint request for proposal for an integral launch and re-entry vehicle to study different configurations for a round-trip vehicle that could reduce costs, increase safety, and carry payloads of up to 22,680 kg (50,000 pounds). This marked the beginning of the design and development of the shuttle.

Maxime Faget, director of engineering and development at the Manned Spacecraft Center in 1969, holding a balsa wood model of his concept of the spaceship that would launch on a rocket and land on a runway.

Four contractors—General Dynamics/Convair, Lockheed, McDonnell Douglas, and North American Rockwell—received 10-month contracts to study different approaches for the integral launch and re-entry vehicle. Experts examined a number of designs, from fully reusable vehicles to the use of expendable rockets. On completion of these studies, NASA determined that a two-stage, fully reusable vehicle met its needs and would pay off in terms of cost savings.

On April 1, 1969, Maxime Faget, director of engineering and development at the Manned Spacecraft Center, asked 20 people to report to the third floor of a building that most thought did not have a third floor. Because of that, many believed it was an April Fool's prank but went anyway. Once there, they spotted a test bay, which had three floors, and that was where they met. Faget then walked through the door with a balsa wood model of a plane, which he glided toward the engineers. "We're going to build America's next spacecraft. And it's going to launch like a spacecraft, it's going to land like a plane," he told the team. America had not yet landed on the moon, but NASA's engineers moved ahead with plans to create a new space vehicle.

As the contractors and civil servants explored various configurations for the next generation of spacecraft, the Space Task Group, appointed by President Richard Nixon, issued its report for future space programs. The committee submitted three options: the first and most ambitious featured a manned Mars landing as early as 1983, a lunar and Earth-orbiting station, and a lunar surface base; the second supported a mission to Mars in 1986; and the third deferred the Mars landing, providing no scheduled date for its completion. Included in the committee's post-Apollo plans were a Space Shuttle, referred to as the Space Transportation System, and a space station, to be developed simultaneously. Envisioned as less costly than the Saturn rocket and Apollo capsules, which were expended after only one use, the shuttle would be reusable and, as a result, make space travel more routine and less costly. The shuttle would be capable of carrying passengers, supplies, satellites, and other equipment—much as an airplane ferries people and their luggage—to and from orbit at least 100 times before being retired. The system would support both the civil and military space programs and be a cheaper way to launch satellites. Nixon, the Space Task Group proposals, and NASA cut the moon and Mars from their plans. This left only the shuttle and station for development, which the agency hoped to develop in parallel.

The decision to build a shuttle was extremely controversial, even though NASA presented the vehicle as economical—a cost-saver for taxpayers—when compared with the large outlays for the Apollo Program. In fact, in 1970 the shuttle was nearly defeated by Congress, which was dealing with high inflation, conflict in Vietnam, spiraling deficits, and an economic recession. In April 1970, representatives in the House narrowly defeated an amendment to eliminate all funding for the shuttle. A similar amendment offered in the Senate was also narrowly defeated. Minnesota Senator Walter Mondale explained that the money NASA requested was simply the "tip of the iceberg." He argued that the $110 million requested for development that year might be better spent on urban renewal projects, veterans' care, or improving the environment. Political support for the program was very tenuous, including poor support from some scientific and aerospace leaders.

To garner support for the shuttle and eliminate the possibility of losing the program, NASA formed a coalition with the US Air Force and established a joint space transportation committee to meet the needs of the two agencies. As an Air Force spokesman explained, given the political and economic realities of the time, "Quite possibly neither NASA nor the DoD could justify the shuttle system alone. But together we can make a strong case."

The Space Shuttle design that NASA proposed did not initially meet the military's requirements. The military needed the ability to conduct a polar orbit with quick return to a military airfield. This ability demanded the now-famous delta wings as opposed to the originally proposed airplane-like straight wings. The Air Force also insisted that it needed a larger payload bay and heavier lift capabilities to carry and launch reconnaissance satellites. A smaller payload bay would require the Air Force to retain their expendable launch vehicles and chip away at the argument forwarded by NASA about the shuttle's economy and utilitarian purpose. The result was a larger vehicle with more cross-range landing capability.

Though the president and Congress had not yet approved the shuttle in 1970, NASA awarded preliminary design contracts to McDonnell Douglas and North American Rockwell, thus beginning the second phase of development. By awarding two contracts for the country's next-generation spacecraft, NASA signaled its decision to focus on securing support for the two-stage reusable space plane over the station, which received little funding and was essentially shelved until 1984 when President Ronald Reagan directed the agency to build a space station within a decade. In fact, when James Fletcher became NASA's administrator in April 1971, he wholeheartedly supported the shuttle and proclaimed, "I don't want to hear any more about a space station, not while I am here."

Fletcher was doggedly determined to see that the federal government funded the shuttle, so he worked closely with the Nixon administration to assure the program received approval. Realizing that the $10.5 billion price tag for the development of the fully reusable, two-stage vehicle was too high, and facing massive budget cuts from the Office of Management and Budget, the administrator had the agency study the use of expendable rockets to cut the high cost and determine the significant cost savings with a partially reusable spacecraft as opposed to the proposed totally reusable one. On learning that use of an expendable External Tank, which would provide liquid oxygen and hydrogen fuel for Orbiter engines, would decrease costs by nearly half, NASA chose that technology—thereby making the program more marketable to Congress and the administration.

Robert Thompson, former Space Shuttle Program manager, believed that the decision to use an expendable External Tank for the Space Shuttle Main Engines was "perhaps the single most important configuration decision made in the Space Shuttle Program," resulting in a smaller, lighter shuttle. "In retrospect," Thompson explained, "the basic decision to follow a less complicated development path at the future risk of possible higher operating costs was, in my judgment, a very wise choice." This decision was one of the program's major milestones, and the decreased costs for development had the desired effect.

Presidential Approval

Nixon made the announcement in support of the Space Shuttle Program at his Western White House in San Clemente, California, on January 5, 1972. Believing that the shuttle was a good investment, he asked the space agency to stress that the shuttle was not an expensive toy. The president highlighted the benefits of the civilian and military applications and emphasized the importance of international cooperation, which would be ushered in with the program. Ordinary people from across the globe, not just American test pilots, could fly on board the shuttle.

From the start, Nixon envisioned the shuttle as a truly international program. Even before the president approved the program, NASA Administrator Thomas Paine, at Nixon's urging, approached other nations about participating. As NASA's budget worsened, partnering with other nations became more appealing to the space agency. In 1973, Europe agreed to develop and build the Spacelab, which

Rollout tests of the Solid Rocket Boosters. Mobile Launcher Platform number 3, with twin Solid Rocket Boosters bolted to it, inches along the crawlerway at various speeds up to 1.6 km (1 mile) per hour in an effort to gather vibration data. The boosters are braced at the top for stability. Data from these tests, completed September 2004, helped develop maintenance requirements on the transport equipment and the flight hardware.

would be housed in the payload bay of the Orbiter and serve as an in-flight space research facility. The Canadians agreed to build the Shuttle Robotic Arm in 1975, making the Space Shuttle Program international in scope.

Having the Nixon administration support the shuttle was a major hurdle, but NASA still had to contend with several members of Congress who disagreed with the administration's decision. In spite of highly vocal critics, both the House and Senate voted in favor of NASA's authorization bill, committing the United States to developing the Space Shuttle and, thereby, marking another milestone for the program.

To further reduce costs, NASA decided to use Solid Rocket Boosters, which were less expensive to build because they were a proven technology used by the Air Force in the Minuteman intercontinental ballistic missile program. As NASA Administrator Fletcher explained, "I think we have made the right decision at the right time. And I think it is the right price." Solids were less expensive to develop and cost less than liquid boosters. To save additional funds, NASA planned to recover the Solid Rocket Boosters and refurbish them for future flights.

Contracting out the Work

Two days after NASA selected the parallel burn Solid Rocket Motor propellant configuration, the agency put out a request for proposal for the development of the Orbiter. Four companies responded. NASA selected North American Rockwell, awarding the company a $2.6 billion contract. The Orbiter that Rockwell agreed to build illustrated the impact the Air Force had on the design. The payload bay measured 18.3 by 4.6 m (60 by 15 ft), to house the military's satellites. The Orbiter also had delta wings and the ability to deploy a 29,483-kg (65,000-pound) payload from a due-east orbit.

As NASA studied alternative concepts for the program, the agency issued a request for proposal for the Space Shuttle Main Engines. In the summer of 1971, NASA selected the Rocketdyne Division of Rockwell. Rocketdyne built the large, liquid fuel rocket engines used on the NASA Saturn V (moon rocket). However, the shuttle engines differed dramatically from their predecessors. As James Kingsbury, the director of Science and Engineering at the Marshall Space Flight Center, explained, "It was an unproven technology. Nobody had ever had a rocket engine that operated at the pressures and temperatures of that engine." Because of the necessary lead time needed to develop the world's first reusable rocket engine, the selection of the Space Shuttle Main Engines contractor preceded other Orbiter decisions, but a contract protest delayed development by 10 months. Work on the engines officially began in April 1972.

Other large companies benefiting from congressional approval of the Space Shuttle Program included International Business Machines, Martin Marietta, and Thiokol. The computer giant International Business Machines would provide five on-board computers, design and maintain their software, and support testing in all ground facilities that used the flight software and general purpose computers, including the Shuttle Avionics Integration Laboratory, the Shuttle Mission Simulator, and other facilities. Thiokol received the contract for the solid rockets, and NASA selected Martin Marietta to build the External Tank. Although Rockwell received the contract for the Orbiter, the corporation parceled out work to other rival aerospace companies: Grumman built the wings; Convair Aerospace agreed to build the mid-fuselage; and McDonnell Douglas managed the Orbiter rocket engines, which maneuvered the vehicle in space.

Delays and Budget Challenges

Although NASA received approval for the program in 1972, inflation and budget cuts continually ate away at funding throughout the rest of the decade. Over time, this resulted in slips in the schedule as the agency had to make do with effectively fewer dollars each year and eventually cut or decrease spending for less-prominent projects, or postpone them. This also led to higher total development costs. Technical problems with the tiles, Orbiter heat shield, and main engines also resulted in delays, which caused development costs to increase. As a result, NASA kept extending the first launch date.

The shuttle continued to evolve as engineers worked to shave weight from the vehicle to save costs. In 1974, engineers decided to remove the shuttle's air-breathing engines, which would have allowed a powered landing of the vehicle. The engines were to be housed in the payload bay and would have cost more than $300 million to design and build, but

The Space Shuttle Main Engines were the first rocket engines to be reused from one mission to the next. This picture is of Engine 0526, tested on July 7, 2003. A remote camera captures a close-up view of a Space Shuttle Main Engine during a test firing at the John C. Stennis Space Center in Hancock County, Mississippi.

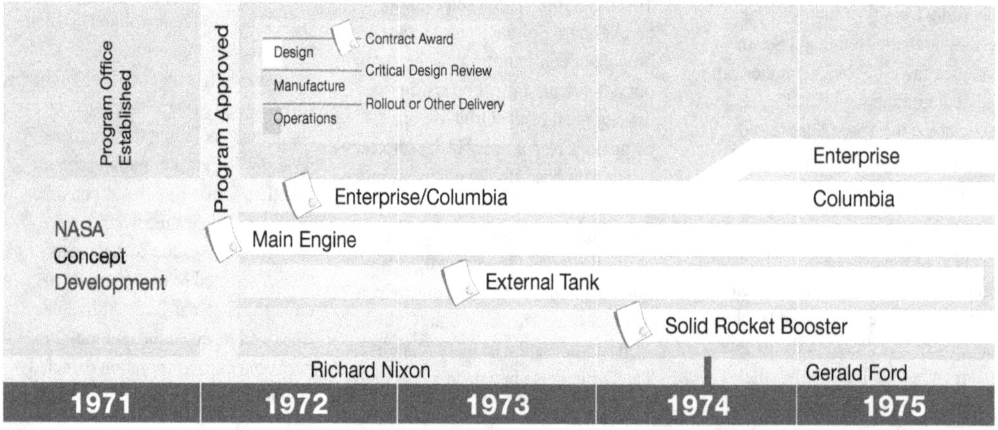

they took up too much space in the bay and added substantial complexity to the design. Thus, the agency decided to go forward with the idea of an unpowered landing to glide the Orbiter and crew safely to a runway.

This decision posed an important question for engineers: how to bring the Orbiter from California, where Rockwell was building it, to the launch sites in Florida, Vandenberg Air Force Base, or test sites in Alabama. NASA considered several options: hanging the Orbiter from a dirigible; carrying the vehicle on a ship; or modifying a Lockheed C-5A or a Boeing 747 to ferry the Orbiter in a piggyback configuration on the back of the plane. Eventually, NASA selected the 747 and purchased a used plane from American Airlines in 1974 to conduct a series of tests before transforming the plane into the Shuttle Carrier Aircraft. Modifications of the 747 began in 1976.

Final Testing

Rollout

On September 17, 1976, Americans got an initial glimpse of NASA's first shuttle, the Enterprise, when a red, white, and blue tractor pulled the glider out of the hangar at the Air Force Plant in Palmdale, California. Enterprise was not a complete shuttle: it had no propellant lines and the propulsion systems (the main engines and orbital maneuvering pods) were mock-ups. Originally, NASA intended to name the vehicle Constitution in honor of the bicentennial of the United States, but fans of the television show *Star Trek* appealed to NASA and President Gerald Ford, who eventually relented and decided to name the shuttle after Captain Kirk's spaceship. Speaking at the unveiling, Fletcher proclaimed that the debut was "a very proud moment" for NASA. He emphasized the dramatic changes brought about by the program: "Americans and the people of the world have made the evolution to man in space—not just astronauts." The rollout of Enterprise marked the beginning of a new era in spaceflight, one in which all could participate.

In fact, earlier that summer, the agency had issued a call for a new class of astronauts, the first to be selected since the late 1960s when nearly all astronauts were test pilots. A few held advanced degrees in science and medicine, but none were women or minorities. Consequently, NASA emphasized its determination to select people from these groups and encouraged women and minorities to apply.

Approach and Landing Tests

In 1977, Enterprise flew the Approach and Landing Tests at Dryden Flight Research Center using Edwards Air Force Base runways in California. The program was a series of ground and flight tests designed to learn more about the landing characteristics of the Orbiter and how the mated shuttle and its carrier operated together. First, crewless high-speed taxi tests proved that the Shuttle Carrier Aircraft, when mated to the Enterprise, could steer and brake with the Orbiter perched on top of the airframe. The pair, then ready for flight, flew five captive inert flights without astronauts in February and March, which qualified the 747 for ferry operations. Captive-active flights followed in June and July and featured two-man crews.

The final phase was a series of free flights (when Enterprise separated from the Shuttle Carrier Aircraft and landed at the hands of the two-man crews) that flew in 1977, from August to October, and proved the flightworthiness of the shuttle and the techniques of unpowered landings. Most important, the Approach and Landing Tests Program pointed out sections of the Orbiter that needed to be strengthened or made of different materials to save weight.

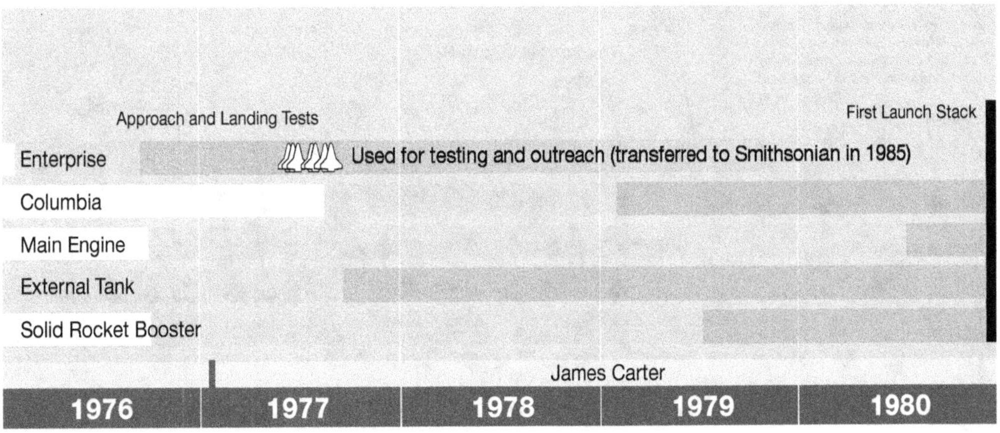

Enterprise atop the Shuttle Carrier Aircraft in a flight above the Mojave Desert, California (1977).

NASA had planned to retrofit Enterprise as a flight vehicle, but that would have taken time and been costly. Instead, the agency selected the other alternative, which was to have the structural test article rebuilt for flight. Eventually called Challenger, this vehicle would become the second Orbiter to fly in space after Columbia. Though Enterprise was no longer slated for flight, NASA continued to use it for a number of tests as the program matured.

Getting Ready to Fly

Concurrent with the Approach and Landing Tests Program, the astronaut selection board in Houston held interviews with 208 applicants selected from more than 8,000 hopefuls. In 1978, the agency announced the first class of Space Shuttle astronauts. This announcement was a historic one. Six women who held PhDs or medical degrees accepted positions along with three African American men and a Japanese American flight test engineer. After completing 1 year of training, the group began following the progress of the shuttle's subsystems, several of which had caused the program's first launch to slip.

The Space Shuttle Main Engines were behind schedule and threatened to delay the first orbital flight, which was tentatively scheduled for March 1979. Problems plagued the engines from the beginning. As early as 1974, the engines ran into trouble as cost overruns threatened the program and delays dogged the modification of facilities in California and the development of key engine components. Test failures occurred at Rocketdyne's California facility and the National Space Technology Laboratory in Mississippi, further delaying development and testing.

Another pacing item for the program was the shuttle's tiles. As Columbia underwent final assembly in California, Rockwell employees began applying the tiles, with the work to be completed in January 1979. Their application was much more time consuming than had been anticipated, and NASA transferred the ship to Kennedy Space Center (KSC) in March, where the task would be completed in the Orbiter Processing Facility and later in the Vehicle Assembly Building. Once in Florida, mating of the tiles to the shuttle ramped up. Unfortunately, engineers found that many of the tiles had to be strengthened. This resulted in many of the 30,000 tiles being removed, tested, and replaced at least once. The bonding process was so time consuming that technicians worked

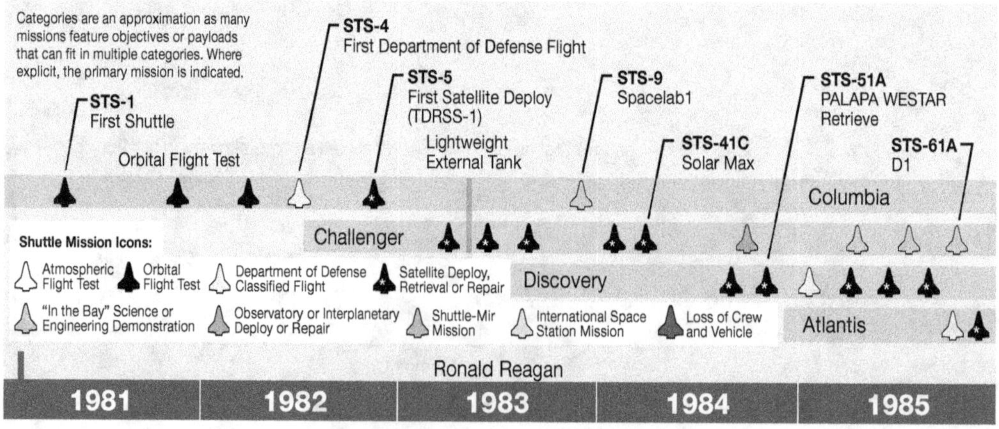

around the clock, 7 days a week at KSC to meet the launch deadline.

Aaron Cohen, former manager for the Space Shuttle Orbiter Project and JSC director, remembered the stress and pressure caused by the delays in schedule. "I really didn't know how we were going to solve the tile problem," he recalled. As the challenges mounted, Cohen, who was under tremendous pressure from NASA, began going gray, a fact that his wife attributed to "every tile it took to put on the vehicle." Eventually, engineers came up with a solution—a process known as densification, which strengthened the tiles and, according to Cohen, "bailed us out of a major, significant problem" and remained the process throughout the program.

After more than 10 years of design and development, the shuttle appeared ready to fly. In 1979 and 1980, the Space Shuttle Main Engines proved their flightworthiness by completing a series of engine acceptance tests. The tile installation finally ended, and the STS-1 crew members, who had been named in 1978, joked that they were "130% trained and ready to go" because of all the time they spent in the shuttle simulators. Young and Crippen's mission marked the beginning of the shuttle flight test program.

Spaceflight Operations

Columbia's First Missions

Columbia flew three additional test flights between 1981 and 1982. These test flights were designed to verify the shuttle in space, the testing and processing facilities, the vehicle's equipment, and crew procedures. Ground testing demonstrated the capability of the Orbiter, as well as of its components and systems. Without flight time, information about these systems was incomplete. The four tests were necessary to help NASA understand heating, loads, acoustics, and other concepts that could not be studied on the ground.

This test program ended on July 4, 1982, when commander Thomas Mattingly landed the shuttle at Dryden Flight Research Center (DFRC) on the 15,000-ft runway at Edwards Air Force Base in California. Waiting at the foot of the steps, President Reagan and First Lady Nancy Reagan congratulated the STS-4 crew on a job well done. Speaking to a crowd of more than 45,000 people at DFRC, the president said that the completion of this task was "the historical equivalent to the driving of the golden spike which completed the first transcontinental railroad. It marks our entrance into a new era."

The operational flights, which followed the flight test program, fell into several categories: DoD missions; commercial satellite deployments; space science flights; notable spacewalks (also called extravehicular activities); or satellite repair and retrieval.

To improve costs, beginning in 1983 all launches and landings at KSC were managed by one contractor, Lockheed Space Operations Company, Titusville, Florida. This consolidated many functions for the entire shuttle processing.

Department of Defense Flights

STS-4 (1982) featured the first classified payload, which marked a fundamental shift in NASA's traditionally open environment. Concerned with national security, the DoD instructed NASA Astronauts Mattingly and Henry Hartsfield to not transmit images of the cargo bay during the flight, lest pictures of the secret payload might inadvertently be revealed. STS-4 did differ somewhat from the other future DoD-dedicated flights: there was no secure communication line, so the crew worked out a system of communicating with the ground.

"We had the checklist divided up in sections that we just had letter names like Bravo Charlie, Tab Charlie, Tab

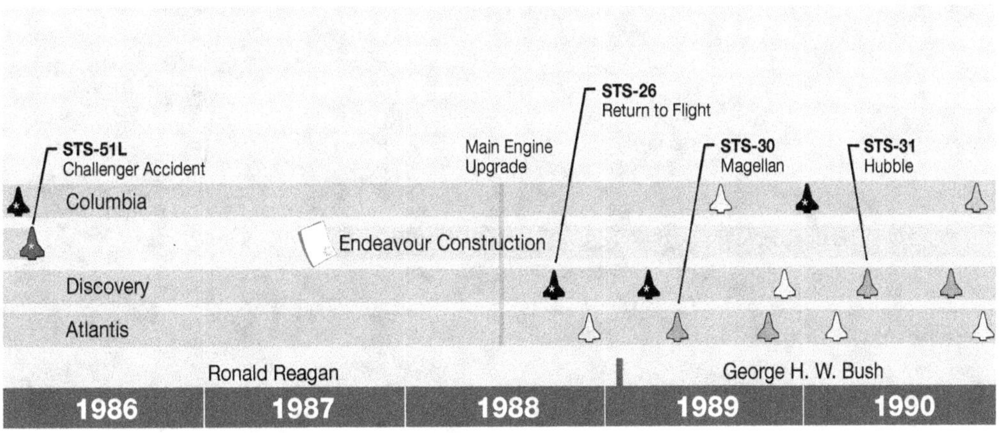

Bravo that they could call out. When we talked to Sunnyvale [California] to Blue Cube out there, military control, they said, 'Do Tab Charlie,' or something. That way it was just unclassified," Hartsfield recalled. Completely classified flights began in 1985.

Even though Vandenberg Air Force Base had been selected as one of the program launch sites in 1972, the California shuttle facilities were not complete when classified flights began. Anticipating slips, the DoD and NASA decided to implement a controlled mode at JSC and KSC that would give the space agency the capability to control classified flights out of the Texas and Florida facilities. Flight controllers at KSC and JSC used secured launch and flight control rooms separate from the rooms used for non-DoD flights. Modifications were also made to the flight simulation facility, and a room was added in the astronaut office, where flight crew members could store classified documents inside a safe and talk on a secure line.

Although the facilities at Vandenberg Air Force Base were nearly complete in 1984, NASA continued to launch and control DoD flights. Two DoD missions flew in 1985: STS-51C and STS-51J. Each flight included a payload specialist from the Air Force. That year, the department also announced the names of the crew of the first Vandenberg flight, STS-62A, which would have been commanded by veteran Astronaut Robert Crippen, but was cancelled in the wake of the Challenger accident (1986).

Flying classified flights complicated the business of spaceflight. For national security reasons, the Mission Operations Control Room at JSC was closed to visitors during simulations and these flights. Launch time was not shared with the press and, for the first time in NASA's history, no astronaut interviews were granted about the flight, no press kits were distributed, and the media were prohibited from listening to the air-to-ground communications.

Shuttle Operations, 1982-1986

STS-5 (1982) marked both the beginning of shuttle operations and another turning point in the history of the Space Shuttle Program. As Astronaut Joseph Allen explained, spaceflight changed "from testing the means of getting into space to using the resources found there." Or, put another way, this four-member crew (the largest space crew up to that point; the flight tests never carried more than two men at a time) was the first to launch two commercial satellites. This "initiated a new era in which the business of spaceflight became business itself." Dubbed the "Ace Moving Company," the crew jokingly promised "fast and courteous service" for its future launch services.

Many of the early shuttle flights were, in fact, assigned numerous commercial satellites, which they launched from the Orbiter's cargo bay. With NASA given a monopoly in the domestic launch market, many flight crews released at least one satellite on each flight, with several unloading as many as three communication satellites for a number of nations and companies. Foreign clients, particularly attracted to NASA's bargain rates, booked launches early in the program.

Another visible change that occurred on this, the fifth flight of Columbia was the addition of mission specialists—scientists and engineers—whose job it was to deploy satellites, conduct spacewalks, repair and retrieve malfunctioning satellites, and work as scientific researchers in space. The first two mission specialists—Joseph Allen, a physicist, and William Lenoir, an electrical engineer—held PhDs in their respective fields and had been selected as astronauts in 1967. Those who followed in their footsteps had similar qualifications, often holding advanced degrees in their fields of study.

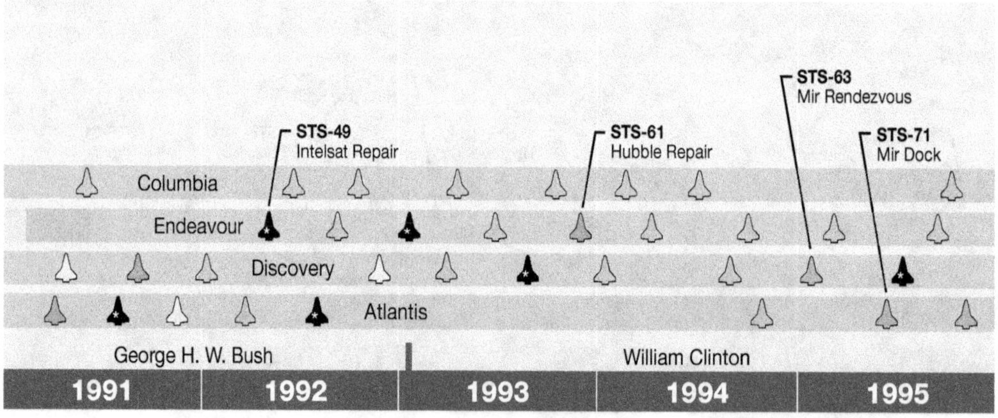

Christopher Kraft
Director of Johnson Space Center during shuttle development and early launches (1972-1982).
Played an instrumental role in the development and establishment of mission control.

"We went through a lot to prove that we should launch STS-1 manned instead of unmanned; it was the first time we ever tried to do anything like that. We convinced ourselves that the reliability was higher and the risk lower, even though we were risking the lives of two men. We convinced ourselves that that was a better way to do it, because we didn't know what else to do. We had done everything we could think of."

With the addition of mission specialists and the beginning of operations, space science became a major priority for the shuttle, and crews turned their attention to research. A variety of experiments made their way on board the shuttle in Get Away Specials, the Shuttle Student Involvement Project, the middeck (crew quarters), pallets (unpressurized platforms designed to support instruments that require direct exposure to space), and Spacelabs. Medical doctors within NASA's own Astronaut Corps studied space sickness on STS-7 (1983) and STS-8 (1983), subjecting their fellow crew members to a variety of tests in the middeck to determine the triggers for a problem that plagues some space travelers. Aside from medical experiments, many of the early missions included a variety of Earth observation instruments. The crews spent time looking out the window, identifying and photographing weather patterns, among other phenomena. A number of flights featured material science research, including STS-61C (1986), which included Marshall Space Flight Center's Material Science Laboratory.

As space research expanded, so did the number of users, and the aerospace industry was not excluded from this list. They were particularly active in capitalizing on the potential benefits offered by the shuttle and its platform as a research facility. Having signed a Joint Endeavor Agreement (a quid pro quo arrangement, where no money exchanged hands) with NASA in 1980, McDonnell Douglas Astronautics flew its Continuous Flow Electrophoresis System on board the shuttle numerous times to explore the capabilities of materials processing in space. The system investigated the ability to purify erythropoietin (a hormone) in orbit and to learn whether the company could mass produce the purified pharmaceutical in orbit. The company even sent one of its employees—who, coincidentally, was the first industrial payload specialist—into space to monitor the experiment on board three flights, including the maiden flight of Discovery. Other companies, like Fairchild Industries and 3M, also signed Joint Endeavor Agreements with NASA.

When the ninth shuttle flight lifted off the pad in November 1983, Columbia had six passengers and a Spacelab in its payload bay. This mission, the first flight of European lab, operated 24 hours a day, featured more than 70 experiments,

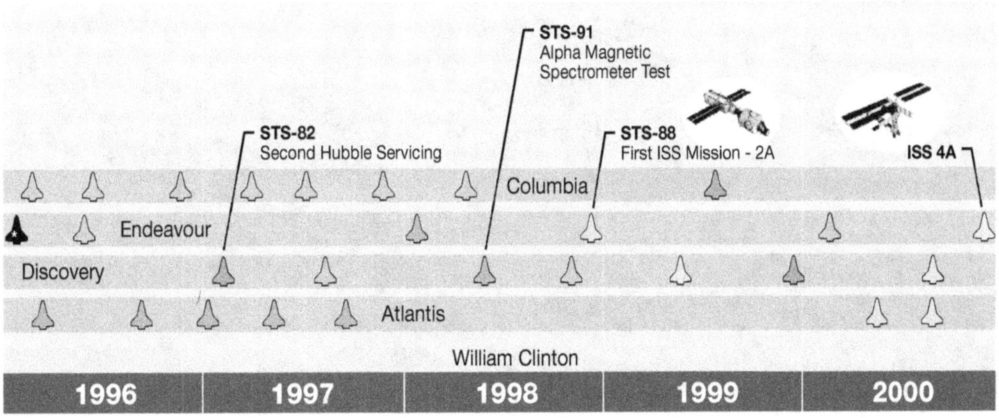

William Lucas, PhD
Former director of Marshall Space Flight Center during shuttle operations until Challenger accident (1974-1986).
Played an instrumental role in Space Shuttle Main Engine, External Tank, and Solid Rocket Booster design, development, and operations.

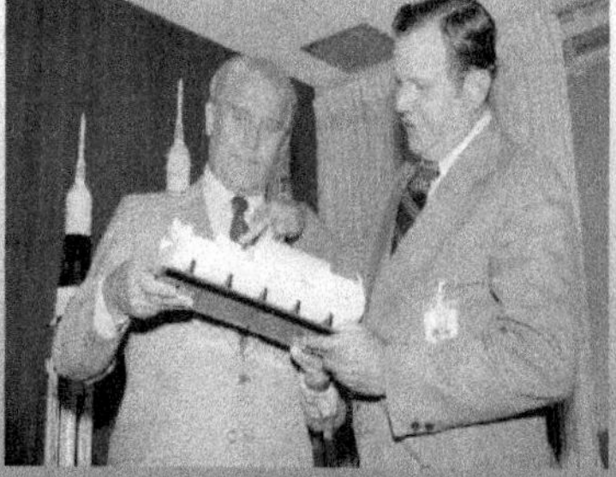

On October 11, 1974, newly appointed Marshall Space Flight Center (MSFC) Director Dr. William Lucas (right) and a former MSFC Director Dr. Wernher von Braun view a model.

"The shuttle was an important part of the total space program and it accomplished, in a remarkable way, the unique missions for which it was designed. In addition, as an element of the continuum from the first ballistic missile to the present, it has been a significant driver of technology for the benefit of all mankind."

and carried the first noncommercial payload specialists to fly in space.

Three additional missions flew Spacelabs in 1985, with West Germany sponsoring the flight of STS-61A, the first mission financed and operated by another nation. One of the unique features of this flight was how control was split between centers. JSC's Mission Control managed the shuttle's systems and worked closely with the commander and pilot while the German Space Operations Center in Oberpfaffenhofen oversaw the experiments and scientists working in the lab.

By 1984, the shuttle's capabilities expanded dramatically when Astronauts Bruce McCandless and Bob Stewart tested the manned maneuvering units that permitted flight crews to conduct untethered spacewalks. At this point in the program, this was by far the most demanding spacewalk conducted by astronauts. The first spacewalk, conducted just months before the flight of STS-41B, tested the suits and the capability of astronauts to work in the payload bay. As McCandless flew the unit out of the cargo bay for the first time, he said, "It may have been one small step for Neil, but it's a heck of a big leap for me." Set against the darkness of space, McCandless became the first human satellite in space. Having proved the capabilities of the manned maneuvering unit, NASA exploited its capabilities and used the device to make satellite retrieval and repair possible without the use of the Shuttle Robotic Arm.

Early Satellite Repair and Retrievals

Between 1984 and 1985, the shuttle flew three complicated satellite retrieval or repair missions. On NASA's 11th shuttle mission, STS-41C, the crew was to capture and repair the Solar Maximum Satellite (SolarMax), the first one built to be serviced and repaired by shuttle astronauts. Riding the manned maneuvering unit, spacewalker George Nelson tried to capture the SolarMax, but neither he nor the Robotic Arm operator Terry Hart was able to do so. Running low on fuel, the crew backed away from the satellite while folks at the Goddard Space Flight Center in

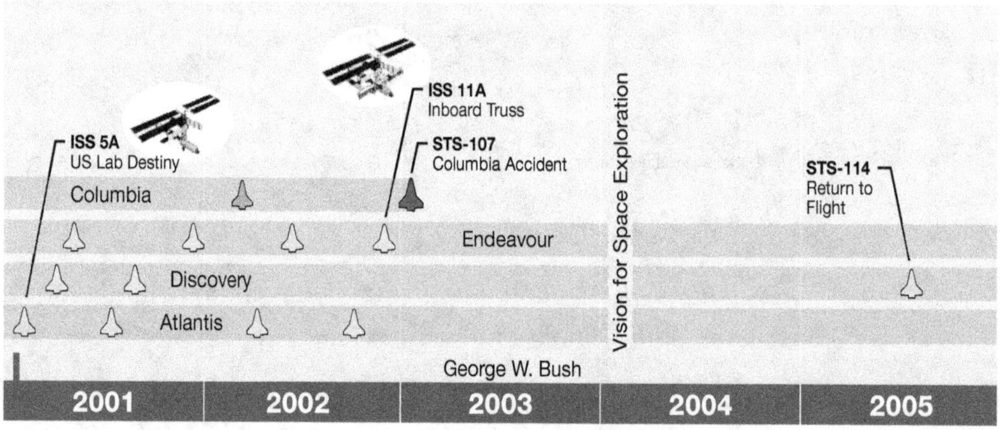

Maryland stabilized the SolarMax. The shuttle had just enough fuel for one more rendezvous with the satellite. Fortunately, Hart was able to grapple the satellite, allowing Nelson and James van Hoften to fix the unit, which was then rereleased into orbit.

The following retrieval mission was even more complex. STS-51A was the first mission to deploy two satellites and then retrieve two others that failed to achieve their desired orbits. Astronauts Joseph Allen and Dale Gardner used the manned maneuvering unit to capture Palapa and Westar, originally deployed on STS-41B 9 months earlier. They encountered problems, however, when stowing the first recovered satellite, forcing Allen to hold the 907-kg (2,000-pound) satellite over his head for an entire rotation of the Earth—90 minutes. When the crew members reported that they had captured and secured both satellites in Discovery's payload bay, Lloyd's of London—one of the underwriters for the satellites—rang the Lutine bell, as they had done since the 1800s, to announce events of importance. As Cohen, former director of JSC, explained, "Historically Lloyd's of London, who would insure high risk adventures, rang a bell whenever ships returned to port with recovered treasure from the sea." He added that the salvage of these satellites in 1984 "was at that time the largest monetary treasure recovered in history."

The program developed a plan for the crew of STS-51I (1985) to retrieve and repair a malfunctioning Hughes satellite that had failed to power up just months before the flight. With only 4 months to prepare, NASA built a number of tools that had not been tested in space to accomplish the crew's goal. In many ways, the crew's flight was a first. Van Hoften, one of the walkers on STS-41C, recalled the difference between his first and second spacewalk: "It wasn't anything like the first one. The first one was so planned out and choreographed. This one, we were winging it, really." Instead of planning their exact moves, crew members focused instead on skills and tasks. Their efforts paid off when the ground activated the satellite.

Space Station Reemerges

As the Space Shuttle Program matured, NASA began working on the Space Station Program, having been directed to do so by President Reagan in his 1984 State of the Union address. The shuttle would play an important role in building the orbiting facility. In the winter of 1985, STS-61B tested structures and assembly methods for the proposed long-duration workshop. Spacewalkers built a 13.8-m (45-ft) tower and a 3.7-m (12-ft) structure, proving that crews could feasibly assemble structures using parts carried into space by the Orbiter. NASA proceeded with plans to build Space Station Freedom, which in the 1990s was transformed to the International Space Station (ISS).

To fund the space station, NASA needed to cut costs for shuttles by releasing requests for proposals for three new contracts. In 1983, the Shuttle Processing Contract integrated all processing at KSC. Lockheed Space Operations Company received this contract. In 1985, the Space Transportation Systems Operations Contract and the Flight Equipment Contract were solicited. The former contract consolidated 22 shuttle operations contracts, while the latter combined 15 agreements involving spaceflight equipment (e.g., food, clothes, and cameras). NASA Administrator James Beggs hoped that by awarding such contracts, he could reduce shuttle costs by as much as a quarter by putting cost incentives into the contracts. Rockwell International won the Space Transportation Systems Operations Contract, and NASA chose Boeing Aerospace Operations to manage the Flight Equipment Processing Contract.

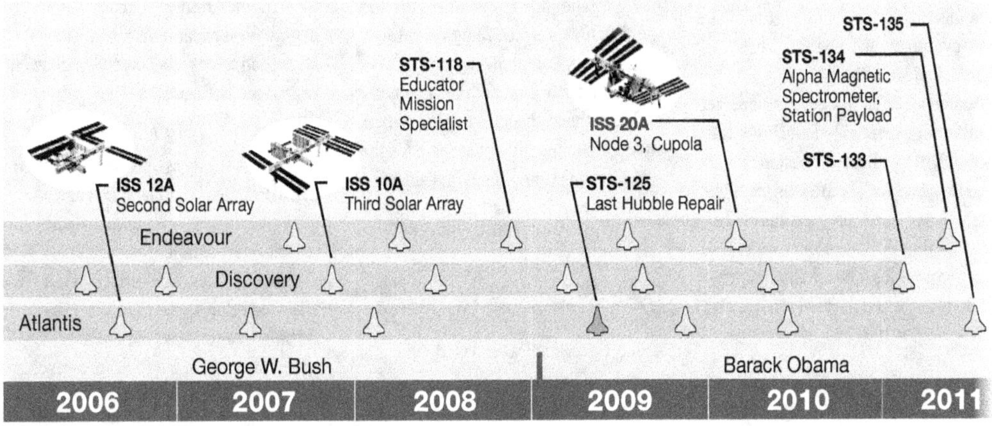

Challenger Accident

In January 1986, NASA suspended all shuttle flights after the Challenger accident in which seven crew members perished. A failure in the Solid Rocket Booster motor joint caused the vehicle to break up. The investigation board was very critical of NASA management, especially about the decision to launch. For nearly 3 years, NASA flew no shuttle flights. Instead, the agency made changes to the shuttle. It added a crew escape system and new brakes, improved the main engines, and redesigned the Solid Rocket Boosters, among other things.

In the aftermath of the accident, the agency made several key decisions, which were major turning points. The shuttle would no longer deliver commercial satellites into Earth orbit unless "compelling circumstances" existed or the deployment required the unique capabilities of the space truck. This decision forced industry and foreign governments who hoped to deploy satellites from the shuttle to turn to expendable launch vehicles. Fletcher, who had returned for a second term as NASA administrator, cancelled the Shuttle/Centaur Program because it was too risky to launch the shuttle carrying a rocket with highly combustible liquid fuel. Plans to finally activate and use the Vandenberg Air Force Base launch site were abandoned, and the shuttle launch site was eventually mothballed. The Air Force decided to launch future payloads on Titan rockets and ordered additional expendable launch vehicles. A few DoD-dedicated missions would, however, fly after the accident. Finally, in 1987, Congress authorized the building of Endeavour as a replacement for the lost Challenger. Endeavour was delivered to KSC in the spring of 1991.

Post-Challenger Accident Return to Flight

STS-26 was the Space Shuttle's Return to Flight. Thirty-two months after the Challenger accident, Discovery roared to life on September 29, 1988, taking its all-veteran crew into space where they deployed the second Tracking and Data Relay Satellite. The crew safely returned home to DFRC 4 days later, and Vice President George H.W. Bush and his wife Barbara Bush greeted the crew. That mission was a particularly significant accomplishment for NASA. STS-26 restored confidence in the agency and marked a new beginning for NASA's human spaceflight program.

Building Momentum

Following the STS-26 flight, the shuttle's launch schedule climbed once again, with the space agency eventually using all three shuttles in the launch processing flow for upcoming missions. The first four flights after the accident alternated between Discovery and Atlantis, adding Columbia to the mix for STS-28 (1989). Even though the flight crews did not launch any commercial satellites from the payload bay, several deep space probes—the Magellan Venus Radar Mapper, Galileo, and Ulysses—required the shuttle's unique capabilities. STS-30 (1989) launched the mapper, which opened a new era of exploration for the agency. This was the first time a Space Shuttle crew deployed an interplanetary probe, thereby interlocking both the manned and unmanned spaceflight programs. In addition, this flight was NASA's first planetary mission of any kind since 1977, when it launched the Voyager spacecraft. STS-34 (1989) deployed the Galileo spacecraft toward Jupiter. Finally, STS-41 (1990) delivered the European Space Agency's Ulysses spacecraft, which would study the polar regions of the sun.

Astronaut James Voss is pictured during an STS-69 (1995) extravehicular activity that was conducted in and around Endeavour's cargo bay. Voss and Astronaut Michael Gernhardt performed evaluations for space station-era tools and various elements of the spacesuits.

Extended Duration Orbiter Program

Before 1988, shuttle flights were short, with limited life science research. NASA thought that if the shuttle could be modified, it could function as a microgravity laboratory for weeks at a time. The first stage was to make modifications to the life support, air, water, and waste management systems for up to a 16-day stay. There were potential drawbacks to extended stays in microgravity. Astronauts were concerned about the preservation of their capability for unaided egress from the shuttle, including the capability for bailout. Another concern was degradation of landing proficiency after such a long stay, as this had never been done before.

Between 1992 (STS-50) and 1995, this program successfully demonstrated that astronauts could land and egress after such long stays, but that significant muscle degradation occurred. The addition of a new pressurized g-suit provided relief to the light-headedness (feeling like fainting) experienced when returning to Earth. Improvements included the addition of a crew transport vehicle that astronauts entered directly from the landed shuttle in which they reclined during medical examination until they were ready to walk. On-orbit exercise was tested to improve their physical capabilities for emergency egress and landing. The research showed that with more than 2 weeks of microgravity, astronauts probably should not land the shuttle as it was too complicated and risky. In the future, shuttle landing would only be performed by a short-duration astronaut.

The Great Observatories

Months before the Ulysses deployment, the crew of STS-31 (1990) deployed the Hubble Space Telescope, which had been slated for launch in August 1986 but slipped to 1990 after the Challenger accident. Weeks before the launch, astronauts and NASA administrators laid out the importance of the flight. Lennard Fisk, NASA's associate administrator for Space Science and Applications, explained, "This is a mission from which (people) can expect very fundamental discoveries. They could begin to understand creation. Hubble could be a turning point in humankind's perception of itself and its place in the universe."

Unfortunately, within just a few short months NASA discovered problems with the telescope's mirror—problems that generated a great deal of controversy. Several in Congress believed that the telescope was a colossal waste of money. Only 4 years after the accident, NASA's morale plunged again. Fortunately, the flight and ground crews, along with employees at Lockheed Martin, took the time to work out procedures to service the telescope in orbit during the flight hiatus. In 1992, NASA named the crew that would take on this challenge. The astronauts assigned to repair the telescope felt pressure to succeed. "Everybody was looking at the servicing and repair of the Hubble Space Telescope as the mission that could prove NASA's worth," Commander Dick Covey recalled. The mission was one of the most sophisticated ever planned at NASA. The spacewalkers rendezvoused for the first time with the telescope, one of the largest objects the shuttle had rendezvoused with at that point, and conducted a record-breaking five spacewalks. The repairs were successful, and the public faith rebounded. Four additional missions serviced the Hubble, with the final launching in 2009.

Two other major scientific payloads, part of NASA's Great Observatories including the Compton Gamma Ray Observatory and the Chandra X-ray Observatory, launched from the Orbiter's cargo bay. When the Compton Gamma Ray Observatory's high-gain antenna failed to deploy, Astronauts Jerry Ross and Jay Apt took the first spacewalk in 6 years (the last walk occurred in 1985) and freed the antenna. The crew of STS-93, which featured NASA's first female mission commander, Eileen Collins, delivered the Chandra X-ray Observatory to Earth orbit in 1999.

Satellite Retrieval and Repair

Satellite retrieval and repair missions all but disappeared from the shuttle manifest after the Challenger accident. STS-49 (1992) was the one exception. An Intelsat was stranded in an improper orbit for several years, and spacewalkers from STS-49 were to attach a new kick-start motor to it. The plan seemed simple enough. After all, NASA had plenty of practice capturing ailing satellites. After two unsuccessful attempts, flight controllers developed a plan that required a three-person spacewalk, a first in the history of NASA's space operations. This finally allowed the crew to repair and redeploy the satellite, which occurred—coincidentally—during Endeavour's first flight.

New Main Engine

STS-70 flew in the summer of 1995 and launched a Tracking and Data Relay Satellite. The shuttle flew the new main engine, which contained an improved high-pressure liquid oxygen turbopump, a two-duct powerhead, and a single-coil heat exchanger. The new pumps were a breakthrough in shuttle reliability and quality, for they were much safer than those previously used on the Orbiter. The turbopumps required less maintenance than those used prior to 1995. Rather than removing each pump after every flight, engineers would only have to conduct detailed inspections of the pumps after six missions. A single-coil heat exchanger eliminated many of the welds that existed in the previous pump, thereby increasing engine reliability, while the powerhead enhanced the flow of fuel in the engine.

Space Laboratories

NASA continued to fly space laboratory missions until 1998, when Columbia launched the final laboratory and crew into orbit for the STS-90 mission. The shuttle had two versions of the payload bay laboratory: European Spacelab and US company Spacehab, Inc. Fifteen years had passed since the flight of STS-9—the first mission—and the project ended with the launch of Neurolab, which measured the impact of microgravity on the nervous system: blood pressure; eye-hand coordination; motor coordination; sleep patterns; and the inner ear. Scientists learned a great deal from Spacelab Life Sciences-1 and -2 missions, which flew in the summer of 1991 and 1993, respectively, and represented a turning point in spaceflight human physiology research. Previous understandings of how the human body worked in space were either incomplete or incorrect. The program scientist for the flight explained that the crew obtained "a significant number of surprising results" from the flight.

Other notable flights included the ASTRO-1 payload, which featured four telescopes designed to measure ultraviolet light from astronomical objects, life sciences missions, the US Microgravity Labs, and even a second German flight called D-2. The day before the crew of D-2 touched down at DFRC on an Edwards Air Force Base runway, the Space Shuttle Program reached a major milestone, having accrued a full year of flight time by May 5, 1993.

Spacehab, a commercially provided series of modules similar to Spacelab and used for science and logistics, was a significant part of the shuttle manifest in the 1990s. One of those Spacehab flights featured the return of Mercury 7 Astronaut and US Senator John Glenn, Jr. Thirty-six years had passed since he had flown in space and had become the first American to fly in Earth orbit. He broke records again in 1998 when he became the oldest person to fly in space. Given his age, researchers hoped to compare the similarities between aging on Earth with the effects of microgravity on the human body. Interest in this historic flight, which also fell on NASA's 40th anniversary, was immense. Not only was Glenn returning to orbit, but Pedro Duque—a European Space Agency astronaut—became the first Spanish astronaut, following in the footsteps of Spanish explorers Hernán Cortés and Francisco Pizarro.

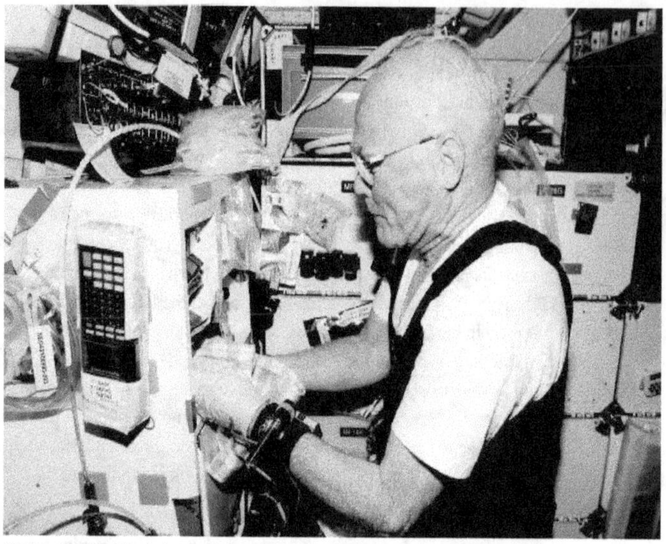

US Senator John Glenn, Jr., payload specialist, keeps up his busy test agenda during Flight Day 7 on board Discovery STS-95 in 1998. This was a Spacehab flight that studied the effect of microgravity on human physiology. He is preparing his food, and on the side is the bar code reader used to record all food, fluids, and drug intakes.

Consolidating Contracts

The Space Shuttle Program seemed to hit its stride in the 1990s. In 1995, NASA decided to consolidate 12 individual contracts under a single prime contractor. United Space Alliance (USA), a hybrid venture between Rockwell International and Lockheed Martin, became NASA's selection to manage the space agency's Space Flight Operations Contract. USA was the obvious choice because those two companies combined held nearly 70% of the dollar value of prime shuttle contracts. Although the idea of handing over all processing and launch operations to a contractor was controversial, NASA Administrator Daniel Goldin, known for his "faster, better, cheaper" mantra, enthusiastically supported the sole source contract as part of President William Clinton's effort to trim the federal budget and increase efficiency within government. NASA awarded USA a $7 billion contract, which went into effect on October 1, 1996. Speaking at JSC about the agreement, Goldin proclaimed, "Today is the first day of a new space program in America. We are opening up the space program to commercial space involving humans. May it survive and get stronger."

STS-80, the first mission controlled by USA, launched in November 1996. The all-veteran crew, on the final flight of the year and the 80th of the program, stayed in space for a record-breaking 17 days. A failure with the hatch prohibited crew members from conducting two scheduled spacewalks, but NASA considered the mission a success because the crew brought home more scientific data than they had expected to gather with the Orbiting and Retrievable Far and Extreme Ultraviolet Spectrometer-Shuttle Pallet Satellite-II.

The Shuttle-Mir Program

As the Cold War (the Soviet-US conflict between the mid 1940s and early 1990s) ended, the George H.W. Bush administration began laying the groundwork for a partnership in space between the United States and the Soviet Union. Following the collapse of the Soviet Union in 1991, President Bush and Russian President Boris Yeltsin signed a space agreement, in June 1992, calling for collaboration between the two countries in space. They planned to place American astronauts on board the Russian space station Mir and to take Russian cosmonauts on board shuttle flights. Noting the historic nature of the agreement, Goldin said, "Our children and their children will look upon yesterday and today as momentous events that brought our peoples together." This agreement brokered a new partnership between the world's spacefaring nations, once adversaries.

Known as the Shuttle-Mir Program, these international flights were the first phase of the ISS Program and marked a turning point in history. The Shuttle-Mir Program—led from JSC, with its director George Abbey—was a watershed and a symbol of the thawing of relations between the United States and Russia.

For more than 4 years, from the winter of 1994 to the summer of 1998, nine shuttle flights flew to the Russian space station, with seven astronauts living on board the Mir for extended periods of time. The first phase began when Cosmonaut Sergei Krikalev flew on board STS-60 (1994).

Twenty years had passed since the Apollo-Soyuz Test Project when, in the summer of 1995, Robert Gibson made history when he docked Atlantis to the much-larger Mir. The STS-71 crew members exchanged gifts and shook hands with the Mir commander in the docking tunnel that linked the shuttle and the Russian station. They dropped off the next Mir crew and picked up two cosmonauts and America's first resident of Mir, Astronaut Norman Thagard. Additional missions ferried crews and necessary supplies to Mir. One of the major milestones of the program was the STS-74 (1995) mission, which delivered and attached a permanent docking port to the Russian space station.

In 1996, Astronaut Shannon Lucid broke all American records for time in orbit and held the flight endurance record for all women, from any nation, when she stayed on board Mir for 188 days. Clinton presented Lucid with the Congressional Space Medal of Honor for her service, representing the first time a woman or scientist had received this accolade. Speaking about the importance of the Shuttle-Mir Program, the president said, "Her mission did much to cement the alliance in space we have formed with Russia. It demonstrated that, as we move into a truly global society, space exploration can serve to deepen our understanding, not only of our planet and our universe, but of those who share the Earth with us."

STS-91 (1998), which ended shuttle visits to Mir, featured the first flight of the super-lightweight External Tank. Made of aluminum lithium, the newly designed tank weighed 3,402 kg (7,500 pounds) less than the previous tank (the lightweight or second-generation tank) used on the previous flight, but its metal was stronger than that flown prior to the summer of 1998. By removing so much launch weight, engineers expanded the shuttle's ability to carry heavier payloads, like the space station modules, into Earth's orbit. Launching with less weight also enabled the crew to fly to a high inclination orbit of 51.6 degrees, where NASA and its partners would build the ISS. STS-91 also carried a prototype of the Alpha Magnetic Spectrometer into space. This instrument was designed to look for dark and missing matter in the universe. The preliminary test flight was in preparation for its launch to the ISS on STS-134. The Alpha Magnetic Spectrometer has a state-of-the-art particle physics detector, and includes the participation of 56 institutions and 16 countries led by Nobel Laureate Samuel Ting. By the end of the Shuttle-Mir Program, the number of US astronauts who visited the Russian space station exceeded the number of Russian cosmonauts who had worked aboard Mir.

The International Space Station

With the first phase completed, NASA began constructing the ISS with the assistance of shuttle crews, who played an integral role in building the outpost. In 1998, 13 years after spacewalker Jerry Ross demonstrated the feasibility of assembling structures in space (STS-61B [1985]), ISS construction began. During three spacewalks, Ross and James Newman connected electrical power and cables between the Russian Zarya module and America's Unity Module, also called Node 1. They installed additional hardware—handrails and antennas—on the station. NASA's dream of building a space station had finally come to fruition.

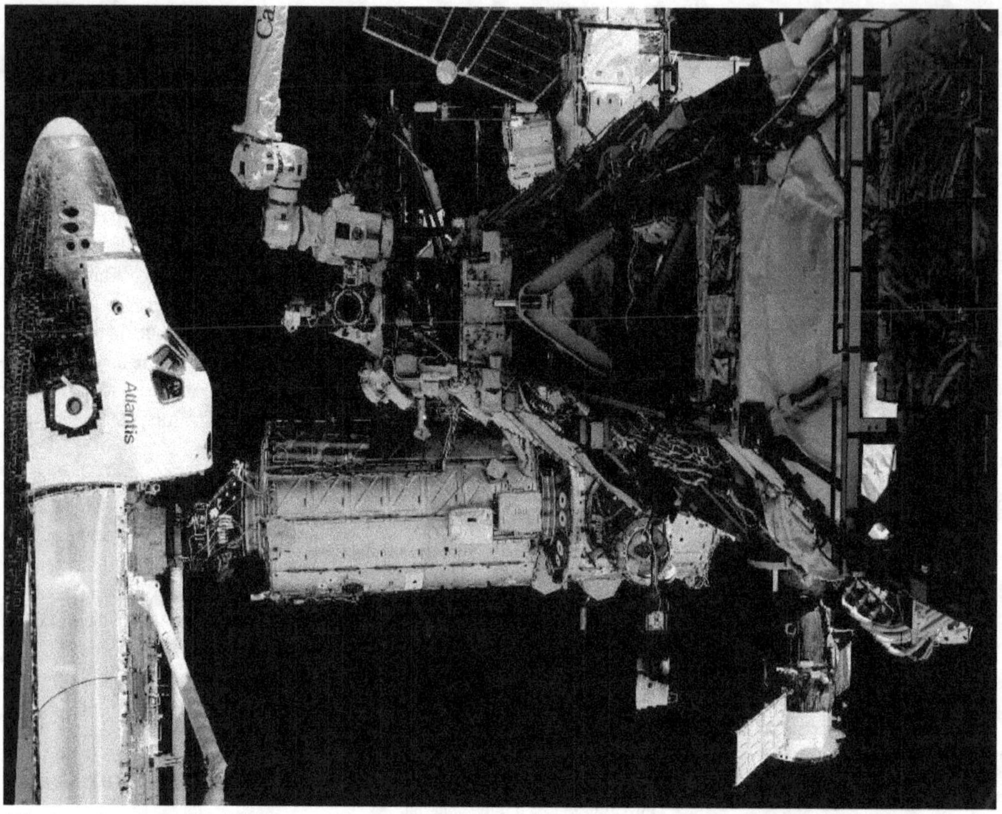

Although no astronauts are visible in this picture, action was brisk outside the Space Shuttle (STS-116)/space station tandem in 2006.

The shuttle's 100th mission (STS-92) launched from KSC in October 2000, marking a major milestone for the Space Shuttle and the International Space Station Programs. The construction crew delivered and installed the initial truss—the first permanent latticework structure—which set the stage for the future addition of trusses. The crew also delivered a docking port and other hardware to the station. Four spacewalkers spent more than 27 hours outside the shuttle as they reconfigured these new elements onto the station. The seven-member crew also prepared the station for the first resident astronauts, who docked with the station 14 days after the crew left the orbital workshop. Of the historic mission, Lead Flight Director Chuck Shaw said, "STS-92/ISS Mission 3A opens the next chapter in the construction of the International Space Station," when human beings from around the world would permanently occupy the space base.

Crews began living and working in the station in the fall of 2000, when the first resident crew (Expedition 1) of Sergei Krikalev, William Shepherd, and Yuri Gidzenko resided in the space station for 4 months. For the next 3 years, the shuttle and her crews were the station's workhorse. They transferred crews; delivered supplies; installed modules, trusses, the Space Station Robotic Arm, an airlock, and a mobile transporter, among other things. By the end of 2002, NASA had flown 16 assembly flights. Flying the shuttle seemed fairly routine until February 2003, when Columbia disintegrated over East Texas, resulting in the loss of the shuttle and her seven-member crew.

Columbia Accident

The cause of the Columbia accident was twofold. The physical cause resulted from the loss of insulating foam from the External Tank, which hit the Orbiter's left wing during launch and created a hole. When Columbia entered the Earth's atmosphere, the left wing leading edge thermal protection (reinforced carbon-carbon panels) was unable to prevent heating due to the breach. This led to the loss of control and disintegration of the shuttle, killing the crew. NASA's flawed culture of complacency also bore responsibility for the loss of the vehicle and its astronauts. All flights were put on hold for more than 2 years as NASA implemented numerous safety improvements, like redesigning the External Tank with an improved bipod fitting that minimized potential foam debris from the tank. Other improvements were the Solid Rocket Booster Bolt Catcher, impact sensors added to the wing's leading edge, and a boom for the shuttle's arm that allowed the crew to inspect the vehicle for any possible damage, among other things.

As NASA worked on these issues, President George W. Bush announced his new Vision for Space Exploration, which included the end of the Space Shuttle Program. As soon as possible, the shuttles would return to flight to complete the ISS by 2010 and then NASA would retire the fleet.

Post-Columbia Accident Return to Flight

In 2005, STS-114 returned NASA to flying in space. Astronaut Eileen Collins commanded the first of two Return to Flight missions, which were considered test flights. The first mission tested and evaluated new flight safety procedures as well as inspection and repair techniques for the vehicle. One of the changes was the addition of an approximately 15-m (50-ft) boom to the end of the robotic arm. This increased astronauts' capabilities to inspect the tile located

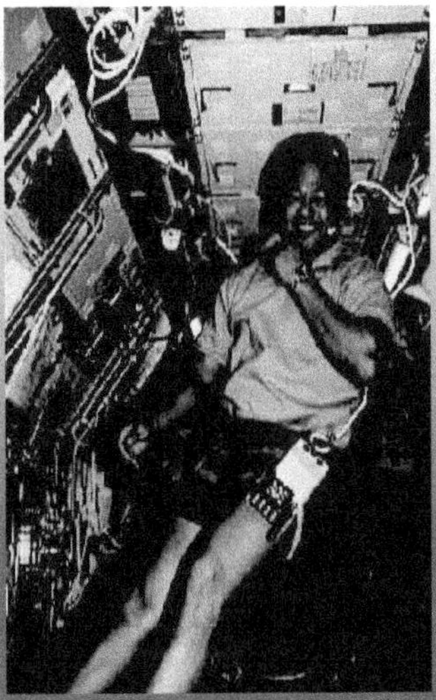

Leroy Chiao, PhD

Astronaut on STS-65 (1994), STS-72 (1996), and STS-92 (2000). Commander and science officer on ISS Expedition 10 (2004-2005).

"To me, the Space Shuttle is an amazing flying machine. It launches vertically as a rocket, turns into an extremely capable orbital platform for many purposes, and then becomes an airplane after re-entry into the atmosphere for landing on a conventional runway. Moreover, it is a reusable vehicle, which was a first in the US space program.

"The Space Shuttle Program presented me the opportunity to become a NASA astronaut and to fly in space. I never forgot my boyhood dream and years later applied after watching the first launch of Columbia. In addition to being a superb research and operations platform, the Space Shuttle also served as a bridge to other nations. Never before had foreign nationals flown aboard US spacecraft. On shuttle, the US had flown representatives from nations all around the world. Space is an ideal neutral ground for cooperation and the development of better understanding and relationships between nations.

"Without the Space Shuttle as an extravehicular activity test bed, we would not have been nearly as successful as we have been so far in assembling the ISS. The Space Shuttle again proved its flexibility and capability for ISS construction missions.

"Upon our landing (STS-92), I realized that my shuttle days were behind me. I was about to begin training for ISS. But on that afternoon, as we walked around and under Discovery, I savored the moment and felt a mixture of awe, satisfaction, and a little sadness. Shuttle, to me, represents a triumph and remains to this day a technological marvel. We learned so much from the program, not only in the advancement of science and international relations, but also from what works and what doesn't on a reusable vehicle. The lessons learned from shuttle will make future US spacecraft more reliable, safer, and cost effective.

"I love the Space Shuttle. I am proud and honored to be a part of its history and legacy."

on the underbelly of the shuttle. When NASA discovered two gap fillers sticking out of the tiles on the shuttle's belly on the first mission, flight controllers and the astronauts came up with a plan to remove the gap fillers—an unprecedented and unplanned spacewalk that they believed would decrease excessive temperatures on re-entry. The plan required Astronaut Stephen Robinson to ride the arm underneath the shuttle and pull out the fillers. In 24 years of shuttle operations, this had never been attempted, but the fillers were easily removed. STS-114 showed that improvements in the External Tank insulation foam were insufficient to prevent dangerous losses during ascent. Another year passed before STS-121 (2006), the second Return to Flight mission, flew after more improvements were made to the foam applications.

Final Flights

Educator Astronaut

Excitement began to build at NASA and across the nation as the date for Barbara Morgan's flight, STS-118 (2007), grew closer. Morgan had been selected as the backup for Christa McAuliffe, NASA's first Teacher in Space in 1985. After the Challenger accident, Morgan became the Teacher in Space Designee and returned to teaching in Idaho. She came back to Houston in 1998 when she was selected as an astronaut candidate. More than 20 years after being selected as the backup Teacher in Space, Morgan fulfilled that dream by serving as the first educator mission specialist. NASA Administrator Michael Griffin praised Morgan "for her interest, her toughness, her resiliency, her persistence in wanting to fly in space and eventually doing so." Adults recalled the Challenger accident and watched this flight with interest. STS-118 drew attention from students, from across America and around the globe, who were curious about the flight.

Return to Hubble

In May 2009, the crew of STS-125 made the final repairs and upgrades to the Hubble Space Telescope to ensure quality science for several more years. This flight was a long time coming due to the Columbia accident, after which NASA was unsure whether it could continue to fly to destinations with no safe haven such as the ISS.

With the ISS, if problems arose, especially with the thermal protection, the astronauts could stay in the space station until either another shuttle or the Russian Soyuz could bring them home. The Hubble orbited beyond the ability for the shuttle to get to the ISS if the shuttle was critically damaged. Thus, for several years, the agency had vetoed any possibility that NASA could return to the telescope.

At that point, the Hubble had been functioning for 12 years in the very hostile environment of space. Not only did its instruments eventually wear out, but the telescope needed important upgrades to expand its capabilities. After the Return to Flight of STS-114 and STS-121, NASA reevaluated the ability to safety return astronauts after launch. The method to ensure safe return in the event of shuttle damage was to have a backup vehicle in place. So in 2009, Atlantis launched to repair the telescope, with Endeavour as the backup.

Improvements on the International Space Station Continued

Discovery flight STS-128, in 2009, provided capability for six crew members for ISS. This was a major milestone for ISS as the station had been operating with two to three crew members since its first occupation in 1999. The shuttle launched most of the ISS, including Canadian, European, and Japanese elements, to the orbiting laboratory. In 2010, Endeavour provided the final large components: European Space Agency Node 3 with additional hygiene compartment; and Cupola with a robotic work station to assist in assembly/maintenance of the ISS and a window for Earth observations.

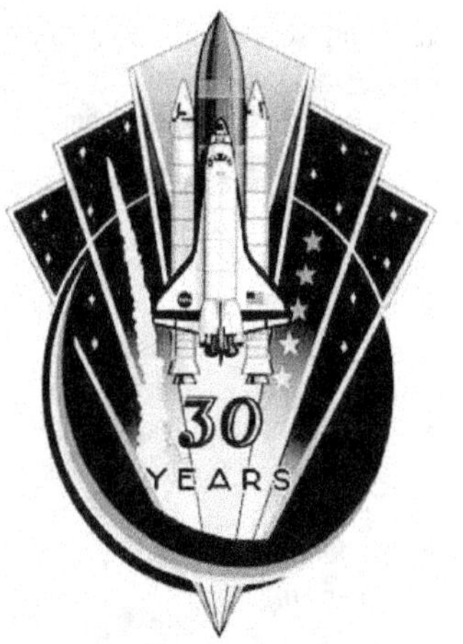

This Commemorative Patch celebrates the 30-year life and work of the Space Shuttle Program. Selected from over 100 designs, this winning patch by Mr. Blake Dumesnil features the historic icon set within a jewel-shape frame. It celebrates the shuttle's exploration within low-Earth orbit, and our desire to explore beyond. Especially poignant are the seven stars on each side of the shuttle, representing the 14 lives lost—seven on Columbia, seven on Challenger—in pursuit of their dream, and this nation's dream of further exploration and discovery. The five larger stars represent the shuttles that made up the fleet—each shuttle a star in its own right.

Changes in Mission Complexity Over Nearly 3 Decades

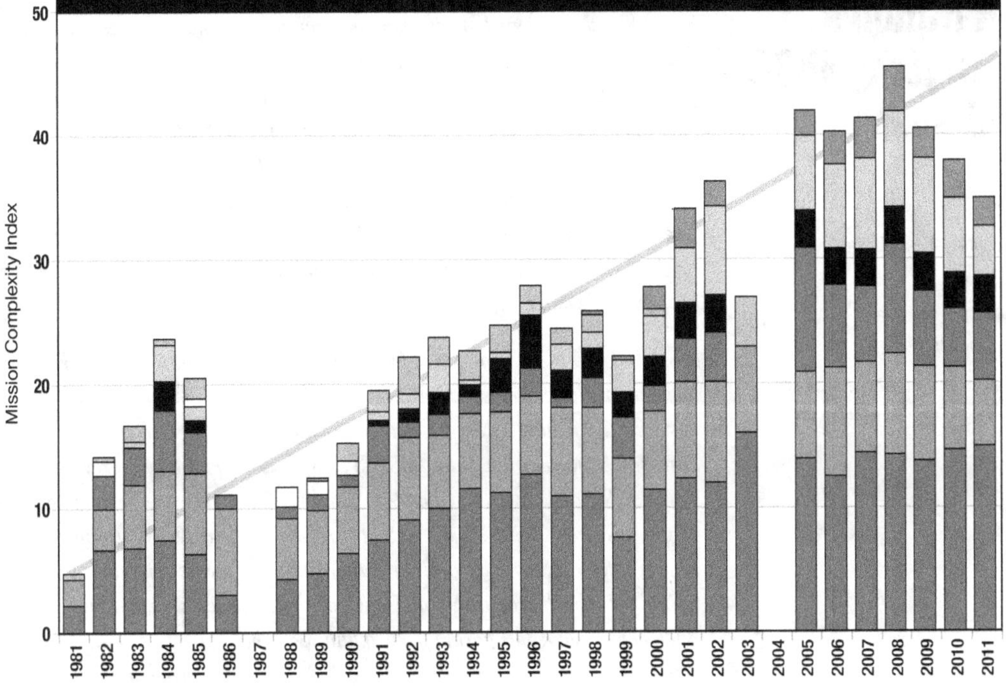

Components of Mission Complexity

- *Length of flight as mission days. Early flights lasted less than 1 week, but, as confidence grew, some flights lasted 14 to 15 days.*
- *Crew size started at two—a commander and pilot—and grew to routine flights with six crew members. During the Shuttle-Mir and International Space Station (ISS) Programs, the shuttle took crew members to the station and returned crew members, for a total of seven crew members.*
- *Deploys occurred throughout the program. During the first 10 years, these were primarily satellites with sometimes more than one per flight. Some satellites, such as Hubble Space Telescope, were returned to the payload bay for repair. With construction of the ISS, several major elements were deployed.*
- *Rendezvous included every time the shuttle connected to an orbiting craft from satellites, to Hubble, Mir, and ISS. Some flights had several rendezvous.*
- *Extravehicular activity (EVA) is determined as EVA crew days. Many flights had no EVAs, while others had one every day with two crew members.*
- *Secret Department of Defense missions were very complex.*
- *Spacelabs were missions with a scientific lab in the payload bay. Besides the complexity of launch and landing, these flights included many scientific studies.*
- *Construction of the ISS by shuttle crew members.*

Over the 30 years of the Space Shuttle Program, missions became more complex with increased understanding of the use of this vehicle, thereby producing increased capabilities. This diagram illustrates the increasing complexity as well as the downtime between the major accidents—Challenger and Columbia.

The Accidents: A Nation's Tragedy, NASA's Challenge

Randy Stone
Jennifer Ross-Nazzal

The Crew
Michael Chandler
Philip Stepaniak

Witness Accounts—Key to Understanding Columbia Breakup
Paul Hill

Who heard the whispers that were coming from the shuttle's Solid Rocket Boosters (SRBs) on a cold January morning in 1986? Who thought the mighty Space Shuttle, designed to withstand the thermal extremes of space, would be negatively affected by launching at near-freezing temperatures? Very few understood the danger, and most of the smart people working in the program missed the obvious signs. Through 1985 and January 1986, the dedicated and talented people at the NASA Human Spaceflight Centers focused on readying the Challenger and her crew to fly a complex mission. Seventy-three seconds after SRB ignition, hot gases leaking from a joint on one of the SRBs impinged on the External Tank (ET), causing a structural failure that resulted in the loss of the vehicle and crew.

Most Americans are unaware of the profound and devastating impact the accident had on the close-knit NASA team. The loss of Challenger and her crew devastated NASA, particularly at Johnson Space Center (JSC) and Marshall Space Flight Center (MSFC) as well as the processing crews at Kennedy Space Center (KSC) and the landing and recovery crew at Dryden Flight Research Center. Three NASA teams were primarily responsible for shuttle safety—JSC for on-orbit operation and crew member issues; MSFC for launch propulsion; and KSC for shuttle processing and launch. Each center played its part in the two failures. What happened to the "Failure is not an option" creed, they asked. The engineering and operations teams had spent months preparing for this mission. They identified many failure scenarios and trained relentlessly to overcome them. The ascent flight control team was experienced with outstanding leadership and had practiced for every contingency. But on that cold morning in January, all they could do was watch in disbelief as the vehicle and crew were lost high above the Atlantic Ocean. Nothing could have saved the Challenger and her crew once the chain of events started to unfold. On that day, everything fell to pieces.

Seventeen years later, in 2003, NASA lost a second shuttle and crew—Space Transportation System (STS)-107. The events that led up to the loss of Columbia were eerily similar to those surrounding Challenger. As with Challenger, the vehicle talked to the program but no one understood. Loss of foam from the ET had been a persistent problem in varying degrees for the entire program. When it occurred on STS-107, many doubted that a lightweight piece of foam could damage the resilient shuttle. It made no sense, but that is what happened. Dedicated people missed the obvious. In the end, foam damaged the wing to such an extent that the crew and vehicle could not safely reenter the Earth's atmosphere. Just as with Challenger, there was no opportunity to heroically "save the day" as the data from the vehicle disappeared and it became clear that friends and colleagues were lost. Disbelief was the first reaction, and then a pall of grief and devastation descended on the NASA family of operators, engineers, and managers.

The Challenger Accident

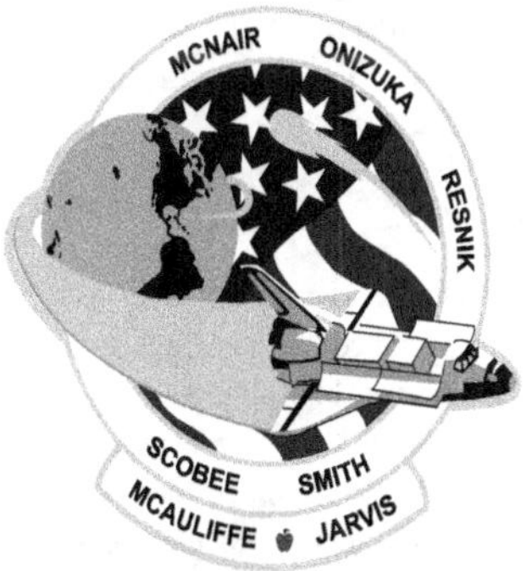

Pressure to Fly

As the final flight of Challenger approached, the Space Shuttle Program and the operations community at JSC, MSFC, and KSC faced many pressures that made each sensitive to maintaining a very ambitious launch schedule. By 1986, the schedule and changes in the manifest due to commercial and Department of Defense launch requirements began to stress NASA's ability to plan, design, and execute shuttle missions. NASA had won support for the program in the 1970s by emphasizing the cost-effectiveness and economic value of the system. By December 1983, 2 years after the maiden flight of Columbia, NASA had flown only nine missions. To make spaceflight more routine and therefore more economical, the agency had to accelerate the number of missions it flew each year. To reach this goal, NASA announced an ambitious rate of 24 flights by 1990.

NASA flew five missions in 1984 and a record nine missions the following year. By 1985, strains in the system were evident. Planning, training, launching, and flying nine flights stressed the agency's resources and workforce, as did the constant change in the flight manifest. Crews scheduled to fly in 1986 would have seen a dramatic decrease in their number of training hours or the agency would have had to slow down its pace because NASA simply lacked the staff and facilities to safely fly an accelerated number of missions.

By the end of 1985, pressure mounted on the space agency as they prepared to launch more than one flight a month the next year. A record four launch scrubs and two launch delays of STS-61C, which finally launched in January 1986, exacerbated tensions. To ensure that no more delays would threaten the 1986 flight rate or schedule, NASA cut the flight 1 day short to make sure Columbia could be processed in time for the scheduled ASTRO-1 science mission in March. Weather conditions prohibited landing that day and the next, causing a slip in the processing schedule. NASA had to avoid any additional delays to meet its goal of 15 flights that year.

The agency needed to hold to the schedule to complete at least three flights that could not be delayed. Two flights had to be launched in May 1986: the Ulysses and the Galileo flights, which were to launch within 6 days of each other. If the back-to-back flights missed their launch window, the payloads could not be launched until July 1987. The delay of STS-61C and Challenger's final liftoff in January threatened the scheduled launch plans of these two flights in particular. The Challenger needed to launch and deploy a second Tracking and Data Relay Satellite, which provided continuous global coverage of Earth-orbiting satellites at various altitudes. The shuttle would then return promptly to be reconfigured to hold the liquid-fueled Centaur rocket in its payload bay. The ASTRO-1 flight had to be launched in March or April to observe Halley's Comet from the shuttle.

On January 28, 1986, NASA launched Challenger, but the mission was never realized. Hot gases from the right-hand Solid Rocket Booster motor had penetrated the thermal barrier and blown by the O-ring seals on the booster field joint. The joints were designed to join the motor segments together and contain the immense heat and pressure of the motor combustion. As the Challenger ascended, the leak became an intense jet of flame that penetrated the ET, resulting in structural failure of the vehicle and loss of the crew.

Prior to this tragic flight, there had been many O-ring problems witnessed as early as November 1981 on the second flight of Columbia. The hot gases had significantly eroded the STS-2 booster right field joint—deeper than on any other mission until the accident—but knowledge was not widespread in mission management. STS-6 (1983) boosters did not have erosion of the O-rings, but heat had impacted them. In addition, holes were blown through the putty in both nozzle joints. NASA reclassified the new field joints Criticality 1, noting that the failure of a joint could result in "loss of life or vehicle if the component fails." Even with this new categorization, the topic of O-ring erosion was not discussed in any Flight Readiness Reviews until March 1984, in preparation for the 11th flight of the program. Time and again these anomalies popped up in other missions flown in 1984 and 1985, with the issue eventually classified as an "acceptable risk" but not desirable. The SRB project manager regularly waived these anomalies, citing them as "repeats of conditions that had already been accepted for flight" or "within their experience base," explained Arnold Aldrich, program manager for the Space Shuttle Program.

Senior leadership like Judson Lovingood believed that engineers "had thoroughly worked that joint problem." As explained by former Chief Engineer Keith Coates, "We knew the gap was opening. We knew the O-rings were getting burned. But there'd been some engineering rationale that said, 'It won't be a failure of the joint.' And I thought justifiably so at the time I was there. And I think that if it hadn't been for the cold weather, which was a whole new environment, then it probably would have continued. We didn't like it, but it wouldn't fail."

Each time the shuttle launched successfully, the accomplishment masked the recurring field joint problems. Engineers and managers were fooled into complacency because they were told it was not a flight safety issue. They concluded that it was safe to fly again because the previous missions had flown successfully. In short, they reached the same conclusion each time—it was safe to fly another mission. "The argument that the same risk was flown before without failure is often accepted as an argument for the safety of accepting it again. Because of this, obvious weaknesses are accepted again and again, sometimes without a sufficiently serious attempt to remedy them or to delay a flight because of their continued presence," wrote Richard Feynman, Nobel Prize winner and member of the presidential-appointed Rogers Commission charged to investigate the Challenger accident.

Operational Syndrome

The Space Shuttle Program was also "caught up in a syndrome that the shuttle was operational," according to J.R. Thompson, former project manager for the Space Shuttle Main Engines. The Orbital Flight Test Program, which ended in 1982, marked the beginning of routine operations of the shuttle, even though there were still problems with the booster joint. Nonetheless, MSFC and Morton Thiokol, the company responsible for the SRBs, seemed confident with the design.

Although the design of the boosters had proven to be a major complication for MSFC and Morton Thiokol, the engineering debate occurring behind closed doors was not visible to the entire Space Shuttle Program preparing for the launch of STS-51L. There had been serious erosions of the booster joint seals on STS-51B (1985) and STS-51C (1985), but MSFC had not pointed out any problems with the boosters right before the Challenger launch. Furthermore, MSFC failed to bring the design issue, failures, or concern with launching in cold temperatures to the attention of senior management. Instead, discussions of the booster engines were resolved at the local level, even on the eve of the Challenger launch. "I was totally unaware that these meetings and discussions had even occurred until they were brought to light several weeks following the Challenger accident in a Rogers Commission hearing at KSC," Arnold Aldrich recalled. He also recalled that he had sat shoulder to shoulder with senior management "in the firing room for approximately 5 hours leading up to the launch of Challenger and no aspect of these deliberations was ever discussed or mentioned."

Even the flight control team "didn't know about what was lurking on the booster side," according to Ascent Flight Director Jay Greene. Astronaut Richard Covey, then working as capsule communicator, explained that the team "just flat didn't have that insight" into the booster trouble. Launch proceeded and, in fewer than 2 minutes, the joint failed, resulting in the loss of seven lives and the Challenger.

Looking back over the decision, it is difficult to understand why NASA launched the Challenger that morning. The history of troublesome technical issues with the O-rings and joint are easily documented. In hindsight, the trends appear obvious, but the data had not been compiled. Wiley Bunn noted, "It was a matter of assembling that data and looking at it [in] the proper fashion. Had we done that, the data just jumps off the page at you."

Devastated

The accident devastated NASA employees and contractors. To this day Aldrich asks himself regularly, "What could we have done to prevent what happened?" Holding a mission management team meeting the morning of launch might have brought up the Thiokol/MSFC teleconference the previous evening. "I wish I had made such a meeting happen," he lamented. The flight control team felt some responsibility for the accident, remembered STS-51L Lead Flight Director Randy Stone. Controllers "truly believed they could handle absolutely any problem that this vehicle could throw at us." The accident, however, "completely shattered the belief that the flight control team can always save the day. We have never fully recovered from that." Alabama and Florida employees similarly felt guilty about the loss of the crew and shuttle, viewing it as a personal failure. John Conway of KSC pointed out that "a lot of the fun went out of the business with that accident."

Rebounded

Over time, the wounds began to heal and morale improved as employees reevaluated the engineering design and process decisions of the program. The KSC personnel dedicated themselves to the recovery of Challenger and returning as much of the vehicle back to the launch site as possible. NASA spent the next 2½ years fixing the hardware and improving processes, and made over 200 changes to the shuttle during this downtime. Working on design changes to improve the vehicle contributed to the healing process for people at the centers.

The Crew

Following the breakup of Challenger (STS-51L) during launch over the Atlantic Ocean on January 28, 1986, personnel in the Department of Defense STS Contingency Support Office activated the rescue and recovery assets. This included the local military search and rescue helicopters from the Eastern Space and Missile Center at Patrick Air Force Base and the US Coast Guard. The crew compartment was eventually located on March 8, and NASA officially announced that the recovery operations were completed on April 21. The recovered remains of the crew were taken to Cape Canaveral Air Force Station and then transported, with military honors, to the Armed Forces Institute of Pathology where they were identified. Burial arrangements were coordinated with the families by the Port Mortuary at Dover Air Force Base, Delaware. Internal NASA reports on the mechanism of injuries sustained by the crew contributed to upgrades in training and crew equipment that supported scenarios of bailout, egress, and escape for Return to Flight.

Following the breakup of Columbia (STS-107) during re-entry over Texas and Louisiana on February 1, 2003, personnel from the NASA Mishap Investigation Team were dispatched to various disaster field offices for crew recovery efforts. The Lufkin, Texas, office served as the primary area for all operations, including staging assets and deploying field teams for search, recovery, and security. Many organizations had operational experience with disaster recovery, including branches of the federal, state, and local governments together with many local citizen volunteers. Remains of all seven crew members were found within a 40- by 3-km (25- by 2-mile) corridor in East Texas. The formal search for crew members was terminated on February 13, 2003. Astronauts, military, and local police personnel transported the crew, with honors, to Barksdale Air Force Base, Louisiana, for preliminary identification and preparation for transport. The crew was then relocated, with military honor guard and protocol, to the Armed Forces Institute of Pathology medical examiner for forensic analysis. Burial preparation and arrangements were coordinated with the families by the Port Mortuary at Dover Air Force Base, Delaware. Additional details on the mechanism of injuries sustained by the crew and lessons learned for enhanced crew survival are found in the Columbia Crew Survival Investigation Report NASA/SP-2008-565.

Reconstruction of the Columbia from parts found in East Texas. From this layout, NASA was able to determine that a large hole occurred in the leading edge of the wing and identify the burn patterns that eventually led to the destruction of the shuttle.

Making the boosters and main engines more robust became extremely important for engineers at MSFC and Thiokol. The engineers and astronauts at JSC threw themselves into developing an escape system and protective launch and re-entry suits and improving the flight preparation process. All of the improvements then had to be incorporated into the KSC vehicle processing efforts.

All NASA centers concentrated on how they could make the system better and safer. For civil servants and contractors, the recovery from the accident was not just business. It was personal. Working toward Return to Flight was almost a religious experience that restored the shattered confidence of the workforce.

NASA instituted a robust flight preparation process for the Return to Flight mission, which focused on safety and included a series of revised procedures and processes at the centers. At KSC, for instance, new policies were instituted for 24-hour operations to avoid the fatigue and excessive overtime noted by the Rogers Commission. NASA implemented the NASA Safety Reporting System. Safety, reliability, maintainability, and quality assurance staff increased considerably.

JSC's Mission Operations Director Eugene Kranz noted that Mission Operations examined "every job we do" during the stand down. They microscopically analyzed their processes and scrutinized those decisions. They learned that the flight readiness process prior to the Challenger accident frequently lacked detailed documentation and was often driven more by personality than by requirements. The process was never identical or exact but unique. Changes were made to institute a more rigorous program, which was well-documented and could be instituted for every flight.

Astronaut Robert Crippen became the deputy director of the National Space Transportation System Operations. He helped to determine and establish new processes for running and operating the flight readiness review and mission management team (headed by Crippen), as well as the launch commit criteria procedures, including temperature standards. He instituted changes to ensure the agency maintained clear lines of responsibility and authority for the new launch decision process he oversaw.

Retired Astronaut Richard Truly also participated in the decision-making processes for the Return to Flight effort. Truly, then working as associate administrator for spaceflight, invited the STS-26 (1988) commander Frederick Hauck to attend any management meetings in relation to the preparation for flight. By attending those meetings, Hauck had "confidence in the fixes that had been made" and "confidence in the team of people that had made those decisions," he remarked.

Return to Flight After Challenger Accident

As the launch date for the flight approached, excitement began to build at the centers. Crowds surrounded the shuttle when it emerged from the Vehicle Assembly Building on July 4, 1988. The Star-Spangled Banner played as the vehicle crawled to the pad, while crew members and other workers from KSC and Headquarters spoke about the milestone. David Hilmers, a member of the crew, tied the milestone to the patriotism of the day. "What more fitting present could we make to our country on the day of its birth than this? America, the dream is still alive," he exclaimed. The Return to Flight effort was a symbol of America's pride and served as a healing moment not only for the agency but also for the country. Tip Talone of KSC likened the event to a "rebirth."

Indeed, President Ronald Reagan, who visited JSC in September 1988, told workers, "When we launch Discovery, even more than the thrust of great engines, it will be the courage of our heroes and the hopes and dreams of every American that will lift the shuttle into the heavens."

Without any delays, the launch of STS-26 went off just a few days after the president's speech, returning Americans to space. The pride in America's accomplishment could be seen across the country. In Florida, the Launch Control Center raised a large American flag at launch time and lowered it when the mission concluded. In California, at Dryden Flight Research Center, the astronauts exited the vehicle carrying an American flag—a patriotic symbol of their flight. Cheering crowds waving American flags greeted the astronauts at the crew return event at Ellington Field in Houston, Texas. The launch restored confidence in the program and the vehicle. Pride and excitement could be found across the centers and at contract facilities around the country.

The Columbia Accident

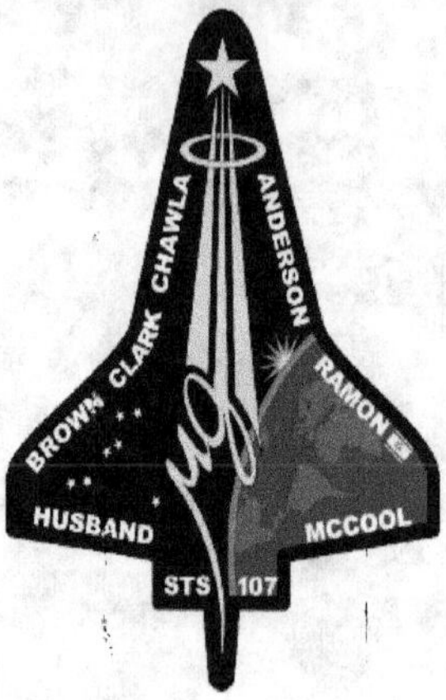

NASA flew 87 successful missions following the Return to Flight effort. As the 1990s unfolded, the post-Challenger political and economic environment changed dramatically.

Environment Changes

As the Soviet Union disintegrated and the Soviet-US conflict that began in the mid 1940s came to an end, NASA (established in 1958) struggled to find its place in a post-Cold War world. Around the same time, the federal deficit swelled to a height that raised concern among economists and citizens. To cut the deficit, Congress and the White House decreased domestic spending, and NASA was not spared from these cuts. Rather than eliminate programs within the agency, NASA chose to become more cost-effective. A leaner, more efficient agency emerged with the appointment of NASA Administrator Daniel Goldin in 1992, whose slogan was "faster, better, cheaper."

The shuttle, the most expensive line item in NASA's budget, underwent significant budget reductions throughout the 1990s. Between 1993 and 2003, the program suffered from a 40% decrease in its purchasing capability (with inflation included in the figures), and its workforce correspondingly decreased. To secure additional cost savings, NASA awarded the Space Flight Operations Contract to United Space Alliance in 1995 to consolidate numerous shuttle contracts into one.

Pressure Leading up to the Accident

As these changes took effect, NASA began working on Phase One of the Space Station Program, called Shuttle-Mir. Phase Two, assembly of the ISS, began in 1998. The shuttle was critical to the building of the outpost and was the only vehicle that could launch the modules built by Europe, Japan, and the United States. By tying the two programs so closely together, a reliable, regular launch schedule was necessary to maintain crew rotations, so the ISS management began to dictate NASA's launch schedule. The program had to meet deadlines outlined in bilateral agreements signed in 1998. Even though the shuttle was not an operational vehicle, the agency worked its schedules as if the space truck could be launched on demand, and there was increasing pressure to meet a February 2004 launch date for Node 2. When launch dates slipped, these delays affected flight schedules.

On top of budget constraints, personnel reductions, and schedule pressure, the program suffered from a lack of vision on replacing the shuttle. There was uncertainty about the program's lifetime. Would the shuttle fly until 2030 or be replaced with new technology? Ronald Dittemore, manager of the Space Shuttle Program from 1999 to 2003, explained, "We had no direction." NASA would "start and stop" funding initiatives, like the shuttle upgrades, and then reverse directions. "Our reputation was kind of sullied there, because we never finished what we started out to do."

This was the environment in which NASA found itself in 2003. On the morning of January 16, Columbia launched from KSC for a lengthy research flight. On February 1, just minutes from a successful landing in Florida, the Orbiter broke up over East Texas and Louisiana. Debris littered its final path. The crew and Columbia were lost.

Recovering Columbia and Her Crew

Recovery of the Orbiter and its crew began at 9:16 a.m., when the ship failed to arrive in Florida. The rapid response and mishap investigation teams from within the agency headed to Barksdale Air Force Base in Shreveport, Louisiana. Hundreds of NASA employees and contractors reported to their centers to determine how they could help bring the crew and Columbia home. Local emergency service personnel were the first responders at the various scenes. By that evening, representatives from local, state, and federal agencies were in place and ready to assist NASA.

The recovery effort was unique, quite unlike emergency responses following other national disasters. David Whittle, head of the mishap investigation team, recalled that there were "130 state, federal, and local agencies" represented in the effort; but as he explained, we "never, ever had a tiff. Matter of fact, the Congressional Committee on Homeland Security sent some people down to interview us to figure out how we did that, because that was not the experience of 9/11." The priority of the effort was the recovery of the vehicle and the astronauts, and all of these agencies came together to see to it that NASA achieved this goal.

While in East Texas and Louisiana, the space agency came to understand how important the Space Shuttle Program was to the area and America. Volunteers traveled from all over the United States to help in the search. People living in the area opened their arms to the thousands of NASA employees who were grieving. They offered their condolences, while some local restaurants provided free food to workers. Ed Mango, KSC launch manager and director of the recovery for approximately 3 months, learned "that people love the space program and want to support it in any way they can." His replacement, Jeff Angermeier, added, "When you work in the program all the time, you care deeply about it, but it isn't glamorous to you. Out away from the space centers, NASA is a big deal."

As volunteers collected debris, it was shipped to KSC where the vehicle was reconstructed. For the center's employees, the fact that Columbia would not be coming back whole was hard to swallow. "I never thought I'd see Columbia going home in a box," said Michael Leinbach of KSC. Many others felt the same way. Working with the debris and reconstructing the ship did help, however, to heal the wounds.

As with the loss of Challenger, NASA employees continue to be haunted by questions of "what if." "I'll bet you a day hardly goes by that we don't think about the crew of Columbia and if there was something we might have been able to do to prevent" the accident, admitted Dittemore. Wayne Hale, shuttle program manager for launch integration at KSC, called the decisions made by the mission management team his "biggest" regret. "We had the opportunity to really save the day, we really did, and we just didn't do it, just were blind to it."

Causes

Foam had detached from the ET since the beginning of the program, even though design requirements specifically prohibited shedding from the tank. Columbia sustained major damage on its maiden flight, eventually requiring the replacement of 300 tiles. As early as 1983, six other missions witnessed the left tank bipod ramp foam loss that eventually led to the loss of the STS-107 crew and vehicle. For more than 20 years, NASA had witnessed foam shedding and debris hits. Just one flight after STS-26 (the Return to Flight after Challenger), Atlantis was severely damaged by debris that resulted in the loss of one tile.

Two flights prior to the loss of Columbia and her crew, STS-112 (2002) experienced bipod ramp loss, which hit both the booster and tank attachment ring. The result was a 10.2-cm- (4-in.)-wide, 7.6-cm- (3 in.)-deep tear in the insulation. The program assigned the ET Project with the task of determining the cause and a solution. But the project failed to understand the severity of foam loss and its impact on the Orbiter, so the due date for the assignment slipped to after the return of STS-107.

Foam loss became an expected anomaly and was not viewed as risky. Instead, the issue became one the program had regularly experienced, and one that engineers believed they understood. It was never seen as a safety issue. The fact that previous missions, which had experienced severe debris hits, had successfully landed only served to reinforce confidence within the program concerning the robustness of the vehicle.

After several months of investigation and speculation about the cause of the accident, investigators determined that a breach in the tile on the left wing led to the loss of the vehicle. Insulation foam from the ET's left bipod ramp, which damaged the wing's reinforced carbon-carbon panel, created the gap. During re-entry, superheated air entered the breach. Temperatures were so extreme that the aluminum in the left wing began to melt, which eventually destroyed it and led to a loss of vehicle control. Columbia experienced aerodynamic stress that the damaged airframe could not withstand, and the vehicle eventually broke up over East Texas and Louisiana.

Senior program management had been alerted to the STS-107 debris strike on the second day of the flight but had failed to understand the risks to the crew or the vehicle. No one thought that foam could create a hole in the leading edge of the wing. Strikes had been within their experience base. In short, management made assumptions based on previous successes, which blinded them to serious problems. "Even in flight when we saw (the foam) hit the wing, it was a failure of imagination that it could cause the damage that it undoubtedly caused," said John Shannon, who later became manager of the Space Shuttle Program. Testing later proved that foam could create cracks in the reinforced carbon-carbon and holes of 40.6 by 43.2 cm (16 by 17 in.).

Aside from the physical cause of the accident, flaws within the decision-making process also significantly impacted the outcome of the STS-107 flight. A lack of effective and clear communication stemmed from organizational barriers and hierarchies within the program. These obstacles made it difficult for engineers with real concerns about vehicle damage to share their views with management. Investigators found that management accepted opinions that mirrored their own and rejected dissent.

Changes

The second Return to Flight effort focused on reducing the risk of failures documented by the Columbia Accident Investigation Board. The focus was on improving risk assessments, making system improvements, and implementing cultural changes in workforce interaction. In the case of improved risk assessments, Hale explained, "We [had] reestablished the old NASA culture of doing it right, relying more on test and less on talk, requiring exacting analysis, doing our homework." As an example, he cited the ET-120, which was to have been the Return to Flight tank for STS-114 and was to be sent to KSC late in 2004. But, he admitted, "We knew there [were] insufficient data to determine the tank was safe to fly." After the Debris Verification Review, management learned that some minor issues still had to be handled before these tanks would be approved for flight.

During the flight hiatus, NASA upgraded many of the shuttle's systems and began the process of changing its culture. Engineers redesigned the boosters' bolt catcher and modified the tank in an attempt to eliminate foam loss from the bipod ramp. Engineers developed an Orbiter Boom Sensor System to inspect the tiles in space, and NASA added a Wing Leading Edge Impact Detection System. NASA also installed a camera on the ET umbilical well to document separation and any foam loss.

Finally, NASA focused on improving communication and listening to dissenting opinions. To help the agency implement plans to open dialogue between managers and engineers, from the bottom up, NASA hired the global safety consulting firm Behavioral Science Technology, headquartered in Ojai, California.

Return to Flight After Columbia Accident

When the crew of STS-114 finally launched in the summer of 2005, it was a proud moment for the agency and the country. President George W. Bush, who watched the launch from the Oval Office's dining room, said, "Our space program is a source of great national pride, and this flight is an essential step toward our goal of continuing to lead the world in space science, human spaceflight, and space exploration." First Lady Laura Bush and Florida

Witness Accounts—Key to Understanding Columbia Breakup

The early sightings assessment team—formed 2 days after the Space Shuttle Columbia accident on February 1, 2003—had two primary goals:

- Sift through and characterize the witness reports during re-entry.
- Obtain and analyze all available data to better characterize the pre-breakup debris and ground impact areas. This included providing the NASA interface to the Department of Defense (DoD) through the DoD Columbia Investigation Support Team.

Of the 17,400 public phone, e-mail, and mail reports received from February 1 through April 4, more than 2,900 were witness reports during re-entry, prior to the vehicle breakup. Over 700 of those included photographs or video. Public imagery provided a near-complete record of Columbia's re-entry and video showed debris being shed from the shuttle. Final analysis revealed 20 distinct debris shedding events and three flashes/flares during re-entry. Analysis of these videos and corresponding air traffic control radar produced 20 pre-breakup search areas, ranging in size from 2.6 to 4,403 square km (1 to 1,700 square miles) extending from the California-Nevada border through West Texas.

To facilitate the trajectory analysis, witness reports were prioritized to process re-entry imagery with precise observer location and time calibration first. The process was to time-synchronize all video, determine the exact debris shedding time, measure relative motion, determine ballistic properties of the debris, and perform trajectory analysis to predict the potential ground impact areas or footprints. Key videos were hand carried, expedited through the photo assessment team, and put into ballistic and trajectory analysis as quickly as possible. The Aerospace Corporation independently performed the ballistic and trajectory analysis for process verification.

The public reports, which at first seemed like random information, were in fact a diamond in the rough. This information became invaluable for the search teams on the ground. The associated trajectory analyses also significantly advanced the study of spacecraft breakup in the atmosphere and the subsequent ground impact footprints.

After the Columbia broke apart over East Texas, volunteers from federal agencies, as well as members of the East Texas First Responders, participated in walking the debris fields, forest, and wetlands to find as many parts as possible. This facilitated in determining the cause of the accident.

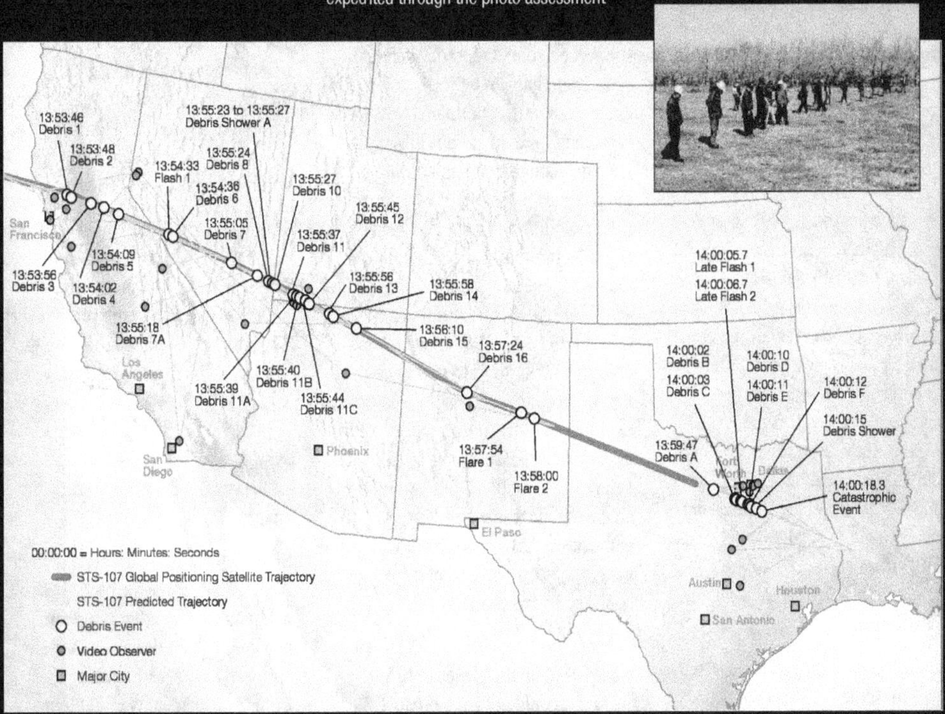

Governor Jeb Bush were among the guests at KSC. Indeed, the Return to Flight mission had been a source of pride for the nation since its announcement. For instance, troops in Iraq sent a "Go Discovery" banner that was hung at KSC. At the landing at Dryden Flight Research Center, the astronauts exited the vehicle carrying an American flag. When the crew returned to Ellington Field, a huge crowd greeted the crew, waving flags as a symbol of the nation's accomplishment. Houston Mayor Bill White declared August 10, 2005, "Discovery STS-114 Day." Standing on a stage, backed by a giant American flag, the crew thanked everyone for their support.

Impact of the Accidents on NASA

The two shuttle tragedies shook NASA's confidence and have significantly impacted the agency in the long term. At the time of both accidents, the Space Shuttle Program office, astronauts, and flight and launch control teams were incredibly capable and dedicated to flying safely. Yet, from the vantage point of hindsight, these teams overlooked the obvious, allowing two tragedies to unfold on the public stage.

Many of the people directly involved in those flights remain haunted by the realization that their decisions resulted in the loss of human lives. NASA was responsible for the safety of the crew and vehicles, and they failed. The flight control teams who worked toward perfection with the motto of "Failure is not an option" felt responsible and hesitant to make hard decisions. Likewise, the engineering communities at JSC and MSFC, and the KSC team that prepared the vehicles, shared feelings of guilt and shaken confidence.

The fact that these tragedies occurred in front of millions of spectators and elected officials made the aftermath even more difficult for the NASA team. The American public and the elected officials expected perfection. When it was not delivered, the outcry of "How could this have happened?" made the headlines of every newspaper and television newscast and became a topic of concern in Congress. The second accident was harder on the agency because the question was now: "How could this have happened *again*?"

Because of the accidents, the agency had a more difficult challenge in convincing Congress of NASA's ability to safely fly people in space. That credibility gap made each NASA administrator's job more difficult and raised doubts in Congress about whether human spaceflight was worth the risk and money. To this day, doubts have not been fully erased on the value of human spaceflight, and the questions of safety and cost are at the forefront of every yearly budget cycle.

In contrast with American politicians, the team of astronauts, engineers, and support personnel that makes human spaceflight happen believes that space exploration must continue. "Yes, there is risk in space travel, but I think that it's safe enough that I'm willing to take the risk," STS-114 (2005) Commander Eileen Collins admitted before her final flight. "I think it's much, much safer than what our ancestors did in traveling across the Atlantic Ocean in an old ship. Frankly, I think they were crazy doing that, but they wanted to do that, and we need to carry on the human exploration of the universe that we live in. I'm honored to be part of that and I'm proud to be part of it. I want to be able to hand on that belief or enthusiasm that I have to the younger generation because I want us to continue to explore."

Without this core belief, the individuals who picked up the pieces after both accidents could not have made it through those terrible times. All of the human spaceflight centers—KSC, MSFC, and JSC—suffered terribly from the loss of Challenger and Columbia. The personnel of all three centers recovered by rededicating themselves to understanding what caused the accidents and how accidents could be prevented in the future. Together, they found the problems and fixed them.

Did the agency change following these two accidents? The answer is absolutely. Following the Challenger accident, the teams looked at every aspect of the processes used to prepare for a shuttle mission. As a result, they went from the mentality that every flight was completely new with a custom solution to a mindset that included a documented production process that was repeatable, flight after flight. The flight readiness process evolved from a process of informally asking each element if all was flight ready to a well-documented set of processes that required specific questions be answered and documented for presentation to management at a formal face-to-face meeting. A rigorous process emerged across the engineering and the operations elements at the centers that made subsequent flights safer.

Yet in spite of all the formal processes put in place, Columbia was still lost. These procedures were not flawed, but the decision-making process was flawed with regard to assessing the loss of foam. Tommy Holloway, who served for several years as the Space Shuttle Program manager, observed that the decision to fly had been based on previous success and not on the analysis of the data.

Since 2003, NASA has gone to great lengths to improve the processes to determine risk and how the team handles difficult decisions. A major criticism of NASA following the Columbia accident was that managers did not always listen to minority and dissenting positions. NASA has since diligently worked toward transforming the culture of its employees to be inclusive of all opinions while working toward a solution.

In hindsight, NASA should not have made an "OK to fly" decision for the final missions of Challenger and Columbia. NASA depended on the requirements that went into the Launch Commit Criteria and Flight Rules to assure that the shuttle was safe to fly. Since neither flight had a "violation" of these requirements, the missions were allowed to proceed even though some people were uncomfortable with the conditions. As a result, NASA has emphasized that the culture should be "prove it is safe" as opposed to "prove it is unsafe" when a concern is raised. The process is better, and the culture is changing as a result of both of these accidents.

As a tribute to the human spirit, teams did not quit or give up after either accident but rather pressed on to Return to Flight each time with a better-prepared and more robust vehicle and team. Some individuals never fully recovered, and they drifted away from human spaceflight. The majority, however, stayed with a renewed vigor to find ways to make spaceflight safer. They still believe in the creed "Failure is not an option" and work diligently to meet the expectation of perfection by the American people and Congress.

NASA has learned from past mistakes and continues on with ventures in space exploration, recognizing that spaceflight is hard, complex, and—most importantly—will always have inherent risk. Accidents will happen, and the teams will have to dig deep into their inner strength to find a way to recover, improve the system, and continue the exploration of space for future generations.

On an Occasion of National Mourning

Howard Nemerov
Poet Laureate of the United States
1963-1964 and 1988-1990

It is admittedly difficult for a whole
Nation to mourn and be seen to do so, but
It can be done, the silvery platitudes
Were waiting in their silos for just such
An emergent occasion, cards of sympathy
From heads of state were long ago prepared
For launching and are bounced around the world
From satellites at near the speed of light,
The divine services are telecast
From the home towns, children are interviewed
And say politely, gravely, how sorry they are,
And in a week or so the thing is done,
The sea gives up its bits and pieces and
The investigating board pinpoints the cause
By inspecting bits and pieces, nothing of the sort
Can ever happen again, the prescribed course
Of tragedy is run through omen to amen
As in a play, the nation rises again
Reborn of grief and ready to seek the stars;
Remembering the shuttle, forgetting the loom.

National Security

Jeff DeTroye
James Armor
Sebastian Coglitore
James Grogan
Michael Hamel
David Hess
Gary Payton
Katherine Roberts
Everett Dolman

To fully understand the story of the development of the Space Shuttle, it is important to consider the national defense context in which it was conceived, developed, and initially deployed.

The Cold War between the United States and the Union of Soviet Socialist Republics (USSR), which had played such a large role in the initiation of the Apollo Program, was also an important factor in the decisions that formed and guided the Space Shuttle Program. The United States feared that losing the Cold War (1947-1991) to the USSR could result in Soviet mastery over the globe. Since there were few direct conflicts between the United States and the USSR, success in space was an indicator of which country was ahead—which side was winning. Having lost the tactical battles of first satellite and first human in orbit, the United States had recovered and spectacularly won the race to the moon. To counter the successful US man-on-the-moon effort, the USSR developed an impressive space station program. By the early 1980s, the USSR had launched a series of space stations into Earth orbit. The Soviets were in space to stay, and the United States could not be viewed as having abdicated leadership in space after the Apollo Program.

The need to clearly demonstrate the continued US leadership in space was an important factor in the formation of the Space Shuttle Program. While several other programs were considered, NASA ultimately directed their planning efforts to focus on a reusable, crewed booster that would provide frequent, low-cost access to low-Earth orbit. This booster would launch all US spacecraft, so there would have to be direct interaction between the open, civilian NASA culture and the Defense-related National Security Space (NSS) programs. Use of the civilian NASA Space Shuttle Program by the NSS programs was controversial, with divergent goals, and many thought it was a relationship made for political reasons only—not in the interest of national security. The relationship between these two very different cultures was often turbulent and each side had to change to accommodate the other. Yet it was ultimately successful, as seen in the flawless missions that followed.

National Security Space Programs

The Department of Defense uses space systems in support of air, land, and sea forces to deter and defend against hostile actions directed at the interests of the United States. The Intelligence community uses space systems to collect intelligence. These programs, as a group, are referred to as National Security Space (NSS). Despite having a single name, the NSS did not have a unified management structure with authority over all programs.

Since the beginning of the space era, these defense-related space missions had been giving the president, as well as defense and intelligence leadership in the United States, critical insights into the actions and intents of adversaries. In 1967, President Lyndon Johnson said, "I wouldn't want to be quoted on this—we've spent $35 or $40 billion on the space program. And if nothing else had come out of it except the knowledge that we gained from space photography, it would be worth 10 times what the whole program has cost. Because tonight we know how many missiles the enemy has and, it turned out, our guesses were way off. We were doing things we didn't need to do. We were building things we didn't need to build. We were harboring fears we didn't need to harbor." Due to these important contributions and others, the NSS programs had a significant amount of political support and funding. As a result, both the NSS program leadership and the NASA program leadership often held conflicting views of which program was more important and, therefore, whose position on a given issue ought to prevail.

These two characteristics of the NSS programs—lack of unified NSS program management and a competing view of priorities—would cause friction between NASA and the NSS programs management throughout the duration of the relationship.

1970-1981: Role of National Security Space Programs in Development of the Shuttle

The National Security Space (NSS) is often portrayed as having forced design requirements on NASA to gain NSS commitment to the Space Shuttle Program. In reality, NASA was interested in building the most capable (and largest) shuttle that Congress and the administration would approve. It is true that NSS leaders argued for a large payload bay and a delta wing to provide a 1,600-km (1,000-mile) cross range for landing. NASA, however, also wanted a large payload bay for space station modules as well as for spacecraft and high-energy stage combinations. NASA designers required the shuttle to be able to land at an abort site, one orbit after launch from the West Coast, which would also require a delta wing. Indeed, NASA cited the delta wing as an essential NASA requirement, even for launches from the East Coast. NASA was offered the chance to build a smaller shuttle when, in January 1972, President Richard Nixon approved the Space Transportation System (STS) for development. The NASA leadership decided to stick with the larger, delta wing design.

National Space Policy: The Shuttle as Sole Access to Space

The Space Shuttle Program was approved with the widely understood but unstated policy that when it became operational it would be used to launch all NSS payloads. The production of all other expendable launch vehicles, like the reliable Titan, would be abandoned. In 1981, shortly after the launch of STS-1, the National Space Transportation Policy signed by President Ronald Reagan formalized this position: "The STS will be the primary space launch system for both United States military and civil government missions. The transition to the shuttle should occur as expeditiously as practical. . . . Launch priority will be provided to national security missions, and such missions may use the shuttle as dedicated mission vehicles."

This mandated dependence on the shuttle worried NSS leaders, with some saying the plan was "seriously deficient, both operationally and economically." In January 1984, Secretary of Defense Caspar Weinberger directed the purchase of additional expendable boosters because "total reliance upon the STS for sole access to space in view of the technical and operational uncertainties, represents an unacceptable national security risk." This action, taken 2 years before the Challenger accident, ensured that expendable launch vehicles would be available for use by the NSS programs in the event of a shuttle accident. Furthermore, by 1982 the full costs of shuttle missions were becoming clearer and the actual per-flight cost of a shuttle mission had risen to over $280 million, with a Titan launch looking cheap in comparison at less than $180 million. With the skyrocketing costs of a shuttle launch, the existence of an expendable launch vehicles option for the NSS programs made the transition from the shuttle inevitable.

Military "Man in Space"

To this day, the US Air Force (USAF) uses flight crews for most of their airborne missions. Yet, there was much discussion within the service about the value of having a military human in space program. Through the 1960s, development of early reconnaissance satellites like Corona demonstrated that long-life electronics and complex systems on the spacecraft and on the ground could be relied on to accomplish the crucial task of reconnaissance. These systems used inexpensive systems on orbit and relatively small expendable launch vehicles, and they proved that human presence in space was not necessary for these missions.

During the early 1960s, NSS had two military man in space programs: first the "Dyna Soar" space plane, and then the Manned Orbiting Laboratory program. Both were cancelled, largely due to skepticism on the part of the Department of Defense (DoD) or NSS leadership that the programs' contributions were worth the expense as well as the unwanted attention that the presence of astronauts would bring to these highly classified missions.

Although 14 military astronauts were chosen for the Manned Orbiting Laboratory program, the sudden cancellation of this vast program in 1969 left them, as well as the nearly completed launch facility at Vandenberg Air Force Base, California, without a mission. With NASA's existing programs ramping down, NASA was reluctant to take the military astronauts into its Astronaut Corps. Eventually, only the seven youngest military astronauts transferred to NASA. The others returned to their military careers. These military astronauts did not fly until the 1980s, with the first being Robert Crippen as pilot on STS-1. The Manned Orbiting Laboratory pad at Vandenberg Air Force Base would lie dormant until the early 1980s when modifications were begun for use with the shuttle.

The Space Shuttle Program plans included a payload specialist selected for a particular mission by the payload sponsor or customer. Many NSS leadership were not enthusiastic about the concept; however, in 1979, a selection board made up of NSS leadership and a NASA representative chose the first cadre of 13 military officers from the USAF and US Navy. These officers were called manned spaceflight engineers. There was considerable friction with the NASA astronaut office over the military payload specialist program. Many of the ex-Manned Orbiting Laboratory astronauts who had been working at NASA and waiting for over a decade to fly in space were not enthusiastic about the NSS plans to fly their own officers as payload specialists. In the long run, NASA astronauts had little to be concerned about. When asked his opinion of the role of military payload specialists in upcoming shuttle missions, General Lew Allen, then chief of staff of the USAF, related a story about when he played a major role in the cancellation of the Manned Orbiting Laboratory Military Man in Space program. In 1984, another NSS senior wrote: "The major driver in the higher STS costs is the cost of carrying man on a mission which does not need man. . . . It is clear that man is not needed on the transport mission. . . ." The NSS senior leadership was still very skeptical about the need for a military man in space. Ultimately, only two NSS manned spaceflight engineers flew on shuttle missions.

Launch System Integration: Preparing for Launch

The new partnership between NASA and the NSS programs was very complex. Launching the national security payloads on the shuttle required the cooperation of two large, proud organizations, each of which viewed their mission as being of the highest national priority. This belief in their own primacy was a part of each organization's culture. From the very beginning, it was obvious that considerable effort would be required by both organizations to forge a true partnership. At the beginning of the Space Shuttle Program, NASA focused on the shuttle, while NSS program leaders naturally focused on the spacecraft's mission. As the partnership developed, NASA had to become more payload focused. Much of the friction was over who was in charge. The NSS programs were used to having control of the launch of their spacecraft. NASA kept firm control of the shuttle missions and struggled with the requests for unique support from each of the many programs using the shuttle.

Launch system integration—the process of launching a spacecraft on the shuttle—was a complex activity that had to be navigated successfully. For an existing spacecraft design, transitioning to fly on the shuttle required a detailed engineering and safety assessment. Typically, some redesign was required to make the spacecraft meet the shuttle's operational and safety requirements, such as making dangerous propellant and explosive systems safe for a crewed vehicle. This effort actually offered an opportunity for growth due to the shuttle payload bay size and the lift capacity from the Kennedy Space Center (KSC) launch site. Typically flying alone on dedicated missions, the NSS spacecraft had all the shuttle capacity to grow into. Since design changes were usually required for structural or safety reasons, most NSS program managers could not resist taking at least some advantage of the available mass or volume. So many NSS spacecraft developed during the shuttle era were much larger than their predecessors had been in the late 1960s.

National Security Space Contributions to the Space Shuttle Program

The NSS programs agreed to provide some of the key capabilities that the Space Shuttle Program would need to achieve all of its goals. As the executive agent for DoD space, the USAF funded and managed these programs.

One of these programs, eventually known as the Inertial Upper Stage, focused on an upper stage that would take a spacecraft from the shuttle in low-Earth orbit to its final mission orbit or onto an escape trajectory for an interplanetary mission. Another was a West Coast launch site for the shuttle, Vandenberg Air Force Base, California. Launching from this site would allow the shuttle to reach high inclination orbits over the Earth's poles. Although almost complete, it was closed after the Challenger accident in 1986 and much of the equipment was disassembled and shipped to KSC to improve or expand its facilities. Another program was a USAF shuttle flight operation center in Colorado. This was intended to be the mission control center for NSS shuttle flights, easing the workload on the control center in Houston, Texas, for these classified missions. USAF built the facility and their personnel trained at Johnson Space Center; however, when the decision was made to remove NSS missions from the shuttle manifest after the Challenger accident, the facility was not needed for shuttle flights and eventually it was used for other purposes.

Space Shuttle Enterprise on Space Launch Complex 6 during pad checkout tests at Vandenberg Air Force Base in 1985. Enterprise was the Orbiter built for the Approach and Landing Tests to prove flightworthiness. It never became part of the shuttle fleet.

Flying National Security Space Payloads on the Shuttle

The NSS program leadership matured during a period when spacecraft and their ground systems were fairly simple and orbital operations were not very complex. In the early 1980s, one senior NSS program director was often heard to say, "All operations needs is a roll of quarters and a phone booth." This was hyperbole, but the point was clear: planning and preparing for orbital operations was not a priority. It wasn't unheard of for an NSS program with budget, schedule, or political pressures to launch a new spacecraft before all the details for how to operate the spacecraft on orbit had been completely worked out.

Early on, NASA flight operations personnel were stunned to see that the ground systems involved in operating the most critical NSS spacecraft were at least a decade behind equivalent NASA systems. Some even voiced concern that, because the NSS systems were so antiquated, they weren't sure the NSS spacecraft could be operated safely with the shuttle. In NASA, flight operations was a major organizational focus and had been since the days of Project Mercury. NASA flight operations leaders such as John O'Neil, Jay Honeycutt, Cliff Charlesworth, and Gene Kranz had an important voice in how the Space Shuttle Program allocated its resources and in its development plans. Line managers in NASA, including Jay Greene, Ed Fendell, and Hal Beck, worked closely with the NSS flight operations people to merge NSS spacecraft and shuttle operations into one seamless activity. Many of the NASA personnel, especially flight directors, had no counterpart on the NSS government team.

To prepare for a mission, NASA flight operations employed a very thorough process that focused on ensuring that flight controllers were ready for anything the mission might throw at them. This included practice sessions in the control centers using spacecraft simulators that were better than anything the NSS personnel had seen. NSS flight operations personnel thought they had died and gone to heaven. Here, finally, was an organization that took "ops" seriously and committed the resources to do it right. As the partnership developed, NASA forced, cajoled, and convinced the NSS programs to adopt a more thorough approach to the shuttle integration and operations readiness processes. Over time, NASA's approach caught on within the NSS. It was simply a best practice worth emulating.

Another component of NASA human spaceflight—the role of the astronaut—was initially very foreign to NSS personnel. Astronauts tended to place a very personal stamp on the plans for "their" mission, which came as a shock to NSS program personnel. Some NSS personnel chafed at the effort required to satisfy the crew member working with their payload. On early missions, the commander or other senior crew members would not start working with the payload until the last 6 months or so prior to launch and would want to make changes in the plans. This caused some friction. The NSS people did not want to deal with last-minute changes so close to launch. After a few missions, as the relationship developed, adjustments were made by both sides to ease this "last-minute effect."

1982-1992: National Security Space and NASA Complete 11 Missions

The first National Security Space (NSS) payload was launched on Space Transportation System (STS)-4 in June 1982. This attached payload (one that never left the payload bay), called "82-1," carried the US Air Force (USAF) Space Test Program Cryogenic Infrared Radiance Instrumentation for Shuttle (CIRRIS) telescope and several other small experiments. This mission was originally scheduled for the 18th shuttle flight; but, as the Space Shuttle Program slipped, NSS program management was able to maintain its schedule and was ready for integration into the shuttle early in 1982. Since the first two shuttle missions had gone so well, NASA decided to allow the 82-1 payload to fly on this flight test mission despite the conflicts this decision would cause with the mission's test goals. This rather selfless act on the part of NASA was characteristic of the positive relationship between NASA and the NSS programs once the shuttle began to fly. For the NSS programs, a major purpose of this mission was to be a pathfinder for subsequent NSS missions. This payload was controlled from the Sunnyvale USAF station in California. This was also the only NSS mission where the NSS flight controllers talked directly to the shuttle crew.

Operational Missions

The next NSS mission, STS-51C, occurred January 1985, 2½ years after STS-4. STS-51C was a classified NSS mission that included the successful use of the Inertial Upper Stage. The

Inertial Upper Stage had experienced a failure during the launch of the first NASA Tracking and Data Relay Satellite mission on STS-6 in 1983. The subsequent failure investigation and redesign had resulted in a long delay in Inertial Upper Stage missions. With the problem solved, the shuttle launched into a 28.5-degree orbit with an altitude of about 407 km (220 nautical miles). The first manned spaceflight engineer, Gary Payton, flew as a payload specialist on this 3-day mission. This was also the first use of the "Department of Defense (DoD) Control Mode"—a specially configured Mission Operations Control Room at Johnson Space Center that was designed and equipped with all the systems required to protect the classified nature of these missions.

Gary Payton, US Air Force (USAF) Lieutenant General (retired), flew on STS-51C (1985) as a payload specialist. He was part of the USAF manned spaceflight engineering program and served as USAF Deputy Under Secretary for Special Programs.

The second and final manned spaceflight engineer, William Pailes, flew on the 4-day flight of STS-51J in October 1985. This shuttle mission deployed a defense communications satellite riding on an Inertial Upper Stage, which took the satellite up to geosynchronous orbit.

The Challenger and her crew were lost in a tragic accident the following January. After launching only three spacecraft payloads on the first 25 missions, the NSS response to the Challenger accident was to move all spacecraft that it could off shuttle flights. The next NSS spacecraft flew almost 2 years after the Challenger accident on the 4-day mission of STS-27 in December 1988. This mission was launched into a 57-degree orbit and had an all-NASA crew, as did the subsequent NSS spacecraft payload missions with only one exception (STS-44 [1991]). No other details on the STS-27 mission have been released.

The launch rate picked up 8 months later with the launch of STS-28 in August and STS-33 in November (both in 1989), followed by STS-36 in February and STS-38 in November (both in 1990). The details of these missions remain classified, but the rapid launch rate—four missions in 15 months—was working off the backlog that had built up during the delays after the Challenger accident. This pace also demonstrated the growing maturity of the NSS/NASA working relationship.

In April 1991, in a departure from the NSS unified approach to classification of its activities on the shuttle, the USAF Space Test Program AFP-675 with the CIRRIS telescope was launched on STS-39. This was the first time in the NSS/NASA relationship that the details of a dedicated DoD payload were released to the world prior to launch. The focus of this mission was Strategic Defense Initiative research into sensor designs and environmental phenomena. The details of this flight and STS-44 in November 1991 were released to the public. Their payloads were from previously publicized USAF programs.

Defense Support Program spacecraft and attached Inertial Upper Stage prior to release from Atlantis on STS-44 (1991). This spacecraft provides warning of ballistic missile attacks on the United States.

STS-44 crew members included an Army payload specialist, Tom Hennan. This mission marked the end of flights on the shuttle for non-NASA military payload specialists. Ironically, Warrant Officer Hennan performed experiments called "Military Man in Space." The spacecraft launched on this mission was the USAF Defense Support Program satellite designed to detect nuclear detonations, missile launches, and space launches from geosynchronous orbit. This satellite program had been in existence for over 20 years. The satellite launched on STS-44 replaced an older satellite in the operational Defense Support Program constellation.

Space Test Program

Another series of experiments, called "M88-1," on STS-44 was announced as an ongoing series of tri-service experiments designed to assess man's visual and communication capabilities from space. The objectives of M88-1

Michael Griffin, PhD
Deputy for technology at the Strategic Defense Initiative Organization (1986-1991).
NASA administrator (2005-2009).

Strategic Defense Initiative Test

"STS-39 was a very complex mission that led to breakthroughs in America's understanding of the characteristics of missile signatures in space. The data we gathered enhanced our ability to identify and protect ourselves from future missile threats. This is one of the most under-recognized achievements of the shuttle era."

View of the Aurora Australis—or Southern Lights—taken by Air Force Program-675 Uniformly Redundant Array and Cryogenic Infrared Radiance Instrumentation during STS-39 (1991). One of the equipment's objectives was to gather data on the Earth's aurora, limb, and airglow.

STS-39's Air Force Program-675 equipment mounted on the experiment support system pallet in Discovery's payload bay.

overlapped those done by Hennan with his experiments; however, NASA Mission Specialist Mario Runco and the rest of the NASA crew performed the M88-1 experiments. This activity used a digital camera to produce images that could be evaluated on orbit. Observations were to be radioed to tactical field users seconds after the observation pass was complete. Emphasis was on coordinating observations with ongoing DoD exercises to fully assess the military benefits of a spaceborne observer. The policy implications of using NASA astronauts to provide input directly to military forces on the ground during shuttle missions have long been debated. This flight and the following mission (STS-53) are the only acknowledged examples of this policy.

A year later in December 1992, STS-53 was launched with a classified payload called "DoD-1" on a 7-day mission. Marty Faga, assistant secretary of the USAF (space), said: "STS-53 marks a milestone in our long and productive partnership with NASA. We have enjoyed outstanding support from the Space Shuttle Program. Although this is the last dedicated shuttle payload, we look forward to continued involvement with the program with DoD secondary payloads."

With the landing of STS-53 at Kennedy Space Center, the NSS/NASA partnership came to an end. During the 10 years of shuttle missions, 11 of the 52 missions were dedicated to NSS programs. The end of NSS-dedicated shuttle missions resulted from the rising costs of shuttle missions and policy decisions made as a result of the Challenger accident. There were few NSS-dedicated missions relative to the enthusiastic plans laid in the late 1970s; however, the Space Shuttle Program had a lasting impact on the NSS programs. While the number of NSS-dedicated missions was small, the partnership between the NSS programs and NASA had a lasting impact.

Legacy of the Space Shuttle Program and National Security Space

The greatest legacy of the NASA/National Security Space (NSS) partnership was at the personal level for NSS engineers and managers. Working on the Space Shuttle Program in the early 1980s was exciting and provided just the sort of motivation that could fuel a career. NSS personnel learned new and different operational and engineering techniques through direct contact with their NASA counterparts. As a result, engineering and operations practices developed by NASA were applied to the future complex NSS programs with great success.

Another significant legacy is that of leadership in the NSS programs. The manned spaceflight engineer program in particular was adept at selecting young officers with potential to be future leaders of the NSS programs. A few examples of current or recent NSS leaders who spent their formative years in the manned spaceflight engineer program include: Gary Payton, Mike Hamel, Jim Armor, Kathy Roberts, and Larry James. Others, such as Willie Shelton, were US Air Force (USAF) flight controllers assigned to work in Houston, Texas.

Many military personnel working with NASA returned to the NSS space programs, providing outstanding leadership to future programs. Several ex-astronauts, such as Bob Stuart, John Fabian, and Kevin Chilton, have held or are now holding senior leadership roles in their respective services.

The role that the NASA/NSS collaboration played in the formation of Space Command also left a legacy. While the formation of the USAF Space Command occurred late in the NASA/NSS relationship, close contact between the NSS programs and the shuttle organizations motivated the Department of Defense to create an organization that would have the organizational clout and budget to deal with the Space Shuttle Program on a more equal basis.

The impact on mission assurance and the rigor in operations planning and

US Air Force Space Test Program—Pathfinder for Department of Defense Space Systems

The US Air Force (USAF) Space Test Program was established as a multiuser space program whose role is to be the primary provider of spaceflight for the entire Department of Defense (DoD) space research community. From as early as STS-4 (1982), the USAF Space Test Program used the shuttle to fly payloads relevant to the military. The goal of the program was to exploit the use of the shuttle as a research and development laboratory. In addition to supplying the primary payloads on several DoD-dedicated missions, more than 250 secondary payloads and experiments flew on 95 shuttle missions. Space Test Program payloads flew in the shuttle middeck, cargo bay, Spacelab, and Spacehab, and on the Russian space station Mir during the Shuttle-Mir missions in the mid 1990s.

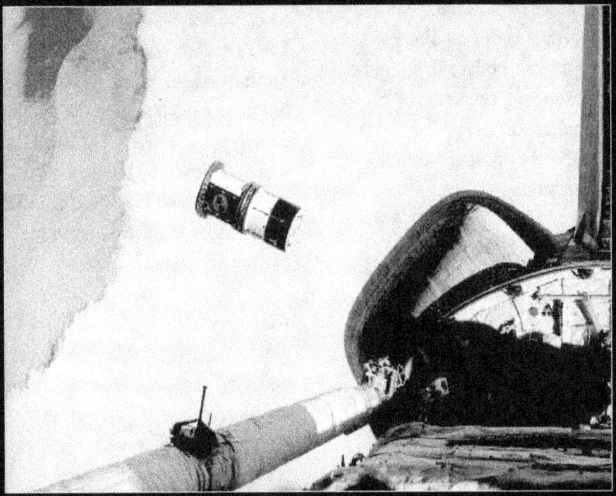

A Department of Defense pico-satellite known as Atmospheric Neutral Density Experiment (ANDE) is released from the STS-116 (2006) payload bay. ANDE consists of two micro-satellites that measure the density and composition of the low-Earth orbit atmosphere while being tracked from the ground. The data are used to better predict the movement of objects in orbit.

preparation could be the most significant technical legacy the Space Shuttle Program left the NSS programs. NASA required participation by the NSS spacecraft operators in the early stages of each mission's planning. NSS operations personnel quickly realized that this early involvement resulted in improved operations or survivability and provided the tools and experience necessary to deal with the new, more complex NSS spacecraft.

The impact of the Space Shuttle Program on the NSS cannot be judged by the small number of NSS-dedicated shuttle missions. The policy decision that moved all NSS spacecraft onto the shuttle formed a team out of the most creative engineering minds in the country. There was friction between the two organizations, but ultimately it was the people on this NSS/NASA team who made it work. It is unfortunate that, as a result of the Challenger accident, the end of the partnership came so soon. The success of this partnership should be measured not by the number of missions or even by the data collected, but rather by the lasting impact on the NSS programs' personnel and the experiences they brought to future NSS programs.

Another Legacy: Relationship with USSR and Its Allies

In 1972, with the US announcement of the Space Shuttle as its primary space transportation system, the USSR quickly adapted to keep pace. "Believing the Space Shuttle to be a military threat to the Soviet Union, officials of the USSR Ministry of Defense found little interest in lunar bases or giant space stations. What they wanted was a parallel deterrent to the shuttle." Premier Leonid Brezhnev, Russian sources reported, was particularly distraught at the thought of a winged spacecraft on an apparently routine mission in space suddenly swooping down on Moscow and delivering an unthinkably dangerous cargo.

Russian design bureaus offered a number of innovative counter-capabilities, but Brezhnev and the Ministry of Defense were adamant that a near match was vital. They may not have known what the American military was planning with the shuttle, but they wanted to be prepared for exactly what it might be. The Soviets were perplexed by the decision to go forward with the Space Shuttle. Their estimates of cost-performance, particularly over their own mass-produced space launch vehicles, were very high. It seemed to make little practical sense until the announcement that a military shuttle launch facility at Vandenberg Air Force Base was planned; according to one Soviet space scientist, "… trajectories from Vandenberg allowed an overflight of the main centers of the USSR on the first orbit. So our hypothesis was that the development of the shuttle was mainly for military purposes." It was estimated that a military payload could reenter Earth's atmosphere from orbit and engage any target within the USSR in 3 to 4 minutes—much faster than the anticipated 10 minutes from launch to detonation by US nuclear submarines stationed off Arctic coastlines. This drastically changed the deterrence calculations of top Soviet decision makers.

Indeed, deterrence was the great game of the Cold War. Each side had amassed nuclear arsenals sufficient to destroy the other side many times over, and any threat to the precarious balance of terror the two sides had achieved was sure to spell doom. The key to stability was the capacity to deny any gain from a surprise or first strike. A guaranteed response in the form of a devastating counterattack was the hole card in this international game of bluff and brinksmanship. Any development that threatened to mitigate a full second strike was a menace of the highest order.

Several treaties had been signed limiting or barring various anti-satellite activities, especially those targeted against nuclear launch detection capabilities (in a brute attempt to blind the second-strike capacity of the other side). The shuttle, with its robotic arm used for retrieving satellites in orbit, could act as an anti-satellite weapon in a crisis, expensive and dangerous as its use might be. Thus, the shuttle could get around prohib itions against anti-satellite capabilities through its public image as a peaceful NASA space plane. So concerned were the Soviets

Buran/Energiya shuttle and heavy-lift booster, built by the USSR, flew once—uncrewed—in 1988.

with the potential capability of the shuttle, they developed designs for at least two orbiting "laser-equipped battle stations" as a counter and conducted more than 20 "test launches" of a massive ground-launched anti-satellite weapon in the 1970s and 1980s.

In the 1978-1979 strategic arms limitation talks, the Soviets asked for a guarantee that the shuttle would not be used for anti-satellite purposes. The United States refused. In 1983, the USSR offered to prohibit the stationing of any weapons in space, if the United States would agree. The catch was the shuttle could not be used for military activities. In exchange, the Soviets would likewise limit the Mir space station from military interaction—an untenable exchange.

So a shuttle-equivalent space plane was bulldozed through the Soviet budget and the result was the Buran/Energiya shuttle and heavy-lift booster. After more than a decade of funding—and, for the cash-strapped Soviet government, a crippling budget—the unmanned Buran debuted and flew two orbits before landing flawlessly in November 1988. Immediately after the impressive proof-of-concept flight, the Soviets mothballed Buran.

James Moltz, professor of national security at the Naval Postgraduate School, commented that the "self-inflicted extreme cost of the Buran/Energiya program did more to destabilize the Soviet economy than any response to the Reagan administration's efforts in the 1980s." If so, the Space Shuttle can be given at least partial credit for winning the Cold War.

The Space Shuttle and Its Operations

The Space Shuttle

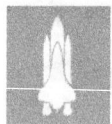

Roberto Galvez
Stephen Gaylor
Charles Young
Nancy Patrick
Dexer Johnson
Jose Ruiz

The Space Shuttle design was remarkable. The idea of "wings in orbit" took concrete shape in the brilliant minds of NASA engineers, and the result was the most innovative, elegant, versatile, and highly functional vehicle of its time. The shuttle was indeed an engineering marvel on many counts. Accomplishing these feats required the design of a very complex system.

In several ways, the shuttle combined unique attributes not witnessed in spacecraft of an earlier era. The shuttle was capable of launching like a rocket, reentering Earth's atmosphere like a capsule, and flying like a glider for a runway landing. It could rendezvous and dock precisely, and serve as a platform for scientific research within a range of disciplines that included biotechnology and radar mapping. The shuttle also performed satellite launches and repairs, bestowing an almost "perpetual youth" upon the Hubble Space Telescope through refurbishments.

The most impressive product that resulted from the shuttle's capabilities and contributions is the International Space Station—a massive engineering assembly and construction undertaking in space.

No other crewed spacecraft to date has replicated these capabilities. The shuttle has left an indelible mark on our society and culture, and will remain an icon of space exploration for decades to come.

What Was the Space Shuttle?

Physical Characteristics

The Space Shuttle was the most complex space vehicle design of its time. It was comprised of four main components: the External Tank (ET); three Space Shuttle Main Engines; two Solid Rocket Boosters (SRBs); and the Orbiter vehicle. It was the first side-mounted space system dictated by the need to have a large winged vehicle for cross-range capability for re-entry into Earth's atmosphere and the ability to land a heavyweight payload.

These four components provided the shuttle with the ability to accomplish a diverse set of missions over its flight history. The Orbiter's heavy cargo/payload carrying capability, along with the crew habitability and flexibility to operate in space, made this vehicle unique. Because of its lift capability and due-East inclination, the shuttle was able to launch a multitude of satellites, Spacelab modules, science platforms, interplanetary probes, Department of Defense payloads, and components/modules for the assembly of the International Space Station (ISS).

The shuttle lift capability or payload decreased with increased operational altitude or orbit inclination because more fuel was required to reach the higher altitude or inclination.

Shuttle lift capability was also limited by total vehicle landing weight—different limits for different cases (nominal or abort landing). An abort landing was required if a system failure

Space Shuttle Launch Configuration

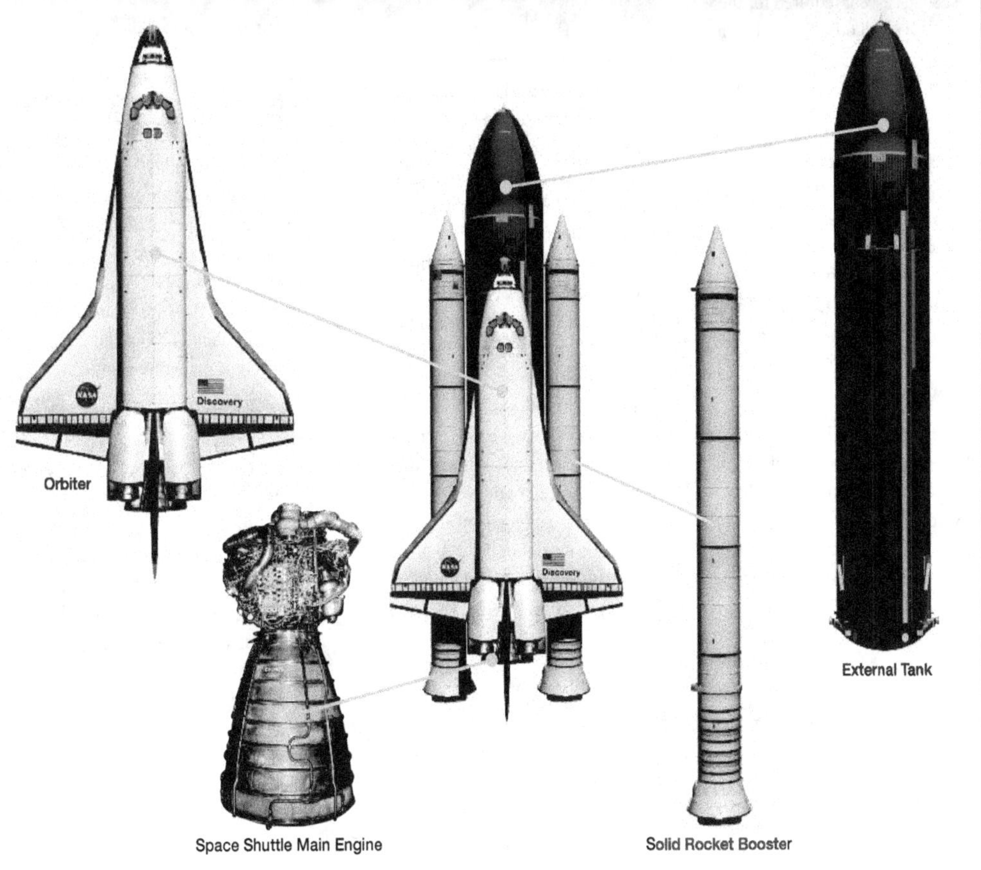

during ascent caused the shuttle not to have enough energy to reach orbit or was a hazard to crew or mission. Abort landing sites were located around the world, with the prime abort landing sites being Kennedy Space Center (KSC) in Florida, Dryden Flight Research Center on the Edwards Air Force Base in California, and Europe.

The entire shuttle vehicle, fully loaded, weighed about 2 million kg (4.4 million pounds) and required a combined thrust of about 35 million newtons (7.8 million pounds-force) to reach orbital altitude. Thrust was provided by the boosters for the first 2 minutes and the main engines for the approximately 8 minutes and 30 seconds ascent required for the vehicle to reach orbital speed at the requisite altitude range of 185 to about 590 km (100 to 320 nautical miles).

Once in orbit, the Orbital Maneuvering System engines and Reaction Control System thrusters were used to perform all orbital operations, Orbiter maneuvers, and deorbit. Re-entry required orbital velocity decelerations of about 330 km/hr (204 mph) depending on orbital altitude, which caused the Orbiter to slow and fall back to Earth.

The Orbiter Thermal Protection System, which covered the entire vehicle, provided the protection needed to survive the extreme high temperatures experienced during re-entry. Primarily friction between the Orbiter and the Earth's atmosphere generated temperatures ranging from 927°C (1,700°F) to 1,600°C (3,000°F). The highest temperatures experienced were on the wing leading edge and nose cone.

The time it took the Orbiter to start its descent from orbital velocity of about 28,160 km/hr (17,500 mph) to a landing speed of about 346 km/hr (215 mph) was 1 hour and 5 minutes.

During re-entry, the Orbiter was essentially a glider. It did not have any propulsion capability, except for the Reaction Control System thrusters required for roll control to adjust its trajectory early during re-entry.

Management of the Orbiter energy from its orbital speed was critical to allow the Orbiter to reach its desired runway target. The Orbiter's limited cross-range capability of about 1,480 km (800 nautical miles) made management of the energy during final phases of re-entry close to the ground—otherwise called terminal area energy management—critical for a safe landing.

The Orbiter performed as a glider during re-entry, thus its mass properties had to be well understood to ensure that the Flight Control System could control the vehicle and reach the required landing site with the right amount of energy for landing. One of the critical components of its aerodynamic flight was to ensure that the Orbiter center of gravity was correctly calculated and entered into the Orbiter flight design process. Because of the tight center of gravity constraints, the cargo bay payloads were placed in the necessary cargo bay location to protect the down weight and center of gravity of the Orbiter for landing. Considering the Orbiter's size, the center of gravity box was only 91 cm (36 in.) long, 5 cm (2 in.) wide, and 5 cm (2 in.) high.

External Tank

The ET was 46.8 m (153.6 ft) in length with a diameter of 8.4 m (27.6 ft), which made it the largest component of the shuttle. The ET contained two internal tanks—one for the storage of liquid hydrogen and the other for the storage of liquid oxygen. The hydrogen tank, which was the bigger of the two internal tanks, held 102,737 kg (226,497 pounds) of hydrogen. The oxygen tank, located at the top of the ET, held 619,160 kg (1,365,010 pounds) of oxygen. Both tanks provided the fuel to the main engines required to provide the thrust for the vehicle to achieve a safe orbit. During powered flight and ascent to orbit, the ET provided about 180,000 L/min (47,000 gal/min) of hydrogen and about 67,000 L/min (18,000 gal/min) of oxygen to all three Space Shuttle Main Engines with a 6-to-1 mixture ratio of liquid hydrogen to liquid oxygen.

Solid Rocket Boosters

The two SRBs provided the main thrust to lift the shuttle off the launch pad. Each booster provided about 14.7 meganewtons (3,300,000 pounds-force) of thrust at launch, and they were only ignited once the three main engines reached the required 104.5% thrust level for launch. Once the SRBs were ignited, they provided about 72% of the thrust required of the entire shuttle at liftoff and through the first stage, which ended at SRB separation.

The SRB thrust vector control system enabled the nozzles to rotate, allowing the entire shuttle to maneuver to the required ascent trajectory during first stage. Two minutes after launch, the spent SRBs were jettisoned, having taken the vehicle to an altitude of about 45 km (28 miles). Not only were the boosters reusable, they were also the largest solid propellant motors in use then. Each measured about 45.4 m (149 ft) long and about 3.6 m (12 ft) in diameter.

External Tank

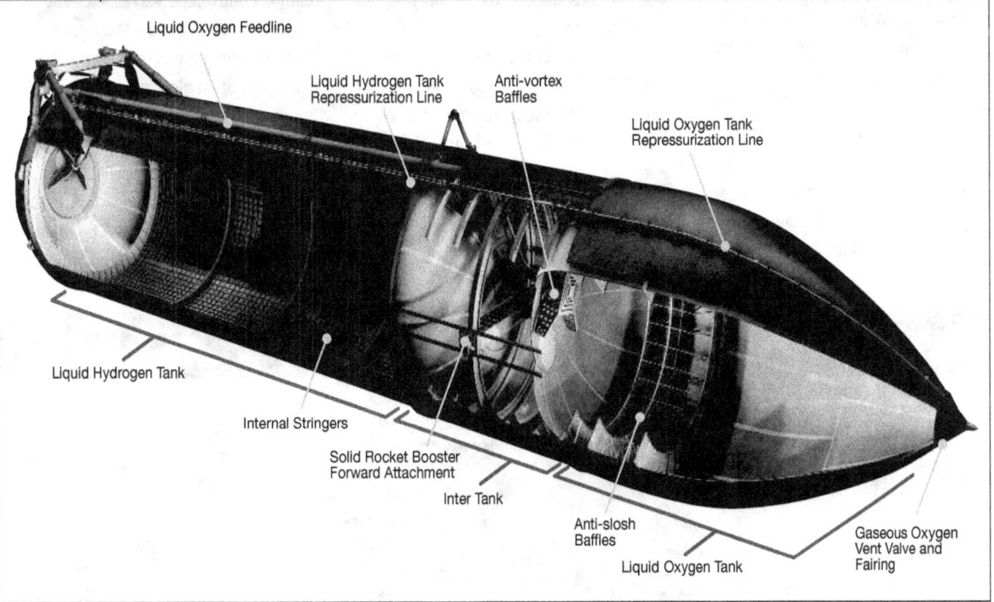

Solid Rocket Boosters

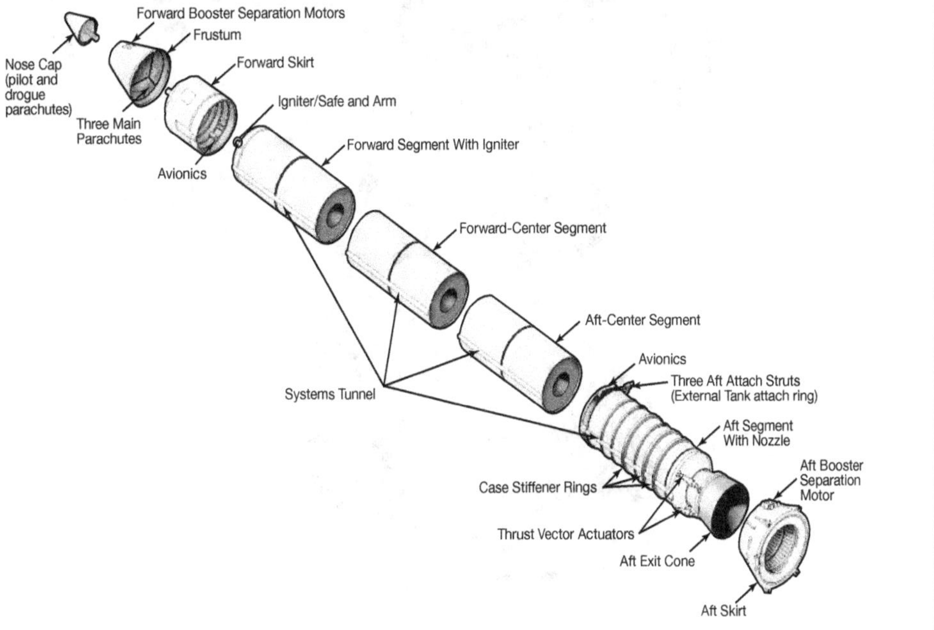

Space Shuttle Main Engines

After SRB separation, the main engines provided the majority of thrust required for the shuttle to reach orbital velocity. Each main engine weighed about 3,200 kg (7,000 pounds). With a total length of 4.3 m (14 ft), each engine, operating at the 104.5% power level, provided a thrust level of about 1.75 meganewtons (394,000 pounds-force) at sea level and about 2.2 meganewtons (492,000 pounds-force) at vacuum throughout the entire 8 minutes and 30 seconds of powered flight. The engine nozzle by itself was 2.9 m (9.4 ft) long with a nozzle exit diameter of 2.4 m (7.8 ft). Due to the high heat generated by the engine thrust, each engine contained 1,082 tubes throughout its entire diameter, allowing circulation of liquid hydrogen to cool the nozzle during powered flight. The main engines were a complex piece of machinery comprised of high- and low-pressure fuel and oxidizer pumps, engine controllers, valves, etc. The engines were under constant control by the main engine controllers. These consisted of an electronics package mounted on each engine to control engine operation under strict and critical performance parameters. The engines ran at 104.5% performance for much of the entire operation, except when they were throttled down to about

Space Shuttle Main Engine

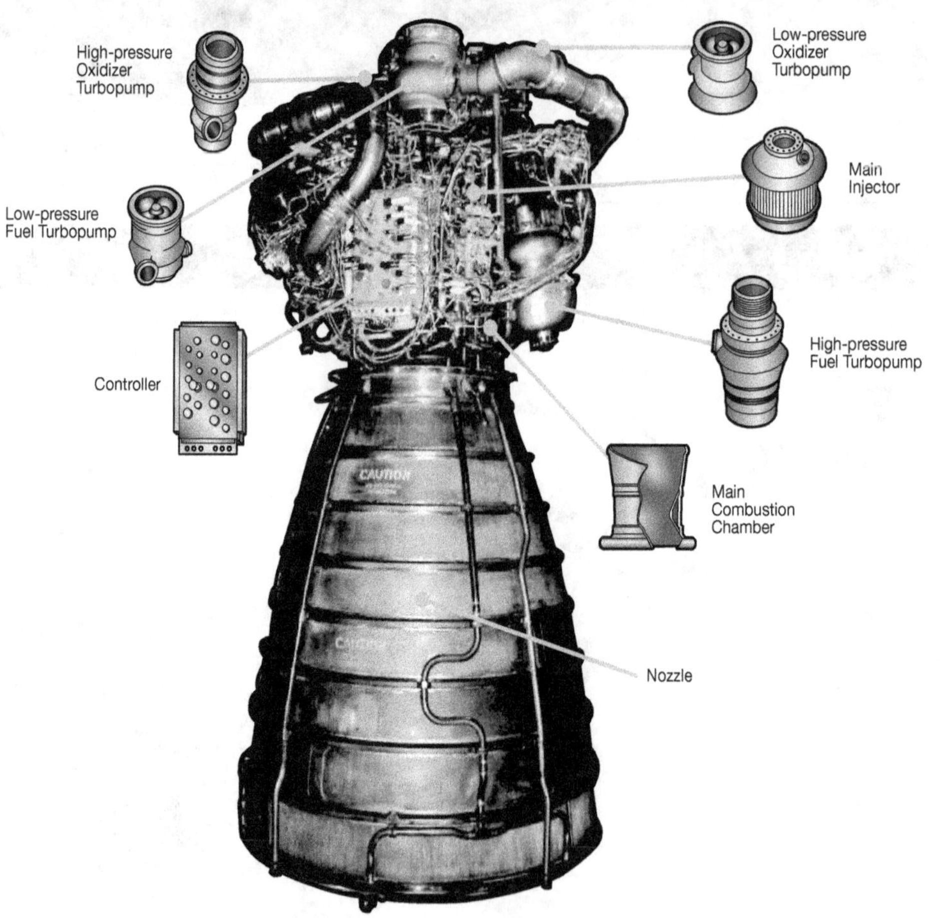

James Thompson, Jr.
Space Shuttle Main Engine project manager (1974-1982).
Deputy NASA administrator (1989-1991).

"A major problem we had to solve for the Space Shuttle Main Engine was called rotary stability subsynchronous whorl. After numerous theories and suggestions on turbo machinery, Joe Stangler at Rocketdyne and his team came up with the vortex and vaporizing theory down in the passages. He proposed putting a paddle in the flow stream and killing the vortex. Even though I thought Joe's theory was crazy, 2 weeks before the program review they had success. Government, industries, and universities all contributed to its success."

72% during first stage to preclude having the vehicle exceed structural limits during high dynamic pressure as well as close to main engine shutdown to preclude the vehicle from exceeding 3 gravitational force (3g) limits.

The only manual main engine control capability available to the crew was the manual throttle control, which allowed the crew to decrease engine performance from 104.5% to a level of 72% if required for vehicle control. The main engines had the capability to gimbal about 10.5 degrees up and down and 8.5 degrees to either side to change the thrust direction required for changes in trajectory parameters.

Orbiter

The Orbiter was the primary component of the shuttle; it carried the crew members and mission cargo/payload hardware to orbit. The Orbiter was about 37.1 m (122 ft) long with a wingspan of about 23.8 m (78 ft). The cargo/payload carrying capacity was limited by the 18.3-m- (60-ft)-long by 4.6-m- (15-ft)-wide payload bay. The cargo/payload weighed up to 29,000 kg (65,000 pounds), depending on the desired orbital inclination. The Orbiter payload bay doors, which were constructed of graphite epoxy composite material, were 18.3 m (60 ft) in length and 4.5 m (15 ft) in diameter and rotated through an angle of 175 degrees. A set of radiator panels, affixed to each door, dissipated heat from the crew cabin avionic systems.

The first vehicle, Columbia, was the heaviest Orbiter fabricated due to the installation of additional test instrumentation required to gather data on vehicle performance. As each Orbiter was fabricated, the test instrumentation was deleted and system changes implemented, resulting in each subsequent vehicle being built lighter.

The Orbiter crew cabin consisted of the flight deck and the middeck and could be configured for a maximum crew size of seven astronauts, including their required equipment to accomplish the mission objectives. The flight deck contained the Orbiter cockpit and aft station where all the vehicle and systems controls were located. The crew used six windows in the forward cockpit, two windows overhead, and two windows looking aft for orbit operations and viewing. The middeck was mostly the crew accommodations area, and it housed all the crew equipment required to live and work in space. The middeck also contained the three avionic bays where the Orbiter electronic boxes were installed. Due to their limited power generation capability, the Orbiter fuel cells consumables (power generation cryogenics) provided mission duration capability on the order of about 12 to 14 days, dependent on vehicle configuration.

In 2006, NASA put into place the Station-to-Shuttle Power Transfer System, which allowed the ISS to provide power to the Orbiter vehicle, thereby allowing the Orbiter to have a total mission duration of about 16 days. The Orbiter configuration (amount of propellant loaded in the forward and aft propellant tanks, payload mounting hardware in the payload bay, loading of cryogenic tanks required for power generation, crew size, etc.) was adjusted and optimized throughout the pre-mission process.

Because of its payload size and robotic arm capability, the Orbiter could be configured to perform as a platform for different cargo/payload hardware configurations. In the total 132 Space Shuttle missions (as of October 2010) over a period of 29 years, the Orbiter deployed a multitude of satellites for Earth observation and telecommunications; interplanetary probes such as Galileo/Jupiter spacecraft and Magellan/Venus Radar Mapper; and great observatories that included the Hubble Space Telescope, Compton Gamma Ray Observatory, and Chandra X-ray Observatory. The Orbiter even functioned as a science platform/laboratory; e.g., Spacelab, Astronomy Ultraviolet Telescope, US Microgravity Laboratory, US Microgravity Payload, etc. Aside from the experiments and satellite deployments the shuttle performed, its most important accomplishment was the delivery and assembly of the ISS.

Space Shuttle Reusability

All components of the Space Shuttle vehicle, except for the ET, were designed to be reusable flight after flight. The ET, once jettisoned from the Orbiter, fell to Earth where atmospheric heating caused the tank to break up over the ocean.

The SRBs, once jettisoned from the tank, parachuted back to the ocean where they were recovered by special ships and brought back to KSC. With their solid propellant spent, the boosters were de-stacked and shipped back to aerospace and defense company Thiokol in Utah for refurbishment and reuse. The SRBs were thoroughly inspected after every mission to ensure that the components were not damaged and could be refurbished for another flight. Any damage found was either repaired or the component was discarded.

The Orbiter was the only fully reusable component of the shuttle system. Each Orbiter was designed and certified for 100 space missions and required about 5 months, once it landed, to service the different systems and configure the payload bay to support requirements for its next mission. NASA replaced the components only when they sustained a system failure and could not be repaired. Even though certified for 100 missions, Discovery, Atlantis, and Endeavour completed 39, 32, and 25 missions, respectively, by October 2010. Challenger flew 10 missions and Columbia flew 28 missions before their loss on January 28, 1986, and February 1, 2003, respectively.

The Orbiter

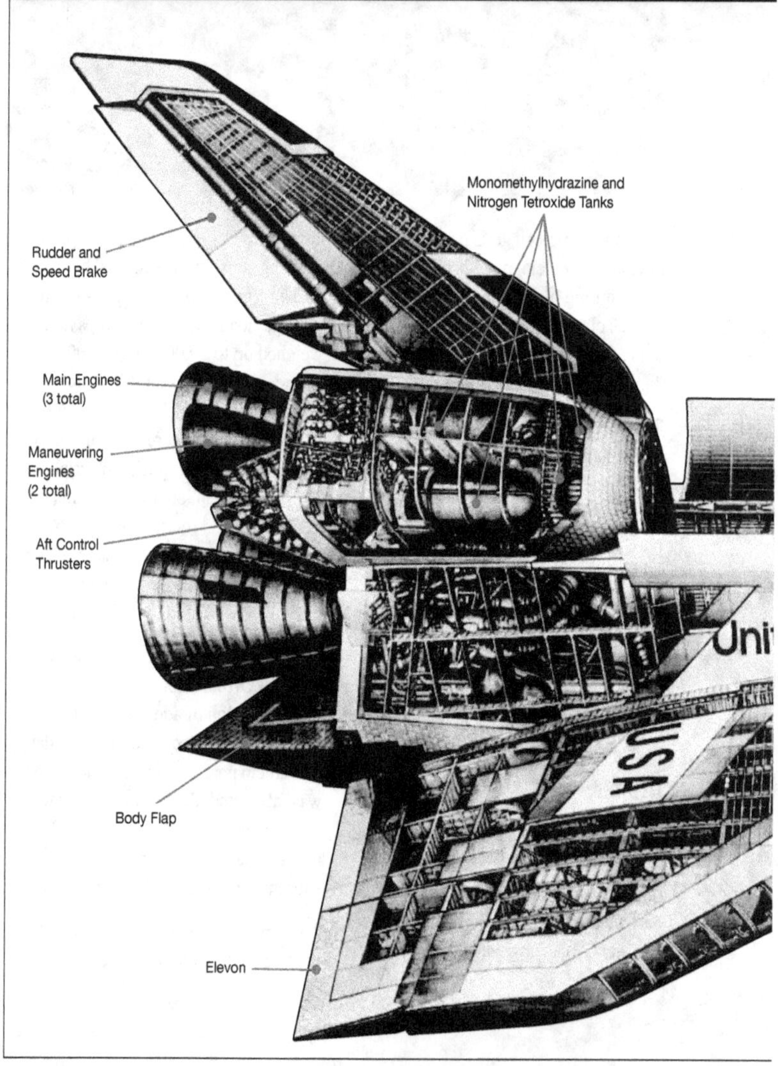

Typical Flight Profile

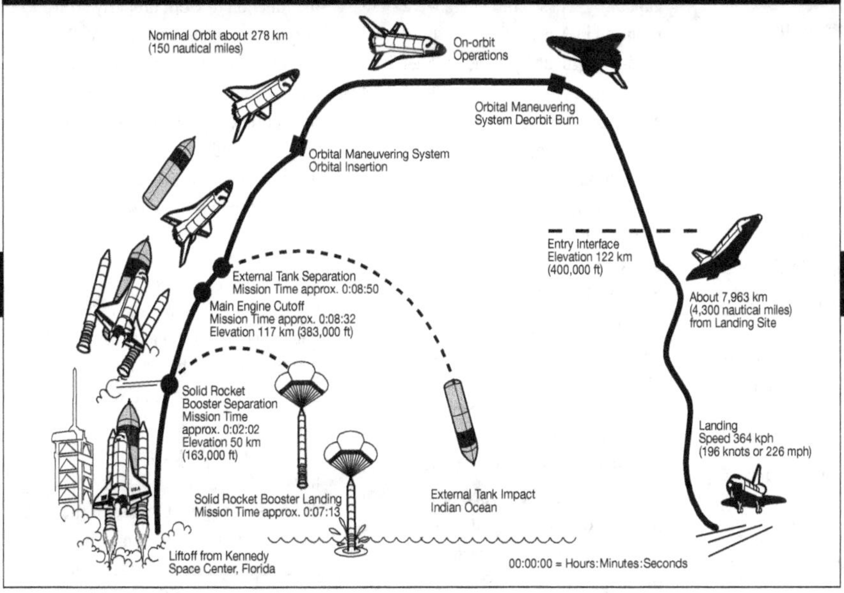

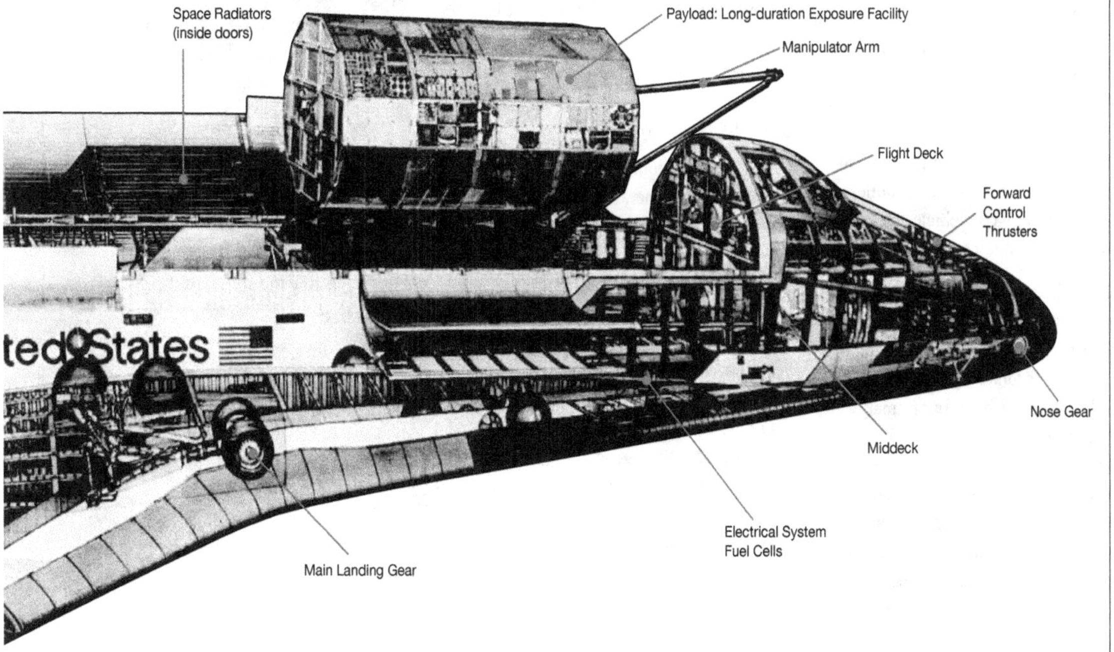

Automation, Autonomy, and Redundancy

The Space Shuttle was the first space vehicle to use the fly-by-wire computerized digital flight control system. Except for manual switch throws for system power-up and certain valve actuations, control of the Orbiter systems was through the general purpose computers installed in the forward avionics bay in the middeck.

Each Orbiter had five hardware-identical general purpose computers; four functioned as the primary means to control the Orbiter systems, and one was used as a backup should a software anomaly or problem cause the loss of the four primary computers. During ascent and re-entry—the critical phases of flight—four general purpose computers were used to control the spacecraft. The primary software, called the Primary Avionics Software System, was divided into two major systems: system software, responsible for computer operation, synchronization, and management of input and output operations; and applications software, which performed the actual duties required to fly the vehicle and operate the vehicle systems.

Even though simple in their architecture compared to today's computers, the general purpose computers had a complex redundancy management scheme in which all four primary computers were tightly coupled together and processed the same information at the same time. This tight coupling was achieved through synchronization steps and cross-check results of their processes about 440 times per second. The original International Business Machines computers had only about 424 kilobytes of memory each. The central processing unit could process about 400,000 instructions per second and did not have a hard disk drive capability. These computers were replaced in April 1991 (first flight was STS-37) with an upgraded model that had about 2.5 times the memory capacity and three times the processor speed. To protect against corrupt software, the general purpose computers had a backup computer that operated with a completely different code independent of the Primary Avionics Software System. This fifth computer, called the Backup Flight System, operated in the background, processing the same critical ascent/re-entry functions in case the four general purpose computers failed or were corrupted by problems with their software. The Backup Flight System could be engaged at any moment only by manual crew command, and it also performed oversight and management of Orbiter noncritical functions. For the first 132 flights of the Space Shuttle Program, the Backup Flight System computer was never engaged and, therefore, was not used for Orbiter control.

The overall avionics system architecture that used the general purpose computer redundancy was developed with a redundancy requirement for fail-operational/fail-safe capability. These redundancy schemes allowed for the loss of redundancy in the avionics systems and still allowed continuation of the mission or safe landing of the Orbiter. All re-entry critical avionics functions, such as general purpose computers, aero surface actuators, rate gyro assemblies, accelerometer assemblies, air data transducer assemblies, etc., were designed with four levels of redundancy. This meant that each of these functions was controlled by four avionic boxes that performed the same specific function. The loss of the first box allowed for safe continuation of the mission. The loss of the second box still allowed the function to work properly with only two remaining boxes, which subsequently allowed for safe re-entry and landing of the Orbiter. Other critical functions were designed with only triple redundancy, which meant that fail-operational/fail-safe reliability allowed the loss of two of the boxes before the function was lost.

The avionics systems redundancy management scheme was essentially controlled via computer software that operated within the general purpose computers. This scheme was to select the middle value of the avionics components when the systems had three or four avionics boxes executing the same function. On loss of the first box, the redundancy management scheme would down mode to the "average value" of the input received from the functioning boxes. Upon the second box failure, the scheme would further down mode to the "use value," which essentially meant that the function was performed by using input data from only one remaining unit in the system. This robust avionics architecture allowed the loss of avionics redundancy within a function without impacting the ability of the Orbiter to perform its required mission.

Maneuverability, Rendezvous, and Docking Capability

Maneuverability

The Orbiter was very maneuverable and could be tightly controlled in its pointing accuracy, depending on the objective it was trying to achieve. The Orbiter controllability and pointing capability was performed by the use of 44 Reaction Control System thrusters installed both in the forward and the aft portions of the

Reaction Control Thrusters and Orbital Maneuvering System

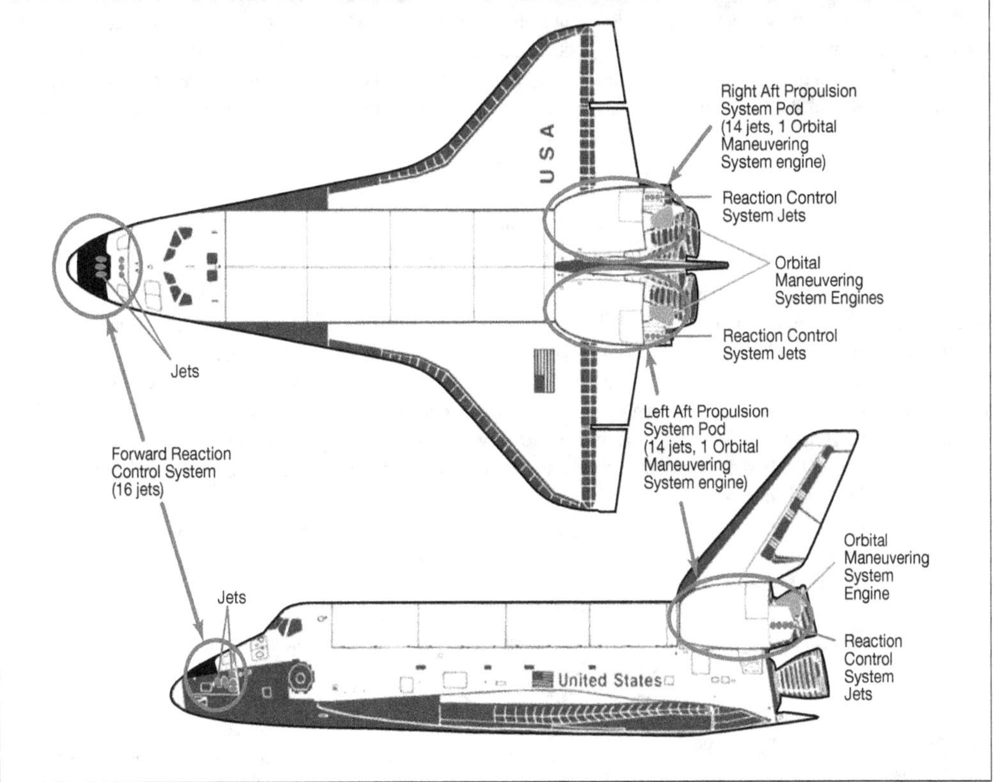

vehicle. Of the 44 thrusters, six were Reaction Control Systems and each had a thrust level of only 111 newtons (25 pounds-force). The remaining 38 thrusters were considered primary thrusters and each had a thrust level of 3,825 newtons (860 pounds-force).

The total thruster complement was divided between the forward thrusters located forward of the crew cabin, and the aft thrusters located on the two Orbital Maneuvering System pods in the tail of the Orbiter. The forward thrusters (total of 16) consisted of 14 primary thrusters and two vernier thrusters. Of the 28 thrusters in the aft, 24 were primary thrusters and four were vernier thrusters. The thrusters were installed on the Orbiter in such a way that both the rotational and the translational control was provided to each of the Orbiter's six axes of control with each axis having either two or three thrusters available for control.

The Orbital Maneuvering System provided propulsion for the shuttle. During the orbit phase of the flight, it was used for the orbital maneuvers needed to achieve orbit after the Main Propulsion System had shut down. It was also the primary propulsion system for orbital transfer maneuvers and the deorbit maneuver.

The general purpose computers also controlled the tight Orbiter attitude and pointing capability via the Orbiter Digital Auto Pilot—a key piece of application software within the computers. During orbit operations, the Digital Auto Pilot was the primary means for the crew to control Orbiter pointing by the selection of different attitude and attitude rate deadbands, which varied between +/-1.0 and 5.0 degrees for attitude and +/-0.02 and 0.2 deg/sec for attitude rate. The Digital Auto Pilot could perform three-axis automatic maneuver, attitude tracking, and rotation about any axis or body vector. Crew interface to the Digital

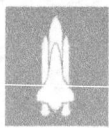

Auto Pilot was via the Orbiter cathode ray tubes/keyboard interface, which allowed the crew to control parameters in the software. With very accurate control of its orientation, the Orbiter could provide a pointing capability to any part of the celestial sky as required to accomplish its mission objectives.

Rendezvous and Docking

The shuttle docked to, grappled, deployed, retrieved, and otherwise serviced a more diverse set of orbiting objects than any other spacecraft in history. It became the world's first general purpose space rendezvous vehicle. Astronauts retrieved payloads no larger than a refrigerator and docked to targets as massive as the ISS, despite the shuttle being designed without specific rendezvous targets in mind. In fact, the shuttle wasn't designed to physically dock with anything; it was intended to reach out and grapple objects with its robotic arm.

A rendezvous period lasted up to 4 days and could be divided into three phases: ground targeted; on-board targeted; and human-piloted proximity operations. The first phase began with launch into a lower orbit, which lagged the target vehicle. The Orbiter phased toward the target vehicle due to the different orbital rates caused by orbital altitude. Mission Control at Johnson Space Center tracked the shuttle via ground assets and computed orbital burn parameters to push the shuttle higher toward the target vehicle. As the shuttle neared the target, it transitioned to on-board targeting using radar and star trackers. These sensors provided navigation data that allowed on-board computers to calculate subsequent orbital burns to reach the target vehicle.

The final stage of rendezvous operations—proximity operations—began with the Orbiter's arrival within thousands of meters (feet) of the target orbital position. During proximity operations, the crew used their highest fidelity sensors (laser, radar, or direct measurement out the window with a camera) to obtain the target vehicle's relative position. The crew then transitioned to manual control and used the translational hand controller to delicately guide the Orbiter in for docking or grappling operations.

The first rendezvous missions targeted satellite objects less massive than the shuttle and grappled these objects with its robotic arm. During the proximity operations phase, the commander only had a docking camera view and accompanying radar information to guide the vehicle. Other astronauts aimed payload bay cameras at the target and recorded elevation angles, which were charted on paper to give the commander awareness of the Orbiter's position relative to the target. Once the commander maneuvered into a position where the target was above the payload bay, a mission specialist grappled the target with the robotic arm. This method proved highly reliable and applicable to a wide array of rendezvous missions.

Shuttle rendezvous needed a new strategy to physically dock with large vehicles: the Russian space station Mir and the ISS. Rendezvous with larger space stations required more precise navigation, stricter thruster plume limitations, and tighter tolerances during docking operations. New tools such as the laser sensors provided highly accurate range and range rate information for the crew. The laser was mounted in the payload bay and its data were routed into the shuttle cabin but could not be incorporated directly into the shuttle guidance, navigation, and control software. Instead, data were displayed on and controlled by a laptop computer mounted in the aft cockpit. This laptop hosted software called the Rendezvous Proximity Operations Program that displayed the Orbiter's position relative to the target for increased crew situational awareness. This display was used extensively by the commander to manually fly the vehicle from 610 m (2,000 ft) to docking.

This assembly of hardware and software aptly met the increased accuracy required by delicate docking mechanisms and enabled crews to pilot the massive shuttle within amazing tolerances. In fact, during the final 0.9 m (3 ft) of docking with the ISS, the Orbiter had to maintain a 7.62-cm (3-in.) lateral alignment cylinder and the closing rate had to be controlled to within 0.02 m/sec (0.06 ft/sec). The commander could control this with incredibly discrete pulses of the Reaction Control System thrusters. Both the commander and the pilot were trained extensively in the art of shuttle proximity operations, learning techniques that allowed them to pilot the Orbiter to meet tolerances. The shuttle was never meant to be piloted to this degree of accuracy, but innovative engineering and training made these dockings uneventful and even routine.

The success of shuttle rendezvous missions was remarkable considering its operational complexity. Spacecraft rendezvous is an art requiring the highly scripted choreography of hardware systems, astronauts, and members of Mission Control. It is a precise and graceful waltz of billions of dollars of hardware and human decision making.

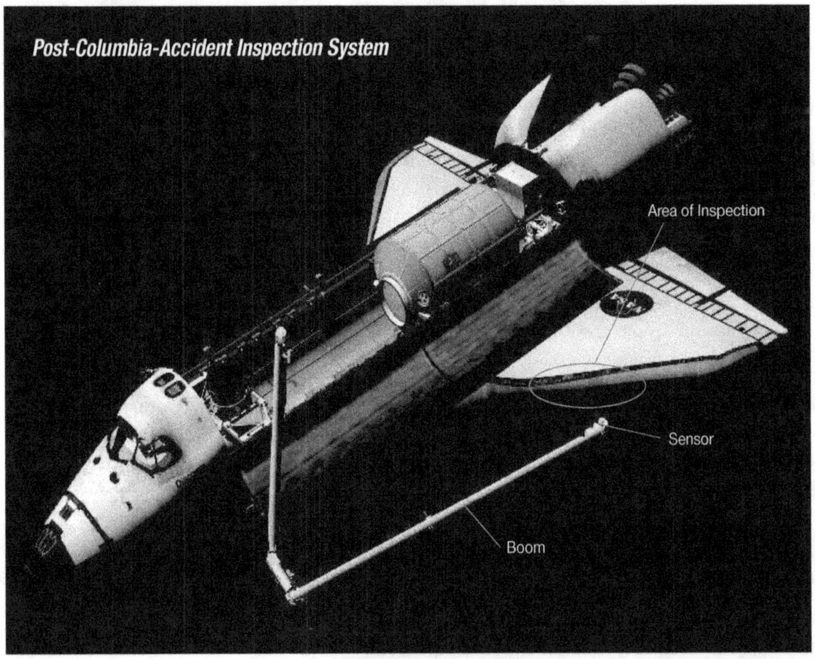

The Orbiter Boom Sensor System inspects the wing leading edge. This system was built for inspections after the Columbia accident (STS-107 [2003]).

STS-88 (1998) Endeavour's Shuttle Robotic Arm grapples the Russian Module Zarya for berthing onto the International Space Station (ISS) Node 1, thus beginning the assembly sequence for the ISS.

Robotic Arm/Operational Capability

The Canadian Space Agency provided the Shuttle Robotic Arm. It was designed, built, and tested by Spar Aerospace Ltd., a Canadian Company. The electromechanical arm measured about 15 m (50 ft) long and 0.4 m (15 in.) in diameter with a six-degree-of-freedom rotational capability, and it consisted of a manipulator arm that was under the control of the crew via displays and control panels located in the Orbiter aft flight deck. The Shuttle Robotic Arm was comprised of six joints that corresponded roughly to the joints of a human arm and could handle a payload weighing up to 29,000 kg (65,000 pounds). An end effector was used to grapple a payload or any other fixture and/or component that had a grapple fixture for handling by the arm.

Even though NASA used the Shuttle Robotic Arm primarily for handling payloads, it could also be used as a platform for extravehicular activity (EVA) crew members to attach themselves via a portable foot restraint. The EVA crew member, affixed to the portable foot restraint grappled by the end effector, could then be maneuvered around the Orbiter vehicle as required to accomplish mission objectives.

Following the Return to Flight after the loss of Columbia, the Shuttle Robotic Arm was used to move around the Orbiter Boom Sensor System, which allowed the flight crew to inspect the Thermal Protection System around the entire Orbiter or the reinforced carbon-carbon panels installed on the leading edge of the wings.

During buildup of the ISS, the Shuttle Robotic Arm was instrumental in the handling of modules carried by the Orbiter—a task that would not have been possible without the use of this robotic capability.

Extravehicular Activity Capability

The Space Shuttle Program provided a dramatic expansion in EVA capability for NASA, including the ability to perform tasks in the space environment and ways to best protect and accommodate a crew member in that environment. The sheer number of EVAs performed during the course of the program resulted in a significant increase in knowledge of how EVA systems and EVA crew members perform.

Prior to the start of the program, a total of 38 EVAs were performed by all US space programs combined, including Gemini, Apollo, and Skylab. During previous programs, EVAs focused primarily on simple tasks, such as the jettison of expended hardware or the collection of geology samples. The Space Shuttle Program advanced EVA capability to construction of massive space structures, high-strength maneuvers, and repair of complicated engineering components requiring a combination of precision and gentle handling of sensitive materials and structures. As of October 2010, the shuttle accomplished about 157 EVAs in 132 flights. Of those EVAs, 105 were dedicated to ISS assembly and repair tasks. Shuttle EVA crews succeeded in handling and manipulating elements as large as 9,000 kg (20,000 pounds); relocating and installing large replacement parts; capturing and repairing failed satellites; and performing surgical-like repairs of delicate solar arrays, rotating joints, and much more.

The Orbiter's EVA capability consisted of several key engineering components and equipment. For a crew member to step out of the shuttle and safely enter the harsh environment of space, that crew member had to use the integrated airlock, an extravehicular mobility unit spacesuit, a variety of EVA tools, and EVA translation and attachment aids attached to the vehicle or payload. EVA tools consisted of a suite of components that assisted in handling and translating cargo, translating and stabilizing at the work site, operating manual mechanisms, and attaching bolts and fasteners, often with relatively precise torque requirements. Photo and television operations provided documentation of the results for future troubleshooting, when necessary.

Extravehicular Mobility Unit

The extravehicular mobility unit was a fully self-sufficient individual spacecraft providing critical life support systems and protection from the harsh space environment. Unlike previous suits, the shuttle suit was designed specifically for EVA and was the cornerstone component for safe conduct of EVA during the shuttle era. It operated at 0.03 kgf/cm^2 (4.3 psi) pressure in the vacuum environment and provided thermal protection for interfacing with environments and components from -73°C (-100°F) to 177°C (350°F). It provided oxygen and removed carbon dioxide during an EVA, and it supplied battery power to run critical life support and ancillary extravehicular mobility unit systems, including support lights, cameras, and radio. The suit, which also provided crew members with critical feedback on system operations during EVA, was the first spacesuit controlled by a computer.

Future space programs will benefit tremendously from NASA's EVA experience during the shuttle flights. To ensure success, the goal has been and always will be to design for EVAs that are as simple and straightforward as possible. Fewer and less-complicated provisions will be required for EVA interfaces on spacecraft, and functions previously thought to require complicated and automated systems can now rely on EVA instead. During the shuttle era, NASA took the training wheels off of EVA capability and now has a fully developed and highly efficient operational resource in support of both scheduled and contingency EVA tasks.

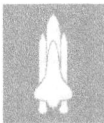

Crew Compartment Accommodation for Crew and Payloads

The Orbiter's crew cabin had a habitable volume of 71.5 m^3 (2,525 ft^3) and consisted of three levels: flight deck, middeck, and utility area. The flight deck, located on the top level, accommodated the commander, pilot, and two mission specialists behind them. The Orbiter was flown and controlled from the flight deck. The middeck, located directly below the flight deck, accommodated up to three additional crew members and included a galley, toilet, sleep locations, storage lockers, and the side hatch for entering and exiting the vehicle. The Orbiter airlock was also located

Flight Deck

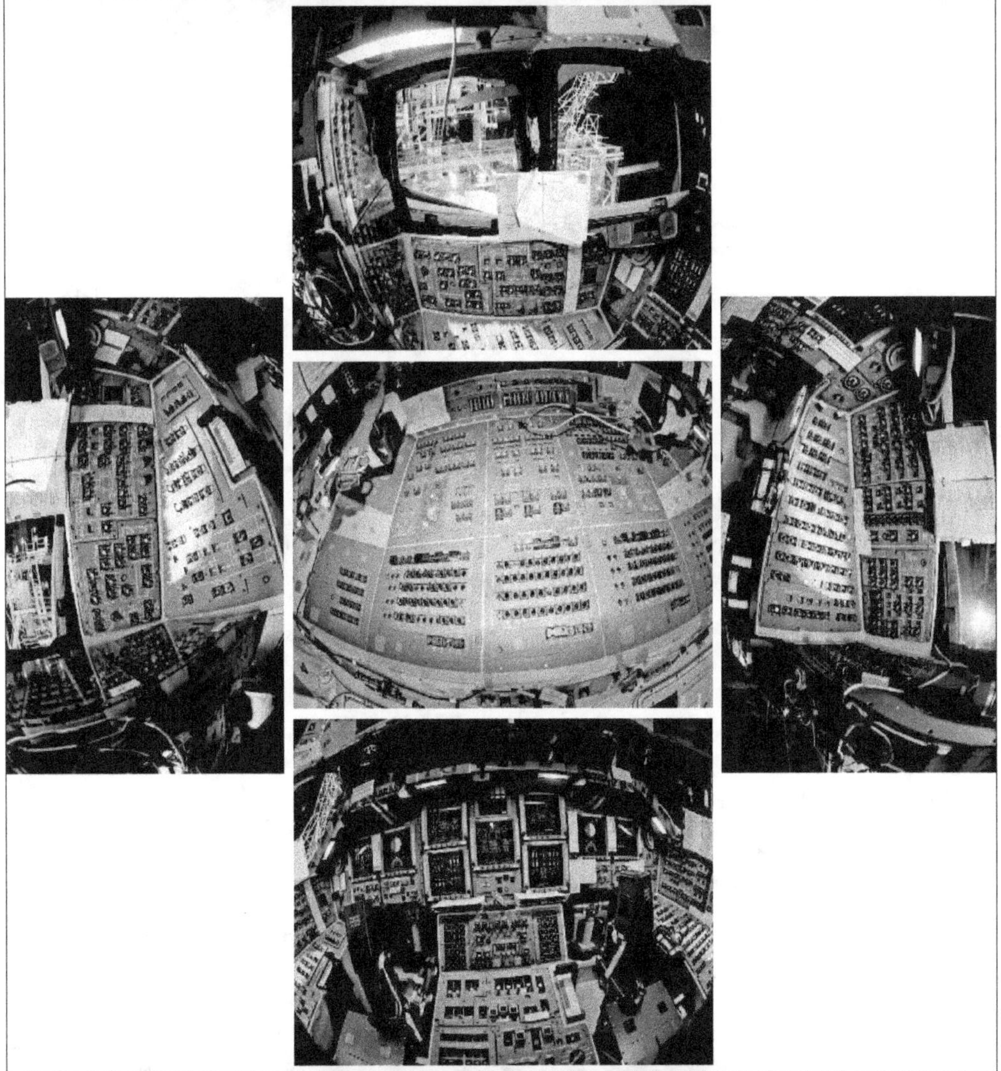

Flight deck showing the commander and pilot seats, along with cockpit controls.

in the middeck area; it allowed up to three astronauts, wearing extravehicular mobility unit spacesuits, to perform an EVA in the vacuum of space. The standard practice was for only two crew members to perform an EVA.

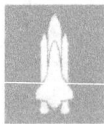

Most of the day-to-day mission operations took place on the middeck. The majority of hardware required for crew members to live, work, and perform their mission objectives was stowed in stowage lockers and bags within the middeck volume. The entire middeck stowage capability was equivalent to 127.5 middeck lockers in which each locker was about 0.06 m^3 (2 ft^3) in volume. This volume could accommodate all required equipment and supplies for a crew of seven for as many as 16 days.

Middeck

Crew compartment middeck configuration showing the forward middeck lockers in Avionics Bay 1 and 2, crew seats, and sleeping bags.

Performance Capabilities and Limitations

Throughout the history of the program, the versatile shuttle vehicle was configured and modified to accomplish a variety of missions, including: the deployment of Earth observation and communication satellites, interplanetary probes, and scientific observatories; satellite retrieval and repair; assembly; crew rotation; science and logistics resupply of both the Russian space station Mir and the ISS, and scientific research and operations. Each mission type had its own capabilities and limitations.

Deploying and Servicing Satellites

The largest deployable payload launched by the shuttle in the life of the program was the Chandra X-ray Observatory. Deployed in 1999 at an inclination of 28.45 degrees and an altitude of about 241 km (130 nautical miles), Chandra—and the support equipment deployed with it—weighed 22,800 kg (50,000 pounds).

In 1990, NASA deployed the Hubble Space Telescope into a 28.45-degree inclination and a 555-km (300-nautical-mile) altitude. Hubble weighed 13,600 kg (30,000 pounds). Five servicing missions were conducted over the next 19 years to upgrade Hubble's science instrumentation, thereby enhancing its scientific capabilities. These subsequent servicing missions were essential in

Maximum Deploy Capability for Each Vehicle Versus Altitude

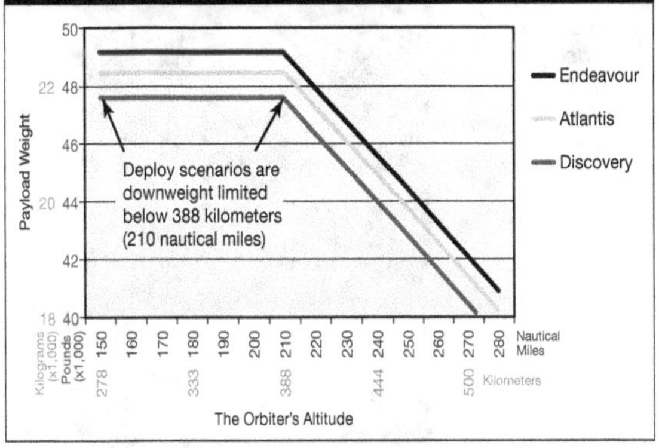

The up-mass capability of the shuttle decreased relative to the orbital altitude.

Atlantis' (STS-125 [2009]) robotic arm lifts Hubble from the cargo bay and is moments away from releasing the orbital observatory to start it on its way back home to observe the universe.

correcting the Hubble mirror spherical aberration, thereby extending the operational life of the telescope and upgrading its science capability.

Assembling the International Space Station

The ISS Node 1/Unity module was launched on STS-88 (1998), thus beginning the assembly of the ISS, which required a total of 36 shuttle missions to assemble and provide logistical support for ISS vehicle operations. As of October 2010, Discovery had flown 12 missions and Atlantis and Endeavour had flown 11 missions to the ISS, with each mission carrying 12,700 to 18,600 kg (28,000 to 41,000 pounds) of cargo in the cargo bay and another 3,000 to 4,000 kg (7,000 to 9,000 pounds) of equipment stowed in the crew cabin. The combined total of ISS structure, logistics, crew, water, oxygen, nitrogen, and avionics delivered to the station for all shuttle visits totaled more than 603,300 kg (1,330,000 pounds). No other launch vehicle in the world could deliver these large 4.27-m- (14-ft)-diameter by 15.24-m- (50-ft)-long structures or have this much capability.

ISS missions required modifications to the three vehicles cited above—Discovery, Atlantis, and Endeavour—to dock to the space station. The docking requirement resulted in the Orbiter internal airlock being moved externally in the payload bay. This change, along with the inclusion of the docking mechanism, added about 1,500 kg (3,300 pounds) of mass to the vehicle weight.

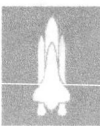

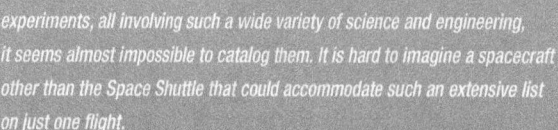

Kenneth Reightler
Captain, US Navy (retired). Pilot on STS-48 (1991) and STS-60 (1994).

"When I think about the legacy of the Space Shuttle Program in terms of scientific and engineering accomplishments, the word that comes to mind is versatility. Each of my flights involved so many projects and experiments, all involving such a wide variety of science and engineering, it seems almost impossible to catalog them. It is hard to imagine a spacecraft other than the Space Shuttle that could accommodate such an extensive list on just one flight.

"The shuttle's large cargo bay could hold large, complex structures or many small experiments, an amazing variety of experiments. We carried big, intricate satellites as well as smaller, simpler ones able to be deployed remotely or using robotic and/or human assistance.

"For me, as an engineer and a pilot, it was an unbelievable experience to now be conducting world-class science in a range of disciplines with the potential to benefit so many people back on Earth, such as experiments designed to help produce vaccines used to eradicate deadly diseases, to produce synthetic hormones, or to develop countermeasures for the effects of aging. I consider it to be a rare honor and privilege to have operated experiments to which so many scientists and engineers had devoted their time, energy, and thought. In some cases, people had spent entire careers preparing for the day when their experiments could be conducted, knowing that they could only work in space and there might be only one chance to try.

"Each of my flights brought moments of pride and satisfaction in such singular experiences."

A Platform for Scientific Research

The Orbiter was configured to accommodate many different types of scientific equipment, ranging from large pressurized modules called Spacelab or Spacehab where the crew conducted scientific research in a shirt-sleeve environment to the radars and telescopes for Earth mapping, celestial observations, and the study of solar, atmospheric, and space plasma physics. The shuttle was often used to deploy and retrieve science experiments and satellites. These science payloads were: deployed using the Shuttle Robotic Arm; allowed to conduct free-flight scientific operations; and then retrieved using the arm for return to Earth for further data analysis. This was a unique capability that only the Orbiter could perform.

The Orbiter was also unique because it was an extremely stable platform on which to conduct microgravity research studies in material, fundamental physics, combustion science, crystal growth, and biotechnology that required minimal movement or disturbance from the host vehicle. NASA studied the effect of space adaptation on both humans and animals. Crews of seven worked around the clock conducting research in these pressurized modules/laboratories that were packed with scientific equipment.

Much research was conducted with the international community. These missions brought together international academic, industrial, and governmental

The crew from the International Space Station captured this view of STS-97 (2000).

Franklin Chang-Díaz, PhD
Astronaut on STS-61C (1986), STS-34 (1989), STS-46 (1992), STS-60 (1994), STS-75 (1996), STS-91 (1998), and STS-111 (2002).

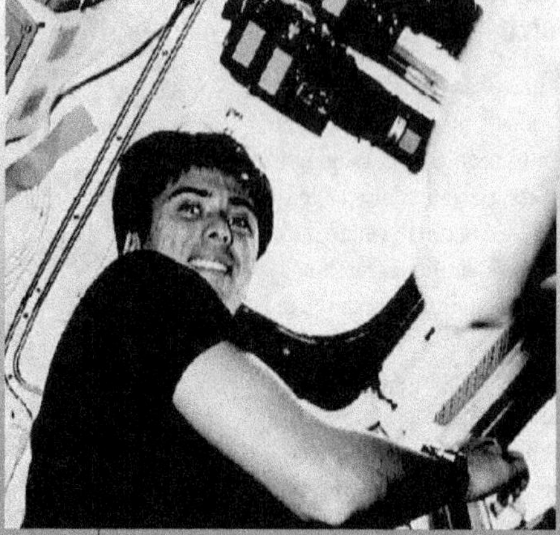

Memories of Wonder

"We have arrived at the base of the launch pad, dressed for the occasion in bright orange pressure suits that fit worse than they look. This is the day! As we enter the service elevator that will take us 193 feet up to the level of the shuttle cabin, we get to appreciate the size of this ship, the mighty solid rockets that hold the gargantuan External Tank and the seemingly fragile shuttle craft, poised on this unlikely contraption like a gigantic moth, gathering strength, for she knows full well where she is going today. One by one, between nervous smiles and sheer anticipation, we climb into our ship, aided by expert technicians who execute their tasks with seamless and clockwork precision, while soothing our minds with carefree conversation. The chatter over the audio channels reverberates, unemotional, precise, relentless, and the countdown clock is our master. We often say that, on launch day, the ship seems alive, hissing and creaking with the flow of the super-cold fluids that give her life. Over the course of 3 hours, waiting patiently for the hour of deliverance, we have each become one with the Orbiter. The chatter has subsided, the technicians have gone. It is just us now, our orange cocoons securely strapped and drawing the sap of the mother ship through multiple hoses and cables. It feels cozy and safe, alas, our comfort is tempered by the knowledge of the machine and the job we are about to do. 'GLS is go for main engine start…' sounds the familiar female voice. The rumbling below signals the beginning of an earthquake. We feel a sudden jolt, the ship is free and she flies! We feel the shaking and vibration and the onset of the 'g' forces that build up uncomfortably, squeezing our chests and immobilizing our limbs as the craft escapes the pull of the Earth. And in less than 9 minutes, we are in space. The view is the most beautiful thing we ever saw and we will see this over and over from what is now our new home in the vacuum of space. The days will pass and this extraordinary vehicle will carry us to our destination…to our destiny. It has learned to dance in space, with exquisite precision and grace, first alone, then with other lonely dates, the Hubble telescope, the Russian Mir station and the International Space Station, and when the job was done, it returned to land softly, majestically, triumphant…and ready to do it all again."

partners to obtain maximum benefits and results. The facilities included middeck glove boxes for conducting research and testing science procedures and for developing new technologies in microgravity. These boxes enabled crew members to handle, transfer, and manipulate experiment hardware and material that were not approved for use in the shuttle. There were furnaces to study diffusion, and combustion modules for conducting research on the single most important chemical process in our everyday lives. The shuttle had freezers for sample return as well as the capability to store large amounts of data for further analysis back on Earth. Scientists used spin tables to conduct biological and physiological research on the crew members.

The Orbiter provided all the power and active cooling for the laboratories. A typical Spacelab was provided approximately 6.3 kW (8.45 hp) of power, with peak power as high as 8.1 kW (10.86 hp). To cool the laboratories' electronics, the modules were tied into the Orbiter's cooling system so thermal control of the payload was the same as thermal control for the Orbiter avionics.

In an effort to share this national resource with industry and academia, NASA developed the Get Away Special Program, designed to provide inexpensive access to space for both novices and professionals to explore new concepts at little risk. In total, over 100 Get Away Special payloads were flown aboard the shuttle, and each payload often consisted of several individual experiments. The cylindrical payload canisters in which these experiments were flown measured 0.91 m (3 ft) in length with a 0.46-m (1.5-ft) diameter. They were integrated into the Orbiter cargo bay on the sill/sidewall and required minimal space and cargo integration engineering. The experiments could be confined inside a sealed canister, or the canister could be configured with a lid that could be opened for experiment pointing or deployment.

The shuttle was also an extremely accurate platform for precise pointing of scientific payloads at the Earth and celestial targets. These unpressurized payloads were also integrated into the cargo bay; however, unlike the Spacelab and Spacehab science modules, these payloads were not accessible by the crew, but rather were exposed to the space environment. The crew activated and operated these experiments from the pressurized confines of the Orbiter flight deck. The Shuttle Radar Topography Mission was dedicated to mapping the Earth's topography between 60° North and 58° South, including the ocean floor. The result of the mission was a three-dimensional digital terrain map of 90% of the Earth's surface. The Orbiter provided about 10 kW (13.4 hp) of power to the Shuttle Radar Topography Mission payload during on-orbit operations and all of the cooling for the payloads' electronics.

An Enduring Legacy

The shuttle was a remarkable, versatile, complex piece of machinery that demonstrated our ingenuity for human exploration. It allowed the United States and the world to perform magnificent space missions for the benefit of all. Its ability to deploy satellites to explore the solar system, carry space laboratories to perform human/biological/material science, and carry different components to assemble the ISS were accomplishments that will not be surpassed for years to come.

Processing the Shuttle for Flight

Steven Sullivan

Preparing the Shuttle for Flight

Ground Processing
Jennifer Hall
Peter Nickolenko
Jorge Rivera
Edith Stull
Steven Sullivan

Space Operations Weather
Francis Merceret
Robert Scully
Terri Herst
Steven Sullivan
Robert Youngquist

When taking a road trip, it is important to plan ahead by making sure your vehicle is prepared for the journey. A typical road trip on Earth can be routine and simple. The roadways are already properly paved, service stations are available if vehicle repairs are needed, and food, lodging, and stores for other supplies can also be found. The same, however, could not be said for a Space Shuttle trip into space. The difficulties associated with space travel are complex compared with those we face when traveling here. Food, lodging, supplies, and repair equipment must be provided for within the space vehicle.

Vehicle preparation required a large amount of effort to restore the shuttle to nearly new condition each time it flew. Since it was a reusable vehicle with high technical performance requirements, processing involved a tremendous amount of "hands-on" labor; no simple tune-up here. Not only was the shuttle's exterior checked and repaired for its next flight, all components and systems within the vehicle were individually inspected and verified to be functioning correctly. This much detail work was necessary because a successful flight was dependent on proper vehicle assembly. During a launch attempt, decisions were made within milliseconds by equipment and systems that had to perform accurately the first time—there was no room for hesitation or error. It has been said that a million things have to go right for the launch, mission, and landing to be a success, but it can take only one thing to go wrong for them to become a failure.

In addition to technical problems that could plague missions, weather conditions also significantly affected launch or landing attempts. Unlike our car, which can continue its road trip in cloudy, windy, rainy, or cold weather conditions, shuttle launch and landing attempts were restricted to occur only during optimal weather conditions. As a result, weather conditions often caused launch delays or postponed landings.

Space Shuttle launches were a national effort. During the lengthy processing procedures for each launch, a dedicated workforce of support staff, technicians, inspectors, engineers, and managers from across the nation at multiple government centers had to pull together to ensure a safe flight. The whole NASA team performed in unison during shuttle processing, with pride and dedication to its work, to make certain the success of each mission.

Preparing the Shuttle for Flight

Ground Processing

Imagine embarking on a one-of-a-kind, once-in-a-lifetime trip. Everything must be exactly right. Every flight of the Space Shuttle was just that way. A successful mission hinged on ground operations planning and execution.

Ground operations was the term used to describe the work required to process the shuttle for each flight. It included landing-to-launch processing—called a "flow"—of the Orbiter, payloads, Solid Rocket Boosters (SRBs), and External Tank (ET). It also involved many important ground systems. Three missions could be processed at one time, all at various stages in the flow. Each stage had to meet critical milestones or throw the entire flow into a tailspin.

Each shuttle mission was unique. The planning process involved creating a detailed set of mission guidelines, writing reference materials and manuals, developing flight software, generating a flight plan, managing configuration control, and conducting simulation and testing. Engineers became masters at using existing technology, systems, and equipment in unique ways to meet the demands of the largest and most complex reusable space vehicle.

The end of a mission set in motion a 4- to 5-month process that included more than 750,000 work hours and literally millions of processing steps to prepare the shuttle for the next flight.

Landing

During each mission, NASA designated several landing sites—three in the Continental United States, three overseas contingency or transatlanic abort landing sites, and various emergency landing sites located in the shuttle's orbital flight path. All of these sites had one thing in common: the commander got one chance to make the runway. The Orbiter dropped like a rock and there were no second chances. If the target was missed, the result was disaster.

Kennedy Space Center (KSC) in Florida and Dryden Flight Research Center (DFRC)/Edwards Air Force Base in California were the primary landing sites for the entire Space Shuttle Program. White Sands Space Harbor in New Mexico was the primary shuttle pilot training site and a tertiary landing site in case of unacceptable weather conditions at the other locations.

The initial six operational missions were scheduled to land at DFRC/Edwards Air Force Base because of the safety margins available on the lakebed runways. Wet lakebed conditions diverted one of those landings—Space Transportation System (STS)-3 (1982)—to White Sands Space Harbor. STS-7 (1983) was the first mission scheduled to land at KSC, but it was diverted to Edwards Air Force Base runways due to unfavorable Florida weather. The 10th shuttle flight—STS-41B (1984)—was the first to land at KSC.

Landing Systems

Similar to a conventional airport, the KSC shuttle landing facility used visual and electronic landing aids both on the ground and in the Orbiter to help direct the landing. Unlike conventional aircraft, the Orbiter had to land perfectly the first time since it lacked propulsion and landed in a high-speed glide at 343 to 364 km/hr (213 to 226 mph).

Following shuttle landing, a convoy of some 25 specially designed vehicles or units and a team of about 150 trained personnel converged on the runway. The team conducted safety checks for explosive or toxic gases, assisted the crew in leaving the Orbiter, and prepared the Orbiter for towing to the Orbiter Processing Facility.

The landing-to-launch ground operations "flow" at Kennedy Space Center prepared each shuttle for its next flight. This 4- to 5-month process required thousands of work hours and millions of individual processing steps.

Space Shuttle Atlantis landing, STS-129 (2009).

After landing, the Orbiter is moved to the Orbiter Processing Facility.

Landing

Orbiter Processing Facility: 120-130 days

Orbiter Processing

The Orbiter Processing Facility was a sophisticated aircraft hangar (about 2,700 m^2 [29,000 ft^2]) with three separate buildings or bays. Trained personnel completed more than 60% of the processing work during the approximately 125 days the vehicle spent in the facility.

Technicians drained residual fuels and removed remaining payload elements or support equipment. More than 115 multilevel, movable access platforms could be positioned to surround the Orbiter and provide interior and exterior access. Engineers performed extensive checkouts involving some 6 million parts. NASA removed and transferred some elements to other facilities for servicing. The Orbiter Processing Facility also contained shops to support Orbiter processing.

Tasks were divided into forward, midbody, and aft sections and required mechanical, electrical, and Thermal Protection System technicians, engineers, and inspectors as well as planners and schedulers. Daily activities included test and checkout schedule meetings that required coordination and prioritization among some 35 engineering systems and 32 support groups. Schedules ranged in detail from minutes to years.

Personnel removed the Orbital Maneuvering System pods and Forward Reaction Control System modules and modified or repaired and retested them in the Hypergolic Maintenance Facility. When workers completed modifications and repairs, they shipped the pods and modules back to the Orbiter Processing Facility for reinstallation.

Johnson Space Center Orbiter Laboratories

Several laboratories at Johnson Space Center supported Orbiter testing and modifications.

The Electrical Power Systems Laboratory was a state-of-the-art electrical compatibility facility that supported shuttle and International Space Station (ISS) testing. The shuttle breadboard, a high-fidelity replica of the shuttle electrical power distribution and control subsystem, was used early in the program for equipment development testing and later for ongoing payload and shuttle equipment upgrade testing. During missions, the breadboard replicated flow problems and worked out solutions.

Engineers also tested spacecraft communications systems at the Electronic Systems Test Laboratory, where multielement, crewed spacecraft communications systems were interfaced with relay satellites and ground elements for end-to-end testing in a controlled radio-frequency environment.

The Avionics Engineering Laboratory supported flight system hardware and software development and evaluation as well as informal engineering evaluation and formal configuration-controlled verification testing of non-flight and flight hardware and software. Its real-time environment consisted of a vehicle dynamics simulation for all phases of flight, including contingency aborts, and a full complement of Orbiter data processing system line replacement units.

The Shuttle Avionics Integration Laboratory was the only program test facility where avionics, other flight hardware (or simulations), software, procedures, and ground support equipment were brought together for integrated verification testing.

Inside the Orbiter Processing Facility, technicians process the Space Shuttle Main Engine and install it into the Orbiter.

Orbiter Processing Facility (continued)

Kennedy Space Center Shuttle Logistics Depot

Technicians at the Shuttle Logistics Depot in Florida manufactured, overhauled and repaired, and procured Orbiter line replacement units. The facility was certified to service more than 85% of the shuttle's approximately 4,000 replaceable parts.

At the Shuttle Logistics Depot, Rick Zeitler assesses the cycling of a main propulsion fill and drain valve after a valve anomoly during launch countdown caused a scrub.

This facility established capabilities for avionics and mechanical hardware ranging from wire harnesses and panels to radar and communications systems, and from ducts and tubing to complex actuators, valves, and regulators. Capability included all aspects of maintenance, repair, and overhaul activities.

Kennedy Space Center Tile Processing

Following shuttle landing, the Thermal Protection System—about 24,000 silica tiles and about 8,000 thermal blankets—was visually inspected in the Orbiter Processing Facility.

Thermal Protection System products included tiles, gap fillers, and insulation blankets to protect the Orbiter exterior from the searing heat of launch, re-entry into Earth's atmosphere, and the cold soak of space. The materials were repaired and manufactured in the Thermal Protection Systems Facility.

Tile technicians and engineers used manual and automated methods to fabricate patterns for areas of the Orbiter that needed new tiles. Engineers used the automotive industry tool Optigo™ to take measurements in tile cavities. Optigo™ used optics to record the hundreds of data points needed to manufacture tile accurate to 0.00254 cm (0.001 in.). Tile and external blanket repair and replacement processing included: removal of damaged tile and preparation of the cavity; machining, coating, and firing the replacement tile; and fit-checking, waterproofing, bonding, and verifying the bond.

Prior to the launch of STS-119 (2009), Discovery gets boundary layer transition tile, which monitors the heating effects of early re-entry at high Mach numbers.

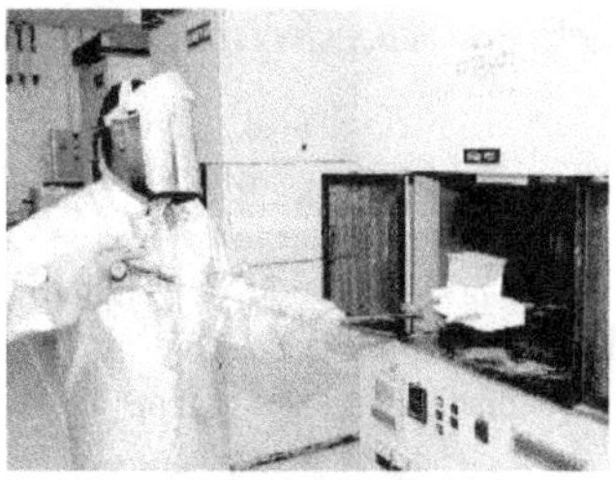

At the Kennedy Space Center tile shop, a worker places a Boeing replacement insulation 18 tile in the oven to be baked at 1,200°C (2,200°F) to cure the ceramic coating.

Solid Rocket Boosters and the External Tank are delivered to Kennedy Space Center and transported to the Vehicle Assembly Building to be readied for the Space Shuttle.

Vehicle Assembly Building: 7-9 days

Space Shuttle Main Engine Processing

Trained personnel removed the three reusable, high-performance, liquid-fueled main engines from the Orbiter following each flight for inspection. They also checked engine systems and performed maintenance. Each engine had 50,000 parts, about 7,000 of which were life limited and periodically replaced.

Solid Rocket Booster Processing

The SRBs were repaired, refurbished, and reused for future missions. The twin boosters were the largest ever built and the first designed for refurbishment and reuse. They provided "lift" for the Orbiter to a distance of about 45 km (28 miles) into the atmosphere.

Booster Refurbishment

Following shuttle launch, NASA recovered the spent SRBs from the Atlantic Ocean, disassembled them, and transported them from Florida to ATK's Utah facilities via specially designed rail cars—a trip that took about 3 weeks.

After refurbishment, the motor cases were prepared for casting. Each motor consisted of nine cylinders, an aft dome, and a forward dome. These elements were joined into four units called casting segments. Insulation was applied to the inside of the cases and the propellant was bonded to this insulation.

The semiliquid, solid propellant was poured into casting segments and cured over 4 days. Approximately forty 2.7-metric-ton (3-ton) mixes of propellant were required to fill each segment.

The nozzle consisted of layers of glass- and carbon-cloth materials bonded to aluminum and steel structures. These materials were wound at specified angles and then cured to form a dense, homogeneous insulating material capable of withstanding temperatures reaching 3,300°C (6,000°F). The cured components were then adhesively bonded to their metal support structures and the metal sections were joined to form the complete nozzle assembly.

Transporting a flight set of two Solid Rocket Motors to KSC required four major railroads, nine railcars, and 7 days.

KSC teams refurbished, assembled, tested, and integrated many SRB elements, including the forward and aft skirts, separation motors, frustum, parachutes, and nose cap.

Technicians at the Rotation Processing and Surge Facility received, inspected, and offloaded the booster segments from rail cars, then rotated the segments from horizontal to vertical and placed them on pallets.

Many booster electrical, mechanical, thermal, and pyrotechnic subsystems were integrated into the flight structures. The aft skirt subassembly and forward skirt assembly were processed and then integrated with the booster aft segments.

After a complete flight set of boosters was processed and staged in the surge buildings, the boosters were transferred to the Vehicle Assembly Building for stacking operations.

External Tank Processing

The ET provided propellants to the main engines during launch. The tank was manufactured at the Michoud Assembly Facility in New Orleans and shipped to Port Canaveral in Florida. It was towed by one of NASA's SRB retrieval ships. At the port, tugboats moved the barge upriver to the KSC turn basin. There, the

Inside the Vehicle Assembly Building, technicians complete the process of stacking the Solid Rocket Booster components.

Vehicle Assembly Building (continued)

tank was offloaded and transported to the Vehicle Assembly Building.

Payload Processing

Payload processing involved a variety of payloads and processing requirements.

The cargo integration test equipment stand simulated and verified payload/cargo mechanical and functional interfaces with the Orbiter before the spacecraft was transported to the launch pad. Payload processing began with power-on health and status checks, functional tests, computer and communications interface checks, and spacecraft command and monitor tests followed by a test to simulate all normal mission functions through payload deployment.

Hubble Space Telescope servicing missions provided other challenges. Sensitive telescope instruments required additional cleaning and hardware handling procedures. Payload-specific ground support equipment had to be installed and monitored throughout the pad flow, including launch countdown.

William Parsons

Space Shuttle program manager (2003-2005) and director of Kennedy Space Center (2007-2008).

In the firing room, William Parsons (left), director of Kennedy Space Center, and Dave King, director of Marshall Space Flight Center, discuss the imminent launch of STS-124 (2008).

"The shuttle is an extremely complex space system. It is surprising how many people and vendors touch the vehicle. At the Kennedy Space Center, it is amazing to me how we are able to move a behemoth space structure, like the Orbiter, and mate to another structure with incredibly precise tolerances."

Following processing, payloads were installed in the Orbiter either horizontally at the Orbiter Processing Facility or vertically at the launch pad.

Space Station Processing Facility Checkout

All space station elements were processed, beginning with Node 1 in 1997.

Most ISS payloads arrived at KSC by plane and were delivered to the Space Station Processing Facility where experiments and other payloads were integrated.

ISS flight hardware was processed in a three-story building that had two processing bays, an airlock, operational control rooms, laboratories, logistic areas, and office space. For all payloads, contamination by even the smallest particles could impair their function in the space environment.

Payloads, including the large station modules, were processed in this

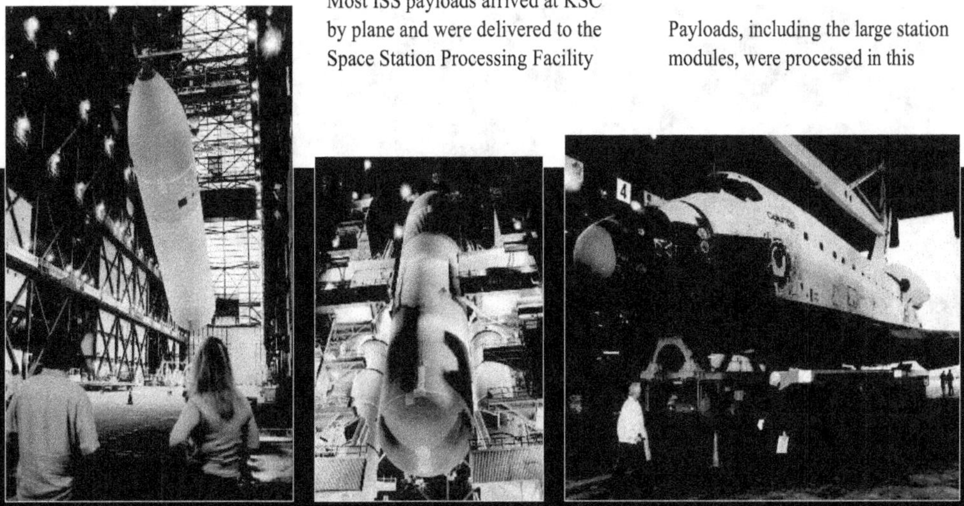

After the External Tank is mated to the Solid Rocket Booster, the Orbiter is brought to the Vehicle Assembly Building.

Vehicle Assembly Building (continued)

state-of-the-art, nonhazardous facility that had a nonconductive, air-bearing pallet compatible floor. This facility had a Class 100K clean room that regularly operated in the 20K range. Class 100K refers to the classification of a clean room environment in terms of the number of particles allowed. In a Class 100K, 0.03 m^3 (1 ft^3) of air is allowed to have 100,000 particles whose size is 0.5 micrometer (0.0002 in.).

Vehicle Assembly Integration for Launch

The SRB, ET, and Orbiter were vertically integrated in the Vehicle Assembly Building.

Mobile Launch Platform

Technicians inside the building stacked the shuttle on one of three mobile launcher platforms originally built in 1964 for the Apollo moon missions. These platforms were modified to accommodate the weight of the shuttle and still be transportable by crawler transporters, and to handle the increased pressure and heat caused by the SRBs. NASA strengthened the platform deck and added an over-pressurization water deluge system. Two additional flame trenches accommodated the SRB exhaust. Tail service masts, also added, enabled cryogenic fueling and electrical umbilical interfaces.

Technology inside the mobile launcher platforms remained basically unchanged for the first half of the program, reusing much of the Apollo-era hardware. The Hazardous Gas Leak Detection System was the first to be updated. It enabled engineers in the firing room to monitor levels of hydrogen gas in and around the vehicle. Many manual systems also were automated and some could be controlled from remote locations other than the firing rooms.

Assembly

Massive Cranes

The size and weight of shuttle components required a variety of lifting devices to move and assemble the vehicle. Two of the largest and most critical were the 295-metric-ton (325-ton) and 227-metric-ton (250-ton) cranes. The 295-metric-ton (325-ton) cranes lifted and positioned the Solid Rocket Motor sections, ET, and Orbiter. The 227-metric-ton (250-ton) cranes were backups.

Both cranes were capable of fine movements, down to 0.003 cm (0.001 in.), even when lifting fully rated loads. The 295-metric-ton (325-ton) cranes used computer controls and graphics and could be set to release the brakes and "float" the load, holding the load still in midair using motor control alone without overloading any part of the crane or its motors.

The cranes were located 140 m (460 ft) above the Vehicle Assembly Building ground floor. Crane operators relied on radio direction from ground controllers at the lift location.

The cranes used two independent wire ropes to carry the loads. Each crane carried about 1.6 km (1 mile) of wire rope that was reeved from the crane to the load block many times. The wire ropes were manufactured at the same time and from the same lot to ensure rope diameters were identical

The Orbiter is then mated with the External Tank and the Solid Rocket Booster.

Vehicle Assembly Building (continued)

and would wind up evenly on the drum as the load was raised.

Stacking the Orbiter, External Tank, and Solid Rocket Booster

SRB segments were moved to the Vehicle Assembly Building. A lifting beam was connected to the booster clevis using the 295-metric-ton (325-ton) crane hook. The segment was lifted off the pallet and moved into the designated high bay, where it was lowered onto the hold-down post bearings on the mobile launcher platform. Remaining segments were processed and mated to form two complete boosters.

Next in the stacking process was hoisting the ET from a checkout cell, lowering into the integration cell, and mating it to the SRBs. Additional inspections, tests, and component installations were then performed.

The Orbiter was towed from the Orbiter Processing Facility to the Vehicle Assembly Building transfer aisle, raised to a vertical position, lowered onto the mobile launcher platform, and mated. Following inspections, tests, and installations, the integrated shuttle vehicle was ready for rollout to the launch pad.

Rollout to Launch Pad

Technicians retracted the access platforms, opened the Vehicle Assembly Building doors, and moved the tracked crawler transporter vehicle under the mobile launcher platform that held the assembled shuttle vehicle.

The transporter lifted the platform off its pedestals and rollout began. The trip to the launch pad took about 6 to 8 hours along the specially built crawlerway—two lanes of river gravel separated by a median strip. The rock surface supported the weight of the crawler and shuttle, and it reduced vibration. The crawler's maximum unloaded speed was 3.2 km/hr (2 mph) and 1.6 km/hr (1 mph) loaded.

Engineers and technicians on the crawler, assisted by ground crews, operated and monitored systems during rollout while drivers steered the vehicle toward the pad. The crawler leveling system kept the top of the shuttle vertical within +/-10 minutes of 1 degree of arc—the diameter of a basketball. The system also provided the leveling required to negotiate the 5% ramp leading to the launch pads and keep the load level when raised and lowered on pedestals at the pad.

Launch Pad Operations

Once the crawler lowered the mobile launcher platform and shuttle onto a launch pad's hold-down posts, a team began launch preparations. These required an average of 21 processing days to complete.

The two steel towers of Launch Pads 39A and 39B stood 105.7 m (347 ft) above KSC's coastline, atop 13-m- (42-ft)-thick concrete pads. Each complex housed a fixed service structure and a rotating service structure that provided access to electrical, pneumatic, hydraulic, hypergolic, and high-pressure gas lines to support vehicle servicing while protecting the shuttle from inclement weather. Pad facilities also included hypergolic propellant storage (nitrogen tetroxide and monomethylhydrazine),

Once the process is complete, the Space Shuttle is transported to the launch pad.

Crawler moving the shuttle stack to the launch pad.

Launch Pad: 28-30 days

cryogenic propellant storage (liquid hydrogen and liquid oxygen), a water tower, a slide wire crew escape system, and a pad terminal connection room.

Liquid Hydrogen/Liquid Oxygen—Tankers, Spheres

Chicago Bridge & Iron Company built the liquid hydrogen and liquid oxygen storage spheres in the 1960s for the Apollo Program. The tanks were two concentric spheres. The inner stainless-steel sphere was suspended inside the outer carbon-steel sphere using long support rods to allow thermal contraction and minimize heat conduction from the outside environment to the propellant. The space between the two spheres was insulated to keep the extremely cold propellants in a liquid state. For liquid hydrogen, the temperature is -253°C (-423°F); for liquid oxygen, the temperature is -183°C (-297°F).

The spheres were filled to near capacity prior to a launch countdown. A successful launch used about 1.7 million L (450,000 gal) of liquid hydrogen and about 830,000 L (220,000 gal) of liquid oxygen. A launch scrub consumed about

Technicians in the Payload Changeout Room at Launch Pad 39B process the Hubble Space Telescope for STS-31 (1990).

380,000 L (100,000 gal) of each commodity. The spheres contained enough propellant to support three launch attempts before requiring additional liquid from tankers.

Pad Terminal Connection Room

The Pad Terminal Connection Room was a reinforced-concrete room located on the west side of the flame trench, underneath the elevated launch pad hardstand. It was covered with about 6 m (20 ft) of dirt fill and housed the equipment that linked elements of the shuttle, mobile launcher platform, and pad with the Launch Processing System in the Launch Control Center. NASA performed and controlled checkout, countdown, and launch of the shuttle through the Launch Processing System.

Payload Changeout Room

Payloads were transported to the launch pad in a payload canister. At the pad, the canister was lifted with a 81,647-kg (90-ton) hoist and its doors were opened to the Payload Changeout Room—an enclosed, environmentally controlled area mated to the Orbiter payload bay. The payload ground-handling mechanism—a rail-suspended, mechanical structure measuring 20 m (65 ft) tall—captured the payload with retention fittings that used a water-based hydraulic system with gas-charged accumulators as a cushion. The mechanism, with the payload, was then moved to the aft wall of the Payload Changeout Room, the main doors were closed, and the canister

The Space Shuttle arrives at the launch pad, where payloads are installed into the Orbiter cargo bay.

Payload Changeout Room at launch pad.

Launch Pad (continued)

was lowered and removed from the pad by the transporter.

Once the rotating service structure was in the mate position and the Orbiter was ready with payload bay doors open, technicians moved the payload ground-handling mechanism forward and installed the payload into the Orbiter cargo bay. This task could take as many as 12 hours if all went well. When installation was complete, the payload was electrically connected to the Orbiter and tested, final preflight preparations were made, and the Orbiter payload bay doors were closed for flight.

Sound Suppression

Launch pads and mobile launcher platforms were designed with a water deluge system that delivered high-volume water flows into key areas to protect the Orbiter and its payloads from damage by acoustic energy and rocket exhaust.

The water, released just prior to main engine ignition, flowed through pipes measuring 2.1 m (7 ft) in diameter for about 20 seconds. The mobile launcher platform deck water spray system was fed from

Water spray at the launch pad was used to suppress the acoustic vibration during launch.

six 3.7-m- (12-ft)-high water spray diffusers nozzles dubbed "rainbirds."

Operational Systems—Test and Countdown

Launch Processing System

Engineers used the Launch Processing System computers to monitor thousands of shuttle measurements and control systems from a remote and safe location. Transducers, built into on-board systems and ground support equipment, measured each important function (i.e., temperature, pressure). Those measurements were converted into engineering data and delivered to the Launch Processing System in the firing rooms, where computer displays gave system engineers detailed views of their systems.

The unique Launch Processing System software was specifically written to process measurements and send commands to on-board computers and ground support equipment to control the various systems. The software reacted either to measurements reaching predefined values or when the countdown clock reached a defined time.

Launch was done by the software. If there were no problems, the button to initiate that software was pushed at the designated period called T minus 9 minutes (T=time). One of the last commands sent to the vehicle was "Go for main engine start," which was sent 10 seconds before launch. From that point on, the on-board computers were in control. They ignited the main engines and the SRBs.

In the firing room at Kennedy Space Center, NASA clears the Space Shuttle for launch.

STS-108 (2001) launch.

Training and Simulations

Launch Countdown Simulation

The complexity of the shuttle required new approaches to launch team training. During Mercury, Gemini, and Apollo, a launch-day rehearsal involving the launch vehicle, flight crew, and launch control was adequate to prepare for launch. The shuttle, however, required more than just one rehearsal.

Due to processing and facility requirements, access to actual hardware in a launch configuration only occurred near the actual launch day after the vehicle was assembled and rolled to the launch pad. The solution was to write a computer program that simulated shuttle telemetry data with a computer math model and fed those data into launch control in place of the actual data sent by a shuttle on the pad.

Terminal Countdown Demonstration Test

The Terminal Countdown Demonstration Test was a dress rehearsal of the terminal portion of the launch countdown that included the flight crew suit-up and flight crew loading into the crew cabin. The Orbiter was configured to simulate a launch-day posture, giving the flight crew the opportunity to run through all required procedures. The flight crew members also was trained in emergency egress from the launch pad, including use of emergency equipment, facility fire-suppression systems, egress routes, slidewire egress baskets, emergency bunker, emergency vehicles, and the systems available if they needed to egress the launch pad.

Space Station Processing Facility for modules and other hardware at Kennedy Space Center.

Special Facilities and Tools

Facility Infrastructure

Although the types of ground systems at KSC were common in many large-scale industrial complexes, KSC systems often were unique in their application, scale, and complexity.

The Kennedy Complex Control System was a custom-built commercial facility control system that included

After launch, Solid Rocket Boosters separate from the Space Shuttle and are recovered in the Atlantic Ocean, close to Florida's East Coast.

Solid Rocket Booster Recovery

about 15,000 monitored parameters, 800 programs, and 300 different displays. In 1999, it was replaced with commercial off-the-shelf products.

The facility heating, ventilating, and air conditioning systems for Launch Pads 39A and 39B used commercial systems in unique ways. During launch operations that required hazard proofing of the mobile launcher platform, a fully redundant fan—149,140 W (200 hp), 1.12 m (44 in.) in diameter—pressurized the mobile launcher platform and used more than 305 m (1,000 ft) of 1.2- by 1.9-m (48- by 75-in.) concrete sewer pipe as ductwork to deliver this pressurization air.

Facility systems at the Orbiter Processing Facility high bays used two fully redundant, spark-resistant air handling units to maintain a Class 100K clean work area in the 73,624-m^3 (2.6-million-ft^3) high bay. During hazardous operations, two spark-resistant exhaust fans, capable of exhausting 2,492 m^3/min (88,000 ft^3/min), worked in conjunction with high bay air handling units and could replace the entire high bay air volume in fewer than 30 minutes.

The launch processing environment included odorless and invisible gaseous commodities that could pose safety threats. KSC used an oxygen-deficiency monitoring system to continuously monitor confined-space oxygen content. If oxygen content fell below 19.5%, an alarm was sounded and beacons flashed, warning personnel to vacate the area.

Communications and Tracking

Shuttle communications systems and equipment were critical to safe vehicle operation. The communications and tracking station in the Orbiter Processing Facility provided test, checkout, and troubleshooting for Orbiter preflight, launch, and landing activities. Communications and tracking supported Orbiter communications and navigations subsystems.

Following landing at KSC, the communications and tracking station monitored the Orbiter and Merritt Island Launch Area communications transmissions during tow and spotting of the vehicle in the Orbiter Processing Facility. In that facility, the station was configured as a passive repeater to route the uplink and downlink radio frequency signals to and from the Orbiter Processing Facility and Merritt Island Launch Area using rooftop antennas.

Operations Planning Tools

Requirements and Configuration Management

Certification of Flight Readiness was the process by which the Space Shuttle Program manager determined the shuttle was ready to fly. This process verified that all design requirements were properly approved, implemented, and closed per the established requirements and configuration management processes in place at KSC.

Requirements and configuration management involved test requirements and modifications. Test requirements ensured shuttle integrity, safety, and performance. Modifications addressed permanent hardware or software changes, which improved the safety of flight or vehicle performance, and mission-specific hardware or software changes required to support the payload and mission objectives.

The recovered Solid Rocket Boosters are returned to Kennedy Space Center for refurbishment and reusability.

NASA generated planning, executing, and tracking products to ensure the completion of all processing flow steps. These included: process and support plans; summary and detailed assessments; milestone, site, maintenance, and mini schedules; and work authorization documents. Over time, many operations tools evolved from pen and paper, to mainframe computer, to desktop PC, and to Web-based applications.

Work authorization documents implemented each of the thousands of requirements in a flow. Documents included standard procedures performed every flow as well as nonstandard documents such as problem and discrepancy reports, test preparation sheets, and work orders.

Kennedy Space Center Integrated Control Schedule

The KSC Integrated Control Schedule was the official, controlling schedule for all work at KSC's shuttle processing sites. This integration tool reconciled conflicts between sites and resources among more than a dozen independent sites and multiple shuttle missions in work simultaneously. Work authorization documents could not be performed unless they were entered on this schedule, which distributed the required work authorization documents over time and sequenced the work in the proper order over the duration of the processing flow. The schedule, published on the Web every workday, contained the work schedule for the following 11 days for each of the 14 shuttle processing sites, including the three Orbiter Processing Facility bays, Vehicle Assembly Building, launch pads, Shuttle Landing Facility, and Hypergolic Maintenance Facility.

Space Shuttle Launch Countdown Operations

Launch countdown operations occurred over a period of about 70 hours during which NASA activated, checked out, and configured the shuttle vehicle systems to support launch. Initial operations configured shuttle data and computer systems. Power Reactant Storage and Distribution System loading was the next major milestone in the countdown operation. Liquid oxygen and liquid hydrogen had to be transferred from tanker trucks on the launch pad surface, up the fixed service structure, across the rotating service structure, and into the on-board storage tanks, thus providing the oxygen and hydrogen gas that the shuttle fuel cells required to supply power and water while on orbit.

The next major milestones were activation of the communication equipment and movement of the rotating service structure from the mate position (next to the shuttle) to the park position (away from the shuttle), which removed much access to the vehicle.

The most hazardous operation, short of launch, was loading the ET with liquid oxygen and liquid hydrogen. This was performed remotely from the Launch Control Center. The Main Propulsion System had to be able to control the flow of cryogenic propellant through a wide range of flow rates. The liquid hydrogen flow through the vehicle was as high as 32,550 L/min (8,600 gal/min). While in stable replenish, flow rates as low as 340 L/min (90 gal/min) had to be maintained with no adverse affects on the quality of the super-cold propellant.

Once the tank was loaded and stable, NASA sent teams to the launch pad. One team inspected the vehicle for issues that would prevent launch, including ice formation and cracks in the ET foam associated with the tank loading. Another team configured the crew cabin and the room used to access the shuttle cabin. Flight crew members, who arrived a short time later, were strapped into their seats and the hatch was secured for launch.

The remaining operations configured the vehicle systems to support the terminal countdown. At that point, the ground launch sequencer sent the commands to perform the remaining operations up to 31 seconds before launch, when the on-board computers took over the countdown and performed the main engine start and booster ignition.

Solid Rocket Booster Recovery

Following shuttle launch, preparations continued for the next mission, beginning with SRB recovery.

Approximately 1 day before launch, the two booster recovery ships—Freedom Star and Liberty Star—left Cape Canaveral Air Force Station and

Port Canaveral to be on station prior to launch to retrieve the boosters from the Atlantic Ocean.

Approximately 6½ minutes after launch, the boosters splashed down 258 km (160 miles) downrange. Divers separated the three main parachutes from each booster and the parachutes were spun onto reels on the decks of each ship. The divers also retrieved drogue chutes and frustums and lifted them aboard the ships.

For the boosters to be towed back to KSC, they were repositioned from vertical to horizontal. Divers placed an enhanced diver-operated plug into the nozzle of the booster, which was 32 m (105 ft) below the ocean surface. Air was pumped into the boosters, displacing the water inside them and repositioning the boosters to horizontal. The boosters were then moved alongside the ships for transit to Cape Canaveral Air Force Station where they were disassembled and refurbished. Nozzles and motor segments were shipped to the manufacturer for further processing.

Following recovery, the segments were taken apart and the joints were inspected to make sure they had performed as expected. Booster components were inspected and hydrolased—the ultimate pressure cleaning—to remove any residual fuel and other contaminants. Hydrolasing was done manually with a gun operating at 103,421 kPa (15,000 psi) and robotically at up to 120,658 kPa (17,500 psi). Following cleaning, the frustum and forward skirt were media-blasted and repainted.

Parachutes

SRB main parachute canopies were the only parachutes in their size class that were refurbished. NASA removed the parachutes from the retrieval ships and transported them to the Parachute Refurbishment Facility.

At the facility, technicians unspooled, defouled, and inspected the parachutes. Following a preliminary damage mapping to assess the scope of repairs required, the parachutes were hung on a monorail system that facilitated movement through the facility. The first stop was a 94,635-L (25,000-gal) horizontal wash tank where each parachute underwent a 4- to 6-hour fresh water wash cycle to remove all foreign material. The parachutes were transferred to the drying room and exposed to 60°C (140°F) air for 10 to 12 hours, after which they were inspected, repaired, and packed into a three-part main parachute cluster and transferred to the Assembly and Refurbishment Facility for integration into a new forward assembly.

Summary

In conclusion, the success of each shuttle mission depended, without exception, on ground processing. The series of planning and execution steps required to process the largest and most complex reusable space vehicle was representative of NASA's ingenuity, dedicated workforce, and unmatched ability, thus contributing immensely to the legacy of the Space Shuttle Program.

Technicians assemble a Solid Rocket Booster parachute at Kennedy Space Center.

Space Operations Weather: How NASA, the National Weather Service, and the Air Force Improved Predictions

Weather was the largest single cause of delays or scrubs of launch, landing, and ground operations for the Space Shuttle.

Rollout of Space Shuttle Discovery, STS-128 (2009), was delayed by onset of lightning in the area of Launch Pad 39A at Kennedy Space Center. Photo courtesy of Environmental Protection Agency.

The Shuttle Weather Legacy

NASA and the US Air Force (USAF) worked together throughout the program to find and implement solutions to weather-related concerns. The Kennedy Space Center (KSC) Weather Office played a key role in shuttle weather operations. The National Weather Service operated the Spaceflight Meteorology Group at Johnson Space Center (JSC) to support on-orbit and landing operations for its direct customers—the shuttle flight directors. At Marshall Space Flight Center, the Natural Environments Branch provided expertise in climatology and analysis of meteorological data for both launch and landing operations with emphasis on support for engineering analysis and design. The USAF 45th Weather Squadron provided the operational weather observations and forecasting for ground operations and launch at the space launch complex. This collaborative community, which worked effectively as a team across the USAF, NASA, and the National Weather Service, not only improved weather prediction to support the Space Shuttle Program and spaceflight worldwide in general, it also contributed much to our understanding of the atmosphere and how to observe and predict it. Their efforts not only enabled safe ground launch and landing, they contributed to atmospheric science related to observation and prediction of lightning, wind, ground and atmosphere, and clouds.

By the late 1980s, 50% of all launch scrubs were caused by adverse weather conditions—especially the destructive effects of lightning, winds, hail, and temperature extremes. So NASA and their partners developed new methods to improve the forecasting of weather phenomena that threatened missions, including the development of technologies for lightning, winds, and other weather phenomena. The Space Shuttle Program led developments and innovations that addressed weather conditions specific to Florida, and largely supported and enhanced launch capability from the Eastern Range. Sensor technologies developed were used by, and shared with, other meteorological organizations throughout the country.

Living With Lightning, a Major Problem at Launch Complexes Worldwide

Naturally occurring lightning activity associated with thunderstorms occurs at all launch complexes, including KSC and Cape Canaveral Air Force Station. Also, the launch itself can trigger lightning—a problem for launch complexes that have relatively infrequent lightning may have a substantial potential for rocket-triggered lightning. The launch complex at Vandenberg Air Force Base, California, is a primary example.

Natural lightning discharges may occur within a single thundercloud, between thunderclouds, or as cloud-to-ground strikes. Lightning may also be triggered by a conductive object, such as a Space Shuttle, flying into a region of atmosphere where strong electrical charge exists but is not strong enough by itself to discharge as a lightning strike.

Natural lightning is hazardous to all aerospace operations, particularly those that take place outdoors and away from protective structures. Triggered lightning is only a danger to vehicles in flight but, as previously described, may occur even when natural lightning is not present.

Lightning Technology at the Space Launch Complex

Crucial to the success of shuttle operations were the activities of the USAF 45th Weather Squadron, which provided all launch and landing orbit weather support for the space launch complex. Shuttle landing support was provided by the National Weather Service Spaceflight Meteorology Group located at JSC. The 45th Weather Squadron operated from Range Weather Operations at Cape Canaveral Air Force Station. The Spaceflight Meteorology Group housed weather system computers for forecast and also analyzed data from the National Centers for Environmental Prediction, weather satellite imagery, and local weather sensors as well as assisted in putting together KSC area weather forecasts.

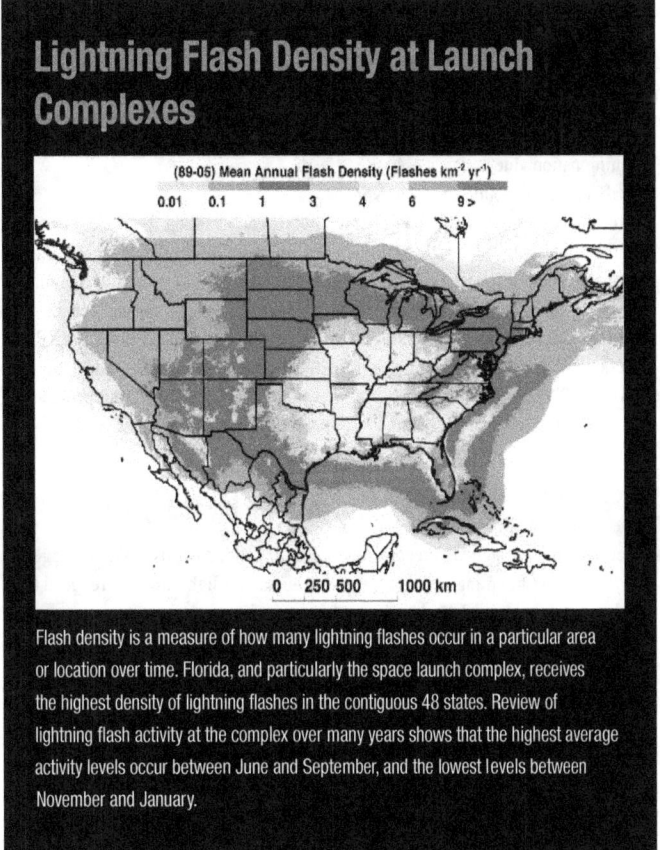

Lightning Flash Density at Launch Complexes

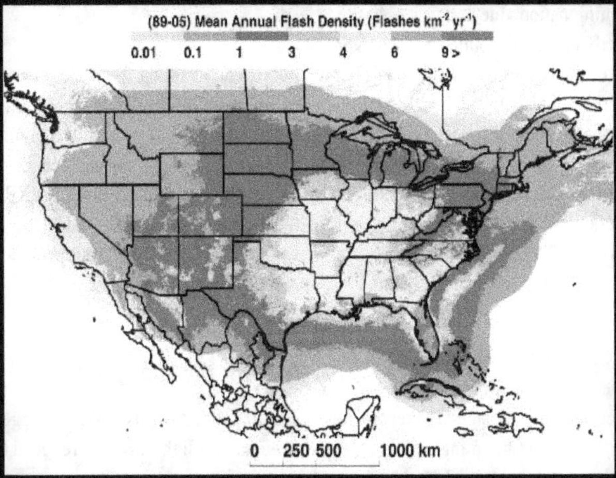

Flash density is a measure of how many lightning flashes occur in a particular area or location over time. Florida, and particularly the space launch complex, receives the highest density of lightning flashes in the contiguous 48 states. Review of lightning flash activity at the complex over many years shows that the highest average activity levels occur between June and September, and the lowest levels between November and January.

Another key component of shuttle operations was the KSC Weather Office, established in the late 1980s. The KSC Weather Office ensured all engineering studies, design proposals, anomaly analyses, and ground processing and launch commit criteria for the shuttle were properly considered. It coordinated all weather research and development, incorporating results into operations.

Lightning Evaluation Tools	System Network
Launch Pad Lightning Warning System	Thirty-one electric-field mills that serve as an early warning system for electrical charges building aloft due to a storm system.
Lightning Detection and Ranging	Nine antennas that detect and locate lightning in three dimensions within 185 km (100 nautical miles) using a "time of arrival" computation on signals.
National Lightning Detection Network	One-hundred ground-based sensing stations that detect cloud-to-ground lightning activity across the continental US. The sensors instantaneously detect the electromagnetic signal given off when lightning strikes the ground.
Cloud-to-Ground Lightning Surveillance System	Six sensors spaced much closer than in the National Lightning Detection Network.
Weather Radar	Two radars that provide rain intensity and cloud top information.

Systems used for weather and thunderstorm prediction and conditions.

Launch Pad Lightning Warning System data helped forecasters determine when surface electric fields may have been of sufficient magnitude to create triggered lightning during launch. The data also helped determine when to issue and cancel lightning advisories and warnings. The original Lightning Detection and Ranging System, developed by NASA at KSC, sensed electric fields produced by the processes of breakdown and channel formation in both cloud lightning and cloud-to-ground flashes. The locational accuracy of this system was on the order of +/-100 m (328 ft). In 2008, a USAF-owned system replaced the original KSC Lightning Detection and Ranging System, which served the space launch complex for about 20 years.

The National Lightning Detection Network plots cloud-to-ground lightning nationwide and was used to identify cloud-to-ground strikes at KSC and to ensure safe transit of the Orbiter atop the Shuttle Carrier Aircraft. A National Lightning Detection Network upgrade in 2002-2003 enabled the system to provide a lightning flash-detection efficiency of approximately 93% of all flashes with a location accuracy on the order of +/-500 to 600 m (1,640 to 1,968 ft).

The Cloud-to-Ground Lightning Surveillance System is a lightning detection system designed to record cloud-to-ground lightning strikes in the vicinity of the space launch complex. A Cape Canaveral Air Force Station upgrade in 1998 enabled the system to provide a lightning flash-detection efficiency within the sensor array of approximately 98% of all flashes and with a location accuracy on the order of +/-250m (820 ft).

The Lightning Detection and Ranging System was completely upgraded during the shuttle era with new sensors positioned in nine locations around the space launch complex proper. Along with a central processor, the system was referred to as the Four-Dimensional Lightning Surveillance System. This new central processor was also capable of processing the Cloud-to-Ground Lightning Surveillance System sensor data at the same time and, moreover, produced full cloud-to-ground stroke data rather than just the first stroke in real time. The synergistic combination of the upgraded Four-Dimensional Lightning Surveillance System and the Cloud-to-Ground Lightning Surveillance System provided a more accurate and timely reporting capability over that of the upgraded Cloud-to-Ground Lightning Surveillance System or the older Lightning Detection and Ranging System individually, and it allowed for enhanced space launch operations support.

Launch and landing forecasters located in Texas, and Cape Canaveral, Florida, accessed displays from two different Florida radar sites—one located at Patrick Air Force Base, and a NEXRAD (next-generation weather radar) Doppler, located in Melbourne at the National Weather Service.

Lightning Operational Impacts; Warning Systems

The likelihood of sustaining damage from natural lightning was reduced by minimizing exposure of personnel and hardware during times when lightning threatened. To accomplish this, it was necessary to have in place a balanced warning system whereby lightning activity could be detected and reported far enough in advance to permit protective action to be taken. Warnings needed to be accurate to prevent harm yet not stop work unnecessarily. Lightning advisories were important for ground personnel, launch systems, and the transport of hardware, including the 6- to 8-hour transport of the Space Shuttle to the launch pad.

The original deployment of the Lightning Detection and Ranging System pioneered a two-phase lightning policy. In Phase I, an advisory was issued that lightning was forecast within 8 km (5 miles) of the designated site within 30 minutes of the effective time of the advisory. The 30-minute warning gave personnel time to get to a protective shelter and gave personnel working on lightning-sensitive tasks time to secure operations in a safe and orderly manner. A Phase II warning was issued when lightning was imminent or occurring within 8 km (5 miles) of the designated site. All lightning-sensitive operations were terminated until the Phase II warning was lifted. This two-phase policy provided adequate lead time for sensitive operations without shutting down less-sensitive operations until the hazard became immediate. Much of this activity was on the launch pads, which were tall, isolated, narrow structures in wide-open areas and were prime targets for lightning strikes. Lightning advisories were critical for the safety of over 25,000 people and resource protection of over $18 billion in facilities. Several more billion dollars could be added to this value, depending on what payloads and rockets were at the launch pads or in transit outside. This policy ultimately reduced ground processing downtime by as much as 50% compared to the older system, saving millions of dollars annually.

Operationally, warnings were sometimes not sufficient, for example during launch operations when real-time decisions had to be made based on varying weather conditions with a potentially adverse effect on flight. Following a catastrophic lightning-induced failure of an Atlas/Centaur rocket in 1987, a blue-ribbon "Lightning Advisory Panel" comprising top American lightning scientists was convened to assist the space program. The panel recommended a set of "lightning launch commit criteria" to avoid launching into an environment conducive to either natural or triggered lightning. These criteria were adopted by NASA for the Space Shuttle Program, and also by the USAF for all military and civilian crewless launches from the Eastern and Western Ranges.

Hail Damage to the External Tank

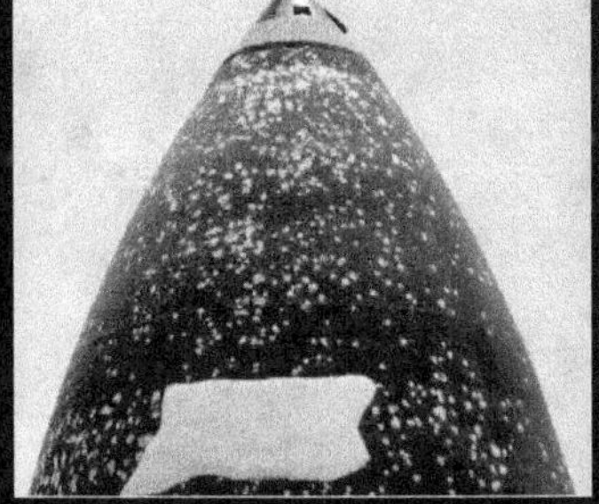
ET-124 damage repairs, post storm.

On the afternoon of February 26, 2007, during STS-117 prelaunch processing at Kennedy Space Center (KSC) Launch Pad A, a freak winter thunderstorm with hail struck the launch complex and severely damaged the External Tank (ET) (ET-124) Thermal Protection System foam insulation. The hail strikes caused approximately 7,000 divots in the foam material. The resulting damage revealed that the vehicle stack would have to be returned to the Vehicle Assembly Building to access the damage. This would be the second time hail caused the shuttle to be returned to the building. To assess the damage, NASA built customized scaffolding. The design and installation of the scaffolding needed to reach the sloping forward section of the tank was a monumental task requiring teams of specialized riggers called "High Crew" to work 24 hours a day for 5 straight days. A hand-picked engineering assessment team evaluated the damage. The ET liquid oxygen tank forward section was the most severely damaged area and required an unprecedented repair effort. There were thousands of damaged areas that violated the ET engineering acceptance criteria for flight. NASA assembled a select repair team of expert technicians, quality inspectors, and engineers to repair the damage. This team was assisted by manufacturing specialists from Lockheed Martin, the ET manufacturer, and Marshall Space Flight Center.

KSC developed an inexpensive, unique hail monitoring system using a piezoelectric device and sounding board to characterize rain and hail. While the shuttle was at the pad, three remote devices constantly monitored the storms for potential damage to the vehicle.

The lightning launch commit criteria, as initially drafted, were very conservative as electrical properties of clouds were not well understood. Unfortunately, this increased the number of launches that had to be postponed or scrubbed due to weather conditions. The program undertook a series of field research initiatives to learn more about cloud electrification in hopes that the criteria could safely be made less restrictive.

These field research initiatives used aircraft instrumented with devices called electric field mills that could measure the strength of the electric field in clouds as the aircraft flew through them. The research program was known as Airborne Field Mill. Data collected by the Airborne Field Mill program were subjected to extensive quality control, time-synchronized, and consolidated into a carefully documented, publicly accessible online archive. This data set is the largest, most comprehensive of its kind.

The Airborne Field Mill science team developed a quantity called Volume Averaged Height Integrated Radar Reflectivity that could be observed with weather radar. This quantity, when small enough, assured safe electric fields aloft. As a result, the Lightning Advisory Panel was able to recommend changes to the lightning launch commit criteria to make them both safer and less restrictive. The new criteria are used by all US Government launch facilities, and the Federal Aviation Administration is including them in its regulations governing the licensing of private spaceports. These criteria were expressed in detailed rules that described weather conditions likely to produce or be associated with lightning activity, the existence of which precluded launch.

Lightning Protection and Instrumentation Systems

Physical lightning protection for the shuttle on the pad was provided by a combination of a large, loose network of wiring known as a counterpoise beneath the pad structure and surrounding environs and a large wire system comprising a 2.5-cm- (1-in.)-, 610-m- (2,000-ft)-long steel cable anchored and grounded at either end and supported in the middle by a 24.4-m- (80-ft)-tall nonconductive mast. The mast also served to prevent currents—from lightning strikes to the wire—from passing into the pad structure. A 1.2-m (4-ft) air terminal, or lightning rod, was mounted atop the mast and electrically connected to the steel cable. The cable arrangement assumed a characteristic curved shape to either side of the pad described mathematically as a catenary and therefore called the Catenary Wire System.

A grounded stainless-steel cable extends from the lightning mast to provide a zone of protection for the launch vehicle.

Additional lightning protection devices at the launch pads included a grounded overhead shield cable that protected the crew emergency egress slide wires attached to the fixed service structure. Grounding points on the pad surface and the mobile launcher platform and electrical connections in contact with the shuttle completed the system that conducted any lightning-related currents safely away from the vehicle. Overhead grid-wire systems protected hypergolic fuel and oxidizer storage areas. The huge 3,407,000-L (900,000-gal) liquid hydrogen and liquid oxygen tanks at each pad were constructed of metal and did not need overhead protection.

The shuttle and its elements were well protected from both inclement weather and lightning away from the pad while in the Vehicle Assembly Building. This 160-m- (525-ft)-high structure had eleven 8-m- (25-ft)-high lightning conductor towers on its roof. When lightning hit the building's air terminal system, wires conducted the charge to the towers, which directed the current down the Vehicle Assembly Building's sides and into bedrock through the building's foundation pilings.

In addition to physical protection features, the Space Shuttle Program employed lightning monitoring systems to determine the effects of lightning strikes to the catenary system, the immediate vicinity of the launch pad, and the shuttle itself. The shuttle used two specific lightning monitoring systems—the Catenary Wire Lightning Instrumentation System and the Lightning Induced Voltage Instrumentation System. The Catenary Wire Lightning Instrumentation System used sensors located at either end of the Catenary Wire System to sense currents in the catenary wire induced by nearby or direct lightning strikes. The data were then used to evaluate the potential for damage to sensitive electrical equipment on the shuttle. The Lightning Induced Voltage Instrumentation System used voltage taps and current sensors located in the shuttle and the mobile launcher platform to detect and record voltage or current transients in the shuttle Electrical Power System.

After STS-115, NASA performed a system review and decided to upgrade the two systems. The Ground Lightning Monitoring System was implemented. It was comprised of both voltage monitoring on the Orbiter power busses and magnetic field sensing internal to the Orbiter middeck, the aft avionics bay, the Payload Changeout Room, and locations on the pad structure. The collected voltage and magnetic field data were used to determine induced current and voltage threats to equipment, allowing direct comparison to known, acceptable maximum levels for the vehicle and its equipment.

The elaborate lightning detection and personnel protection systems at KSC proved their worth the hard way. The lightning masts at Launch Pads 39A and 39B were struck many times with a shuttle on the pad, with no damage to equipment. No shuttle was endangered during launch, although several launches were delayed due to reported weather conditions.

Ultimately, one of the biggest contributions to aerospace vehicle design for lightning protection was the original standard developed by NASA for the shuttle. New standards developed by the Department of Defense, the Federal Aviation Administration, and

Lightning Delays Launch

In August 2006, while STS-115 was on the pad, the lightning mast suffered a 50,000-ampere attachment, much stronger than the more typical 20,000- to 30,000-ampere events, resulting in a 3-day launch delay while engineers and managers worked feverishly to determine the safety of flight condition of the vehicle. The vehicle, following extensive data review and analysis, was declared safe to fly.

commercial organizations over the years have leveraged this pioneering effort, and the latest of these standards is now applicable for design of the new spacecraft.

Hurricane Damage

Space Shuttle processing during Florida's hurricane season was a constant challenge to ground processing. Hurricane weather patterns were constantly monitored by the team. If the storms could potentially cause damage to the vehicle, the stack was rolled back to the Vehicle Assembly Building for protection. During Hurricane Frances in September 2004, Kennedy Space Center suffered major damage resulting from the storm. The Vehicle Assembly Building lost approximately 820 aluminum side panels and experienced serious roof damage.

Damage to Vehicle Assembly Building at Kennedy Space Center during Hurricane Frances.

Working With Winds

Between the Earth's surface and about 18 km (10 nautical miles) altitude, the Earth's atmosphere is dense enough that winds can have a big effect on an ascending spacecraft. Not only can the wind blow a vehicle toward an undesirable direction, the force of the wind can cause stress on the vehicle. The steering commands in the vehicle's guidance computer were based on winds measured well before launch time. If large wind changes occurred between the time the steering commands were calculated and launch time, it was difficult for the vehicle to fly the desired trajectory or the vehicle would be stressed beyond its limits and break up. Therefore, frequent measurements of wind speed and direction as a function of height were made during countdown.

The Space Shuttle Program measured upper air winds in two ways: high-resolution weather balloons and a Doppler radar wind profiler. Both had a wind speed accuracy of about 1 m/sec (3.3 ft/sec). Balloons had the advantage of being able to detect atmospheric features as small as 100 m (328 ft) in vertical extent, and have been used since the beginning of the space program. Their primary disadvantages were that they took about 1 hour to make a complete profile from the surface to 18 km (11 miles), and they blew downwind. In the winter at KSC, jet stream winds could blow a balloon as much as 100 km (62 miles) away from the launch site before the balloon reached the top of its trajectory.

The wind profiler was located near the Shuttle Landing Facility, close to the launch pad. The profiler scattered radar waves off turbulence in the atmosphere and measured their speed in a manner similar to a traffic policeman's radar gun. It produced a complete profile of wind speed and direction every 5 minutes. This produced profiles 12 times faster than a balloon and much closer to the flight path of the vehicle. Its only technical disadvantage was that the smallest feature in the atmosphere it could distinguish was 300 m (984 ft) in vertical extent. The Doppler radar wind profiler was first installed in the late 1980s.

When originally delivered, the profiler was equipped with commercial software that provided profiles with unknown accuracy every 30 minutes. For launch support, NASA desired a higher rate of measurement and accuracy as good as the high-resolution balloons. Although the Median Filter First Guess software, used in a laboratory to evaluate the potential value of the Doppler radar wind profiler, significantly outperformed any commercially available signal processing methodology for wind profilers, it was sufficiently complex and its run time too long for operational use to be practical.

To use wind profiler data, NASA developed algorithms for wind profiles that included the ground wind profile, high-altitude weather balloons, and Doppler radar. This greatly enhanced the safety of space launches.

Landing Weather Forecasts

The most important shuttle landing step occurred just prior to the deorbit burn decision. The National Weather Service Spaceflight Meteorology Group's weather prediction was provided to the JSC flight director about 90 minutes prior to the scheduled landing. This forecast supported the Mission Control Center's "go" or "no-go" deorbit burn decision. The deorbit burn occurred about 60 minutes prior to landing. The shuttle had to land at the specified landing site. The final 90-minute landing forecast had to be precise, accurate, and clearly communicated for NASA to make a safe landing decision.

Flight Operations

Jack Knight
Gail Chapline
Marissa Herron
Mark Kelly
Jennifer Ross-Nazzal

For nearly 3 decades, NASA's Johnson Space Center (JSC) Mission Operations organization planned, trained, and managed the on-orbit operations of all Space Shuttle missions. Every mission was unique, and managing a single mission was an extremely complex endeavor. At any one time, however, the agency simultaneously handled numerous flights (nine in 1985 alone). Each mission featured different hardware, payloads, crew, launch date, and landing date. Over the years, shuttle missions became more complicated—even more so when International Space Station (ISS) assembly flights began. Besides the JSC effort, Kennedy Space Center managed all launches while industry, the other centers, and other countries managed many of the payloads.

NASA defined the purpose of each mission several years before the mission's flight. Types of missions varied from satellite releases, classified military payloads, science missions, and Hubble Space Telescope repair and upgrades to construction of the ISS. In addition to completion of the primary mission, all flights had secondary payloads such as education, science, and engineering tests. Along with executing mission objectives, astronauts managed Orbiter systems and fulfilled the usual needs of life such as eating and sleeping. All of these activities were integrated into each mission.

This section explains how NASA accomplished the complicated tasks involved in flight operations. The Space Transportation System (STS)-124 (2008) flight provides examples of how mission operations were conducted.

Plan, Train, and Fly

Planning the Flight Activities

NASA's mission operations team planned flight activities to assure the maximum probability of safe and complete success of mission objectives for each shuttle flight. The planning process encompassed all aspects of preflight assessments, detailed preflight planning and real-time replanning, and postflight evaluations to feed back into subsequent flights. It also included facility planning and configuration requirements. Each vehicle's unique characteristics had to be considered in all flight phases to remain within defined constraints and limitations. The agency made continual efforts to optimize each flight's detailed execution plan, including planning for contingencies to maximize safety and performance margins as well as maximizing mission content and probability of mission success.

During the initial planning period, NASA selected the flight directors and determined the key operators for the Mission Control Team. This team then began planning and training. The flight crew was named 1 to 1½ years prior to launch. The commander acted as the leader for the flight crew through all planning, training, and execution of the mission while the flight directors led the mission operations team.

Approximately 14 months before launch, the mission operations team developed a detailed flight plan. To create the comprehensive timeline, team members worked closely with technical organizations like engineering, the astronaut office, specific NASA contractors, payload suppliers, government agencies, international partners, and other NASA centers including Kennedy Space Center (KSC) and Marshall Space Flight Center (MSFC). Crew timeline development required balancing crew task completion toward mission objectives and the individual's daily life needs, such as nutrition, sleep, exercise, and personal hygiene. The timeline was in 5-minute increments to avoid overextending the crew, which could create additional risks due to crew fatigue. Real-time changes to the flight plan were common; therefore, the ground team had to be prepared to accommodate unexpected deviations. Crew input was vital to the process.

Initial Planning: Trajectory Profile

Planning included the mission's trajectory profile. This began with identifying the launch window, which involved determining the future time at which the planes from the launch site and the targeted orbit intersect. The latitude of the launch site was important in determining the direction of launch because it defined the minimum inclination that could be achieved, whereas operational maximum inclinations were defined by range safety limits to avoid landmass. For International Space Station (ISS) missions, the shuttle launched from the

Collaboration Paved the Way for a Successful Mission… of International Proportions

In 2000, Mission Operations Directorate worked with Japan in preparation for the flight of STS-124 in 2008. To integrate Japan Aerospace Exploration Agency (JAXA) into the program, the US flight team worked closely with the team from Japan to assimilate JAXA's Japanese Experiment Module mission with the requirements deemed by the International Space Station Program. The team of experts taught Japanese flight controllers how Mission Operations Directorate handled flight operations—the responsibilities of mission controllers, dealing with on-orbit failures, writing mission rules and procedures, structuring flight control teams—to help them determine how to plan future missions and manage real-time operations. The downtime created by the Columbia accident (2003) provided additional time to the Japanese to develop necessary processes, since this was the first time JAXA commanded and controlled a space station module.

In addition to working closely with Japan on methodology and training, flight designers integrated the international partners (Russian Federal Space Agency, European Space Agency, Canadian Space Agency, and JAXA) in their planning process. The STS-124 team worked closely with JAXA's flight controllers in the Space Station Integration and Promotion Center at Tsukuba, Japan, to decide the sequence of events—from unberthing the module to activating the science lab. Together, they determined plans and incorporated these plans into the extensive timeline.

launch site's 28.5-degree latitude into a 51.6-degree inclination orbit, so the launch ground track traveled up the East Coast. For an orbit with a lower inclination, the shuttle headed in a more easterly direction off the launch pad. Imagine that, as the ISS approached on an ascending pass, the shuttle launched along a path that placed it into an orbit just below and behind the ISS orbit. NASA optimized the fuel usage (for launch and rendezvous) by selecting an appropriate launch time. The optimal time to launch was when the ISS orbit was nearest the launch site. Any other time would have resulted in an inefficient use of expensive fuel and resources; however, human factors and mission objectives also influenced mission design and could impose additional requirements on the timing of key mission events. The availability of launch days was further constrained by the angle between the orbital plane and the sun vector. That angle refers to the amount of time the spacecraft spends in sunlight. When this angle exceeded 60 degrees, it was referred to as a "beta cutout." This variable, accounted for throughout a shuttle mission, limited the availability of launch days.

Operational Procedures Development

NASA developed crew procedures and rules prior to the first shuttle flight—Space Transportation System (STS)-1 in 1981—and refined and modified them after each flight, as necessary. A basic premise was that the crew should have all requisite procedures to operate the vehicle safely with respect to the completion of launch, limited orbit operations, and deorbit without ground involvement in the event of a loss of communication. This was not as simple as it might sound. Crew members had no independent knowledge of ground site

During the early flights, NASA established the core elements of the mission operations shuttle processes. The emblem for Johnson Space Center Mission Operations included a sigma to indicate that the history of everything learned was included in planning for the next missions.

status, landing site weather, or on-board sensor drift, and they had considerably less insight into the total set of vehicle telemetry available to the ground.

Each flight increased NASA's experience base with regard to actual vehicle, crew, and ground operations performance. Each mission's operational lessons learned were incorporated into the next mission's crew procedures, flight team training, Flight Rules modifications, and facilities modifications (mostly software).

Flight Control Team

Flight controllers were a vital part of every mission. For each flight control position in the flight control room, one or more supporting positions were in the back room, or the multipurpose support room. For example, the flight dynamics officer and the guidance procedures officer, located in "the trench" of the flight control room, relied on a team of flight controllers sitting just a few feet away in the multipurpose support room to provide them with recommendations. These back room flight controllers provided specialized support in areas such as aborts, navigation, and weather as well as communications with external entities (i.e., Federal Aviation Administration, US State Department).

Back room support had more time and capabilities to perform quick analyses while front room flight controllers were working higher level issues and communicating with the other front room controllers (i.e., propulsion engineer, booster engineer) and the flight director. This flow of communications enabled analyses to be performed in real time, with appropriate discussions among all team players to result in a recommended course of action that was then passed on to the front room. The front room remained involved in back room discussions when feasible and could always redirect their support if they received new information from another front room flight controller, the flight director, or the capsule communicator (responsible for all communications with the on-orbit crew).

It can easily be surmised that being a flight controller required a quick and decisive mindset with an equally important team player attitude. The pressure to make immediate decisions was greatest during the launch phase and similarly so during the re-entry phase. During those times, flight controllers worked under a high level of pressure and had to trust their counterparts to work together through any unplanned challenges that may have occurred.

Flight Controller Preparation

Preparations for any off-nominal situations were regularly practiced prior to any mission through activities that simulated a particular phase of flight and any potential issue that could occur during that timeframe. These simulated activities, simply referred to as "Sims," involved both the front room

Flight Rules

Part of the planning process included writing Flight Rules. Flight Rules were a key element of the real-time flight control process and were predefined actions to be taken, given certain defined circumstances. This typically meant that rules were implemented, as written, during critical phases such as launch and re-entry into Earth's atmosphere. Generally, during the orbit phase, there was time to evaluate exact circumstances. The Flight Rules defined authorities and responsibilities between the crew and ground, and consisted of generic rules, such as system loss definition, system management, and mission consequence (including early mission termination) for defined failures.

For each mission, lead flight directors and their teams identified flight-specific mission rules to determine how to proceed if a failure occurred. These supplemented the larger book of generic flight rules. For instance, how would the team respond if the payload bay doors failed to open in orbit? The rules minimized real-time rationalization because the controllers thoroughly reviewed and simulated requirements and procedures before the flight.

and the back room flight controllers, just as if the Sim were the real thing. Sims allowed the flight control team and the astronauts to familiarize themselves with the specifics of the missions and with each other. These activities were just as much team-building exercises as they were training exercises in what steps to take and the decisions required for a variety of issues, any of which could have had catastrophic results. Of course, the best part of a simulation was that it was not real. So if a flight controller or an astronaut made a mistake, he or she could live and learn while becoming better prepared for the real thing.

Training to become a flight controller began long before a mission flew. Flight controllers had to complete a training flow and certification process before being assigned to a mission. The certification requirements varied depending on the level of responsibility of the position. Most trainees began by reading technical manuals related to their area of flight control (i.e., electrical, environmental, consumables manager or guidance, navigation, and controls system engineer), observing currently certified flight controllers during simulations, and performing other hands-on activities appropriate to their development process. As the trainee became more familiar with the position, he or she gradually began participating in simulations until an examination of the trainee's performance was successfully completed to award formal certification. Training and development was a continually improving process that all flight controllers remained engaged in whether they were assigned to a mission or maintaining proficiency. A flight controller also had the option to either remain in his or her current position or move on to a more challenging flight control position with increased responsibilities, such as those found in the front room. An ascent phase, front room flight control position was typically regarded as having the greatest level of responsibility because this flight controller was responsible for the actions of his or her team in the back room during an intense and time-critical phase of flight. Similarly, the flight director was responsible for the entire flight control team.

Flight Techniques

The flight techniques process helped develop the procedures, techniques, and rules for the vehicle system, payload, extravehicular activities (EVAs), and robotics for the flight crew, flight control team, flight designers, and engineers. NASA addressed many topics over the course of the Space Shuttle Program, including abort modes and techniques, vehicle power downs, system loss integrated manifestations and responses, risk assessments, EVA and robotic procedures and techniques, payload deployment techniques, rendezvous and docking or payload capture procedures, weather rules and procedures, landing site selection criteria, and others. Specific examples involving the ISS were the development of techniques to rendezvous, conduct proximity operations, and dock the Orbiter while minimizing plume impingement contamination and load imposition.

Crew Procedures

Prior to the first shuttle flight, NASA developed and refined the initial launch, orbit, and re-entry crew procedures, as documented in the Flight Data File. This document evolved and expanded over time, especially early in the program, as experience in the real operational environment increased rapidly.

A "fish-eye" lens on a digital still camera was used to record this image of the STS-124 and International Space Station (ISS) Expedition 17 crew members as they share a meal on the middeck of the Space Shuttle Discovery while docked with the ISS. Pictured counterclockwise (from the left bottom): Astronaut Mark Kelly, STS-124 commander; Russian Federal Space Agency Cosmonaut Sergei Volkov, Expedition 17 commander; Astronaut Garrett Reisman; Russian Federal Space Agency Cosmonaut Oleg Kononenko, Astronaut Gregory Chamitoff, Expedition 17 flight engineers; Astronaut Michael Fossum, Japan Aerospace Exploration Agency Astronaut Akihiko Hoshide, Astronaut Karen Nyberg; and Astronaut Kenneth Ham, pilot.

The three major flight phases—ascent, orbit, and re-entry—often required different responses to the same condition, many of which were time critical. This led to the development of different checklists for these phases. New vehicle features such as the Shuttle Robotic Arm and the airlock resulted in additional Flight Data File articles. Some of these, such as the malfunction procedures, did not change unless the underlying system changed or new knowledge was gained, while flight-specific articles, such as the flight plan, EVA, and payload operations checklists, changed for each flight. The Flight Data File included in-flight maintenance

Commander Mark Kelly's personal crew notebook from STS-124.

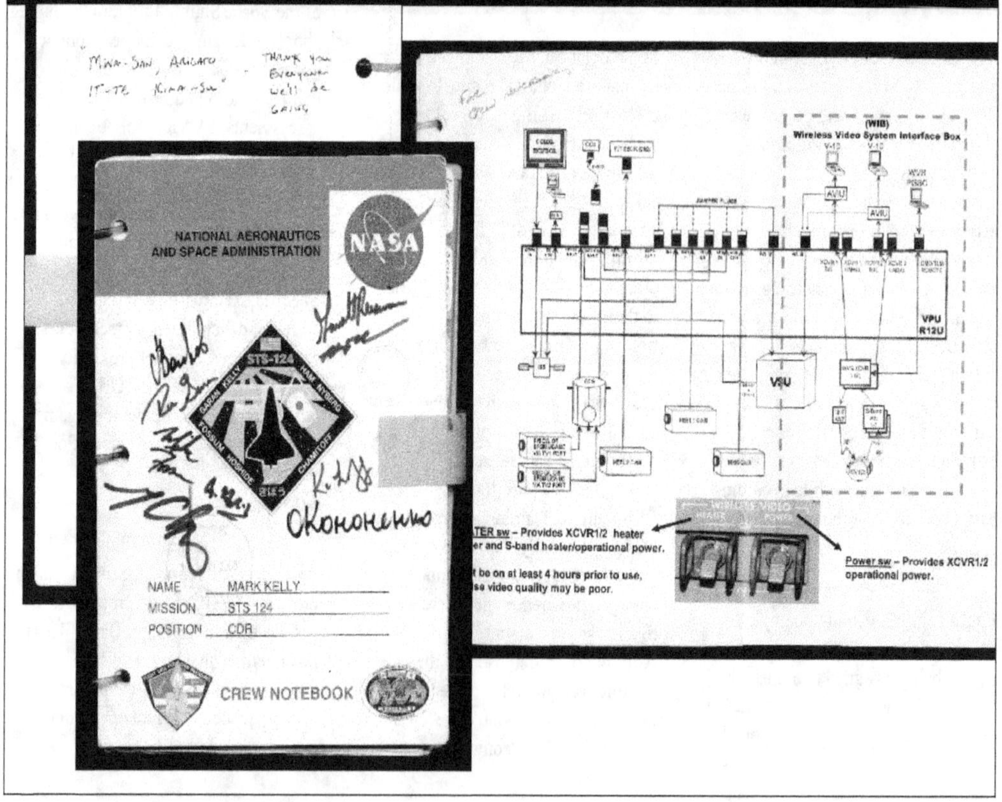

procedures based on experience from the previous programs. Checklist formats and construction standards were developed and refined in consultation with the crews. NASA modeled the pocket checklists, in particular, after similar checklists used by many military pilots for their operations. Flight versions of the cue cards were fitted with Velcro® tabs and some were positioned in critical locations on the various cockpit panels for instantaneous reference.

In addition, the crew developed quick-reference, personal crew notebooks that included key information the crew member felt important, such as emails or letters from individuals or organizations. During ISS missions, the crews established a tradition where the shuttle crew and the ISS crew signed or stamped the front of each other's notebook.

Once the official Flight Data File was completed, crew members reviewed the flight version one last time and often added their own notes on various pages. All information was then copied and the flight versions of the Flight Data File were loaded on the shuttle. Multiple copies of selected Flight Data File books were often flown to enhance on-board productivity.

All flight control team members and stakeholders, including the capsule communicator and flight director, had nearly identical copies of the Flight Data File at their consoles. This was to ensure the best possible communications between the space vehicle and the flight control team. The entire flown Flight Data File with crew annotations, both preflight and in-flight, was recovered Postflight and archived as an official record.

Detailed Trajectory Planning

Trajectory planning efforts, both preflight and in real time, were major activities. Part of the preflight effort involved defining specific parameters called I-loads, which defined elements of the ascent trajectory control software, some of which were defined and loaded on launch day via the Day-of-Launch I-Load Update system. The values of these parameters were uniquely determined for each flight based on the time of year, specific flight vehicle, specific main engines, mass properties including the specific Solid Rocket Boosters (SRBs), launch azimuth, and day-of-launch wind measurements. It was a constant optimization process for each flight to minimize risk and maximize potential success. Other constraints were space radiation events, predictable conjunctions, and predictable meteoroid events, such as the annual Perseid meteor shower period in mid August. The mission operations team developed the Flight Design Handbook to document, in detail, the process for this planning.

Re-entry trajectory planning was initially done preflight and was continuously updated during a mission. NASA evaluated daily landing site opportunities for contingency deorbit purposes, and continuously tracked mass properties and vehicle center of gravity to precisely predict deorbit burn times and re-entry maneuvers. After the Columbia accident (STS-107) in 2003, the agency established new ground rules to minimize the population overflown for normal entries.

Planning also involved a high level of NASA/Department of Defense coordination, particularly following the Challenger accident (STS-51L) in 1986. This included such topics as threat and warning, orbital debris, and search and rescue.

Orbiter and Payload Systems Management

Planning each mission required management of on-board consumables for breathing oxygen, fuel cell reactants, carbon dioxide, potable water and wastewater, Reaction Control System and Orbital Maneuvering System propellants, Digital Auto Pilot, attitude constraints, thermal conditioning, antenna pointing, Orbiter and payload data recording and dumping, power downs, etc. The ground team developed and validated in-flight maintenance activities, as required, then put these activities in procedure form and uplinked the activity list for crew execution. There was an in-flight maintenance checklist of predefined procedures as well as an in-flight maintenance tool kit on board for such activities. Unique requirements for each flight were planned preflight and optimized during the flight by the ground-based flight control team and, where necessary, executed by the crew on request.

Astronaut Training

Training astronauts is a continually evolving process and can vary depending on the agency's objectives. Astronaut candidates typically completed 1 year of basic training, over half of which was on the shuttle. This initial year of training was intended to create a strong foundation on which the candidates would build for future mission assignments. Astronaut candidates learned about the shuttle systems, practiced operation of the shuttle in hands-on mock-ups, and trained in disciplines such as space

Shuttle Training Aircraft

Commanders and pilots used the Shuttle Training Aircraft—a modified Gulfstream-2 aircraft—to simulate landing the Orbiter, which was often likened to landing a brick, especially when compared with the highly maneuverable high-speed aircraft that naval aviators and pilots had flown. The Shuttle Training Aircraft mimicked the flying characteristics of the shuttle, and the left-hand flight deck resembled the Orbiter. Trainers even blocked the windows to simulate the limited view that a pilot experienced during the landing. During simulations at the White Sands Space Harbor in New Mexico, the instructor sat in the right-hand seat and flew the plane into simulation. The commander or pilot, sitting in the left-hand seat, then took the controls. To obtain the feel of flying a brick with wings, he or she lowered the main landing gear and used the reverse thrusters. NASA requirements stipulated that commanders complete a minimum of 1,000 Shuttle Training Aircraft approaches before a flight. Even Commander Mark Kelly—a pilot for two shuttle missions, a naval aviator, and a test pilot with over 5,000 flight hours—recalled that he completed at least "1,600 approaches before [he] ever landed the Orbiter." He conceded that the training was "necessary because the Space Shuttle doesn't have any engines for landing. You only get one chance to land it. You don't want to mess that up."

Two aircraft stationed at Ellington Air Force Base for Johnson Space Center are captured during a training and familiarization flight over White Sands, New Mexico. The Gulfstream aircraft (bottom) is NASA's Shuttle Training Aircraft and the T-38 jet serves as a chase plane.

Flight Simulation Training

For every hour of flight, the STS-124 crew spent 6 hours training on the ground for a total of about 1,940 hours per crew member. This worked out to be nearly a year of 8-hour workdays.

Commander Mark Kelly and Pilot Kenneth Ham practiced rendezvousing and docking with the space station on the Shuttle Engineering Simulator, also known as the dome, numerous times (on weekends and during free time) because the margin of error was so small.

and life sciences, Earth observation, and geology. These disciplines helped develop them into "jacks-of-all-trades."

Flight assignment typically occurred 1 to 1½ years prior to a mission. Once assigned, the crew began training for the specific objectives and specialized needs for that mission. Each crew had a training team that ensured each crew member possessed an accurate understanding of his or her assignments. Mission-specific training was built off of past flight experience, if any, and basic training knowledge. Crew members also received payload training at the principal investigator's facility. This could be at a university, a national facility, an international facility, or another NASA facility. Crew members were the surrogates for the scientists and engineers who designed the payloads, and they trained extensively to ensure a successfully completed mission. As part of their training for the payloads, they may have actually spent days doing the operations required for each day's primary objectives.

Crew members practiced mission objectives in simulators both with and without the flight control teams in Mission Control. Astronauts trained in Johnson Space Center's (JSC's) Shuttle Mission Simulator, shuttle mock-ups, and the Shuttle Engineering Simulator. The Shuttle Mission Simulator contained both a fixed-base and a motion-based high-fidelity station. The motion-based simulator duplicated, as closely as possible, the experience of launch and landing, including the release of the SRBs and External Tank (ET) and the views seen out the Orbiter windows. Astronauts practiced aborts and disaster scenarios in this simulator. The fixed-base simulator included a flight deck and middeck, where crews practiced on-orbit activities. To replicate the feeling of space, the simulator featured views of space and Earth outside the mock-up's windows. Astronauts used the full-fuselage mock-up trainer for a number of activities, including emergency egress practice and EVA training. Crew compartment trainers (essentially the flight deck and the middeck) provided training on Orbiter stowage and related subsystems.

A few months before liftoff, the crew began integrated simulations with the flight control teams in the Mission Control Center. These simulations prepared the astronauts and the flight control teams assigned to the mission to safely execute critical aspects of the mission. They were a crucial step in flight preparation, helping to identify any problems in the flight plan.

With the exception of being in Earth environment, integrated simulations were designed to look and feel as they would in space, except equipment did not malfunction as frequently in space as it did during simulations. Elaborate scripts always included a number of glitches, anomalies, and failures. Designed to bring the on-orbit and Mission Control teams together to work toward a solution, integrated simulations tested not only the crews and controllers but also the mission-specific Flight Rules.

An important part of astronaut crew training was a team-building activity completed through the National Outdoor Leadership School. This involved a camping trip that taught astronaut candidates how to be leaders as well as followers. They had to learn to depend on one another and balance each other's strengths and weaknesses. The astronaut candidates needed to learn to work together as a crew and eventually recognize that their crew was their family. Once a crew was assigned to a mission, these team-building

Team Building

Commander Mark Kelly took his crew and the lead International Space Station flight director to Alaska for a 10-day team-building exercise in the middle of mission training. These exercises were important, Kelly explained, as they provided crews with the "opportunity to spend some quality time together in a stressful environment" and gave the crews an opportunity to develop leadership skills. Because shuttle missions were so compressed, Kelly wanted to determine how his crew would react under pressure and strain. Furthermore, as a veteran, he knew the crew members had to work as a team. They needed to learn more about one another to perform effectively under anxious and stressful circumstances. Thus, away from the conveniences of everyday life, STS-124's crew members lived in a tent, where they could "practice things like team building, Expedition behavior, and working out conflicts." Building a team was important not only to Kelly, but also to the lead shuttle flight director who stressed the importance of developing "a friendship and camaraderie with the crew." To build that support, crew members frequently gathered together for social events after work. A strong relationship forged between the flight control team and crews enabled Mission Control to assess how the astronauts worked and how to work through stressful situations.

The STS-124 crew members celebrate the end of formal crew training with a cake-cutting ceremony in the Jake Garn Simulation and Training Facility at Johnson Space Center. Pictured from the left: Astronauts Mark Kelly, commander; Ronald Garan, mission specialist; Kenneth Ham, pilot; Japan Aerospace Exploration Agency Astronaut Akihiko Hoshide, Astronauts Michael Fossum, Karen Nyberg, and Gregory Chamitoff, all mission specialists. The cake-cutting tradition shows some of the family vibe between the training team and crew as they celebrate key events in an assigned crew training flow.

Prior to launch, astronauts walk around their launch vehicle at Kennedy Space Center.

activities became an important part of the mission-specific training flow. Teamwork was key to the success of a shuttle mission.

When basic training was complete, astronauts received technical assignments; participated in simulations, support boards, and meetings; and made public appearances. Many also began specialized training in areas such as EVA and robotic operations. Extensive preflight training was performed when EVAs were required for the mission.

Each astronaut candidate completed an EVA skills program to determine his or her aptitude for EVA work. Those continuing on to the EVA specialty completed task training and systems training, the first of which was specific to the tasks completed by an astronaut during an EVA while the latter focused on suit operations. Task training included classes on topics such as the familiarization and operation of tools. For their final EVA training, the astronauts practiced in a swimming pool that produced neutral buoyancy, which mimicked some aspect of microgravity. Other training included learning about their EVA suits, the use of the airlock in the Orbiter or ISS, and the medical requirements to prevent decompression sickness.

Mission-specific EVA training typically began 10 months before launch. An astronaut completed seven neutral buoyancy training periods for each spacewalk that was considered complex, and five training periods for noncomplex or repeat tasks. The last training runs before launch were usually completed in the order in which they would occur during the mission. Some astronauts found that the first EVA was more intimidating than the others simply because it represented that initial hurdle to overcome before gaining their rhythm. This concern was eased by practicing an additional Neutral Bouyancy Laboratory training run for their first planned spacewalk as the very last training run before launch.

EVA and robotic operations were commonly integrated, thereby creating the need to train both specialties together and individually. The robotic arm operator received specialized training with the arm on the ground using skills to mimic microgravity and coordination through a closed-circuit television.

EVA training was also accomplished in the Virtual Reality Laboratory, which was similarly used for robotic training. The Virtual Reality Laboratory complemented the underwater training with a more comfortable and flexible environment for reconfiguration changes. Virtual reality software was also used to increase an astronaut's situational awareness and develop effective verbal commands as well as to familiarize him or her with mass handling on the arm and r-bar pitch maneuver photography training.

T-38 aircraft training was primarily used to keep astronauts mentally conditioned to handle challenging, real-time situations. Simulators were an excellent training tool, but they were limited in that the student had the comfort of knowing that he or she was safely on the ground. The other benefit of T-38 training was that the aircraft permitted frequent and flexible travel, which was necessary to accommodate an astronaut's busy training schedule.

In Need of a Plumber

Just a few days before liftoff of STS-124, the space station's toilet broke. This added a wrinkle to the flight plan redrafted earlier. Russia delivered a spare pump to Kennedy Space Center, and the part arrived just in time to be added to Discovery's middeck. Storage space was always at a premium on missions. The last-minute inclusion of the pump involved some shifting and the removal of 15.9 kg (35 pounds) of cargo, including some wrenches and air-scrubber equipment. This resulted in changes to the flight plan—Discovery's crew and the station members would use the shuttle's toilet until station's could be used. If that failed, NASA packed plenty of emergency bags typically used by astronauts to gather in-flight urine specimens for researchers.

When the crew finally arrived and opened the airlock, Commander Mark Kelly joked, "Hey, you looking for a plumber?" The crews, happy to see each other, embraced one another.

Crew Prepares for Launch

With all systems "go" and launch weather acceptable, STS-124 launched on May 31, 2008, marking the 26th shuttle flight to the International Space Station. Three hours earlier, technicians had strapped in seven astronauts for NASA's 123rd Space Shuttle mission. Commander Mark Kelly was a veteran of two shuttle missions. By contrast, the majority of his crew consisted of rookies—Pilot Kenneth Ham along with Astronauts Karen Nyberg, Ronald Garan, Gregory Chamitoff, and Akihiko Hoshide of the Japan Aerospace Exploration Agency. Although launch typically represented the beginning of a flight, more than 2 decades of work went into the coordination of this single mission.

After suiting up, STS-124 crew members exited the Operations and Checkout Building to board the Astrovan, which took them to Launch Pad 39A for the launch of Space Shuttle Discovery. On the right (front to back): Astronauts Mark Kelly, Karen Nyberg, and Michael Fossum. On the left (front to back): Astronauts Kenneth Ham, Ronald Garan, Akihiko Hoshide, and Gregory Chamitoff.

There were roughly two dozen T-38 aircraft at any time, all of which were maintained and flown out of Ellington Field in Houston, Texas. As part of astronaut candidate training, they received T-38 ground school, ejection seat training, and altitude chamber training. Mission specialists frequently did not have a military flying background, so they were sent to Pensacola, Florida, to receive survival training from the US Navy. As with any flight certification, currency requirements were expected to be maintained. Semiannual total T-38 flying time minimum for a pilot was 40 hours. For a mission specialist, the minimum flight time was 24 hours. Pilots were also required to meet approach and landing minimum flight times.

Launching the Shuttle

Launch day was always exciting. KSC's firing room controlled the launch, but JSC's Mission Operations intently watched all the vehicle systems. The Mission Control Center was filled with activity as the flight controllers completed their launch checklists. For any shuttle mission, the weather was the most common topic of discussion

The Countdown Begins

The primary objective of the STS-124 mission was to deliver Japan's Kibo module to the International Space Station. As Commander Mark Kelly said, "We're going to deliver Kibo, or hope, to the space station, and while we tend to live for today, the discoveries from Kibo will certainly offer hope for tomorrow." The Japanese module is an approximately 11-m (37-ft), 14,500-kg (32,000-pound) pressurized science laboratory, often referred to as the Japanese Pressurized Module. This module was so large that the Orbiter Boom Sensor System had to be left on orbit during STS-123 (2008) to accommodate the extra room necessary in Discovery's payload bay.

During the STS-124 countdown, the area experienced some showers. By launch time, however, the sea breeze had pushed the showers far enough away to eliminate any concerns. The transatlantic abort landing weather proved a little more challenging, with two of the three landing sites forecasted to have weather violations. Fortunately, Moron Air Base, Spain, remained clear and became the chosen transatlantic abort landing site.

Space Shuttle Discovery and its seven-member STS-124 crew head toward low-Earth orbit and a scheduled link-up with the International Space Station.

and the most frequent reason why launches and landings were delayed. Thunderstorms could not occur too close to the launch pad, crosswinds had to be sufficiently low, cloud decks could not be too thick or low, and visibility was important. Acceptable weather needed to be forecast at the launch site and transatlantic abort landing sites as well as for each ascent abort option.

Not far from the launch pad, search and rescue forces were always on standby for both launch and landing. This included pararescue jumpers to retrieve astronauts from the water if a bailout event were to occur. The more well-known assets were the support ships, which were also supported by each of the military branches and the US Coast Guard. This team of search-and-rescue support remained on alert throughout a mission to ensure the safe return of all crew members.

Shortly before a launch, the KSC launch director polled the KSC launch control room along with JSC Mission Control for a "go/no go" launch decision. The JSC front room flight controllers also polled their back room flight controllers for any issues. If no issues were identified, the flight controllers, representing their specific discipline, responded to the flight director with a "go." If an issue was identified, the flight controller was required to state "no go" and why. Flight Rules existed to identify operational limitations, but even with these delineations the decision to launch was never simple.

Fly

Ground Facilities Operations

The Mission Control Center relied on the NASA network, managed by Goddard Space Flight Center (GSFC), to route the spacecraft downlink telemetry, tracking, voice, and television and uplink voice, data, and command. The primary in-flight link was to/from the Mission Control Center to the White Sands Ground Terminal up to the tracking and data relay satellites and then to/from the Orbiter. In addition, there were still a few ground sites with a direct linkage to/from the Orbiter as well as specific C-band tracking sites for specific phases as needed. The preflight planning function included arranging for flight-specific support from all these ground facilities and adjusting them, as necessary, based on in-flight events. The readiness of all these support elements for each flight was certified by the GSFC network director at the Mission Operations Flight Readiness Review.

The Mission Control Center was the focus of shuttle missions during the flight phase. Control of the mission and communication with the crew transferred from the KSC firing room to the JSC Mission Control Center at main engine ignition. Shuttle systems data, voice communications, and television were relayed almost instantaneously to the Mission Control Center through the NASA ground and space networks. In many instances, external facilities such as MSFC and GSFC as well as US Air Force and European Space Agency facilities also provided support for specific payloads. The facility support effort, the responsibility of the operations support team, ensured the Mission Control Center and all its interfaces were ready with the correct software, hardware, and interfaces to support a particular flight.

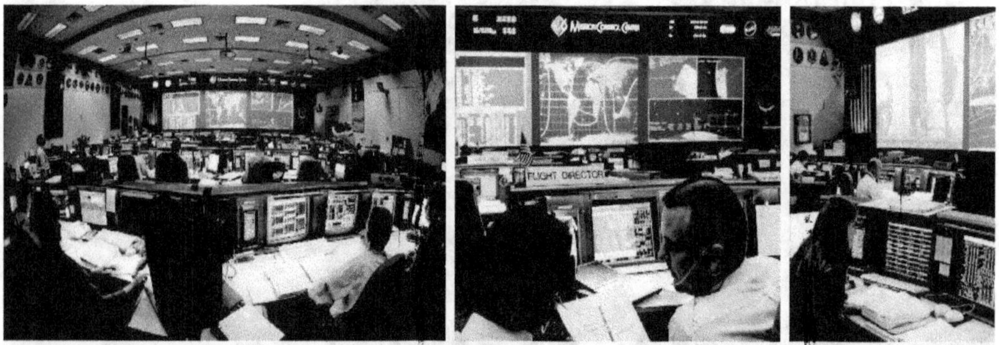

The Mission Control Center front room houses the capsule communicator, flight director and deputy, and leads for all major systems such as avionics, life support, communication systems, guidance and navigation, extravehicular activity lead and robotic arm, propulsion and other expendables, flight surgeon, and public affairs officer. These views show the extensive support and consoles. ***Left photo:*** *At the front of the operations center are three screens. The clocks on the left include Greenwich time, mission elapsed time, and current shuttle commands. A map of the world with the shuttle position-current orbit is in the center. The right screen shows shuttle attitude.* ***Center photo:*** *Flight Director Norman Knight (right) speaks with one of the leads at the support console.* ***Right photo:*** *Each console in the operations center has data related to the lead's position; e.g., the life support position would have the data related to Orbiter air, water, and temperature readings and the support hardware functions.*

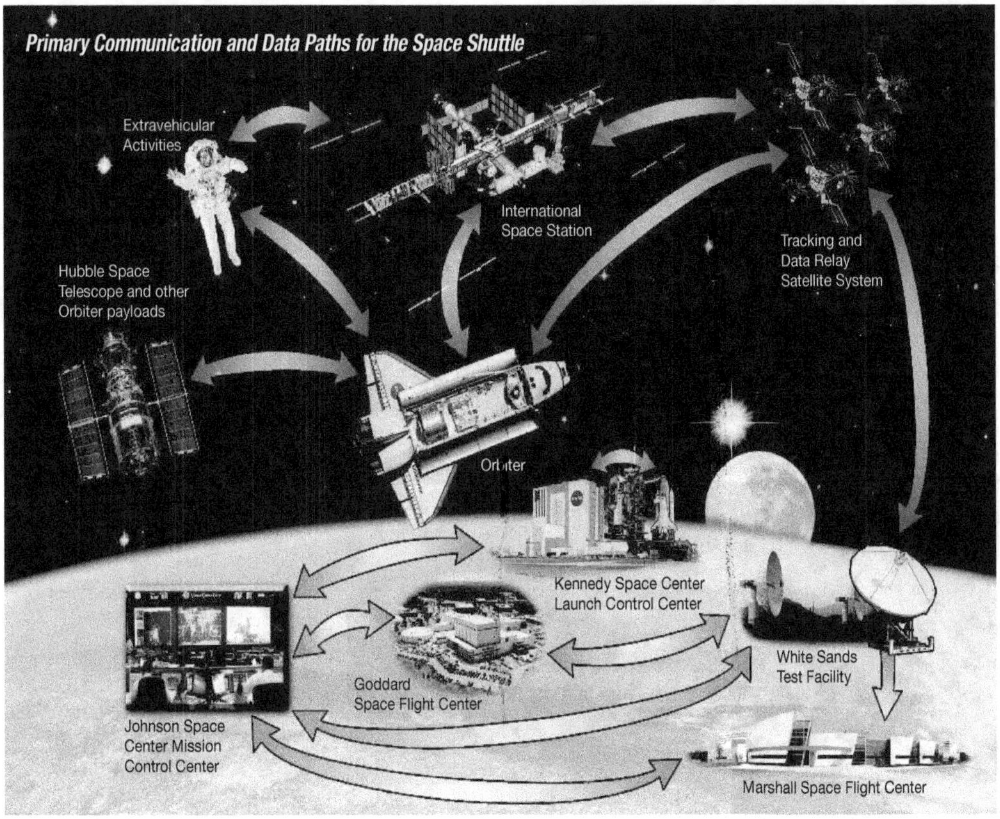

Just before shuttle liftoff, activity in the Mission Control Center slowed and the members of the flight control team became intently focused on their computer screens. From liftoff, the performance of the main engines, SRBs, and ET were closely observed with the team ready to respond if anything performed off-nominally. If, for example, a propulsion failure occurred, the flight control team would identify a potential solution that may or may not require the immediate return of the Orbiter to the ground. If the latter were necessary, an abort mode (i.e., return to launch site, transatlantic abort landing) and a landing site would be selected. The electrical systems and the crew environment also had to function correctly while the Orbiter was guided into orbit. For the entire climb to orbit, personnel in the Mission Control Center remained intensely focused. Major events were called out during the ascent. At almost 8½ minutes, when target velocity was achieved, main engine cutoff was commanded by the on-board computers and flight controllers continued verifying system performance. Every successful launch was an amazing accomplishment.

Before and after a shuttle launch, KSC personnel performed walkdowns of the launch pad for a visual inspection of any potential debris sources. Shuttle liftoff was a dynamic event that could cause ice/frost or a loose piece of hardware to break free and impact the Orbiter. Finding these debris sources and preventing potential damage was important to the safety of the mission.

Debris Impact on the Orbiter

Debris from launch and on orbit could make the Orbiter unable to land. The Orbiter could also require on-orbit repair.

Ascent Inspection

After the Columbia accident (2003), the shuttle was closely observed during the shuttle launch and for the duration of the ascent phase by a combination of ground and vehicle-mounted cameras, ground Radio Detection and Ranging, and the Wing Leading Edge Impact Detection System. The ground cameras were located on the fixed service structure, the mobile launch platform, around the perimeter of the launch pad, and on short-, medium-, and long-range trackers located along the Florida coast. The ground cameras

Orbiter Survey

The Orbiter survey included the Orbiter's crew cabin Thermal Protection System and the wing leading edge and nose cap reinforced carbon-carbon using the Shuttle Robotic Arm and the Orbiter Boom Sensor System. The survey involved detailed scanning in a specified pattern and required most of the day to complete. A focused inspection was only performed when a suspect area was identified and more detailed information was required to determine whether a repair or alternative action was necessary.

Astronaut Karen Nyberg, STS-124, works the controls on the aft flight deck of Space Shuttle Discovery during Flight Day 2 activities.

Due to the unique nature of the STS-124 mission, the Shuttle Robotic Arm was used instead of the Orbiter Boom Sensor System. Astronaut Karen Nyberg operated the robotic arm for the inspection of the Thermal Protection System. The nose cap and wing leading edge reinforced carbon-carbon survey was scheduled for post undock after the Orbiter Boom Sensor System had been retrieved during a Flight Day 4 extravehicular activity.

provided high-resolution imagery of liftoff and followed the vehicle through SRB separation and beyond. The vehicle-mounted cameras were strategically placed on the tank, boosters, and Orbiter to observe the condition of specific areas of interest and any debris strikes. The crew took handheld video and still imagery of the tank following separation when lighting conditions permitted. This provided another source of information to confirm a clean separation or identify any suspect areas on the tank that might potentially represent a debris concern for the Orbiter Thermal Protection System. The Wing Leading Edge Impact Detection System used accelerometers mounted within the Orbiter's wing leading edge to monitor for impacts throughout the ascent and orbit phases, power permitting.

The world's largest C-band radar and two X-band radars played an integral role in the ascent debris observation through a valuable partnership with the US Navy. The C-band radar watched for falling debris near the Orbiter, and the X-band radar further interpreted the velocity characteristics of any debris events with respect to the vehicle's motion. The X-band radars were on board an SRB recovery ship located downrange of the launch site and a US Army vessel south of the groundtrack. The US Navy C-band radar sat just north of KSC.

Data collected from ground and vehicle-mounted cameras, ground radar, and the Wing Leading Edge Impact Detection System created a comprehensive set of ascent data. Data were sent to the imagery analysis teams at JSC, KSC, and MSFC for immediate review. Each team had its area of specialty; however, intentional overlap of the data analyses existed as a conservative measure. As early as 1 hour after launch, these teams of imagery specialists gathered in a dark room with a large screen and began reviewing every camera angle captured. They watched the videos in slow motion, forward, and backward as many times as necessary to thoroughly analyze the data. The teams were looking for debris falling off the vehicle stack or even the pad structure that may have impacted the Orbiter. If the team observed or even suspected a debris strike on the Orbiter, the team reported the location to the mission management team and the Orbiter damage assessment team for on-orbit inspection. The damage assessment team oversaw the reported findings of the on-orbit imagery analysis and delivered a recommendation to the Orbiter Project Office and the mission management team stating the extent of any damage and the appropriate forward action. This cycle of obtaining imagery, reviewing imagery, and recommending forward actions continued throughout each phase of the mission.

On-orbit Inspections

The ISS crew took still images of the Orbiter as it approached the station and performed maneuvers, exposing the underside tiles. Pictures were also taken of the ET umbilical doors to verify proper closure as well as photos of the Orbiter's main engines, flight deck windows, Orbital Maneuvering System pods, and vertical stabilizer. The shuttle crew photographed the pods and the leading edge of the vertical stabilizer from the windows of the flight deck. The ISS crew took still images of the Orbiter. All images were downlinked for review by the damage assessment team.

For all missions to the ISS that took place after the Columbia accident, late inspection was completed after the Orbiter undocked. This activity included a survey of the reinforced carbon-carbon to look for any micrometeoroid orbital debris damage that may have occurred during the time on orbit. Since the survey was only of the reinforced carbon-carbon, it took less time to complete than did the initial on-orbit survey. As with the Flight Day 2 survey, the ground teams compared the late inspection imagery to Flight Day 2 imagery and either cleared the Orbiter for re-entry or requested an alternative action.

On-orbit Activities

Extravehicular Activity Preparation

For missions that had EVAs, the day after launch was reserved for extravehicular mobility unit checkout and the Orbiter survey. EVA suit checkout was completed in the airlock where the suit systems were verified to be operating correctly. Various procedures developed over the nearly 30-year history for an EVA mission were implemented to prevent decompression sickness and ensure the crew and all the hardware were ready. The day of the EVA, both crew members suited up with the assistance of the other crew members and then left the airlock. EVAs involving the Shuttle Robotic Arm required careful coordination between crew members. This was when the astronauts applied the meticulously practiced verbal commands.

For missions to the ISS, the primary objective of Flight Day 3 was to rendezvous and dock with the ISS. As the Orbiter approached the ISS, it performed a carefully planned series of burns to adjust the orbit for a smooth approach to docking.

On-orbit Operations

Within an hour of docking with the ISS, the hatch opened and the shuttle crew was welcomed by the ISS crew. For missions consisting of a crew change, the first task was to transfer the custom Soyuz seat liners to crew members staying on station. Soyuz is the Russian capsule required for emergency return to Earth and for crew rotations. Completion of this task marked the formal change between the shuttle and ISS crews.

A Flawless Rendezvous

On day three, STS-124 rendezvoused and docked with the space station. About 182 m (600 ft) below the station, Commander Mark Kelly flipped Discovery 360 degrees so that the station crew members could photograph the underbelly of the shuttle. Following the flip, Kelly conducted a series of precise burns with the Orbital Maneuvering System, which allowed the shuttle—flying about 28,200 km/hr (17,500 mph)—to chase the station, which was traveling just as fast. Kelly, who had twice flown to the station, described the moment: "It's just incredible when you come 610 m (2,000 ft) underneath it and see this giant space station. It's just an amazing sight." Once the Orbiter was in the same orbit with the orbiting lab, Kelly nudged the vehicle toward the station. As the vehicle moved, the crew encountered problems with the Trajectory Control System, a laser that provided range and closure rates. This system was the primary sensor, which the crew members used to gauge how far they were from the station. Luckily, the crew had simulated this failure numerous times, so the malfunction had no impact on the approach or closure. The lead shuttle flight director called the rendezvous "absolutely flawless." Upon docking ring capture, the crew congratulated Kelly with a series of high fives.

Trust and Respect Do Matter

During activation of the Japanese Experiment Module, the flight controllers in Japan encountered a minor hiccup. As the crew attached the internal thermal control system lines, ground controllers worried that there was an air bubble in the system's lines, which could negatively impact the pump's performance. Controllers in Houston, Texas, and Tsukuba, Japan, began discussing options. The International Space Station (ISS) flight director noticed that the relationship she had built with the Japanese "helped immensely." The thermal operations and resource officer had spent so many years working closely with his Japan Aerospace Exploration Agency counterpart that, when it came time to decide to use the nominal plan or a different path, "the respect and trust were there," and the Japanese controllers agreed with his recommendations to stay with the current plan. "I think," the ISS flight director said, "that really set the mission on the right course, because then we ended up proceeding with activation."

Every mission included some housekeeping and maintenance. New supplies were delivered to the station and old supplies were stowed in the Orbiter for return to Earth. Experiments that completed their stay on board the ISS were also returned home for analyses of the microgravity environment's influence.

Returning Home

If necessary, a flight could be extended to accommodate extra activities and weather delays. The mission management team decided on flight extensions for additional activities where consideration was given for impacts to consumables, station activities, schedule, etc. Landing was typically allotted 2 days with multiple opportunities to land. NASA's preference was always to land at KSC since the vehicle could be processed at that facility; however, weather would sometimes push the landing to Dryden Flight Research Center/Edwards Air Force Base. If the latter occurred, the Orbiter was flown back on a modified Boeing 747 in what was referred to as a "ferry flight."

Once the Orbiter landed and rolled to a stop, the Mission Control Center turned control back to KSC. After landing, personnel inspected the Orbiter for any variations in Thermal Protection System and reinforced carbon-carbon integrity. More imagery was taken for comparison to on-orbit imagery. Once the Orbiter was at the Orbiter Processing Facility, its cameras were removed for additional imagery analysis and the repairs began in preparation for another flight.

Returning to Earth

Space Shuttle Discovery's drag chute is deployed as the spacecraft rolls toward a stop on runway 15 of the Shuttle Landing Facility at Kennedy Space Center, concluding the 14-day STS-124 mission to the International Space Station.

After nearly 9 days at the space station, the crew of STS-124 undocked and said farewell to Gregory Chamitoff, who would be staying on as the flight engineer for the Expedition crew, and the two other crew members. When watching the goodbyes on video, it appeared as if the crew said goodbye, closed the hatch, and dashed away from the station. "It's more complicated than that," Commander Mark Kelly explained. "You actually spend some time sitting on the Orbiter side of the hatch." About 1 hour passed before the undocking proceeded. Afterward, the crew flew around the station and then completed a full inspection of the wing's leading edge and nose cap with the boom.

The crew began stowing items like the Ku-band antenna in preparation for landing on June 15. On the day of landing, the crew suited up and reconfigured the Orbiter from a spaceship to an airplane. The re-entry flight director and his team worked with the crew to safely land the Orbiter, and continually monitored weather conditions at the three landing sites. With no inclement weather at Kennedy Space Center, the crew of STS-124 was "go" for landing. The payload bay doors were closed several minutes before deorbit burn. The crew then performed checklist functions such as computer configuration, auxiliary power unit start, etc. Sixty minutes before touchdown the deorbit burn was performed. After the Columbia accident, the re-entry profiles for the Orbiter changed so that the crew came across the Gulf of Mexico, rather than the United States. As the Orbiter descended, the sky turned from pitch black to red and orange. Discovery hit the atmosphere at Mach 25 and a large fireball surrounded the glider. It rapidly flew over Mexico. By the time it passed over Orlando, Florida, the Orbiter slowed. As they approached the runway, Kelly pulled the nose up and lowered the landing gear. On touchdown—after main gear touchdown but before nose gear touchdown—he deployed a parachute, which helped slow the shuttle as it came to a complete stop.

The Shuttle Carrier Aircraft transported the Space Shuttle Endeavour from Dryden Research Center, California, back to Kennedy Space Center, Florida.

Solid Foundations Assured Success

Two pioneers of flight operations, Christopher Kraft and Gene Kranz, established the foundations of shuttle mission operations in the early human spaceflight programs of Mercury, Gemini, and Apollo. Their "plan, train, fly" approach made controllers tough and competent, "flexible, smart, and quick on their feet in real time," recalled the lead flight director for STS-124 (2008). That concept, created in the early 1960s, remained the cornerstone of mission operations throughout the Space Shuttle Program, as exemplified by the flight of STS-124.

Endeavour touches down at Dryden Flight Research Center located at Edwards Air Force Base in California to end the STS-126 (2008) mission.

Extravehicular Activity Operations and Advancements

Nancy Patrick
Joseph Kosmo
James Locke
Luis Trevino
Robert Trevino

A dramatic expansion in extravehicular activity (EVA)—or "spacewalking"—capability occurred during the Space Shuttle Program; this capability will tremendously benefit future space exploration. Walking in space became almost a routine event during the program—a far cry from the extraordinary occurrence it had been. Engineers had to accommodate a new cadre of astronauts that included women, and the tasks these spacewalkers were asked to do proved significantly more challenging than before. Spacewalkers would be charged with building and repairing the International Space Station. Most of the early shuttle missions helped prepare astronauts, engineers, and flight controllers to tackle this series of complicated missions while also contributing to the success of many significant national resources—most notably the Hubble Space Telescope. Shuttle spacewalkers manipulated elements up to 9,000 kg (20,000 pounds), relocated and installed large replacement parts, captured and repaired failed satellites, and performed surgical-like repairs of delicate solar arrays, rotating joints, and sensitive Orbiter Thermal Protection System components. These new tasks presented unique challenges for the engineers and flight controllers charged with making EVAs happen.

The Space Shuttle Program matured the EVA capability with advances in operational techniques, suit and tool versatility and function, training techniques and venues, and physiological protocols to protect astronauts while providing better operational efficiency. Many of these advances were due to the sheer number of EVAs performed. Prior to the start of the program, 38 EVAs had been performed by all prior US spaceflights combined. The shuttle astronauts accomplished 157 EVAs.

This was the primary advancement in EVA during the shuttle era—an expansion of capability to include much more complicated and difficult tasks, with a much more diverse Astronaut Corps, done on a much more frequent basis. This will greatly benefit space programs in the future as they can rely on a more robust EVA capability than was previously possible.

Spacewalking: Extravehicular Activity

If We Can Put a Human on the Moon, Why Do We Need to Put One in the Payload Bay?

The first question for program managers at NASA in regard to extravehicular activities (EVAs) was: Are they necessary? Managers faced the challenge of justifying the added cost, weight, and risk of putting individual crew members outside and isolated from the pressurized cabin in what is essentially a personal spacecraft. Robotics or automation are often considered alternatives to sending a human outside the spacecraft; however, at the time the shuttle was designed, robotics and automation were not advanced enough to take the place of a human in all required external tasks. Just as construction workers and cranes are both needed to build skyscrapers, EVA crew members and robots are needed to work in space.

Early in the Space Shuttle Program, safety engineers identified several shuttle contingency tasks for which EVA was the only viable option. Several shuttle components could not meet redundancy requirements through automated means without an untenable increase in weight or system complexity. Therefore, EVA was employed as a backup. Once EVA capability was required, it became a viable and cost-effective backup option as NASA identified other system problems. Retrieval or repair of the Solar Maximum Satellite (SolarMax) and retrieval of the Palapa B2 and Westar VI satellites were EVA tasks identified very early in the program. Later, EVA became a standard backup option for many shuttle payloads, thereby saving cost and resolving design issues.

Gregory Harbaugh
Astronaut on STS-39 (1991), STS-54 (1993), and STS-82 (1997).
Manager, Extravehicular Activity (EVA) Office (1997-2001).

"In my opinion, one of the major achievements of the Space Shuttle era was the dramatic enhancement in productivity, adaptability, and efficiency of EVA, not to mention the numerous EVA-derived accomplishments. At the beginning of the shuttle era, the extravehicular mobility unit had minimal capability for tools, and overall utility of EVA was limited. However, over the course of the program EVA became a planned event on many missions and ultimately became the fallback option to address a multitude of on-orbit mission objectives and vehicle anomalies. Speaking as the EVA program manager for 4 years (1997-2001), this was the result of incredible reliability of the extravehicular mobility unit thanks to its manufacturers (Hamilton Sundstrand and ILC Dover), continuous interest and innovation led by the EVA crew member representatives, and amazing talent and can-do spirit of the engineering/training teams. In my 23 years with NASA, I found no team of NASA and contractor personnel more technically astute, more dedicated, more innovative, or more ultimately successful than the EVA team.

EVA became an indispensible part of the Space Shuttle Program. EVA could and did fix whatever problems arose, and became an assumed tool in the holster of the mission planners and managers. In fact, when I was EVA program manager we had shirts made with the acronym WOBTSYA—meaning 'we've only begun to save your Alpha' (the ISS name at the time). We knew when called upon we could handle just about anything that arose."

Automation and Extravehicular Activity

EVA remained the preferred method for many tasks because of its efficiency and its ability to respond to unexpected failures and contingencies. As amazing and capable as robots and automation are, they are typically efficient for anticipated tasks or those that fall within the parameters of known tasks. Designing and certifying a robot to perform tasks beyond known requirements is extremely costly and not yet mature enough to replace humans.

Robots and automation streamlined EVA tasks and complemented EVA, resulting in a flexible and robust capability for building, maintaining, and repairing space structures and conducting scientific research.

Designing the Spacesuit for the Space Shuttle

Once NASA established a requirement for EVA, engineers set out to design and build the hardware necessary to provide this capability. Foremost, a spacesuit was required to allow a crew member to venture outside the pressurized cabin. The Gemini and Apollo spacesuits were a great starting point; however, many changes were needed to create a workable suit for the shuttle. The shuttle suit had to be reusable, needed to fit many different crew members, and was required to last for many years of repeated use. Fortunately, engineers were able to take advantage of advanced technology and lessons learned from earlier programs to meet these new requirements.

Contingency extravehicular activity: Astronaut Scott Parazynski, atop the Space Station Robotic Arm and the Shuttle Robotic Arm extension, the Orbiter Boom Sensor System, approaches the International Space Station solar arrays to repair torn sections during STS-120 (2009).

The cornerstone design requirement for any spacesuit is to protect the crew member from the space environment.

Suit Environment as Compared to Space Environment

Atmosphere	Suit Environment Requirements	Space Environment
Pressure:	23.44 kPa-27.57 kPa (3.4-4.4 psi)	1 Pa (1.45 x 10^{-4} psi)
Oxygen:	100%	0%
Temperature:	10°C-27°C (50°F-80°F)	-123°C-+232°C (-190°F-+450°F)

The target suit pressure was an exercise in balancing competing requirements. The minimum pressure required to sustain human life is 21.4 kPa (3.1 psi) at 100% oxygen. Higher suit pressure allows better oxygenation and decreases the risk of decompression sickness to the EVA crew member. Lower suit pressure increases crew member flexibility and dexterity, thereby reducing crew fatigue. This is similar to a water hose. A hose full of water is difficult to bend or twist, while an empty hose is much easier to move around. Higher suit pressures also require more structural stiffening to maintain suit integrity (just as a thicker balloon is required to hold more air). This further exacerbates the decrease in flexibility and dexterity. The final suit pressure selected was 29.6 kPa (4.3 psi), which has proven to be a reasonable compromise between these competing constraints.

The next significant design requirements came from the specific mission applications: what EVA tasks were required, who would perform them, and to what environmental conditions the spacewalkers would be exposed. Managers decided that the shuttle spacesuit would only be required to perform in microgravity and outside the shuttle cabin. This customized requirement allowed designers to optimize the spacesuit. The biggest advantage of this approach was that designers didn't have to worry as much about the mass of the suit.

Improving mobility was also a design goal for the shuttle extravehicular mobility unit (i.e., EVA suit). Designers added features to make it more flexible and allow the crew member greater range of motion than with previous suits. Bearings were included in the shoulder, upper arm, and waist areas to provide a useful range of mobility. The incorporation of the waist bearing enabled the EVA crew member to rotate.

Shuttle managers decided that, due to the duration of the program, the suit should also be reusable and able to fit many different crew members. Women were included as EVA crew members for the first time, necessitating unique accommodations and expanding the size range required. The range had to cover from the 5% American Female to the 95% American Male with variations in shoulders, waist, arms, and legs.

A modular "tuxedo" approach was used to address the multi-fit requirement. Tuxedos use several different pieces, which can be mixed and matched to best fit an individual—one size of pants can be paired with a different size shirt, cummerbund, and shoes to fit the individual. The EVA suit used a modular design, thereby allowing various pieces of different sizes to achieve a reasonably good fit. The design also incorporated a custom-tailoring capability using inserts, which allowed a reasonably good fit with minimal modifications.

Crew Member Size Variations and Ranges

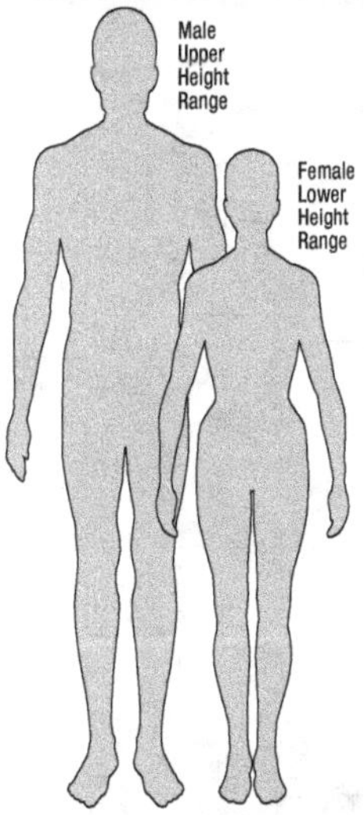

Critical Body Dimension	5th % Female cm (in.)	95th % Male cm (in.)	Max. Size Variation cm (in.)
Standing Height	152.1 (59.9)	188.7 (74.3)	36.6 (14.4)
Chest Breadth	25.1 (9.9)	36.6 (14.4)	11.7 (4.6)
Chest Depth	20.8 (8.2)	27.7 (10.9)	6.9 (2.7)
Chest Circumference	82.3 (32.4)	109.7 (43.2)	27.4 (10.8)
Shoulder Circumference	95.5 (36.7)	128.5 (50.6)	35.3 (13.9)
Shoulder Breadth	38.6 (15.2)	46.7 (18.4)	8.1 (3.2)
Shoulder Height	122.9 (48.4)	156.7 (61.7)	33.8 (13.3)
Fingertip Span	152.4 (60.0)	195.6 (77.0)	43.2 (17.0)
Torso Length	56.1 (22.1)	70.4 (27.7)	14.2 (5.6)
Hip Breadth	31.5 (12.4)	38.9 (15.3)	7.4 (2.9)
Crotch Height	60.1 (26.8)	93.5 (36.8)	25.4 (10.0)
Knee Height	38.1 (15.0)	54.1 (21.3)	16.0 (6.3)

While the final design didn't accommodate the entire size range of the Astronaut Corps, it was flexible enough to allow for a wide variety of crew members to perform spacewalks, especially those crew members who had the best physical attributes for work on the International Space Station (ISS).

One notable exception to this modular approach was the spacesuit gloves. Imagine trying to assemble a bicycle while wearing ski gloves that are too large and are inflated like a balloon. This is similar to attempting EVA tasks like driving bolts and operating latches while wearing an ill-fitting glove. Laser-scanning technology was used to provide a precise fit for glove manufacture patterns. Eventually, it became too expensive to maintain a fully customized glove program. Engineers were able to develop a set of standard sizes with adjustments at critical joints to allow good dexterity at a much lower cost. In contrast, a single helmet size was deemed sufficient to fit the entire population without compromising a crew member's ability to perform tasks.

The responsibility for meeting the reuse requirement was borne primarily by the Primary Life Support System, or "backpack," which included equipment within the suit garment to control various life functions. The challenge

Extravehicular Mobility Unit

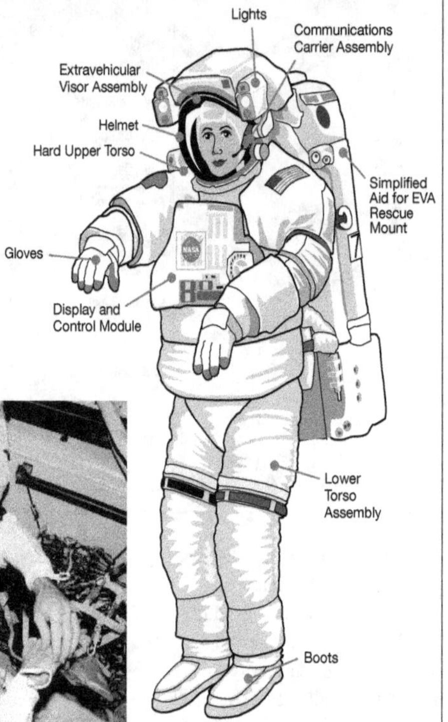

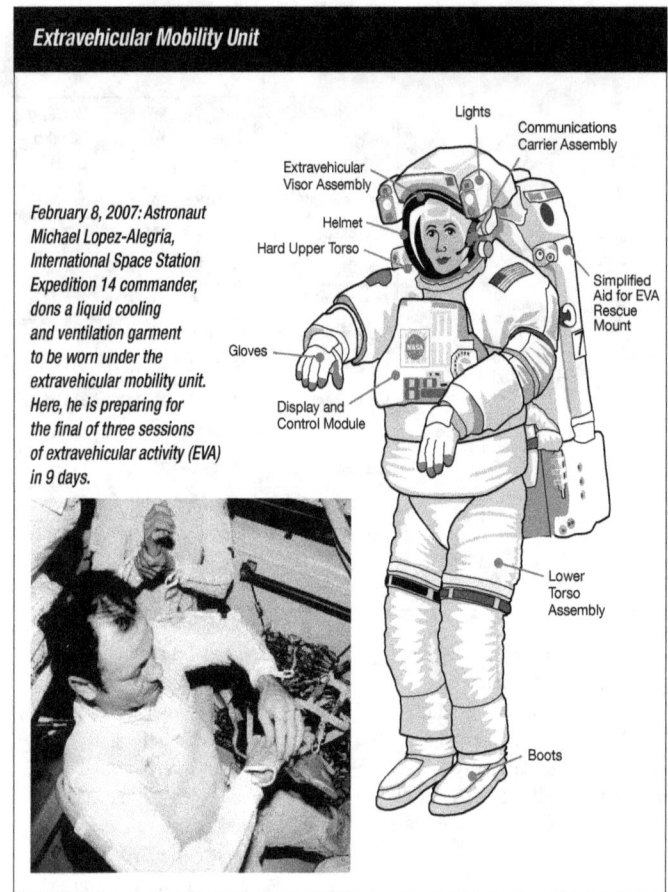

February 8, 2007: Astronaut Michael Lopez-Alegria, International Space Station Expedition 14 commander, dons a liquid cooling and ventilation garment to be worn under the extravehicular mobility unit. Here, he is preparing for the final of three sessions of extravehicular activity (EVA) in 9 days.

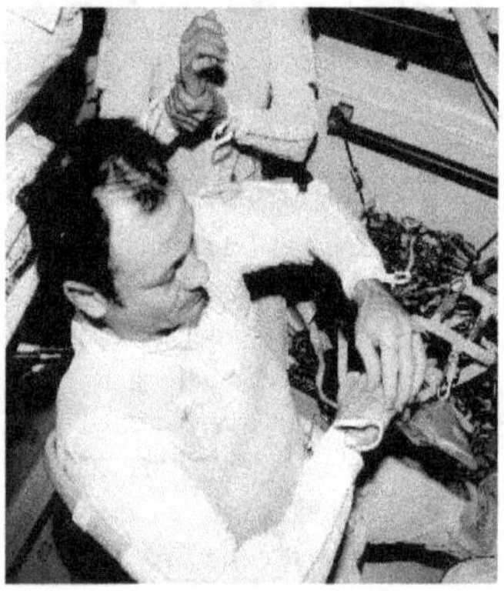

for Primary Life Support System designers was to provide a multiyear, 25-EVA system. This design challenge resulted in many innovations over previous programs.

One area that had to be improved to reduce maintenance was body temperature control. Both the Apollo and the shuttle EVA suit used a water cooling system with a series of tubes that carried chilled water and oxygen around the body to cool and ventilate the crew member. The shuttle EVA suit improved on the Apollo design by removing the water tubes from the body of the suit and putting them in a separate garment—the liquid cooling ventilation garment. This garment was a formfitting, stretchable undergarment (think long johns) that circulated water and oxygen supplied by the Primary Life Support System through about 91 m (300 ft) of flexible tubing. This component of the suit was easily replaceable, inexpensive, easy to manufacture, and available in several sizes.

Materials changes in the Primary Life Support System also helped to reduce maintenance and refurbishment requirements. Shuttle designers replaced the tubing in the liquid cooling ventilation garment with ethylene vinyl acetate to reduce impurities carried by the water into the system. The single change that likely contributed the most toward increasing component life and reducing maintenance requirements was the materials selection for the Primary Life Support System water tank bladder. The water tank bladder expanded and contracted as the water quantity changed during the EVA, and functioned as a barrier between the water and the oxygen system. Designers replaced the molded silicon bladder material with Flourel™, which leached fewer and less-corrosive effluents and was half as permeable to water, resulting in dryer bladder cavities. This meant less corrosion and cleaner filters—all resulting in longer life and less maintenance.

Using the Apollo EVA suit as the basis for the shuttle EVA suit design saved time and money. It also provided a better chance for success by using proven design. The changes that were incorporated, such as using a modular fit approach, including more robust materials, and taking advantage of advances in technology, helped meet the challenges of the Space Shuttle Program. These changes also resulted in a spacesuit that allowed different types of astronauts to perform more difficult EVA tasks over a 30-year program with very few significant problems.

Extravehicular Activity Mission Operations and Training—All Dressed Up, Time to Get to Work

If spacesuit designers were the outfitters of spacewalks, flight controllers, who also plan the EVAs and train the crew members, were the choreographers. Early in the program, EVAs resembled a solo dancer performing a single dance. As flights became more complicated, the choreography became more like a Broadway show—several dancers performing individual sequences, before coming together to dance in concert. On Broadway, the individual sequences have to be choreographed so that dancers come together at the right time. This choreography is similar to developing EVA timelines for a Hubble repair or an ISS assembly mission. The tasks had to be scheduled so that crew members could work individually when only one person was required for a task, but allow them to come together when they had a jointly executed task.

The goal was to make timelines as efficient as possible, accomplish as many tasks as possible, and avoid one crew member waiting idle until the other crew member finished a task. The most significant contribution of EVA operations during the shuttle era was the development of this ability to plan and train for a large number of interdependent and challenging EVA tasks during short periods of time. Over time, the difficulty increased to require interdependent spacewalks within a flight and finally interdependent spacewalks between flights. This culminated in the assembly and maintenance of the ISS, which required the most challenging series of EVAs to date.

The first shuttle EVAs were devoted to testing the tools and suit equipment that would be used in upcoming spacewalks. After suit/airlock problems scrubbed the first attempt, NASA conducted the first EVA since 1974 during Space Transportation System (STS)-6 on April 7, 1983. This EVA practiced some of the shuttle contingency tasks and exercised the suit and tools. The goal was to gain confidence and experience with the new EVA hardware. Then on STS-41B (1984), the second EVA flight tested some of the critical tools and techniques that would be used on upcoming spacewalks to retrieve and repair satellites. One of the highlights was a test of the manned maneuvering unit, a jet pack designed to allow EVA crew members to fly untethered, retrieve satellites, and return with the satellite to the payload bay for servicing. The manned maneuvering unit allowed an EVA crew member to perform precise maneuvering around a target and dock to a payload in need of servicing.

Shuttle Robotic Arm

Another highlight of the STS-41B EVAs was the first demonstration of an EVA crew member performing tasks while positioned at the end of the

Astronaut Bruce McCandless on STS 41B (1984) in the nitrogen-propelled manned maneuvering unit, completing an extravehicular activity. McCandless is floating without tethers attaching him to the shuttle.

Shuttle Robotic Arm. This capability was a major step in streamlining EVAs to come as it allowed a crew member to be moved from one worksite to another quickly. This capability saved the effort required to swap safety tethers during translation and set up and adjust foot restraints—sort of like being able to roll a chair to move around an office rather than having to switch from chair to chair. It was also a first step in evaluating how an EVA crew member affected the hardware with which he or she interacted.

The concern with riding the Shuttle Robotic Arm was ensuring that the EVA crew member did not damage the robotic arm's shoulder joint by imparting forces and moments at the end of the 15-m (50-ft) boom that didn't have much more mass than the crew member. Another concern was the motion that the Shuttle Robotic Arm could experience under EVA loads—similar to how a diving board bends and flexes as a diver bounces on its end. Too much motion could make it too difficult to perform EVA tasks and too time consuming to wait until the motion damps out. Since the arm joints were designed to slip before damage could occur and crew members would be able to sense a joint slip, the belief was that the arm had adequate safeguards to preclude damage.

Allowing a crew member to work from the end of the arm required analysis of the arm's ability to withstand EVA crew member forces. Since both the Shuttle Robotic Arm and the crew member were dynamic systems, the analysis could be complicated; however, experts agreed that any dynamic EVA load case with a static Shuttle Robotic Arm would be enveloped by the case of applying brakes to the arm at its worst-case runaway speed with a static EVA crew member on the end. After this analysis demonstrated that the Shuttle Robotic Arm would not be damaged, EVA crew members were permitted to work on it. Working from the Shuttle Robotic Arm became an important technique for performing EVAs.

Satellite Retrieval and Repair

Once these demonstrations and tests of EVA capabilities were complete, the EVA community was ready to tackle satellite repairs. The first satellite to be repaired was SolarMax, on STS-41C (1984), 1 year after the first shuttle EVA. Shortly after STS-41B landed, NASA decided to add retrieval of Palapa B2 and Westar VI to the shuttle manifest, as the satellites had failed shortly after their deploy on that flight. While these early EVAs were ultimately successful, they did not go as originally planned.

NASA developed several new tools to assist in the retrieval. For SolarMax, the trunnion pin attachment device was built to attach to the manned maneuvering unit on one side and then mate to the SolarMax satellite on the other side to accommodate the towing of SolarMax back to the payload bay. Similarly, an apogee kick motor capture device (known as the "stinger") was built to attach to the manned maneuvering unit to mate with the Palapa B2 and Westar VI satellites. An a-frame was also provided to secure the Palapa B and Westar satellites in the payload bay. All was ready for the first operational EVAs; however, engineers, flight controllers, and managers would soon have their first of many experiences demonstrating the value of having a crew member in the loop.

When George Nelson flew the manned maneuvering unit to SolarMax during STS-41C, the trunnion pin attachment device jaws failed to close on the service module docking pins. After several attempts to mate, the action induced a slow spin and eventually an unpredictable tumble. SolarMax was stabilized by ground commands from Goddard Space Flight Center during the crew sleep period. The next day, Shuttle Robotic Arm operator Terry Hart grappled and berthed the satellite—a procedure that flight controllers felt was too risky preflight. EVA crew members executed a second EVA to complete the planned repairs.

The STS-51A (1984) Palapa B2/Westar VI retrieval mission was planned, trained, and executed within 10 months of the original satellite failures. In the wake of the problem retrieving SolarMax, flight planners decided to develop backup plans in case the crew had problems with the stinger or a-frame. Joseph Allen flew the manned maneuvering unit/stinger and mated it to the Palapa B2 satellite; however, Dale Gardner, working off the robotic arm, was unable to attach the a-frame device designed to assist in handling the satellite. The crew resorted to a backup plan, with Gardner grasping the satellite then slowly bringing it down and securing it for return to Earth. On a subsequent EVA, Gardner used the manned maneuvering unit and stinger to capture the Westar VI satellite, and the crew used the Shuttle Robotic Arm to maneuver it to the payload bay where the EVA crew members secured it.

Although the manned maneuvering unit was expected to be used extensively, the Shuttle Robotic Arm proved more

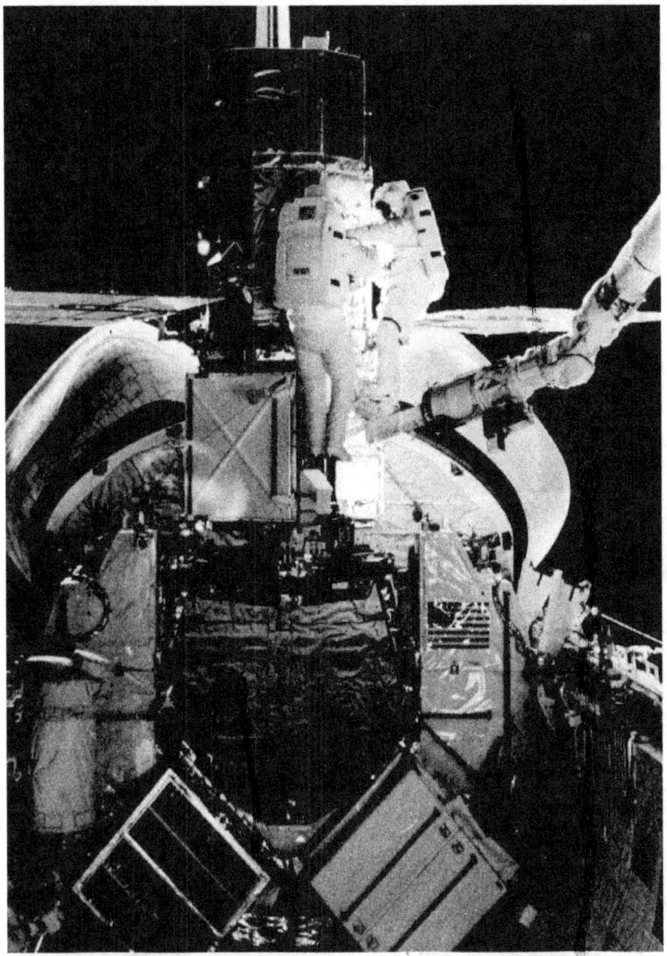

Astronauts George Nelson (right) and James van Hoften captured Solar Maximum Satellite in the aft end of the Challenger's cargo bay during STS-41C (1984). The purpose was to repair the satellite. They used the mobile foot restraint and the robotic arm for moving about the satellite.

efficient because it had fewer maintenance costs and less launch mass.

The next major EVA missions were STS-51D and STS-51I, both in 1985. STS-51D launched and deployed Syncom-IV/Leasat 3 satellite, which failed to activate after deployment. The STS-51D crew conducted the first unscheduled shuttle EVA. The goal was to install a device on the Shuttle Robotic Arm that would be used to attempt to flip a switch to activate the satellite. Although the EVA was successful, the satellite did not activate and STS-51I was replanned to attempt to repair the satellite. STS-51I was executed within 4 months of STS-51D, and two successful EVAs repaired it.

These early EVA flights were significant because they established many of the techniques that would be used throughout the Space Shuttle Program. They also helped fulfill the promise that the shuttle was a viable option for on-orbit repair of satellites. EVA flight controllers, engineers, and astronauts proved their ability to respond to unexpected circumstances and still accomplish mission objectives. EVA team members learned many things that would drive the program and payload customers for the rest of the program. They learned that moving massive objects was not as difficult as expected, and that working from the Shuttle Robotic Arm was a stable way of positioning an EVA crew member. Over the next several years, EVA operations were essentially a further extension of the same processes and operations developed and demonstrated on these early flights.

During the early part of the Space Shuttle Program, EVA was considered to be a last resort because of inherent risk. As the reliability and benefits of EVA were better understood, however, engineers began to have more confidence in it. They accepted that EVA could be employed as a backup means, be used to make repairs, or provide a way to save design complexity. Engineers were able to take advantage of the emerging EVA capability in the design of shuttle payloads. Payload designers could now include manual EVA overrides on deployable systems such as antennas and solar arrays instead of adding costly automated overrides. Spacecraft subsystems such as batteries and scientific instruments were designed to be repaired or replaced by EVA. Hubble and the Compton Gamma Ray Observatory were two notable science satellites that were able to use a significant number of EVA-serviceable components in their designs.

EVA flight controllers and engineers began looking ahead to approaching missions to build the ISS. To prepare for this, program managers approved a test program devoted to testing tools, techniques, and hardware design concepts for the ISS. In addition to direct feedback to the tool and station hardware designs, the EVA community gained valuable experience in planning, training, and conducting more frequent EVAs than in the early part of the program.

Hubble Repair

As NASA had proven the ability to execute EVAs and accomplish some remarkable tasks, demand for the EVA resource increased sharply on the agency. One of the most dramatic and demanding EVA flights began development shortly after the deployment of Hubble in April 1990. NASA's reputation was in jeopardy from the highly publicized Hubble failure, and the scientific community was sorely disappointed with the capability of the telescope. Hubble was designed with several servicing missions planned, but the first mission—to restore its optics to the expected performance—took on greater significance. EVA was the focal point in recovery efforts. The mission took nearly 3 years to plan, train, and develop the necessary replacement parts.

The Hubble repair effort required significant effort from most resources in the EVA community. Designers from Goddard Space Flight Center, Johnson Space Center, Marshall Space Flight

Three Spacewalkers Capture Satellite

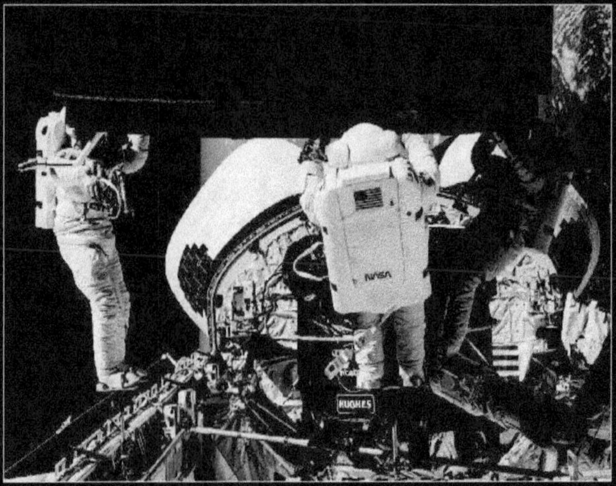

Astronauts Rick Hieb on the starboard payload bay mounted foot restraint work station, Bruce Melnick with his back to the camera, and Tom Akers on the robotic arm mounted foot restraint work station—on the backside of the Intelsat during STS-49 (1992).

STS-49 significantly impacted planning for future EVAs. It was the most aggressive EVA flight planned, up to that point, with three EVAs scheduled. Engineers designed a bar with a grapple fixture to capture Intelsat and berth it in the payload bay. The data available on the satellite proved inadequate and it was modeled incorrectly for ground simulations. After two EVA attempts to attach the capture bar, flight controllers looked at other options.

The result was an unprecedented three-man EVA using space hardware to build a platform for the crew members, allowing them to position themselves in a triangle formation to capture the Intelsat by hand. This required an intense effort by ground controllers to verify that the airlock could fit three crew members, since it was only designed for two, and that there were sufficient resources to service all three. Additional analyses looked at whether there were sufficient handholds to grasp the satellite, that satellite temperatures would not exceed the glove temperature limits, and that structural margins were sufficient. Practice runs on the ground convinced ground operators that the operation was possible. The result was a successful capture and repair during the longest EVA in the shuttle era.

Center, and the European Space Agency delivered specialized tools and replacement parts for the repair. Approximately 150 new tools and replacement parts were required for this mission. Some of these tools and parts were the most complicated ones designed to date. Flight controllers concentrated on planning and training the unprecedented number of EVA tasks to be performed—a number that continued to grow until launch. What started as a three-EVA mission had grown to five by launch date. The EVA timeliners faced serious challenges in trying to accomplish so many tasks, as precious EVA resources were stretched to the limit.

New philosophies for managing EVA timelines developed in response to the growing task list. Until then, flight controllers included extra time in timelines to ensure all tasks would be completed, and crews were only trained in the tasks stated in those timelines. For Hubble, timelines included less flexibility and crews were trained on extra tasks to make sure they could get as much done as possible. With the next servicing mission years away, there was little to lose by training for extra tasks. To better ensure the success of the aggressive timelines, the crew logged more than twice the training time as on earlier flights.

When astronauts were sent to the Hubble to perform its first repair, engineers became concerned that the crew members would put unacceptable forces on the great observatory. Engineers used several training platforms to measure forces and moments from many different crew members to gain a representative set of both normal and contingency EVA tasks. These cases were used to analyze Hubble for structural integrity and to sensitize EVA crew members to where and when they needed to be careful to avoid damage.

EVA operators also initiated three key processes that would prove very valuable both for Hubble and later for ISS. Operators and tool designers requested that, during Hubble assembly, all tools be checked for fit against all Hubble components and replacement parts. They also required extensive photography of all Hubble components and catalogued the images for ready access to aid in real-time troubleshooting. Finally, engineers analyzed all the bolts that would be actuated during the repair

Fatigue—A Constant Concern During Extravehicular Activity

Why are extravehicular activities (EVAs) so fatiguing if nothing has any weight in microgravity?

Lack of suit flexibility and dexterity forces the wearer to exert more energy to perform tasks. With the EVA glove, the fingers are fixed in a neutral position. Any motion that changes the finger/hand position requires effort.

Lack of gravity removes leverage. Normally, torque used to turn a fastener is opposed by a counter-torque that is passively generated by the weight of the user. In weightlessness, a screwdriver user would spin aimlessly unless the user's arm and body were anchored to the worksite, or opposed the torque on the screwdriver with an equal muscular force in the opposite direction. Tool use during EVAs is accomplished by direct muscle opposition with the other arm, locking feet to the end of a robotic arm, or rigidly attaching the suit waist to the worksite. EVA tasks that require many hand/arm motions over several hours lead to significant forearm fatigue.

The most critical tasks—ingressing the airlock, shutting the hatch, and reconnecting the suit umbilical line— occur at the end of an EVA. Airlocks are cramped and tasks are difficult, especially when crew members are fatigued and overheated. Overheating occurs because the cooling system must be turned off before an astronaut can enter the airlock. The suit does not receive cooling until the airlock umbilical is connected. The helmet visor can fog over at this point, making ingress even more difficult.

Along with crew training, medical doctors and the mission control team monitor exertion level, heart rate, and oxygen usage. Communication between ground personnel and astronauts is essential in preventing fatigue from having disastrous consequences.

Astronaut John Grunsfeld, working from the end of the Shuttle Robotic Arm, installs replacement parts on the Hubble Space Telescope during the final repair mission, STS-125 (2009).

to provide predetermined responses to problems operating bolts—data like the maximum torque allowed across the entire thermal range. Providing these data and fit checks would become a standard process for all future EVA-serviceable hardware.

The first Hubble repair mission was hugely successful, restoring Hubble's functionality and NASA's reputation. The mission also flushed out many process changes that the EVA community would need to adapt as the shuttle prepared to undertake assembly of the ISS. What had been a near disaster for NASA when Hubble was deployed turned out to be a tremendous opportunity for engineers, flight controllers, and mission managers to exercise a station-like EVA mission prior to when such missions would become routine. This mission helped demonstrate NASA's ability to execute a complex mission while under tremendous pressure to restore a vital international resource.

Flight Training

Once NASA identified the tasks for a shuttle mission, the crew had to be trained to perform them. From past programs, EVA instructors knew that the most effective training for microgravity took place under water, where hardware and crew members could be made neutrally buoyant. The Weightless Environment Training Facility— a swimming pool that measured 23 m (75 ft) long, 15 m (50 ft) wide, and 8 m (25 ft) deep—was the primary location for EVA training early in the Space Shuttle Program. The Weightless Environment Training Facility contained a full-size mock-up of the shuttle payload bay with all EVA interfaces represented. In the same manner that scuba divers use buoyancy compensation vests and weights, crew members and their tools were configured to be neutrally buoyant through the use of air, foam inserts, and weights. This enabled them to float suspended at the worksite, thus simulating a weightless environment. Crew members trained an average of 10 hours in the Weightless Environment Training Facility for every 1 hour of planned on-orbit EVA. For complicated flights, as with the first Hubble repair mission, the training ratio was increased. Later, EVA training moved to a new, larger, and more updated water tank—the Neutral Buoyancy Laboratory—to accommodate training on the ISS.

A few limitations to the neutral buoyancy training kept it from being a perfect zero-gravity simulation. The water drag made it less accurate for simulating the movement of large objects. And since they were still in a gravity environment, crew members had to maintain a "heads-up" orientation most of the time to avoid blood pooling in the head. So mock-ups had to be built and oriented to allow crew members to maintain this position.

The gravity environment of the water tank also contributed to shoulder injuries—a chronic issue, especially in the latter part of the program. Starting in the mid 1990s, several crew members experienced shoulder injuries during the course of their EVA training. This was due to a design change made at that time to the extravehicular mobility unit shoulder joint. The shoulder joint was optimized for mobility, but designers noticed wear in the fabric components of the original joint. To avoid the risk of a catastrophic suit depressurization, NASA replaced the joint with a scye bearing that was much less subject to wear but limited to rotation in a single plane, thus reducing the range of motion. The scye bearing had to be placed to provide good motion for work and allow the wearer to don the extravehicular mobility unit through the waist ring (like putting on a shirt),

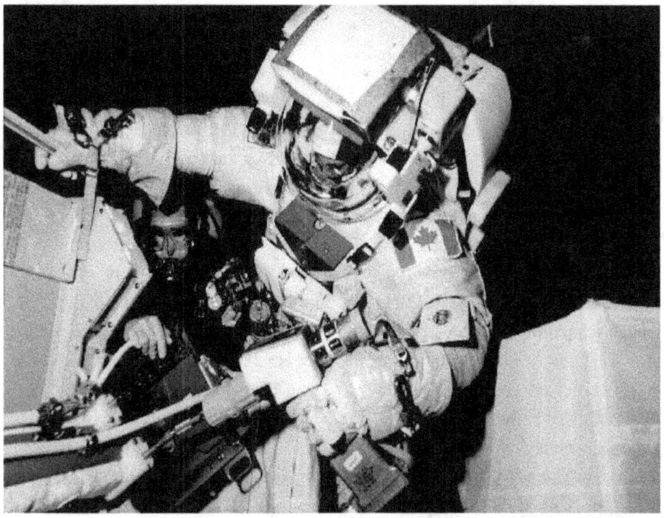

Astronaut Dafydd Williams, STS-118, representing the Canadian Space Agency, is wearing a training version of the extravehicular mobility unit spacesuit while participating in an underwater simulation of extravehicular activity in the Neutral Buoyancy Laboratory near Johnson Space Center. Scuba-equipped divers are in the water to assist Williams in his rehearsal, intended to help prepare him for work on the exterior of the International Space Station. Observe Williams holding the Pistol Grip Tool in his left hand with his shoulder extended. This position causes shoulder pain during training in neutral bouyancy.

which placed the arms straight up alongside the head. Placement of the shoulder joint was critical to a good fit, but there were only a few sizes of upper torsos for all crew members. Some crew members had reasonably good fit with the new joint, but others suffered awkward placement of the ring, which exerted abnormal forces on the shoulders. This was more a problem during training, when stress on the shoulder joint was increased due to gravity.

On Earth, the upper arm is held fairly close to the body during work activities. The shoulder joint is least prone to injury in this position under gravity. In space, the natural position of the arms is quite different, with arms extended in front of the torso. Shoulders were not significantly stressed by EVA tasks performed in microgravity. In ground training, however, it was difficult to make EVA tools and equipment completely neutrally buoyant, so astronauts often held heavy tools with their shoulders fully extended for long periods. Rotator cuff injuries, tendonitis, and other shoulder injuries occurred despite best efforts to prevent them. The problem was never fully resolved during the shuttle era, given the design limitations of the EVA suit and the intensity of training required for mission success.

The Precision Air Bearing Floor, also used for EVA training, is a 6-m (20-ft) by 9-m (30-ft), highly polished steel floor that works on the same principles as an air hockey table. Large mock-ups of flight hardware were attached to steel plates that had high-pressure air forced through tubes that ran along the bottom and sides. These formed a cushion under the mock-up that allowed the mock-up to move easily in the horizontal plane, simulating zero-gravity mass handling. Despite the single plane limitation of the Precision Air Bearing Floor, when combined with neutral buoyancy training the two facilities provided comprehensive and valuable training of moving large objects.

Another training and engineering platform was the zero-gravity aircraft. This specially outfitted KC-135 (later replaced by a DC-9) aircraft was able to fly a parabolic trajectory that provided approximately 20 seconds of microgravity on the downward slope, similar to the brief periods experienced on a roller coaster. This platform was not limited by water drag as was the Weightless Environment Training Facility, or to single plane evaluations as was the Precision Air Bearing Floor; however, it was only effective for short-duration tasks. Therefore, the zero-gravity aircraft was only used for short events that required a high-fidelity platform.

Extravehicular Activity Tools

EVA tools and support equipment are the Rodney Dangerfield of spacewalks. When they work, they are virtually unnoticed; however, when they fail to live up to expectations, everyone knows. Looking at the cost of what appear to be simple tools, similar to what might be found at the local hardware store, one wonders why they cost so much and don't always work. The reality is that EVA tool engineers had a formidable task—to design tools that could operate, in vacuum, in temperatures both colder than the Arctic and as hot as an oven, and be operable by someone wearing the equivalent of several pairs of ski gloves, in vacuum, while weightless. These factors combined to produce a set of competing constraints that was difficult to balance. When adding that the complete space environment cannot be simulated on the ground, the challenge for building specialized tools that perform in space became clear. Any discussion of tools invariably involves the reasons why they fail and the lessons learned from those failures.

EVA tools are identified from two sources: the required EVA tasks, and engineering judgment on what general tools might be useful for unplanned events. Many of the initial tools were fairly simple—tethers, foot restraints, sockets, and wrenches. There were also specialized tools devoted to closing and latching the payload bay doors. Many tools were commercial tools available to the public but that were modified for use in space. This was thought to be a cost savings since they were designed for many of the same functions. These tools proved to be adequate for many uses; however, detailed information was often unavailable for commercial tools and they did not generally hold up to the temperature extremes of space. Material impurities made them unpredictable at cold temperatures and lubricants became too runny at high temperatures, causing failures. Therefore, engineers moved toward custom tools made with high-grade materials that were reliable across the full temperature range.

Trunnion pin attachment device, a-frame, and capture bar problems on the early satellite repair flights were found to be primarily due to incorrect information on the satellite interfaces. Engineers determined that interfering objects weren't represented on satellite design drawings. After these events, engineers stepped up efforts to better document EVA interfaces, but it is never possible to fully document the precise configuration of any individual spacecraft. Sometimes drawings include a range of options for components for which many units will be produced, and that will be manufactured over a long period of time. Designers must also have the flexibility to perform quick fixes to minor problems to maintain launch schedules. The balance between providing precise documentation and allowing design and processing flexibility will always be a judgment call and will, at times, result in problems.

Engineers modified tools as they learned about the tools' performance in space. White paint was originally used as a thermal coating to keep tools from getting too hot. Since tools bump against objects and the paint tends to chip, the paint did not hold up well under normal EVA operations. Engineers thus switched to an anodizing process (similar to electroplating) to make the tools more durable. Lubricants were also a problem. Oil-based lubricants would get too thick in cold temperatures and inhibit moving parts from operating. In warm environments, the lubricants would become too thin. Dry-film lubricants (primarily Braycote®, which acts like Teflon® on frying pans) became the choice for almost all EVA tools because they are not vulnerable to temperature changes in the space environment.

Pistol Grip Tool

Some of the biggest problems with tools came from attempting to expand their use beyond the original purpose. Sometimes new uses were very similar to the original use, but the details were different—like trying to use a hacksaw to perform surgery. The saw is designed for cutting, but the precision required is extremely different. An example is the computerized Pistol Grip Tool, which was developed to actuate bolts while providing fairly precise torque information. This battery-operated tool was similar to a powered screwdriver, but had some sophisticated features to allow flexibility in applying and measuring different levels of torque or angular rotation. The tool was designed for Hubble, and the accuracy was more than adequate for Hubble. When ISS required a similar tool, the program chose to purchase several units of the Hubble power tool rather than design a new tool specific to ISS requirements. The standards for certification and documentation were different for Hubble. ISS had to reanalyze bolts, provide for additional ground and on-orbit processing of the Pistol Grip Tool to meet ISS accuracy needs, and provide additional units on orbit to meet fault tolerance requirements and maintain calibration.

The use of the Pistol Grip Tool for ISS assembly also uncovered another shortcoming with regard to using a tool developed for a different spacecraft. The Pistol Grip Tool was advertised as having an accuracy of 10% around the selected torque setting. This accuracy was verified by setting the Pistol Grip Tool in a fixed test stand on the ground where it was held rigidly in place. This was a valid characterization when used on Hubble where EVA worksites were designed to be easily accessible and where the Pistol Grip Tool was used directly on the bolts. It was relatively easy for crew members to center the tool and hold it steady on any bolt. ISS worksites were not as elegant as Hubble worksites, however, since ISS is such a large vehicle and the Pistol Grip Tool

Extravehicular Activity Tools

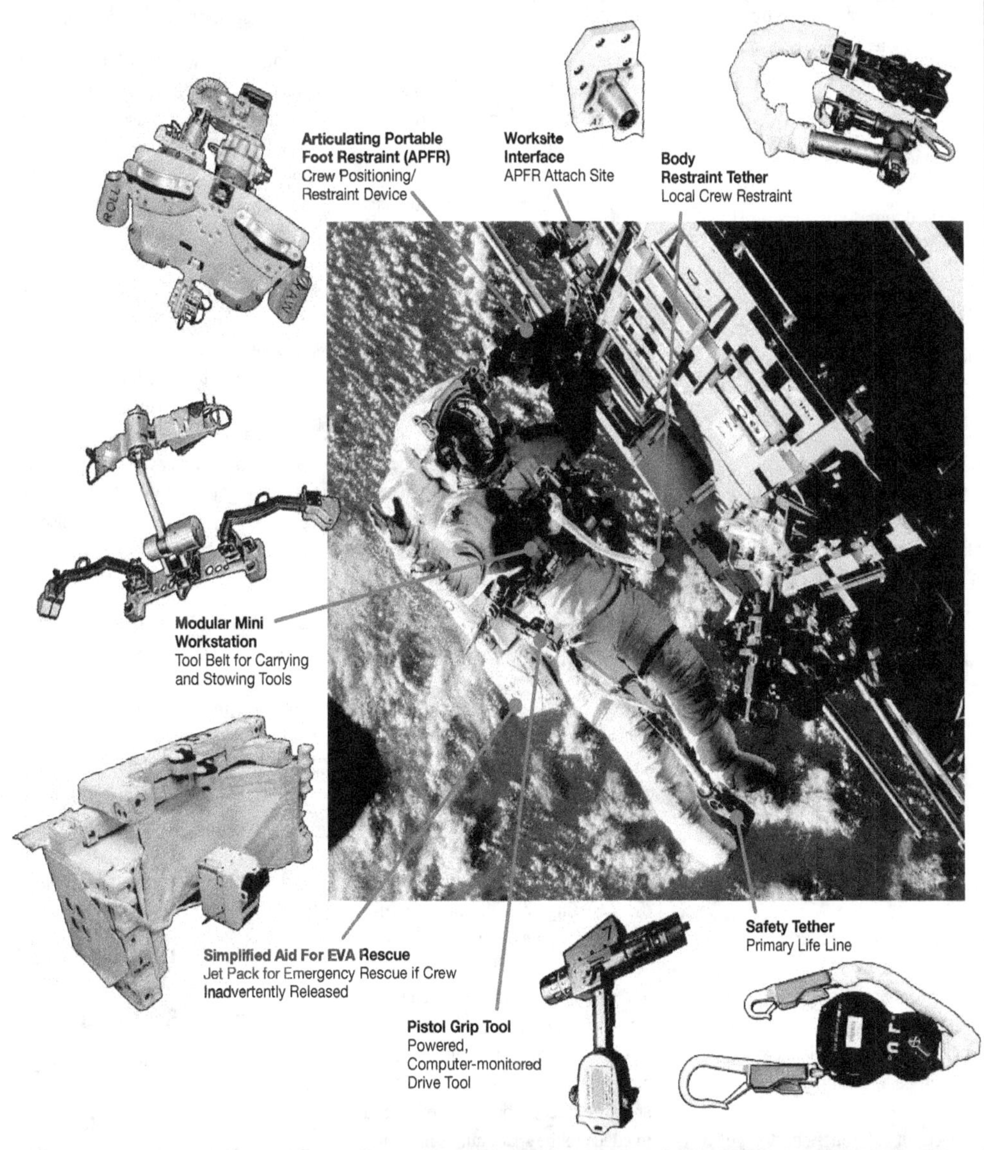

Astronaut Rick Mastracchio, STS-118 (2007), is shown using several extravehicular activity (EVA) tools while working on construction and maintenance of the International Space Station during the shuttle mission's third planned EVA activity.

often had to be used with socket extensions and other attachments that had inaccuracies of their own. Crew members often had to hold the tool off to the side with several attachments, and the resulting side forces could cause the torque measured by the tool to be very different than the torque actually applied. Unfortunately, ISS bolts were designed and analyzed to the advertised torque accuracy for Hubble and they didn't account for this "man-in-the-loop" effect. The result was a long test program to characterize the accuracy of the Pistol Grip Tool when used in representative ISS worksites, followed by analysis of the ISS bolts to this new accuracy.

To focus only on tool problems, however, is a disservice. It's like winning the Super Bowl and only talking about the fumbles. While use of the Pistol Grip Tool caused some problems as NASA learned about its properties, it was still the most sophisticated tool ever designed for EVA. It provided a way to deliver a variety of torque settings and accurately measure the torque delivered. Without this tool, the assembly and maintenance of the ISS would not have been possible.

Other Tools

NASA made other advancements in tool development as well. Tools built for previous programs were generally simple tools required for collecting geology samples. While there weren't many groundbreaking discoveries in the tool development area, the advances in tool function, storage, and transport greatly improved EVA efficiency during the course of the program. The fact that Henry Ford didn't invent the internal combustion engine doesn't mean he didn't make tremendous contributions to the automobile industry.

One area where tool engineers expanded EVA capabilities was in astronaut translation and worksite restraint. Improvements were made to the safety tether to include a more reliable winding device and locking crew hooks to prevent inadvertent release. Engineers developed portable foot restraints that could be moved from one location to another, like carrying a ladder from site to site. The foot restraints consisted of a boot plate to lock the crew member's feet in place and an adjustment knob to adjust the orientation of the plate for better positioning. The foot restraint had a probe to plug into a socket at the worksite. These foot restraints gave crew members the stability to work in an environment where unrestrained crew members would have otherwise been pushed away from the worksite whenever they exerted force.

The portable foot restraints were an excellent starting point, but they required a fair amount of time to move. They also became cumbersome when crew members had to work in many locations during a single EVA (as with the ISS). Engineers developed tools that could streamline the time to stabilize at a new location. The Body Restraint Tether is one of these tools. This tool consists of a stack of balls connected through its center by a cable with a clamp on one end to attach to a handrail and a bayonet probe on the other end to attach to the spacesuit. Similar to flexible shop lights, the Body Restraint Tether can be bent and twisted to the optimum position, then locked in that position with a knob that tightens the cable. The Body Restraint Tether is a much quicker way for crew members to secure themselves for lower-force tasks.

Another area where tool designers made improvements was tool stowage and transport. Crew members had to string tools to their suits for transport until designers developed sophisticated tool bags and boxes that allowed crews to carry a large number of tools and use the tools efficiently at a worksite. The Modular Mini Workstation—the EVA tool belt—was developed to attach to the extravehicular mobility unit and has become invaluable to conducting spacewalks. Specific tools can be attached to the arms on the workstation, thereby allowing ready access to the most-used tools. Various sizes of tool caddies and bags also help to transport tools and EVA "trash" (e.g., launch restraints).

Space Shuttle Program tool designers expanded tool options to include computer-operated electronics and improved methods for crew restraint, tool transport, and stowage. While there were hiccups along the way, the EVA tools and crew aids performed admirably and expanded NASA's ability to perform more complicated and increasingly congested EVAs.

Extravehicular Activity During Construction of the International Space Station

From 1981 through 1996, the Space Shuttle Program accomplished 33 EVAs. From 1997 through 2010, the program managed 126 EVAs devoted primarily to ISS assembly and maintenance, with several Hubble

Space Telescope repair missions also included. Assembly and maintenance of the ISS presented a series of challenges for the program. EVA tools and suits had to be turned around quickly and flawlessly from one flight to the next. Crew training had to be streamlined since several flights would be training at the same time and tasks were interdependent from one flight to the next. Plans for one flight, based on previous flight results, could change drastically just months (or weeks) before launch. Sharing resources with the International Space Station Program was also new territory—the same tools, spacesuits, and crew members would serve both programs after the ISS airlock was installed.

Medical Risks of Extravehicular Activity—Decompression Sickness

One risk spacewalkers share with scuba divers is decompression sickness, or "the bends." "The bends" name came from painful contortions of 19th-century underwater caisson workers suffering from decompression sickness, which occurs when nitrogen dissolves in blood and tissues while under pressure, and then expands when pressure is lowered. Decompression sickness can occur when spacewalkers exit the pressurized spacecraft into vacuum in a spacesuit

Decompression sickness can be prevented if nitrogen tissue concentrations are lowered prior to reducing pressure. Breathing 100% oxygen causes nitrogen to migrate from tissues into the bloodstream and lungs, exiting the body with exhaling. The first shuttle-based extravehicular activities used a 4-hour in-suit oxygen prebreathe. This idle time was inefficient and resulted in too long a crew day. New solutions were needed.

One solution was to lower shuttle cabin pressure from its nominal pressure of 101.2 kPa (14.7 psi) to 70.3 kPa (10.2 psi) for at least 12 hours prior to the EVA. This reduced cabin pressure protocol was efficient and effective, with only 40 minutes prebreathe.

Shuttle EVA crew members working International Space Station (ISS) construction required a different approach. It is impossible to reduce large volume ISS pressure to 70.3 kPa (10.2 psi). To increase the rate of nitrogen release from tissues, crew members exercised before EVA while breathing 100% oxygen. This worked, but it added extra time to the packed EVA day and exhausted the crew. Planners used the reduced cabin pressure protocol by isolating EVA crew members in the ISS airlock the night before the EVA and lowering the pressure to 70.3 kPa (10.2 psi). This worked well for the remainder of ISS EVAs, with no cases of decompression sickness throughout the Space Shuttle Program.

Extravehicular Loads for Structural Requirements

The EVA loads development program, first started for the Hubble servicing missions, helped define the ISS structural design requirements. ISS was the first program to have extensive EVA performed on a range of structural interfaces. The load cases for Hubble repair had to protect the telescope for a short period of EVA operations and for a finite number of well-known EVA tasks.

ISS load cases had to have sufficient margin for tasks that were only partially defined at the time the requirements were fixed, to protect for hundreds of EVAs over the planned life of the ISS. The size of ISS was also a factor. An EVA task on one end of the truss structure could be much more damaging than the same task closer to the center (just like bouncing on the end of a diving board creates more stress at the base than bouncing on the base itself). EVA loads had to account for intentional tasks (e.g., driving bolts) and unintentional events (e.g., pushing away from a rotating structure to avoid collision). Engineers had to protect for a reasonable set of EVA scenarios without overly restricting the ISS design to protect against simultaneous low-probability events. This required an iterative process that included working with ISS structures experts to zero in on the right requirements.

A considerable test program—using a range of EVA crew members executing a variety of tasks in different ground venues—characterized the forces and moments that an EVA crew member could impart. The resulting cases were used throughout the programs to evaluate new tasks when the tasks were needed. While the work was done primarily for ISS, the loads that had been developed were used extensively in the post-Columbia EVA inspection and repair development.

Rescue From Inadvertent Release

NASA always provided for rescue of an accidentally released EVA crew member by maintaining enough fuel to fly to him or her. Once ISS assembly began, however, the Orbiter was docked during EVAs and would not have been able to detach and pursue an EVA crew member in time. The ISS Program required a self-rescue jet pack for use during ISS EVAs. The Simplified Aid for EVA Rescue was designed to meet this requirement. Based on the manned maneuvering unit design but greatly simplified, the Simplified Aid for EVA Rescue was a reliable, nitrogen-propelled backpack that provided limited capability for a crew member to stop and fly back to the station or Orbiter. It was successfully tested on two shuttle flights when shuttle rescue was still possible if something went wrong. Fortunately, the Simplified Aid for EVA Rescue never had to be employed for crew rescue.

Extravehicular Activity Suit Life Extension and Multiuse Certification for International Space Station Support

A significant advancement for the EVA suit was the development of a regenerable carbon dioxide removal system. Prior to the ISS, NASA used

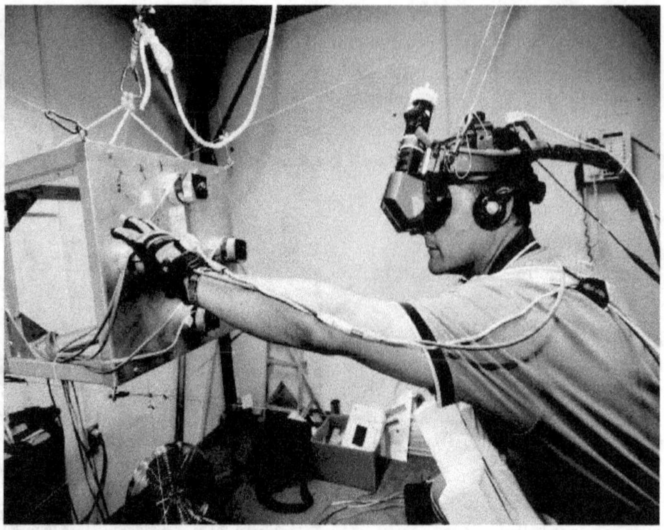

Astronaut Douglas Wheelock, STS-120 (2007), uses virtual reality hardware in the Space Vehicle Mockup Facility at Johnson Space Center to rehearse some of his duties on the upcoming mission to the International Space Station.

single-use lithium hydroxide canisters for scrubbing carbon dioxide during an EVA. Multiple EVAs were routine during flights to the ISS. Providing a regenerative alternative using silver oxide produced significant savings in launch weight and volume. These canisters could be cleaned in the ISS airlock regenerator, thereby allowing the canisters to be left on orbit rather than processed on the ground and launched on the shuttle. This capability saved approximately 164 kg (361 pounds) up-mass per year.

Training Capability Enhancements

During the early shuttle missions, the Weightless Environment Training Facility and Precision Air Bearing Facility were sufficient for crew training. To prepare for space station assembly, however, virtually every mission would include training for three to five EVAs—often with two EVA teams—with training for three to five flights in progress simultaneously.

To do this, NASA built the Neutral Buoyancy Laboratory to accommodate EVA training for both the Space Shuttle and ISS Programs. At 62 m (202 ft) long, 31 m (102 ft) wide, and 12 m (40 ft) deep, the Neutral Buoyancy Laboratory is more than twice the size of the previous facility, and it dramatically increased neutral buoyancy training capability. It also allowed two simultaneous simulations to be conducted using two separate control rooms to manage each individual event.

Trainers took advantage of other resources not originally designed for EVA training. The Virtual Reality Laboratory, which was designed primarily to assist in robotic operations, became a regular EVA training venue. This lab helped crew members train in an environment that resembled the space environment, from a crew member's viewpoint, by using payload and vehicle engineering models working with computer software to display a view that changed as the crew member "moved" around the space station.

The Virtual Reality Laboratory also provided mass simulation capability by using a system of cables and pulleys controlled by a computer as well as special goggles to give the right visual cues to the crew member, thus allowing him or her to get a sense of moving a large object in a microgravity environment. Most of the models used in the Virtual Reality Laboratory were actually built for other engineering facilities, so the data were readily available and parameters could be changed relatively quickly to account for hardware or environment changes. This gave the lab a distinct advantage over other venues that could not accommodate changes as quickly.

In addition to the new training venues, changes in training philosophy were required to support ISS assembly. Typically, EVA crew training began at least 1 year prior to the scheduled launch. Therefore, crew members for four to five missions would have to train at the same time, and the tasks required were completely dependent on the previous flights' accomplishments. A hiccup in on-orbit operations could cascade to all subsequent flights, changing the tasks that were currently in training. In addition, on-orbit ISS failures often resulted in changes to the tasks, as repair of those components may have taken a higher priority.

To accommodate late changes, flight controllers concentrated on training individual tasks rather than timelines early in the training schedule. They also engaged in skills training—training the crew on general skills required to perform EVAs on the ISS rather than individual tasks. Flight controllers still developed timelines, but they held off training the timelines until closer to flight. Crews also trained on "get-ahead" tasks—those tasks that did not fit into the pre-mission timelines but that could be added if time became available. This flexibility provided time to allow for real-time difficulties.

Extravehicular Activity Participation in Return to Flight After Space Shuttle Columbia Accident

One other significant EVA accomplishment was the development of a repair capability for the Orbiter Thermal Protection System after the Space Shuttle Columbia accident in 2003. This posed a significant challenge for EVA for several reasons. The Thermal Protection System was a complex design that was resistant to high temperatures but was also delicate. It was located in areas under the fuselage that was inaccessible to EVA crew members. The materials used for repair were a challenge to work with, even in an Earth environment, since they did not adhere well to the damage. Finally, the repair had to be smooth since even very small rough edges or large surface deviations could cause turbulent airflow behind the repair, like rocks disrupting flow in a stream. Turbulent flow increased surface heating dramatically, with potentially disastrous results. These challenges, along with the schedule pressure to resume building and resupplying the ISS, made Thermal Protection System repair a top priority for EVA for several years.

The process included using repair materials that engineers originally began developing at the beginning of the program that now had to be refined and certified for flight. Unique tools and equipment, crew procedures, and methods to ensure stabilizing the crew member at the worksite were required to apply the material. The tools mixed the two-part silicone rubber repair material but also kept it from hardening until it was dispensed in

Astronauts Robert Curbeam (foreground) and Rex Walheim (background) simulate tile repair, using materials and tools developed after the Space Shuttle Columbia accident, on board the zero-gravity training aircraft KC-135.

Astronaut Piers Sellers, STS-121 (2006), wearing a training version of the extravehicular mobility unit, participates in an extravehicular activity simulation while anchored on the end of the training version of the Shuttle Robotic Arm in the Space Vehicle Mockup Facility at Johnson Space Center (JSC). The arm has an attached 15-m (50-ft) boom used to reach underneath the Orbiter to access tiles. Lora Bailey (right), manager, JSC Engineering Tile Repair, assisted Sellers.

the damage area. The tools also maintained the materials within a fairly tight thermal range to keep them viable. Engineers were able to avoid the complexity of battery-powered heaters by selecting materials and coatings to passively control the material temperature. The reinforced carbon-carbon Thermal Protection System (used on the wing leading edge) repair required an additional set of tools and techniques with similar considerations regarding precision application of sensitive materials.

Getting a crew member to the worksite proved to be a unique challenge. NASA considered several options, including using the Simplified Aid for EVA Rescue with restraint aids attached by adhesives. Repair developers determined, however, that the best option was to use the new robotic arm extension boom provided for Orbiter inspection. The main challenge to using the extension boom was proving that it was stable enough to conduct repairs, and that the forces the EVA crew member imparted on the boom would not damage the boom or the arm. These concerns were similar to those involved with putting a crew member on a robotic arm, but the "diving board" was twice as long. The EVA loads work performed earlier provided a foundation for the process by which EVA loads could be determined for this situation; however, the process had to be modified since the work platform was much more flexible.

Previous investigations into EVA loads usually involved a crew member imparting loads into a fixed platform. When the loads were continuously applied to the boom/arm configuration, they resulted in a large (about 1.2 m [4 ft]) amount of sway as well as structural concerns for the arm and boom. Engineers knew that the boom/arm configuration was more like a diving board than a floor, meaning that the boom would slip away as force was applied, limiting the force a crew member could put into the system. Engineers developed a sophisticated boom/arm simulator and used it on the precision air bearing floor to measure EVA loads. These tests provided the data for analysis of the boom/arm motion. The work culminated in a flight test on STS-121 (2006), which demonstrated that the boom/arm was stable enough for repair and able to withstand reasonable EVA motions without damage.

Although the repair capability was never used, both the shuttle and the space station benefited from the repair development effort. Engineers made several minor repairs to the shuttle Thermal Protection System that would not have been possible without demonstrating that the EVA crew member could safely work near the fragile system. The boom was also used on the Space Station Robotic Arm to conduct a successful repair of a damaged station solar array wing that was not reachable any other way.

Major Extravehicular Activity Milestones

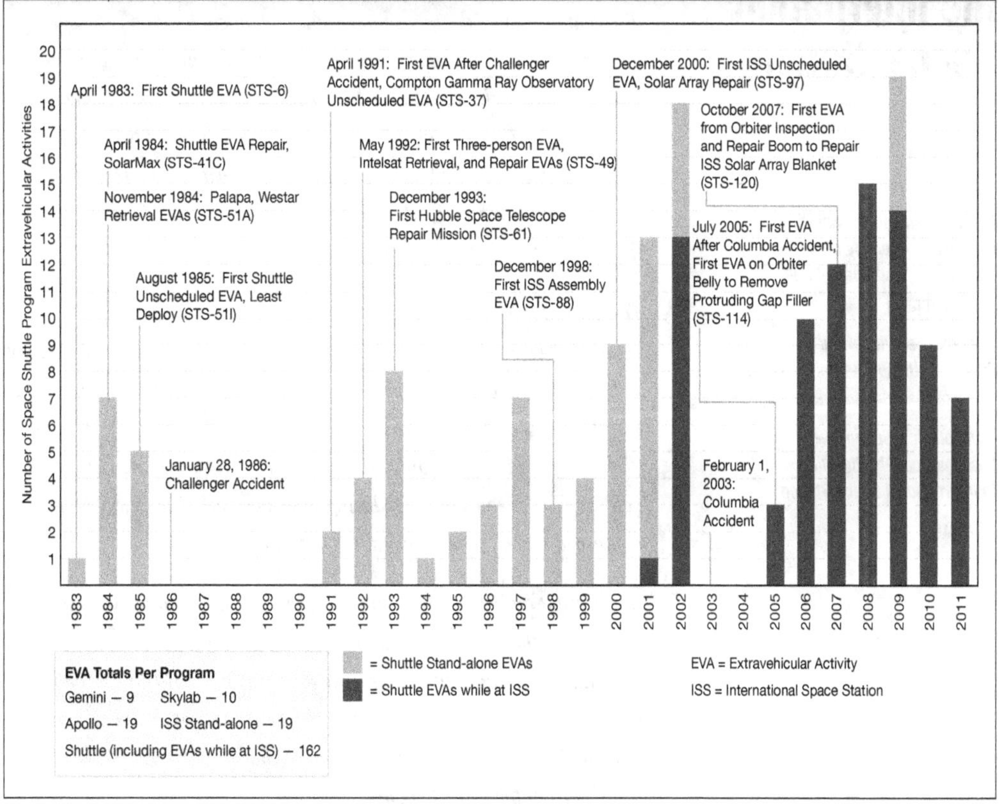

Summary

The legacy of EVA during the Space Shuttle Program consists of both the actual work that was done and the dramatic expansion of the EVA capability. EVA was used to successfully repair or restore significant national resources to their full capacity, such as Hubble, communications satellites, and the Orbiter, and to construct the ISS. EVA advanced from being a minor capability used sparingly to becoming a significant part of almost every shuttle mission, with an increasing list of tasks that EVA crew members were able to perform. EVA tools and support equipment provided more capability than ever before, with battery-powered and computer-controlled tools being well understood and highly reliable.

Much was learned about what an EVA crew member needs to survive and work in a harsh environment as well as how an EVA crew member affected his or her environment. This tremendous expansion in EVA capability will substantially benefit the future exploration of the solar system as engineers design vehicles and missions knowing that EVA crew members are able to do much more than they could at the beginning of the Space Shuttle Program.

Shuttle Builds the International Space Station

John Bacon
Melanie Saunders

Improvements to the Shuttle Facilitated Assembly of the International Space Station
Lee Norbraten

Financial Benefits of the Space Shuttle for the United States
Melanie Saunders

Psychological Support—Lessons from Shuttle-Mir to International Space Station
Albert Holland

Since its inception, the International Space Station (ISS) was destined to have a close relationship with the Space Shuttle. Conceived for very different missions, the two spacecraft drew on each other's strengths and empowered each other to achieve more than either could alone. The shuttle was the workhorse that could loft massive ISS elements into space. It could then maneuver, manipulate, and support these pieces with power, simple data monitoring, and temperature control until the pieces could be assembled. The ISS gradually became the port of call for the shuttles that served it.

The idea of building a space station dates back to Konstantin Tsiolkovsky's writings in 1883. A space station would be a small colony in space where long-term research could be carried out. Visionaries in many nations offered hundreds of design concepts over the next century and a half, and a few simple outposts were built in the late 20th century. The dreams of an enduring international space laboratory coalesced when the shuttle made it a practical reality.

As a parent and child grow, so too did the relationship between the shuttle and the ISS as the fledgling station grew out of its total dependence on the shuttle to its role as a port of call. The ISS soon became the dominant destination in the heavens, hosting vehicles launched from many spaceports in four continents below, including shuttles from the Florida coast.

Creating the International Space Station Masterpiece—in Well-planned Increments

Building this miniature world in the vacuum of space was to be the largest engineering challenge in history. It was made possible by the incomparable capabilities of the winged fleet of shuttles that brought and assembled the pieces. The space station did not spring into being "out of thin air." Rather, it made use of progressively sophisticated engineering and operations techniques that were matured by the Space Shuttle Program over the preceding 17 years. This evolution began before the first International Space Station (ISS) assembly flight ever left the ground—or even the drawing board.

Early Tests Form a Blueprint

NASA ran a series of tests beginning with a deployable solar power wing experiment on Discovery's first flight (Space Transportation System [STS]-41D in 1984) to validate the construction techniques that would be used to build the ISS. On STS-41G (1984), astronauts demonstrated the safe capability for in-space resupply of dangerous rocket propellants in a payload bay apparatus. Astronauts practiced extravehicular activity (EVA) assembly techniques for space-station-sized structures in experiments aboard STS-61B (1985). Several missions tested the performance of large heat pipes in space. NASA explored mobility aids and EVA handling limits during STS-37 (1991).

In April 1984, STS-41C deployed one of the most important and comprehensive test programs—the Long Duration Exposure Facility. STS-32 retrieved the facility in January 1990, giving critical evidence of the performance and degradation timeline of materials in the low-Earth environment. It was a treasure trove of data about the micrometeoroid orbital debris threat that the ISS would face. NASA's ability to launch such huge test fixtures and to examine them back on Earth after flight added immensely to the engineers' understanding of the technical refinements that would be necessary for the massively complicated ISS construction.

The next stage in the process would involve an international connection and the coming together of great scientific and engineering minds.

Spacelab and Spacehab Flights

Skylab had been an interesting first step in research but, after the Saturn V production ceased, all US space station designs would be limited to something similar to the Orbiter's 4.6-m (15-ft.) payload bay diameter. The shuttle had given the world ample ways to evolve concepts of space station modules, including a European Space Agency-built Spacelab and an American-built Spacehab. Each module rode in the payload bay of the Orbiter. These labs had the same outer diameter as subsequent ISS modules.

The shuttle could provide the necessary power, communications, cooling, and life support to these laboratories. Due to consumables limits, the shuttle could only keep these labs in orbit for a maximum of 2 weeks at a time. Through the experience, however,

Space Shuttle Atlantis (STS-71) is docked with the Russian space station Mir (1995). At the time, Atlantis and Mir had formed the largest spacecraft ever in orbit. Photo taken from Russian Soyuz vehicle as shuttle begins undocking from Mir. Photo provided to NASA by Russian Federal Space Agency.

astronaut crews and ground engineers discovered many issues of loading and deploying real payloads, establishing optimum work positions and locations, clearances, cleanliness, mobility, environmental issues, etc.

Shuttle-Mir

In 1994, the funding of the Space Station Program passed the US Senate by a single vote. Later that year, Vice President Al Gore and Russian Deputy Premier Viktor Chernomyrdin signed the agreement that redefined both countries' space station programs. That agreement also directed the US Space Shuttle Program and the Russian space program to immediately hone the complex cooperative operations required to build the new, larger-than-dreamed space station. That operations development effort would come through a series of increasingly complex flights of the shuttle to the existing Russian space station Mir. George Abbey, director of Johnson Space Center, provided the leadership to ensure the success of the Shuttle-Mir Program.

The Space Shuttle Program immediately engaged Mir engineers and the Moscow Control Center to begin joint operations planning. Simultaneously, engineers working on the former US-led Space Station Program, called Freedom, went to work with their counterparts who had been designing and building Mir's successor—Mir-II. The new joint program was christened the ISS Program. Although NASA's Space Shuttle and ISS Programs emerged as flagships for new, vigorous international cooperation with the former Soviet states, the immediate technical challenges were formidable. The Space Shuttle Program had to surmount many of these challenges on shorter notice than did the ISS Program.

Astronaut Shannon Lucid floats in the tunnel that connects Atlantis' (STS-79 [1996]) cabin to the Spacehab double module in the cargo bay. Lucid and her crew mates were already separated from the Russian space station Mir and were completing end-of-mission chores before their return to Earth.

Striving for Lofty Heights—And Reaching Them

The biggest effect on the shuttle in this merged program was the need to reach a higher-inclination orbit that could be accessed from Baikonur Cosmodrome in Kazakhstan. At an inclination of 51.6 degrees to the equator, this new orbit for the ISS would not take as much advantage of the speed of the Earth's rotation toward the East as had originally been planned. Instead of launching straight eastward and achieving nearly 1,287 km/hour (800 mph) from Earth's rotation, the shuttle now had to aim northward to meet the vehicles launched from Baikonur, achieving a benefit of only 901 km/hour (560 mph). The speed difference meant that each shuttle could carry substantially less mass to orbit for the same maximum propellant load. The Mir was already in such an orbit, so the constraint was in place from the first flight (STS-63 in 1995).

The next challenge of the 51.6-degree orbit was a very narrow launch window each day. In performing a rendezvous, the shuttle needed to launch close to the moment when the shuttle's launch pad was directly in the same flat plane as the orbit of the target spacecraft. Typically, there were only 5 minutes when the shuttle could angle enough to meet the Russian orbit.

Thus, in a cooperative program with vehicles like Mir (and later the ISS), the shuttle had only a tiny "window" each day when it could launch. The brief chance to beat any intermittent weather meant that the launch teams and Mission Control personnel often had to wait days for acceptable weather during the launch window. As a result of the frequent launch slips, the Mir and ISS control teams had to learn to pack days with spontaneous work schedules for the station crew on a single day's notice. Flexibility grew to become a high art form in both programs.

Once the shuttle had launched into the orbit plane of the Mir, it had to catch up to the station before it could dock and begin its mission at the outpost. Normally, rendezvous and docking would be completed 2 days after launch, giving the shuttle time to make up any differences between its location around the orbit compared to where the Mir or ISS was positioned at the time of launch, as well as time for ground operators to create the precise maneuvering plan that could only be perfected after the main engines cut off 8½ minutes after launch.

Generally, the plan was to launch then execute the lengthy rendezvous preparation the day after launch. The shuttle conducted the last stages of the rendezvous and docking the next morning so that a full day could be devoted to assembly and cargo transfer. This 2-day process maximized the available work time aboard the station before the shuttle consumables gave out and the shuttle had to return to Earth. The Mir and ISS teams worked in the months preceding launch to place their vehicles in the proper phase in their respective orbits, such that this 2-day rendezvous was always possible.

Arriving at the rendezvous destination was only the first step of the journey. The shuttle still faced a formidable hurdle: docking.

Docking to Mir

The American side had not conducted a docking since the Apollo-Soyuz Test Project of 1975. Fortunately, Moscow's Rocket and Space Corporation Energia had further developed the joint US-Russian docking system originally created for the Apollo-Soyuz Test Project in anticipation of their own shuttle—the Buran. Thus, the needed mechanism was already installed on Mir.

The Russians had a docking mechanism on their space station in a 51.6-degree orbit, awaiting a shuttle. That mechanism had a joint US-Russian design heritage. The Americans had a fleet of shuttles that needed to practice servicing missions to a space station in a 51.6-degree orbit. In a surprisingly rapid turn of events, the US shuttle's basic design began to include a sophisticated Russian mechanism. That mechanism would remain a part of most of the shuttle's ensuing missions.

The mechanism—called an Androgynous Peripheral Docking System—became an integral part of the shuttle's future. It looked a little like a three-petal artichoke when seen from the side. US engineers were challenged to work scores of details and unanticipated challenges to incorporate this exotic Russian apparatus in the shuttle. The bolts that held the Androgynous Peripheral Docking System to the shuttle were manufactured according to *Système International* (SI, or metric) units whereas all other shuttle hardware and tools were English units. For the first time, the US space program began to create hardware and execute operations in SI units—a practice that would become the norm during the ISS era.

All connectors in the cabling were of Russian origin and were unavailable in the West. Electrical and data interfaces had to be made somewhere. The obvious solution would be to put a US connector on the "free" end of each cable that led to the docking system. Each side could engineer from there to its own standards and hardware. Yet, even that simple plan had obstacles. Whose wire would be in the cable?

The Russian wires were designed to be soldered into each pin and socket while the US connector pins and sockets were all crimped under pressure to their wires in an exact fit. US wire had nickel plating, Russian wire did not. US wire could not be easily soldered into Russian connector pins, and Russian wire could not be reliably crimped into American connector pins. Ultimately, unplated Russian wire was chosen and new techniques were certified to assure a reliable crimped bond at each American pin. Even though the Russian system and the shuttle were both designed to operate at 28 volts, direct current (Vdc), differences in the grounding strategy required extensive discussions and work.

The Space Shuttle Atlantis (STS-71) arrived at the Mir on June 29, 1995, with the international boundary drawn at the crimped interface to a Russian wire in every US connector pin and socket. US 28-Vdc power flowed in every Russian Androgynous Peripheral Docking System electronic component, beginning a new era in international cooperation. And this happened just in time, as the US and partners were poised to begin work on a project of international proportions.

View of the Orbiter Docking System that allowed the shuttle to attach to the International Space Station. This close-up image shows the payload bay closeout on STS-130 (2010).

Construction of the International Space Station Begins

The International Space Station (ISS) was a new kind of spacecraft that would have been impossible without the shuttle's unique capabilities; it was the first spacecraft designed to be assembled in space from components that could not sustain themselves independently. The original 1984 International Freedom Space Station—already well along in its manufacture—was reconfigured to be the forward section of the ISS. The Freedom heritage was a crucial part of ISS plans, as its in-space construction was a major goal of the program. All previous spacecraft had either been launched intact from the ground (such as the shuttle itself, Skylab, or the early Salyut space stations) or made of fully functional modules, each launched intact from the ground and hooked together in a cluster of otherwise independent spacecraft.

This timeline represents the Space Shuttle fleet's delivery and attachment of several major components to the International Space Station. The specific components are outlined in red in each photo.

The Mir and the late-era Salyut stations were built from such self-contained spacecraft linked together. Although these Soviet stations were big, they were somewhat like structures built primarily out of the trucks that brought the pieces and were not of a monolithic design. Only about 15% of each module could be dedicated to science. The rest of the mass was composed of the infrastructure needed to get the mass to the station.

The ISS would take the best features of both the merged Mir-II and the Freedom programs. It would use proven Russian reliability in logistics, propulsion, and basic life support and enormous new capabilities in US power, communications, life support, and thermal control. The integrated Russian modules helped to nurture the first few structural elements of the US design until the major US systems could be carried to the station and activated. These major US systems were made possible by assembly techniques enabled by the shuttle. The United States could curtail expensive and difficult projects in both propulsion and crew rescue vehicles and stop worrying about the problem of bootstrapping their initial infrastructure, while the Russians would be able to suspend sophisticated-but-expensive efforts in in-space construction techniques, power systems, large gyroscopes, and robotics. What emerged out of the union of the Freedom and the Mir-II programs was a space station vastly larger and more robust (and more complicated) than either side had envisioned.

The Pieces Begin to Come Together

Although the ISS ultimately included several necessary Mir-style modules in the Russian segment, the other partner elements from the United States, Canada, European Space Agency, Italy, and Japan were all designed with the shuttle in mind. Each of these several dozen components was to be supported by the shuttle until each could be supported by the ISS infrastructure. These major elements typically required power, thermal control, and telemetry support from the shuttle. Not one of these chunks could make it to the ISS on its own, nor could any be automatically assembled into the ISS by itself. Thus, the shuttle enabled a new era of unprecedented *in situ* construction capability.

Because it grew with every mission, the ISS presented new challenges to

Endeavour (STS-88) brought US-built Unity node, which attached to Russian-built Zarya.

1998

Discovery (STS-92) delivered Z1 truss and antenna (top) and one of the mating adapters.

2000

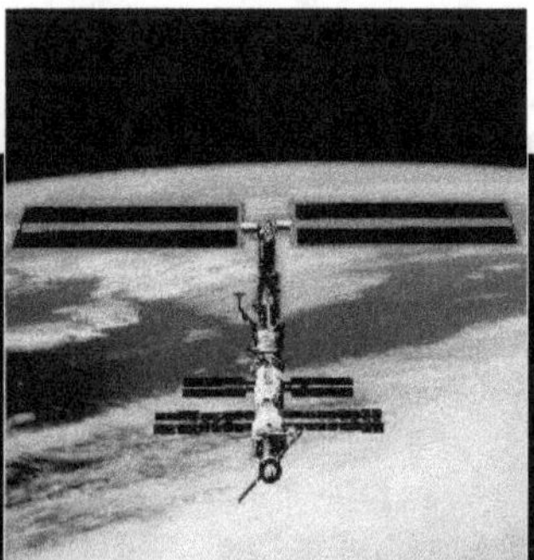

Endeavour (STS-97) delivered new solar array panels.

spacecraft engineering in general and to the shuttle in particular. With each new module, the spacecraft achieved more mass, a new center of mass, new antenna blockages, and some enhanced or new capability and constraints.

During the assembly missions, the shuttle and the ISS would each need to reconfigure the guidance, navigation, and control software to account for several different configurations. Each configuration needed to be analyzed for free flight, initial docked configuration with the arriving element still in the Orbiter payload bay, and final assembled and mated configuration with the element in its ISS position. There were usually one or two intermediate configurations with the element robotically held at some distance between the cargo bay and its final destination.

Consequently, crews had to update a lot of software many times during the mission. At each step, both the ISS and the shuttle experienced a new and previously unflown shape and size of spacecraft.

Even the most passive cargos involved active participation from the shuttle. For example, in the extremely cold conditions in space, most cargo elements dramatically cooled throughout the flight to the ISS. On previous space station generations like Skylab, Salyut, and Mir, such modules needed heaters, a control system to regulate them, and a power supply to run them both. These functions all passed to the shuttle, allowing an optimized design of each ISS element.

Each mission, therefore, had a kind of special countdown called the "Launch to Activation" timeline. This unique timeline for every cargo considered how long it would take before such temperature limits were reached. Sometimes, the shuttle's ground support systems would heat the cargo in the payload bay for hours before the launch to gain some precious time in orbit. Other times electric heaters were provided to the cargo element at the expense of shuttle power. At certain times the shuttle would spend extra time pointing the payload bay intentionally toward the sun or the Earth during the long rendezvous with the ISS. All these activities led to a detailed planning process for every flight that involved thermal systems, attitude control, robotics, and power.

The growth of the ISS did not come at the push of a button or even solely at the tip of a remote manipulator. The assembly tasks in orbit involved a combination of docking, berthing, automatic capture, automatic deployment, and good old-fashioned elbow grease.

The shuttle had mastered the rendezvous and docking issues in a high-inclination orbit during the Mir Phase 1 Program. However, just getting there and getting docked would not assemble the ISS. Berthing and several other attachment techniques were required.

Docking and Berthing

Docking

Docking and berthing are conceptually similar methods of connecting a pressurized tunnel between two objects in space. The key differences arise from the dynamic nature of the docking process with potentially large residual motions. In addition, under docking there is a need to complete the rigid structural mating quickly. Such constraints are not imposed on the slower, robotically controlled berthing process.

Docking spacecraft need to mate quickly so that attitude control can be restored. Until the latches are secured,

Atlantis (STS-98) brought Destiny laboratory.

Endeavour (STS-100) delivered and attached Space Station Robotic Arm.

Atlantis (STS-104) delivered Quest airlock.

2001

there is very little structural strength at the interface. Therefore, neither vehicle attempts to fire any thrusters or exert any control on "the stack." During this period of free drift, there is no telling which attitude can be expected. The sun may consequently end up pointing someplace difficult, such as straight onto a radiator or edge-on to the arrays. Thus, it pays to get free-flying vehicles latched firmly together as quickly as possible.

Due to the large thermal differences—up to 400° C (752°F) between sun-facing metal and deep-space-facing metal—the thermal expansion of large metal surfaces can quickly make the precise alignment of structural mating hooks or bolts problematic, unless the metal surfaces have substantial time to reach the same temperature. As noted, time is of the essence. Hence, docking mechanisms were forced to be small—about the size of a manhole—due to this need to rapidly align in the presence of large thermal differences.

A docking interface is a sophisticated mechanism that must accomplish many difficult functions in rapid succession. It must mechanically guide the approaching spacecraft from its first contact into a position where a "soft capture" can be engaged. Soft capture is somewhat akin to the moment when a large ship first tosses its shore lines to dock hands on the pier; it serves only to keep the two vehicles lightly connected while the next series of functions is completed.

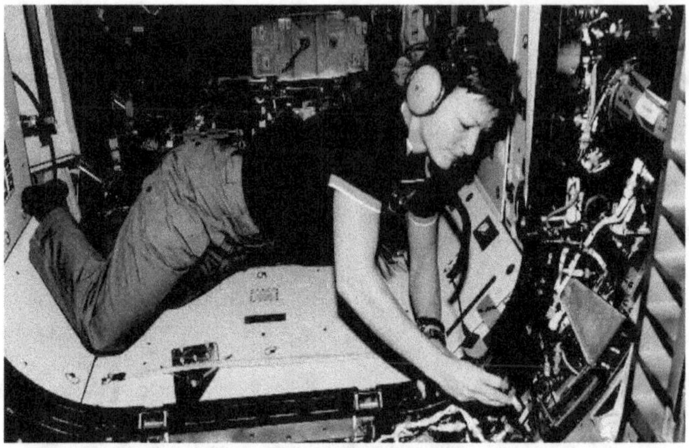

Astronaut Peggy Whitson, Expedition 16 commander, works on Node 2 outfitting in the vestibule between the Harmony node and Destiny laboratory of the International Space Station in November 2007.

The mechanism must next damp out leftover motions in X, Y, and Z axes as well as damp rotational motions in pitch, yaw, and roll while bringing the two spacecraft into exact alignment. This step was a particular challenge for shuttle dockings. For the first time in space history, the docking mechanism was placed well away from the vehicle's center of gravity. Sufficient torque had to be applied at the interface to overcome the large moment of the massive shuttle as it damped its motion.

Next, the mechanism had to retract, pulling the two spacecraft close enough together that strong latches could engage. The strong latches clamped the two halves of the mechanism together with enough force to compress the seals. These latches held the halves together against the huge force of pressure that would try to push them apart once the hatches were opened inside. While this final cinching of

Atlantis (STS-110) delivered S0 truss.

Atlantis (STS-112) brought S1 truss.

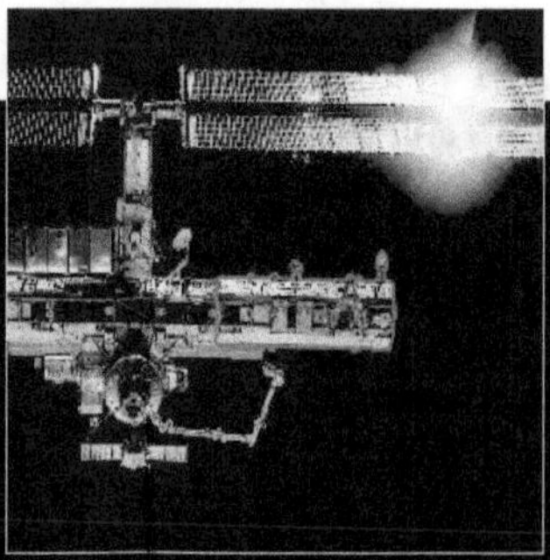

Endeavour (STS-113) delivered P1 truss.

2002

the latches happened, hundreds of electrical connections and even a few fluid transfer lines had to be automatically and reliably connected. Finally, there had to be a means to let air into the space between the hatches, and all the hardware that had been filling the tunnel area had to be removed before crew and cargo could freely transit between the spacecraft.

Berthing

Once docked, the shuttle and station cooperated in a gentler way called berthing, which led to much larger passageways.

Berthing was done under the control of a robotic arm. It was the preferred method of assembling major modules of the ISS. The mechanism halves could be held close to each other indefinitely to thermally equilibrate. The control afforded by the robotic positioning meant that the final alignment and damping system in berthing could be small, delicate, and lightweight while the overall tunnel could be large.

In the case of the ISS, the berthing action only completed the hard structural mating and sealing, unlike docking, where all utilities were simultaneously mated. All berthing interface utilities were subsequently hooked between the modules in the pressurized tunnel (i.e., in a "shirtsleeve" environment). During extravehicular activities (EVAs), astronauts connected major cable routes only where necessary.

The interior cables and ducts connected in a vestibule area inside the sealing rings and around the hatchways. This arrangement allowed thousands of wires and ducts to course through the shirtsleeve environment where they could be easily accessed and maintained while allowing the emergency closure of any hatch in seconds. This hatch closure could be done without the need to clear or cut cables that connected the modules. This "cut cable to survive" situation occurred, at great peril to the crew, for several major power cables across a docking assembly during the Mir Program.

Robotic Arms Provide Necessary Reach

The assembly of the enormous ISS required that large structures were placed with high precision at great distance from the shuttle's payload bay. As the Shuttle Robotic Arm could only reach the length of the payload bay, the ISS needed a second-generation arm to position its assembly segments and modules for subsequent hooking, berthing, and/or EVA bolt-downs.

The Unity connecting module is being put into position to be mated to Endeavour's (STS-88 [1998]) docking system in the cargo bay. This mating was the first link in a long chain of events that led to the eventual deployment of the connected Unity and Zarya modules.

Building upon the lessons learned from the shuttle experience, the same Canadian Space Agency and contractor team created the larger, stiffer, and more nimble Space Station Robotic

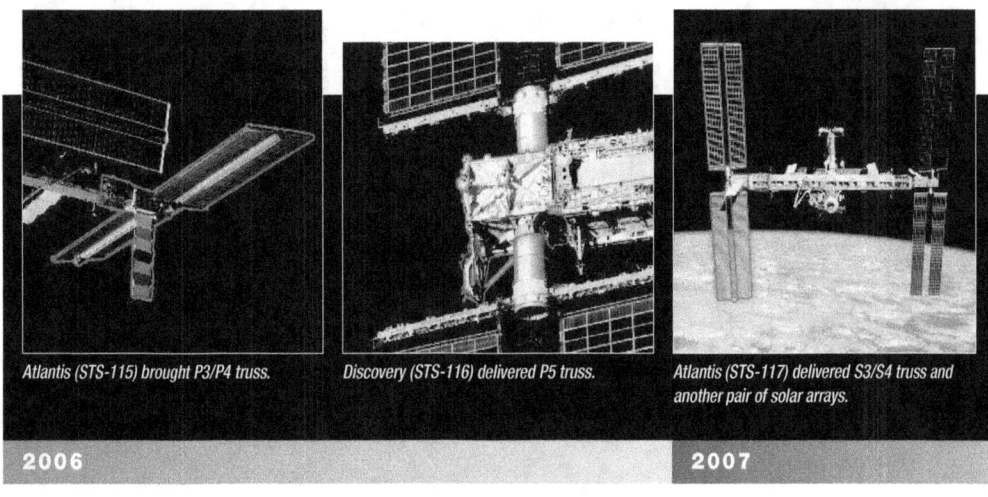

Atlantis (STS-115) brought P3/P4 truss.

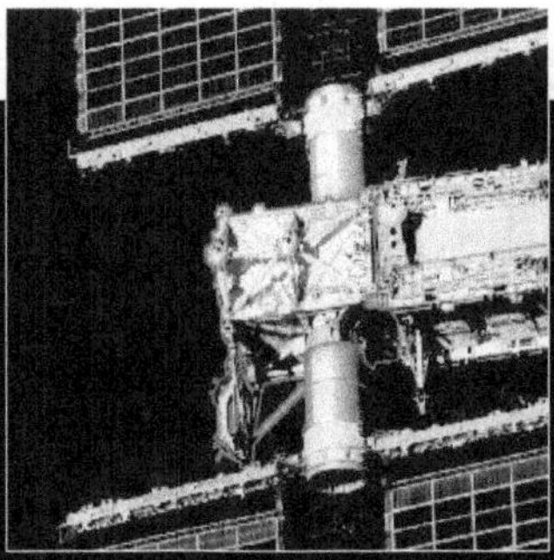

Discovery (STS-116) delivered P5 truss.

Atlantis (STS-117) delivered S3/S4 truss and another pair of solar arrays.

2006

2007

Arm, also known as the "big arm." The agency and team created a 17-m (56-ft) arm with seven joints. The completely symmetric big arm was also equipped with the unique ability to use its end effector as a new base of operations, walking end-over-end around the ISS. Together with a mobile transporter that could carry the new arm with a multiton cargo element at its end, the ISS robotics system worked in synergy with the Shuttle Robotic Arm to maneuver all cargos to their final destinations.

The Space Station Robotic Arm could grip nearly every type of grapple fixture that the shuttle's system could handle, which enabled the astounding combined robotic effort to repair a torn outboard solar array on STS-120 (2007). On that memorable mission, the Space Station Robotic Arm "borrowed" the long Orbiter Boom Sensor System, allowing an unprecedented stretch of 50 m (165 ft) down the truss and 27 m (90 ft) up to reach the damage.

The Space Station Robotic Arm was robust. Analysis showed that it was capable of maneuvering a fully loaded Orbiter to inspect its underside from the ISS windows.

The robotic feats were amazing indeed—and unbelievable at times—yet successful construction of the ISS depended on a collaboration of human efforts, ingenuity, and a host of other "nuts-and-bolts" mechanisms and techniques.

Other Construction Mechanisms

The many EVA tests conducted by shuttle crews in the 1980s inspired ISS designers to create several simplifying construction techniques for the enormous complex. While crews assembled the pressurized modules using the Common Berthing Mechanism, they had to assemble major external structures using a simple large-hook system called the Segment-to-Segment Attachment System designed for high strength and rapid alignment.

The Segment-to-Segment Attachment System had many weight and reliability enhancements resulting from the lack of a need for a pressurized seal. Such over-center hooks were used in many places on the ISS exterior. In major structural attachments (especially between segments of the 100-m [328-ft] truss), the EVA crew additionally drove mechanical bolts between the segments. The crew then attached major appendages and payloads with a smaller mechanism called a Common Attachment System.

Where appropriate, major systems were automatically deployed or retracted from platforms that were pre-integrated to the delivered segment before launch. The solar array wings were deployed by swinging two half-blanket boxes open from a "folded hinge" launch position and then deploying a collapsible mast to extend and finally to stiffen the blankets. Like the Russian segment's smaller solar arrays, the tennis-court-sized US thermal radiators deployed automatically with an extending scissor-like mechanism.

Meanwhile, the ISS design had to accommodate the shuttle. It needed to provide a zigzag tunnel mechanism (the Pressurized Mating Adapter) to optimize the clearance to remove payloads from the bay after the shuttle had docked. ISS needed to withstand the shuttle's thruster plumes for heating, loads, contamination, and erosion. It also had to provide the proper electrical grounding path for shuttle electronics, even though the ISS operated at a significantly higher voltage.

Endeavour (STS-118) delivered the S5 truss segment.

Discovery (STS-120) brought Harmony Node 2 module.

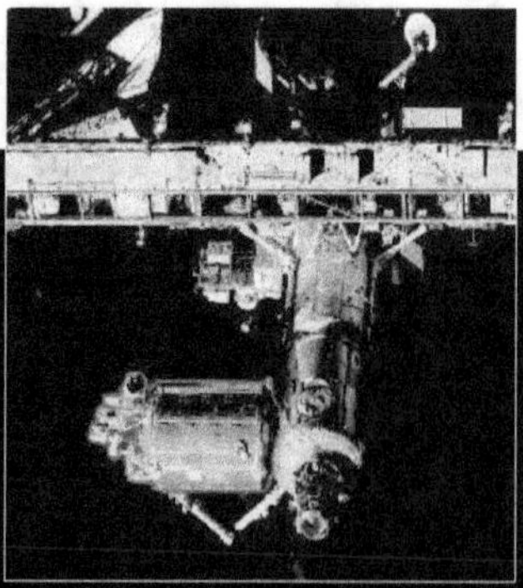

Atlantis (STS-122) delivered European Space Agency's Columbus laboratory.

2007

2008

Improvements to the Shuttle Facilitated Assembly of the International Space Station

NASA had to improve Space Shuttle capability before the International Space Station (ISS) could be assembled. The altitude and inclination of the ISS orbit required greater lift capability by the shuttle, and NASA made a concerted effort to reduce the weight of the vehicle. Engineers redesigned items such as crew seats, storage racks, and thermal tiles. The super lightweight External Tank allowed the larger ISS segments to be launched and assembled. Modifications to the ascent flight path and the firing of Orbital Maneuvering System engines alongside the main engines during ascent provided a more efficient use of propellant.

Launch reliability was another concern. For the shuttle to rendezvous with the ISS, the launch window was limited to a period of about 5 minutes, when the launch pad on the rotating Earth was aligned with the ISS orbit. By rearranging the prelaunch checklist to complete final tests earlier and by adding planned hold periods to resolve last-minute technical concerns, the 5-minute launch window could be met with high reliability.

Finally, physical interfaces between the shuttle and the ISS needed to be coordinated. NASA designed docking fixtures and transfer bags to accommodate the ISS. The agency modified the rendezvous sequence to prevent contamination of the ISS by the shuttle thrusters. In addition, NASA could transfer electrical power from the ISS to the shuttle. This allowed the shuttle to remain docked to the ISS for longer periods, thus maximizing the work that could be accomplished.

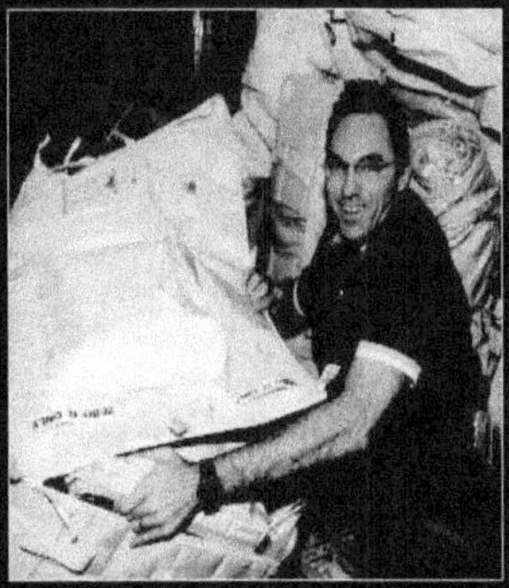

Astronaut Carl Walz, Expedition 4 flight engineer, stows a small transfer bag into a larger cargo transfer bag while working in the International Space Station Unity Node 1 during joint docked operations with STS-111 (2002).

Further Improvements Facilitate Collaboration Between Shuttle and Station

The ISS needed a tiny light source that could be seen at a distance of hundreds of miles by the shuttle's star tracker so that rendezvous could be conducted. The ISS was so huge that in sunlight it would saturate the star trackers of the shuttle, which were accustomed to seeking vastly dimmer points of light. Thus, the shuttle's final rendezvous with the ISS involved taking a relative navigational "fix" on the ISS at night, when the ISS's small light bulb approximated the light from a star.

Endeavour (STS-123) brought Kibo Japanese Experiment Module.

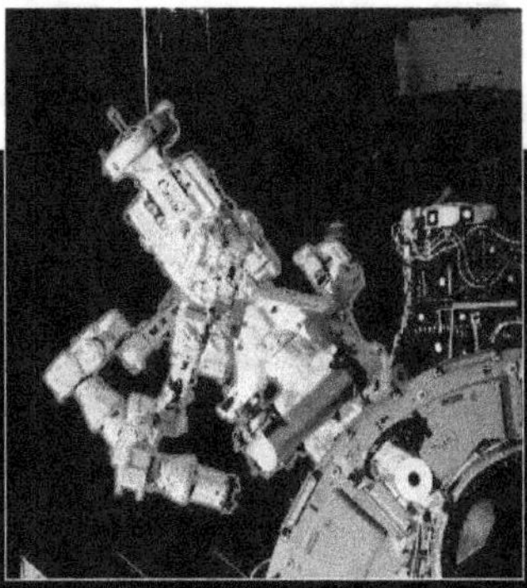

Endeavour (STS-123) also delivered Canadian-built Special Purpose Dextrous Manipulator.

Discovery (STS-124) brought Pressurized Module and robotic arm of Kibo Japanese Experiment Module.

2008 *continued*

Other navigational aids were mounted on the ISS as well. These aids included a visual docking target that looked like a branding iron of the letter "X" erected vertically from a background plate in the center of the hatch. Corner-cube glass reflectors were provided to catch a laser beam from the shuttle and redirect it straight back to the shuttle. This remarkable optical trick is used by several alignment systems, including the European Space Agency's rendezvous system that targeted other places on the ISS. Thus, it was necessary to carefully shield the different space partners' reflectors from the beams of each other's spacecraft during their respective final approaches to the ISS. Otherwise a spacecraft might "lock on" to the wrong place for its final approach.

As the station grew, it presented new challenges to the shuttle's decades-old control methods. The enormous solar arrays, larger than America's Cup yacht sails, caught the supersonic exhaust from the shuttle's attitude control jets and threatened to either tear or accelerate the station in some strange angular motion. Thus, when the shuttle was in the vicinity of or docked to the ISS, a careful ballet of shuttle engine selection and ISS array positions was always necessary to keep the arrays from being damaged.

This choreography grew progressively more worrisome as the ISS added more arrays. It was particularly difficult during the last stages of docking and in the first moments of a shuttle's departure, when it was necessary to fire thrusters in the general direction of the station.

There were also limits as to how soon a shuttle might be allowed to fire an engine after it had just fired one. It was possible that the time between each attitude correction pulse could match the natural structural frequency of that configuration of the ISS. This pulsing could amplify oscillations to the point where the ISS might break if protection systems were not in place. Of course, this frequency changed each time the ISS configuration changed. Thus, the shuttle was always loading new "dead bands" in its control logic to prevent it from accidentally exciting one of these large station modes.

In all, the performances of all the "players" in this unfolding drama were stellar. The complexity of challenges required flexibility and tenacity. The shuttle not only played the lead in the process, it also served in supporting roles throughout the entire construction process.

The Roles of the Space Shuttle Program Throughout Construction

Logistics Support—Expendable Supplies

The shuttle was a workhorse that brought vast quantities of hardware and supplies to the International Space Station (ISS). Consumables and spare parts were a key part of that manifest, with whole shuttle missions dedicated to resupply. These missions were called "Utilization and Logistics Flights." All missions—even the assembly flights—contributed to the return of trash, experiment samples, completed experiment apparatus, and other items.

Unique Capacity to Return Hardware and Scientific Samples

Perhaps the greatest shuttle contribution to ISS logistics was its unsurpassed capability to return key systems and components to Earth. Although most of the ISS worked perfectly from the start, the shuttle's ability to bring components and systems back was essential in rapidly advancing NASA's engineering

Discovery (STS-119) brought S6 truss segment.

Endeavour (STS-127) delivered Kibo Japanese Experiment Module Exposed Facility and Experiment Logistics Module Exposed Section.

Endeavour (STS-130) delivered Node 3 with Cupola.

2009

2010

knowledge in many key areas. This allowed ground engineers to thoroughly diagnose, repair, and sometimes redesign the very heart of the ISS.

The shuttle upmass was a highly valued financial commodity within the ISS Program, but its recoverable down-mass capability was unique, hotly pursued, and the crown jewel at the negotiation table. As it became clear that more and more partners would have the capability to deliver cargo to the ISS but only NASA retained any significant ability to return cargo intact to Earth, the cachet only increased. Even the Russian partner—with its own robust resupply capabilities and long, proud history in human spaceflight—was seduced by the lure of recoverable down mass and agreed that its value was twice that of 1 kg (2.2 pounds) of upmass. NASA negotiators had a particular fondness for this one capability that the Russians seemed to value higher than their own capabilities.

Mass Transported by Space Shuttle to Station

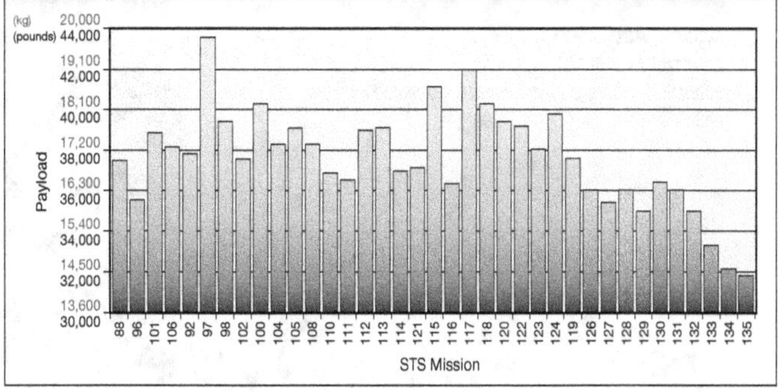

Symbiotic Relationship Between Shuttle and the International Space Station

Over time the two programs developed several symbiotic logistic relationships. The ISS was eager to take the pure-water by-product of the shuttle's fuel cell power generators because water is the heaviest and most vital consumable of the life support system. The invention of the Station to Shuttle Power Transfer System allowed the shuttle to draw power from the ISS solar arrays, thereby conserving its own oxygen and hydrogen supplies and extending its stay in orbit.

The ISS maintained the shared contingency supply of lithium hydroxide canisters for carbon dioxide scrubbing by both programs, allowing more cargo to ride up with the shuttle on every launch in place of such canisters. The shuttle would even carry precious ice cream and frozen treats for the ISS crews in freezers needed for the return of frozen medical samples.

The shuttle would periodically reboost the ISS, as needed, using any leftover propellant that had not been required for contingencies. The shuttle introduced air into the cabin and transferred compressed oxygen and nitrogen to the ISS tanks as its unused reserves allowed. ISS crews even encouraged shuttle crews to use their toilet so that the precious water could be later recaptured from the wastes for oxygen generation.

The ISS kept stockpiles of food, water, and essential consumables that were collectively sufficient to keep a guest crew of seven aboard for an additional 30 days—long enough for a rescue shuttle to be prepared and launched to the ISS in the event a shuttle already at the station could not safely reenter the Earth's atmosphere.

Extravehicular Activity by Space Shuttle Crews

Even with all of the automated and robotic assembly, a large and complex vehicle such as the ISS requires an enormous amount of manual assembly—much of it "hands on"—in the harsh environment of space. Spacewalking crews assembled the ISS in well over 100 extravehicular activity (EVA) sessions, usually lasting 5 hours or more. EVA is tiring, time consuming, and more dangerous than routine cabin flight. It is also exhilarating to all involved. Despite the dangers of EVA, the main role for shuttle in the last decade of flight was to assemble the ISS. Therefore, EVAs came to dominate the shuttle's activities during most station visits.

These shuttle crew members were trained extensively for their respective missions. NASA scripted the shuttle flights to achieve ambitious assembly objectives, sometimes requiring four EVAs in rapid succession. The level of proficiency required for such long, complicated tasks was not in keeping with the ISS training template. Therefore, the shuttle crews handled most of the burden. They trained until mere days before launch for the marathon sessions that began shortly after docking.

Shuttle Airlock

Between assembly flights STS-97 (2000) and STS-104 (2001)—the first time a crew was already aboard the ISS to host a shuttle and the flight when

Clayton Anderson
Astronaut on STS-117 (2007) and STS-131 (2010). Spent 152 days on the International Space Station before returning on STS-120 (2007).

"Life was good on board the International Space Station (ISS). Time typically passed quickly, with much to do each day. This was especially true when an ISS crew prepared to welcome 'interplanetary guests'...or more specifically, a Space Shuttle crew! During my 5-month ISS expedition, our 'visitors from another planet' included STS-117 (my ride up), STS-118, and STS-120 (my ride down).

"While awaiting a shuttle's arrival, ISS crews prepared in many ways. We may have said goodbye to 'trash-collecting tugs' or welcomed replacement ships (Russian Progress, European Space Agency Automated Transfer Vehicle, and the Japanese Aerospace Exploration Agency H-II Transfer Vehicle) fully stocked with supplies. Just as depicted in the movies, life on the ISS became a little bit like Grand Central Station!

"Prepping for a shuttle crew was not trivial. It was reminiscent of work you might do when guests are coming to your home! ISS crews 'pre-packed...,' gathering loads of equipment and supplies no longer needed that must be disposed of or may be returned to Earth...like cleaning house! This wasn't just 'trash disposal'—sending a vehicle to its final rendezvous with the fiery friction of Earth's atmosphere. Equipment could be returned on shuttle to enable refurbishment for later use or analyzed by experts to figure out how it performed in the harsh environment of outer space. It was also paramount to help shuttle crews by prepping their spacewalking suits and arranging the special tools and equipment that they would need. This allowed them to 'jump right in' and start their work immediately after crawling through the ISS hatch! Shuttle flights were all about cramming much work into a short timeframe! The station crew did their part to help them get there!

Astronaut Clayton Anderson, Expedition 15 flight engineer, smiles for a photo while floating in the Unity node of the International Space Station.

"The integration of shuttle and ISS crews was like forming an 'All-Star' baseball team. In this combined form, wonderful things happened. At the moment hatches swung open, a complicated, zero-gravity dance began in earnest and a well-oiled machine emerged from the talents of all on board executing mission priorities flawlessly!

"Shuttle departure was a significant event. I missed my STS-117 and STS-118 colleagues as soon as they left! I wanted them to stay there with me, flying through the station, moving cargo to and fro, knocking stuff from the walls! The docked time was grand...we accomplished so much. To build onto the ISS, fly the robotic arm, perform spacewalks, and transfer huge amounts of cargo and supplies, we had to work together, all while having a wonderfully good time. We talked, we laughed, we worked, we played, and we thoroughly enjoyed each other's company. That is what camaraderie and 'crew' was all about. I truly hated to see them go. But then they were home...safe and sound with their feet firmly on the ground. For that, I was always grateful, yet I must admit that when a crew departed I began to think more of the things that I did not have in orbit, some 354 km (220 miles) above the ground.

"Life was good on board the ISS...I cherished every single minute of my time in that fantastic place."

the ISS Quest airlock was activated, respectively—the shuttle crews were hampered by a short-term geometry problem. The shuttle's airlock was part of the docking tunnel that held the two spacecraft together, so in that period the shuttle crew had to be on its side of the hatch during all such EVAs in case of an emergency departure. Further, the preparations for EVA required that the crew spend many hours at reduced pressure, which was accomplished prior to Quest by dropping the entire shuttle cabin pressure. Since the ISS was designed to operate at sea-level atmosphere, it was necessary to keep the shuttle and station separated by closed hatches while EVAs were in preparation or process. This hampered the transfer of internal cargos and other intravehicular activities.

International Space Station Airlock

On assembly flight 7A (STS-104), the addition of the joint airlock Quest allowed shuttle crews to work in continuous intravehicular conditions while their EVA members worked outside. Even in this airlock, shuttle crews continued to conduct the majority of ISS EVAs and shuttles provided the majority of the gases for this work. Docked shuttles could replenish the small volume of unrecoverable air that could not be compressed from the airlock. The prebreathe procedure of pure oxygen to the EVA crew also was supported by shuttle reserves through a system called Recharge Oxygen Orifice Bypass Assembly. This system was delivered on STS-114 (2005) and used for the first time on STS-121 (2006). Finally, the shuttle routinely repressurized the ISS high-pressure oxygen and nitrogen tanks and/or the cabin itself prior to leaving. The ISS rarely saw net losses in its on-board supplies, even in the midst of such intense operations. Fewer ISS consumables were thus used whenever a shuttle could support the EVAs.

The Shuttle as Crew Transport

Although many crews came and went aboard the Russian Soyuz rescue craft, the shuttle assisted the ISS crew rotations at the station during early flights. This shuttle-based rotation of ISS crew had several significant drawbacks, however, and the practice was abandoned in later flights. Launch and re-entry suits needed to be shared or, worse, spared on the Orbiter middeck to fit the arriving and departing crew member. Different Russian suits were used in the Soyuz rescue craft, so those suits had to make the manifest somewhere. Further, a special custom-fit seat liner was necessary to allow each crew member to safely ride the Soyuz to an emergency landing. This seat liner had to be ferried to the ISS with each new crew member who might use the Soyuz as a lifeboat. Thus, a lot of duplication occurred in the hardware required for shuttle-delivered crews.

Shuttle Launch Delays

As a shuttle experienced periodic delays of weeks or even months from its original flight plan, it was necessary to replan the activities of ISS crews who were expecting a different crew makeup. Down-going crews sometimes found their "tours of duty" had been extended. Arriving crews found their tours of duty shortened and their work schedule compressed. As the construction evolved, the shuttle carried a smaller fraction of the ISS crew.

***Left photo:** Astronauts John Olivas (top) and Christer Fuglesang pose for a photo in the STS-128 (2009) Space Shuttle airlock.*
***Right photo:** Astronauts Garrett Reisman (left) and Michael Good—STS-132 (2010)—pose for a photo between two extravehicular mobility units in the International Space Station (ISS) Quest airlock. By comparison, the Quest airlock is much larger and thus allows enough space for the prebreathe needed to prevent decompression sickness to occur in the airlock, isolated from the ISS.*

Michael Foale, PhD

Astronaut on STS-45 (1992), STS-56 (1993), STS-63 (1995), STS-84 (1997), and STS-103 (1999). Spent 145 days on Russian space station Mir before returning on STS-86 (1997). Spent 194 days as commander of Expedition 8 on the International Space Station (2003-2004).

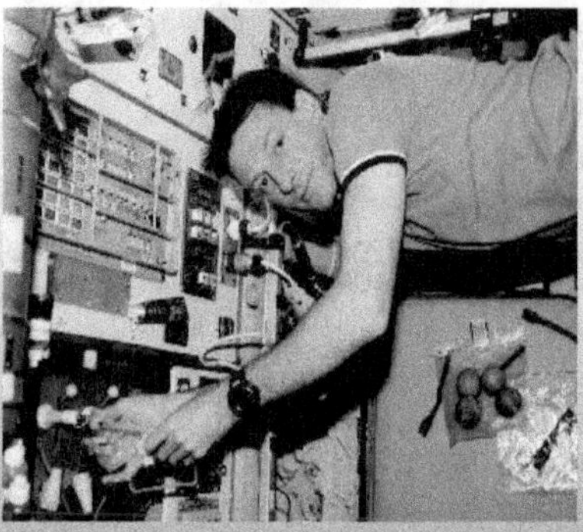

On board the International Space Station, Astronaut Michael Foale fills a water microbiology bag for in-flight analysis.

"When we look back 50 years to this time, we won't remember the experiments that were performed, we won't remember the assembly that was done. What we will know was that countries came together to do the first joint international project, and we will know that that was the seed that started us off to the moon and Mars."

Whenever NASA scrubbed a launch attempt for even 1 day, the scrub disrupted the near-term plan on board the ISS. Imagine the shuttle point of view in such a scrub scenario: "We'll try again tomorrow and still run exactly the script we know."

Now imagine the ISS point of view in the same scenario: "We've been planning to take 12 days off from our routine to host seven visitors at our home. These visitors are coming to rehab our place with a major new home addition. We need to wrap up any routine life we've established and conclude our special projects and then rearrange our storage to let these seven folks move back and forth, start packing things for the visitors to take with them, and reconfigure our wiring and plumbing to be ready for them to do their work. Then we must sleep shift to be ready for them at the strange hour of the day that orbital mechanics says that they can dock. Two days before they are to get here, they tell us that they're not coming on that day. For the next week or so of attempts, they will be able to tell us only at the moment of launch that they will in fact be arriving 2 days later."

At that juncture, did ISS crew members sleep shift? Did they shut down the payloads and rewire for the shuttle's arrival? Did they try to cram in one more day of experiments while they waited? Did they pack anything at all? This was the type of dilemma that crews and planners faced leading up to every launch. Therefore, a few weeks before each launch, ISS planners polled the technical teams for the tasks that could be put on the "slip schedule," such as small tasks or day-long procedures that could be slotted into the plan on very short notice. Some of these tasks were complex, like tearing down a piece of exercise equipment and then refurbishing it; not the sort of thing they could just dive in and do without reviewing the procedures.

Shuttle Helps Build International Partnerships

Partnering With the Russians

It is hard to overstate the homogenizing but draconian effect that the shuttle initially had on all the original international partners who had joined the Freedom Space Station Program or who took part in other cooperative spaceflights and payloads. The shuttle was the only planned way to get their hardware and astronauts to orbit. Thus, "international integration" was decidedly one-sided as NASA engineers and operators worked with existing partners to meet shuttle standards.

Such standards included detailed specifications for launch loads capability, electrical grounding and power quality, radio wave emission and susceptibility limits, materials outgassing limits, flammability limits, toxicity, mold resistance, surface temperature limits, and tens of thousands of other shuttle standards. The Japanese H-II Transfer Vehicle and European Space Agency's (ESA's) Automated Transfer Vehicle were not expected until nearly a decade after shuttle began assembly of the ISS. Neither could carry crews, so all astronauts, cargoes, supplies, and structures had to play by shuttle's rules.

Then the Earth Moved

The Russians and Americans started working together with a series of shuttle visits to the Russian space station Mir. There was more at stake than technical standards. Leadership roles were more equitably distributed and cooperation took on a new diplomatic flavor in a true partnership.

In the era following the fall of the Berlin Wall (1989) along with the end of Soviet communism and the Soviet Union itself, the US government seized the possibility of achieving two key goals—the seeding of a healthy economy in Russia through valuable western contracts, and the prevention of the spread of the large and now-saleable missile and weapon technology to unstable governments from the expansive former Soviet military-industrial complex that was particularly cash-strapped. The creation of a joint ISS was a huge step toward each of those goals, while providing the former Freedom program with an additional logistics and crew transport path. It also provided the Russian government a huge boost in prestige as a senior partner in the new worldwide partnership. That critical role made Russian integration the dominant focus of shuttle integration, and it subsequently changed the entire US perspective on international spaceflight.

Two existing spacecraft were about to meet, and engineers in each country had to satisfy each other that it was safe for each vehicle to do so. Neither side could be compelled to simply accept the other's entire system of standards and practices. The two sides certainly could not retool their programs, even if they had wanted to accept new standards. Tens of thousands of agreements and compromises had to be reached, and quickly. Only where absolutely necessary did either side have to retest its hardware to a new standard. During the Mir Phase 1 Program, the shuttle encountered the new realities of cooperative spaceflight and set about the task of defining new ways of doing business.

It was difficult but necessary to compare every standard for mutual acceptability. In most cases, the intent of the constraint was instantly compatible and the implementation was close enough to sidestep an argument. The standards compatibility team worked tirelessly for 4 years to allow cross certification. This was an entirely new experience for the Americans.

As difficult as the technical requirements were, an even more fundamental issue existed in the documents themselves. The Russians had never published in English and, similarly, the United States had not published in Cyrillic, the alphabet of the Russian language. Chaos might immediately ensue in the computers that tracked each program's data.

Communicating With Multiple Alphabets

The space programs needed something robust to handle multiple alphabets, and they needed it soon. In other words, the programs needed more bytes for every character. Thus, the programs became early adopters of the system that several Asian nations had been forced to adopt as a national standard to capture the 6,000+ characters of kanji—pictograms of Chinese origin used in modern Japanese writing. The Universal Multiple-Octet Coded Character Set—known in one ubiquitous word processing environment as "Unicode" and standardized worldwide as International Standards Organization (ISO) Standard 10646—allowed all character sets of

Financial Benefits of the Space Shuttle for the United States

Just as the International Space Station (ISS) international agreements called for each partner to meet its obligations to share in common operations costs such as propellant delivery and reboost, the agreements also required each partner to bear the cost of delivering its contributions and payloads to orbit and encouraged use of barter. As a result, the European Space Agency (ESA) and the Japan Aerospace Exploration Agency (JAXA) took on the obligation to build some of the modules within NASA's contribution as payment in kind for the launch of their laboratories. In shifting the cost of development and spares for these modules to the international partners—and without taking on any additional financial obligation for the launch of the partner labs—NASA was able to provide much-needed fiscal relief to its capped "build-to-cost" development budget in the post-redesign years. The Columbus laboratory took a dedicated shuttle flight to launch. In return, ESA built Nodes 2 and 3 and some research equipment. The Japanese Experiment Module that included Kibo would take 2.3 shuttle flights to place in orbit. JAXA paid this bill by building the Centrifuge Accommodation Module (later deleted from the program by NASA after the Vision for Space Exploration refocused research priorities on the ISS) and by providing other payload equipment and a non-ISS launch.

the world to be represented in all desired fonts. Computers in space agencies around the world quickly modified to accept the new character ISO Standard, and instantly the cosmos was accessible to the languages of all nations. This also allowed a common lexicon for acronyms.

National Perceptions

The Russians had a highly "industrial" approach to operating a spacecraft. Their cultural view of a space station appeared to most Americans to be more as a facility for science, not necessarily a scientific wonder unto itself. Although the crews continued to be revered as Russian national heroes, the spacecraft on which they flew never achieved the kind of iconic status that the Space Shuttle or the ISS achieved in the United States. By contrast, the American public was more likely to know the name of the particular one of four Orbiters flying the current mission than the names of the crew members aboard.

Although the Soyuz was reliable, it was a small capsule—so small that it limited the size of crews that could use it as a lifeboat. All crew members required long stays in Russia to train for Soyuz and many Russian life-critical systems. This was in addition to their US training and short training stays with the other partners. Overall, however, the benefits of having this alternate crew and supply launch capability were abundantly clear in the wake of the Columbia (STS-107) accident in 2003. The Russians launched a Progress supply ship to the ISS within 24 hours and then launched an international crew of Ed Lu and Yuri Malenchenko exactly 10 weeks after the accident. Both crew members wore the STS-107 patch on their suits in tribute to their fallen comrades. After the Columbia accident, the Russians launched 14 straight uncrewed and crewed missions to continue the world's uninterrupted human presence in space before the shuttle returned to share in those duties.

Other Faces on the International Stage

All the while, teams of specialists from the Canadian Space Agency, Japanese Space Exploration Agency, Italian Space Agency, and ESA each worked side-by-side with NASA shuttle and station specialists at Kennedy Space Center to prepare their modules for launch aboard the shuttle. Shortly after the delivery of the ESA Columbus laboratory on STS-122 (2008) and the Japanese Kibo laboratory on STS-124 (2008), each agency's newly developed visiting cargo vehicle joined the fleet.

The Europeans had elected to dock their Automated Transfer Vehicle at the Russian end of the station, whereas the Japanese elected to berth their vehicle—the H-II Transfer Vehicle—to the station. The manipulation of the H-II Transfer Vehicle and its berthing to the ISS were similar to the experience of all previous modules that the shuttle had brought to the space station. The big change was that the vehicle had to be grabbed in free flight by the station arm—a trick previously only performed by the much more nimble shuttle arm. NASA ISS engineers and Japanese specialists worked for years with shuttle robotics veterans to develop this exotic procedure for the far-more-sluggish ISS.

The experience paid off. In the grapple of H-II Transfer Vehicle 1 in 2009, and following the techniques first pioneered by shuttle, the free-flight grapple and berth emerged as the attachment technique for the upcoming fleet of commercial space transports expected at the ISS.

"For Shuttle ESA was a junior partner, but now with ISS we are equal partners" **—Volker Damann, ESA**

CSA ASC
Canadian Space Agency

esa
European Space Agency

JAXA
Japan Aerospace Exploration Agency

NASA
National Aeronautics and Space Administration

РОСКОСМОС
Russian Federal Space Agency

From Shuttle-Mir to International Space Station—Crews Face Additional Challenges

The Shock of Long-Duration Spaceflights

NASA had very little experience with the realities of long-term flight. Since the shuttle's inception, the shuttle team had been accustomed to planning single-purpose missions with tight scripts and well-identified manifests. The shuttle went through time-critical stages of ascent and re-entry into Earth's atmosphere on every flight, with limited life-support resources aboard. Thus, the overall shuttle culture was that every second was crucial and every step was potentially catastrophic. It took a while for NASA to become comfortable with the concept of "time to criticality," where systems aboard a large station did not necessarily have to have immediate consequences. These systems often didn't even have immediate failure recovery requirements.

For instance, the carbon dioxide scrubber or the oxygen generator could be off for quite some time before the vast station atmosphere had to be adjusted. What mattered most was flexibility in the manifest to get needed parts up to space. The shuttle's self-contained missions with well-defined manifests were not the best experience base for this pipeline of supplies.

New Realities

Russia patiently guided shuttle and then International Space Station (ISS) teams through these new realities. The delivery of parts, while always urgent, was handled in stride and with great flexibility. Their flexible manifesting practices were a shock to veteran shuttle planners. The Soyuz and the uncrewed Progress were particularly reliable at getting off the pad on time, come rain, sleet, wind, or clouds. This reliability came from the Russians' simple capsule-on-a-missile heritage, and allowed mission planners to pinpoint spacecraft arrivals and departures months in advance. The cargos aboard the Progress, however, were tweaked up until the final day as dictated by the needs at the destination, just as overnight packages are identified and manifested until the final minutes aboard a regularly scheduled airline flight. In contrast, the shuttle's heritage was one of well-defined cargos with launch dates that were weather-dependent.

Prior to the Mir experience, the shuttle engineers had maintained stringent manifesting deadlines to keep the weight and balance of the Orbiter within tight constraints and to handle the complex task of verifying the structural loads during ascent for the unique mix of items bolted to structures that would press against their fittings in the payload bay in nonlinear ways. Nonlinearity was a difficult side effect of the way that heavy loads had to be distributed. The load that each part of the structure would see was completely dependent on the history of the loads it

Unheeded Skylab Lesson: Take a Break!

The US planners might be applauded for their optimism and ambition in scheduling large workloads for the crew, but they had missed the lesson of a previous generation of planners resulting from the "Skylab Rebellion." This rebellion occurred when the Skylab-4 crew members suddenly took a day off in response to persistent over-tasking by the ground planners during their 83-day mission. From "Challenges of Space Exploration" by Marsha Freeman:

"At the end of their sixth week aboard Skylab, the third crew went on strike. Commander Carr, science pilot Edward Gibson, and Pogue stopped working, and spent the day doing what they wanted to do. As have almost all astronauts before and after them, they took the most pleasure and relaxation from looking out the windows at the Earth, taking a lot of photographs. Gibson monitored the changing activity of the Sun, which had also been a favourite pastime of the crew."

It is both ironic and instructive to note that during the so-called "rebellion," the crew members actually filled their day off with intellectually stimulating activities that were also of scientific use. Although these activities of choice were not the ones originally scripted, they were a form of mental relaxation for these exhausted but dedicated scientists. The crew members of Skylab-4 just needed some time to call their own.

had seen recently. If a load was moved, removed, or added to any of the cargo, it could invalidate the analysis.

This was an acceptable way of operating a stand-alone mission until one faced a manifesting crisis such as the loss of an oxygen generator or a critical computer on the space station.

Shortly after starting the Mir Phase I Program, the pressures of emergency manifest demands led to a new suite of tools and capabilities for the shuttle team. Engineers developed new computer codes and modeling techniques to rapidly reconfigure the models of where the masses were attached and to show how the shuttle would respond as it shook during launch. Items as heavy as 250 kg (551 pounds) were swapped out in the cargo within months or weeks of launch. In some cases, items as large as suitcases were swapped out within hours of launch.

During the ISS Program, Space Transportation System (STS)-124 carried critical toilet repair parts that had been hand-couriered from Russia during the 3-day countdown. The parts had to go in about the right place and weigh about the same amount as parts removed from the manifest for the safety analysis to be valid. Nevertheless, on fewer than 72 hours' notice, the parts made it from Moscow to space aboard the shuttle.

Training

The continuous nature of space station operations led to significant philosophical changes in NASA's training and operations. A major facet of the training adjustment had to do with the emotional nature of long-duration activities. Short-duration shuttle missions could draw on the astronauts' emotional "surge" capability to conduct operations for extended hours, sleep shift as necessary, and develop proficiency in tightly scripted procedures. It was like asking performers to polish a 15-day performance, with up to 2 years of training to perfect the show. Astronauts spent about 45 days of training for each day on orbit. They would have time to rest before and after the mission, with short breaks, if any, included in their timeline.

That would be a lot of training for a half-year ISS expedition. The crew would have to train for over 22 years under that model. One way to put the training issue into perspective is to realize that most ISS expedition members expect to remain about 185 days in orbit. This experience, per crew member, is equal to the combined Earth orbital, lunar orbital, and trans-lunar experience accumulated by all US astronauts until the moment the United States headed to the moon on Apollo 11. Thus, each such Mir (or ISS) crew member matched the accumulated total crew experience of the first 9 years of the US space effort.

With initially three and eventually six long-duration astronauts permanently aboard the ISS, the US experience in space grew at a rapidly expanding rate. By the middle of ISS Expedition 5

Posing in Node 2 during STS-127 (2009)/Expedition 20 Joint Operations: ***Front row (left to right):*** *Expedition 20 Flight Engineer Robert Thirsk (Canadian Space Agency); STS-127 Commander Mark Polansky; Expedition 19/20 Commander Gennady Padalka (Cosmonaut); and STS-127 Mission Specialist David Wolf.* ***Second row (left to right):*** *Astronaut Koichi Wakata (Japanese Aerospace Exploration Agency); Expedition 19/20 Flight Engineer Michael Barratt; STS-127 Mission Specialist Julie Payette (Canadian Space Agency); STS-127 Pilot Douglas Hurley; and STS-127 Mission Specialist Thomas Marshburn.* ***Back row (left to right):*** *Expedition 20/21 Flight Engineer Roman Romanenko (Cosmonaut); STS-127 Mission Specialist Christopher Cassidy; Expedition 20 Flight Engineer Timothy Kopra; and Expedition 20 Flight Engineer Frank De Winne (European Space Agency).*

(2002), only 2½ years into the ISS occupation, the ISS expedition crews had worked in orbit longer than crews had worked aboard all other US-operated space missions in the previous 42 years, including the shuttle's 100+ flights. Clearly, the training model had to change.

Shuttle operations were like a decathlon of back-to-back sporting events—all intense, all difficult, and all in a short period of time—while space station operations were more like an ongoing trek of many months, requiring a different kind of stamina. ISS used the "surge" of specialized training by the shuttle crews to execute most of the specialized extravehicular activities (EVAs) to assemble the vehicle. The station crew training schedule focused on the necessary critical-but-general skills to deal with general trekking as well as a few planned specific tasks for that expedition. Only rarely did ISS crews take on major assembly tasks in the period between shuttle visits (known in the ISS Program as "the stage").

Another key in the mission scripting and training problem was to consider when and how that "surge capability" could be requested of the ISS crew. That all depended on how long that crew would be expected to work at the increased pace, and how much rest the crew members had had before that period. Nobody can keep competing in decathlons day after day; however, such periodic surges were needed and would need to be compensated by periodic holidays and recovery days.

Humans need a balanced workday with padding in the schedule to freshen up after sleep, read the morning news, eat, exercise, sit back with a good movie, write letters, create, and generally relax before sleep, which should be a minimum of 8 hours per night for long-term health. The Russians had warned eager US mission planners that their expectations of 10 hours of productive work from every crew member every day, 6 days per week was unrealistic. A 5-day workweek with 8-hour days (with breaks), plus periodic holidays, was more like it.

Different Attitude and Planning of Timelines

The ISS plan eventually settled in exactly as the veteran Russian planners had recommended. That is not to say that ISS astronauts took all the time made available to them for purely personal downtime. These are some of the galaxy's most motivated people, so several "unofficial" ways evolved to let them contribute to the program beyond the scripted activities, but only on a voluntary basis.

The ISS planners ultimately learned one productivity technique from the Russians and the crews invented another. At the Russians' suggestion, the ground added a "job jar" of tasks with no particular deadline. These tasks could occupy the crew's idle hours. If a job-jar item had grown too stale and needed doing soon, it found its way onto the short-term plan. Otherwise, the job jar (in reality, a computer file of good "things to do") was a useful means to keep the crew busy during off-duty time. The crew was inventive, even adding new education programs to such times.

Tasks vs. Skills

Generally, training for both the ground and the crew was skills oriented for station operations and task oriented for shuttle operations. The trainers grew to rely on electronic file transfers of intricate procedures, especially videos, to provide specialized training on demand. These were played on on-board notebook computers for the station crew but occasionally for the shuttle crews as well. This training was useful in executing large tasks on the slip schedule, unscheduled maintenance, or on contingency EVAs scheduled well after the crew arrival on station.

Station crews worked on generic EVA skills, component replacement techniques, maintenance tasks, and general robotic manipulation skills. Many systems-maintenance skills needed to be mastered for such a huge "built environment." The station systems needed to closely replicate a natural existence on Earth, including air and water revitalization, waste management, thermal and power control, exercise, communications and computers, and general cleaning and organizing.

The 363-metric-ton (400-ton) ISS had a lot of hardware in need of routine inspection and maintenance that, in shuttle experience, was the job of ground technicians—not astronauts. These systems were the core focus of ISS training. There were multiple languages and cultures to consider (most crew members were multilingual) and usually two types of everything: two oxygen generators; two condensate collectors; two carbon dioxide separators; multiple water systems; different computer architectures; and even different food rations. Each ISS crew member then trained extensively for the specific payloads that would be active during his or her stay on orbit. Scores of payloads needed operators and human subjects. Thus, it took about 3 years to prepare an astronaut for long-duration flight.

Major Missions of Shuttle Support

By May 2010, the shuttle had flown 34 missions to the International Space Station (ISS). Although no human space mission can be called "routine," some missions demonstrated particular strengths of the shuttle and her crews—sometimes in unplanned heroics. A few such missions are highlighted to illustrate the high drama and extraordinary achievement of the shuttle's 12-year construction of the ISS.

STS-88—The First Big Step

The shuttle encountered the full suite of what would soon be routine challenges during its first ISS assembly mission—Space Transportation System (STS)-88 (1998). The narrow launch window required a launch in the middle of the night. This required a huge sleep shift. The cargo element (Node 1 with two of the three pressurized mating adapters already attached) needed to be warmed in the payload bay for hours before launch to survive until the heaters could be activated after the first extravehicular activity (EVA). The rendezvous was conducted with the cargo already erected in a 12-m (39-ft) tower above the Orbiter docking mechanism. This substantially changed the flight characteristics of the shuttle and blocked large sections of the sky as seen from the Orbiter's high-gain television antenna.

The rendezvous required the robotic capture of the Russian-American bridge module: the FGB named Zarya. (Zarya is Russian for "sunrise." "FGB" is a Russian acronym for the generic class of spacecraft—a Functional Cargo Block—on which the Zarya had been slightly customized.) Due to the required separation of the robotic capture of the FGB from the shuttle's cargo element, Space Shuttle Endeavour needed to extend its arm nearly to its limit just to reach the free-flying FGB. Even so, the arm could only touch Zarya's forward end.

In the shuttle's first assembly act of the ISS Program, Astronaut Nancy Currie grappled the heaviest object the Shuttle Robotic Arm had ever manipulated, farther off-center than any object had ever been manipulated. Because of the blocked view of the payload bay (obstructed by Node 1 and the Pressurized Mating Adapter 2), she completed this grapple based on television cues alone—another first.

After the FGB was positioned above the top of the cargo stack, the shuttle used new software to accommodate the large oscillations that resulted from the massive off-center object as it moved. Next, the shuttle crew reconnected the Androgynous Peripheral Docking System control box to a second Androgynous Peripheral Docking System cable set and prepared to drive the interface between the Pressurized Mating Adapter 1 and the FGB. Finally, Currie limped the manipulator arm while Commander Robert Cabana engaged Endeavour's thrusters and flew the Androgynous Peripheral Docking System halves together. The successful mating was followed by a series of three EVAs to link the US and Russian systems together and to deploy two stuck Russian antennas.

This process required continuous operation from two control centers, as had been practiced during the Mir Phase I Program.

Before departing, the shuttle (with yet another altitude-control software configuration) provided a substantial reboost to the fledgling ISS. At a press conference prior to the STS-88 mission, Lead Flight Director Robert Castle called it "…the most difficult mission the shuttle has ever had to fly, and the simplest of all the missions it will have to do in assembling the ISS." He was correct. The shuttle began an ambitious series of firsts, expanding its capabilities with nearly every assembly mission.

STS-97—First US Solar Arrays

STS-97 launched in November 2000 with one of its heaviest cargos: the massive P6 structural truss; three radiators; and two record-setting solar array wings. At nearly 300 m^2 (3,229 ft^2) each, the solar wings could each generate more power than any spacecraft in history had ever used.

After docking in an unusual-but-necessary approach corridor that arrived straight up from below the ISS, Endeavour and her US/Canadian crew gingerly placed the enormous mast high above the Orbiter and seated it with the first use of the Segment-to-Segment Attachment System.

The first solar wing began to automatically deploy as scheduled, just as the new massive P6 structure began to block the communications path to the Tracking and Data Relay Satellites. The software dutifully switched off the video broadcast so as not to beam high-intensity television signals into the structure. When the video resumed, ground controllers saw a disturbing "traveling wave" that violently shook the thin wing as it unfolded. Later, it was determined that lubricants intended to assist in deployment instead added enough surface tension to act as a delicate adhesive. This subtle sticking kept the fanfolds together in irregular clumps rather than letting them gracefully unfold out of the storage box. The clumps would be carried outward in the blanket and then would release rapidly when tension built up near the final tensioning of the array.

Robert Cabana

Colonel, US Marine Corps (retired).
Pilot on STS-41 (1990) and STS-53 (1992).
Commander on STS-65 (1994) and STS-88 (1998).

Reflections on the International Space Station

"Of all the missions that have been accomplished by the Space Shuttle, the assembly of the International Space Station (ISS) certainly has to rank as one of the most challenging and successful. Without the Space Shuttle, the ISS would not be what it is today. It is truly a phenomenal accomplishment, especially considering the engineering challenge of assembling hardware from all parts of the world, on orbit, for the first time and having it work. Additionally, the success is truly amazing when one factors in the complexity of the cultural differences between the European Space Agency and all its partners, Canada, Japan, Russia, and the United States.

"When the Russian Functional Cargo Block, also known as Zarya, which means sunrise in Russian, launched on November 20, 1998, it paved the way for the launch of Space Shuttle Endeavour carrying the US Node 1, Unity. The first assembly mission had slipped almost a year, but in December 1998, we were ready to go. Our first launch attempt on December 3 was scrubbed after counting down to 18 seconds due to technical issues with the Auxiliary Power Units. It was a textbook count for the second attempt on the night of December 4, and Endeavour performed flawlessly.

"Nancy Currie carefully lifted Unity out of the bay and we berthed it to Endeavour's docking system with a quick pulse of our engines once it was properly positioned. With that task complete, we set off for the rendezvous and capture of Zarya. The handling qualities of the Orbiter during rendezvous and proximity operations are superb and amazingly precise. Once stabilized and over a Russian ground site, we got the 'go' for grapple, and Nancy did a great job on the arm capturing Zarya and berthing it to Unity high above the Orbiter. This was the start of the ISS, and it was the shuttle, with its unique capabilities, that made it all possible.

"On December 10, Sergei Krikalev and I entered the ISS for the first time. What a unique and rewarding experience it was to enter this new outpost side by side. It was a very special 2 days that we spent working inside this fledgling space station.

Robert Cabana (left), mission commander, and Sergei Krikalev, Russian Space Agency mission specialist, helped install equipment aboard the Russian-built Zarya module and the US-built Unity module.

"We worked and talked late into the night about what this small cornerstone would become and what it meant for international cooperation and the future of exploration beyond our home planet. I made the first entry into the log of the ISS that night, and the whole crew signed it the next day. It is an evening I'll never forget.

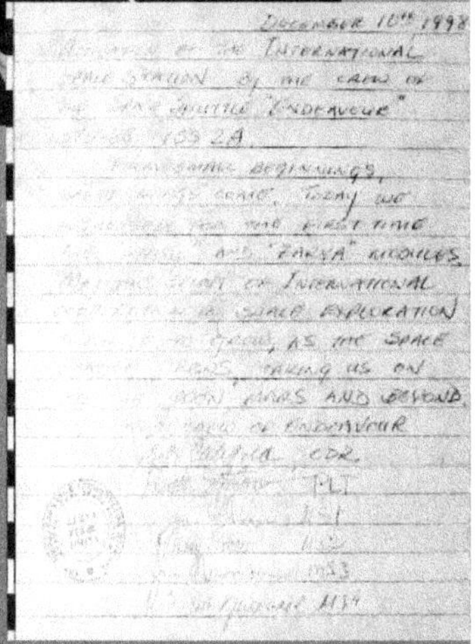

"Since that flight, the ISS has grown to reach its full potential as a world-class microgravity research facility and an engineering proving ground for operations in space. As it passes overhead, it is the brightest star in the early evening and morning skies and is a symbol of the preeminent and unparalleled capabilities of the Space Shuttle."

Psychological Support—

Lessons From Shuttle-Mir to International Space Station

Using crew members' experiences from flying on Mir long-duration flights, NASA's medical team designed a psychological support capability. The Space Shuttle began carrying psychological support items to the International Space Station (ISS) from the very beginning. Prior to the arrival of the Expedition 1 crew, STS-101 (2000) and STS-106 (2000) pre-positioned crew care packages for the three crew members. Subsequently, the shuttle delivered 36 such packages to the ISS. The shuttle transported approximately half of all the packages that were sent to the ISS during that era. The contents were tailored to the individual (and crew). Packages contained music CDs, DVDs, personal items, cards, pictures, snacks, specialty foods, sauces, holiday decorations, books, religious supplies, and other items.

The shuttle delivered a guitar (STS-105 [2001]), an electronic keyboard (STS-108 [2001]), a holiday tree (STS-112 [2002]), external music speakers (STS-116 [2006]), numerous crew personal support drives, and similar nonwork items. As communications technology evolved, the shuttle delivered key items such as the Internet Protocol telephones.

The shuttle also brought visitors and fellow space explorers to the dinner table of the ISS crews. In comparison to other vehicles that visited the space station, the shuttle was self-contained. It was said that when the shuttle visited, it was like having your family pull up in front of your home in their RV—they arrived with their own independent sleeping quarters, galley, food, toilet, and electrical power. This made a shuttle arrival a very welcome thing.

The deployment was stopped and a bigger problem became apparent. The wave motion had dislodged the key tensioning cable from its pulley system and the array could not be fully tensioned. The scenario was somewhat like a huge circus tent partially erected on its poles, with none of the ropes pulled tight enough to stretch the tent into a strong structure. The whole thing was in danger of collapsing, particularly if the shuttle fired jets to leave. Rocket plumes would certainly collapse the massive wings. If Endeavour left without tensioning the array, another shuttle might never be able to arrive unless the array was jettisoned.

Within hours, several astronauts and engineers flew to Boeing Rocketdyne in Canoga Park, California, to develop special new EVA techniques with the spare solar wing. A set of tools and at least three alternate plans were conceived in Houston, Texas, and in California. By the time the crew woke up the next morning, a special EVA had been scripted to save the array. Far beyond the reach of the Shuttle Robotic Arm, astronauts Joseph Tanner and Carlos Noriega crept slowly along the ISS to the array base and gently rethreaded the tension cable back onto the pulleys. They used techniques developed overnight in California that were relayed in the form of video training to the on-board notebook computers.

Meanwhile, engineers rescripted the deployment of the second wing to minimize the size of the traveling waves. The new procedures worked. As STS-97 departed, the ISS had acquired more electric power than any prior spacecraft and was in a robust configuration, ready to grow.

STS-100—An Ambitious Agenda, and an Unforeseen Challenge

STS-100 launched with a four-nation crew in April 2001 to deliver the Space Station Robotic Arm and the Raffaello Italian logistics module with major experiments and supplies for the new US Destiny laboratory, which had been delivered in February. The Space Station Robotic Arm deployed worked well, guided by Canada's first spacewalker, Chris Hadfield. Hadfield reconnected a balky power cable at the base of the Space Station Robotic Arm to give the arm the required full redundancy.

Raffaello, the Italian logistics module, flies in the payload bay on STS-100 in 2001.

Raffaello was successfully berthed and the mission went smoothly until a software glitch in the evolving ISS computer architecture brought all ISS communications to a halt, along with the capability of the ground to command and control the station. Coordinating through the shuttle's communications systems, the station, shuttle, and ground personnel organized a dramatic restart of the ISS.

A major control computer was rebuilt using a payload computer's hard drive, while the heartbeat of the station was maintained by a tiny piece of rescue software—appropriately called "Mighty Mouse"—in the lowest-level computer on the massive spacecraft. Astronaut Susan Helms directly commanded the ISS core computers through a notebook computer. That job was normally assigned to Mission Control. Having rescued the ISS computer architecture, the ISS crew inaugurated the new Space Station Robotic Arm by using it to return its own delivery pallet to Endeavour's cargo bay. Through a mix of intravehicular activity, EVA, and robotic techniques shared across four space agencies, the ISS and Endeavour each ended the ambitious mission more capable than ever.

STS-120—Dramatic Accomplishments

By 2007, with the launch of STS-120, ISS construction was in its final stages. Crew members encountered huge EVA tasks in several previous flights, usually dealing with further problems in balky ISS solar arrays. A severe Russian computer issue had occurred during flight STS-117 in June of that year, forcing an international problem resolution team to spring into action while the shuttle took over attitude control of the station.

STS-120, however, was to be one for the history books. It was already historic in that by pure coincidence both the shuttle and the station were commanded by women. Pamela Melroy commanded Space Shuttle Discovery and Peggy Whitson commanded the ISS. Further, the Harmony connecting node would need to be relocated during the stage in a "must succeed" EVA. During that EVA, the ISS would briefly be in an interim configuration where the shuttle could not dock to the ISS. On this flight, the ISS would finally achieve the full complement of solar arrays and reach its full width.

Shortly after the shuttle docked, the ISS main array joint on the starboard side exhibited a problem that was traced to crushed metal grit from improperly treated bearing surfaces that fouled the whole mechanism. While teams worked to replan the mission to clean and lubricate this critical joint, a worse problem came up. The outermost solar array ripped while it was being deployed. The wing could not be retracted or further deployed without sustaining greater damage. It would be destroyed if the shuttle tried to leave. The huge Space Station Robotic Arm could not reach the distant tear, and crews could not safely climb on the 160-volt array to reach the tear.

In an overnight miracle of cooperation, skill, and ingenuity, ISS and shuttle engineers developed a plan to extend the Space Station Robotic Arm's reach using the Orbiter Boom Sensor System with an EVA astronaut on the end. The use of the boom on the shuttle's arm for contingency EVA had been

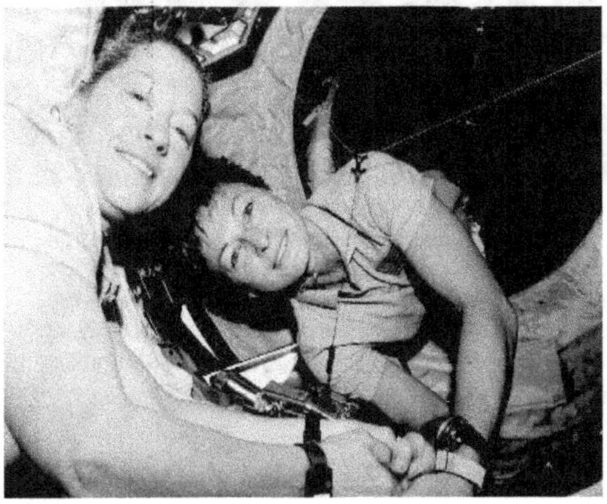

Astronaut Pamela Melroy (left), STS-120 (2007) commander, and Peggy Whitson, Expedition 16 commander, pose for a photo in the Pressurized Mating Adapter of the International Space Station as the shuttle crew members exit the station to board Discovery for their return trip home.

While anchored to a foot restraint on the end of the Orbiter Boom Sensor System, Astronaut Scott Parazynski, STS-120 (2007), assesses his repair work as the solar array is fully deployed during the mission's fourth session of extravehicular activity while Discovery is docked with the International Space Station. During the 7-hour, 19-minute spacewalk, Parazynski cut a snagged wire and installed homemade stabilizers designed to strengthen the damaged solar array's structure and stability in the vicinity of the damage. Astronaut Douglas Wheelock (not pictured) assisted from the truss by keeping an eye on the distance between Parazynski and the array.

validated on the previous flight. The new technique using the Space Station Robotic Arm and boom would barely reach the damaged area with the tallest astronaut in the corps—Scott Parazynski—at its tip in a portable foot restraint. This technique came with the risk of potential freezing damage to some instruments at the end of the Orbiter Boom Sensor System. Overnight, Commander Whitson and STS-120 Pilot George Zamka manufactured special wire links that had been specified to the millimeter in length by ground crews working with a spare array.

In one of the most dramatic repairs (and memorable images) in the history of spaceflight, Parazynski, surrounded by potentially lethal circuits, rode the boom and arm combination on a record-tying fifth single-mission EVA to the farthest edge of the ISS. Once there, he carefully "stitched" the vast array back into perfect shape and strength with the five space-built links.

These few selected vignettes cannot possibly capture the scope of the ISS assembly in the vacuum of space. Each shuttle mission brought its own drama and its own major contributions to the ISS Program, culminating in a new colony in space, appearing brighter to everyone on Earth than any planet. This bright vision would never have been possible without the close relationship—and often unprecedented cooperative problem solving—that ISS enjoyed with its major partner from Earth.

The International Space Station and Space Shuttle Endeavour, STS-135 (2011)—as photographed by European Space Agency astronaut Paolo Nespoli from aboard the Russian Soyuz spacecraft—following completion of space station assembly.

Summary

When humans learn how to manipulate any force of nature, it is called "technology," and technology is the fabric of the modern world and its economy. One such force—gravity—is now known to affect physics, chemistry, and biology more profoundly than the forces that have previously changed humanity, such as fire, wind, electricity, and biochemistry. Humankind's achievement of an international, permanent platform in space will accelerate the creation of new technologies for the cooperating nations that may be as influential as the steam engine, the printing press, and fire. The shuttle carried the modules of this engine of invention, assembled them in orbit, provided supplies and crews to maintain it, and even built the original experience base that allowed it to be designed.

Over the 12 years of coexistence, and even further back in the days when the old Freedom design was first on the drawing board, the International Space Station (ISS) and Space Shuttle teams learned a lot from each other, and both teams and both vehicles grew stronger as a result. Like a parent and child, the shuttle and station grew to where the new generation took up the journey while the accomplished veteran eased toward retirement.

The shuttle's true legacy does not live in museums. As visitors to these astounding birds marvel up close at these engineering masterpieces, they need only glance skyward to see the ongoing testament to just a portion of the shuttles' achievements. In many twilight moments, the shuttle's greatest single payload and partner—the stadium-sized ISS—flies by for all to see in a dazzling display that is brighter than any planet.

Engineering Innovations

Propulsion

Thermal Protection Systems

Materials and Manufacturing

Aerodynamics and Flight Dynamics

Avionics, Navigation, and Instrumentation

Software

Structural Design

Robotics and Automation

Systems Engineering for Life Cycle of Complex Systems

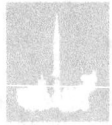

Propulsion

Introduction
Yolanda Harris

Space Shuttle Main Engine
Fred Jue

Chemochromic Hydrogen Leak Detectors
Luke Roberson
Janine Captain
Martha Williams
Mary Whitten

The First Human-Rated Reuseable Solid Rocket Motor
Fred Perkins
Holly Lamb

Orbital Propulsion Systems
Cecil Gibson
Willard Castner
Robert Cort
Samuel Jones

Pioneering Inspection Tool
Mike Lingbloom

Propulsion Systems and Hazardous Gas Detection
Bill Helms
David Collins
Ozzie Fish
Richard Mizell

The launch of the Space Shuttle was probably the most visible event of the entire mission cycle. The image of the Main Propulsion System—the Space Shuttle Main Engine and the Solid Rocket Boosters (SRBs)—powering the Orbiter into space captured the attention and the imagination of people around the globe. Even by 2010 standards, these main engines' performance was unsurpassed compared to any other engines. They were a quantum leap from previous rocket engines. The main engines were the most reliable and extensively tested rocket engine before and during the shuttle era.

The shuttle's SRBs were the largest ever used, the first reusable rocket, and the only solid fuel certified for human spaceflight. This technology, engineering, and manufacturing may remain unsurpassed for decades to come.

But the shuttle's propulsion capabilities also encompassed the Orbiter's equally important array of rockets—the Orbital Maneuvering System and the Reaction Control System—which were used to fine-tune orbits and perform the delicate adjustments needed to dock the Orbiter with the International Space Station. The design and maintenance of the first reusable space vehicle—the Orbiter—presented a unique set of challenges. In fact, the Space Shuttle Program developed the world's most extensive materials database for propulsion. In all, the shuttle's propulsion systems achieved unprecedented engineering milestones and launched a 30-year era of American space exploration.

Space Shuttle Main Engine

NASA faced a unique challenge at the beginning of the Space Shuttle Program: to design and fly a human-rated reusable liquid propulsion rocket engine to launch the shuttle. It was the first and only liquid-fueled rocket engine to be reused from one mission to the next during the shuttle era. The improvement of the Space Shuttle Main Engine (SSME) was a continuous undertaking, with the objectives being to increase safety, reliability, and operational margins; reduce maintenance; and improve the life of the engine's high-pressure turbopumps.

The reusable SSME was a staged combustion cycle engine. Using a mixture of liquid oxygen and liquid hydrogen, the main engine could attain a maximum thrust level (in vacuum) of 232,375 kg (512,300 pounds), which is equivalent to greater than 12,000,000 horsepower (hp). The engine also featured high-performance fuel and oxidizer turbopumps that developed 69,000 hp and 25,000 hp, respectively. Ultra-high-pressure operation of the pumps and combustion chamber allowed expansion of hot gases through the exhaust nozzle to achieve efficiencies never previously attained in a rocket engine.

Requirements established for Space Shuttle design and development began in the mid 1960s. These requirements called for a two-stage-to-orbit vehicle configuration with liquid oxygen (oxidizer) and liquid hydrogen (fuel) for the Orbiter's main engines. By 1969, NASA awarded advanced engine studies to three contractor firms to further define designs necessary to meet the leap in performance demanded by the new Space Transportation System (STS).

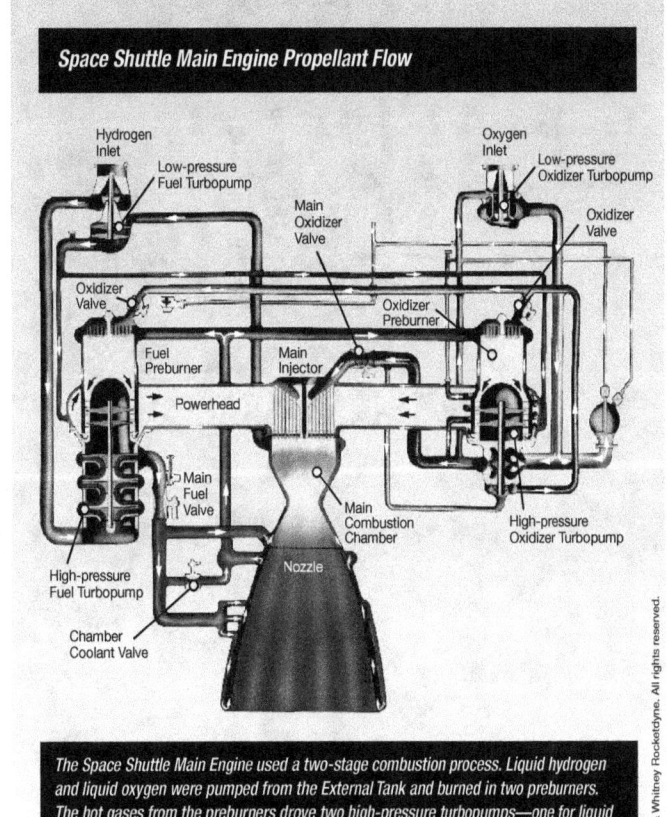

Space Shuttle Main Engine Propellant Flow

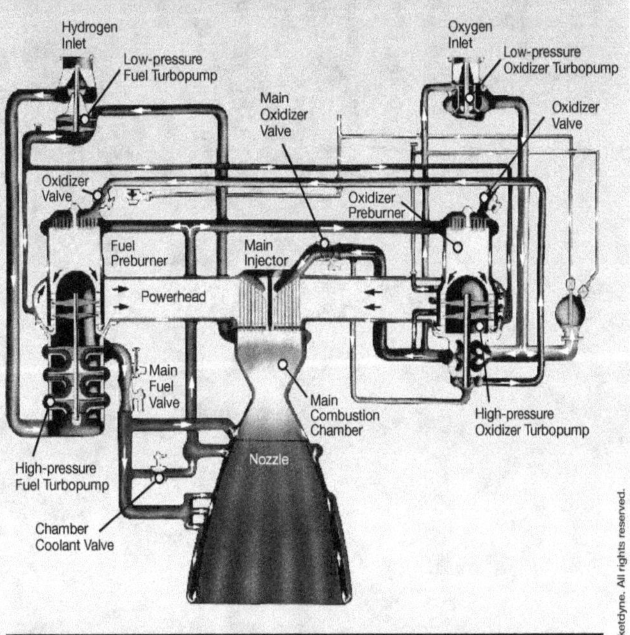

The Space Shuttle Main Engine used a two-stage combustion process. Liquid hydrogen and liquid oxygen were pumped from the External Tank and burned in two preburners. The hot gases from the preburners drove two high-pressure turbopumps—one for liquid hydrogen (fuel) and one for liquid oxygen (oxidizer).

In 1971, the Rocketdyne division of Rockwell International was awarded a contract to design, develop, and produce the main engine.

The main engine would be the first production-staged combustion cycle engine for the United States. (The Soviet Union had previously demonstrated the viability of staged combustion cycle in the Proton vehicle in 1965.) The staged combustion cycle yielded high efficiency in a technologically advanced and complex engine that operated at pressures beyond known experience.

The design team chose a dual-preburner powerhead configuration to provide precise mixture ratio and throttling control. A low- and high-pressure turbopump, placed in series for each of the liquid hydrogen and liquid oxygen loops, generated high pressures across a wide range of power levels.

A weight target of 2,857 kg (6,300 pounds) and tight Orbiter ascent envelope requirements yielded a compact design capable of generating a nominal chamber pressure of 211 kg/cm^2 (3,000 pounds/in^2)—about four times that of the Apollo/Saturn J-2 engine.

Michael Coats
Pilot on STS-41D (1984). Commander on STS-29 (1989) and STS-39 (1991).

A Balky Hydrogen Valve Halts Discovery Liftoff

"I had the privilege of being the pilot on the maiden flight of the Orbiter Discovery, a hugely successful mission. We deployed three large communications satellites and tested the dynamic response characteristics of an extendable solar array wing, which was a precursor to the much-larger solar array wings on the International Space Station.

"But the first launch attempt did not go quite as we expected. Our pulses were racing as the three main engines sequentially began to roar to life, but as we rocked forward on the launch pad it suddenly got deathly quiet and all motion stopped abruptly. With the seagulls screaming in protest outside our windows, it dawned on us we weren't going into space that day. The first comment came from Mission Specialist Steve Hawley, who broke the stunned silence by calmly saying 'I thought we'd be a lot higher at MECO (main engine cutoff).' So we soon started cracking lousy jokes while waiting for the ground crew to return to the pad and open the hatch. The joking was short-lived when we realized there was a residual fire coming up the left side of the Orbiter, fed from the same balky hydrogen valve that had caused the abort. The Launch Control Center team was quick to identify the problem and initiated the water deluge system designed for just such a contingency. We had to exit the pad elevator through a virtual wall of water. We wore thin, blue cotton flight suits back then and were soaked to the bone as we entered the air-conditioned astronaut van for the ride back to crew quarters. Our drenched crew shivered and huddled together as we watched the Discovery recede through the rear window of the van, and as Mike Mullane wryly observed, 'This isn't exactly what I expected spaceflight to be like.' The entire crew, including Commander Henry Hartsfield, the other Mission Specialists Mike Mullane and Judy Resnik, and Payload Specialist Charlie Walker, contributed to an easy camaraderie that made the long hours of training for the mission truly enjoyable."

For the first time in a boost-to-orbit rocket engine application, an on-board digital main engine controller continuously monitored and controlled all engine functions. The controller initiated and monitored engine parameters and adjusted control valves to maintain the performance parameters required by the mission. When detecting a malfunction, it also commanded the engine into a safe lockup mode or engine shutdown.

Design Challenges

Emphasis on fatigue capability, strength, ease of assembly and disassembly, maintainability, and materials compatibility were all major considerations in achieving a fully reusable design.

Specialized materials needed to be incorporated into the design to meet the severe operating environments. NASA successfully adapted advanced alloys, including cast titanium, Inconel® 718 (a high-strength, nickel-based superalloy used in the main combustion chamber support jacket and powerhead), and NARloy-Z (a high-conductivity, copper-based alloy used as the liner in the main combustion chamber). NASA also oversaw the development of single-crystal turbine blades for the high-pressure turbopumps. This innovation essentially eliminated the grain boundary separation failure mechanism (blade cracking) that had limited the service life of the pumps. Nonmetallic materials such as Kel-F® (a plastic used in turbopump seals), Armalon® fabric (turbopump bearing cage material), and P5N carbon-graphite seal material were also incorporated into the design.

Material sensitivity to oxygen environment was a major concern for compatibility due to reaction and

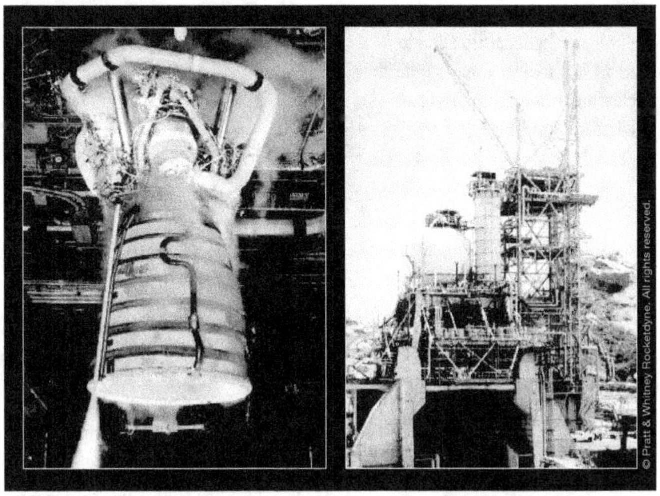

A 1970s-era Space Shuttle Main Engine undergoes testing at Rocketdyne's Santa Susana Field Laboratory near Los Angeles, California.

ignition under the high pressures. Mechanical impact testing had vastly expanded in the 1970s to accommodate the shuttle engine's varied operating conditions. This led to a new class of liquid oxygen reaction testing up to 703 kg/cm^2 (10,000 pounds/in^2).

Engineers also needed to understand long-term reaction to hydrogen effects to achieve full reusability. Thus, a whole field of materials testing evolved to evaluate the behavior of hydrogen charging on all affected materials.

NASA developed new tools to accomplish design advancements. Engineering design tools advanced along with the digital age as analysis migrated from the mainframe platform to workstations and desktop personal computers. Fracture mechanics and fracture control became critical tools for understanding the characteristics of crack propagation to ensure design reusability. As the analytical tools and processor power improved over the decades, cycle time for engineering analysis such as finite element models, computer-aided design and manufacturing, and computational fluid dynamics dropped from days to minutes. Real-time engine performance analyses were conducted during ground tests and flights at the end of the shuttle era.

Development and Certification

The shuttle propulsion system was the most critical system during ascent; therefore, a high level of testing was needed prior to first flight to demonstrate engine maturity. Component-level testing of the preburners and thrust chamber began in 1974 at Rocketdyne's Santa Susana Field Laboratory in Southern California.

The first engine-level test of the main engine—the Integrated Subsystem Test Bed—occurred in 1975 at the NASA National Space Technology Laboratory (now Stennis Space Center) in Mississippi and relied on facility controls, as the main engine controller was not yet available.

NASA and Rocketdyne pursued an aggressive test schedule at their respective facilities. Stennis Space Center with three test stands and Rocketdyne with one test stand completed 152 engine tests in 1980 alone—a record that has not been exceeded since. This ramp-up to 100,000 seconds represented a team effort of personnel and facilities to overachieve a stated development goal of 65,000 seconds set by then-Administrator John Yardley as the maturity level deemed flightworthy. NASA verified operation at altitude conditions and also demonstrated the rigors of sea-level performance and engine gimballing for thrust vector control. The Rocketdyne laboratory supplemented sea-level testing as well as deep throttling by using a low 33:1 expansion ratio nozzle. This testing was crucial in identifying shortcomings related to the initial design of the high-pressure turbopumps, powerhead, valves, and nozzles.

Extensive margin testing beyond the normal flight envelope—including high-power, extended-duration tests and near-depleted inlet propellant conditions to simulate the effects of microgravity—provided further confidence in the design. Engineers subjected key components to a full series of design verification tests, some with intentional hardware defects, to validate safety margins should the components develop undetected flaws during operation.

NASA and Rocketdyne also performed system testing to replicate the three engine cluster interactions with the Orbiter. The Main Propulsion Test Article consisted of an Orbiter aft fuselage, complete with full thrust structure, main propulsion electrical and system plumbing, External Tank, and three main engines. To validate that the Main Propulsion System was ready for launch, engineers completed 18 tests at the National Space Technology Laboratory by 1981.

The completion of the main engine preliminary flight certification in March 1981 marked a major milestone in clearing the initial flights at 100% rated power level.

Design Evolutions

A major requirement in engine design was the ability to operate at various power levels. The original engine life requirement was 100 nominal missions and 27,000 seconds (7.5 hours) of engine life. Nominal thrust, designated as rated power level, was 213,189 kg (470,000 pounds) in vacuum. The life requirement included six exposures at the emergency power level of 232,375 kg (512,300 pounds), which was designated 109% of rated power level. To maximize the number of missions possible at emergency power level, an assessment of the engine capability resulted in reducing the number of nominal missions per engine to 55 missions at 109%. Emergency power level was subsequently renamed full power level.

Ongoing ascent trajectory analysis determined 65% of rated power level to be sufficient to power the vehicle through its period of maximum aerodynamic pressure during ascent. Minimum power level was later refined upward to 67%.

On April 12, 1981, Space Shuttle Columbia lifted off Launch Pad 39A from Kennedy Space Center in Florida on its maiden voyage. The first flight configuration engines were aptly named the First Manned Orbital Flight SSMEs. These engines were flown during the initial five shuttle development missions at 100% rated power level thrust. Work done to prepare for the next flight validated the ability to perform routine engine maintenance without removing them from the Orbiter.

The successful flight of STS-1 initiated the development of a full-power (109% rated power level) engine. The higher thrust capability was needed to support an envisioned multitude of NASA, commercial, and Department of Defense payloads, especially if the shuttle was launched from the West Coast. By 1983, however, test failures demonstrated the basic engine lacked margin to continuously operate at 109% thrust, and full-power-level development was halted. Other engine improvements were implemented into what was called the Phase II engine. During this period, the engine program was restructured into two programs—flight and development.

Post-Challenger Return to Flight

The 1986 Challenger accident provoked fundamental changes to the shuttle, including an improved main engine called Phase II. This included changes to the high-pressure turbopumps and main combustion chamber, avionics, valves, and high-pressure fuel duct insulation. An additional 90,241 seconds of engine testing accrued, including recertification to 104% rated power level.

The new Phase II engine continued to be the workhorse configuration for shuttle launches up to the late 1990s while additional improvements envisioned during the 1980s were undergoing development and flight certification for later incorporation. NASA targeted five major components for advanced development to further enhance safety and reliability, lower recurring costs, and increase performance capability. These components included the powerhead, heat exchanger, main combustion chamber, and high-pressure oxidizer and fuel turbopumps.

These major changes would later be divided into two "Block" configuration upgrades, with Rocketdyne tasked to improve the powerhead, heat exchanger, and main combustion chamber while Pratt & Whitney was selected to design, develop, and produce the improved high-pressure turbopumps.

Pratt & Whitney Company of United Technologies began the effort in 1986 to provide alternate high-pressure turbopumps as direct line replaceable units for the main engines. Pratt & Whitney used staged combustion experience from its development of the XLR-129 engine for the US Air Force and cryogenic hydrogen experience from the RL-10 (an upper-stage engine used by NASA, the military, and commercial enterprises) along with SSME lessons learned to design the new pumps. The redesign of the components eliminated critical failure modes and increased safety margins.

Next Generation

The Block I configuration became the successor to the Phase II engine. A new Pratt & Whitney high-pressure oxygen turbopump, an improved two-duct engine powerhead, and a single-tube heat exchanger were introduced that collectively used new design and production processes to eliminate failure causes. Also it increased the inherent reliability and operating margin and reduced production cycle time and costs. This Block I engine first flew on STS-70 (1995).

The powerhead redesign was less risky and was chosen to proceed ahead of the main combustion chamber.

The Technology Test Bed Space Shuttle Main Engine test program was conducted at Marshall Space Flight Center, Alabama, between September 1988 and May 1996. The program demonstrated the ability of the main engine to accommodate a wide variation in safe operating ranges.

The two-duct powerhead eliminated 74 welds and had 52 fewer parts. This improved design led to production simplification and a 40% cost reduction compared to the previous three-duct configuration. The two-duct configuration provided an improvement to the hot gas flow field distribution and reductions in dynamic pressures. The improved heat exchanger eliminated all inter-propellant welds, and its wall thickness was increased by 25% for added margin against penetration by unexpected foreign debris impact.

The new high-pressure oxygen turbopump eliminated 293 welds, added improved suction performance, and introduced a stiff single-piece disk/shaft configuration and thin-cast turbine blades. The oxygen turbopump incorporated silicon nitride (ceramic) ball bearings in a rocket engine application and could be serviced without removal from the engine. Initial component-level testing occurred at the Pratt & Whitney West Palm Beach, Florida, testing facilities. Testing then graduated to the engine level at Stennis Space Center as well as at Marshall Space Flight Center's (MSFC's) Technology Test Bed test configuration.

The large-throat main combustion chamber began prototype testing at Rocketdyne in 1988. But it was not until 1992, after a series of combustion stability tests at the MSFC Technology Test Bed facility, that concerns regarding combustion stability were put to rest. The next improved engine—Block II—incorporated the new high-pressure fuel turbopump, modified low-pressure turbopumps, software operability enhancements, and previous Block I upgrades. These upgrades were needed to support International Space Station (ISS) launches with their heavy payloads beginning in 1998.

As Block II development testing progressed, the engineering accomplishments on the large-throat main combustion chamber matured more rapidly than the high-pressure fuel turbopump.

By February 1997, NASA had decided to go forward with an interim configuration called the Block IIA. Using the existing Phase II high-pressure fuel pump, this configuration would allow early implementation of the large-throat main combustion chamber to support ISS launches. The large-throat main combustion chamber was simpler and producible. The new chamber lowered the engine's operating pressures and temperatures while increasing the engine's operational safety margin. Changes to the low-pressure turbopumps to operate in this derated environment, along with further avionics improvements, were flown in 1998 on STS-89.

The large-throat main combustion chamber became one of the most significant safety improvements for the main engine by effectively reducing operating pressures and temperatures up to 10% for all subsystems. This design also incorporated improved cooling capability for longer life and used high-strength castings, thus eliminating 50 welds.

By the time the first Block IIA flew on STS-89 in January 1998, the large-throat main combustion chamber design had accumulated in excess of 100,000 seconds of testing time. By late 1999, the Block II high-pressure fuel turbopump had progressed into certification testing. The design philosophy mirrored those proven successful in the high-pressure oxidizer turbopump and included the elimination of 387 welds

The Improved Space Shuttle Main Engine Powerhead Component Arrangement for Block II Engines

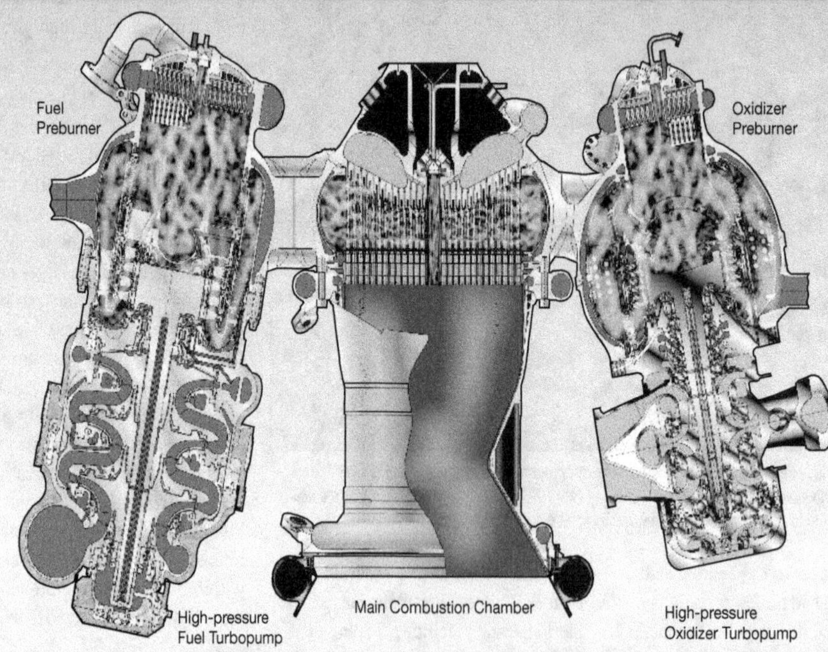

The Block II engine combined a new high-pressure fuel turbopump with the previously flown redesigned high-pressure oxygen turbopump. Risk analysis showed that the Block II engine was twice as safe as the 1990s-era engine. Beginning with STS-110 in April 2002, all shuttle flights were powered by the improved Space Shuttle Main Engine.

and incorporation of a stiff single-piece disk/shaft, thin-cast turbine blades, and a cast pump inlet that improved the suction performance and robustness against pressure surges. As with the high-pressure oxidizer turbopump, the high-pressure fuel turbopump turbine inlet did not require off-engine inspections, which contributed significantly to improving engine turnaround time. The high-pressure fuel turbopump also demonstrated that a turbine blade failure would result in a contained, safe engine shutdown. By introducing the added operational margin of the large-throat main combustion chamber with the new turbopumps, quantitative risk analysis projected that the Block II engine was twice as safe as the Phase II engine.

The first two single-engine flights of Block II occurred on STS-104 and STS-108 in July 2001 and December 2001, respectively, followed by the first three-engine cluster flight on STS-110 in April 2002. The high-pressure fuel turbopump had accumulated 150,843 seconds of engine test maturity at the time of the first flight.

The Block II engine also incorporated the advanced health management system on STS-117 in 2007. This on-board system could detect and mitigate anomalous high-pressure turbopump vibration behavior, and the system further improved engine ascent safety by an additional 23%.

Summary

Another major SSME milestone took place in 2004 when the main engine passed 1,000,000 seconds in test and operating time. This unprecedented level of engine maturity over the preceding 3 decades established the main engine as one of the world's most reliable rocket engines, with a 100% flight safety record and a demonstrated reliability exceeding 0.9996 in over 1,000,000 seconds of hot-fire experience.

Chemochromic Hydrogen Leak Detectors

The Chemochromic Point Detector for sensing hydrogen gas leakage is useful in any application in which it is important to know the presence and location of a hydrogen gas leak.

This technology uses a chemochromic pigment and polymer that can be molded or spun into a rigid or pliable shape useable in variable-temperature environments including atmospheres of inert gas, hydrogen gas, or mixtures of gases. A change in the color of detector material reveals the location of a leak. Benefits of this technology include: temperature stability, from -75°C to 100°C (-103°F to 212°F); use in cryogenic applications; ease of application and removal; lack of a power requirement; quick response time; visual or electronic leak detection; nonhazardous qualities, thus environmentally friendly; remote monitoring capability; and a long shelf life. This technology is also durable and inexpensive.

The detector can be fabricated into two types of sensors—reversible and irreversible. Both versions immediately notify the operator of the presence of low levels of hydrogen; however, the reversible version does not require replacement after exposure. Both versions were incorporated into numerous polymeric materials for specific applications including: extruded tapes for wrapping around valves and joints suspected of leaking; injection-molded parts for seals, O-rings, pipe fittings, or plastic piping material; melt-spun fibers for clothing applications; and paint for direct application to ground support equipment. The versatility of the sensor for several different applications provides the operator with a specific-use safety notification while working under hazardous operations.

Hydrogen-sensing tape applied to the Orbiter midbody umbilical unit during fuel cell loading for STS-118 through STS-123 at Kennedy Space Center, Florida.

Hydrogen-sensing tape application at liquid hydrogen cross-country vent line flanges on the pad slope.

The First Human-Rated Reusable Solid Rocket Motor

The Space Shuttle reusable solid rocket motors were the largest solid rockets ever used, the first reusable solid rockets, and the only solids ever certified for crewed spaceflight. The closest solid-fueled rival—the Titan IV Solid Rocket Motor Upgrade—was known for boosting heavy payloads for the US Air Force and National Reconnaissance Organization. The motors were additionally known for launching the 5,586-kg (12,220-pound) Cassini mission on its 7-year voyage to Saturn. By contrast, the Titan booster was 76 cm (30 in.) smaller in diameter and 4.2 m (14 ft) shorter in length, and held only two-thirds of the amount of propellant.

In a class of its own, the Reusable Solid Rocket Motor Program was characterized from its inception by four distinguishing traits: hardware reusability, postflight recovery and analysis, a robust ground-test program, and a culture of continual improvement via process control.

The challenge NASA faced in developing the first human-rated solid rocket motor was to engineer a pair of solid-fueled rocket motors capable of meeting the rigorous reliability requirements associated with human spaceflight. The rocket motors would have to be powerful enough to boost the shuttle system into orbit. The motors would also need to be robust enough to meet stringent reliability requirements and survive the additional rigors of re-entry into Earth's atmosphere and subsequent splashdown, all while being reusable. The prime contractor—Morton Thiokol, Utah—completed its

The two shuttle reusable solid rocket motors, which stood more than 38 m (126 ft) tall, harnessed 29.4 meganewtons (6.6 million pounds) of thrust. The twin solid-fueled rockets provided 80% of the thrust needed to achieve liftoff.

first full-scale demonstration test within 3 years.

NASA learned a poignant lesson in the value of spent booster recovery and inspection with the Challenger tragedy in January 1986. The postflight condition of the hardware provided valuable information on the health of the design and triggered a redesign effort that surpassed, in magnitude and complexity, the original development program.

For the substantial redesign that occurred between 1986 and 1988, engineers incorporated lessons learned from the first 25 shuttle flight booster sets. More than 100 tests, including five full-scale ground tests, were conducted to demonstrate the strength of the new design. Flaws were deliberately manufactured into the final test motor to check redundant systems.

The redesigned motors flew for the first time in September 1988 and performed flawlessly.

A Proven Design

To construct the reusable solid rocket motor, four cylindrical steel segments—insulated and loaded with a high-performance solid propellant—were joined together to form what was essentially a huge pressure vessel and combustion chamber. The segmented design provided maximum flexibility in motor fabrication, transportation, and handling. Each segment measured 3.7 m (12 ft) in diameter and was forged from D6AC steel measuring approximately 1.27 cm (0.5 in.) in thickness.

Case integrity and strength were maintained during flight by insulating the case interior. The insulating liner was a fiber-filled elastomeric (rubber-like) material applied to the interior of the steel cylinders. A carefully formulated tacky rubber bonding layer—or "liner"—was applied to the rubber insulator surface to facilitate a strong bond with the propellant.

Producing an accurate insulating layer was critical. Too little insulation, and the steel could be heated and melted by the 2,760°C (5,000°F) combustion gases. Too much insulation, and weight requirements were exceeded. Engineers employed sophisticated design analysis and testing to optimize this balance between protection and weight. By design, much of the insulation was burned away during the 2 minutes of motor operation.

The propellant was formulated from three major ingredients: aluminum powder (fuel); ammonium perchlorate (oxidizer); and a synthetic polymer binding agent. The ingredients were batched, fed into large 2,600-L (600-gal) mix bowls, mixed, and tested before being poured into the insulated and lined segments. Forty batches were produced to fill each case segment. The propellant mixture had an initial consistency similar to that of peanut butter, but was cured to a texture and color that resembled a rubber pencil eraser—strong, yet pliable. The propellant configuration or "shape" inside each segment was carefully designed and cast to yield the precise thrust trace upon ignition.

Once each segment was insulated and cast with propellant and finalized, the segments were shipped from ATK's manufacturing facility in Utah to Kennedy Space Center (KSC) in Florida, on specially designed, heavy-duty covered rail cars. At KSC, they were stacked and assembled into the flight configuration.

The segments were joined together with tang/clevis joints pinned in 177 locations and sealed with redundant O-rings. Each joint, with its redundant seals and multiple redundant seal protection features, was pressure checked during assembly to ensure a good pressure seal.

Reusable Solid Rocket Motor Propellant Configuration

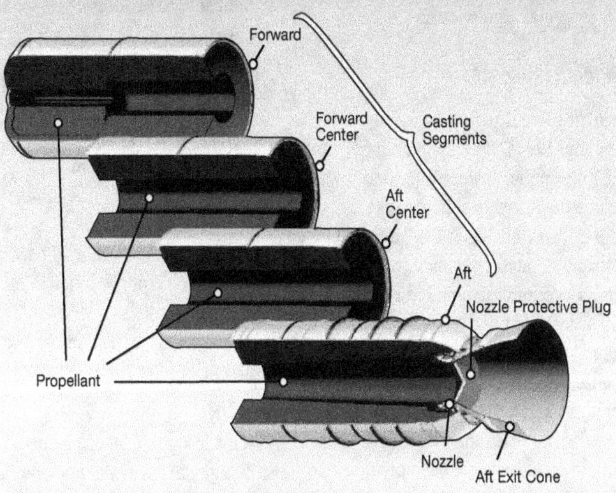

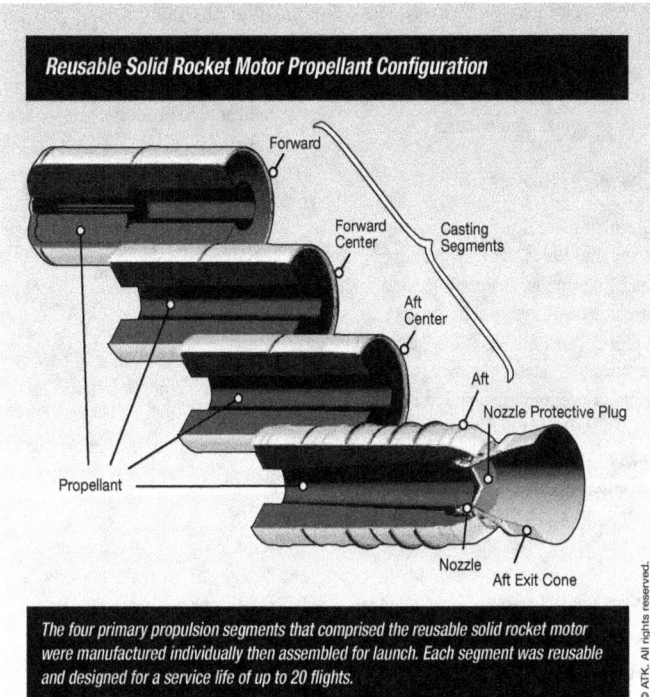

The four primary propulsion segments that comprised the reusable solid rocket motor were manufactured individually then assembled for launch. Each segment was reusable and designed for a service life of up to 20 flights.

Forward Segment Propellant Grain Configuration

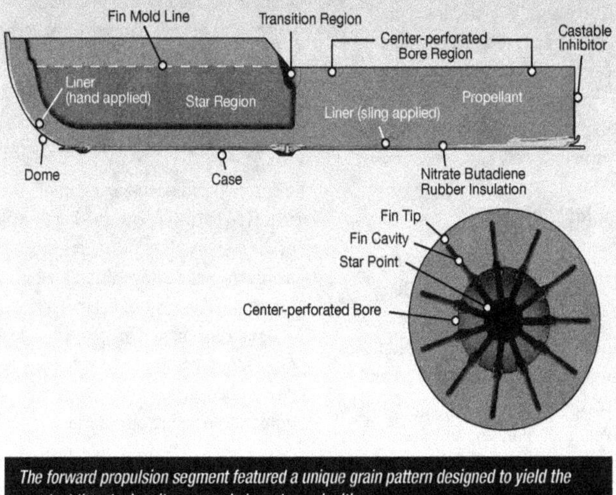

The forward propulsion segment featured a unique grain pattern designed to yield the greatest thrust when it was needed most—on ignition.

An igniter was installed in the forward end of the forward segment—at the top of the rocket. The igniter was essentially a smaller rocket motor that fired into the solid rocket motor to ignite the main propellant grain. Design and manufacture closely mirrored the four main segments.

The nozzle was installed at the aft end of the aft segment, at the bottom of the rocket. The nozzle was the "working" component of the rocket in which hot exhaust gases were accelerated and directed to achieve performance requirements and vehicle control.

The nozzle structure consisted of metal housings over which were bonded layers of carbon/phenolic and silica/phenolic materials that protected the metal structure from the searing exhaust gases by partially decomposing and ablating. A flexible bearing, formed with vulcanized rubber and steel, allowed for nozzle maneuverability up to 8 degrees in any direction to steer the shuttle during the first minutes of flight.

Engineers employed significant analysis and testing to develop a reliable and efficient nozzle capable of being manufactured. The nozzle flexible bearing—measuring up 2.35 m (92.4 in.) at its outside diameter—was an example of one component that required multiple processing iterations to ensure the manufactured product aligned with design requirements.

NASA enhanced the nozzle design following the Challenger accident when severe erosion on one section of the nozzle on one motor was noted through postflight analysis. While the phenolic liners were designed to erode smoothly and predictably, engineers found—at certain ply orientations—that internal stresses resulting from exposure to hot gases exceeded the material strength. Under such stress, the hot charred material had the potential to erode erratically and jeopardize component integrity. Engineers modified nozzle ply angels to reduce material stress, and this condition was successfully eliminated on all subsequent flights.

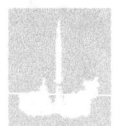

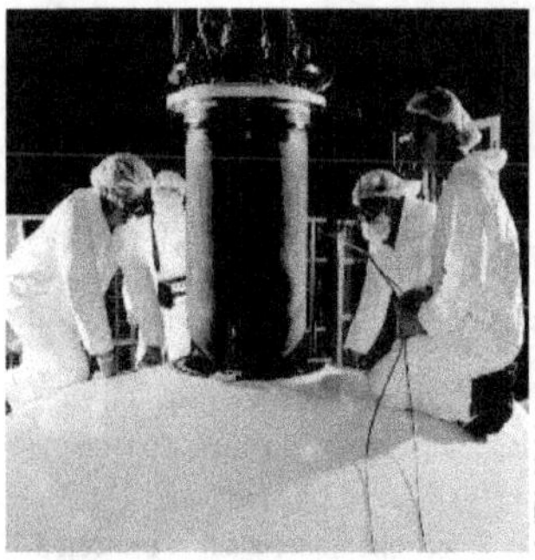

Technicians shown installing igniter used to initiate the propellant burn in a forward motor segment. The Igniter was a small rocket motor loaded with propellant that propagated flame down the bore of the motor.

The Reusable Rocket

All metal hardware—including structures from the case, igniter, safe-and-arm device, and nozzle—were designed to support up to 20 shuttle missions. This was unique to the reusable solid rocket motor. Besides the benefits of conservation and affordability, the ability to recover the motors allowed NASA to understand exactly how the components performed in flight. This performance analysis provided a wealth of valuable information and created a synergy to drive improvements in motor performance, implemented through motor manufacturing and processing.

This recovery and postflight capability was particularly important for the long-term Space Shuttle Program since, over time, changes were inevitable. Change to design or process became mandatory as a result of factors such as material/vendor obsolescence or new environmental regulations.

Changing Processes

During a 10-year period beginning in the mid 1990s, for example, more than 100 supplier materials used to produce the reusable solid rocket motor became obsolete. The largest contributing factor stemmed from supplier economics, captured in three main scenarios. First, suppliers changed their materials or processes. Second, suppliers consolidated operations and either discontinued or otherwise modified their materials. Third, the materials were simply no longer available from subtier vendors.

US environmental regulations, such as the requirement to phase out the use of ozone-depleting chemicals, were an additional factor. Methyl chloroform, for example, was a solvent used extensively in hardware processing. A multimillion-dollar effort was launched within NASA and ATK to eventually eliminate methyl chloroform use altogether in motor processing. Eight alternate materials were selected following thorough testing and analysis to ensure program performance was not compromised.

New Technology

Advancements in technology that occurred during the decades-long program were a further source of change. Engineers incorporated new technologies into motor design and processing as the technology could be proven. Incorporating braided carbon fiber material as a thermal barrier in the nozzle-to-case joint is one example.

Postflight Analysis

The ability to closely monitor flight performance through hands-on postflight analysis—after myriad material, design, and process changes—was only possible by virtue of the motor's reusable nature.

Developing methods to scrutinize and recertify spent rocket motor hardware that had raced through the stratosphere at supersonic speeds was new. NASA had the additional burden of working with components that had experienced splashdown loads and were subsequently soaked in corrosive saltwater prior to retrieval.

In the early days of the program, NASA made significant efforts in identifying relevant evaluation criteria and establishing hardware assessment methods. A failure to detect hardware stresses and material weaknesses could result in an unforgivable catastrophic event later on. The criteria used to evaluate the first motors and the accompanying data collected would also become the benchmark from which future flights would be measured. Included in the evaluation criteria were signs of case damage or material loss caused by external debris; integrity of major components such as case segments, nozzle and igniter; and fidelity of insulation, seals, and joints.

Inspection and documentation of retrieved hardware occurred in two parts of the country: Florida, where the hardware was retrieved; and Utah, where it underwent in-depth inspection and refurbishment. On recovery, a team of 15 motor engineers conducted what was termed an "open assessment," primarily focusing on exterior components. After retrieval, teams of specialists rigorously dissected, measured, sampled, and assessed joints,

Field Joint Comparison for Use on Reusable Solid Rocket Motor

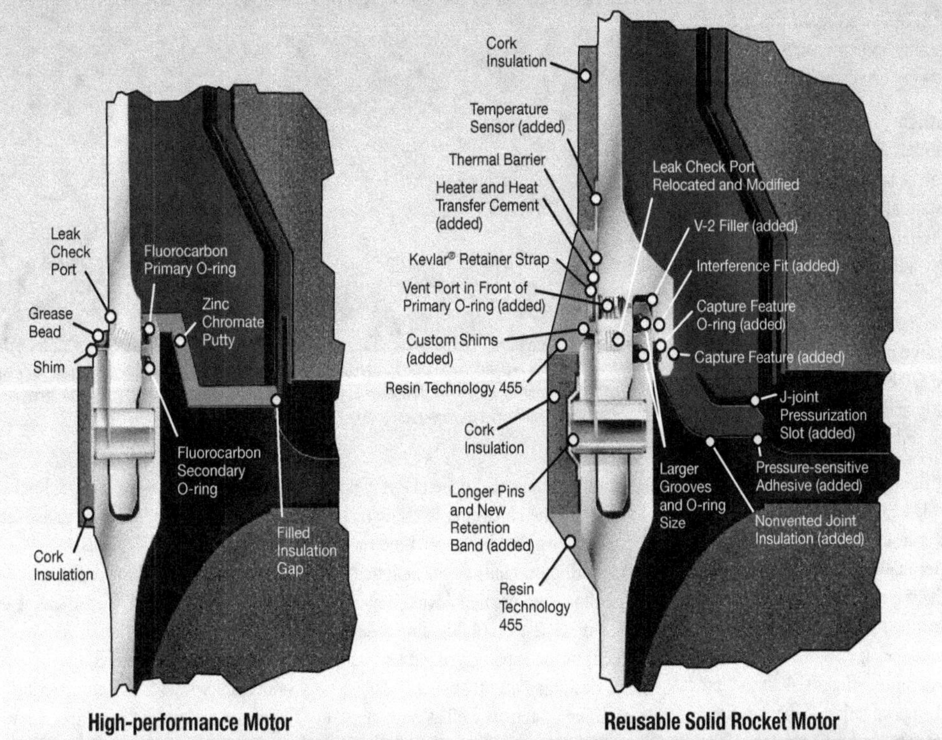

Reusable solid rocket motors incorporated significant improvements over the earlier shuttle motors in the design of the joints between the main segments. Redesign of this key feature was part of the intensive engineering redesign and demonstration feat accomplished following the Challenger accident. The result was a fail-safe joint/seal configuration that, with continued refinement, had a high demonstrated reliability. Each joint, with its redundant seals and multiple redundant seal protection features, could be pressure checked during assembly to ensure a good pressure seal was achieved. A similar design approach was implemented on the igniter joints during that same time period.

bondlines, ablatives, fasteners, and virtually all remaining flight hardware. Engineers promptly evaluated any significant observations that could affect the orbiting vehicle or the next motor launch sets.

Before the motor was returned to the flight inventory, the recovered metal parts were inspected for corrosion, deformations, cracks, and other potential damage. Dimensional measurement data were fed into a system-wide database containing documentation dating back to the program's inception. The wealth of information available for performance trend analysis was unmatched by any other solid rocket motor manufacturing process in the world. Gates and checks within the system ensured the full investigation of any anomalies to pinpoint root cause and initiate corrective action.

The postflight analysis program collected the actual flight performance data—most of which would not have been available if the motors had not been recovered.

Through this tightly defined process, engineers were able to address the subtle effects that are often a result of an unintended drift in the manufacturing process or new manufacturing materials introduced into the process. The process addressed these concerns in the incipient phase rather than allowing for a potentially serious issue to escalate undetected. The ultimate intangible benefit of this program was greater reliability, as demonstrated by the following two examples.

Postflight assessment of nozzle bondlines was a catalyst to augment adhesive bonding technology and substantially improve hardware quality and reliability. Storage controls for epoxy adhesives were established in-house and with adhesive suppliers. Surface preparation, cleanliness, adhesive primer, and process timelines were established. Adhesive bond quality and robustness were increased by an order of magnitude.

Postflight inspections also occasionally revealed gas paths through the nozzle-to-case joint polysulfide thermal barrier that led to hot gas impingement on the wiper O-ring—a structure protecting the primary O-ring from thermal damage. While this condition did not pose a flight risk, it did indicate performance failed to meet design intent. The root cause: a design that was impossible to manufacture perfectly every time. Engineers resolved this concern by implementing a nozzle-to-case joint J-leg design similar to that successfully used on case field joints and igniters.

In Utah, rigorous test program included 53 reusable solid rocket motor ground tests between 1977 and 2010. Spectators flocked by the thousands to witness firsthand the equivalent of 15 million horsepower safely unleashed from a vantage point of 2 to 3 km (1 to 2 miles) away.

Robust Systems Testing

The adage "test before you fly," adopted by the Space Shuttle Program, was the standard for many reusable solid rocket motor processes and material, hardware, and design changes. What ATK, the manufacturer, was able to learn from the vast range of data collected and processed through preflight and ground testing ensured the highest levels of dependability and safety for the hardware. Immediate challenges posed by the 570-metric-ton (1.2-million-pound) motor included handling, tooling, and developing a 17.8-meganewton (4,000,000-pound-force) thrust-capable ground test stand; and designing a 1,000-channel data handling system as well as new support systems, instrumentation capability, data acquisition, and countdown procedures.

Hot-fire testing of full-scale rocket motors in the Utah desert became a hallmark of the reusable solid rocket motor development and sustainment program. Individual motor rockets were fired horizontally, typically once or twice a year, lighting up the mountainside with the brightness of a blazing sun, even in broad daylight.

Following a test firing, quick-look data were available within hours. Full data analyses required several months.

On average, NASA collected between 400 and 700 channels of data for each test. Instrumentation varied according to test requirements but typically included a suite of sensors not limited to accelerometers, pressure transducers, calorimeters, strain gauges, thermocouples, and microphones. Beyond overall system assessment and component qualification, benefits of full-scale testing included the opportunity to enhance engineering expertise and predictive skills, improve engineering techniques, and conduct precise margin testing. The ability to tightly measure margins for many motor process, material, components, and design parameters provided valuable verification data to demonstrate whether even the slightest modification was safe for flight.

Quick-look data revealed basic ballistics performance—pressure and thrust measurements—that could be compared with predicted performance and historic data for an initial assessment.

Full analysis included scrutiny of all data recorded during the actual test as well as additional data gathered from visual inspections and measurements of disassembled hardware, similar to that of postflight inspection. Engineers assessed specific data tied to test objectives. When qualifying a new motor insulation, for example, posttest inspection would additionally include measurements of remaining insulation material to calculate the rate of material loss.

Subscale propellant batch ballistics tests, environmental conditioning testing, vibration tests, and custom sensor development and data acquisition were also successful components of the program to provide specific reliability data.

Culture of Continual Improvement

The drive to achieve 100% mission success, paired with the innovations of pre- and postflight testing that allowed performance to be precisely quantified, resulted in an operating culture in which the bar was continually raised.

Design and processing improvements were identified, pursued, and implemented through the end of the program to incrementally reduce risk and waste. Examples of relatively late program innovations included: permeable carbon fiber rope as a thermal protection element in various nozzle and nozzle/case joints; structurally optimized bolted joints; reduced stress forward-grain fin transition configuration; and improved adhesive bonding systems.

This culture, firmly rooted in the wake of the Challenger accident, led to a comprehensive process control program with systems and tools to ensure processes were appropriately defined, correctly performed, and adequately maintained to guarantee reliable and repeatable product performance.

Noteworthy elements of the motor process control program included an extensive chemical fingerprinting program to analyze and monitor the quality of vendor-supplied materials, the use of statistical process control to better monitor conditions, and the comprehensive use of witness panels—product samples captured from the live manufacturing process and analyzed to validate product quality.

With scrupulous process control, ATK and NASA achieved an even greater level of understanding of the materials and processes involved with reusable solid rocket motor processing. As a result, product output became more consistent over the life of the program. Additionally, partnerships with vendors and suppliers were strengthened as increased performance measurement and data sharing created a win-win situation.

An Enduring Legacy

The reusable solid rocket motor was more than an exceptional rocket that safely carried astronauts and hundreds of metric tons of hardware into orbit for more than 25 years. Throughout the Reusable Solid Rocket Motor Program, engineers and scientists generated the technical know-how in design, test, analysis, production, and process control that is essential to continued space exploration. The legacy of the first human-rated reusable solid rocket motor will carry on in future decades. In the pages of history, the shuttle reusable solid rocket motor will be known as more than a stepping-stone. It will also be regarded as a benchmark by which future solid-propulsion systems will be measured.

Orbital Propulsion Systems—Unique Development Challenges

Until the development of the Space Shuttle, all space vehicle propulsion systems were expendable. Influenced by advances in technologies and materials, NASA decided to develop a reusable propulsion system. Although reusability saved overall costs, maintenance and turnaround costs offset some of those benefits.

NASA established a general redundancy requirement of fail operational/fail safe for these critical systems: Orbital Maneuvering System, Reaction Control System, and Auxiliary Power Unit. In addition, engineers designed the propulsion systems for a life of 100 missions or 10 years combined storage and operations. Limited refurbishment was permitted at the expense of higher operational costs.

Orbital Maneuvering System

The Orbital Maneuvering System provided propulsion for the Orbiter during orbit insertion, orbit circularization, orbit transfer, rendezvous, and deorbit. NASA faced a major challenge in selecting the propellant. The agency originally chose liquid oxygen and liquid hydrogen propellants. However, internal volume constraints could not be met for a vehicle configuration that provided a payload of 22,680 kg (50,000 pounds) in a bay measuring 4.6 m (15 ft) in diameter and 18.3 m (60 ft) in length. This, coupled with concerns regarding complexity of cryogenic propellants, led to the consideration of storable hypergolic propellants.

Orbital Maneuvering System/Reaction Control System

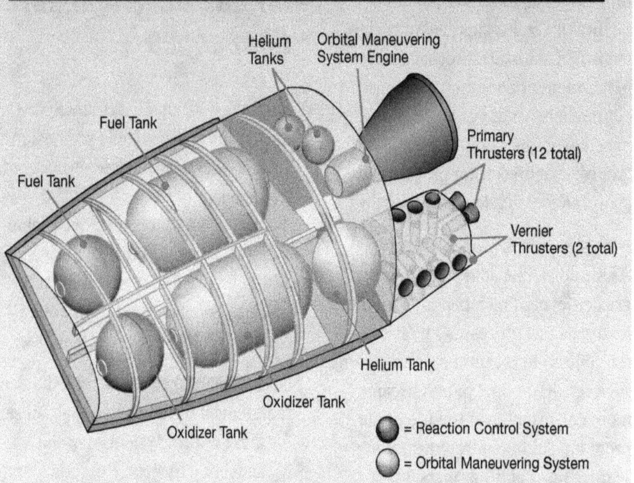

Orbital Maneuvering System/Reaction Control System pods viewed from the underside.

NASA ultimately selected monomethylhydrazine as the fuel and nitrogen tetroxide as the oxidizer for this system. As these propellants were hypergolic—they ignited when coming into contact with each other—no ignition device was needed. Both propellants remained liquid at the temperatures normally experienced during a mission. Electrical heaters prevented freezing during long periods in orbit when the system was not in use.

Modular Design Presents Obstacles for Ground Support

Trade studies and design approach investigations identified challenges and solutions. For instance, cost and weight could be reduced with a common integrated structure for the Orbital Maneuvering System and Reaction Control System. This integrated structure was combined with the selection of nitrogen tetroxide and monomethylhydrazine propellants. Thus, NASA adopted an interconnect system in which the Reaction Control System used Orbital Maneuvering System propellants because of cost, weight, and lower development risk.

Disadvantages of a storable propellant system were higher maintenance requirements resulting from their corrosive nature and hazards to personnel exposed to the toxic propellants. NASA partially addressed these considerations by incorporating the Orbital Maneuvering System into a removable modular pod. This allowed maintenance and refurbishment of those components exposed to hypergols to be separated from other turnaround activities.

For ground operations, it was not practical to remove modules for each turnaround activity, and sophisticated equipment and processes were required for servicing between flights. Fluid and gas connections to the propellants and pressurants used quick disconnects to allow servicing on the launch pad, in Orbiter processing facilities, and in the hypergolic maintenance facility. However, quick disconnects occasionally caused problems, including leakage that damaged Orbiter thermal tiles.

Engineers tested and evaluated many ground support equipment design concepts at the White Sands Test Facility (WSTF). In particular, they tested, designed, and built the equipment used to test and evaluate the propellant acquisition screens inside the propellant tanks before shipment to Kennedy Space Center for use on flight vehicles. The Orbital Maneuvering System/Reaction Control System Fleet Leader Program used existing qualification test articles to detect and evaluate "life-dependent" problems before these problems affected the shuttle fleet. This program provided a test bed for developing and evaluating ground support equipment design changes and improving processes and procedures. An example of this was the Reaction Control System Thruster Purge System, which used low-pressure nitrogen to prevent propellant vapors from accumulating in the thruster chamber. This WSTF-developed ground support system proved beneficial in reducing the number of in-flight thruster failures.

Henry Pohl
Director of Engineering at Johnson Space Center (1986-1993).

"To begin to understand the challenges of operating without gravity, imagine removing the commode from your bathroom floor, bolting it to the ceiling. And then try to use it. You would then have a measure of the challenges facing NASA."

Additional Challenges

Stable combustion was a concern for NASA. In fact, stable combustion has always been the most expensive schedule-constraining development issue in rocket development. For the Orbital Maneuvering System engine, engineers investigated injector pattern designs combined with acoustic cavity concepts. In propulsion applications with requirements for long-duration firings and reusability, cavities had an advantage because they were easy to cool and therefore less subject to failure from either burnout or thermal cycling.

To accomplish precise injector fabrication, engineers implemented platelet configuration. The fuel and oxidizer flowed through the injector and impinged on each other, causing mixing and combustion. Platelet technology, consisting of a series of thin plates manufactured by photo etching and diffusion bonded together, eliminated mechanical manufacturing errors and increased injector life and combustion efficiency.

The combustion chamber was regenerative-cooled by fuel flowing in a single pass through non-tubular coolant channels. The chamber was composed of a stainless-steel liner, an electroformed nickel shell, and an aft flange and fuel inlet manifold assembly. Its structural design was based on life

Formation of Metal Nitrates Caused Valve Leaks

Being the first reusable spacecraft—and in particular, the first to use hypergolic propellants—the shuttle presented technical challenges, including leaky and sticky propellant valves in the Reaction Control System thrusters. Early in the program, failures in this system were either an oxidizer valve leak or failure to reach full chamber pressure within an acceptable amount of time after the thruster was commanded on. NASA attributed both problems to the buildup of metal nitrates on and around the valve-sealing surfaces.

Metal nitrates were products of iron dissolved in the oxidizer when purchased and iron and nickel that were leached out of the ground and flight fluid systems. When the oxidizer was exposed to reduced pressure or allowed to evaporate, metal nitrates precipitated out of solution and contaminated the valve seat.

Subsequent valve cycling caused damage to the Teflon® valve seat, further exacerbating the leakage until sufficient nitrate deposition resulted in "gumming" up the valve. At that point, the valve was either slow to operate or failed to operate.

Multiple changes reduced the metal nitrate problem but may have contributed to fuel valve seat extrusion, which manifested years later. The fuel valve extrusion was largely attributed to the use of throat plugs. These plugs trapped oxidizer vapor leakage in the combustion chamber, which subsequently reacted at a low level of fuel that had permeated the Teflon® fuel valve seat. This problem was successfully addressed with the implementation of the NASA-developed thruster nitrogen purge system, which kept the thruster combustion chamber relatively free of propellant vapors.

cycle requirements, mechanical loads, thrust and aerodynamic loading on the nozzle, ease of fabrication, and weight requirements.

The nozzle extension was radiation cooled and constructed of columbium metal consistent with experience gained during the Apollo Program. The mounting flange consisted of a bolt ring, made from a forging and a tapered section, that could either be spun or made from a forging. The forward and aft sections were made from two panels each. This assembly was bulge formed to the final configuration and the stiffening rings were attached by welding. The oxidation barrier diffusion operation was done after machining was completed.

A basic design challenge for the bipropellant valve was the modular valve. The primary aspect of the assembly design was modularization, which reduced fabrication problems and development time and allowed servicing and maintenance goals to be met with lower inventory.

NASA Seeks Options as Costs Increase

The most significant lesson learned during Orbital Maneuvering System development was the advantage of developing critical technologies before initiating full-scale hardware designs. The successful completion of predevelopment studies not only reduced total costs, also it minimized schedule delays.

In the 1980s, NASA began looking for ways to decrease the cost of component refurbishment and repair. NASA consolidated engineering, evaluation, and repair capabilities for many components, and reduced overall costs. Technicians serviced, acceptance

An Ordinary Solution to the Extraordinary Challenge of Rain Protection

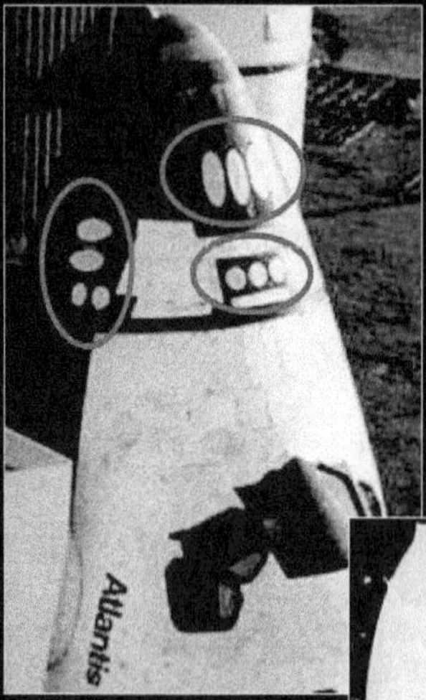

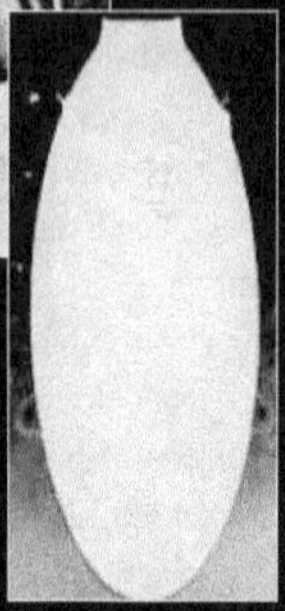

Tyvek® covers shown installed on forward Reaction Control System thrusters (top) and a typical cover (right). Note that the covers were designed to fit certain thruster exit plane configurations.

During operations, Orbiter engines needed rain protection after the protective structure was moved away and protective ground covers were removed. This requirement protected the three upward-facing engines and eight of the left-side engines from rainwater accumulation on the launch pad. The up-firing engine covers had to prevent water accumulation that could freeze in the injector passages during ascent. The side-firing engine covers prevented water from accumulating in the bottom of the chamber and protected the chamber pressure sensing ports. Freezing of accumulated water during ascent could block the sensing port and cause the engine to be declared "failed off" when first used. The original design concept allowed for Teflon® plugs installed in the engine throats and a combination of Teflon® plugs tied to a Teflon® plate that covered the nozzle exit. This concept added vehicle weight, required special procedures to eject the plugs in flight, and risked accidental ejection in ascent that could damage tiles. The solution used ordinary plastic-coated freezer paper cut to fit the exit plane of the nozzle. Tests proved this concept could provide a reliable seal under all expected rain and wind conditions. The covers were low cost, simple, and added no significant weight. The thruster rain cover material was changed to Tyvek® when NASA discovered pieces of liberated plastic-coated paper beneath the cockpit window pressure seals. The new Tyvek® covers were designed to release at relatively low vehicle velocity so that the liberated covers did not cause impact damage to windows, tile, or any other Orbiter surface.

Forward Reaction Control System

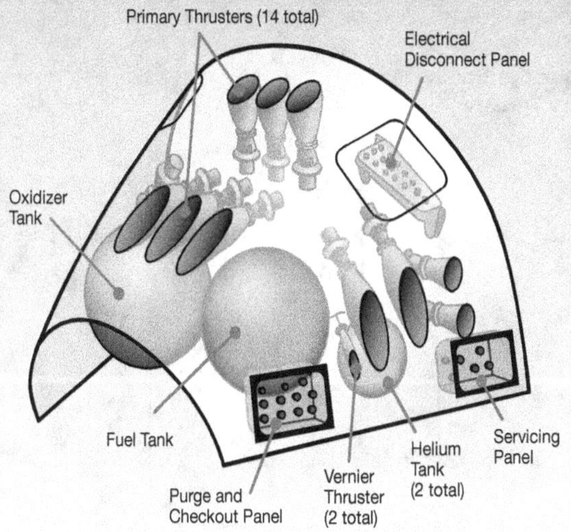

Forward Reaction Control System on Discovery.

tested, and prepared all hypergolic wetted components for reinstallation on the vehicles.

Reaction Control System

The Reaction Control System provided propulsive forces to control the motion of the Orbiter for attitude control, rotational maneuvers, and small velocity changes along the Orbiter axes. The requirement of a fail-operational/fail-safe design introduced complexity of additional hardware and a complex critical redundancy management system. The reuse requirement posed problems in material selection and compatibility, ground handling and turnaround procedures, and classical wear-out problems. The requirement for both on-orbit operations and re-entry into Earth's atmosphere complicated propellant tank acquisition system design because of changes in the gravitational environment.

NASA Makes Effective Selections

As with the Orbital Maneuvering System, propellant selection was important for the Reaction Control System. NASA chose a bipropellant of monomethylhydrazine and nitrogen

Low Temperatures, Increased Leakage, and a Calculated Solution

Some primary thruster valves could leak when subjected to low temperature. NASA discovered this problem when they observed liquid dripping from the system level engines during a cold environment test. The leakage became progressively worse with increased cycling. Continued investigation indicated that tetrafluoroethylene Teflon® underwent a marked change in the thermal expansion rate in a designated temperature range. Because machining, done as a part of seat fabrication, was accomplished in this temperature range, some parts had insufficient seat material exposed at reduced temperatures.

To reduce susceptibility to cold leakage, engineers machined Teflon® at 0°C (32°F) to ensure uniform dimensions with adequate seat material exposed at reduced temperatures and raised the thruster heater set points to maintain valve temperature above 16°C (60°F).

Cracks Prompt Ultrasonic Inspection

Late in the Space Shuttle Program, NASA discovered cracks in a thruster injector. The thruster was being refurbished at White Sands Test Facility (WSTF) during the post-Columbia accident Return to Flight time period. The cracks were markedly similar to those that had occurred in injectors in 1979 and again in 1982.

These earlier cracks were discovered during manufacturing of the thrusters and occurred during the nozzle insulation bake-out process. Results from the laboratory testing indicated that cracks were developed due to chemical processing and manufacturing.

In addition to using leak testing to screen for injector cracking, NASA engineers developed and implemented an ultrasonic inspection procedure to screen for cracks that measured less than the injector wall thickness.

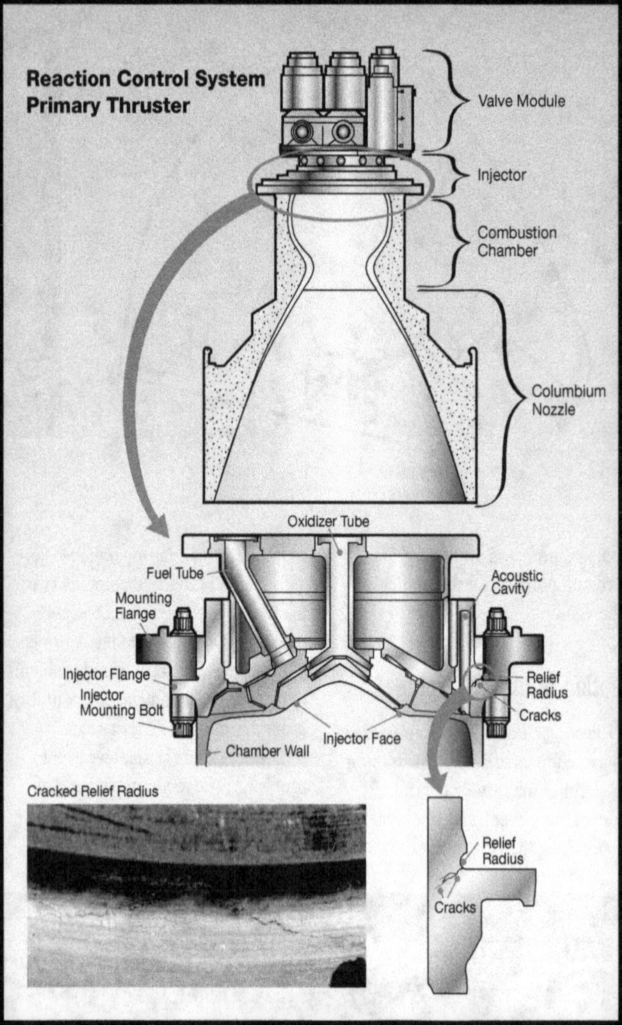

Reaction Control System thruster cross sections showing the crack location and its actual surface appearance.

The marked similarity of the crack location and crack surface appearance strongly suggested the WSTF-discovered cracks were due to the original equipment manufacturing process and were not flight induced or propagated. Laboratory tests and analyses confirmed that those cracks were induced in manufacturing. The cracks had not grown significantly over the years of the thruster's use and its many engine firings. Laboratory nondestructive testing showed that the original ultrasonic inspection process was not very reliable and it was possible that manufacturing-induced cracks could escape detection and cracked thrusters could have been placed in service. The fact that there was no evidence of crack growth associated with the WSTF-discovered cracks due to the service environment was a significant factor in the development of flight rationale for the thrusters.

tetroxide system, which allowed for integration of this system with the Orbital Maneuvering System. This propellant combination offered a favorable weight tradeoff, reasonable development cost, and minimal development risk.

NASA selected a screen tank as a reusable propellant supply system to provide gas-free propellants to the thrusters. Screen tanks worked by using the surface tension of the liquid to form a barrier to the pressurant gas. The propellant acquisition device was made of channels covered with a finely woven steel mesh screen. Contact with liquid wetted the screen and surface tension of the liquid prevented the passage of gas. The strength of the liquid barrier was finite. The pressure differential at which gas would be forced through the wetted screen was called the "bubble point." When the bubble point was exceeded, the screen broke down and gas was transferred. If the pressure differential was less than the bubble point, gas could not penetrate the liquid barrier and only liquid was pulled through the channels. NASA achieved their goal in designing the tank to minimize the pressure loss while maximizing the amount of propellant expelled.

Several Reaction Control System component failures were related to nitrate contamination. Storage of oxidizer in tanks and plumbing that contained iron caused contamination in the propellant. This contamination formed a nitrate that could cause valve leakage, filter blockage, and interference in sliding fits. The most prominent incident was the failure of a ground half-quick disconnect to close, resulting in an oxidizer spill on the launch pad. NASA implemented a program to determine the parameters that caused the iron nitrate formation and implement procedures to prevent its formation in the future. This resulted in understanding the relationship between iron, water, nitric oxide content, and nitrate formation. The agency developed production and storage controls as well as filtration techniques to remove the iron, which resolved the iron nitrate problem.

Auxiliary Power Unit

The Auxiliary Power Unit generated power to drive hydraulic pumps that produced pressure for actuators to control the main engines, aero surfaces, landing gear, brakes, and nose wheel steering. The Auxiliary Power Unit shared common hardware and systems with the Hydraulic Power Unit used on the solid rocket motors. The shuttle needed a hydraulic power unit that could operate from zero to three times gravity, at vacuum and sea-level pressures, from -54°C to 107°C (-65°F to 225°F), and be capable of restarting. NASA took the basic approach of using a small, high-speed, monopropellant-fuel, turbine-powered unit to drive a conventional aircraft-type hydraulic pump.

If the Auxiliary Power Unit was restarted before the injector cooled to less than 204°C to 232°C (400°F to 450°F), the fuel would thermally decompose behind the injector panels and damage the injector and the Gas Generator Valve Module. Limited hot-restart capability was achieved by adding an active water cooling system to the gas generator to be used only for hot restarts. This system injected water into a cavity within the injector. The steam generated was vented overboard. Use of this system enabled restarts at any time after the cooling process, which required a 210-second delay.

Improved Machining and Manufacturing Solves Valve Issue

Development of a reliable valve to control fuel flow into the gas generator proved to be one of the most daunting tasks of the propulsion systems. The valve was required to pulse fuel into the gas generator at frequencies of 1 to 3 hertz. Problems with the valve centered on leakage and limited life due to wear and breakage of the tungsten carbide seat. NASA's considerable effort in redesigning the seat and developing manufacturing processes resulted in an intricate seat design with concentric dual sealing surfaces and redesigned internal flow passages. The seat was diamond-slurry honed as part of the manufacturing process to remove the recast layer left by the electro-discharge machining. This recast layer was a source of stress risers and was considered one of the primary factors causing seat failure. The improved design and machining and manufacturing processes were successful.

Additional Challenges and Subsequent Solutions

During development testing of the gear box, engineers determined that the oil pump may not funtion satisfactorily on orbit due to low pressure. It became necessary to provide a fluid for the pump to displace to assure the presence of oil at the inlet and to have a mechanism to provide needed minimum pressure at startup and during operation.

The Auxiliary Power Unit was designed with a turbine wheel radial containment ring and a blade tip seal and rub ring to safely control failures of the high-speed assembly. The containment ring was intended to keep any wheel fragments from leaving the Auxiliary Power Unit envelope. NASA provided safety features that would allow operation within the existing degree of containment. The agency used an over-speed safety circuit to automatically shut down a unit at 93,000 revolutions per minute. To provide further insurance against wheel failure, NASA imposed stringent flaw detection inspections. With these controls, results of fracture mechanics analyses showed the theoretical life to be 10 times the 100-mission requirement.

With these improvements, the Auxiliary Power Unit demonstrated success of design and exhibited proven durability, performance, and reusability.

NASA Encounters Obstacle Course in Turbine Wheel Design

The space agency faced multiple challenges with the development of the turbine wheel. Aerodynamically induced high-cycle fatigue caused cracking. Analysis indicated this part of the blade could be removed with a small chamfer at the blade tip without significant effect on performance. This cracking problem was resolved by careful design and control of electromechanical machining.

The shroud cracking problem was related to material selection and the welding process. Increased strength and weld characteristics were achieved by changing the shroud material. Engineers developed a controlled electron beam weld procedure to ensure no overheating of the shroud. These actions eliminated the cracking problem.

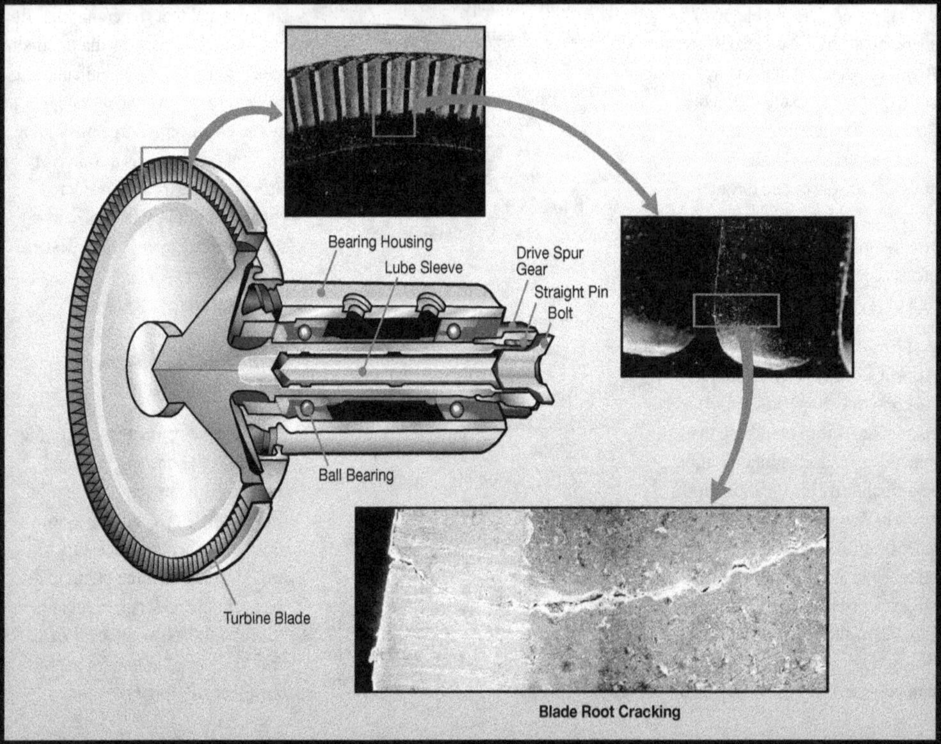

Stress Corrosion and Propellant Ignition

One of the most significant Auxiliary Power Unit problems occurred during the STS-9 (1983) mission when two of the three units caught fire and detonated. Postflight analysis indicated the presence of hydrazine leaks in Auxiliary Power Units 1 and 2 when they were started for re-entry while still in orbit. The leaking hydrazine subsequently ignited and the resulting fire overheated the units, causing the residual hydrazine to detonate after landing. The fire investigation determined the source of the leaks to be nearly identical cracks in the gas generator injector tubes in both units. Laboratory tests further determined that the injector tube cracks were due to stress corrosion from ammonium hydroxide vapors generated by decomposition of hydrazine in the catalyst bed after Auxiliary Power Unit shutdown.

Initial corrective actions included removal of the electrical machined recast layer on the tube inside diameter and an improved assembly of the injector tube. Later, resistance to stress corrosion and general corrosion was further improved by chromizing the injector tubes.

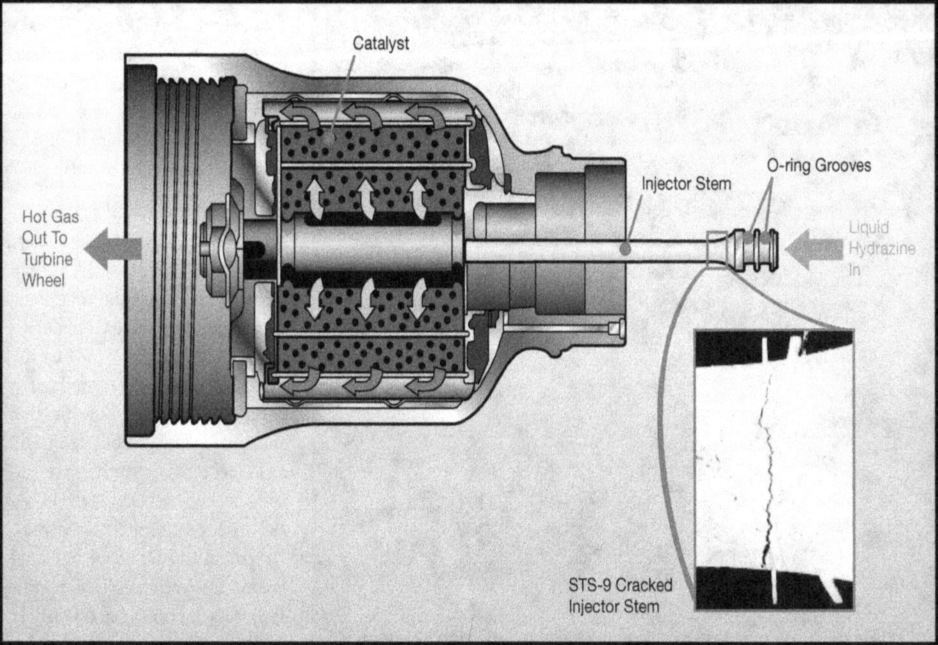

Summary

The evolution of orbital propulsion systems for the Space Shuttle Program began with Apollo Program concepts, expanded with new technologies required to meet changing requirements, and continued with improvements based on flight experience. The design requirements for 100 missions, 10 years, and reuse presented challenges not previously encountered. In addition, several problems were not anticipated. NASA met these challenges, as demonstrated by the success of these systems.

Pioneering Inspection Tool

Contamination Scanning of Bond Surfaces

Inspection technology capitalizing on the photoelectric effect provided significant benefits over the traditional method of visual inspection using handheld black lights. The technology was developed through a NASA/industry partnership managed by Marshall Space Flight Center. Specific benefits included increased accuracy in contamination detection and an electronic data record for each hardware inspection.

Bonding thermal insulation to metal case surfaces was a critical process in solid rocket motor manufacturing during the Space Shuttle Program. Surfaces had to be immaculately clean for proper adherence. The steel alloy was susceptible to corrosion and was coated with grease for protection during storage. That grease, and the solvents to remove it, became potential contaminants.

The improvement of contamination inspection techniques was initiated in the late 1980s. The development of a quantitative and recordable inspection technique was based on the physics of optically stimulated electron emission (photoelectric effect) technology being developed at NASA's Marshall Space Flight Center at the time.

Fundamentally, incident ultraviolet light excites and frees electrons from the metal surface. The freed electrons having a negative charge are attracted to a positively charged collector ring in the "Con Scan" (short for Contamination Scanning) sensor. When contamination exists on a metal surface, the amount of ultraviolet radiation that reaches the surface is reduced. In turn, the current is reduced, confirming the presence of a contaminant.

Approximately 90% of each reusable solid rocket motor barrel assembly was inspected using automated Con Scan before bond operations. Technicians mounted the sensor on a robotic arm, which allowed longitudinal translation of the sensor as the barrel assembly rotated on a turntable. Inspection results were mapped, showing color-coded contamination levels (measured current) vs. axial and circumferential locations on the case inner diameter. Color coding made acceptable and rejected areas visually apparent.

By pioneering optically stimulated electron emission technology, which was engineered into a baseline inspection tool, the Space Shuttle Program significantly improved contamination control methods for critical bonding applications.

Propulsion Systems and Hazardous Gas Detection

Shuttle propulsion had hazardous gases requiring development of detection systems including purged compartments. This development was based on lessons learned from the system first used during Saturn I launches.

NASA performed an exhaustive review of all available online monitoring mass spectrometry technology for the shuttle. The system the agency selected for the prototype Hazardous Gas Detection System had an automated high-vacuum system, a built-in computer control interface, and the ability to meet all program-anticipated detection limit requirements.

The instrument arrived at Kennedy Space Center (KSC) in December 1975 and was integrated into the sample delivery subsystem, the control and data subsystem, and the remote control subsystem designed by KSC. Engineers extensively tested the unit for functionality, detection limits and dynamic range, long-term drift, and other typical instrumental performance characteristics. In May 1977, KSC shipped the prototype Hazardous Gas Detection System to Stennis Space Center to support the shuttle main propulsion test article engine test firings. The system remained in use at Stennis Space Center for 12 years and supported the testing of upgraded engines.

The first operational Hazardous Gas Detection System was installed for the system on the Mobile Launch

Platform-1 during the late summer of 1979. Checkout and operations procedure development and activation required almost 1 year, but the system was ready to support initial purge activation and propellant loading tests in late 1980. A special test in which engineers introduced simulated leaks of hydrogen and oxygen into the Orbiter payload bay, lower midbody, aft fuselage, and the External Tank intertank area represented a significant milestone. The system accurately detected and measured gas leaks.

After the new system's activation issues were worked out, it could detect and measure small leaks from the Main Propulsion System. The Hazardous Gas Detection System did not become visible until Space Transportation System (STS)-6—the first launch of the new Orbiter Challenger—during a flight readiness test. In this test, the countdown would proceed normally to launch time, the Orbiter main engines would ignite, but the Solid Rocket Booster engines would not ignite and the shuttle would remain bolted to the launch pad during a 20-second firing of the main engines. The STS-1 firing test for Columbia had proceeded normally, but during Challenger's firing test, the Hazardous Gas Detection System detected a leak exceeding 4,000 parts per million. Rerunning the firing test and performing further leak hunting and analysis revealed a number of faults in the main engines. The manager for shuttle operation propulsion stated that all the money spent on the Hazardous Gas Detection System, and all that would ever be spent, was paid for in those 20 seconds when the leak was detected.

Originally, NASA declined to provide redundancy for the Hazardous Gas Detection System due to a lack of a launch-on-time requirement; however, the agency subsequently decided that redundancy was required. After a detailed engineering analysis followed by lab testing of candidate mass spectrometers, the space agency selected the PerkinElmer MGA-1200 as the basis of the backup Hazardous Gas Detection System. This backup was an ion-pumped, magnetic-sector, multiple-collector mass spectrometer widely used in operating rooms and industrial plants. Although the first systems were delivered in late 1985, full installation on all mobile launch platforms did not occur until NASA completed the Return to Flight activities following the Challenger accident in 1986.

In May 1990, the Hazardous Gas Detection System gained attention once again when NASA detected a hydrogen leak in the Orbiter aft fuselage on STS-35. The space agency also detected a hydrogen leak at the External Tank to Orbiter hydrogen umbilical disconnect and thought that the aft fuselage leakage indication was due to hydrogen from the external leak migrating inside the Orbiter. Workers rolled STS-35 back into the Vertical Assembly Building and replaced the umbilical disconnect. Meanwhile, STS-38 had been rolled to the pad and leakage was again detected at the umbilical disconnect, but not in the aft fuselage. STS-38 was also rolled back, and its umbilical disconnect was replaced. The ensuing investigation revealed that manufacturing defects in both units caused the leaks, but not before STS-35 was back on the pad.

During launch countdown, NASA detected the aft fuselage hydrogen leak. It was then apparent that STS-35 had experienced two separate leaks. The Space Shuttle Program director appointed a special tiger team to investigate the leak problem. This team suspected that the Hazardous Gas Detection System was giving erroneous data, and brought 10 experts from Marshall Space Flight Center to assess the system design. KSC design engineering provided an in-depth, 2-week description of the design and performance details of both the Hazardous Gas Detection System and the backup system. The most compelling evidence of the validity of the readings was that both systems, which used different technology, had measured identical data, and both systems had recorded accurate calibration data before and after leakage detection. After a series of mini-tanking tests—each with increased temporary instrumentation—engineers located and repaired the leak, and STS-35 lifted off for a successful mission on December 9, 1990.

The Hazardous Gas Detection System and backup Hazardous Gas Detection System continued to serve the shuttle until 2001, when both systems were replaced with Hazardous Gas Detection System 2000—a modern state-of-the-art system with a common sampling system and identical twin quadrupole mass spectrometers from Stanford Research Institute. The Hazardous Gas Detection System served for 22 years and the backup Hazardous Gas Detection System served for 15 years.

Thermal Protection Systems

Introduction
Gail Chapline

Orbiter Thermal Protection System
Alvaro Rodriguez
Cooper Snapp
Geminesse Dorsey
Michael Fowler
Ben Greene
William Schneider
Carl Scott

External Tank Thermal Protection System
Myron Pessin
Jim Butler
J. Scott Sparks

Solid Rocket Motor Joint—An Innovative Solution
Paul Bauer
Bruce Steinetz

Ice Detection Prevents Catastrophic Problems
Charles Stevenson

Aerogel-based Insulation System
Charles Stevenson

The Space Shuttle design presented many thermal insulation challenges. The system not only had to perform well, it had to integrate with other subsystems. The Orbiter's surfaces were exposed to exceedingly high temperatures and needed reusable, lightweight, low-cost thermal protection. The vehicle also required low vulnerability to orbital debris and minimal thermal conductivity. NASA decided to bond the Orbiter's thermal protection directly to its aluminum skin, which presented an additional challenge.

The External Tank required insulation to maintain the cryogenic fuels, liquid hydrogen, and liquid oxygen as well as to provide additional structural integrity through launch and after release from the Orbiter.

The challenge and solutions that NASA discovered through tests and flight experience represent innovations that will carry into the next generation of space programs.

Orbiter Thermal Protection System

Throughout the design and development of the Space Shuttle Orbiter Thermal Protection System, NASA overcame many technical challenges to attain a reusable system that could withstand the high-temperature environments of re-entry into Earth's atmosphere. Theodore von Karman, the dean of American aerodynamicists, wrote in 1956, "Re-entry is perhaps one of the most difficult problems one can imagine. It is certainly a problem that constitutes a challenge to the best brains working in these domains of modern aerophysics." He was referring to protecting the intercontinental ballistic missile nose cones. Fifteen years later, the shuttle offered considerably greater difficulties. It was vastly larger. Its thermal protection had to be reusable, and this thermal shield demanded both light weight and low cost. The requirement for a fully reusable system meant that new thermal protection materials would have to be developed, as the technology from the previous Mercury, Gemini, and Apollo flights were only single-mission capable.

Engineers embraced this challenge by developing rigid silica/alumina fibrous materials that could meet the majority of heating environments on windward surfaces of the Orbiter. On the nose cap and wing leading edge, however, the heating was even more extreme. In response, a coated carbon-carbon composite material was developed to

Thermal Protection System Could Take the Heat

Orbiter remained protected during catalytic heating.

While the re-entry surface heating of the Orbiter was predominantly convective, sufficient energy in the shock layer dissociated air molecules and provided the potential for additional heating. As the air molecules broke apart and collided with the surface of the vehicle, they recombined in an exothermic reaction. Since the surface acted as a catalyst, it was important that the interfacing material/coating have a low propensity to augment the reaction. Atomic recombination influenced NASA's selection of glass-type materials, which have low catalycity and allowed the surface of the Orbiter to reject a majority of the chemical energy. Engineers performed precise arc jet measurements to quantify this effect over a range of surface temperatures for both oxygen and nitrogen recombination. This resulted in improved confidence in the Thermal Protection System.

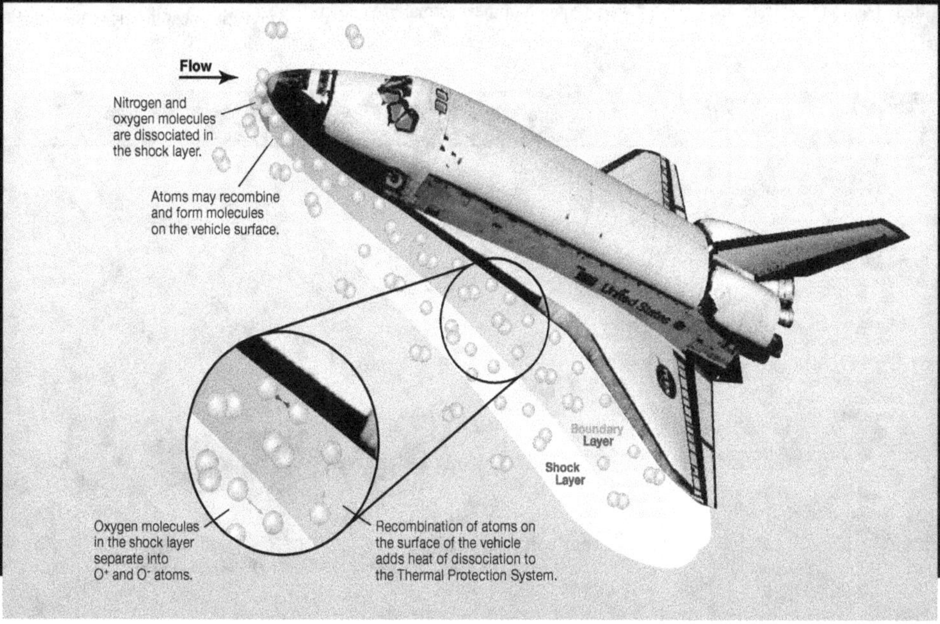

form the contours of these structural components. NASA made an exhaustive effort to ensure these materials would operate over a large spectrum of environments during launch, ascent, on-orbit operations, re-entry, and landing.

Environments

During re-entry, the Orbiter's external surface reached extreme temperatures—up to 1,648°C (3,000°F). The Thermal Protection System was designed to provide a smooth, aerodynamic surface while protecting the underlying metal structure from excessive temperature. The loads endured by the system included launch acoustics, aerodynamic loading and associated structural deflections, and on-orbit temperature variations as well as natural environments such as salt fog, wind, and rain. In addition, the Thermal Protection System had to resist pyrotechnic shock loads as the Orbiter separated from the External Tank (ET).

The Thermal Protection System consisted of various materials applied externally to the outer structural skin of the Orbiter to passively maintain the skin within acceptable temperatures, primarily during the re-entry phase of the mission. During this phase, the Thermal Protection System materials protected the Orbiter's outer skin from exceeding temperatures of 176°C (350°F). In addition, they were reusable for 100 missions with refurbishment and maintenance. These materials performed in temperatures that ranged from -156°C (-250°F) in the cold soak of space to re-entry temperatures that reached nearly 1,648°C (3,000°F). The Thermal Protection System also withstood the forces induced by deflections of the Orbiter airframe as it responded to various external environments.

At the vehicle surface, a boundary layer developed and was designed to be laminar—smooth, nonturbulent fluid flow. However, small gaps and discontinuities on the vehicle surface could cause the flow to transition from laminar to turbulent, thus increasing the overall heating. Therefore, tight fabrication and assembly tolerances were required of the Thermal Protection System to prevent a transition to turbulent flow early in the flight when heating was at its highest.

Requirements for the Thermal Protection System extended beyond the nominal trajectories. For abort scenarios, the systems had to continue to perform in drastically different environments. These scenarios included: Return-to-Launch Site; Abort Once Around; Transatlantic Abort Landing; and others. Many of these abort scenarios increased heat load to the vehicle and pushed the capabilities of the materials to their limits.

Thermal Protection System Materials

Several types of Thermal Protection System materials were used on the Orbiter. These materials included tiles, advanced flexible reusable surface insulation, reinforced carbon-carbon, and flexible reusable surface insulation. All of these materials used high-emissivity coatings to ensure the maximum rejection of incoming convective heat through radiative heat

Orbiter Tile Placement System Configuration

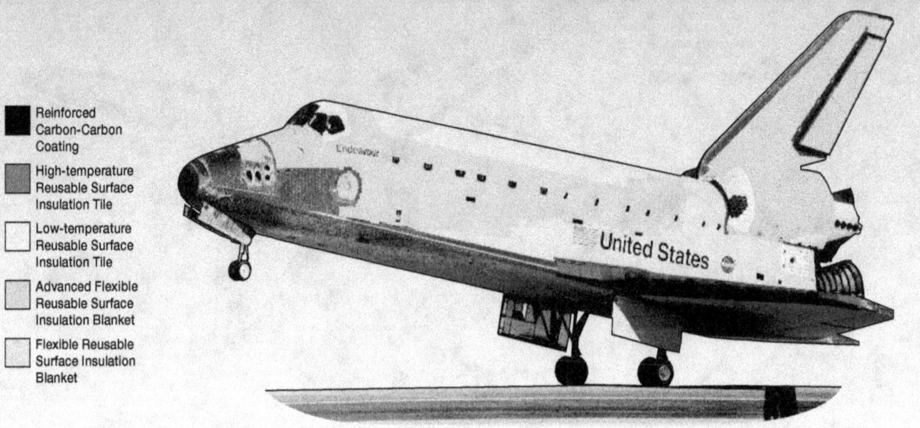

Orbiter Tile Attachment System High-temperature Reusable Surface Insulation

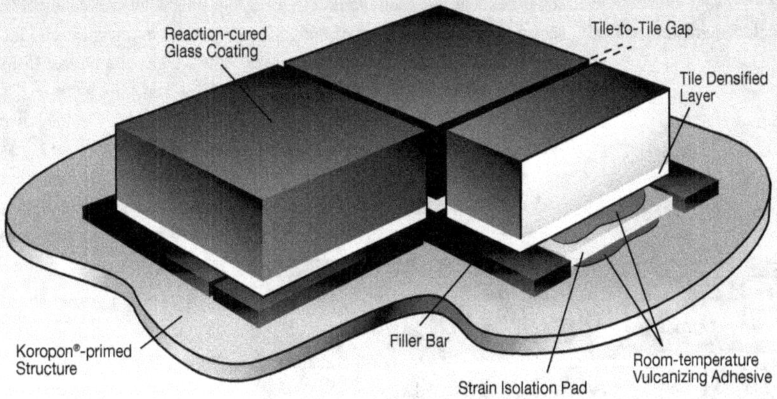

transfer. Selection was based on the temperature on the vehicle. In areas in which temperatures fell below approximately 1,260°C (2,300°F), NASA used rigid silica tiles or fibrous insulation. At temperatures above that point, the agency used reinforced carbon-carbon.

Tiles

The background to the shuttle's tiles lay in work dating to the early 1960s at Lockheed Missiles & Space Company. A Lockheed patent disclosure provided the first description of a reusable insulation made of ceramic fibers for use as a re-entry vehicle heat shield. In other phased shuttle Thermal Protection System development efforts, ablatives and hot structures were the early competitors. However, tight cost constraints and a strong desire to build the Orbiter with an aluminum airframe pointed toward the innovative, lightweight, and reusable insulation material that could be bonded directly to the airframe skin.

NASA used two categories of Thermal Protection System tiles on the Orbiter—low- and high-temperature reusable surface insulation. Surface coating constituted the primary difference between these two categories. High-temperature reusable surface insulation tiles used a black borosilicate glass coating that had an emittance value greater than 0.8 and covered areas of the vehicle in which temperatures reached up to 1,260°C (2,300°F). Low-temperature reusable surface insulation tiles contained a white coating with the proper optical properties needed to maintain the appropriate on-orbit temperatures for vehicle thermal control purposes. The low-temperature reusable surface insulation tiles covered areas of the vehicle in which temperatures reached up to 649°C (1,200°F).

The Orbiter used several different types of tiles, depending on thermal requirements. Over the years of the program, the tile composition changed with NASA's improved understanding of thermal conditions. The majority of these tiles, manufactured by Lockheed Missiles & Space Company, were LI-900 (bulk density of 144 kg/m^3 [9 pounds/ft^3]) and LI-2200 (bulk density of 352 kg/m^3 [22 pounds/ft^3]). Fibrous Refractory Composite Insulation tiles helped reduce the overall weight and later replaced the LI-2200 tiles used around door penetrations. Alumnia Enhanced Thermal Barrier was used in areas in which small particles would damage fragile tiles. As part of the post-Columbia Return to Flight effort, engineers developed Boeing Rigidized Insulation. Overall, the major improvements included reduced weight, decreased vulnerability to orbital debris, and minimal thermal conductivity.

Orbiter tiles were bonded using strain isolation pads and room-temperature vulcanizing silicone adhesives. The inner mold line of the tile was densified prior to the strain isolation pad bond, which aided in the uniform distribution of the stress concentration loads at the tile-to-strain isolation pad interface. The structure beneath the tile-to-tile gaps was protected by filler bar that prevented gas flow from penetrating into the tile bond line. NASA used gap fillers (prevented hot air intrusion and tile-to-tile contact) in areas of high differential pressures, extreme

aero-acoustic excitations and to passivate over-tolerance step and gap conditions. The structure used for the bonding surface was, for the most part, aluminum; however, several other substrates used included graphite epoxy, beryllium, and titanium.

Design Challenges

Determining the strength properties of the tile-to-strain isolation pad interface was no small feat. The allowable strength for the interface was approximately 50% less than the LI-900 tile material used on the Orbiter. This reduction was caused by stress concentrations in the reusable surface insulation because of the formation of "stiff spots" in the strain isolation pad by the needling felting process. Accommodating these stiff spots for the more highly loaded tiles was met by locally densifying the underside of the tile. NASA applied a solution of colloidal silica particles to the non-coated tile underside and baked in an oven at 1,926°C (3,500°F) for 3 hours. The densified layer produced measured about 0.3 cm (0.1 in.) in thickness and increased the weight of a typical 15-by-15-cm (6-by-6-in.) tile by only 27 grams (0.06 pounds). For load distribution, the densified layer served as a structural plate that distributed the concentrated strain isolation pad loads evenly into the weaker, unmodified reusable surface insulation tiles.

NASA faced a greater structural design challenge in the creation of numerous unique tiles. It was necessary to design thousands of these tiles that had compound curves, interfaced with thermal barriers and hatches, and had penetrations for instrumentation and structural access. The overriding challenge was to ensure the strength integrity of the tiles had a probability of tile failure of no greater than $1/10^8$. To accomplish this magnitude of system reliability and still minimize the weight, it was necessary to define the detailed loads and environments on each tile. To verify the integrity of the Thermal Protection System tile design, each tile experienced stresses induced by the following combined sources:

- Substrate or structure out-of-plane displacement
- Aerodynamic loads on the tile
- Tile accelerations due to vibration and acoustics
- Mismatch between tile and structure at installation
- Thermal gradients in the tile
- Residual stress due to tile manufacture
- Substrate in-plane displacement

Other Thermal Protection System Materials? NASA had it Covered.

Flexible Reusable Surface Insulation

White blankets made of coated Nomex® Felt Reusable Surface Insulation protected areas where surface temperatures fell below 371°C (700°F). The blankets were used on the upper payload bay doors, portions of the mid-fuselage, and on the aft fuselage sides.

Advanced Flexible Reusable Surface Insulation

After initial delivery of Columbia to the assembly facility, NASA developed an advanced flexible reusable surface insulation consisting of composite quilted fabric insulation batting sewn between two layers of white fabric. The insulation blankets provided improved producibility and durability, reduced fabrication and installation time and costs, and reduced weight. This insulation replaced the majority of low-temperature reusable surface insulation tiles on two of the shuttles: Discovery and Atlantis. Following Columbia's seventh flight, the shuttle was modified to replace most of the low-temperature reusable surface insulation tiles on portions of the upper wing. For Endeavour, the advanced flexible reusable surface insulation was directly built into the shuttle.

Additional Materials

NASA used additional materials in other areas of the Orbiter, such as in thermal glass for the windows, Inconel® for the forward Reaction Control System fairings, and elevon seal panels on the upper wing. Engineers employed a combination of white and black pigmented silica cloth for thermal barriers and gap fillers around operable penetrations such as main and nose landing gear doors, egress and ingress flight crew side hatch, umbilical doors, elevon cove, forward Reaction Control System, Reaction Control System thrusters, mid-fuselage vent doors, payload bay doors, rudder/speed brake, and gaps between Thermal Protection System tiles in high differential pressure areas.

Reinforced Carbon-Carbon

The temperature extremes on the nose cap and wing leading edge of the Orbiter required a more sophisticated material that would operate over a large spectrum of environments during launch, ascent, on-orbit operations, re-entry, and landing. Developed by the Vought Corporation, Dallas, Texas, in collaboration with NASA, reinforced carbon-carbon formed the contours of the nose cap and wing leading edge structural components.

Reinforced carbon-carbon is a composite made by curing graphite fabric that has been pre-impregnated with phenolic resin laid up in complex shaped molds. After the parts are rough trimmed, the resin polymer is converted to carbon by pyrolysis—a chemical change brought about by the action of heat. The part is then impregnated with furfuryl alcohol and pyrolyzed multiple times to increase its density with a resultant improvement in its mechanical properties.

Since carbon oxidizes at elevated temperatures, a silicon carbide coating is used to protect the carbon substrate. Any oxidation of the substrate directly affects the strength of the material and, therefore—in the case of the Orbiter—had to be limited as much as possible to ensure high performance over multiple missions. Silicon carbide is formed by converting the outer two plies of the carbon-carbon material through a diffusion coating process, resulting in a stronger coating-to-substrate interlaminar strength.

As a result of the silicon carbide formation, which occurs at temperatures of 1,648°C (3,000°F), craze cracks develop in the coating on cool-down as the carbon substrate and coating have a different coefficient of thermal expansion. Impregnating the carbon part with tetraethyl orthosilicate and applying a brush-on sealant provides additional protection against oxygen paths to the carbon from the craze cracks.

The tetraethyl orthosilicate is applied via a vacuum impregnation with the intent of filling any remaining porosity within the part. Once the tetraethyl orthosilicate has cured, a silicon dioxide residue coats the pore walls throughout the part, thus inhibiting oxidation. After the tetraethyl orthosilicate process is complete, a sodium silicate sealant is brushed onto the surface of the reinforced carbon-carbon. The sealant fills in the craze cracks and, once cured, forms a glass. The craze cracks close at high temperatures and the sealant will flow onto the surface; however, since there is sufficient viscosity, the sealant remains on the part. When the reinforced carbon-carbon cools down, the glass fills back into the craze crack.

Orbiter Wing Panel Assembly

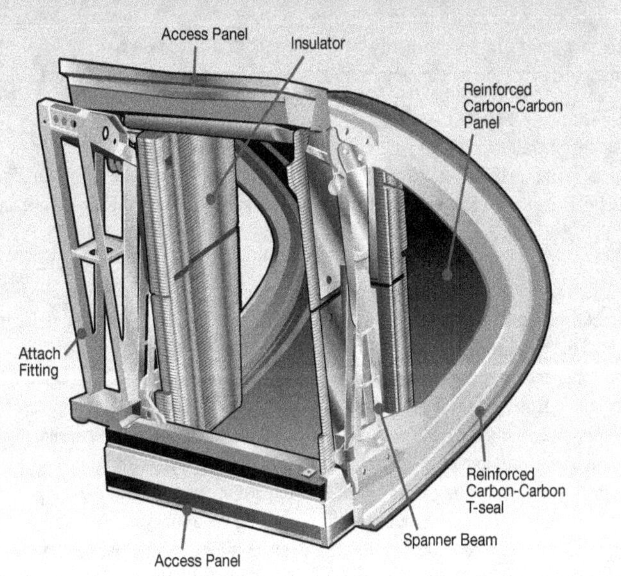

Why Reinforced Carbon-Carbon?

The functionality of the reinforced carbon-carbon is largely due to its ability to reject heat by external radiation (i.e., giving off heat from surface to the surroundings) and cross-radiation, which is the internal reinforced carbon-carbon heat transfer between the lower and upper structures. Reinforced carbon-carbon has an excellent surface emissivity and can reject heat by radiating to space similar to the other Thermal Protection Systems. It is designed as a shell section with an open interior cavity that promotes cross-radiation.

Since the highest heating is biased toward the lower surface, heat can be cross-radiated to the cooler upper surfaces, thus reducing temperatures of the lower windward surface. Another benefit is that the thermal gradients across the part are minimized.

While reinforced carbon-carbon is designed to withstand high temperatures and maintain its structural shape, the material has a relatively high thermal conductivity so it did not significantly inhibit the heat flow to reach the internal Orbiter wing structure. The metallic attachments that mated the reinforced carbon-carbon to the wing structure were crucial for accommodating the thermal expansion of reinforced carbon-carbon and maintaining a smooth outer mold line of the vehicle. Protecting these attachments and the spar structure itself required internal insulation. Incoflex®, an insulative batting encased by a thin Inconel® foil, protected the metal structural components from the internal cavity radiation environment.

Certification

Prior to the Orbiter's first flight, NASA performed extensive test and analysis to satisfy all requirements related to the natural and induced environments. The space agency accomplished certification of the wing leading edge subsystem for flight by analyses verified with development and qualification tests conducted on full-scale hardware. Engineers performed subscale testing to establish thermal and mechanical properties, while full-scale testing ensured the system performance and provided the necessary data to correlate analytical models. This included a full-scale nose cap test article and twin wing leading edge panel configuration tested through multiple environments (i.e., acoustic/vibration, static loads, and radiant testing). Full-scale testing ensured that the metallic mechanisms worked in concert with the hot structure as a complete system in addition to meeting the multi-mission requirements.

Reinforced Carbon-Carbon Flight Experience Lessons Learned

While NASA confirmed the fundamental concepts and design sufficiency through the wing leading edge subsystem certification work and early flight test phase of the Space Shuttle Program, the agency also identified design deficiencies. In most cases, modifications rectified those deficiencies. These modifications included addressing the gap heating between the reinforced carbon-carbon and reusable surface insulation to inhibit hot gas flow-through and retrofitting hardware to the wing leading edge subsystem design to account for a substantial increase in the predicted airloads. With increasing design environment maturity, temperature predictions on the attach fittings were significantly lowered, which allowed a design change from steel to titanium and a weight reduction of 136 kg (300 pounds).

Over the 30 years of flight, the shuttle encountered many anomalies that required investigative testing and analysis. Inspections revealed several cracks in the T-seals—i.e., components made of reinforced carbon-carbon that fit between reinforced carbon-carbon panels that allowed for thermal expansion of those components while keeping a smooth outer mold line. The cracks were later found to be caused by convoluted plies from the original layup of the T-seals. NASA corrected the cracking by modifying the manufacturing techniques and implementing additional inspections. In 1993, the agency identified small pinholes that went down to the carbon substrate and were subsequently traced to a change in maintenance of the launch pad structure. Engineers altered the silica/cement topcoat over the zinc primer such that zinc particles were able to come into contact with the wing leading edge and react with the silicon carbide coating during re-entry, thereby forming pinholes. NASA developed criteria for the pinholes as well as vacuum heat clean and repair methods.

Improved Damage Assessment and Repair With Return to Flight After Columbia Accident

NASA performed rigorous testing and analysis on the Thermal Protection System materials to adequately identify risks and to mitigate failure as much as practical. Engineers developed impact testing, damage-tolerance assessments, and inspection and repair capabilities as part of the Return to Flight effort.

Impact Testing

The greatest lesson learned was that failure of the reinforced carbon-carbon and the catastrophic loss of the vehicle was caused by a large piece of foam debris that was liberated from the ET.

While modifications to the thermal protection foam on the tank reduced the risk of shedding large debris during launch, NASA still expected smaller-sized debris shedding. It was critical that engineers understand the impact of foam shedding on the Orbiter's wing leading edge and tiles. The Southwest Research Institute, San Antonio, Texas, conducted many of these impact tests to understand the important parameters that governed structural failure of reinforced carbon-carbon and tile materials. Additionally, NASA developed finite element modeling capabilities to derive critical-damage thresholds.

Tile Repair—A Critical Capability Was Developed

Prior to the first shuttle launch, NASA recognized the need for a capability to repair tiles on orbit. The loss of a tile during launch due to an improper bond posed the greatest threat. In response, NASA prioritized the development of an ablative material, MA-25S, for repairs of missing or damaged tiles. The biggest obstacle, however, was finding a stable work platform. Thus, NASA cancelled the early repair effort in 1979.

After the Columbia accident in 2003, NASA prioritized tile repair capability. Prior to the Columbia accident, the inspections after every flight revealed damage greater than 2.5 cm (1 in.) in approximately 50 to 100 locations. The original ablative material formed the basis for the repair material developed in the Return to Flight effort.

Some reformulation of MA-25S began in 2003. At that time, NASA changed the name of the material to Shuttle Tile Ablator, 865 kg/m^3 (54 pounds/ft^3) (STA-54). This material decreased the amount of swell during re-entry while maintaining a low enough viscosity to dispense with the extravehicular activity hardware. The material did not harden and would remain workable for approximately 1 hour but still cured within 24 hours in the on-orbit environments.

Simulating a damaged shuttle tile created dust that prevented the STA-54 from penetrating the surface of the tiles. This led to the development of additional materials: a gel cleaning brush that was coated with a sticky silicone substance used to clean tile dust from the repair cavity prior to filling; and primer material that provided a contact surface to which the STA-54 could adhere. Once the primer was cured, the bond strength was stronger than the shuttle tile.

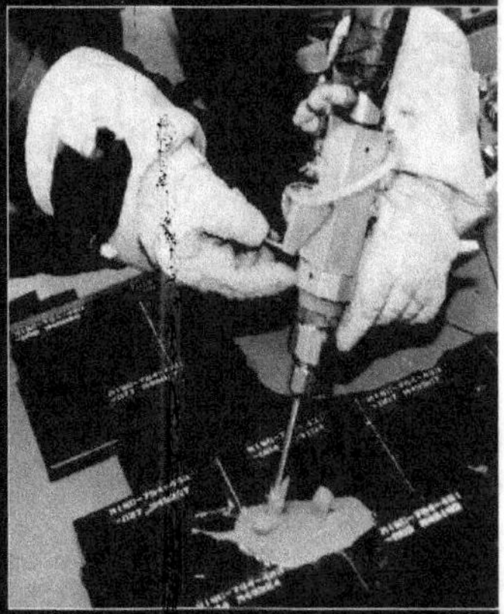

Ground test of Orbiter tile repair.

Finally, NASA performed an on-orbit experiment during STS-123 (2008). Crew member Michael Foreman dispensed STA-54 into several damaged tile specimens. The on-orbit experiment was a success, showing that the material behaved exactly as it had during vacuum dispenses on the ground.

Damage Tolerance Criteria

To make use of the inspection data, NASA developed criteria for critical damage. Damage on reinforced carbon-carbon ranged from spallation (i.e., breaking up or reducing) of the silicon carbide coating to complete penetration of the substrate. Tiles could be gouged by ascent debris to varying depths with a wide variety of cavity shapes. The seriousness of any given damage was highly dependent on local temperature and pressure environments. NASA initiated an extensive Arc Jet test program during Return to Flight activities to characterize the survivability of multiple damage configurations in different environments. Testing in an Arc Jet facility provided the closest ground simulation for the temperature and chemical constituents of re-entry. Engineers performed numerous tests for both reinforced carbon-carbon and tile to establish damage criteria and verify newly developed thermal math models used for real-time mission support.

Inspection Capability

NASA developed an inspection capability to survey the reinforced carbon-carbon and tile surfaces. This capability provided images to assess any potential impact damages from ascent and orbital debris. A boom with an imagery sensor package attached to the Shuttle Robotic Arm was used to perform the inspection. The sensor package contained two laser imaging systems and a high-resolution digital camera. Additionally, astronauts residing on the International Space Station (ISS) photographed the entire Orbiter as it executed an aerial maneuver, similar to a backflip, 182 m (600 ft) from the ISS. The crew transmitted photographs to Houston, Texas, where engineers on the ground evaluated the images for any potential damage.

NASA employed an additional detection system to gauge threats from ascent and on-orbit impacts to the wing leading edge. As part of preparing the

Reinforced Carbon-Carbon Repair—Damage Control in the Vacuum of Space

Following the Space Shuttle Columbia accident in 2003, a group of engineers and scientists gathered at Johnson Space Center to discuss concepts for the repair of damaged reinforced carbon-carbon in the weightless vacuum environment of space. Few potential repair materials could withstand the temperatures and pressures on the surface. Of those materials, few were compatible with the space environment and none had been tested in this type of application. Thus, the team developed two repair systems that were made available for contingency use on the next flight.

The first system—Non-Oxide Adhesive Experimental—was designed to repair coating damage or small cracks in reinforced carbon-carbon panels. This pre-ceramic polymer had the consistency of a thick paste. COI Ceramics, Inc., headquartered in San Diego, California, developed this system and the NASA repair team slightly modified it to optimize its material properties for use in space. Technicians used a modified commercial caulk gun to apply the material to the damaged wing. The material was spread out over the damage using spatulas similar to commercial trowels. Once dried and cured by the sun, Non-Oxide Adhesive Experimental used the heat of re-entry to convert the material into a ceramic, which protected exposed damage from extreme temperatures and pressures.

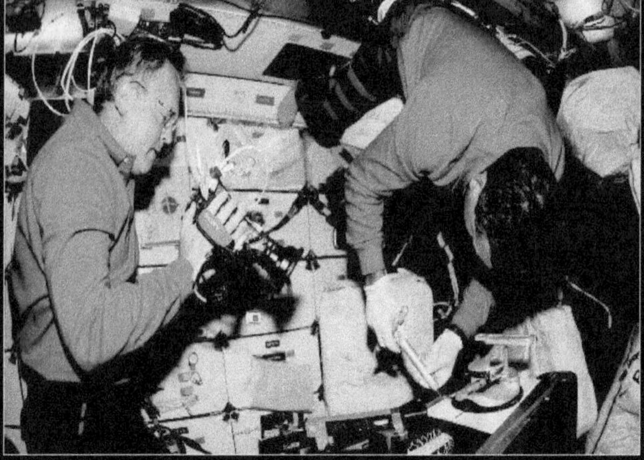

Astronaut Andrew Thomas (left) watches as Charles Camarda tests the reinforced carbon-carbon plug repair (STS-114 [2005]).

For larger damages, a plug repair system protected the reinforced carbon-carbon using a series of thin, flexible composite discs designed to fit securely against the curvature of the surface. Engineers developed 19 geometric shapes, which were flown to provide contingency repair capability. An attach mechanism held the plugs in place. The anchor was made up of a refractory alloy called titanium zirconium molybdenum that was capable of withstanding the 1,648°C (3,000°F) re-entry temperature.

Orbiter for launch, technicians placed accelerometers on the spar aluminum structure behind the reinforced carbon-carbon panels at the attachment locations. Forty-four sensors across both wings detected accelerations from potential impacts and relayed the data to on-board laptops, which could be transmitted to ground engineers. Using test-correlated dynamic models, engineers assessed suspected impacts for their level of risk based on accelerometer output.

Conclusion

The Orbiter Thermal Protection Systems on the shuttle proved to be effective, with the exception of STS-107 (2003). On that flight, the catastrophic loss was caused by a large piece of foam debris that was liberated from the ET. Advanced materials and coatings were key in enabling the success of the shuttle in high-temperature environments. Experience gathered over many shuttle missions led the Thermal Protection Systems team to modify and upgrade both design and materials, thus increasing the robustness and safety of these critical systems during the life of the program. Through the tragedy of the Columbia accident, NASA developed new inspection and repair techniques as protective measures to ensure the success and safety of subsequent shuttle missions.

External Tank Thermal Protection System

The amount of Thermal Protection System material on the shuttle's External Tank (ET) could cover an acre. NASA faced major challenges in developing and improving tank-insulating materials and processes for this critical feature. Yet, the space agency's solutions were varied and innovative. These solutions represented a significant advance in understanding the use of Thermal Protection System materials as well as the structures, aerodynamics, and manufacturing processes involved.

The tanks played two major roles during launch: containing and delivering cryogenic propellants to the Space Shuttle Main Engines, and serving as the structural backbone for the attachment of the Orbiter and Solid Rocket Boosters. The Thermal Protection System, composed of spray-on foam and hand-applied insulation and ablator, was applied primarily to the outer surfaces of the tank. It was designed to maintain the quality of the cryogenic propellants, protect the tank structure from ascent heating, prevent the formation of ice (a potential impact debris source), and stabilize tank internal temperature during re-entry into Earth's atmosphere, thus helping to maintain tank structural integrity prior to its breakup within a predicted landing zone.

Basic Configuration

NASA applied two basic types of Thermal Protection System materials to the ET. One type was a low-density, rigid, closed-cell foam. This foam was sprayed on the majority of the tank's "acreage"—larger areas such as the liquid hydrogen and liquid oxygen tanks as well as the intertank—also referred to as the tank "sidewalls." The other major component was a composite ablator material (a heat shield material designed to burn away) made of silicone resins and cork.

NASA oversaw the development of the closed-cell foam to keep propellants at optimum temperature—liquid hydrogen fuel at -253°C (-423°F) and liquid oxygen oxidizer at -182°C (-296°F)—while preventing a buildup of ice on the outside of the tank, even as the tank remained on the launch pad under the hot Florida sun.

The foam insulation had to be durable enough to endure a 180-day stay at the launch pad, withstand temperatures up to 46°C (115°F) and humidity as high as 100%, and resist sand, salt fog, rain, solar radiation, and even fungus. During launch, the foam had to tolerate temperatures as high as 649°C (1,200°F) generated by aerodynamic friction and rocket exhaust. As the tank reentered the atmosphere approximately 30 minutes after launch, the foam helped hold the tank together as temperatures and internal pressurization worked to break it up, allowing the tank to disintegrate safely over a remote ocean location.

Though the foam insulation on the majority of the tank was only about 2.5 cm (1 in.) thick, it added approximately 1,700 kg (3,800 pounds) to the tank's weight. Insulation on the liquid hydrogen tank was somewhat thicker—between 3.8 and 5 cm (1.5 to 2 in.). The foam's density varied with the type, but an average density was 38.4 kg/m^3 (2.4 pounds/ft^3).

The tank's spray-on foam was a polyurethane material composed of five primary ingredients: an isocyanate and a polyol (both components of the polymeric backbone); a flame retardant; a surfactant (which controls surface tension and bubble or cell formation); and a catalyst (to enhance the efficiency and speed of the polymeric reaction). The blowing agent—originally chlorofluorocarbon (CFC)-11, then hydrochlorofluorocarbon (HCFC)-141b—created the foam's cellular structure, making millions of tiny bubble-like foam cells.

NASA altered the Thermal Protection System configuration over the course of the Space Shuttle Program; however, by 1995, ET performance requirements led the program to baseline four specially engineered closed-cell foams. The larger sections were covered in polyisocyanurate (an improved version of polyurethane) foam (NCFI 24-124) provided by North Carolina Foam Industries. NCFI 24-124 accounted for 77% of the total foam used on the tank and was sprayed robotically. A similar foam, NCFI 24-57, was sprayed robotically on the aft dome of the liquid hydrogen tank. Stepanfoam® BX-265 was sprayed manually on closeout areas, exterior tank feedlines, and internal tank domes. The tank's ablator, Super-Lightweight Ablator (SLA)-561, was sprayed onto areas subjected to extreme heat, such as brackets and other protuberances, and the exposed, exterior lines that fed the liquid oxygen and liquid hydrogen to the shuttle's main engines. NASA used Product Development Laboratory-1034, a hand-poured foam, for filling odd-shaped cavities.

Application Requirements

Application of the foam, whether automated or hand-sprayed, was designed to meet NASA's requirements for finish, thickness, roughness, density, strength, adhesion, and size and frequency of voids within the foam. The foam was applied in

External Tank Thermal Protection Systems Materials

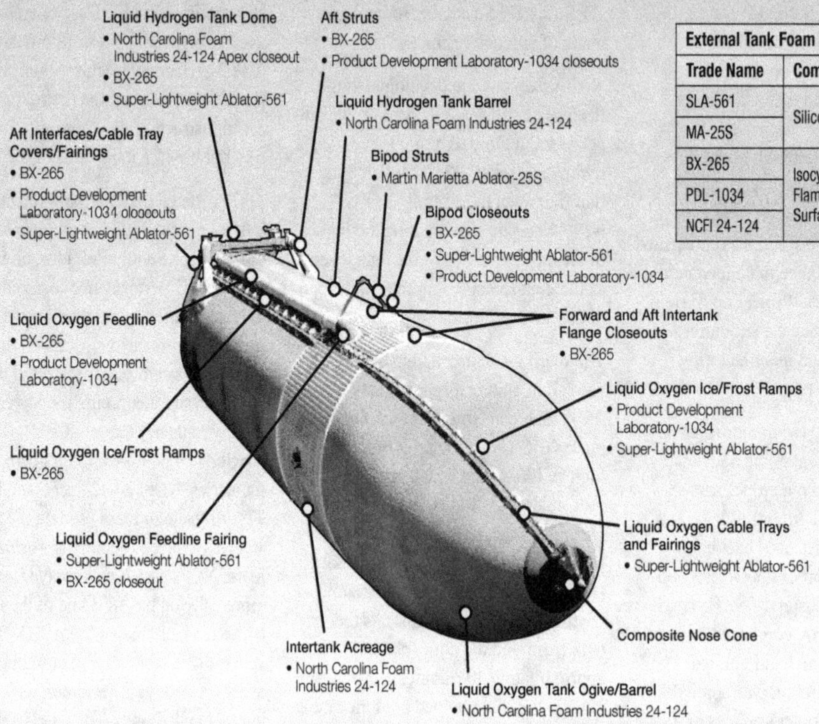

External Tank Foam Material	
Trade Name	**Composition**
SLA-561	Silicone Resin, Cork
MA-25S	
BX-265	Isocyanate Polyol, Flame-Retardant, Surfactant Catalyst
PDL-1034	
NCFI 24-124	

The External Tank's Thermal Protection System consisted of a number of different foam formulations displayed here. NASA selected materials for their insulating properties, and for their ability to withstand ascent aerodynamic forces.

specially designed, environmentally controlled spray cells and sprayed in several phases, often over a period of several weeks. Prior to spraying, engineers tested the foam's raw material and mechanical properties to ensure the materials met NASA specifications. After the spraying was complete, NASA performed multiple visual inspections of all foam surfaces as well as tests of "witness" specimens in some cases.

More than 90% of the foam was sprayed onto the tank robotically, leaving 10% to be applied by manual spraying or by hand. Most foam was applied at Lockheed Martin's Michoud Assembly Facility in New Orleans, Louisiana, where the tank was manufactured. Some closeout Thermal Protection System was applied either by hand or manual spraying at the Kennedy Space Center (KSC) in Florida.

Design and Testing

In the early 1970s, NASA developed a spiral "barber pole" Thermal Protection System application technique that was used through the end of the program. This was an early success for the ET Program, but many challenges soon followed. As the ET was the only expendable part of the shuttle, NASA placed particular emphasis on keeping tank manufacturing costs at a minimum. To achieve this objective, the agency based its original design and manufacturing plans on the use of existing, well-proven materials and processes with a planned evolution to newer products as they became available.

The original baseline Thermal Protection System configuration called for the sprayable Stepanfoam® BX-250 foam (used on the Saturn S-II stage) on the liquid hygrogen sidewalls (acreage)

Solid Rocket Motor Joint—An Innovative Solution

Alliant Techsystems (ATK) Aerospace Systems, in partnership with NASA Glenn Research Center, developed a solution for protecting the temperature-sensitive O-rings used to seal the shuttle reusable solid rocket motor nozzle segments. The use of a carbon fiber material promoted safety and enabled joint assembly in a fraction of the time required by previous processes, with enhanced reproducibility.

The reusable solid rocket motors were fabricated in segments and pinned together incorporating O-ring seals. Similarly, nozzles consisted of multiple components joined and sealed at six joint locations using O-rings. A layer of rubber insulation, referred to as "joint fill" compound, kept the 3,038°C (5,500°F) combustion gases a safe distance away from these seals. In a few instances, however, hot gases breached the compound, leaving soot within the joint. NASA modified the compound installation process and instituted reviews of postflight conditions. Although the modifications proved effective, damage was still possible in the unlikely event that gases breached the compound.

ATK chose an innovative approach through emerging technologies. Rather than attempt to prevent gas intrusion with manually applied rubber fill compound, the heat energy from internal gases would be extracted with a special joint filler and the O-ring seals would be pressurized with the cooled gas.

ATK's solution was based on a pliable, braided form of high-performance carbon material able to withstand harsh temperature environments. The braided design removed most of the thermal energy from the gas and inhibited flow induced by pressure fluctuations. The carbon fiber thermal barrier was easier to install and significantly reduced motor assembly time.

In a rocket environment, carbon fibers withstood temperatures up to 3,816°C (6,900°F). The braided structure and high surface area-to-mass ratio made the barrier an excellent heat exchanger while allowing a restricted yet uniform gas flow. The weave

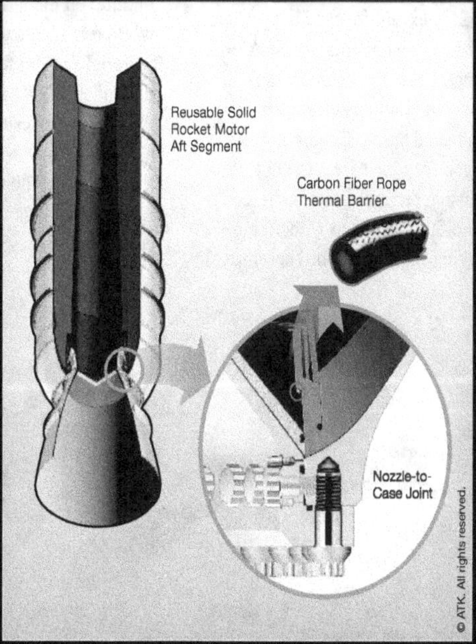

Using carbon fiber rope instead of rubber insulation in solid rocket motor nozzle joints simplified the joint assembly process and improved shuttle safety margins.

structure allowed it to conform to tolerance assembly conditions. The thermal barrier provided flexibility and resiliency to accommodate joint opening or closing during operation. Upon pressurization, the thermal barrier seated itself in the groove to obstruct hot gas flow from bypassing the barrier.

The carbon fiber solution increased Space Shuttle safety margins. Carbon fibers are suited to a nonoxidizing environment, withstanding high temperatures without experiencing degradation. The barrier provided a temperature drop across a single diameter, reducing gas temperature to O-rings well below acceptable levels. The thermal barrier also kept molten alumina slag—generated during solid fuel burn—from contacting and affecting O-rings.

and forward dome, and SLA-561 (used on the Viking Mars Lander) on the aft dome, intertank, and liquid oxygen tank in the areas of high heating. In the late 1970s, however, design of the Orbiter tiles advanced to the point where it became apparent that they were susceptible to damage from ice detaching from the ET. This caused a reassessment of the Thermal Protection System design to prevent the formation of ice anywhere on the tank forward of the liquid hydrogen tank aft-end structural ring frame. The Orbiter/ice issue drove the requirement to cover the entire tank with Stepanfoam® BX-250, except for the high-heating aft dome, which remained SLA-561. Ice was to be prevented on tank pressurization lines through the use of a heated purge. Certain liquid oxygen feedline brackets, subject to extensive thermal contraction, could not be fully insulated without motion breaking the insulation. Therefore, NASA accepted ice formation on these brackets as unavoidable.

While attempting to prevent ice buildup on the tank, NASA also worked to characterize both the ablator material and the foams for expected heating rates. NASA worked with Arnold Engineering Development Center in Tennessee to modify its wind tunnel to provide the capability to test foam materials under realistic flight conditions. SLA-561 was tested in the plasma arc facility at NASA's Ames Research Center in California, which could deliver the required high heating rates. Better understanding of ablation rates and the flow fields around ET protuberances permitted refinement of the Thermal Protection System configuration.

A secondary function of the Thermal Protection System was to stabilize tank internal temperature during re-entry into Earth's atmosphere, thus helping to maintain tank structural integrity prior to its breakup over a remote ocean location.

The key to the External Tank's foam Thermal Protection System insulating properties was its cellular structure, creating millions of tiny bubble-like foam cells. The sprayed foam (NCFI 24-124) can be seen here after application to an area of the tank's aluminum "acreage," consisting of the liquid oxygen tank, liquid hydrogen tank, and intertank.

Another unique project was the testing of spray-on foam insulation on a subscale tank, measuring 3 m (10 ft) in diameter, in the environmental hanger at Eglin Air Force Base, Florida. The insulated tank was filled with liquid nitrogen and subjected to various rain, wind, humidity, and temperature conditions to determine the rate of ice growth. These data were then converted to a computer program known as Surfice, which was used at KSC to predict whether unacceptable ice would form prior to launch.

To provide information on application techniques, the agency ran cryogenic flexure tests that verified substrate adhesion and strength as well as crush tests on the Thermal Protection System materials.

In a continuous search for optimum Thermal Protection System performance, NASA—still in the Thermal Protection System design and testing phase—decided to use Chemical Products Research (CPR)-421, a commercial foam insulation with good high-heating capability. Lockheed Martin developed a sprayable Thermal Protection System to apply to tank sidewalls and aft dome. Application needed a relative humidity of less than 30%, which resulted in the addition of a chemical dryer at Michoud. Also, the tank wall had to be heated to 60°C (140°F). This required passing hot gas through the tank while it was being rotated for the "barber pole" foam application mode.

Ice Detection Prevents Catastrophic Problems

NASA had a potentially catastrophic problem with ice that formed on the cryogenic-filled Space Shuttle External Tank. Falling ice could have struck and damaged the crew compartment windows, reinforced carbon-carbon panels on the wing leading edge of the Orbiter, or its thermal protection tiles, thus placing the crew and vehicle at risk.

Kennedy Space Center and the US Army Tank Automotive and Armaments Research, Development and Engineering Center confirmed that a proof-of-concept system, tested by MacDonald, Dettwiler and Associates Ltd. of Canada, offered potential to support cryogenic tanking tests and ice debris team inspections on the launch pad. NASA and its partners initiated a program to develop a system capable of detecting ice on the External Tank spray-on foam insulation surfaces. This system was calibrated for those surfaces and used an infrared strobe, a focal plane sensor array, and a filter wheel to collect successive images over a number of sub-bands. The camera processed the images to determine whether ice was present, and it also computed ice thickness. The system was housed in nitrogen-purged enclosures that were mounted on a two-wheeled portable cart. It was successfully applied to the inspection of the External Tank on STS-116 (2006), where the camera detected thin ice/frost layers on two umbilical connections.

Robert Speece, NASA engineer, is shown operating the ice detection system at the pad, prior to shuttle launch.

The system can be used to detect ice on any surface. It can also be used to detect the presence of water.

First Flight Approaches

As the Space Shuttle Program moved toward the first shuttle flight in 1981, NASA faced another challenge. Approximately 37 m^2 (400 ft^2) of ablator became debonded from the tank's aluminum surface the first time a tank was loaded with liquid hydrogen. While the failure analysis was inconclusive, it appeared that the production team had tried to bond too large an area and did not get the ablator panels under the required vacuum before the adhesive pot life ran out. Technicians at Michoud Assembly Facility reworked the application process for the ET at their facility and the first tank at KSC.

Following the ablator bonding problem, NASA intensified its analysis of the ablator/aluminum bond line. This analysis showed that the higher coefficient of thermal expansion of the ablator binder, as compared to the aluminum, would cause the ablator to shrink. This would introduce biaxial tension in the ablator and corresponding shear forces at the bond line near any edges, discontinuities, or cracks. Then, when the tank was pressurized, tank expansion from pressure would compound this shear force, possibly causing the bond line to fail. NASA decided to pre-pressurize the liquid hydrogen tank with helium gas prior to filling the tank for launch—and to pressures higher than flight pressures—to stretch the ablator when it was warm and elastic.

Because early test data showed the tank insulation could be adversely affected by ultraviolet light, NASA painted the first several tanks white, using a fire-retardant latex paint. Exposure testing of foam samples on the roof of the Michoud Assembly Facility, however, showed the damage to be so shallow that it was insignificant. NASA decided not to paint the tanks, resulting in a weight savings of about 260 kg (580 pounds), lowered labor costs, and the introduction of the "orange" tank.

Environmental Challenges

Knowledge of toxic properties and environmental contaminations increased over the 30 years of the Space Shuttle Program. Federal laws reflected these changes. For instance, ozone-depleting substances, including some Freon® compounds, reduced the protecting atmospheric ozone layer. NASA worked with its contractors to reduce both toxicity and environmental consequences for the cooling agents and the foam compounds.

During the 1990s, the University of Utah published data showing that CPR-421 was potentially toxic. Based on this analysis, Chemical Products Research withdrew CPR-421 from the market. NASA's ET office had Chemical Products Research reformulate this foam, with the new product identified as CPR-488.

New challenges arose related to emerging environmental policies that necessitated changes to Thermal Protection System foam formulations. In 1987, the United States adopted the Montreal Protocol on Substances that Deplete the Ozone Layer, which provided for the eventual international elimination of ozone-depleting substances. The United States implemented the protocol by regulations under the Clean Air Act. Ozone-depleting substances, including CFC-11—the Freon® blowing agent used in the production of the Thermal Protection System sprayable foams for the tanks—were scheduled to be phased out of production. After the phaseout, CFC-11 would only be available for such uses through a rigorous exemption process.

To prepare for the upcoming obsolescence of the foam blowing agent, Marshall Space Flight Center (MSFC) along with Lockheed Martin tracked and mitigated the effect of emerging environmental regulations. After extensive research and testing of potential substitutes, NASA proposed that HCFC-141b replace the CFC-11 blowing agent. NASA continued to use stockpiled supplies of CFC-11-blown foam until the HCFC-141b foam was certified for tank use and phased in beginning in 1996.

NASA undertook the development and qualification of a foam to be phased in as a replacement for the tank sidewall foam, CPR-488. North Carolina Foam Industries reformulated CPR-488 and developed a new product.

As part of qualifying this new product, Lockheed Martin, Wyle Laboratories, and MSFC developed an environmental test. This test used a flat aluminum plate machined to match aft dome stress levels. The plate was attached to a cryostat filled with liquid helium and then strained with hydraulic jacks

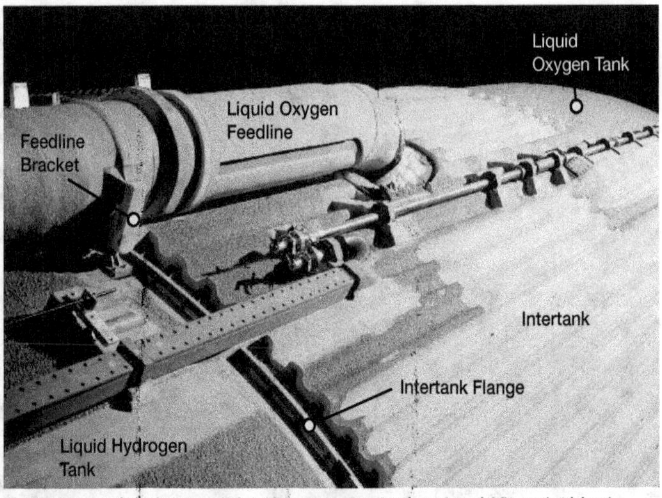

The foam's approximately 2.5-cm (1-in.) thickness borders the circumferential flange that joins the intertank with the liquid hydrogen tank. The ribbed area is the intertank, that, like the liquid oxygen tank in the background and the liquid hydrogen tank in the foreground, was robotically sprayed with NCFI 24-124 foam. The flange would later be hand-sprayed with Stepanfoam® BX-265. The liquid oxygen feedline at the top of the tank and a feedline bracket have been hand-sprayed with BX-265 foam.

A technician at NASA's Michoud Assembly Facility sprays the flange that connects the intertank and liquid hydrogen tank. Stepanfoam® BX-265 was sprayed manually on closeout areas, exterior tank feedlines, internal tank domes, closeout areas of mating External Tank subcomponent surfaces, and small subcomponents.

to the flight biaxial stress levels. Radiant heat lamps were installed to match the radiant heating from the solid rocket motor plumes, and an acoustic horn blasted the test. This simulated the aft dome ascent environment as well as possible. The test results indicated the need to spray ablator on the aft dome. To provide the capability to spray the ablator, personnel at Michoud Assembly Facility built two spray cells, with an additional cell to clean and prime the liquid hydrogen tank before ablator application.

To save the weight of this ablator and its associated cost, NASA had North Carolina Foam Industries develop a foam adequate for the aft dome environment without ablator. The foam was phased in on the aft dome, flying first on Space Transportation System (STS)-79 in 1996. The first usage of the new foam on the tank sidewalls was phased in over three tanks starting with STS-85 in August 1997.

Environmental Protection Agency regulations also required NASA to replace Stepanfoam® BX-250, which was sprayed manually—with a CFC-11 blowing agent—on the tank's "closeout" areas. During STS-108 (2001), Stepanfoam® BX-265—with HCFC-141b as its blowing agent—first flew as a replacement for BX-250. BX-250 continued to be flown in certain applications as BX-265 was phased into the manufacturing process.

The use of HCFC-141b as a foam blowing agent, however, was also problematic. It was classified as a Class II ozone-depleting substance and was subject to phaseout under the

Aerogel-based Insulation System Precluded Hazardous Ice Formation

Testing of gaseous hydrogen vent arm umbilical disconnect equipment at Kennedy Space Center.

During the STS-114 (2005) tanking test, the External Tank Gaseous Hydrogen Vent Arm Umbilical Quick Disconnect formed ice and produced liquid nitrogen/air. The phenomenon was repeated during subsequent testing and launch. For the shuttle, ice presented a debris hazard to the Orbiter Thermal Protection System and was unacceptable at this umbilical location. The production of uncontrolled liquid nitrogen/air presented a hazard to the shuttle, launch pad, and ground support equipment.

NASA incorporated a fix into the existing design to preclude ice formation and the uncontrolled production of liquid nitrogen/air. The resolution was accomplished with two changes to the umbilical purge shroud. First, the space agency improved the shroud purge gas flow to obtain the desired purge cavity gas concentrations. Second, technicians wrapped multiple layers of aerogel blanket material directly onto the quick disconnect metal surfaces within the purged shroud cavity.

NASA tested the design modifications at the Kennedy Space Center Cryo Test Lab. Tests showed that the outer surface of the shroud was maintained above freezing with no ice formation and that no nitrogen penetrated into the shroud purge cavity. NASA used the modified design on STS-121 (2006) and all subsequent flights.

Aerogel insulation is a viable alternative to the current technology for quick disconnect shrouds purged with helium or nitrogen to preclude the formation of ice and liquid nitrogen/air. In most cases, aerogel insulation eliminates the need for active purge systems.

Clean Air Act effective January 2003. NASA was granted exemptions permitting the use of HCFC-141b in foams for specific shuttle applications. These exemptions applied until the end of the program.

Post-Columbia Accident Advances in Thermal Protection

Following the loss of Space Shuttle Columbia in 2003, NASA undertook the redesign of some tank components to reduce the risk of ice and foam debris coming off the tank. These hardware changes drove the need to improve the application of Thermal Protection System foam that served as an integral part of the components' function. The major hardware addressed included the ET/Orbiter attach bipod closeout, protuberance air load ramps, ice frost ramps, and the liquid hydrogen tank-to-intertank flange area.

The ET bipod attached the Orbiter to the tank. The redesign removed the foam ramps that had covered the bipod attach fittings, and which had been designed to prevent the formation of ice when the ET was filled with cold liquid hydrogen and liquid oxygen on the launch pad. This left the majority of each fitting exposed. NASA installed heaters as part of the bipod configuration to prevent ice formation on the exposed fittings.

NASA developed a multistep process to improve the manual bipod Thermal Protection System spray technique. Validation of this process was accomplished on a combination of high-fidelity mock-ups and a full-scale ET test article in a production environment. Wind tunnel tests demonstrated Thermal Protection System closeout capability to withstand maximum aerodynamic loads without generating debris.

The ET protuberance air load ramps were manually sprayed wedge-shaped layers of insulating foam insulation along the pressurization lines and cable tray on the side of the tank. They were designed as a safety precaution to protect the tank's cable trays and pressurization lines from airflow that could potentially cause instability in these attached components. Foam loss from the ramps during ascent, however, drove NASA to remove them from the tank. This required extensive engineering. NASA created enhanced structural dynamics math models to better define the characteristics of this area of the tank and performed numerous wind tunnels tests.

The ET fuel tank Main Propulsion System pressurization lines and cable trays were attached along the length of the tank at multiple locations by metal support brackets. These were protected from forming ice and frost during tanking operations by foam protuberances called ice frost ramps. The feedline bracket configuration had the potential for foam and ice debris loss. Redesign changes were

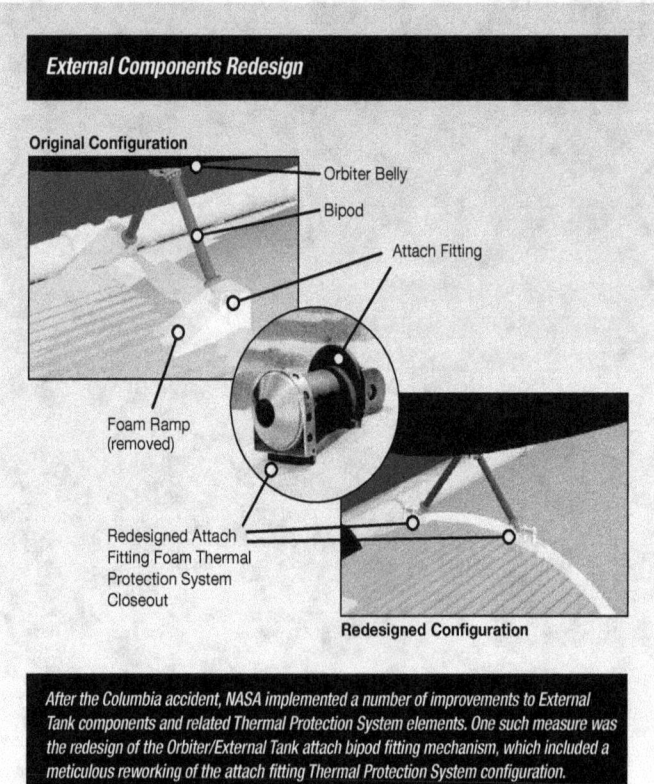

External Components Redesign

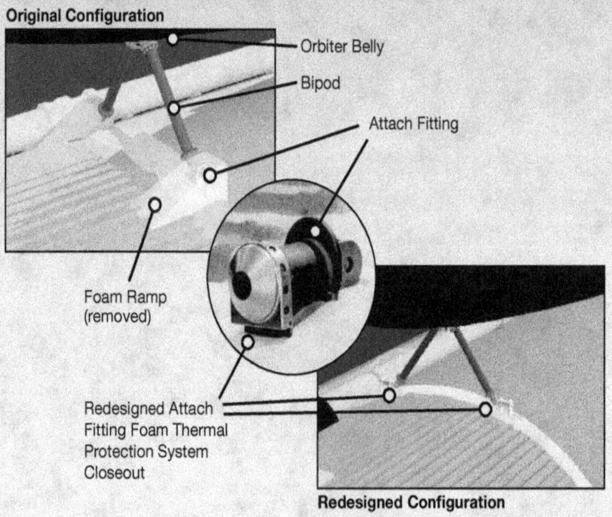

After the Columbia accident, NASA implemented a number of improvements to External Tank components and related Thermal Protection System elements. One such measure was the redesign of the Orbiter/External Tank attach bipod fitting mechanism, which included a meticulous reworking of the attach fitting Thermal Protection System configuration.

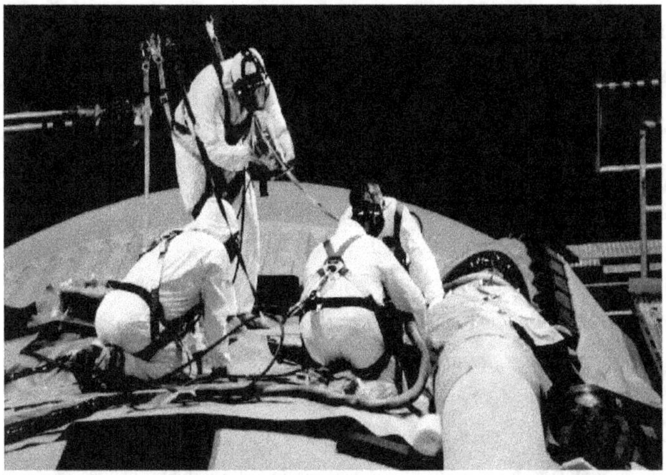

In what used to be a one-person operation, a team of technicians at NASA's Michoud Assembly Facility prepares to hand-spray BX-250 foam on the bipod attach fittings. The videographer (standing) records the process for later review and verification. A quality control specialist (left) witnesses the operation, while two spray technicians make preparations.

incorporated into the 17 ice frost ramps on the liquid hydrogen tank to reduce foam loss. BX-265 manual spray foam replaced foam in the ramps' closeout areas to reduce debonding and cracking.

The NASA/Lockheed Martin team also developed an enhanced three-part procedure to improve the Thermal Protection System closeout process on the liquid hydrogen tank-to-intertank flange area.

In all post-Columbia Thermal Protection System enhancement efforts, NASA modified process controls to ensure that defects were more tightly kept within the design envelope. The space agency simplified application techniques and spelled out instructions in more detail, and technicians had the opportunity to practice their application skills on high-fidelity component models. MSFC and Lockheed Martin also developed an electronic database to store information for each spray. New application certification requirements were added. Improvements included the forward bellows heater, the liquid oxygen feedlines, and titanium brackets. Improved imagery analysis and probabilistic risk assessments also allowed NASA to better track and predict foam loss. Thermal protection debris could never be completely eliminated, but NASA had addressed a complex and unprecedented set of problems with determination and innovation.

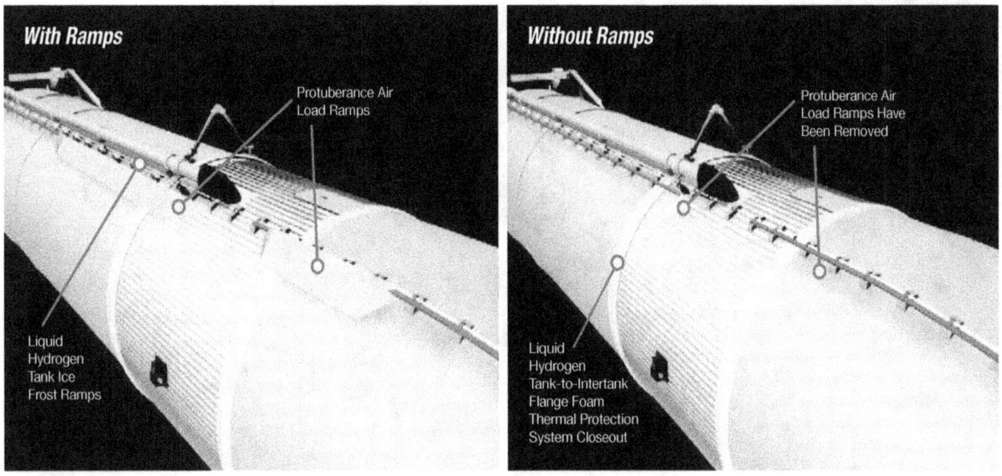

NASA decided to delete the tank's protuberance air load ramps and implement design changes to the 17 ice frost ramps on the liquid hydrogen tank. Both these measures required adjustments in the components' Thermal Protection System configuration and application processes. Materials and techniques were also altered to improve the Thermal Protection System closeout of the flange joining the liquid hydrogen tank with the intertank.

Materials and Manufacturing

Introduction
Gail Chapline

Nondestructive Testing Innovations
Willard Castner
Patricia Howell
James Walker

Friction Stir Welding Advancements
Robert Ding
Jim Butler

Characterization of Materials in the Hydrogen Environment
Jon Frandsen
Jonathan Burkholder
Gregory Swanson

Space Environment: It's More Than a Vacuum
Lubert Leger
Steven Koontz

Chemical Fingerprinting
Michael Killpack

Environmental Assurance
Anne Meinhold

Unprecedented Accomplishments in the Use of Aluminum-Lithium Alloy
Preston McGill
Jim Butler
Myron Pessin

Orbiter Payload Bay Door
Lubert Leger
Ivan Spiker

To build a spacecraft, we must begin with materials. Sometimes the material choice is the solution. Other times, the design must accommodate the limitations of materials properties. The design of the Space Shuttle systems encountered many material challenges, such as weight savings, reusability, and operating in the space environment. NASA also faced manufacturing challenges, such as evolving federal regulations, the limited production of the systems, and maintaining flight certification. These constraints drove many innovative materials solutions. Innovations such as large composite payload bay doors, nondestructive materials evaluation, the super lightweight tank, and the understanding of hydrogen effects on materials were pathfinders used in today's industry. In addition, there were materials innovations in engineering testing, flight analysis, and manufacturing processes. In many areas, materials innovations overcame launch, landing, and low-Earth orbit operational challenges as well as environmental challenges, both in space and on Earth.

Nondestructive Testing Innovations

Have you ever selected a piece of fruit based on its appearance or squeezed it for that certain feel? Of course you have. We all have. In a sense, you performed a nondestructive test. Actually, we perform nondestructive testing every day. We visually examine or evaluate the things we use and buy to see whether they are suitable for their purpose. In most cases, we give the item just a cursory glance or squeeze; however, in some cases, we give it a conscious and detailed examination. We don't think of these routine examinations as nondestructive tests, but they are, and they give us a sense of what nondestructive testing is about.

Nondestructive testing is defined as the inspection or examination of materials, parts, and structures to determine their integrity and future usefulness without compromising or affecting their usefulness. The most fundamental nondestructive test of all is visual inspection. In the industrial world, visual examination can be quite formal, with complex visual aids, pass/fail criteria, training requirements, and written procedures.

Nondestructive testing depends on incident or input energy that interacts with the material or part being examined. The incident or input energy can be modified by reflection from interaction within or transmission through the material or part. The process of detection and interpretation of the modified energy is how nondestructive testing provides knowledge about the material or part. Tests range from the simple detection and interpretation of reflected visible light by the human eye (visual examination) to the complex electronic detection and mathematical reconstruction of through-transmitted x-radiation (computerized axial tomography [CAT] scan). From a nondestructive testing perspective, the similarity between the simple visual examination and the complex CAT scan is the input energy (visible light vs. x-rays) and the modified energy (detected by the human eye vs. an electronic x-ray detector).

Nondestructive testing is a routine part of a spacecraft's life cycle. For the reusable shuttle, nondestructive testing began during the manufacturing and test phases and was applied throughout its service life. NASA performed many such nondestructive tests on the shuttle vehicles and developed most nondestructive testing innovations in response to shuttle problems.

Quantitative Nondestructive Testing of Fatigue Cracks

One of the most significant nondestructive testing innovations was quantifying the flaw sizes that conventional nondestructive testing methods could reliably detect. NASA used artificially induced fatigue cracks to make the determination because such flaws were relatively easy to grow and control, hard to detect, and tended to bound the population of flaws of interest. The need to quantify the reliably detectable crack sizes was

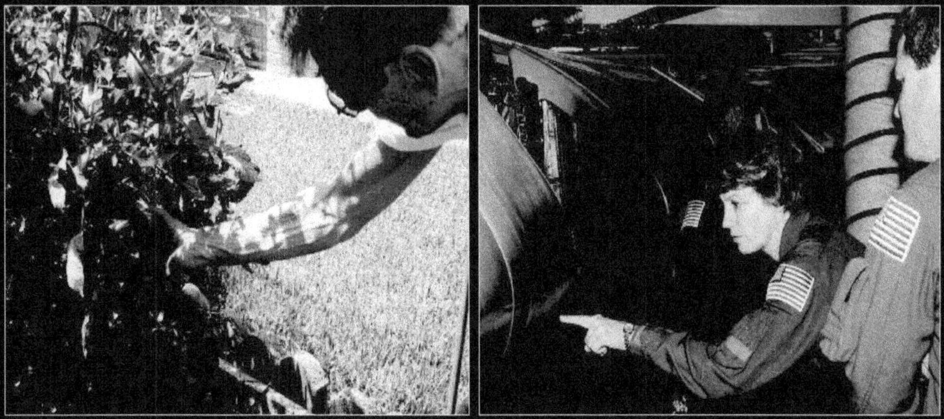

Two examples of the most basic nondestructive testing:
Left, a gardener checks ripening vegetables. Right, Astronaut Eileen Collins, STS-114 (2005) mission commander, looks closely at a reinforced carbon-carbon panel on one of the wings of the Space Shuttle Atlantis in the Orbiter Processing Facility at Kennedy Space Center (KSC). Collins and the other crew members were at KSC to take part in hands-on equipment and Orbiter familiarization.

mandated by a fracture control interest in having confidence in the starting crack size that could be used in fracture and life calculations. Although there was no innovation of any specific nondestructive testing method, quantifying—in a statistical way—the reliably detectable crack sizes associated with the conventional nondestructive evaluation methods was innovative and led the way to the adoption of similar quantitative nondestructive evaluation practices in other industries.

The quantification of nondestructive testing methods is commonly referred to today as probability of detection. The Space Shuttle Program developed some of the earliest data for the penetrant, x-ray, ultrasonic, and eddy current nondestructive testing methods—the principal nondestructive testing methods used to inspect shuttle components during manufacturing. Data showed that inspectors certified to aerospace inspection standards could, on average, perform to a certain probability of detection level defined as standard nondestructive evaluation.

Beyond standard nondestructive evaluation, NASA introduced a special nondestructive evaluation level of probability of detection wherein the detection of cracks smaller than the standard sizes had to be demonstrated by test. Engineers fabricated fatigue-cracked specimens that were used over many years to certify and recertify, by test, the inspectors and their nondestructive evaluation processes to the smaller, special nondestructive evaluation crack size. The size of the fatigue cracks in the specimens was targeted to be a surface-breaking semicircular crack 0.127 cm (0.050 in.) long by 0.063 cm (0.025 in.) deep, a size that was significantly smaller than the standard nondestructive evaluation crack size of 0.381 cm (0.150 in.) long by 0.19 cm (0.075 in.) deep.

Quantitative Nondestructive Testing

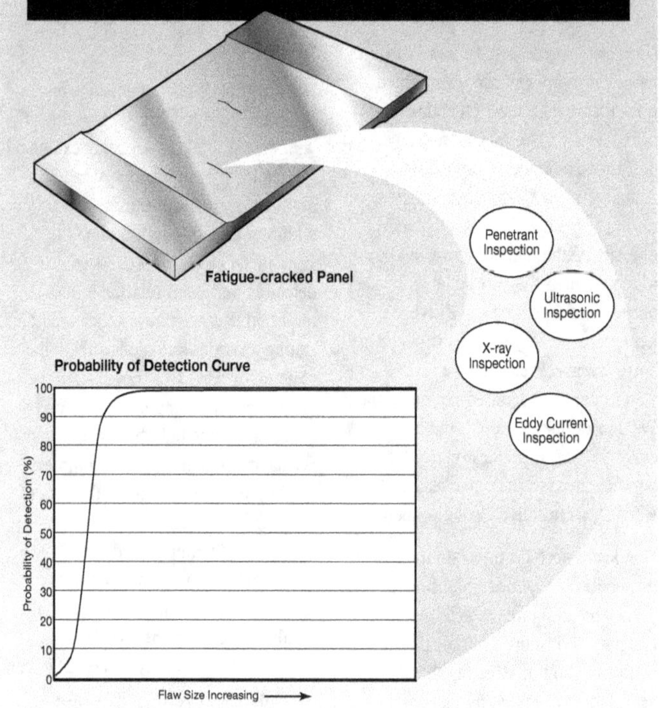

The special probability of detection specimen sets typically consisted of 29 randomly distributed cracks of approximately the same size. By detecting all 29 cracks, the inspector and the specific nondestructive evaluation process were considered capable of detecting the crack size to a 90% probability of detection with 95% confidence.

Nondestructive Testing of Thermal Protection System Tiles

The development of Thermal Protection System tiles was one of the most unique and difficult developments of the program. Because of this material's "unknowns," the tile attachment scheme, and their extremely fragile nature, NASA examined a number of nondestructive testing methods.

Acoustic Emission Monitoring

Late in the development of the shuttle Thermal Protection System and just before the first shuttle launch, NASA encountered a major problem with the attachment of the tiles to the Orbiter's exterior skin. The bond strength of the tile system was lower than the already-low strength of the tile material, and this was not accounted for in the design. The low bond strength was due to stress concentrations at the tile-to-strain isolation pad bond line interface. A Nomex® felt strain isolation pad was bonded between each tile and the Orbiter skin to minimize the lateral strain input to the tile from the aluminum skin. These stress concentrations led to early and progressive failures of the tile material at the tile-to-strain isolation pad bond line interface when the tile was loaded.

To determine whether low bond strengths existed, engineers resorted to proof testing for each tile. This required thousands of individual tile proof tests prior to first flight. Space Shuttle Columbia (Space Transportation System [STS]-1) was at Kennedy Space Center being readied for first flight when NASA decided that proof testing was necessary. Since proof testing was not necessarily nondestructive and tiles could be damaged by the test, NASA sought a means of monitoring potential damage; acoustic emission nondestructive testing was an obvious choice. The acoustic signatures of a low bond strength tile or a tile damaged during proof test were determined through laboratory proof testing of full-size tile arrays.

To say that the development and implementation of acoustic emission monitoring during tile proof testing was done on a crash basis would be an understatement. The fast pace was dictated by a program that was already behind schedule, and the tile bond strength problem threatened significant additional delay. At the height of the effort, 18 acoustic emission systems with fully trained three-person crews were in operation 24 hours a day, 7 days a week. The effort was the largest single concentration of acoustic emission equipment at a single job site. As often happens with such problems, where one solution can be overtaken and replaced by another, a tile densification design fix for the low-strength bond was found and implemented prior to first flight, thus obviating the need for continued acoustic emission monitoring. By the time the acoustic emission monitoring was phased out, NASA had performed 20,000 acoustic emission monitored proof tests.

Acoustic Emission Monitoring of Tiles During Proof Test

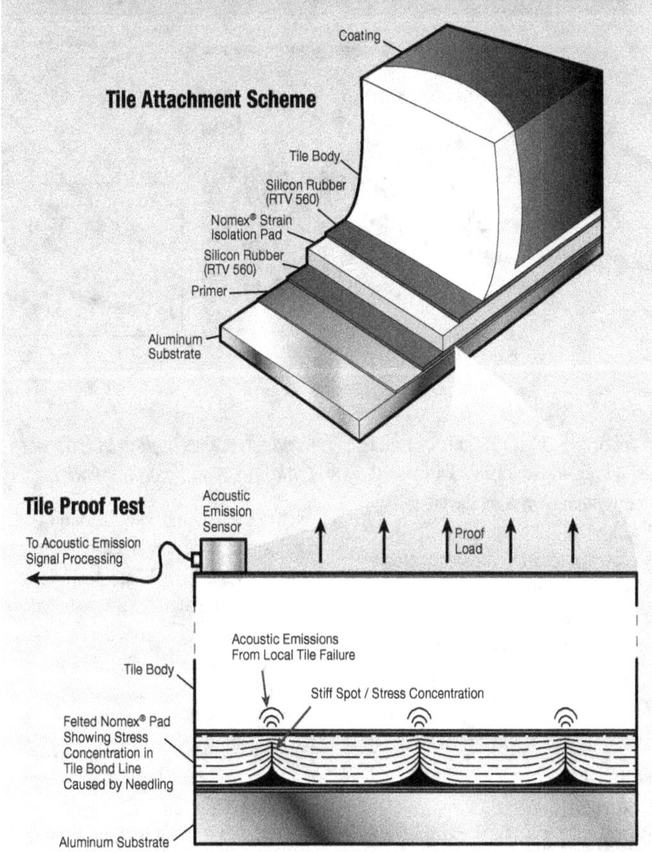

Sonic Velocity Testing

Another early shuttle nondestructive testing innovation was the use of an ultrasonic test technique to ensure that the Thermal Protection System tiles were structurally sound prior to installation. Evaluation of pulse or sonic velocity tests showed a velocity relationship with respect to both tile density and strength. These measurements could be used as a quality-control tool to screen tiles for low density and low strength and could also determine the orientation of the tile.

The sonic velocity technique input a short-duration mechanical impulse into the tile. A transmitting transducer and a receiving transducer, placed on opposite sides of the tile, measured the pulse's transit time through the tile. For the Lockheed-provided tile material, LI-900 (with bulk density of 144 kg/m^3 [9 pounds/ft^3]), the average through-the-thickness sonic velocity was on the

Sonic Velocity Testing of Tiles at Kennedy Space Center Thermal Protection System Facility
The speed of sound through the tile is related to density and strength.

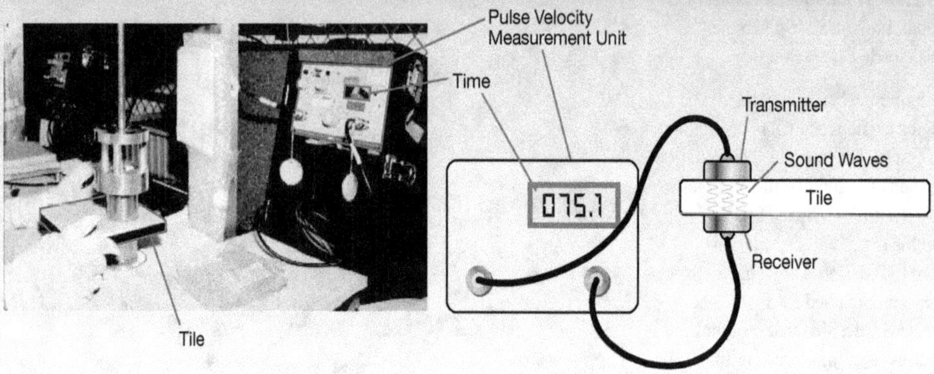

order of 640 m/sec (2,100 ft/sec), and the through-the-thickness flat-wise tensile strength was on the order of 1.69 kg/cm^2 (24 pounds/in^2). The LI-900 acceptance criterion for sonic velocity was set at 518 m/sec (1,700 ft/sec), which corresponded to a minimum strength of 0.91 kg/cm^2 (13 pounds/in^2). Sonic velocity testing was phased out in the early 1990s.

Post-Columbia Accident Nondestructive Testing of External Tank

A consequence of the Columbia (STS-107) accident in 2003 was the development of several nondestructive innovations, including terahertz imaging and backscatter radiography of External Tank foam and thermography of the reinforced carbon-carbon—both on orbit and on the ground—during vehicle turnaround. The loss of foam, reinforced carbon-carbon impact damage, and on-orbit inspection of Thermal Protection System damage were all problems that could be mitigated to some extent through the application of nondestructive testing methods.

Nondestructive Testing of External Tank Spray-on Foam Insulation

Prior to the Columbia accident, no nondestructive testing methods were available for External Tank foam inspection, although NASA pursued development efforts from the early 1980s until the early 1990s. The foam was effectively a collection of small air-filled bubbles with thin polyurethane membranes, making the foam a thermal and electrical insulator with very high acoustic attenuation. Due to these properties, it was not feasible to inspect the foam with conventional methods such as eddy current, ultrasonics, or thermography. In addition, since the foam was considered nonstructural, problems of delaminations occurring during foam application and foam popping off ("popcorning") during ascent were considered manageable through process control.

After the Columbia accident, NASA focused on developing nondestructive testing methods for finding voids and delaminations in the thick, hand-sprayed foam applications around protuberances and closeout areas. The loss of foam applied to the large areas of the tank was not as much of concern because the automated acreage spray-on process was better controlled, making it more unlikely to come off. In the event it did come off, the pieces would likely be small because acreage foam was relatively thin. NASA's intense focus resulted in the development and implementation of two methods for foam inspection—terahertz imaging and backscatter radiography—that represented new and unique application of nondestructive inspection methods.

Terahertz Imaging

Terahertz imaging is a method that operates in the terahertz region of the electromagnetic spectrum between microwave frequencies and far-infrared frequencies. Low-density hydrocarbon materials like External Tank foam were relatively transparent to terahertz radiation. Terahertz imaging used a pulser to transmit energy into a structure and a receiver to record the energy reflected off the substrate or internal defects. As the signal traveled through the structure, its basic wave

Terahertz Imaging System

This system uses high-frequency electromagnetic pulses.

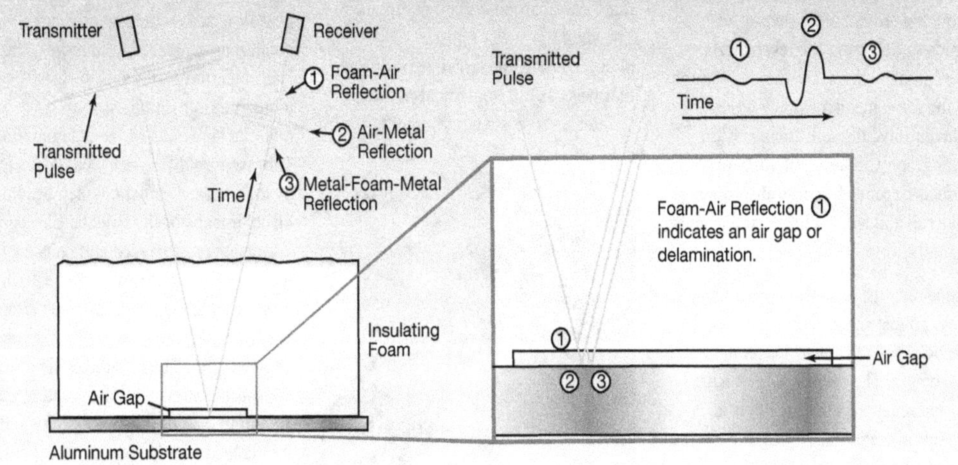

properties were altered by the attenuation of the material and any internal defects. An image was made by scanning the pulser/receiver combination over the foam surface and displaying the received signal.

Probability of detection studies of inserted artificial voids showed around 90% detection of the larger voids in simple geometries, but less than 90% detection in the more-complicated geometries of voids around protrusions. Further refinements showed that delaminations were particularly difficult to detect. The detection threshold for a 2.54-cm- (1-in.)-diameter laminar defect was found to be a height of 0.508 cm (0.2 in.), essentially meaning delaminations could not be detected. The terahertz inspection method was used for engineering evaluation, and any defects found were dealt with by an engineering review process.

Backscatter Radiography

Backscatter radiography uses a conventional industrial x-ray tube to generate a collimated beam of x-rays that is scanned over the test object. The backscattering of x-rays results from the Compton effect—or scattering—in which absorption of the incident or primary x-rays by the atoms of the

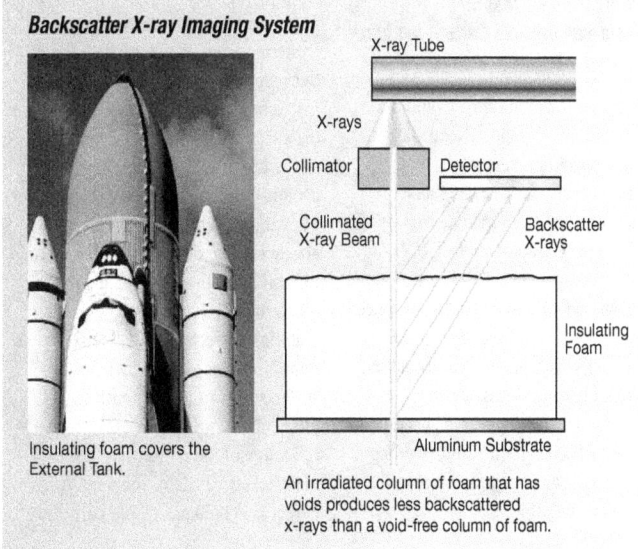

Backscatter X-ray Imaging System

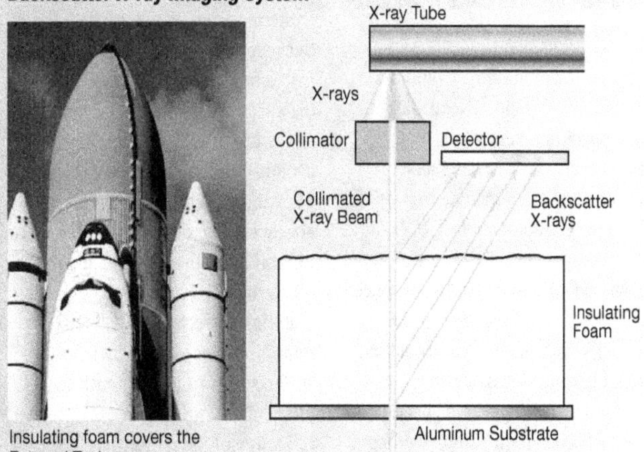

Insulating foam covers the External Tank.

An irradiated column of foam that has voids produces less backscattered x-rays than a void-free column of foam.

test material are reradiated at a lower energy as secondary x-rays in all directions. The reradiated or backscattered x-rays were collected in collimated radiation detectors mounted around the x-ray source. Voids or defects in the test material were imaged in backscatter radiography in the same manner as they were in conventional through-transmission radiography. Imaging of voids or defects depended on less absorbing material and less backscattered x-rays from the void.

Since only the backscattered x-rays were collected, the technique was single sided and suited for foam inspection. The foam was well suited for backscatter radiography since Compton scattering is greater from low atomic number materials. The technique was more sensitive to near surface voids but was unable to detect delaminations. Like terahertz imaging, backscatter radiography was used for engineering evaluation, and defects found were dealt with by an engineering review process.

Nondestructive Testing of Reinforced Carbon-Carbon System Components

A recommendation of the Columbia Accident Investigation Board stated: "Develop and implement a comprehensive inspection plan to determine the structural integrity of all Reinforced Carbon-Carbon (RCC) system components. This inspection plan should take advantage of advanced non-destructive inspection technology." To comply with this recommendation, NASA investigated advanced inspection technology for inspection of the reinforced carbon-carbon leading edge panels during ground turnarounds and while on orbit.

Ground Turnaround Thermography

NASA selected infrared flash thermography as the method to determine the structural integrity of the reinforced carbon-carbon components. Thermography was a fast, noncontacting, one-sided application that was easy to implement in the Orbiter's servicing environment.

Infrared thermography inspection of the Orbiter nose captured at the instant of the xenon lamp flash. Kennedy Space Center Orbiter Processing Facility.

The Thermographic Inspection System was an active infrared flash thermogaphy system. Thermographic inspection examined and recorded the surface temperature transients of the test article after application of a short-duration heat pulse. The rate of heat transfer away from the test article surface depended on the thermal diffusivity of the material and the uniformity and integrity of the test material. Defects in the material would retard the heat flow away from the surface, thus producing surface temperature differentials that were reflective of the uniformity of the material and its defect content. A defect-free material would uniformly transfer heat into the underlying material, and the surface temperature would appear the same over the entire test surface; however, a delamination would prevent or significantly retard heat flow across the gap created by the delamination, resulting in more-local heat retention and higher surface temperature in comparison to the material surrounding the delamination. Temperature differences were detected by the infrared camera, which provided visual images of the defects. Electronic signals were processed and enhanced for easier interpretation. The heat pulse was provided by flashing xenon lamps in a hooded arrangement that excluded ambient light. The infrared camera was transported along a floor-mounted rail system in the Orbiter Processing Facility for the leading edge panel inspections, allowing full and secure access to all of the leading edge surfaces. After the transport cart was positioned, the camera was positioned manually via a grid system that allowed the same areas to be compared from flight to flight.

The thermography system was validated on specimens containing flat bottom holes of different diameters and depths. Validation testing confirmed the ability of the flash thermography system to detect the size holes that needed to be detected.

After the first Return to Flight mission—STS-114 (2005)—the postflight thermography inspection discovered a suspicious indication in the joggle area of a panel. Subsequent investigation showed that the indication was a delamination. This discovery set in motion an intense focus on joggle-area delaminations and their characterization and consequence. Many months of further tests, development, and refinement of the thermography methodology determined that critical delaminations would be detected and sized by flash thermography and provided the basis for flightworthiness.

On-orbit Thermography

The success of infrared thermography for ground-based turnaround inspection of the wing leading edge panels and the extensive use of thermography during Return to Flight impact testing made it the choice for on-orbit inspection of the leading edge reinforced carbon-carbon material. A thermal gradient through the material must exist to detect subsurface reinforced carbon-carbon damage with infrared thermography. A series of ground tests demonstrated that sunlight or solar heating and shadowing could be used to generate the necessary thermal gradient, which significantly simplified the camera development task.

With the feasibility of on-orbit thermography demonstrated and with the spaceflight limitations on weight and power taken into account, NASA selected a commercial off-the-shelf microbolometer camera for modification and development into a space-qualified infrared camera for inspecting the reinforced carbon-carbon for impact damage while on orbit.

The extravehicular activity infrared camera operated successfully on its three flights. Two reinforced carbon-carbon test panels with simulated damage were flown and inspected on STS-121 (2006). The intentional impact damage in one panel and the flat bottom holes in the other panel were clearly imaged. Engineers also performed a similar on-orbit test on two other intentionally damaged reinforced carbon-carbon test panels during a space station extravehicular activity with the

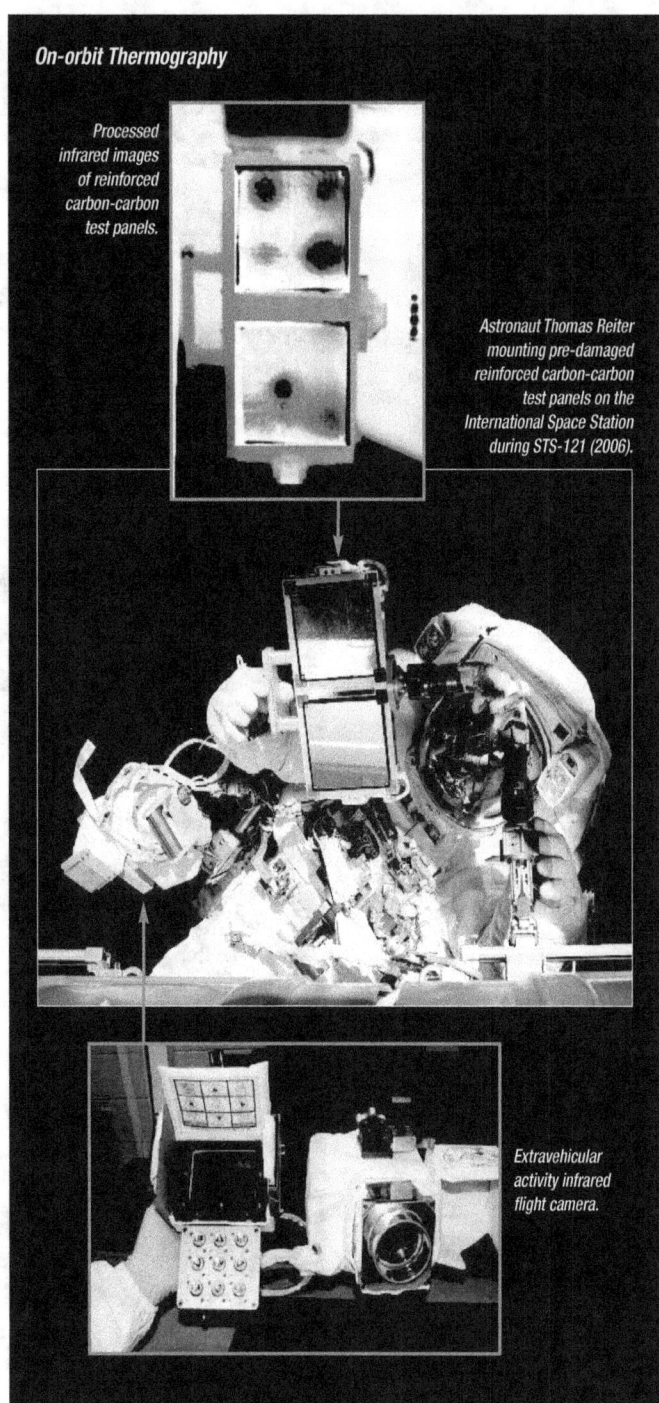

On-orbit Thermography

Processed infrared images of reinforced carbon-carbon test panels.

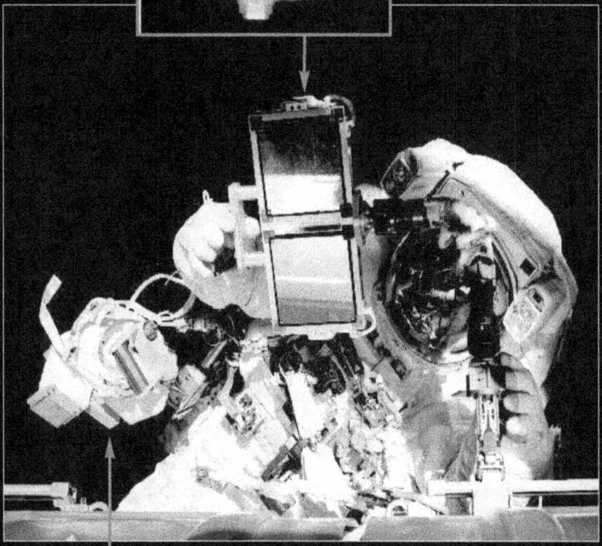

Astronaut Thomas Reiter mounting pre-damaged reinforced carbon-carbon test panels on the International Space Station during STS-121 (2006).

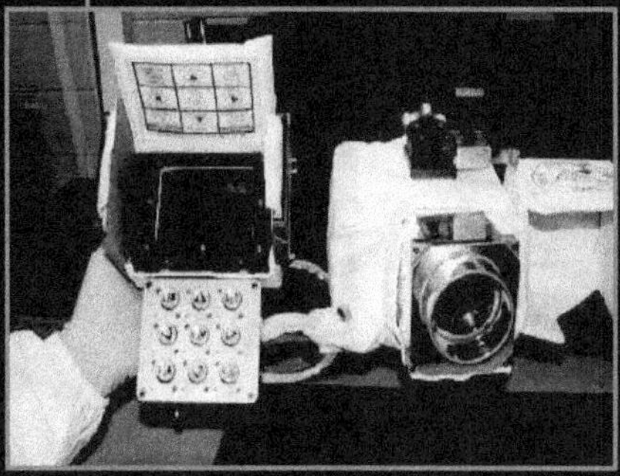

Extravehicular activity infrared flight camera.

same result of clearly imaging the damage. The end result of these efforts was a mature nondestructive inspection technique that was transitioned and demonstrated as an on-orbit nondestructive inspection technique.

Additional Nondestructive Testing

Most nondestructive testing innovations resulted from problems that the shuttle encountered over the years, where nondestructive testing provided all or part of the solution. Other solutions worth mentioning include: ultrasonic extensometer measurements of critical shuttle bolt tensioning; terahertz imaging of corrosion under tiles; phased array ultrasonic testing of the External Tank friction stir welds and the shuttle crawler-transporter shoes; thermographic leak detection of the main engine nozzle; digital radiography of Columbia debris; surface replication of flow liner cracks; and the on-board wing leading edge health monitoring impact system.

Friction Stir Welding Advancements

NASA invents welding fixture.

Friction stir welding units, featuring auto-adjustable pin tools, welded External Tank barrel sections at NASA's Michoud Assembly Facility in New Orleans, Louisiana. The units measured 8.4 m (27.5 ft) in diameter and approximately 7.6 m (25 ft) tall to accommodate the largest barrel sections.

In the mid 1990s, NASA pursued the implementation of friction stir welding technology—a process developed by The Welding Institute of Cambridge, England—to improve External Tank welds. This effort led to the invention of an auto-adjustable welding pin tool adopted by the Space Shuttle Program, the Ares Program (NASA-developed heavy launch vehicles), and industry.

Standard fusion-welding techniques rely on torch-generated heat to melt and join the metal. Friction stir welding does not melt the metal. Instead, it uses a rotating pin and "shoulder" to generate friction, stir the metal together, and forge a bond. This process results in welds with mechanical properties superior to fusion welds.

Standard friction stir welding technology has drawbacks, however; namely, a non-adjustable pin tool that leaves a "keyhole" at the end of a circular weld and the inability to automatically adjust the pin length for materials of varying thickness. NASA's implementation of friction stir welding for the External Tank resulted in the invention and patenting of an auto-adjustable pin tool that automatically retracts and extends in and out of the shoulder. This feature provides the capability to make 360-degree welds without leaving a keyhole, and to weld varying thicknesses.

During 2002-2003, NASA and the External Tank prime contractor, Lockheed Martin, implemented auto-adjustable pin tool friction stir welding for liquid hydrogen and liquid oxygen tank longitudinal welds. Since that time, these friction stir welds have been virtually defect-free. NASA's invention was being used to weld Ares upper-stage cryogenic hardware. It has also been adopted by industry and is being used in the manufacturing of aerospace and aircraft frames.

Characterization of Materials in the Hydrogen Environment

From the humid, corrosion-friendly atmosphere of Kennedy Space Center, to the extreme heat of ascent, to the cold vacuum of space, the Space Shuttle faced one hostile environment after another. One of those harsh environments—the hydrogen environment—existed within the shuttle itself. Liquid hydrogen was the fuel that powered the shuttle's complex, powerful, and reusable main engine. Hydrogen provided the high specific impulse—the bang per pound of fuel needed to perform the shuttle's heavy-lifting duties. Hydrogen, however, was also a potential threat to the very metal of the propulsion system that used it.

The diffusion of hydrogen atoms into a metal can make it more brittle and prone to cracking—a process called hydrogen embrittlement. This effect can reduce the toughness of carefully selected and prepared materials. A concern that exposure to hydrogen might encourage crack growth was present from the beginning of the Space Shuttle Program, but the rationale for using hydrogen was compelling.

The Challenge of the Hydrogen Environment

Hydrogen embrittlement posed more than a single engineering problem for the Space Shuttle. This was partly because hydrogen embrittlement can occur in three different ways. The most common mode occurs when hydrogen is absorbed by a material that is relatively unstressed, such as the components of the shuttle's main engines before they experienced the extreme loads of liftoff and flight; this is called internal hydrogen embrittlement. Under the right conditions, internal hydrogen embrittlement has the potential to render materials too weak and brittle to survive high stresses applied later.

Alternatively, embrittlement can affect a material that is immersed in hydrogen while the material is being stressed and deformed. This phenomenon is called hydrogen environment embrittlement, which can occur in pressurized hydrogen storage vessels. These vessels are constantly stressed while in contact with hydrogen. Hydrogen environment embrittlement can potentially reduce ductility over time and enable cracking, or hydrogen may simply reduce the strength of a vessel until it is too weak to bear its own pressure.

Finally, hydrogen can react chemically with elements that are present in a metal, forming inclusions that can degrade the properties of that metal or even cause blisters on the metal's surface. This effect is called hydrogen reaction embrittlement. In the shuttle's main engine components, the reaction between hydrogen and the titanium alloys occurred to internally form brittle titanium hydrides, which was most likely to occur at locations where there were high tensile stresses in the part. Hydrogen reaction embrittlement can affect steels when hydrogen atoms combine with the carbon atoms dissolved in the metal. Hydrogen reaction embrittlement can also blister copper when hydrogen reacts with the internal oxygen in a solid copper piece, thereby forming steam blisters.

Insights on Hydrogen Environment Embrittlement

NASA studied the effects of hydrogen embrittlement in the 1960s. In the early 1970s, the scope of NASA-sponsored research broadened to include hydrogen environment embrittlement effects on fracture and fatigue. Engineers immersed specimens in hydrogen and performed a battery of tests. They applied repeated load cycles to specimens until they fatigued and broke apart; measured crack growth rates in cyclic loading and under a constant static load; and tested materials in high-heat and high-pressure hydrogen environments. Always, results were compared for each material to its performance in room-temperature air.

During the early years of the Space Shuttle Program, NASA and contractor engineers made a number of key discoveries regarding hydrogen environment embrittlement. First, cracks were shown to grow faster when loaded in a hydrogen environment. This finding would have significant implications for the shuttle design, as fracture assessments of the propulsion system would have to account for accelerated cracking. Second, scientists observed that hydrogen environment embrittlement could result in crack growth under a constant static load. This behavior was unusual for metals. Ductile materials such as metals tend to crack in alternating stress fields, not in fixed ones, unless a chemical or an environmental cause is present. Again, the design of the shuttle would have to account for this effect. Finally, hydrogen environment embrittlement was shown to have more severe effects at higher pressures. Intriguingly, degradation of tensile properties was found to be proportional to the square root of pressure.

The overall approach to hydrogen environment embrittlement research was straightforward. As a matter of common practice, NASA characterized the strength and fracture behavior of its alloys. To determine how these alloys would tolerate hydrogen, engineers simply adapted their tests to include a high-pressure hydrogen environment. After learning that high pressure exacerbates hydrogen environment embrittlement, they further adapted the tests to include a hydrogen pressure of 703 kg/cm^2 (10,000 psi). Later in the program, materials being considered for use in the main engine were tested at a reduced pressure of 492 kg/cm (7,000 psi) to be more consistent with operation conditions. The difference between room-temperature air material property data and these new results was a measurable effect of hydrogen environment embrittlement. Now that these effects could be quantified, the next step was to safeguard the shuttle.

Making Parts Resistant to Hydrogen Environment Embrittlement

One way to protect the main engines from hydrogen environment embrittlement was through materials selection. NASA chose naturally resistant materials when possible. There were, however, often a multitude of conflicting demands on these materials: they had to be lightweight, strong, tough, well suited for the manufacturing processes that shaped them, weldable, and able to bear significant temperature swings. The additional constraint of imperviousness to hydrogen environment embrittlement was not always realistic, so engineers experimented with coatings and plating processes. The concept was to shield vulnerable metal from any contact with hydrogen. A thin layer of hydrogen environment embrittlement-resistant metal would form a barrier that separated at-risk material from hydrogen fuel.

Engineers concentrated their research on coatings that had low solubility and low-diffusion rates for hydrogen at room temperature. Testing had demonstrated that hydrogen environment embrittlement is worst at near-room temperature, so NASA selected coatings based on their effectiveness in that range. The most efficient barrier to hydrogen, engineers found, was gold plating; however, the cost of developing gold plating processes was a significant factor. Engineers observed that copper plating provided as much protection as gold, as long as a thicker and heavier layer was applied.

Protecting weld surfaces was often more challenging. The weld surfaces exposed to hydrogen fuel during flight were typically not accessible to plating after the weld was complete. Overcoming this problem required a more time-consuming and costly approach. Engineers developed weld overlays, processes in which hydrogen environment embrittlement-resistant filler metals were added during a final welding pass. These protective fillers sealed over the weld joints and provided the necessary barrier from hydrogen. NASA used overlays in combination with plating of accessible regions to prevent hydrogen environment embrittlement in engine welds.

These approaches—a combination of two or more hydrogen environment embrittlement prevention methods—were the practical solution for many of the embrittlement-vulnerable parts of the engines. For example, the most heavily used alloy in the engines was Inconel® 718, an alloy known to be affected by hydrogen environment embrittlement. Engineers identified an alternative heat treatment, different from the one typically used, which limited embrittlement. But this alone was insufficient. In the most critical locations, the alternative heat treatment was combined with copper plating and weld overlays.

A unique processing approach was also used to prevent embrittlement in the engine's main combustion chamber. This chamber was made with a highly conductive copper alloy. Its walls contained cooling channels that circulated cold liquid hydrogen and kept the chamber from melting in the extreme heat of combustion. But the hydrogen-filled channels became prone to hydrogen environment embrittlement. These liquid hydrogen channels were made by machining slots in the copper and then plated with nickel, which closed out the open slot and formed a coolant channel. The nickel plate cracked in the hydrogen environment and reduced the pressure capability of the channels. Engineers devised a two-part solution. First, they developed an alternative heat treatment to optimize nickel's performance in hydrogen. Next, they coated the nickel with a layer of copper to isolate it from the liquid hydrogen. This two-pronged strategy worked, and liquid hydrogen could be safely used as the combustion chamber coolant.

Addressing Internal Hydrogen Embrittlement

Whereas hydrogen environment embrittlement was of great concern at NASA in the 1960s, internal hydrogen embrittlement was largely dismissed even through the early years of the Space Shuttle Program. Internal hydrogen embrittlement had never been a significant problem for the types of materials used in spaceflight hardware. The superalloys and particular stainless steels selected by NASA were thought to be resistant to internal hydrogen embrittlement. Engineers thought the face-centered, cubic, close-packed crystal structure would leave too little room for hydrogen to permeate and diffuse.

Recall that internal hydrogen embrittlement occurs when hydrogen is absorbed before high operational stresses. Hydrogen enters into the metal and remains there, making it more brittle and likely to crack when extreme service loads are applied later. It is the accumulation of absorbed hydrogen, rather than the immediate exposure at the moment of high stress, that compromises an internal hydrogen embrittlement-affected material. When NASA initially designed the main engine, engineers accounted for hydrogen absorbed during manufacturing. Engineers, however, thought that the materials that were formed and processed without collecting a significant amount of hydrogen were not in danger of absorbing considerable amounts later.

This notion about internal hydrogen embrittlement was challenged during the preparation of an engine failure analysis document in 1988. The engine was repeatedly exposed to hydrogen in flight and after flight, at high temperatures and extreme pressure. The report suggested that in these exceptional heat and pressure conditions some engine materials might, in fact, gather small amounts of hydrogen with each flight. Gradually, over time, these materials could accumulate enough hydrogen to undermine ductility.

Engineers developed a special test regimen to screen materials for high-temperature, high-pressure hydrogen accumulation. Test specimens were "charged" with hydrogen at 649°C (1,200°F) and 351.6 kg/cm^2 (5,000 psi). They were then quickly cooled and tested for strength and ductility under normal conditions. Surprisingly, embrittlement by internal hydrogen embrittlement was observed to be as severe as by hydrogen environment embrittlement. As a subsequent string of fatigue tests confirmed this comparison, NASA had to reevaluate its approach to preventing hydrogen embrittlement. The agency's focus on hydrogen environment embrittlement had been a near-total focus. Now, a new awareness of internal hydrogen embrittlement would drive a reexamination.

Fortunately, the process for calculating design properties from test data had been conservative. The margins of safety were wide enough to bound the combined effects of internal hydrogen embrittlement and hydrogen environment embrittlement. The wealth of experience gained in studying hydrogen environment embrittlement and mitigating its effects also worked in NASA's favor. Some of the same methodologies could now be applied to internal hydrogen embrittlement. For instance, protective plating would operate on the same principle—the creation of a barrier between hydrogen and a vulnerable alloy—whether hydrogen environment embrittlement or internal hydrogen embrittlement was the chief worry. Continued testing of "charged" specimens would allow quantification of internal hydrogen embrittlement damage, just as hydrogen immersion testing had enabled measurement of hydrogen environment embrittlement effects.

Taking strategies generated to avoid hydrogen environment embrittlement and refitting them to prevent internal hydrogen embrittlement, however, often required additional analysis. For example, from the beginning of the Space Shuttle Program NASA used coatings to separate at-risk metals from hydrogen. The agency intentionally chose these coatings for their performance at near-room temperature, when hydrogen environment embrittlement is most aggressive. Tests showed the coatings were less effective in the high heat that promotes internal hydrogen embrittlement. New research and experimentation was required to prove that these protective coatings were adequate—that, although they didn't completely prevent the absorption of hydrogen when temperatures and pressures were extreme, they did reduce it to safe levels.

Special Cases: High-Pressure Fuel Turbopump Housing

NASA encountered a unique hydrogen embrittlement issue during development testing of the main engine high-pressure fuel turbopump.

High-Pressure Fuel and Oxidizer Turbopump Turbine Blade Cracks

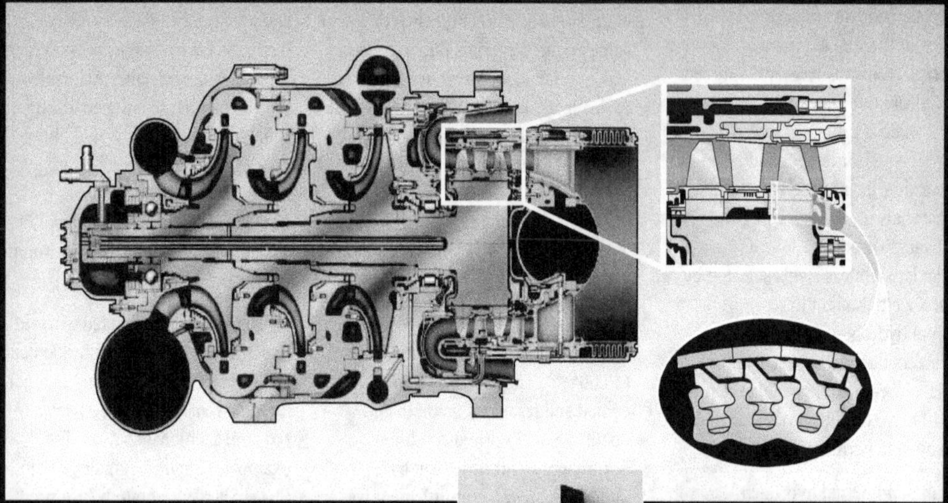

After observing cracks on polycrystalline turbine blades, NASA redesigned the blades as single-crystal parts. When tested in hydrogen, cracks were detected. Scientists used a Brazilian disc test to create the tensile and shear stresses that had caused growth. NASA resolved cracking in the airfoil with changes that eliminated stress concentrations and smoothed the flow of molten metal during casting. To assess cracking at damper contacts, scientists extracted test specimens from single crystal bars, machined contact pins from the damper material, and loaded two specimens. This contact fixture was supported in a test rig that allowed the temperature, loads, and load cycle rate to be varied. Specimens were pre-charged with hydrogen, tested at elevated temperatures, and cycled at high frequency to actual operating conditions.

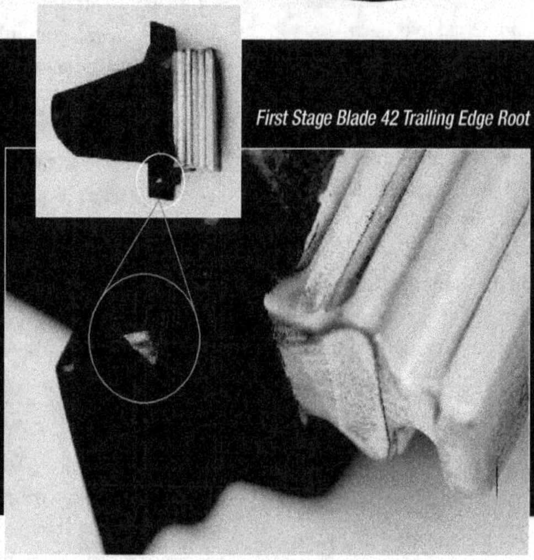

First Stage Blade 42 Trailing Edge Root

Schematic of Test Rig

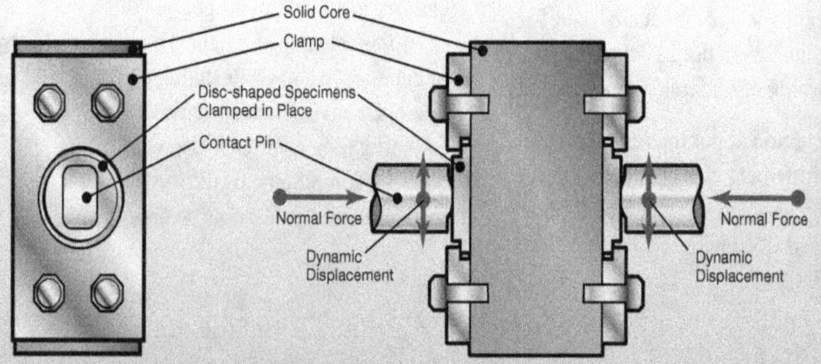

A leak developed during the test; this leak was traced to cracks in the mounting flange of the turbopump's housing. The housing was made from embrittlement-prone nickel-chromium alloy Inconel® 718, and the cracks were found to originate in small regions of highly concentrated stress. So, engineers changed the material to a more-hydrogen-tolerant alloy, Inconel® 100, and they redesigned the housing to reduce stress concentrations. This initially appeared to solve the problem. Then, cracks were discovered in other parts of the housing. Structural and thermal analysis could not explain this cracking. The locations and size of the cracks did not fit with existing fatigue and crack-growth data.

To resolve this inconsistency, engineers considered the service conditions of the housing. The operating environment of the cracked regions was a mixture of high-pressure hydrogen and steam at 149°C to 260°C (300°F to 500°F). Generally, hydrogen environment embrittlement occurs near room temperature and would not be a significant concern at that level of heat; however, because of the unexplained cracking, a decision was made to test Inconel® 100 at elevated temperatures in hydrogen and hydrogen mixed with steam. Again, the results were unexpected. Engineers observed a pronounced reduction in strength and ductility in these environments at elevated temperatures. Crack growth occurred at highly accelerated rates—as high as two orders of magnitude above room-temperature air when the crack was heavily loaded to 30 ksi $\sqrt{\text{in}}$ (33 MPa $\sqrt{\text{m}}$) and held for normal engine operating time. Moreover, crack growth was driven by both the number of load cycles and the duration of each load cycle. Crack growth is typically sensitive to the number and magnitude of load cycles but not to the length of time for each cycle.

Clearly, the combination of the hydrogen and steam mixture and the uncommonly high stress concentrations was promoting hydrogen environment embrittlement in Inconel® 100 at high temperatures. Resolving this issue required three modifications. First, detailed changes to the shape of the housing were made, further reducing stress concentrations. Second, gold plating was added to shield the Inconel® 100 from the hot hydrogen and steam mixture. Finally, a manufacturing process called "shot peening" was used to fortify the surface of the housing against tensile stresses by impacting it with shot, determined to be promoting fracture, and therefore eliminated.

Summary

The material characterization done in the design phase of the main engine, and the subsequent anomaly resolution during its development phase, expanded both the material properties database and the understanding of hydrogen embrittlement. The range of hydrogen embrittlement data has been broadened from essentially encompassing only steels to now including superalloys. It was also extended from including primarily tensile properties to including extensive low-cycle fatigue and fracture-mechanics testing in conditions favorable to internal hydrogen embrittlement or hydrogen environment embrittlement. The resultant material properties database, now approaching 50 years of maturity, is valuable not only because these materials are still being used, but also because it serves as a foundation for predicting how other materials will perform under similar conditions—and in the space programs of the future.

Space Environment: It's More Than a Vacuum

We know that materials behave differently in different environments on Earth. For example, aluminum does not change on a pantry shelf for years yet rapidly corrodes or degrades in salt water.

One would think that such material degradation effects would be eliminated by going to the near-perfect vacuum of space in low-Earth orbit. In fact, many of these effects are eliminated. However, Orbiter systems produced gas, particles, and light when engines, overboard dumps, and other systems operated, thereby creating an induced environment in the immediate vicinity of the spacecraft. In addition, movement of the shuttle through the tenuous upper reaches of Earth's atmosphere (low-Earth orbit) at orbital velocity produced additional contributions to the induced environment in the form of spacecraft glow and atomic oxygen effects on certain materials. The interactions of spacecraft materials with space environment factors like solar ultraviolet (UV) light, atomic oxygen, ionizing radiation, and extremes of temperature can actually be detrimental to the life of materials used in spacecraft systems.

For the Orbiter to perform certain functions and serve as a platform for scientific measurements, the effects of natural and Orbiter-induced environments had to be evaluated and controlled. Payload sensitivities to these environmental effects varied, depending on payload characteristics. Earth-based observatories and other instruments are affected by the Earth's atmosphere in terms of producing unwanted light background and other contamination effects. Therefore, NASA developed essential analytical tools for environment prediction as well as measurement systems for environment definition and performance verification, thus enabling a greater understanding of natural and induced environment effects for space exploration.

Induced Environment Characterization

NASA developed mathematical models to assess and predict the induced environment in the Orbiter cargo bay during the design and development phase of the Space Shuttle Program. Models contained the vehicle geometry, vehicle flight attitude, gas and vapor emission source characteristics, and used low-pressure gas transport physics to calculate local gas densities, column densities (number of molecular species seen along a line of sight), as well as contaminant deposition effects on functional surfaces. Gas transport calculations were based on low-pressure molecular flow physics and included scattering from Orbiter surfaces and the natural low-Earth orbit environment.

The Atlantic Ocean southeast of the Bahamas is in the background as Columbia's Shuttle Robotic Arm and end effector grasp a multi-instrument monitor for detecting contaminants. The experiment, called the Induced Environment Contaminant Monitor, was flown on STS-4 (1982). The tail of the Orbiter can be seen below.

The Induced Environment Contamination Monitor measured the induced environment on three missions—Space Transportation System (STS)-2 (1981), STS-3 (1982), and STS-4 (1982)—and was capable of being moved using the Shuttle Robotic Arm to various locations for specific measurements. Most measurements were made during the on-orbit phase. This measurement package was flown on the three missions to assess shuttle system performance. Instruments included a humidity monitor, an air sampler for gas collection and analysis after return, a cascade impactor for particulate measurement, passive samples for optical degradation of surfaces, quartz-crystal microbalances for deposited mass measurement, a camera/photometer pair for particle measurement in the field of view, and a mass spectrometer. Additional flight measurements made on STS-52 (1992) and many payloads provided more data.

Before the induced environment measurements could be properly interpreted, several on-orbit operational aspects needed to be understood. Because of the size of the vehicle and its payloads, desorption of adsorbed gases such as water, oxygen, and nitrogen (adsorbed on Earth) took a fairly long time, the induced environment on the first day of a mission was affected more than on subsequent days. Shuttle flight attitude requirements could affect the cargo bay gaseous environment via solar heating effects as well as the gases produced by engine firings. These gases could reach the payload bay by direct or scattered flow. Frequently, specific payload or shuttle system attitude or thermal control requirements conflicted with the quiescent induced environment required by some payloads.

With the above operational characteristics, data collected with the monitor and subsequent shuttle operations showed that, in general, the measured data either met or were close to the requirements of sensitive payloads during quiescent periods. A large qualification to this statement had to be made based on a new understanding of the interaction of the natural environment with vehicle surfaces. This interaction resulted in significantly more light emissions and material surface effects than originally expected. Data also identified an additional problem of recontact of particles released from the shuttle during water dumps with surfaces in the payload bay. The induced environment control program instituted for the Space Shuttle Program marked a giant step from the control of small free-flying instrument packages to the control of a large and complex space vehicle with a mixed complement of payloads. This approach helped develop a system with good performance, defined the vehicle associated environment, and facilitated effective communication between the program and users.

The induced environment program also showed that some attached payloads were not compatible with the shuttle system and its associated payloads because of the release of water over long periods of time. Other contamination-sensitive payloads such as Hubble Space Telescope, however, were not only successfully delivered to space but were also repaired in the payload bay.

Unique Features Made It Possible

The Orbiter was the first crewed vehicle to provide protection of instrumentation and sensitive surfaces in the payload bay during ascent and re-entry and allow exposure to the low-Earth orbit environment. Effects were observed without being modified by flight heating or gross contamination. Also, as part of the induced environment control program, the entire payload bay was examined immediately on return. Because of these unique aspects, NASA was able to discover and quantify unexpected interactions between the environment of low-Earth and the vehicle.

Discovery of Effects of Oxygen Atoms

After STS-1 (1981) returned to Earth, researchers visually examined the material surfaces in the payload bay for signs of contamination effects. Most surfaces appeared pristine, except for the exterior of the television camera thermal blankets and some painted surfaces. The outside surface of the blankets consisted of an organic (polyimide) film that, before flight, appeared gold colored and had a glossy finish. After flight, most films were altered to a yellow color and no longer had a glossy finish but, rather, appeared carpet-like under high magnification. Only the surfaces of organic materials were affected; bulk properties remained unchanged.

Patterns on modified surfaces indicated directional effects and, surprisingly, the flight-exposed surfaces were found to have receded rather than having deposited contaminants. The patterns on the surfaces were related to the

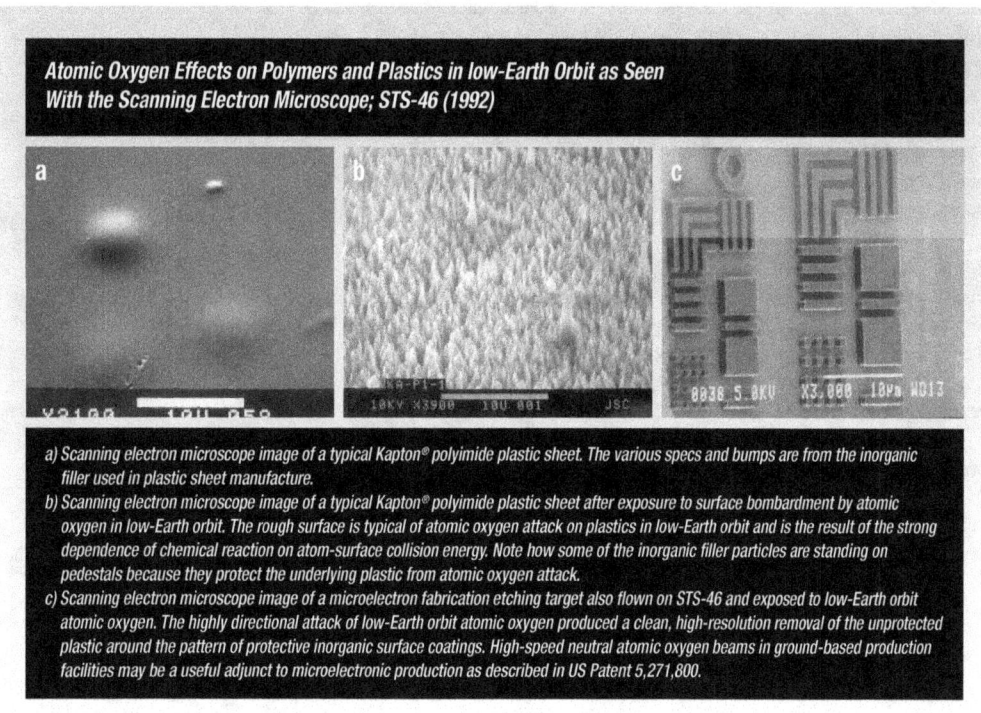

Atomic Oxygen Effects on Polymers and Plastics in low-Earth Orbit as Seen With the Scanning Electron Microscope; STS-46 (1992)

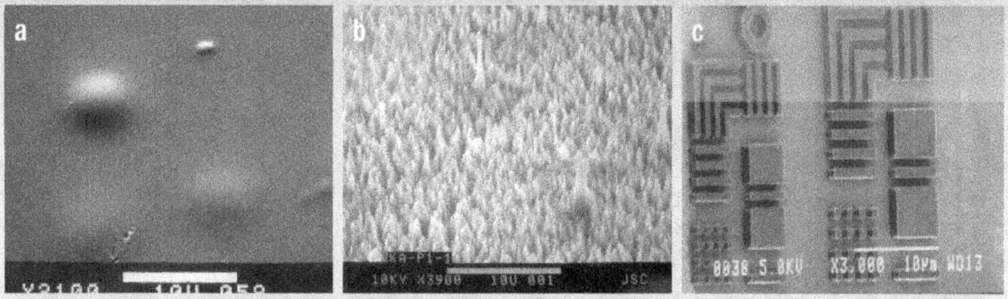

a) Scanning electron microscope image of a typical Kapton® polyimide plastic sheet. The various specs and bumps are from the inorganic filler used in plastic sheet manufacture.
b) Scanning electron microscope image of a typical Kapton® polyimide plastic sheet after exposure to surface bombardment by atomic oxygen in low-Earth orbit. The rough surface is typical of atomic oxygen attack on plastics in low-Earth orbit and is the result of the strong dependence of chemical reaction on atom-surface collision energy. Note how some of the inorganic filler particles are standing on pedestals because they protect the underlying plastic from atomic oxygen attack.
c) Scanning electron microscope image of a microelectron fabrication etching target also flown on STS-46 and exposed to low-Earth orbit atomic oxygen. The highly directional attack of low-Earth orbit atomic oxygen produced a clean, high-resolution removal of the unprotected plastic around the pattern of protective inorganic surface coatings. High-speed neutral atomic oxygen beams in ground-based production facilities may be a useful adjunct to microelectronic production as described in US Patent 5,271,800.

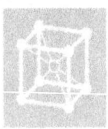

vehicle velocity vector. When combining these data with the atmospheric composition and densities, the material surface recession was caused by the high-velocity collision of oxygen atoms with forward-facing Orbiter surfaces leading to surface degradation by oxidation reactions. Oxygen atoms are a major constituent of the natural low-Earth orbit environment through which the shuttle flew at an orbital velocity of nearly 8 km/sec (17,895 mph). The collision energy of oxygen atoms striking forward-facing shuttle surfaces in low-Earth orbit was extremely high—on the order of 5 electron volts (eV)—100 times greater than the energy of atoms in typical low-pressure laboratory oxygen atom generators. The high collision energy of oxygen atoms in low-Earth orbit plays an important role in surface reactivity and surface recession rates.

Material recession rates are determined by normalizing the change in sample mass to the number of oxygen atoms reaching the surface over the exposure time (atoms/cm^2, fluence). Atom density is obtained from the standard atmospheric density models used by NASA and the Department of Defense. Since oxygen atoms travel much slower than the Orbiter, they impacted the surfaces in question only when facing toward the vehicle velocity vector and had to be integrated over time and vehicle orientation. STS-1 recession data were approximate because they had to be integrated over changing vehicle attitude; had limited atom flux, uncontrolled surface temperatures and solar UV exposure; and predicted atom densities. Recession rates determined from material samples exposed during the STS-5 (1982) mission and Induced Environmental Contamination Monitor flights had the same limitations but supported the STS-1 data. Extrapolation of these preliminary recession data to longer-term missions showed the potential for significant performance degradation of critical hardware, so specific flight experiments were carried out to quantify the recession characteristics and rates for materials of interest.

On-orbit Materials Behavior

Fifteen organizations participated in a flight experiment on STS-8 (1983) to understand materials behavior in the low-Earth orbit environment. The objective was to control some of the parameters to obtain more-accurate recession rates. The mission had a dedicated exposure to direct atom impact (payload bay pointing in the velocity direction) of 41.7 hours at an altitude of 225 km (121 nautical miles) resulting in the largest fluence of the early missions (3.5 x 10^{20} atoms/cm^2). Temperature control at two set points was provided as well as instruments to control UV and exposure to electrically charged ionospheric plasma species.

The STS-8 experiment provided significant insight into low-Earth orbit environment interactions with materials. Researchers established quantitative reaction rates for more than 50 materials, and were in the range of 2-3 x 10^{-24} cm^3/atom for hydrocarbon-based materials. Perfluorinated organic materials were basically nonreactive and silicone-based materials stopped reacting after formation of a protective silicon oxide surface coating. Material reaction rates, as a first approximation, were found to be independent of temperature, material morphology, and exposure to solar radiation or electrically charged ionspheric species. Researchers also evaluated coatings that could be used to protect surfaces from interaction with the environment.

Reaction rates were based on atomic oxygen densities determined from long-term atmospheric density models, potentially introducing errors in short-term experiment data. In addition, researchers obtained very little insight into the reaction mechanism(s).

An additional flight experiment—Evaluation of Oxygen Interaction with Materials III—addressing both of these questions was flown on STS-46 (1992). The primary objective was to produce benchmark atomic oxygen reactivity data by measuring the atom flux during material surface exposure. Secondary experiment objectives included: characterizing the induced environment near several surfaces; acquiring basic chemistry data related to reaction mechanism; determining the effects of temperature, mechanical stress, atom fluence, and solar UV radiation on material reactivity; and characterizing the induced and contamination environments in the shuttle payload bay. This experiment was a team effort involving NASA centers, US Air Force, NASA Space Station Freedom team, Aerospace Corporation, University of Alabama in Huntsville, National Space Agency of Japan, European Space Agency, and the Canadian Space Agency.

STS-46 provided an opportunity to make density measurements at several altitudes: 427, 296, and 230 km (231, 160, and 124 nautical miles). However, the vehicle flew for 42 hours at 230 km (124 nautical miles) with the payload bay surfaces pointed into the velocity vector during the main portion of the mission to obtain high fluence. The mass spectrometer provided by the

US Air Force was the key component of the experiment and was capable of sampling both the direct atomic oxygen flux as well as the local neutral environment created by interaction of atomic oxygen with surfaces placed in a carousel. Five carousel sections were each coated with a different material to determine the material effects on released gases. Material samples trays, which provided temperature control plus instruments to control other exposure conditions, were placed on each side of the mass spectrometer/carousel.

NASA achieved all of the Evaluation of Oxygen Interaction with Materials III objectives during STS-46. A well-characterized, short-term, high-fluence atomic oxygen exposure was provided for a large number of materials, many of which had never been exposed to a known low-Earth orbit atomic oxygen environment. The data provided a benchmark reaction rate database, which has been used by the International Space Station, Hubble, and others to select materials and coatings to ensure long-term durability.

Reaction rate data for many of the materials from earlier experiments were confirmed, as was the generally weak dependence of these reaction rates on temperature, solar UV exposure, oxygen atom flux, and exposure to charged ionospheric species. The role of surface collision energy on oxygen atom reactivity was quantified by comparing flight reaction rates of key Evaluation of Oxygen Interaction with Materials III experiment materials with reactivity measurements made in well-characterized laboratory oxygen atom systems with lower surface collision energies. This evaluation also provided an important benchmark point for understanding the role of solar extreme UV radiation damage in increasing the generally low surface reactivity of perfluorinated organic materials. The mass spectrometer/carousel experiment produced over 46,000 mass spectra providing detailed characterization of both the natural and the induced environment. The mass spectrometer database provided a valuable resource for the verification of various models of rarified gas and ionospheric plasma flow around spacecraft.

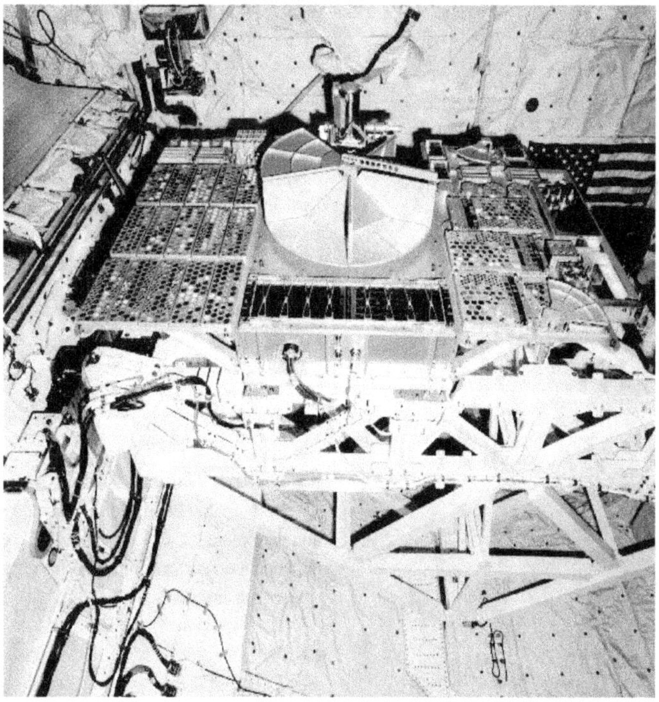

Evaluation of Oxygen Interaction with Materials III flight experiment in the Orbiter payload bay of STS-46 (1992). Material exposure samples are located on both sides of the mass spectrometer gas evolution measurement assembly in the center.

Intelsat Satellite

Knowledge gained from atomic oxygen reactivity studies played a key role in the STS-49 (1992) rescue of the communications satellite Intelsat 603 that was used to maintain communications from a geosynchronous orbit. Failure of the Titan-3 upper stage left Intelsat 603 marooned in an unacceptable low-Earth orbit and subject to the effects of atomic oxygen degradation of its solar panels, which could have rendered the satellite useless. NASA quickly advised the International Telecommunications Satellite Organization (Intelsat) Consortium of the atomic oxygen risk to Intelsat 603, leading to the decision to place the satellite in a configuration that was expected to minimize atomic oxygen damage to the silver interconnects on the solar panels. This was accomplished by raising the satellite altitude and changing its flight attitude so that atomic oxygen fluence was minimized.

The Intelsat Solar Array Coupon flight experiment shown mounted on the Shuttle Robotic Arm lower arm boom and exposed to space environment conditions during STS-41 (1990).

To provide facts needed for a final decision about a rescue flight, NASA designed and executed the Intelsat Solar Array Coupon flight experiment on STS-41 (1990). The experiment results, in combination with ground-based testing, supported the decision to conduct the STS-49 satellite rescue mission. On this mission, Intelsat 603 was captured and equipped with a solid re-boost motor to carry it to successful geosynchronous orbit.

NASA Discovers Light Emissions

On the early shuttle flights, NASA observed another effect caused by the interaction between spacecraft surfaces and the low-Earth orbit environment. Photographs obtained by using intensified cameras and conducted from the Orbiter cabin windows showed light emissions (glow) from the Orbiter surfaces when in forward-facing conditions.

The shuttle provided an excellent opportunity to further study this phenomenon. On STS-41D (1984), astronauts photographed various material samples using a special glow spectrometer to obtain additional data and determine if the glow was dependent on surface composition. These measurements, along with the material recession effects and data obtained on subsequent flights, led to a definition of the glow mechanism.

Spacecraft glow is caused by the interaction of high-velocity oxygen atoms with nitrous oxide absorbed on the surfaces, which produces nitrogen dioxide in an electronically excited state. The excited nitrogen dioxide is released from the surfaces and emits light as it moves away and decays from its excited state. Some nitrous oxide on the surface and some of the released nitrogen dioxide result from the natural environment. The light emission occurs on any spacecraft operating in low-Earth orbit; however, the glow could be enhanced by operation of the shuttle attitude control engines, which produced nitrous oxide and nitrogen dioxide as reaction products. These findings led to a better understanding of the behavior of spacecraft operating in low-Earth orbit and improved accuracy of instrument measurements.

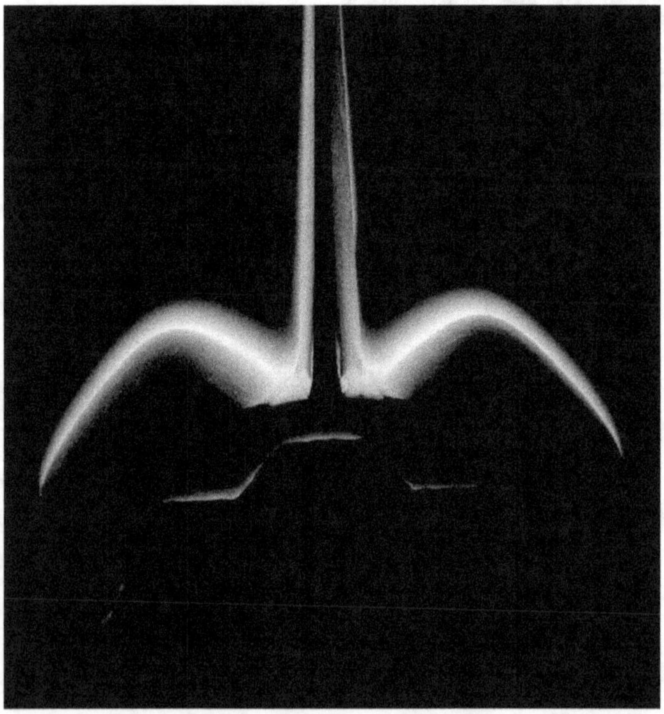

STS-62 (1994) orbits Earth during a "night" pass, documenting the glow phenomenon surrounding the vertical stabilizer and the Orbital Maneuvering System pods of the spacecraft.

Chemical Fingerprinting

Comprehensive Electronic System for Greater Flight Safety

A critical concern for all complex manufacturing operations is that contaminants and material changes over time can creep into the production environment and threaten product quality. This was the challenge for the solid rocket motors, which were in production for 30 years.

It is possible that vendor-supplied raw materials appear to meet specifications from lot to lot and that supplier process changes or even contaminated material can appear to be "in spec" but actually contain subtle, critical differences. This situation has the potential to cause significant problems with hardware performance.

NASA needed a system to readily detect those subtle yet potentially detrimental material variances to ensure the predictability of material properties and the reliability of shuttle reusable solid rocket motors. The envisioned solution was to pioneer consistent and repeatable analytical methods tailored to specific, critical materials that would yield accurate assessments of material integrity over time. Central to the solution was both a foolproof analysis process and an electronic data repository for benchmarking and monitoring.

Environmental Assurance

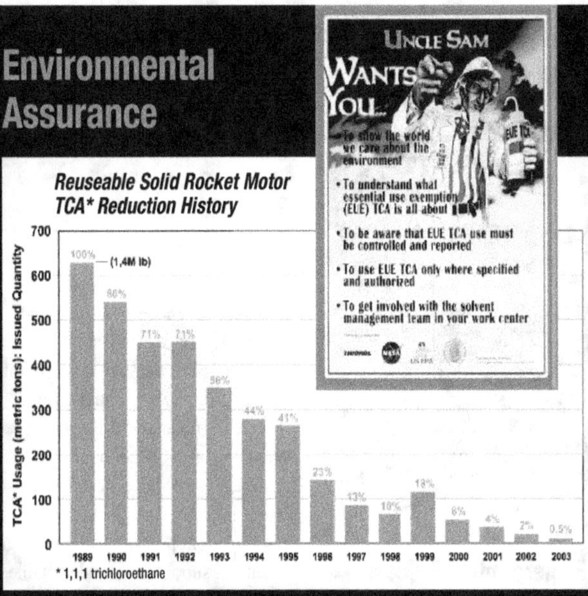

During the Space Shuttle Program's operation, issues arose regarding the use of substances that did not meet emerging environmental regulations and current industry standards. NASA worked to develop chemicals, technologies, and processes that met regulatory requirements, and the agency strove to identify, qualify, and replace materials that were becoming obsolete as a result of environmental issues. The stringent demands of human spaceflight required extensive testing and qualification of these replacement materials.

A Chemical "Fingerprint"

Just as fingerprints are a precise method to confirm an individual's identity, the solid rocket motor project employed chemical "fingerprints" to verify the quality of an incoming raw material. These fingerprints comprised a detailed spectrum of a given material's chemical signature, which could be captured digitally and verified using a combination of sophisticated laboratory equipment and custom analytical methods.

The challenge was to accurately establish a baseline chemical fingerprint of each material and develop reproducible analytical test methods to monitor lot-to-lot material variability. A further objective was to gain a greater understanding of critical reusable solid rocket motor materials, such as insulation and liner ingredients, many of which were the same materials used since the Space Shuttle Program's inception. New analytical techniques such as the atomic force microscope were used to assess materials at fundamental chemical, molecular, and mechanical levels. These new techniques provided the high level of detail sought. Because of unique attributes inherent in each material, a one-size-fits-all analysis method was not feasible.

To facilitate documentation and data sharing, the project team envisioned a comprehensive electronic database to provide ready access to all relevant data. The targeted level of background detail included everything from where and how a material was properly used to details of chemical composition.

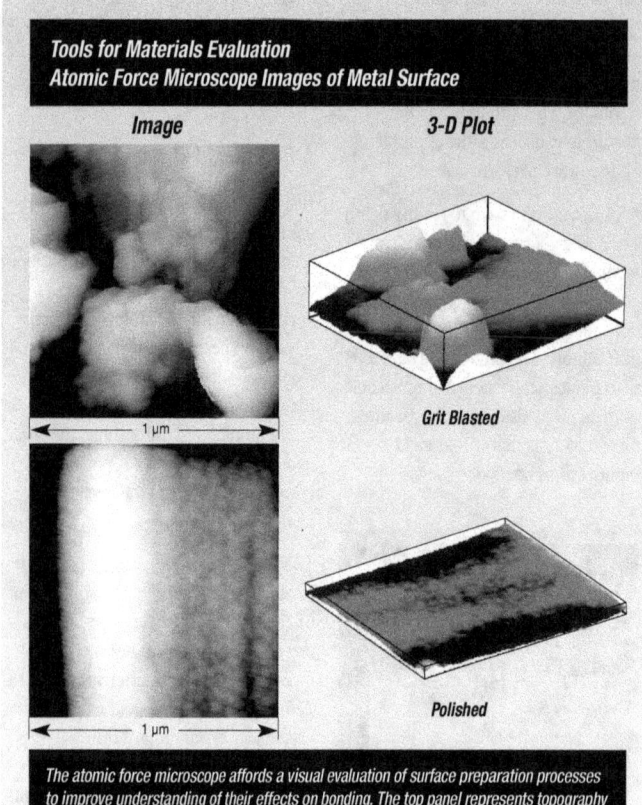

Tools for Materials Evaluation
Atomic Force Microscope Images of Metal Surface

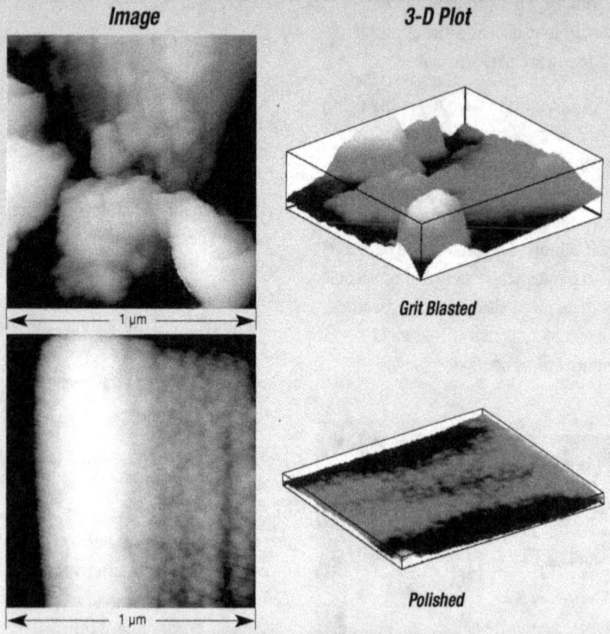

The atomic force microscope affords a visual evaluation of surface preparation processes to improve understanding of their effects on bonding. The top panel represents topography of a grit blast surface for comparison to a highly polished one. The atomic force microscope uses an extremely fine probe to measure minute interactions with surface features even down to an atomic scale. The maps at left are scaled from black at the bottom of valleys to white at the tops of peaks within the scanned area. The 3-D projections at right are on a common height scale. The grit blast surface clearly offers greatly increased surface area and mechanical interlocking for enhanced bonding. Beyond simple topography, the probe interactions with atomic forces can also measure and map properties such as microscopic hardness or elastic modulus on various particles and/or phase transitions in a composite material, which in turn can be correlated with chemical and physical properties.

The ideal system would enable a qualified chemist to immediately examine original chemical analysis data for the subtle yet significant differences between the latest lot of material and previous good or bad samples.

To develop such a system, commercially available hardware and software were used to the greatest extent possible. Since an electronic framework to tie the data together did not exist, one was designed in-house.

The Fingerprinting Process

The chemical fingerprinting program, which began in 1998 with a prioritized list of 14 critical materials, employed a team approach to quantify and document each material. The interdisciplinary team included design engineering, materials and processes engineering, procurement quality engineering, and analytical chemistry. Each discipline group proposed test plans that included the types of testing to be developed. Following approval, researchers acquired test samples (usually three to five lots of materials) and developed reliable test methods. Because of the unique nature of each material, test methods were tailored to each of the 14 materials.

A "material" site in the project database was designed to ensure all data were properly logged and critical reports were written and filed. Once the team agreed sufficient data had been generated, a formal report was drafted and test methods were selected to develop new standard acceptance procedures that would ultimately be used by quality control technicians to certify vendor materials.

The framework developed to package the wide-ranging data was termed the Fingerprinting Viewer. Program data were presented through a series of cascading menu pages, each with increasing levels of detail.

The Outcomes

Beyond meeting the primary program objectives, a number of resulting benefits were noted. First, through increased data sharing, employees communicated more effectively, both internally and with subtier suppliers. The powerful analytical methods employed also added to the suppliers' materials knowledge base. Subtle materials changes that possibly resulted from process drift or changes at subtier suppliers were detectable. Eight subtier suppliers subsequently implemented their own in-house chemical fingerprinting programs to improve product consistency, recertify material after production changes, or even help develop key steps in the manufacturing process to ensure repeatable quality levels.

Additionally, engineers could now accurately establish shelf-life extensions and storage requirements

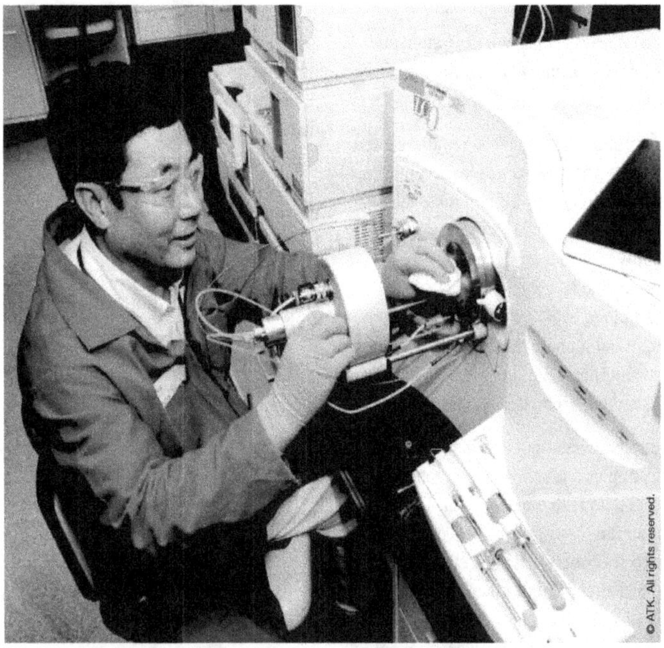

This high-performance liquid chromatography/mass spectrometry is employed to document minute details of a material's chemical and molecular composition. Through the chemical fingerprinting system, seemingly minuscule discrepancies raise red flags that trigger investigations and preclude defective materials from reaching the production floor. Dr. Ping Li shown here at ATK in Utah.

for stockpiled materials. The ability to store greater amounts of materials over longer periods of time was valuable in cases where new materials needed to be certified to replace existing materials that had become obsolete.

Finally, investigators were able to solve production issues with greater efficiency. Comprehensive database features, including standardized test methods and the extensive online reference database, provided resources needed to resolve production issues in a matter of days or even hours—issues that otherwise would have required major investigations. In some cases, fingerprinting was also used to indicate that a suspect material was actually within required specifications. These materials may have been rejected in previous cases but, by using the fingerprinting database to assess the material, the team could look deeper to find the true root cause and implement proper corrective actions.

From Fingerprints to Flight Safety

The overarching value of the chemical fingerprinting program was that it provided greater assurance of the safety and reliability of critical shuttle flight hardware. The fundamental understanding of critical reusable solid rocket motor materials and improved communications with vendors reduced the occurrence of raw materials issues. NASA will implement chemical fingerprinting methods into the acceptance testing of raw materials used in future human space exploration endeavors. The full benefits of the program will continue to be realized in years to come.

Unprecedented Accomplishments in the Use of Aluminum-Lithium Alloy

NASA was the first to use welded aluminum-lithium alloy Al 2195 at cryogenic temperatures, incorporating it into the External Tank under circumstances that demanded innovation.

From the beginning of the Space Shuttle Program's launch phase, NASA sought to reduce the weight of the original tank, thereby increasing payload capacity. Since the tank was carried nearly to orbit, close to 100% of the weight trimmed could be applied to the payload. NASA succeeded in implementing numerous weight-saving measures, but the biggest challenge was to incorporate a lightweight aluminum alloy—aluminum-lithium Al 2195—into the tank structure. This alloy had never been used in welded cryogenic environments prior to NASA's initiative. Several challenges needed to be overcome, including manufacturing the aluminum-lithium tank components, welding the alloy, and repairing the welds. NASA and the External Tank prime contractor broke new ground in the use of aluminum-lithium to produce the "super lightweight tank."

The original tank weighed 34.500 metric tons (76,000 pounds) dry. By the sixth shuttle mission, the tank's weight had been reduced to 29.900 metric tons (66,000 pounds). This configuration was referred to as the "lightweight tank."

The real challenge, however, was still to come. In 1993, the International Space Station Program decided to change the station's orbital inclination to 57 degrees (a "steeper" launch inclination), allowing Russian vehicles to fly directly to the station. That change cost the shuttle 6,123 kg (13,500 pounds) of payload capacity. The External Tank project office proposed to reduce the dry weight of the tank by 3,402 kg (7,500 pounds).

The Space Shuttle Program sought to incorporate lightweight aluminum-lithium Al 2195 into the majority of the tank structure, replacing the original aluminum-copper alloy Al 2219; however, NASA first needed to establish requirements for manufacturing, welding, and repairing aluminum-lithium weld defects.

NASA started the super lightweight tank program in 1994. During the early phase, advice was sought from welding experts throughout the United States and the United Kingdom. The consensus: it was virtually impossible to perform repairs on welded aluminum-lithium.

The aluminum-lithium base metal also presented challenges. Lockheed Martin worked with Reynolds Aluminum to produce the aluminum-lithium base metal. One early problem was related to aluminum-lithium material's fracture toughness—a measure of the ability of material with a defect to carry loads. Although material was screened, flight hardware requirements dictated that structures must have the ability to function in the event a defect was missed by the screening process. The specific difficulty with the aluminum-lithium was that the cryogenic fracture toughness of the material showed little improvement over the room-temperature fracture toughness. Since the two propellant tanks were proof tested at room temperature and flown cryogenically, this fracture toughness ratio was a crucial factor.

A simulated service test requirement was imposed as part of lot acceptance for all aluminum-lithium material used on the tank. The test consisted of applying room temperature and cryogenic load cycles to a cracked sample to evaluate the ability of the material to meet the fracture toughness requirements. Failure resulted in the plate being remelted and reprocessed.

Implementation of simulated service testing as a lot acceptance requirement was unique to the aluminum-lithium material. Testing consisted of cropping two specimens from the end of each plate. Electrical discharge machining (a process that removes metal by discharging a spark between the tool and the test sample) was used to introduce a fine groove in each sample. The samples were then cyclically loaded at low stresses to generate a sharp fatigue crack that simulated a defect in the material.

The first sample was stressed to failure; the second sample was stressed to near failure and then subjected to cyclic loading representative of load cycles the tank would see on the launch pad during tanking and during flight.

In the second sample, initial loading was conducted at room temperature. This simulated the proof test done on the tank. Next, the sample was stressed 13 times (maximum tanking requirement) to the level expected during loading of propellants at cryogenic temperatures and, finally, stressed to maximum expected flight stress at cryogenic temperature. This cycle was repeated three more times to meet a four-mission-life program requirement with the exception that, on the fourth cycle, the sample was stressed to failure and had to exceed a predetermined percent of the flight stress. Given the size of the barrel plates for the liquid hydrogen and liquid oxygen tanks, only one barrel plate could be made from each lot of material. As a result, this process was adopted for every tank barrel plate—32 in each liquid hydrogen tank and four in each liquid oxygen tank—and implemented for the life of the program.

Another challenge was related to the aluminum-lithium weld repair process on compound curvature parts. The effect of weld shrinkage in the repairs caused a flat spot, or even a reverse curvature, in the vicinity of the repairs and contributed to significant levels of residual stress in the repair. Multiple weld repairs, in proximity, showed the propensity for severe cracking. After examination of the repaired area, it was found that welding aluminum-lithium resulted in a zone of brittle material surrounding the weld. Repeated repairs caused this zone to grow until the residual stress from the weld shrinkage exceeded the strength of the weld repair, causing it to crack.

The technique developed to repair these cracks was awarded a US Patent. The repair approach consisted of alternating front-side and back-side grinds as needed to remove damaged microstructure. It was also found that aluminum-lithium could not tolerate as much heating as the previous aluminum-copper alloy. This required increased torch speeds and decreased

The use of aluminum-lithium Al 2195 in manufacturing major External Tank components, such as the liquid hydrogen tank structure shown above, allowed NASA to reduce the overall weight of the External Tank by 3,402 kg (7,500 pounds). The liquid hydrogen tank measured 8.4 m (27.5 ft) in diameter and 29.4 m (96.6 ft) in length. Photo taken at NASA's Michoud Assembly Facility in New Orleans, Louisiana.

fill volumes to limit the heat to which the aluminum-lithium was subjected.

Additional challenges in implementing effective weld repairs caused NASA to reevaluate the criteria for measuring the strength of the welds. In general, weld repair strengths can be evaluated by excising a section of the repaired material and performing a tensile test. The strength behavior of the repaired material is compared to the strength behavior of the original weld material. In the case of the aluminum-copper alloy Al 2219, the strengths were comparable; however, in the case of the aluminum-lithium alloy repair, the strengths were lower.

Past experience and conventional thinking was that in the real hardware, where the repair is embedded in a long initial weld, the repaired weld will yield and the load will be redistributed to the original weld, resulting in higher capability. To demonstrate this assumption, a tensile test was conducted on a 43-cm-(17-in.)-wide aluminum-lithium panel that was fabricated by welding two aluminum-lithium panels together and simulating a weld repair in the center of the original weld. The panel was then loaded to failure. The test that was supposed to indicate better strength behavior than the excised repair material actually failed at a lower stress level.

To understand this condition, an extensive test program was initiated to evaluate the behavior of repairs on a number of aluminum-copper alloy (Al 2219) and aluminum-lithium alloy (Al 2195) panels.

Orbiter Payload Bay Door

One of the largest aerospace composite applications of its time.

With any space vehicle, minimum weight is of critical importance. Initial trade studies indicated that using a graphite/epoxy structure in place of the baselined aluminum structure provided significant weight savings of about 408 kg (900 pounds [4,000 newtons]), given the large size and excellent thermal-structural stability. Two graphite/epoxy composite materials and four structural concepts—full-depth honeycomb sandwich, frame-stiffened thin sandwich, stiffened skin with frames and stringers, and stiffened skin with frames only—were considered for weight savings and manufacturing producibility efficiency. These studies resulted in the selection of the frame-stiffened thin sandwich configuration, and component tests of small specimens finalized the graphite fiber layup, matrix material, and honeycomb materials. Graphite/epoxy properties at elevated temperatures are dependent on moisture content and were taken into account in developing mechanical property design allowables. Additionally, NASA tracked the moisture content through all phases of flight to predict the appropriate properties during re-entry when the payload bay doors encountered maximum temperatures of 177°C (350°F).

Payload bay doors were manufactured in 4.57-m (15-ft) sections, resulting in two 3 x 18.3 m (10 x 60 ft) doors. The panel face sheets consisted of a ± 45-degree fabric ply imbedded between two 0-degree tape plies directed normal to the frames and were pre-cured prior to bonding to the Nomex® honeycomb core. A lightweight-aluminum wire mesh bonded to the outside of face sheets provided lightning-strike protection. Frames consisted primarily of fabric plies with the interspersions of 0-degree plies dictated by strength and/or stiffness. Mechanical fasteners were used for connection of major subassemblies as well as final assembly of the doors.

All five Orbiter vehicles used graphite/epoxy doors, one of the largest aerospace composite applications at the time, and performance was excellent throughout all flights. Not only was the expected weight saving achieved and thermal-structural stability was acceptable, NASA later discovered that the graphite/epoxy material showed an advantage in ease of repair. Ground handling damage occurred on one section of a door, resulting in penetration of the outer skin of the honeycomb core. The door damage was repaired in 2 weeks, thereby avoiding significant schedule delay.

Test panels were covered with a photo-stress coating that, under polarized light, revealed the strain pattern in the weld repair. The Al 2219 panel behaved as expected: the repair yielded, the loads redistributed, and the panel pulled well over the minimum allowable value. In aluminum-lithium panels, however, the strains remained concentrated in the repair. Instead of the 221 MPa (32,000 pounds/in^2) failure stress obtained in the initial welds, the welds were failing around 172 MPa (18,000 pounds/in^2). These lower failure stress values were problematic due to a number of flight parts that had already been sized and machined for the higher 221 MPa (32,000 pounds/in^2) value.

Based on this testing, it was determined that weld shrinkage associated with the repair resulted in residual stresses in the joint, reducing the joint capability. To improve weld repair strengths, engineers developed an approach to planish (lightly hammer) the weld bead, forcing it back into the joint and spreading the joint to redistribute and reduce the residual stresses due to shrinkage. This required scribing and measuring the joint before every repair, making the repair, and then planishing the bead to restore the weld to its previous dimensions. Wide panel test results and photo-stress evaluation of planished repairs revealed that the newly devised repair procedure was effective at restoring repair strengths to acceptable levels.

Testing also revealed that planishing of weld beads is hard to control precisely, resulting in the process frequently forming other cracks, thus leading to additional weld repairs. Because of the difficulty in making and planishing multiple repairs, a verification ground rule was established that every "first repair of its kind" had to be replicated on three wide tensile panels, which were then tested either at room temperature or in a cryogenic environment, depending on the in-flight service condition expected for that part of the tank.

All these measures combined accomplished the first-ever use of welded aluminum-lithium at cryogenic temperatures, meeting the strict demands of human spaceflight. The super lightweight tank incorporated 20 aluminum-lithium ogive gores (the curved surfaces at the forward end of the liquid oxygen tank), four liquid oxygen barrel panels, 32 liquid hydrogen barrel panels, 12 liquid oxygen tank aft dome gores, 12 liquid hydrogen tank forward dome gores, and 11 liquid hydrogen aft dome gores.

Through this complex and innovative program, NASA reduced the 29,937-kg (66,000-pound) lightweight tank by another 3,401.9 kg (7,500 pounds). The 26,560-kg (58,500-pound) super lightweight tank was first flown on Space Transportation System (STS)-91 (1998), opening the door for the shuttle to deliver the heavier components needed for construction of the International Space Station.

Aerodynamics and Flight Dynamics

Introduction
Aldo Bordano

Aeroscience Challenges
Gerald LeBeau
Pieter Buning
Peter Gnoffo
Paul Romere
Reynaldo Gomez
Forrest Lumpkin
Fred Martin
Benjamin Kirk
Steve Brown
Darby Vicker

Ascent Flight Design
Aldo Bordano
Lee Bryant
Richard Ulrich
Richard Rohan

Re-entry Flight Design
Michael Tigges
Richard Rohan

Boundary Layer Transition
Charles Campbell
Thomas Horvath

The shuttle vehicle was uniquely winged so it could reenter Earth's atmosphere and fly to assigned nominal or abort landing strips. The wings allowed the spacecraft to glide and bank like an airplane during much of the return flight phase. This versatility, however, did not come without cost. The combined ascent and re-entry capabilities required a major government investment in new design, development, verification facilities, and analytical tools. The aerodynamic and flight control engineering disciplines needed new aerodynamic and aerothermodynamic physical and analytical models. The shuttle required new adaptive guidance and flight control techniques during ascent and re-entry. Engineers developed and verified complex analysis simulations that could predict flight environments and vehicle interactions. The shuttle design architectures were unprecedented and a significant challenge to government laboratories, academic centers, and the aerospace industry. These new technologies, facilities, and tools would also become a necessary foundation for all post-shuttle spacecraft developments. The following section describes a US legacy unmatched in capability and its contribution to future spaceflight endeavors.

Aeroscience Challenges

One of the first challenges in the development of the Space Shuttle was its aerodynamic design, which had to satisfy the conflicting requirements of a spacecraft-like re-entry into the Earth's atmosphere where blunt objects have certain advantages, but it needed wings that would allow it to achieve an aircraft-like runway landing. It was to be the first winged vehicle to fly through the hypersonic speed regime, providing the first real test of experimental and theoretical technology for high-speed flight. No design precedents existed to help establish necessary requirements. The decision that the first flight would carry a crew further complicated the challenge. Other than approach and landing testing conducted at Dryden Flight Research Center, California, in 1977, there would be no progressive "envelope" expansion as is typically done for winged aircraft. Nor would there be successful uncrewed launch demonstrations as had been done for all spacecraft preceding the shuttle. Ultimately, engineers responsible for characterizing the aeroscience environments for the shuttle would find out if their collective predictions were correct at the same moment as the rest of the world: during the launch and subsequent landing of Space Transportation System (STS)-1 (1981).

Aeroscience encompasses the engineering specialties of aerodynamics and aerothermodynamics. For the shuttle, each specialty was primarily associated with analysis of flight through the Earth's atmosphere.

Aerodynamics involves the study of local pressures generated over the vehicle while in flight and the resultant integrated forces and

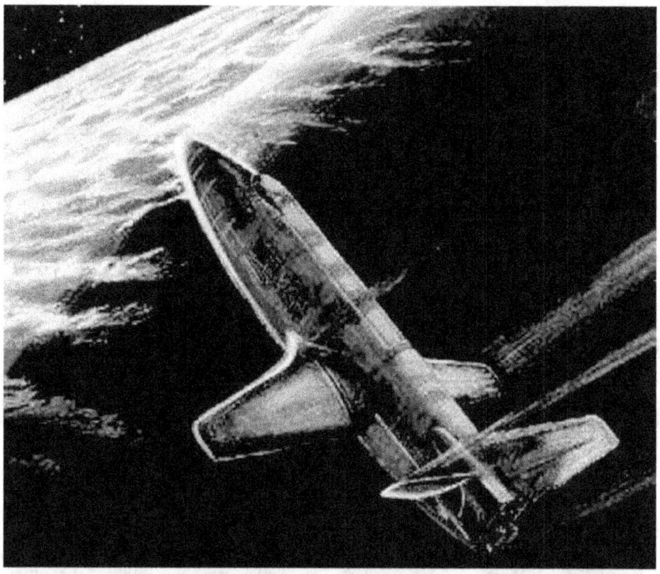

Early conceptual designs for the Orbiter looked much like a traditional airplane with a fairly sharp nose, straight wings, and common horizontal and vertical stabilizers, as shown in this artist's rendering. As a result of subsequent aerodynamic and aerothermodynamic testing and analysis, NASA made the nose more spherical to reduce heating and used a double delta wing planform due to the severe heating encountered by straight wings and the horizontal stabilizer.

moments that, when coupled with forces such as gravity and engine thrust, determine how a spacecraft will fly. Aerothermodynamics focuses on heating to the spacecraft's surface during flight. This information is used in the design of the Thermal Protection System that shields the underlying structure from excessive temperatures. The design of the shuttle employed state-of-the-art aerodynamic and aerothermodynamic prediction techniques of the day and subsequently expanded them into previously uncharted territory.

The historical precedent of flight testing is that it is not possible to "validate"—or prove—that aerodynamic predictions are correct until vehicle performance is measured at actual flight conditions. In the case of the shuttle, preflight predictions needed to be accurate enough to establish sufficient confidence to conduct the first orbital flight with a crew on board. This dictated that the aerodynamic test program had to be extremely thorough. Further complicating this goal was the fact that much of the expected flight regime involved breaking new ground, and thus very little experimental data were available for the early Space Shuttle studies.

Wind tunnel testing—an experimental technique used to obtain associated data—forces air past a scaled model and measures data of interest, such as local pressures, total forces, or heating rates. Accomplishing the testing necessary to cover the full shuttle flight profile required the cooperation of most of the major wind tunnels in North America. The Space Shuttle effort was the largest such program ever undertaken by the United States. It involved a traditional phased approach in the programmatic design evolution of the shuttle configuration.

The shuttle started on the launch pad composed of four primary aerodynamic elements: the Orbiter; External Tank; and two Solid Rocket Boosters (SRBs). It built speed as it rose through the atmosphere. Aeronautical and aerospace engineers often relate to speed in terms of Mach number—the ratio of the speed of an object relative to the speed of sound in the gas through which the object is flying. Anything traveling at less than Mach 1 is said to be subsonic and greater than Mach 1 is said to be supersonic. The flow regime between about Mach 0.8 and Mach 1.2 is referred to as being transonic.

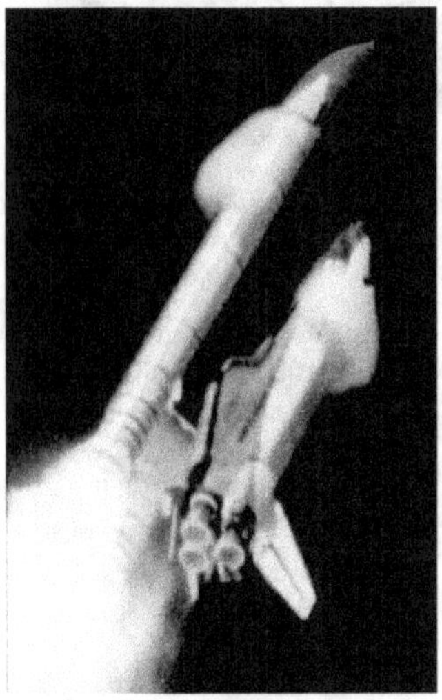

This photo shows clouds enveloping portions of the vehicle (STS-34 [1987]) during ascent. When the launch vehicle was in the transonic regime, shocks formed at various positions along the vehicle to recompress the flow, which greatly impacted the structural loads and aerodynamics. Such shocks, which abruptly transition the flow from supersonic to subsonic flow, were positioned at the trailing edge of the condensation "clouds" that could be seen enveloping portions of the vehicle during ascent. These clouds were created in localized areas of the flow where the pressure and temperature conditions caused the ambient moisture to condense.

Aerodynamic loads decreased to fairly low levels as the shuttle accelerated past about Mach 5 and the atmospheric density decreased with altitude, thus the aerodynamic testing for the ascent configuration was focused on the subsonic through high supersonic regimes.

Other aspects of the shuttle design further complicated the task for engineers. Aerodynamic interference existed between the shuttle's four elements and altered the resultant pressure loads and aerodynamics on neighboring elements. Also, since various shuttle elements were designed to separate at different points in the trajectory, engineers had to consider the various relative positions of the elements during separation. Yet another complication was the effect of plumes generated by SRBs and Space Shuttle Main Engines (SSMEs). The plume flow fields blocked and diverted air moving around the spacecraft, thus influencing pressures on the aft surfaces and altering the vehicle's aerodynamic characteristics.

Unfortunately, wind tunnel testing with gas plumes was significantly more expensive and time consuming than "standard" aerodynamic testing. Thus, the approach implemented was to use the best available testing techniques to completely characterize the basic "power-off" (i.e., no plumes) database. "Power-on" (i.e., with plumes) effects were then measured from a limited number of exhaust plume tests and added to the power-off measurements for the final database.

The re-entry side of the design also posed unique analysis challenges. During ascent, the spacecraft continued

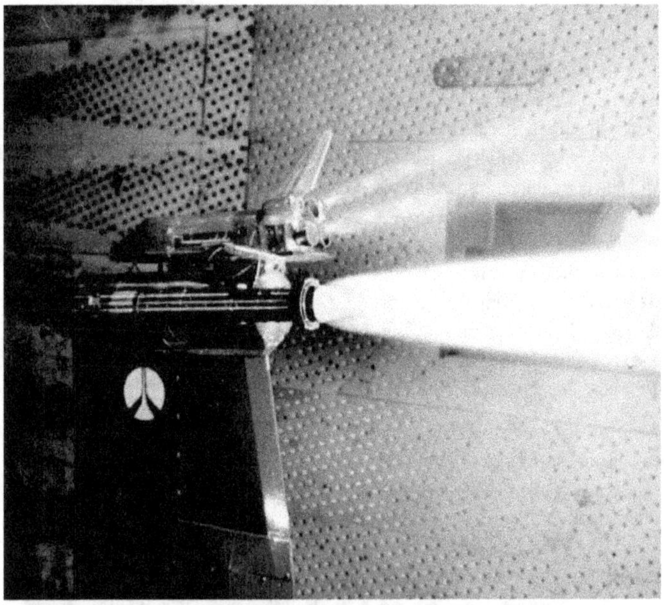

While it may be intuitive to include the major geometric elements of the launch vehicle (Orbiter, External Tank, and two Solid Rocket Boosters) in aerodynamic testing, it was also important to include the plumes eminating from the three main engines on the Orbiter as well as the boosters. The tests were conducted in the 4.9-m (16-ft) Transonic Wind Tunnel at the US Air Force Arnold Engineering and Development Center, Tennessee.

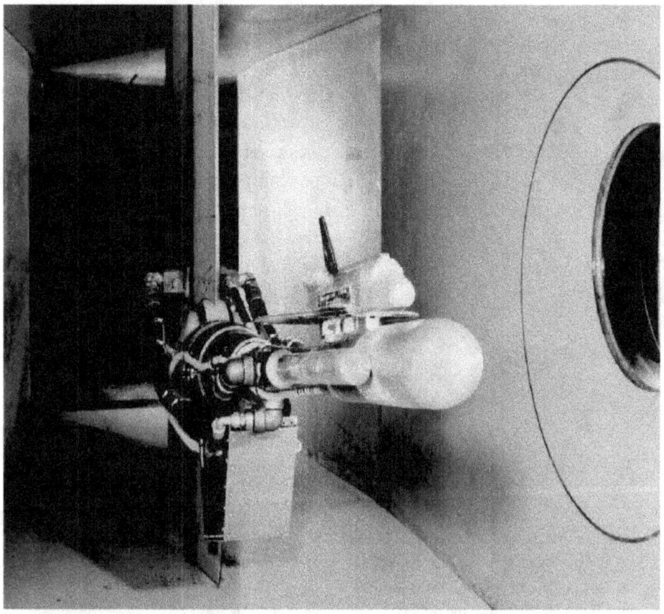

Every effort was made to accurately predict a vehicle's aerodynamic characteristics using wind tunnel testing. Engineers also had to be aware of anything that could adversely affect the results. This image is of the NASA Ames Research Center 2.4 x 2.1 m (8 x 7 ft) Unitary Wind Tunnel, California.

to accelerate past the aerodynamically relevant portion of the ascent trajectory. During re-entry, this speed was carried deep into the atmosphere until there was sufficient atmospheric density to measurably dissipate the related kinetic energy. Therefore, the aerodynamics of the Orbiter were critical to the design of the vehicle from speeds as high as Mach 25 down through the supersonic and subsonic regimes to landing, with the higher Mach numbers being characterized by complex physical gas dynamics that greatly influenced the aerodynamics and heating on the vehicle compared to lower supersonic Mach numbers.

Challenges associated with wind tunnel testing limited direct applicability to the actual flight environment that engineers were interested in simulating, such as: subscale modeling of the vehicle necessary to fit in the wind tunnel and the effect on flow-field scaling; the support structure used to hold the aerodynamic model in the wind tunnel test section, which can affect the flow on the model itself; and any influence of the wind tunnel walls. To protect against any inaccuracies in the database, each aerodynamic coefficient was additionally characterized by an associated uncertainty. Great care had to be taken to not make the uncertainties too large due to the adverse effect an uncertainty would have on the design of the flight control system and the ultimate performance of the spacecraft.

In the end, given the 20,000 hours of wind tunnel test time consumed during the early design efforts and the 80,000 hours required during the final phases, a total of 100,000 hours of wind tunnel testing was conducted for aerodynamic, aerothermodynamic, and structural dynamic testing to characterize the various shuttle system elements.

Initial Flight Experience

Traditionally, a flight test program was used to validate and make any necessary updates to the preflight aerodynamic database. While flight test programs use an incremental expansion of the flight envelope to demonstrate the capabilities of an aircraft, this was not possible with the shuttle. Once launched, without initiation of an abort, the shuttle was committed to flight through ascent, orbital operations, re-entry, and landing. NASA placed a heavy emphasis on comparison of the predicted vehicle performance to the observed flight performance during the first few shuttle missions, and those results showed good agreement over a majority of flight regimes. Two prominent areas, however, were deficient: predictions of the launch vehicle's ascent performance, and the "trim" attitude of the Orbiter during the early phase of re-entry.

On STS-1, the trajectory was steeper than expected, resulting in an SRB separation altitude about 3 km (1.9 miles) higher than predicted. Postflight analysis revealed differences between preflight aerodynamic predictions and actual aerodynamics observed by the shuttle elements due to higher-than-predicted pressures on the shuttle's aft region. It was subsequently determined that wind tunnel predictions were somewhat inaccurate because SRB and SSME plumes were not adequately modeled. This issue also called into question the structural assessment of the wing, given the dependence on the preflight prediction of aerodynamic loads. After additional testing and cross checking with flight data, NASA was able to verify the structural assessment.

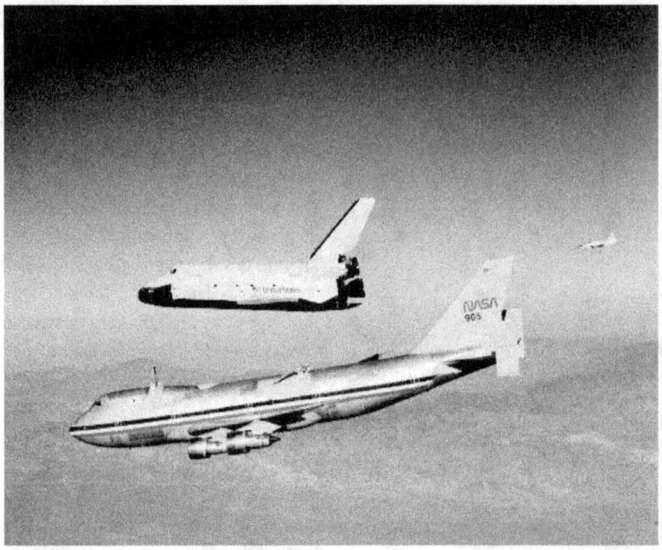

The Space Shuttle Enterprise was used to conduct approach and landing testing (1977) at the Dryden Flight Research Center, California. In the five free flights, the astronaut crew separated the spacecraft from the Shuttle Carrier Aircraft and maneuvered to a landing. These flights verified the Orbiter's pilot-guided approach and landing capability and verified the Orbiter's subsonic airworthiness in preparation for the first crewed orbital flight.

Another discrepancy occurred during the early re-entry phase of STS-1. Nominally, the Orbiter was designed to reenter in an attitude with the nose of the vehicle inclined 40 degrees to the oncoming air. In aeronautical terms, this is a 40-degree angle of attack. To aerodynamically control this attitude, the Orbiter had movable control surfaces on the trailing edge of its wings and a large "body flap." To maintain the desired angle of attack, the Orbiter could adjust the position of the body flap up out of the flow or down into the flow, accordingly. During STS-1, the body flap deflection was twice the amount than had been predicted would be required and was uncomfortably close to the body flap's deployment limit of 22.5 degrees. NASA determined that the cause was "real gas effects"—a phenomenon rooted in high-temperature gas dynamics.

During re-entry, the Orbiter compressed the air of the atmosphere as it smashed into the atmosphere at hypersonic speed, causing the temperature of the air to heat up thermodynamically. The temperature rise was so extreme that it broke the chemical bonds that hold air molecules together, fundamentally altering how the flow around the Orbiter compressed and expanded. These high-temperature gas dynamic effects influenced the pressure distribution on the aft portion of the heat shield, thus affecting its nominal trim condition. The extent to which this effect affected the Orbiter had not been observed before; thus, it was not replicated in the wind tunnel testing used during the design phase. NASA researchers developed an experimental technique to simulate this experience using a special test gas that mimicked the behavior of high-temperature air at the lower temperatures achieved during wind tunnel testing.

Advances in Computational Aerosciences

The use of computational fluid dynamics was eventually developed as a complementary means of obtaining aeroscience information. Engineers used computers to calculate flow-field properties around the shuttle vehicle for a given flight condition. This included pressure, shear stress, or heating on the vehicle surface, as well as density, velocity, temperature, and pressure of the air away from the vehicle. This was accomplished by numerically solving a complex set of nonlinear partial differential equations that described the motion of the fluid and satisfied a fundamental requirement for conservation of mass, momentum, and energy everywhere in the flow field.

Given its relative lack of sophistication and maturity, coupled with the modest computational power afforded by computers in the 1970s, computational fluid dynamics played almost no role in the development of the Space Shuttle aerodynamic database. In the following decades, bolstered by exponential increases in computer capabilities and continuing research, computational fluid dynamics took on a more prominent role. As with any tool, demonstrated validation of results with closely related experimental or flight data was an essential step prior to its use.

The most accurate approach for using wind tunnel data to validate computational fluid dynamics predictions was to directly model the wind tunnel as closely as possible, computationally. After results were validated at wind tunnel conditions, the computational fluid dynamics tool could be run at the flight conditions and used directly, or the difference between the computed flight and wind tunnel predictions could be added to the baseline experimental wind tunnel measured result.

Because different flight regimes have unique modeling challenges, NASA developed separate computational fluid dynamics tools that were tuned to specific flight regimes. This allowed the computational algorithms employed to be optimized for each regime. Although not available during the preflight design of the Space Shuttle, several state-of-the-art computational tools were created that contributed significantly to the subsequent success of the shuttle, providing better understanding of control surface effectiveness, aerodynamic interference effects, and damage assessment. The examples of OVERFLOW and Langley Aerothermodynamic Upwind Relaxation Algorithm (LAURA) software packages were both based on traditional computational fluid dynamics methods while the digital to analog converter (DAC) software employed special-purpose algorithms that allowed it to simulate rarefied, low-density flows.

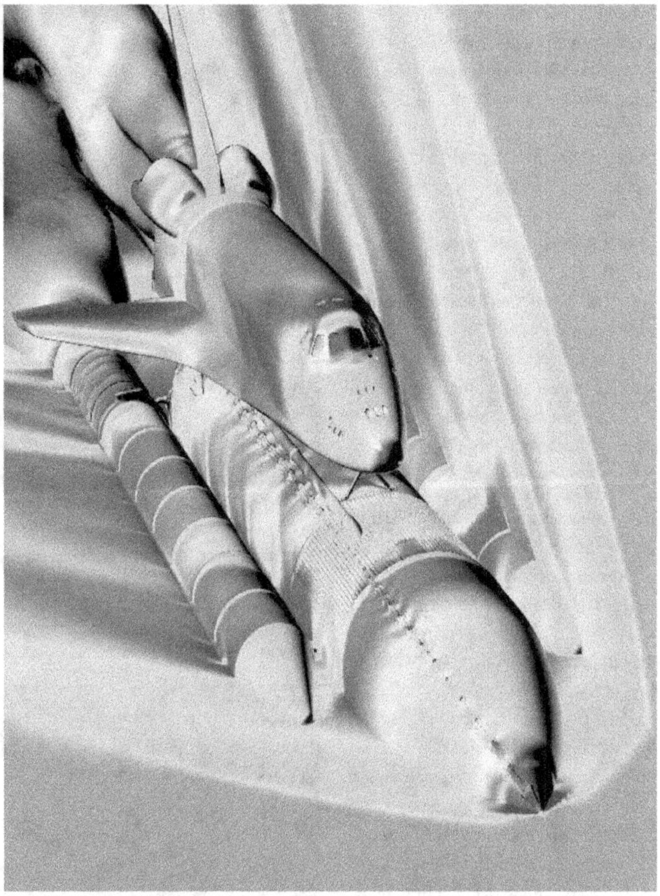

This image depicts the geometric detail included in this high-fidelity modeling capability, as well as some representative results produced by the OVERFLOW tool. The OVERFLOW computational fluid dynamics tool was optimized for lower Mach number subsonic, transonic, and supersonic flows. The surface pressure is conveyed by a progressive color scale that corresponds to the pressure magnitude. A similar color scale with a different range is used to display Mach number in the flow field. OVERFLOW provided extremely accurate predictions for the launch vehicle aerodynamic environments. Color contouring depicts the nominal heating distribution on the Orbiter, where hotter colors represent higher values and cooler colors represent lower values.

The OVERFLOW computational fluid dynamics tool was optimized for lower Mach number subsonic, transonic, and supersonic flows. It was thus most applicable for ascent and late re-entry simulations. Additionally, its underlying methodology was based on an innovative and extremely flexible approach for discretization of the domain around the vehicle. This was especially beneficial for analysis of a complex geometry like the shuttle.

The development of this computational fluid dynamics tool allowed engineers to effectively model the requisite geometric detail of the launch vehicle, as well as the plumes. OVERFLOW was subsequently used to investigate the effect of design changes to the shuttle's aerodynamic performance. Some of these directly impacted shuttle operations, including all of the changes made to the tank after the Columbia accident in 2003 to help minimize the debris. Additionally, OVERFLOW solutions became a key element in the program's risk assessment for ascent debris, as the detailed flow-field information it provided was used to predict trajectories of potential debris sources. OVERFLOW became a key tool for commercial and military transport analyses and was heavily used by industry as well as other NASA programs.

The LAURA package was another traditional computational fluid dynamics code, but designed specifically to predict hypersonic flows associated with re-entry vehicles. It incorporated physical models that account for chemical reactions that take place in air at the extremely high temperatures produced as a spacecraft reenters an atmosphere, as well as the temporal speed at which these reactions take place. This was essential, as the "resident" time a fluid element was in the vicinity of the Orbiter was extremely short given that the vehicle traveled more than 20 times the speed of sound and the chemical reactions taking place in the surrounding fluid occurred at a finite rate.

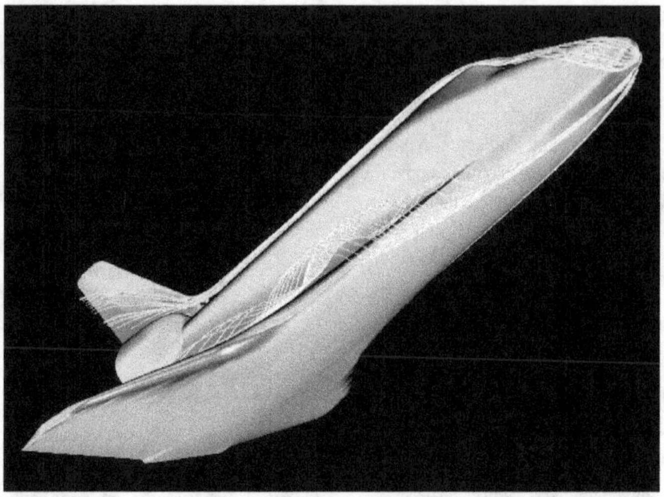

Special computational fluid dynamics programs appropriately model the complex chemically reacting physics necessary to accurately predict a spacecraft's aerodynamic characteristics and the aerothermodynamic heating it will experience. Heating information was needed to determine the appropriate materials and thickness of the Thermal Protection System that insulated the underlying structure of the vehicle from hot gases encountered during re-entry into Earth's atmosphere. Color contouring depicts the nominal heating distribution on the Orbiter, where hotter colors represent higher values and cooler colors represent lower values.

LAURA underwent extensive validation through comparisons to a wide body of experimental and flight data, and it was also used to investigate, reproduce, and answer questions associated with the Orbiter body flap trim anomaly. LAURA was used extensively during the post-Columbia accident investigation activities and played a prominent role in supporting subsequent shuttle operations. This included assessing damaged or repaired Orbiter Thermal Protection System elements, as well as providing detailed flow field characteristics. These characteristics were assessed to protect against dangerous early transitioning of the flow along the heat shield of the Orbiter from smooth laminar flow to turbulent conditions, and thus

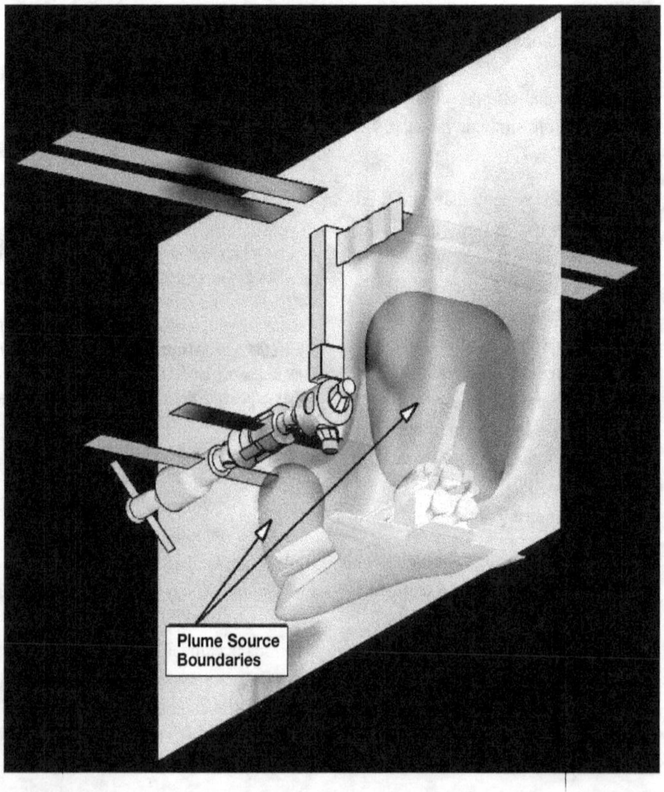

NASA used the Direct Simulation Monte Carlo method to simulate low-density flows, such as those created by maneuvering thrusters during orbital rendezvous and docking of the shuttle to the space station. While the method made use of a distincly different modeling technique to make its predictions, it produced the same detailed information about the flow field as would a traditional computational fluid dynamics technique.

greatly elevated heating that would have endangered the vehicle and crew.

While traditional computational fluid dynamics tools proved extremely useful, their applicability was limited to denser portions of the atmosphere. NASA recognized the need to also be able to perform accurate analysis of low-density flows. Subsequently, the agency invested in the development of a state-of-the-art computer program that would be applicable to low-density rarefied flows. This program was based on the Direct Simulation Monte Carlo (DSMC) method—which is a simulation of a gas at the molecular level that tracks molecules though physical space and their subsequent deterministic collisions with a surface and representative collisions with other molecules. The resulting software, named the DSMC Analysis Code, was used extensively in support of shuttle missions to the Russian space station Mir and the International Space Station, as well as Hubble Space Telescope servicing missions. It also played a critical role in the analysis of the Mars Global Surveyor (1996) and the Mars Odyssey (2001) missions.

Leveraging the Space Shuttle Experience

Never before in the history of flight had such a complex vehicle and challenging flight regime been characterized.
As a result of this challenge, NASA developed new and improved understanding of the associated physics, and subsequently techniques and tools to more accurately simulate them. The aeroscience techniques and technologies that successfully supported the Space Shuttle are useful for exporation of our solar system.

Ascent Flight Design

NASA's challenge was to put wings on a vehicle and have that vehicle survive the atmospheric heating that occurred during re-entry into Earth's atmosphere. The addition of wings resulted in a much-enhanced vehicle with a lift-to-drag ratio that allowed many abort options and a greater cross-range capability, affording more return-to-Earth opportunities. This Orbiter capability did, however, create a unique ascent flight design challenge. The launch configuration was no longer a smooth profiled rocket. The vehicle during ascent required new and complex aerodynamic and structural load relief capabilities.

The Space Shuttle ascent flight design optimized payload to orbit while operating in a constrained environment. The Orbiter trajectory needed to restrict wing and tail structural loading during maximum dynamic pressure and provide acceptable first stage performance. This was achieved by flying a precise angle of attack and sideslip profile and by throttling the main engines to limit dynamic pressure to five-times-gravity loads. The Solid Rocket Boosters (SRBs) had a built-in throttle design that also minimized the maximum dynamic pressure the vehicle would encounter and still achieve orbital insertion.

During the first stage of ascent, the vehicle angle of attack and dynamic pressure produced a lift force from the wings and produced vehicle structural loading. First stage guidance and control algorithms ensured that the angle of attack and sideslip did not vary significantly and resulted in flying through a desired keyhole. The keyhole was defined by the product of dynamic pressure and angle of attack. The product of dynamic pressure and sideslip maintained the desired loading on the vehicle tail.

Varying Throttle to Meet Dynamic Pressure Constraints During Ascent

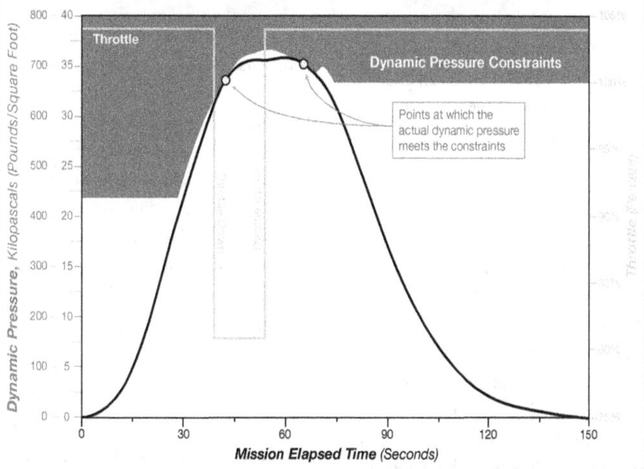

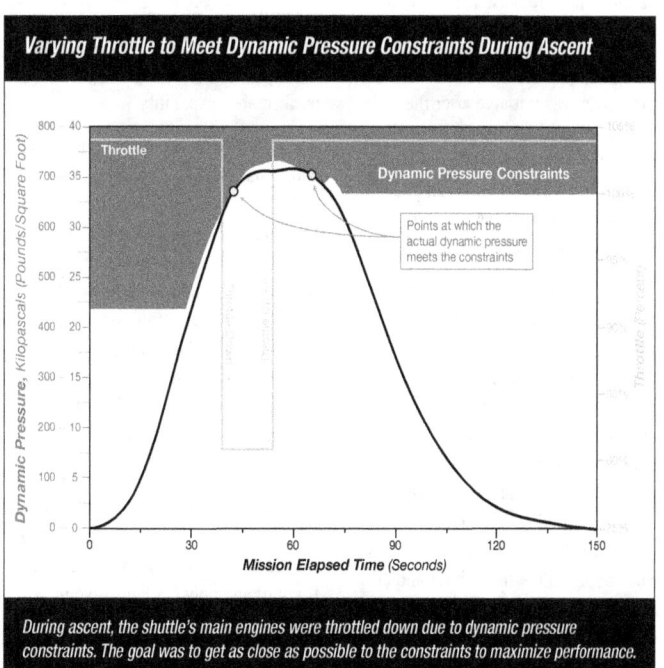

During ascent, the shuttle's main engines were throttled down due to dynamic pressure constraints. The goal was to get as close as possible to the constraints to maximize performance.

Flying Through a Keyhole

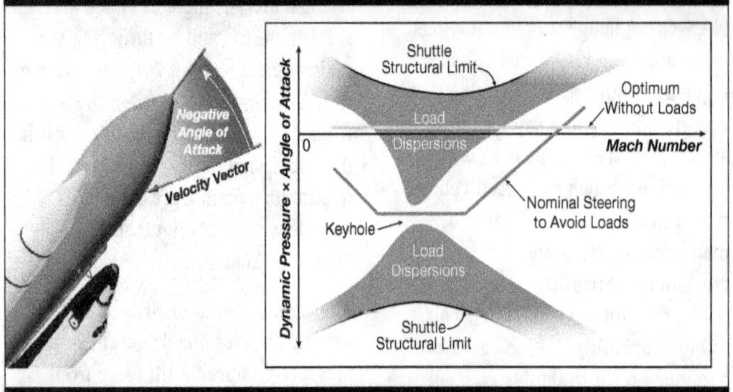

Load dispersions, which are mostly due to atmospheric and thrust variations, added further constraints to the shuttle's flight. To avoid the various load dispersions at certain Mach numbers, the shuttle had to deviate from its optimum angle of attack.

Because day-of-launch winds aloft significantly altered vehicle angle of attack and sideslip during ascent, balloon measurements were taken near liftoff and in proximity of the launch site. Based on these wind measurements, Orbiter guidance parameters were biased and updated via telemetry.

Also during first stage, a roll maneuver was initiated after the vehicle cleared the tower. This roll maneuver was required to achieve the desired orbital inclination and put the vehicle in a heads-down attitude during ascent.

Vehicle performance was maximized during second stage by a linear steering law called powered explicit guidance. This steering law guided the vehicle to orbital insertion and provided abort capability to downrange abort sites or return to launch site. Ascent performance was maintained. If one main engine failed, an intact abort could be achieved to a safe landing site. Such aborts allow the Orbiter and crew to either fly at a lower-than-planned orbit or land.

Ascent flight design was also constrained to dispose the External Tank (ET) in safe waters—either the Indian Ocean or the Pacific Ocean—or in a location where tank debris was not an issue.

After main engine cutoff and ET separation, the remaining main engine fuel and oxidizer were dumped. This event provided some additional performance capability.

After the shuttle became operational, additional ascent performance was added to provide safe orbit insertion for some heavy payloads. Many guidance and targeting algorithm additions provided more payload capability. For example, standard targets were replaced by direct targets, resulting in one Orbital Maneuvering System maneuver instead of two. This saved propellant and resulted in more payload to orbit.

The ascent flight design algorithms and techniques that were generated for the shuttle will be the foundation for ascent flight of any new US launch vehicle.

Ascent Abort

During ascent, a first stage Orbiter main engine out required the shuttle to return to the launch site. The on-board guidance adjusted the pitch profile to achieve SRB staging conditions while satisfying structural and heating constraints. For a side Orbiter main engine out, the vehicle was rolled several degrees so that the normal aerodynamic force canceled the side force induced by the remaining good side engine. Also, vehicle sideslip was maintained near zero to satisfy structural constraints.

After the SRBs were safely separated, second stage guidance commanded a fixed pitch attitude around 70 degrees to minimize vehicle heating and burn the fuel no longer required. This was called the fuel dissipation phase and lasted until approximately 2% of the fuel remained. At this point, guidance commanded the vehicle to turn around and fly back to the launch site using the powered explicit guidance algorithm. As the vehicle returned, it was pitched down so the ET could be safely separated. Dynamic pressure was also minimized so a safe re-entry could occur.

During second stage ascent, a main engine failure usually required the vehicle to abort to a transatlantic landing site. An abort to a downrange landing site was preferred to a return to launch site to reduce complex trajectory targeting and minimize the loads and heating environments, therefore increasing abort success. If a main engine failure occurred late during second stage, an abort to a safe orbit was possible. Abort to orbit was preferred over an abort to a transatlantic landing site. Once the shuttle was in a safe orbit, the vehicle could perform a near nominal re-entry and return to the planned US landing strip.

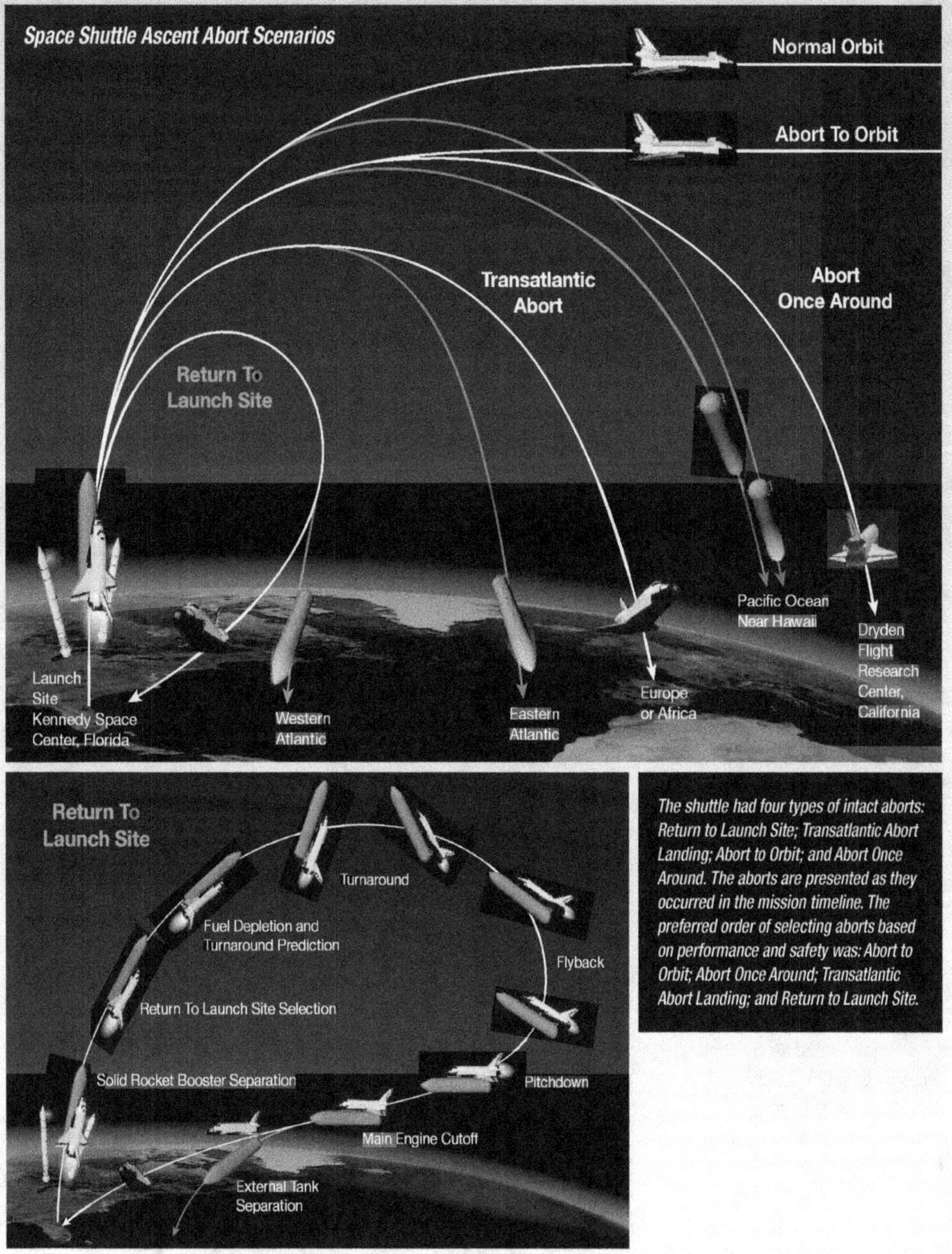

The shuttle had four types of intact aborts: Return to Launch Site; Transatlantic Abort Landing; Abort to Orbit; and Abort Once Around. The aborts are presented as they occurred in the mission timeline. The preferred order of selecting aborts based on performance and safety was: Abort to Orbit; Abort Once Around; Transatlantic Abort Landing; and Return to Launch Site.

If more than one main engine failed during ascent, a contingency abort was required. If a contingency abort was called during first stage, guidance would pitch the vehicle up to loft the trajectory, thereby minimizing dynamic pressure and allowing safe separation of the SRBs and ET. After these events, a pullout maneuver would be performed to bring the vehicle to a gliding flight so a crew bailout could occur.

Two engines out early during second stage allowed the crew to attempt a landing along the US East Coast at predefined landing strips. Two engines out late in second stage allowed an abort to a transatlantic site or abort to safe orbit, depending on the time of the second failure.

In general, Mission Control used vehicle telemetry and complex vehicle performance predictor algorithms to assist the crew in choosing the best abort guidance targets and a safe landing site. The Abort Region Determinator was the primary ground flight design tool that assisted Mission Control in making abort decisions. If communication with the ground was lost, the crew would use on-board computer data and cue cards to assist in selecting the abort mode.

Summary

The shuttle ascent and ascent flight design were complex. NASA developed and verified many innovative guidance algorithms to accomplish mission objectives and maintain vehicle and crew safety. This legacy of flight techniques and computer tools will prove invaluable to all new spacecraft developments.

Re-entry Flight Design

The shuttle vehicle reentered the Earth's atmosphere at over 28,000 km per hour (kph) (17,400 mph)—about nine times faster than the muzzle speed of an M16 bullet. Designing a guidance system that safely decelerated this rapidly moving spacecraft to runway landing speeds while respecting vehicle and crew constraints was a daunting challenge, one that the shuttle re-entry guidance accomplished.

The shuttle re-entry guidance provided steering commands from initial re-entry at a speed of 28,000 kph (17,400 mph), an altitude of 122 km (76 miles), and a distance of 7,600 km (4,722 miles) from the runway until activation of terminal area guidance (a distance of about 90 km [56 miles] and 24 km [15 miles] altitude from the runway). During this interval, a tremendous amount of kinetic energy was transferred into heat energy as the vehicle slowed down. This was all done while the crew experienced only about 1.5 times the acceleration of gravity (1.5g). As a comparison, 1g acceleration is what we feel while sitting on a chair at sea level.

Entry Guidance Drag Velocity Profile

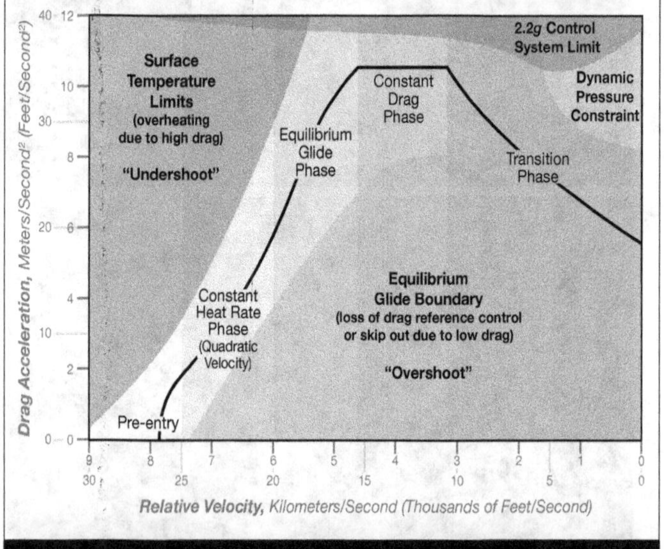

Shuttle re-entry guidance was segmented into several phases—each designed to satisfy unique constraints during flight. The narrow region of acceptable flight conditions was called the "flight corridor." The surface temperature constraints resided at the lower altitude and high drag "undershoot" side of the flight corridor. In contrast, if the vehicle flew too close to the "overshoot" boundary, it would not have enough drag acceleration to reach the landing site and could possibly skip back into orbit. As the vehicle penetrated deeper into the atmosphere, the undershoot corridor was redefined by the vehicle control system and dynamic pressure constraints.

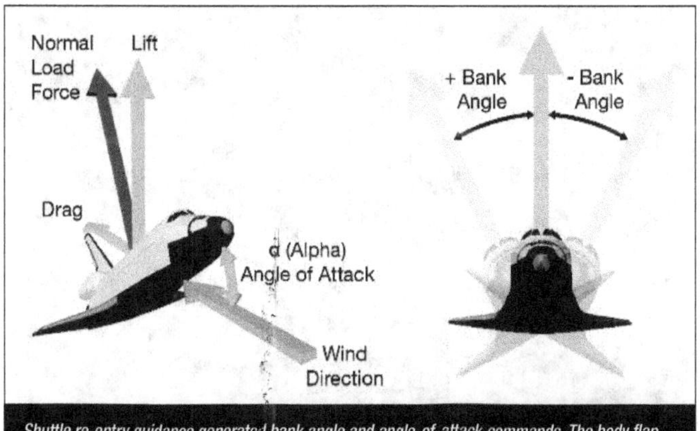

Shuttle re-entry guidance generated bank angle and angle-of-attack commands. The body flap was used to control the angle of attack by balancing the aerodynamic forces and moments about the vehicle center of gravity. The bank angle controlled the direction of the lift vector about the wind velocity vector at a fixed angle of attack. Drag, which was opposite to the wind-relative velocity, slowed the vehicle down. Lift was normal to the drag vector and was used to change the rate at which the vehicle reentered the atmosphere. The total normal load force was the sum of the lift acceleration and drag acceleration and resulted in the force felt by the crew.

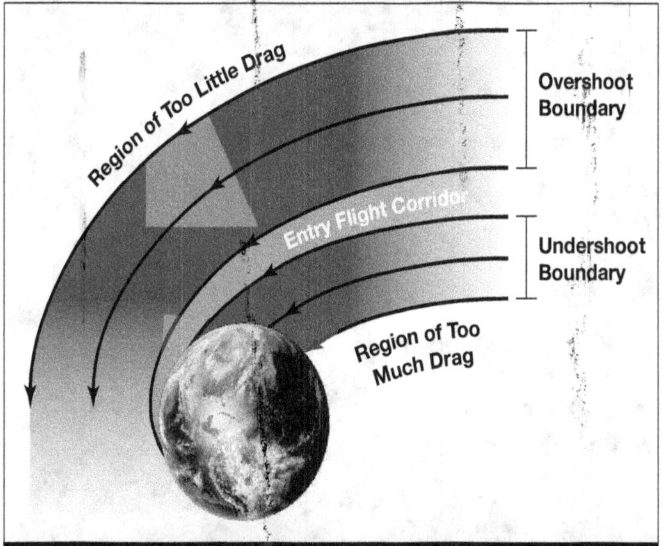

The Entry Flight Corridor defined the atmospheric re-entry angles required for safe re-entry flight. Before any successful re-entry from low-Earth orbit could occur, the shuttle needed to fire engines to place the vehicle on a trajectory that intercepted the atmosphere. This deorbit maneuver had to be executed precisely. With too steep of a re-entry, the guidance could not compute steering commands that would stop the vehicle from overheating. With too shallow of a re-entry, the guidance could not adequately control the trajectory or, for very shallow trajectories, even stop the vehicle from skipping back out into space. The area between these two extremes was called the Entry Flight Corridor.

How did Space Shuttle Guidance Accomplish This Feat?

First, it's important to understand how the shuttle was controlled. Air molecules impacting the vehicle's surface imparted a pressure or force over the vehicle's surface. The shuttle used Reaction Control System jets initially to control the attitude of the vehicle; however, as the dynamic pressure increased on entering denser atmosphere, the position of the body flap was used to control the angle of attack and the ailerons were used to control bank.

Changing the angle of attack had an immediate effect on the drag acceleration of the vehicle, whereas changing the bank angle had a more gradual effect. It took time for the vehicle to decelerate into different portions of the atmosphere where density and speed affected drag. Controlling the direction of the vehicle lift vector by banking the vehicle was the primary control mechanism available to achieve the desired landing target. The vehicle banked about the relative velocity vector using a combination of aft yaw Reaction Control System jets and aileron deflection. The lift vector moved with the vehicle as it banked about the wind vector. The angle of attack was maintained constant during these maneuvers by the balanced aerodynamic forces at a given body flap trim position. The vehicle banked around this wind vector, keeping the blunt side of the shield facing against the flow of the atmosphere. Banking about the wind vector until the lift pointing down accelerated the vehicle into the atmosphere. Over time, this increased drag caused the vehicle to decelerate quickly. Banking about the wind vector until the lift vector pointed up accelerated the vehicle out of the atmosphere. Over time, this decreased the drag acceleration and caused the vehicle to decelerate gradually. Control of the vehicle lift-and-drag acceleration by bank angle and angle-of-attack modulation were the two primary control parameters used to fly the desired range and cross range during re-entry. These concepts had to be clearly grasped before it was possible to understand the operation of the guidance algorithm.

Within each guidance phase, it was possible to use simple equations to analytically compute how much range was flown. As long as the shuttle trajectory stayed "close" to reference profiles, the guidance algorithm could analytically predict how far the vehicle would fly.

By piecing together all of the guidance segments, the total range flown from the current vehicle position all the way to the last guidance phase could be predicted and compared to the actual range required to reach the target. Any difference between the analytically computed range and the required range would trigger an adjustment in the drag-velocity/energy references to remove that range error. The analytic reference profiles were computed every guidance step (1.92 seconds) during flight. In this manner, any range error caused by variations in the environment, navigated state, aerodynamics, or mass properties was sensed and compensated for with adjustments to the real-time computed drag-velocity or drag-energy reference profiles.

In fact, the entire shuttle re-entry guidance system could be described as a set of interlocked drag-velocity or drag-energy pieces that would fly the required range to target and maintain the constraints of flight.

Boundary Layer Transition

Accurate characterization of the aerothermodynamic heating experienced by a spacecraft as it enters an atmosphere is of critical importance to the design of a Thermal Protection System. More intense heating typically requires a thicker Thermal Protection System, which increases a vehicle's weight. During the early phase of entry, the flow near the surface of the spacecraft—referred to as the boundary layer—has a smooth laminar profile. Later in the trajectory, instabilities develop in the boundary layer that cause it to transition to a turbulent condition that can increase the heating to the spacecraft by up to a factor of 4 over the laminar state. Subsequently, a Boundary Layer Transition Flight Experiment was conceived and implemented on Space Shuttle Discovery's later flights. This experiment employed a fixed-height protuberance (speed bump) on the underside of the wing to perturb and destabilize the boundary layer. NASA used instrumentation to measure both the elevated heating on the protuberance as well as the downstream effect so that the progression of the transition could be captured. The experiment provided foundational flight data that will be essential for the validation of future ground-based testing techniques or computational predictions of this flow phenomenon, thus helping improve the design of all future spacecraft.

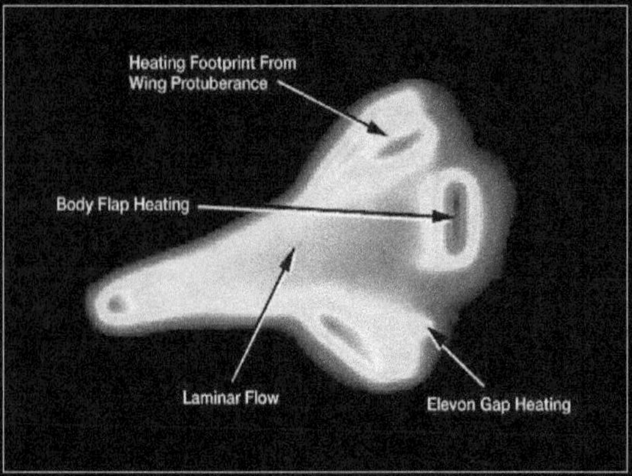

A NASA team—via a US Navy aircraft—captured high-resolution, calibrated infrared imagery of Space Shuttle Discovery's lower surface in addition to discrete instrumentation on the wing, downstream, and on the Boundary Layer Transition Flight Experiment protuberance. In the image, the red regions represent higher surface temperatures.

Constant Heat-rate Phase

The guidance phase was required to protect the structure and interior from the blast furnace of plasma building up outside of the vehicle. That blast furnace was due to the high-velocity impact of the vehicle with the air in the atmosphere.

Typical Angle-of-Attack Profile

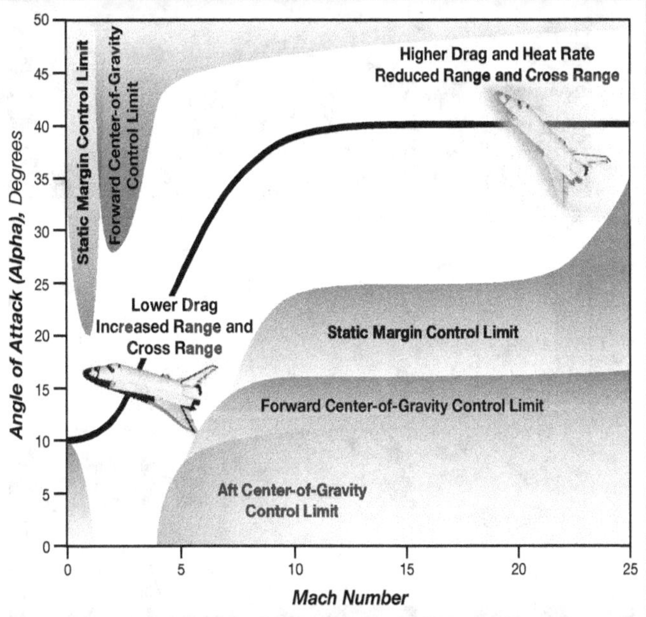

The shuttle guidance was forced to balance conflicting trades to minimize the weight, cost, and complexity of the required subsystems, maximize re-entry performance (range and cross-range capability), and maintain constraint margins. An ideal example was the selection of a constant angle-of-attack (Alpha) profile with a linear-velocity ramp transition. It was known that a high heat-rate trajectory would minimize the tile thickness required to protect the substructure. An initially high Alpha trim (40 degrees) was therefore selected to reduce Thermal Protection System mass and quickly dissipate energy. The 40-degree profile helped shape the forward center-of-gravity control boundaries and define the hypersonic static margin control limits provided by the body flap and ailerons. A linear ramp in the Alpha profile was then inserted to increase the lift-to-drag and cross-range capability and improve the static and dynamic stability of the vehicle.

The Thermal Protection System surface was designed to withstand extremely high temperatures before the temperature limits of the material were exceeded. Even after a successful landing, structural damage from heating could make the vehicle un-reuseable; therefore, it was essential that the surface remain within those limits. To accomplish this, different parts of the vehicle were covered with different types of protective material, depending on local heating.

The objective of the re-entry guidance design during this phase was to ensure that the heat-rate constraints of the Thermal Protection System were not compromised. That is why the constant heat-rate phase used quadratic drag-velocity segments. A vehicle following a drag acceleration profile that was quadratic in velocity experienced a constant rate of heating on the Thermal Protection System. Because the shuttle tile system was designed to radiate heat, the quadratic profiles in shuttle guidance were designed to provide an equilibrium heating environment where the amount of heat transferred by the tiles and to the substructure was balanced by the amount of heat radiated. This meant that there was a temperature at which the radiant heat flux away from the surface matched the rate of atmospheric heating. Once the vehicle Thermal Protection System reached this equilibrium temperature, there would no longer be a net heat flow into the vehicle.

The existence of a temperature limit on the Thermal Protection System material implied the existence of a maximum heat rate the vehicle could withstand. As long as guidance commanded the vehicle to achieve a quadratic velocity reference that was at or below the surface temperature constraint boundaries, the vehicle substructure was maintained at a safe temperature. The Thermal Protection System would be undamaged and reusable, and the crew would be comfortable.

During flight, if the vehicle was too close to the landing site target, the velocity and reference drag profiles were automatically shifted upward, causing an increase in the rate energy is dissipated. The vehicle would, as a result, fly a shorter range. If the vehicle was too far away from the landing site, the combined velocity and reference drag profiles were automatically shifted downward, causing a reduction in the rate at which energy was dissipated. The vehicle would, as a result, fly a longer range.

Lateral Deadband

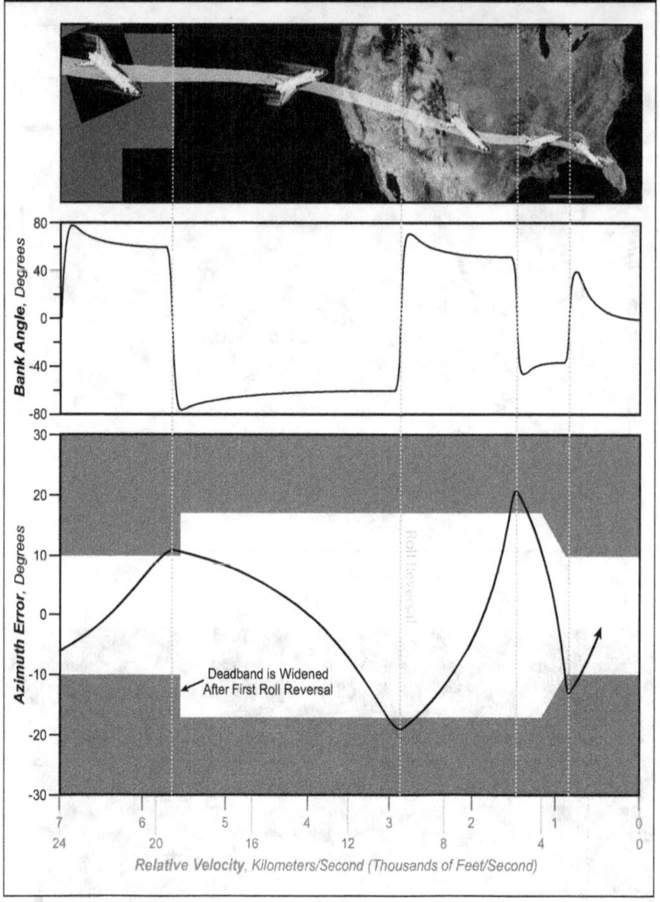

Azimuth Error

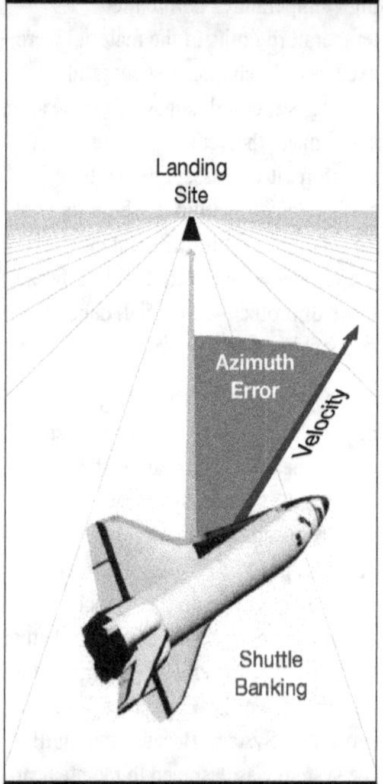

The Space Shuttle removed azimuth errors during flight by periodically executing roll reversals. These changes in the sign (plus or minus) of the vehicle bank command would shift the lift acceleration vector to the opposite side of the current orbit direction and slowly rotate the direction of travel back toward the desired target.

Equilibrium Glide Phase

As the speed of the shuttle dropped below about 6,200 m/s (20,500 ft/s), the constant heat-rate phase ended and the equilibrium glide phase began. This was an intermediate phase between high heating and the rapidly increasing deceleration that occurred as the vehicle penetrated deeper into the atmosphere. This phase determined the drag-velocity reference required to balance gravitational and centrifugal forces on the vehicle. During this phase, only the reference drag profile in the equilibrium glide phase was modified to correct range errors. All future phases were left at their nominal setting. This ranging approach was designed into the shuttle re-entry guidance to reserve ranging capability. This enabled the vehicle to accommodate large navigation errors post ionization blackout (ground communication and tracking loss due to plasma shield interference) and also change runway landing direction due to landing wind changes.

Constant Drag Phase

The constant drag phase began and the equilibrium glide phase ended when either the desired constant drag acceleration target of 10 m/s^2 (33 ft/s^2)

occurred or the transition phase velocity of about 3,200 m/s (10,500 ft/s) was achieved.

During the constant drag phase, the drag-velocity reference was computed to maintain constant drag acceleration on the vehicle. This constrained the accelerations on the vehicle structure and crew. It also constrained maximum load accelerations for crew members confined to a sitting position during re-entry with normal accelerations directed along their spine. For the shuttle, the normal force constraint was set at 2.5*g* maximum; however, typical normal force operational design was set at 1.5*g*. The form of the drag-velocity reference during this phase was particularly simple since the drag accelerations were held constant. Operationally, shuttle guidance continued to command a high 40-degree angle of attack during this phase while the velocity was rapidly reduced and kinetic energy was rapidly removed from the vehicle. Guidance commanded higher drag levels to remove extra energy from the vehicle and to attain a target site that was closer than the nominal prediction. Guidance commanded lower drag levels to reduce the rate energy removed from the vehicle and to attain a target site that was farther away than the nominal prediction.

Transition Phase

When the velocity dropped below approximately 3,200 m/s (10,500 ft/s), the transition phase of guidance was entered and the constant drag phase was terminated. It was during this phase that the guidance system finally began to modulate the energy-vs.-drag reference to remove final trajectory-range errors and issued a command to begin reducing the angle of attack. This pitch-down maneuver prepared the vehicle for transonic and subsonic flight. During the transition phase, the angle of attack was reduced and the vehicle transitioned from flying on the "back side" to the "front side" of the lift-to-drag (lift acceleration divided by drag acceleration) vs. angle-of-attack curve. A vehicle flying on the back side (at a higher angle of attack) was in an aerodynamic posture where increasing the angle of attack decreased the lift-to-drag. In this orientation, the drag on the vehicle was maximized and the vehicle dissipated a great deal of energy, which was highly desirable in the early phases of re-entry flight. A vehicle flying on the front side of the lift-to-drag curve (or at a lower angle of attack) was in an aerodynamic posture where increasing the angle of attack increased the lift-to-drag. In this front-side orientation, the drag was reduced and the vehicle sliced through the air more efficiently. Most airplanes fly on the front side of the lift-to-drag curve, and it was during the transition phase that shuttle guidance began commanding the vehicle to a flying orientation that mimicked the flight characteristics of an airplane.

It was also during the transition phase that the flight-path angle became significantly steeper. This happened naturally as the vehicle began to dig deeper into the atmosphere. A steeper angle was what influenced the formulation of the shuttle guidance to switch from velocity to energy as the independent variable in the reference drag formulation. The linear drag-energy reference acceleration did not use a shallow flight-path angle approximation as was done in the previous guidance phases, and a concise closed-form solution for the range flown at higher flight-path angles was obtained. At the end of transition phase, the vehicle was about 90 km (56 miles) from the runway, flying at an altitude of 24 km (15 miles) and a speed of 750 m/s (2,460 ft/s).

Summary

At this point, the "unique" phase of re-entry required to direct the shuttle from low-Earth orbit was complete. Although other phases of guidance were initiated following the transition phase, these flight regimes were well understood and the guidance formulation was tailored directly for airplane flight.

Avionics, Navigation, and Instrumentation

Introduction
Gail Chapline

Reconfigurable Redundancy
Paul Sollock

Shuttle Single Event Upset Environment
Patrick O'Neill

Development of Space Shuttle Main Engine Instrumentation
Arthur Hill

Unprecedented Rocket Engine Fault-Sensing System
Tony Fiorucci

Calibration of Navigational Aides Using Global Positioning Computers
John Kiriazes

The Space Shuttle faced many vehicle control challenges during ascent, as did the Orbiter during on-orbit and descent operations. Such challenges required innovations such as fly-by-wire, computer redundancy for robust systems, open-loop main engine control, and navigational aides. These tools and concepts led to groundbreaking technologies that are being used today in other space programs and will be used in future space programs. Other government agencies as well as commercial and academic institutions also use these analysis tools. NASA faced a major challenge in the development of instruments for the Space Shuttle Main Engines—engines that operated at speeds, pressures, vibrations, and temperatures that were unprecedented at the time. NASA developed unique instruments and software supporting shuttle navigation and flight inspections. In addition, the general purpose computer used on the shuttle had static random access memory, which was susceptible to memory bit errors or bit flips from cosmic rays. These bit flips presented a formidable challenge as they had the potential to be disastrous to vehicle control.

Reconfigurable Redundancy—The Novel Concept Behind the World's First Two-Fault-Tolerant Integrated Avionics System

Space Shuttle Columbia successfully concluded its first mission on April 14, 1981, with the world's first two-fault-tolerant Integrated Avionics System—a system that represented a curious dichotomy of past and future technologies. On the one hand, many of the electronics components, having been selected before 1975, were already nearing technical obsolescence. On the other hand, it used what were then-emerging technologies; e.g., time-domain-multiplexed data buses, fly-by-wire flight control, and digital autopilots for aircraft, which provided a level of functionality and reliability at least a decade ahead of the avionics in either military or commercial aircraft. Beyond the technological "nuts and bolts" of the on-board system, two fundamental yet innovative precepts enabled and shaped the actual implementation of the avionics system. These precepts included the following:

STS-1 launch (1981) from Kennedy Space Center, Florida. First crewed launch using two-fault-tolerant Integrated Avionics System.

- The entire suite of avionics functions, generally referred to as "subsystems"—data processing (hardware and software), navigation, flight control, displays and controls, communications and tracking, and electrical power distribution and control—would be programmatically and technically managed as an integrated set of subsystems. Given that new and unique types of complex hardware and software had to be developed and certified, it is difficult to overstate the role that approach played in keeping those activities on course and on schedule toward a common goal.
- A digital data processing subsystem comprised of redundant central processor units plus companion input/output units, resident software, digital data buses, and numerous remote bus terminal units would function as the core subsystem to interconnect all avionics subsystems. It also provided the means for the crew and ground to access all vehicle systems (i.e., avionics and non-avionics systems). There were exceptions to this, such as the landing gear, which was lowered by the crew via direct hardwired switches.

Avionics System Patterned After Apollo; Features and Capabilities Unlike Any Other in the Industry

The preceding tenets were very much influenced by NASA's experience with the successful Apollo primary navigation, guidance, and control system. The Apollo-type guidance computer, with additional specialized input/output hardware, an inertial reference unit, a digital autopilot, fly-by-wire thruster control, and an alphanumeric keyboard/display unit represented a nonredundant subset of critical functions for shuttle avionics to perform. The proposed shuttle avionics represented a challenge for two principal reasons: an extensive redundancy scheme and a reliance on new technologies.

Shuttle avionics required the development of an overarching and extensive redundancy management scheme for the entire integrated avionics system, which met the shuttle requirement that the avionics system be "fail operational/fail safe"—i.e., two-fault tolerant with reaction times capable of maintaining safe computerized flight control in a vehicle traveling at more than 10 times the speed of high-performance military aircraft.

Shuttle avionics would also rely on new technologies—i.e., time-domain data buses, digital fly-by-wire flight control, digital autopilots for aircraft, and a sophisticated software operating system that had very limited application in the aerospace industry of that time, even for noncritical applications, much less for "man-rated" usage. Simply put, no textbooks were available to guide the design, development, and flight certification of those technologies and only a modicum of off-the-shelf equipment was directly applicable.

Why Fail Operational/Fail Safe?

Previous crewed spacecraft were designed to be fail safe, meaning that after the first failure of a critical component, the crew would abort the mission by manually disabling the primary system and switching over to a backup system that had only the minimum capability to return the vehicle safely home. Since the shuttle's basic mission was to take humans and payloads safely to and from orbit, the fail-operational requirement was intended to ensure a high probability of mission success by avoiding costly, early termination of missions.

Early conceptual studies of a shuttle-type vehicle indicated that vehicle atmospheric flight control required full-time computerized stability augmentation. Studies also indicated that in some atmospheric flight regimes, the time required for a manual switchover could result in loss of vehicle. Thus, fail operational actually meant that the avionics had to be capable of "graceful degradation" such that the first failure of a critical component did not compromise the avionic system's capability to maintain vehicle stability in any flight regime.

The graceful degradation requirement (derived from the fail-operational/fail-safe requirement) immediately provided an answer to how many redundant computers would be necessary. Since the computers were the only certain way to ensure timely graceful degradation—i.e., automatic detection and isolation of an errant computer—some type of computerized majority-vote technique involving a minimum of three computers would be required to retain operational status and continue the mission after one computer failure. Thus, four computers were required to meet the fail-operational/fail-safe requirement. That level of redundancy applied only to the computers. Triple redundancy was deemed sufficient for other components to satisfy the fail-operational/fail-safe requirement.

Central Processor Units Were Available Off the Shelf—Remaining Hardware and Software Would Need to be Developed

The next steps included: selecting computer hardware that was for military use yet commercially available; choosing the actual configuration, or architecture, of the computer(s), data bus network, and bus terminal units; and then developing the unique hardware and software to implement the world's first two-fault-tolerant avionics.

In 1973, only two off-the-shelf computers available for military aircraft offered the computational capability for the shuttle. Both computers were basic processor units—termed "central processor units"—with only minimal input/output functionality. NASA selected a vendor to provide the central processor units plus new companion input/output processors that would be developed to specifications provided by architecture designers. At the time, no proven best practices existed for interconnecting multiple computers, data buses, and bus terminal units beyond the basic active/standby manual switchover schemes.

The architectural concept figured heavily in the design requirements for the input/output processor and two other new types of hardware "boxes" as

Interconnections Were Key to Avionics Systems Success

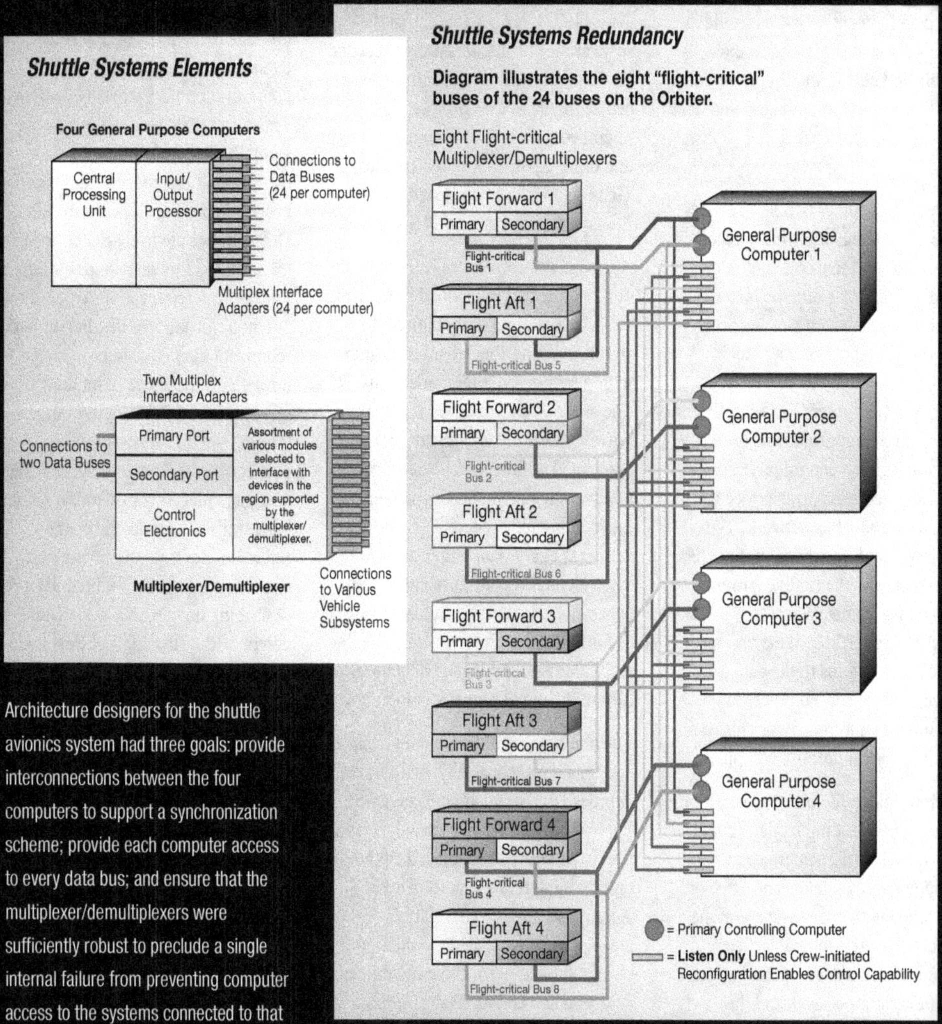

Architecture designers for the shuttle avionics system had three goals: provide interconnections between the four computers to support a synchronization scheme; provide each computer access to every data bus; and ensure that the multiplexer/demultiplexers were sufficiently robust to preclude a single internal failure from preventing computer access to the systems connected to that multiplexer/demultiplexer.

To meet those goals, engineers designed the input/output processor to interface with all 24 data buses necessary to cover the shuttle. Likewise, each multiplexer/demultiplexer would have internal redundancy in the form of two independent ports for connections to two data buses. The digital data processing subsystem possessed eight flight-critical data buses and the eight flight-critical multiplexer/demultiplexers. They were essential to the reconfiguration capability. The total complement of such hardware on the vehicle consisted of 24 data buses, 19 multiplexer/demultiplexers, and an almost equal number of other types of specialized bus terminal units.

well as the operating system software, all four of which had to be uniquely developed for the shuttle digital data processing subsystem. Each of those four development activities would eventually result in products that established new limits for the so-called "state of the art" in both hardware and software for aerospace applications.

In addition to the input/output processor, the other two new devices were the data bus transmitter/receiver units—referred to as the multiplex interface adapter—and the bus terminal units, which was termed the "multiplexer/demultiplexer." NASA designated the software as the Flight Computer Operating System. The input/output processors (one paired with each central processor unit) was necessary to interface the units to the data bus network. The numerous multiplexer/demultiplexers would serve as the remote terminal units along the data buses to effectively interface all the various vehicle subsystems to the data bus network. Each central processor unit/input/output processor pair was called a general purpose computer.

The multiplexer/demultiplexer was an extraordinarily complex device that provided electronic interfaces for the myriad types of sensors and effectors associated with every system on the vehicle. The multiplex interface adaptors were placed internal to the input/output processors and the multiplexer/demultiplexers to provide actual electrical connectivity to the data buses. Multiplex interface adaptors were supplied to each manufacturer of all other specialized devices that interfaced with the serial data buses. The protocol for communication on those buses was also uniquely defined.

The central processor units later became a unique design for two reasons: within the first several months in the field, their reliability was so poor that they could not be certified for the shuttle "man-rated" application; and following the Approach and Landing Tests (1977), NASA found that the software for orbital missions exceeded the original memory capacity. The central processor units were all upgraded with a newer memory design that doubled the amount of memory. That memory flew on Space Transportation System (STS)-1 in 1981.

Although the computers were the only devices that had to be quad redundant, NASA gave some early thought to simply creating four identical strings with very limited interconnections. The space agency quickly realized, however, that the weight and volume associated with so much additional hardware would be unacceptable. Each computer needed the capability to access every data bus so the system could reconfigure and regain capability after certain failures. NASA accomplished such reconfiguration by software reassignment of data buses to different general purpose computers.

The ability to reconfigure the system and regain lost capability was a novel approach to redundancy management. Examination of a typical mission profile illustrates why NASA placed a premium on providing reconfiguration capability. Ascent and re-entry into Earth's atmosphere represented the mission phases that required automatic failure detection and isolation capabilities, while the majority of on-orbit operations did not require full redundancy when there was time to thoroughly assess the implications of any failures that occurred prior to re-entry. When a computer and a critical sensor on another string failed, the failed computer string could be reassigned via software control to a healthy computer, thereby providing a fully functional operational configuration for re-entry.

The Costs and Risks of Reconfigurable Redundancy

The benefits of interconnection flexibility came with costs, the most obvious being increased verification testing needed to certify each configuration performed as designed. Those activities resulted in a set of formally certified system reconfigurations that could be invoked at specified times during a mission. Other less-obvious costs stemmed from the need to eliminate single-point failures. Interconnections offered the potential for failures that began in one redundant element and propagated throughout the entire redundant system—termed a "single-point failure"—with catastrophic consequences. Knowing such, system designers placed considerable emphasis on identification and elimination of failure modes with the potential to become single-point failures. Before describing how NASA dealt with potential catastrophic failures, it is necessary to first describe how the redundant digital data processing subsystem was designed to function.

Establishing Synchronicity

The fundamental premise for the redundant digital data processing subsystem operation was that all four general purpose computers were executing identical software in a time-synchronized fashion such that all received the exact same data, executed the same computations, got the same results, and then sent the exact same time-synchronized commands and/or data to other subsystems.

Maintenance of synchronicity between general purpose computers was one of the truly unique features of the newly developed Flight Computer Operating System. All four general purpose computers ran in a synchronized fashion that was keyed

Shuttle Single Event Upset Environment

Five general purpose computers—the heart of the Orbiter's guidance, navigation, and flight control system—were upgraded in 1991. The iron core memory was replaced with modern static random access memory transistors, providing more memory and better performance. However, the static random access memory computer chips were susceptible to single event upsets: memory bit flips caused by high-energy nuclear particles. These single event upsets could be catastrophic to the Orbiter because general purpose computers were critical to flights since one bit flip could disable the computer.

An error detection and correction code was implemented to "fix" flipped bits in a computer word by correcting any single erroneous bit. Whenever the system experienced a memory bit flip fix, the information was downlinked to flight controllers on the ground in Houston, Texas. The event time and the Orbiter's ground track resulted in the pattern of bit flips around the Earth.

The bit flips correlated with the known space radiation environment. This phenomena had significant consequences for error detection and correction codes, which could only correct one error in a word and would be foiled by a multi-bit error. In response, system architects selected bits for each word from different chips, making it almost impossible for a single particle to upset more than one bit per word.

In all, the upgraded Orbiter general purpose computers performed flawlessly in spite of their susceptibility to ionizing radiation.

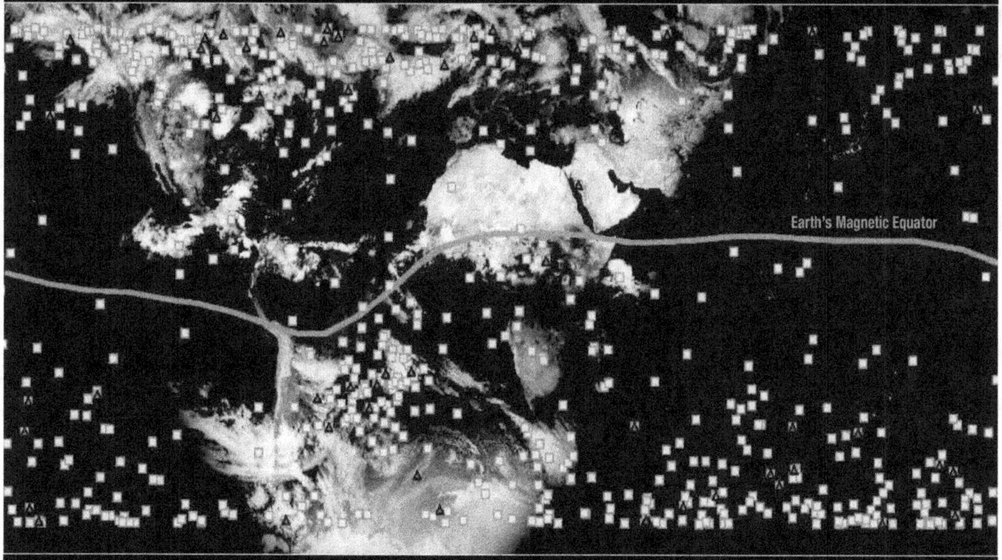

Single event upsets are indicated by yellow squares. Multi-bit single event upsets are indicated by red triangles. In these single events, anywhere from two to eight bits were typically upset by a single charged particle.

to the timing of the intervals when general purpose computers were to query the bus terminal units for data, then process that data to select the best data from redundant sensors, create commands, displays, etc., and finally output those command and status data to designated bus terminal units. That sequence (input/process/output) repeated 25 times per second. The aerodynamic characteristics of the shuttle dictated the 25-hertz (Hz) rate. In other words, the digital autopilot had to generate stability augmentation commands at that frequency for the vehicle to retain stable flight control. The four general purpose computers exchanged synchronization status approximately 350 times per second. The typical failure resulted in the computer halting anything resembling normal operation.

A fish-eye view of the multifunction electronic display subsystem—or "glass cockpit"—in the fixed-base Space Shuttle mission simulator at Johnson Space Center, Texas.

Early Detection of Failure

NASA designed the four general purpose computer redundant set to gracefully degrade from either four to three or from three to two members. Engineers tailored specific redundancy management algorithms for dealing with failures in other redundant subsystems based on knowledge of each subsystem's predominant failure modes and the overall effect on vehicle performance.

NASA paid considerable attention to means of detecting subtle latent failure modes that might create the potential for a simultaneous scenario. Engineers scrutinized sensors such as gyros and accelerometers in particular for null failures. During orbital operation, the vehicle typically spent the majority of time in a quiescent flight control profile such that those sensors were operating very near their null points. Prior to re-entry, the vehicle executed some designed maneuvers to purposefully exercise those devices in a manner to ensure the absence of permanent null failures. The respective design teams for the various subsystems were always challenged to strike a balance between early detection of failures vs. nuisance false alarms, which could cause the unnecessary loss of good devices.

Decreasing Probability of Pseudo-simultaneous Failures

There was one caveat regarding the capability to be two-fault tolerant—the system was incapable of coping with simultaneous failures since such failures obviously defeat the majority-voting scheme. A nuance associated with the practical meaning of "simultaneous" warranted significant attention from the designers. It was quite possible for internal circuitry in complex electronics units to fail in a manner that wasn't immediately apparent because the circuitry wasn't used in all operations. This failure could remain dormant for seconds, minutes, or even longer before normal activities created conditions requiring use of the failed devices; however, should another unrelated failure occur that created the need for use of the previously failed circuitry, the practical effect was equivalent to two simultaneous failures.

To decrease the probability of such pseudo-simultaneous failures, the general purpose computers and multiplexer/demultiplexers were designed to constantly execute cyclic background self-test operations and cease operations if internal problems were detected.

Ferreting Out Potential Single-point Failures

Engineering teams conducted design audits using a technique known as failure modes effects analysis to identify types of failures with the potential to propagate beyond the bounds of the fault-containment region in which they originated. These studies led to the conclusion that the digital data processing subsystem was susceptible to two types of hardware failures with the potential to create a catastrophic condition, termed a "nonuniversal input/output error." As the name implies, under such conditions a majority of general purpose computers may not have received the same data and the redundant set may have diverged into a two-on-two configuration or simply collapsed into four disparate members.

Engineers designed and tested the topology, components, and data encoding of the data bus network to ensure that robust signal levels and data integrity existed throughout the network. Extensive laboratory testing confirmed, however, that the two types of failures would likely create conditions resulting in eventual loss of all four computers.

The first type of failure and the easiest to mitigate was some type of physical failure causing either an open or a short circuit in a data bus. Such a condition would create an impedance mismatch along the bus and produce classic transmission line effects; e.g., signal reflections and standing waves with the end result being unpredictable signal levels at the receivers of any given general purpose computer. The probability of such a failure was deemed to be extremely remote given the robust mechanical and electrical design as well as detailed testing of the hardware, before and after installation on the Orbiter.

The second type of problem was not so easily discounted. That problem could occur if one of the bus terminal units failed, thus generating unrequested output transmissions. Such transmissions, while originating from only one node in the network, would nevertheless propagate to each general purpose computer and disrupt the normal data bus signal levels and timing as seen by each general purpose computer. It should be mentioned that no amount of analysis or testing could eliminate the possibility of a latent, generic software error that could conceivably cause all

Loss of Two General Purpose Computers Tested Resilience

Space Shuttle Columbia (STS-9) makes a successful landing at Dryden Flight Research Center on Edwards Air Force Base runway, California, after reaching a fail-safe condition while on orbit.

Shuttle avionics never encountered any type (hardware or software) of single-point failure in nearly 3 decades of operation, and on only one occasion did it reach the fail-safe condition. That situation occurred on STS-9 (1983) and demonstrated the resiliency afforded by reconfiguration.

While on-orbit, two general purpose computers failed within several minutes of each other in what was later determined to be a highly improbable, coincidental occurrence of a latent generic hardware fault. By definition, the avionics was in a fail-safe condition and preparations were begun in preparation for re-entry into Earth's atmosphere. Upon cycling power, one of the general purpose computers remained failed while the other resumed normal operation. Still, with that machine being suspect, NASA made the decision to continue preparation for the earliest possible return. As part of the preparation, sensors such as the critical inertial measurement unit, which were originally assigned to the failed computer, were reassigned to a healthy one. Thus, re-entry occurred with a three-computer configuration and a full set of inertial measurement units, which represented a much more robust and safe configuration.

The loss of two general purpose computers over such a short period was later attributed to spacelight effects on microscopic debris inside certain electronic components. Since all general purpose computers in the inventory contained such components, NASA delayed subsequent flights until sufficient numbers of those computers could be purged of the suspect components.

four computers to fail. Thus, the program deemed that a backup computer, with software designed and developed by an independent organization, was warranted as a safeguard against that possibility.

This backup computer was an identical general purpose computer designed to "listen" to the flight data being collected by the primary system and make independent calculations that were available for crew monitoring. Only the on-board crew had the switches, which transferred control of all data buses to that computer, thereby preventing any "rogue" primary computers from "interfering" with the backup computer.

Its presence notwithstanding, the backup computer was never considered a factor in the fail-operational/fail-safe analyses of the primary avionics system, and—at the time of this publication—had never been used in that capacity during a mission.

Summary

The shuttle avionics system, which was conceived during the dawn of the digital revolution, consistently provided an exceptional level of dependability and flexibility without any modifications to either the basic architecture or the original innovative design concepts. While engineers replaced specific electronic boxes due to electronic component obsolescence or to provide improved functionality, they took great care to ensure that such replacements did not compromise the proven reliability and resiliency provided by the original design.

Development of Space Shuttle Main Engine Instrumentation

The Space Shuttle Main Engine operated at speeds and temperatures unprecedented in the history of spaceflight. How would NASA measure the engine's performance?

NASA faced a major challenge in the development of instrumentation for the main engine, which required a new generation capable of measuring—and surviving—its extreme operating pressures and temperatures. NASA not only met this challenge, the space agency led the development of such instrumentation while overcoming numerous technical hurdles.

Initial Obstacles

The original main engine instrumentation concept called for compact flange-mounted transducers with internal redundancy, high stability, and a long, maintenance-free life. Challenges presented themselves immediately, however. Few instrumentation suppliers were interested in the limited market projected for the shuttle. Moreover, early engine testing disclosed that standard designs were generally incapable of surviving the harsh environments. Although the "hot side" temperatures were within the realm of jet engines, no sort of instrumentation existed that could handle both high temperatures and cryogenic environments down to minus -253°C (-423°F). Vibration environments with high-frequency spectrums extending beyond commercially testable ranges of 2,000 hertz (Hz) experienced several hundred times the force of gravity over almost 8 hours of an engine's total planned operational exposure. For these reasons, the endurance requirements of the instrumentation constituent materials were unprecedented.

Engine considerations such as weight, concern for leakage that might be caused by mounting bosses, and overall system fault tolerance prompted the need for greater redundancy for each transducer. Existing supplier designs, where available, were single-output devices that provided no redundancy. A possible solution was to package two or more sensors within a single transducer. But this approach required special adaptation to achieve the desired small footprint and weight.

NASA considered the option of strategically placing instrumentation devices and closely coupling them to the desired stimuli source. This approach prompted an appreciation of the inherent simplicity and reliability afforded by low-level output devices. The avoidance of active electronics tended to minimize electrical, electronic, and electromechanical part vulnerability to hostile environments. Direct mounting of transducers also minimized the amount of intermediate hardware capable of producing a catastrophic system failure response. Direct mounting, however, came at a price. In some situations, it was not possible to design transducers capable of surviving the severe environments, making it necessary to off-mount the device. Pressure measurements associated with the combustion process suffered from icing or blockage issues when hardware temperatures dropped below freezing. Purging schemes to provide positive flow in pressure tubing were necessary to alleviate this condition.

Several original system mandates were later shown to be ill advised, such as an early attempt to achieve some measure of standardization through the use of bayonet-type electrical connectors. Early engine-level and laboratory testing revealed the need for threaded connectors since the instrumentation components could not be adequately shock-isolated to prevent failures induced by excessive relative connector motion. Similarly, electromagnetic interference assessments and observed deficiencies resulted in a reconsideration of the need for cable overbraiding to minimize measurement disruption.

Problems also extended to the sensing elements themselves. The lessons of material incompatibilities or deficiencies were evident in the area of resistance temperature devices and thermocouples. The need for the stability of temperature measurements led to platinum-element resistance temperature devices being baselined for all thermal measurements.

Aggressive engine performance and weight considerations also compromised the optimal sensor mountings. For example, it was not practical to include the prescribed straight section of tubing upstream from measuring devices, particularly for flow. This resulted in the improper loading of measuring devices, primarily within the propellant oxygen ducting. The catastrophic failure risks finally prompted the removal or relocation of all intrusive measuring devices downstream of the high-pressure oxygen turbopump. Finally, the deficiencies of vibration redline systems were overcome as processing hardware and algorithms matured to the point where a real-time synchronous vibration redline system could be adopted, providing a significant increase in engine reliability.

Weakness Detection and Solutions

In some instances, the engine environment revealed weaknesses not normally experienced in industrial or aerospace applications. Some hardware successfully passed component-level testing only to experience problems at subsystem or engine-level testing. Applied vibration spectrums mimicked test equipment limitations where frequency ranges typically did not extend beyond 2,000 Hz. The actual engine recognized no limits and continued to expose the hardware to energy above even 20,000 Hz. Therefore, a critical sensor resonance condition might only be excited during engine-level testing. Similarly, segmenting of component testing into separate vibration, thermal, and fluid testing deprived the instrumentation of experiencing the more-severe effect of combined exposures.

The shuttle's reusability revealed failure modes not normally encountered, such as those ascribed to the differences between flight and ground test environments. It was subsequently found that the microgravity exposure of each flight allowed conductive particles within instruments to migrate in a manner not experienced with units confined to terrestrial applications. Main engine pressure transducers experienced electrical shorts only during actual engine performance. During the countdown of Space Transportation System (STS)-53 (1992), a high-pressure oxidizer turbopump secondary seal measurement output pressure transducer data spike almost triggered an on-pad abort. Engineers used pressure transducers screened

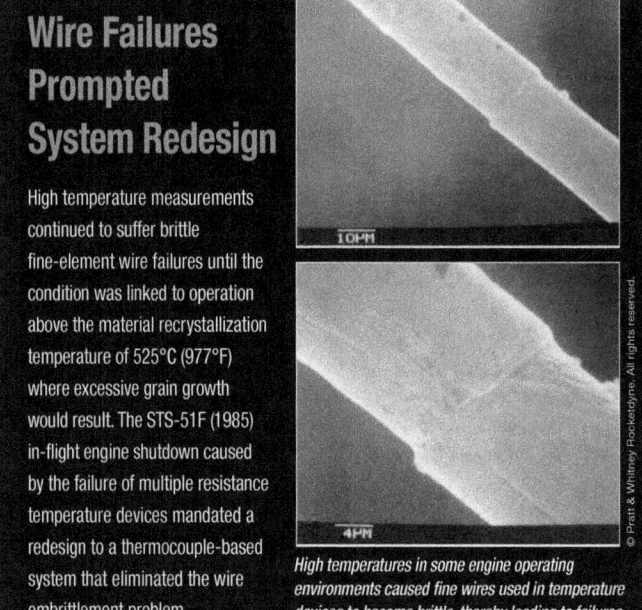

Wire Failures Prompted System Redesign

High temperature measurements continued to suffer brittle fine-element wire failures until the condition was linked to operation above the material recrystallization temperature of 525°C (977°F) where excessive grain growth would result. The STS-51F (1985) in-flight engine shutdown caused by the failure of multiple resistance temperature devices mandated a redesign to a thermocouple-based system that eliminated the wire embrittlement problem.

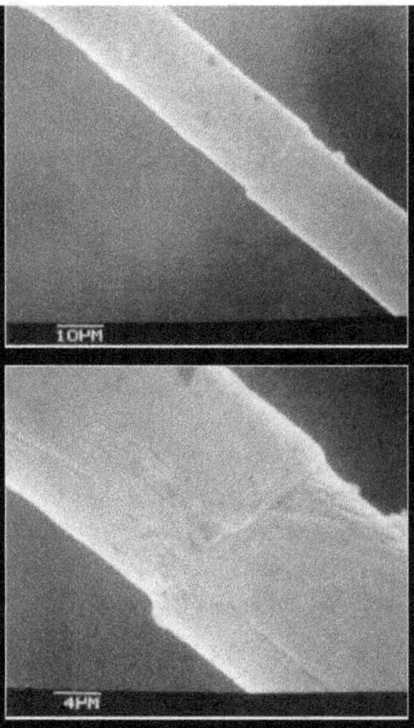

High temperatures in some engine operating environments caused fine wires used in temperature devices to become brittle, thereby leading to failures.

by particle impact noise detection and microfocus x-ray examination on an interim basis until a hardware redesign could be qualified.

Effects of Cryogenic Exposure on Instrumentation

Cryogenic environments revealed a host of related material deficiencies. Encapsulating materials—necessary to provide structural support for fine wires within speed sensors—lacked resiliency at extreme low temperatures. The adverse effects of inadvertent exposure to liquefied gases within the shuttle's aft compartment produced functional failures due to excessively cold conditions. In April 1991, STS-37 was scrubbed when the high-pressure oxidizer turbopump secondary seal pressure measurement became erratic due to the damaging effects of cryogenic exposure of a circuit board.

Problems with cryogenics also extended to the externals of the instrumentation. Cryopumping—the condensation-driven pumping mechanism of inert gases such as nitrogen—severely compromised the ability of electrical connectors to maintain continuity. The normally inert conditions maintained within the engine system masked a problem with residual contamination of glassed resistive temperature devices used for cryogenic propellant measurements. Corrosive flux left over from the manufacturing process remained dormant for years until activated during extended exposures to the humid conditions at the launch site. STS-50 (1992) narrowly avoided a launch delay when a resistive temperature device had to be replaced just days before the scheduled launch date.

Expectations Exceeded

As the original main engine design life of 10 years was surpassed, part obsolescence and aging became a concern. Later designs used more current parts such as industry-standard electrical connectors. Some suppliers chose to invest in technology driven by the shuttle, which helped to ease the program's need for long-term part availability.

The continuing main engine ground test program offered the ability to use ongoing hot-fire testing to ensure that all flight hardware was sufficiently enveloped by older ground test units. Tracking algorithms and extensive databases permitted such comparisons.

Industry standards called for periodic recalibration of measuring devices. NASA excluded this from the Space Shuttle Main Engine Program at its inception to reduce maintenance for hardware not projected for use beyond 10 years. In practice, the hardware life was extended to the point that some engine components approached 40 years of use before the final shuttle flight. Aging studies validated the stable nature of instruments never intended to fly so long without recalibration.

Summary

While initial engine testing disclosed that instrumentation was a weak link, NASA implemented innovative and successful solutions that resulted in a suite of proven instruments capable of direct application on future rocket engines.

Unprecedented Rocket Engine Fault-Sensing System

The Space Shuttle Main Engine (SSME) was a complex system that used liquid hydrogen and liquid oxygen as its fuel and oxidizer, respectively. The engine operated at extreme levels of temperature, pressure, and turbine speed. At these levels, slight material defects could lead to high vibration in the turbomachinery. Because of the potential consequences of such conditions, NASA developed vibration monitoring as a means of monitoring engine health.

The main engine used both low- and high-pressure turbopumps for fuel and oxidizer propellants. Low-pressure turbopumps served as propellant boost pumps for the high-pressure turbopumps, which in turn delivered fuel and oxidizer at high pressures to the engine main combustion chamber.

The high-pressure pumps rotated at speeds reaching 36,000 rpm on the fuel side and 24,000 rpm on the oxidizer side. At these speeds, minor faults were exacerbated and could rapidly propagate to catastrophic engine failure.

During the main engine's 30-year ground test program, more than 40 major engine test failures occurred. High-pressure turbopumps were the source of a large percentage of these failures. Posttest analysis revealed that the vibration spectral data contained potential failure indicators in the form of discrete rotordynamic spectral signatures. These signatures were prime indicators of turbomachinery health and could potentially be used to mitigate

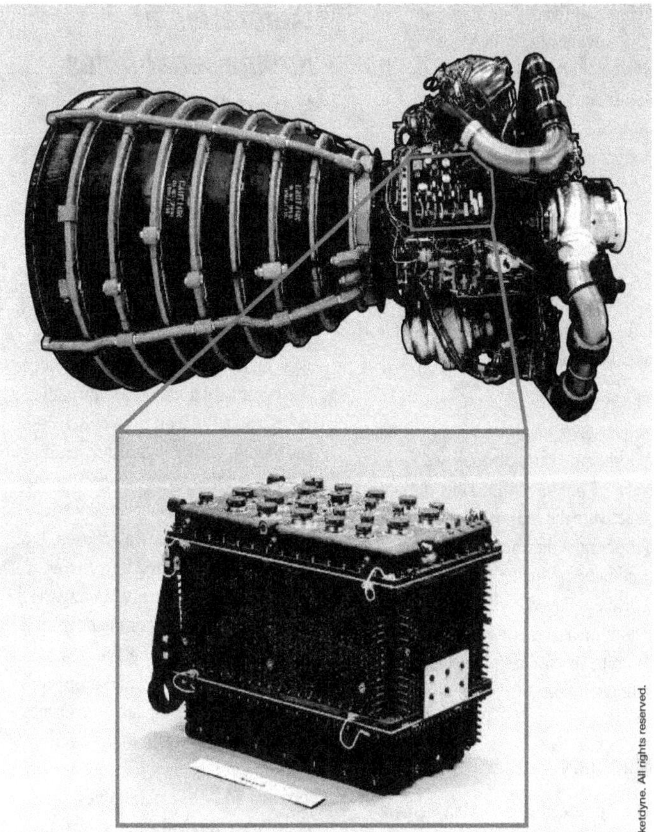

NASA's Advanced Health Monitoring System software was integrated with the Space Shuttle Main Engine controller (shown by itself and mounted on the engine) in 2007.

catastrophic engine failures if assessed at high speeds and in real time.

NASA recognized the need for a high-speed digital engine health management system. In 1996, engineers at Marshall Space Flight Center (MSFC) developed the Real Time Vibration Monitoring System and integrated the system into the main engine ground test program. The system used data from engine-mounted accelerometers to monitor pertinent spectral signatures. Spectral data were produced and assessed every 50 milliseconds to determine whether specific vibration amplitude thresholds were being violated.

NASA also needed to develop software capable of discerning a failed sensor from an actual hardware failure. MSFC engineers developed the sensor validation algorithm—a software algorithm that used a series of rules and threshold gates based on actual vibration spectral signature content to evaluate the quality of sensor data every 50 milliseconds.

Outfitted with the sensor validation algorithm and additional software, the Real Time Vibration Monitoring System could detect and diagnose pertinent indicators of imminent main engine turbomachinery failure and initiate a shutdown command within 100 milliseconds.

The Real Time Vibration Monitoring System operated successfully on more than 550 main engine ground tests with no false assessments and a 100% success rate on determining and disqualifying failed sensors from its vibration redlines. This, the first high-speed vibration redline system developed for a liquid engine rocket system, supported the main engine ground test program throughout the shuttle era.

To prove that a vibration-based, high-speed engine health management system could be used for flight operations, NASA included a subscale version of the Real Time Vibration Monitoring System on Technology Flight Experiment 2, which flew on STS-96 (1999). This led to the concept of the SSME Advanced Health Management System as a means of extending this protection to the main engine during ascent.

The robust software algorithms and redline logic developed and tested for the Real Time Vibration Monitoring System were directly applied to the Advanced Health Management System and incorporated into a redesigned version of the engine controller. The Advanced Health Management System's embedded algorithms continuously monitored the high-pressure turbopump vibrations generated by rotation of the pump shafts and assessed rotordynamic performance every 50 milliseconds. The system was programmed to initiate a shutdown command in fewer than 120 milliseconds if vibration patterns indicated an instability that could lead to catastrophic failure.

The system also used the sensor-validation algorithm to monitor sensor quality and could disqualify a failed sensor from its redline suite or deactivate the redline altogether. Throughout the shuttle era, no other liquid engine rocket system in the world employed a vibration-based health management system that used discrete spectral components to verify safe operation.

Summary

The Advanced Health Management System, developed and certified by Pratt & Whitney Rocketdyne (Canoga Park, California) under contract to NASA, flew on numerous shuttle missions and continued to be active on all engines throughout the remainder of the shuttle flights.

Calibration of Navigational Aides Using Global Positioning Computers

The crew members awakened at 5:00 a.m. After 10 days in orbit, they were ready to return to Earth. By 7:45 a.m., the payload bay doors were closed and they were struggling into their flight suits to prepare for descent. The commander called for a weather report and advice on runway selection. The shuttle could be directed to any one of three landing strips depending on weather at the primary landing site. Regardless of the runway chosen, the descent was controlled by systems capable of automatically landing the Orbiter. The Orbiter commander took cues from these landing systems, controlled the descent, and dropped the landing gear to safely land the Orbiter. During their approach to the landing site, the Orbiter crew depended on a complex array of technologies, including a Tactical Air Navigation System and the Microwave Scanning Beam Landing System, to provide precision navigation. These systems were located at each designated landing site and had to be precisely calibrated to ensure a safe and smooth landing.

Touchdown Sites

Shuttle runways were strategically located around the globe to serve several purposes. After a routine mission, the landing sites included Kennedy Space Center (KSC) in Florida, Dryden Flight Research Center in California, and White Sands Test Facility in New Mexico. The transoceanic abort landing sites—intended for emergencies when the shuttle lost a main engine during ascent and could not return to KSC—were located in Zaragoza and Moron in Spain and in Istres in France. Former transoceanic abort landing sites included: Dakar, Senegal; Ben Guerir, Morocco; Banjul, The Gambia; Honolulu, Hawaii; and Anderson Air Force Base, Guam. NASA certified each site.

Error Sources

Because the ground portion of the Microwave Scanning Beam Landing and Tactical Air Navigation Systems contained moving mechanical components and depended on microwave propagation, inaccuracies could develop over time that might prove detrimental to a shuttle landing. For example, antennas could drift out of mechanical adjustment. Ground settling and external environmental factors could also affect the system's accuracy. Multipath and refraction errors could result from reflections off nearby structures, terrain changes, and day-to-day atmospheric variations.

Flight inspection data gathered by the NASA calibration team could be used to determine the source of these errors. Flight inspection involved flying an aircraft through the landing system coverage area and receiving time-tagged data from the systems under test. Those data were compared to an accurate aircraft positioning reference to determine error. Restoring integrity was easily achieved through system adjustment.

Global Positioning Satellite Position Reference for Flight Inspection

Technologies were upgraded several times since first using the Global Positioning Satellite (GPS)-enabled flight inspection system. The flight inspection system used an aircraft GPS receiver as a position reference. Differences between the system under test and the position reference were recorded, processed, and displayed in real time on board the aircraft. An aircraft position reference used for flight inspection had to be several times more accurate than the system under test. Stand-alone commercial GPS systems did not have enough accuracy for this purpose. Several techniques could be used to improve GPS positioning. Differential GPS used a ground GPS receiver installed over a known surveyed benchmark. Common mode error corrections to the GPS position were calculated and broadcast over a radio data link to the aircraft. After the received corrections were applied, the on-board GPS position accuracy was within 3 m (10 ft). A real-time accuracy within 10 cm (4 in.) was achieved by using a carrier-phase technique and tracking cycles of the L-band GPS carrier signal.

NASA built several versions of the flight inspection system customized to different aircraft platforms. Different NASA aircraft were used based on aircraft availability. These aircraft include NASA's T-39 jet (Learjet), a NASA P-3 turboprop, several C-130 aircraft, and even NASA's KC-135. Each aircraft was modified with shuttle landing system receivers and antennas. Several pallets of equipment were configured and tested to reduce the installation time on aircraft to one shift.

Summary

NASA developed unique instrumentation and software supporting the shuttle navigation aids flight inspection mission. The agency developed aircraft pallets to operate, control, process, display, and archive data from several avionics receivers. They acquired and synchronized measurements from shuttle-unique avionics and aircraft platform avionics with precision time-tagged GPS position. NASA developed data processing platforms and software algorithms to graphically display and trend landing system performance in real time. In addition, a graphical pilot's display provided the aircraft pilot with runway situational awareness and visual direction cues. The pilot's display software, integrated with the GPS reference system, resulted in a significant reduction in mission flight time.

Synergy With the Federal Aviation Administration

In early 2000, NASA and the Federal Aviation Administration (FAA) entered into a partnership for flight inspection. The FAA had existing aircraft assets to perform its mission to flight-inspect US civilian and military navigation aids. The FAA integrated NASA's carrier-phase GPS reference along with shuttle-unique avionics and software algorithms into its existing control and display computers on several flight-inspection aircraft.

The NASA/FAA partnership produced increased efficiency, increased capability, and reduced cost to the government for flight inspection of the shuttle landing aids.

Software

Software was an integral part in the Space Shuttle hardware systems and it played a vital role in the design and operations of the shuttle. The longevity of the program demanded the on-orbit performance of the vehicle to be flexible under new and challenging environments. Because of the flexibility required, quick-turnaround training, simulations, and virtual reality tools were invaluable to the crew for new operational concepts. In addition, ground operations also benefited from software innovations that improved vehicle processing and flight-readiness testing. The innovations in software occurred throughout the life of the program. The topics in this chapter include specific areas where engineering innovations in software enabled solutions to problems and improved overall vehicle and process performance, and have carried over to the next generation of space programs.

Introduction
Gail Chapline
Steven Sullivan

Primary Software
Aldo Bordano
Geminesse Dorsey
James Loveall

Personal Computer Ground Operations Aerospace Language Offered Engineers a "View"
Avis Upton

The Ground Launch Sequencer Orchestrated Launch Success
Al Folensbee

Integrated Extravehicular Activity/Robotics Virtual Reality Simulation
David Homan
Bradley Bell
Jeffrey Hoblit
Evelyn Miralles

Integrated Solutions for Space Shuttle Management...and Future Endeavors
Samantha Manning
Charles Hallett
Dena Richmond
Joseph Schuh

Three-Dimensional Graphics Provide Extraordinary Vantage Points
David Homan
Bradley Bell
Jeffrey Hoblit
Evelyn Miralles

Primary Software

NASA faced notable challenges in the development of computer software for the Space Shuttle in the early 1970s. Only two avionics computers were regarded as having the potential to perform the complex tasks that would be required of them. Even though two options existed, these candidates would require substantial modification. To further compound the problem, the 1970s also suffered a noticeable absence of off-the-shelf microcomputers. Large-scale, integrated-circuit technology had not yet reached the level of sophistication necessary for Orbiter use. This prompted NASA to continue its search for a viable solution.

NASA soon concluded that core memory was the only reasonable choice for Orbiter computers, with the caveat that memory size was subject to power and weight limitations as well as heat constraints. The space agency still faced additional obstacles: data bus technology for real-time avionics systems was not yet fully operational; the use of tape units for software program mass storage in a dynamic environment was limited and unsubstantiated; and a high-order language tailored specifically for aerospace applications was nonexistent. Even at this early juncture, however, NASA had begun developing a high-order software language—HAL/S—for the shuttle. This software would ultimately become the standard for Orbiter operations during the Space Shuttle Program.

Software Capability Beyond Technology Limits

NASA contemplated the number of necessary computer configurations during the early stages of Space Shuttle development. It took into consideration the segregation of flight control from guidance and navigation, as well as the relegation of mechanized aerodynamic ascent/re-entry and spaceflight functions to different machines.

These considerations led to a tightly coupled, synchronized fail-operational/fail-safe computation requirement for flight control and sequencing functions that drove the system toward a four-machine computer complex. In addition, the difficulties NASA faced in attempting to interconnect and operate multiple complexes of machines led to the development of a single complex with central integrated computation.

NASA added a fifth machine for off-loading nonessential mission applications, payload, and system-management tasks from the other four machines. Although this fifth computer was also positioned to handle the additional computation requirements that might be placed on the system, it eventually hosted the backup system flight software.

The space agency had to determine the size of the Orbiter computer memory to be baselined and do so within the constraints of computer design and vehicle structure. Memory limitations posed a formidable

Personal Computer Ground Operations Aerospace Language Offered Engineers a "View"

Personal Computer Ground Operations Aerospace Language (PCGOAL) was a custom, PC-based, certified advisory system that provided engineers with real-time data display and plotting. The enhanced situational awareness aided engineers with the decision-making process and troubleshooting during test, launch, and landing operations.

When shuttle landings first began at Dryden Flight Research Center (DFRC), California, Kennedy Space Center (KSC) engineers had limited data-visualization capability. The original disk operating system (DOS)-based PCGOAL first supported KSC engineers during the STS-34 (1989) landing at DFRC. Data were sent from KSC via telephone modem and engineers had visibility to the Orbiter data on site at DFRC. Firing room console-like displays provided engineers with a familiar look of the command and control displays used for shuttle processing and launch countdown, and the application offered the first high-resolution, real-time plotting capability.

PCGOAL evolved with additional capabilities. After design certification review in 1995, the application was considered acceptable for decision making in conjunction with the command and control applications in the firing rooms and DFRC. In 2004, the application was given a new platform to run on a Windows 2000 operating system.

As the Windows-based version of PCGOAL was being deployed, work had already begun to add visualization capabilities. The upgraded application and upgraded editor were deployed in December 2005 at KSC first and later at DFRC and Marshall Space Flight Center/ Huntsville Operations Support Center.

challenge for NASA early in the development phase; however, with the technological advancements that soon followed came the ability to increase the amount of memory.

NASA faced much skepticism from within its organization, regarding the viability of using a high-order language. Assembly language could be used to produce compact, efficient, and fast software code, but it was very similar in complexity to the computer's machine language and therefore required the programmer to understand the intricacies of the computer hardware and instruction set. For example, assembly language addressed the machine's registers directly and operations on the data in the registers directly.

While it might not result in as fast and efficient a code, using a high-order programming language would provide abstraction from the details of the computer hardware, be less cryptic and closer to natural language, and therefore be easier to develop and maintain. As the space agency contracted for the development of HAL/S, program participants questioned the software's ability to produce code with the size, efficiency, and speed comparable to those of an assembly language program. All participants, however, supported a top-down structured approach to software design.

To resolve the issue and quell any fears as to the capability of HAL/S, NASA tested both options and discovered that the nominal loss in efficiency of the high-order language was insignificant when compared to the advantages of increased programmer productivity, program maintainability, and visibility into the software. Therefore, NASA selected HAL/S for all but one software module (i.e., operating system software), thus fulfilling the remaining baselined requirements and approach.

Operating Software for Avionics System

The Orbiter avionics system operation required two independent software systems with a distinct hierarchy and clear delegation of responsibilities. The Primary Avionics Software System was the workhorse of the two systems. It consisted of several memory loads and performed mission and system functions. The Backup Flight System software was just that: a backup. Yet, it played a critical role in the safety and function of the Orbiter. The Backup Flight System software was composed of one memory load and worked only during critical mission phases to provide an alternate means of orbital insertion or return to Earth in the event of a Primary Avionics Software System failure.

Primary Avionics Software System

The Primary Avionics Software System performed three major functions: guidance, navigation, and control of the vehicle during flight; the systems management involved in monitoring and controlling vehicle subsystems; and payload—later changed to vehicle utility—involving preflight checkout functions.

The depth and complexity of Orbiter requirements demanded more memory capacity than was available from a general purpose computer. As a solution, NASA structured each of the major functions into a collection of programs and capabilities needed to conduct a mission phase or perform an integrated function. These collections were called "operational sequences," and they formed memory configurations that were loaded into the general purpose computers from on-board tape units. Memory overlays were inevitable; however, to a great extent NASA structured these overlays only in quiescent, non-dynamic periods.

The substructure within operational sequences was a choreographed network consisting of major modes, specialist functions, and display functions. Major modes were substructured into blocks that segmented the processes into steps or sequences. These blocks were linked to cathode ray tube display pages so the crew could monitor and control the function. The crew could initiate sequencing through keyboard entry. In certain instances, sequencing could be initiated automatically by the software. Blocks within the specialist functions, initiated by keyboard entry, were linked to cathode ray tube pages. These blocks established and presented valid keyboard entry options available to the crew for controlling the operation or monitoring the process. Major modes accomplished the primary functions within a sequence, and specialist functions were used for secondary or background functions. The display functions, also initiated by keyboard input, contained processing necessary to produce the display and were used only for monitoring data processing results.

Backup Flight System

The Backup Flight System remained poised to take over primary control in the event of Primary Avionics Software System failure, and NASA thoroughly prepared the backup system for this potential problem. The system consisted of the designated general purpose computer, three backup flight controllers, the backup software, and associated switches and displays.

As far as designating a specific general purpose computer, NASA did not favor any particular one over the others—any of the five could be designated the backup machine by appropriate keyboard entry. The designated computer would request the backup

Mission Phase With Corresponding Operational Sequences and Major Modes

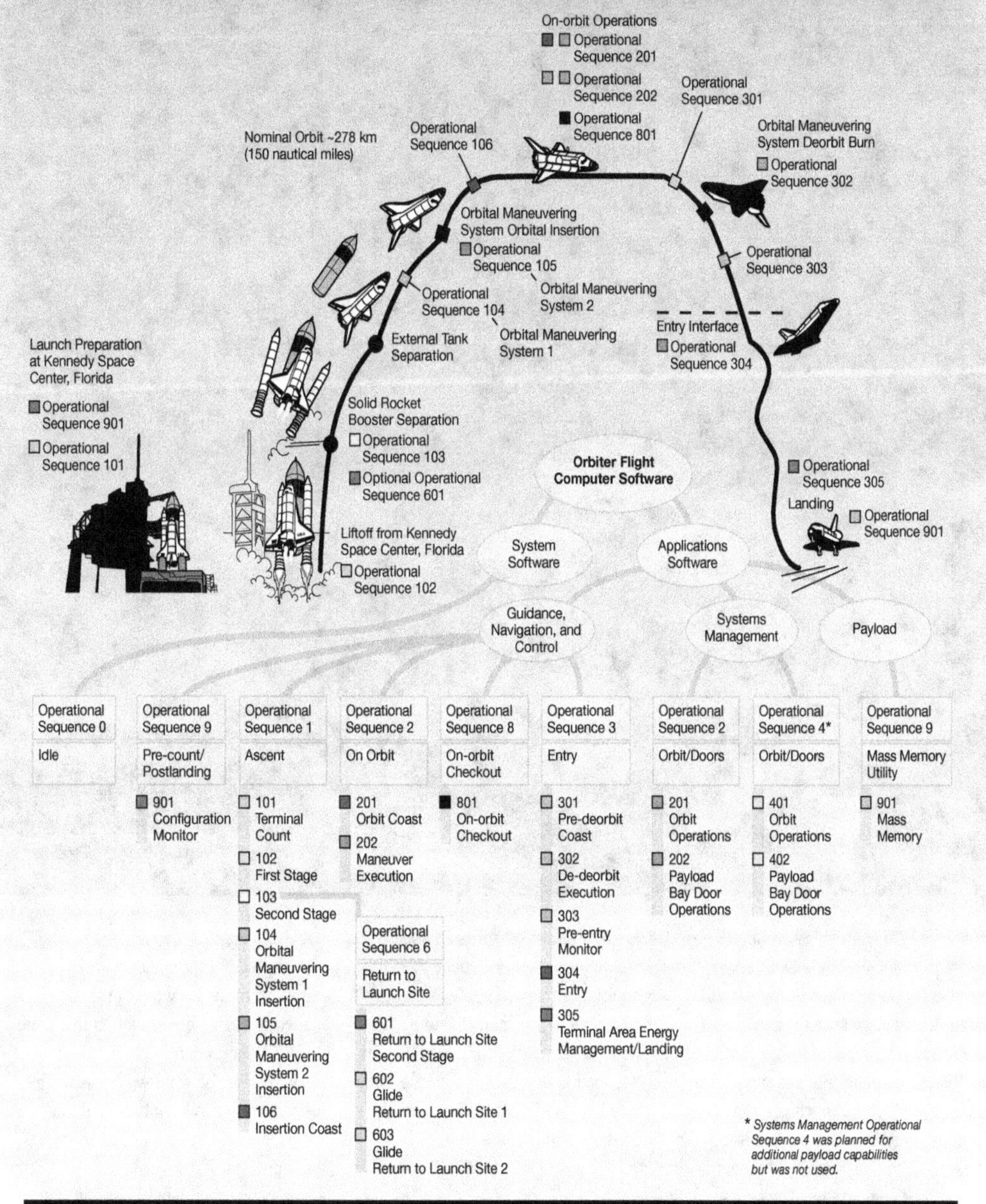

Due to computer memory limitations, the flight software was divided into a number of separate programs called operational sequences. Each sequence provided functions specific to a particular mission phase and were only loaded into memory during that phase of flight.

The Ground Launch Sequencer Orchestrated Launch Success

During launch countdown, the ground launch sequencer was like an orchestra's conductor. Developed in 1978, the sequencer was the software supervisor of critical command sequencing and measurement verification from 2 hours before launch time to launch time and through safing, thus assuring a steady and an appropriate tempo for a safe and successful launch.

Engineered to expedite and automate operations and maximize automatic error detection and recovery, the ground launch sequencer focused on "go/no-go" criteria. Responding to a no-go detection, it could initiate a countdown hold, abort, or recycle or contingency operations. While controlling certain monitoring aspects, the sequencer did not reduce the engineer's capability to monitor his or her system's health/integrity; however, by assuming command responsibility, it integrated launch requirements and activities, and reduced communication traffic and required hardware. Manual intervention was available for off-nominal conditions.

The four ground launch sequencer components included: exception monitoring; sequencer; countdown clock control; and safing. For exception monitoring, the sequencer continuously monitored more than 1,200 measurements. If a measurement violated its expected value, the sequencer checked whether the measurement was part of a voting logic group. If voting failed, it automatically caused the countdown to hold at the next milestone or abort the countdown.

The sequencer provided a single point of control during countdown, issuing all commands to ground and flight equipment from the designated period called T minus 9 minutes (T=time) through liftoff. It verified events required for liftoff. If an event wasn't completed, an automated hold/recycle was requested.

Clock control provided the required synchronization between ground and vehicle systems and managed countdown holds/recycles. Clock control allowed the sequencer to resume the countdown after a problem was resolved. The safing component halted the Orbiter's on-board software and, based on the progression of the sequencer, commanded ground and flight systems into a safe configuration for crew egress.

Launch countdown operations in Firing Room 4 at Kennedy Space Center, Florida.

software load from mass memory. The backup computer would then remain on standby. During normal operations, when the primary system controlled the Orbiter, the backup system operated in "listen" mode to monitor and obtain data from all prime machines and their assigned sensors. By acquiring these data, the Backup Flight System maintained computational currency and, thus, the capability to assume control of the Orbiter at any time.

NASA independently developed and coded the software package for the Backup Flight System as an added level of protection to reduce the possibility of generic software errors common to the primary system. The entire Backup Flight System was contained in one memory configuration, loaded before liftoff, and normally maintained in that machine.

Success—On Multiple Levels

NASA overcame the obstacles it faced in creating the shuttle's Primary Avionics Software System through ingenuity and expertise. Even technology that was current during the initial planning stages did not impose limits on what the space agency could accomplish in this area. NASA succeeded in pushing the boundaries for what was possible by structuring a system that could handle multiple functions within very real parameters. It also structured a backup support system capable of handling the demands of spaceflight at a critical moment's notice.

Integrated Extravehicular Activity/Robotics Virtual Reality Simulation

As the Space Shuttle Program progressed into the 1990s, the integration of extravehicular activity (EVA) and robotics took on a whole new importance when Hubble Space Telescope servicing/repair (first flight 1993) and space-based assembly of the International Space Station (ISS) tasks were realistically evaluated.

Two motivating factors influenced NASA's investigation into the potential use of virtual reality technology that was barely in its infancy at that time. The first factor was in response to a concern that once Hubble was deployed on orbit future astronauts and flight controllers would not have easy access to the telescope to familiarize themselves with the actual hardware configuration to plan, develop, and review servicing procedures.

The second factor was based on previous on-orbit experience with the interaction and communication between EVA crew members and Shuttle Robotic Arm operators. NASA discovered that interpreting instructions given by a crew member located in a foot restraint on the end of the robotic arm was not as intuitive to the arm operator as first thought, especially when both were not in the same body orientation when giving or receiving commands. The EVA crew member could, for example, be upside down with respect to the robotic arm operator in microgravity. Therefore, the command to "Move me up" left the arm operator in a quandary trying to decide what "up" actually meant.

NASA Embraces Advances in Virtual Reality

It was at this same time in the early 1990s that virtual reality hardware started to enter the commercial world in the form of head-mounted displays, data gloves, motion-tracking instruments, etc.

In the astronaut training world, no facility allowed an EVA crew member to ride on a robotic arm operated by another crew member in a realistic space environment. The Water Emersion Test Facility at Johnson Space Center (JSC) in Houston, Texas, provided a training arena for EVA crew members, but the confined space and the desire to not require subjects to be heads down for more than very short periods of time did not allow for suitable integrated training between the EVA crew and the robotic arm operators. Likewise, the Manipulator Development Facility's hydraulic arm and the computer graphic-based robotic arm simulators at JSC were not conducive for EVA crew interaction.

Virtual reality provided a forum to actually tie those two training scenarios together in one simulation. Working closely with the astronaut office, NASA engineers took commercially available virtual reality hardware and developed the computer graphic display software and across-platform communications software that linked into existing "man-in-the-loop" robotic arm computer simulations to produce an integrated EVA/robotics training capability.

Virtual Reality Is Put to the Test

The first use of these new capabilities was in support of crew training for Space Transportation System (STS)-61 (1993)—the Hubble Space Telescope servicing mission. The virtual reality simulation provided a flight-like environment in which the crew was able to develop and practice the intricate choreography between the Shuttle Robotic Arm operator and the EVA crew member affixed to the end of that arm. The view in the head-mounted display was as it would be seen by the astronaut working around the Hubble berthed in the shuttle payload bay at an orbital altitude of 531 km (330 miles) above the Earth.

The next opportunity to take advantage of the virtual reality software involved EVA crew members training to perform the first engineering test flights of the

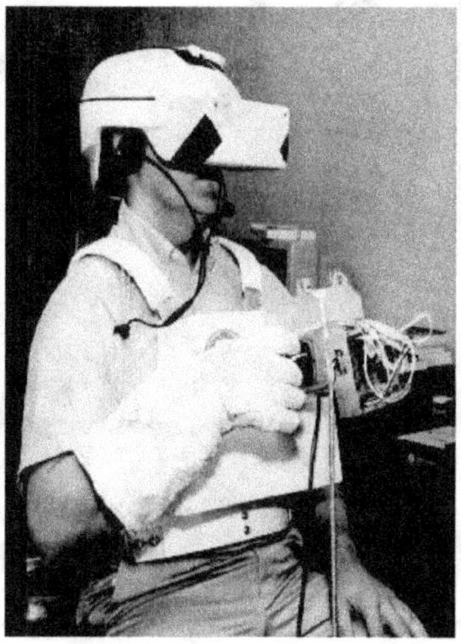

Astronaut Mark Lee trains for his Simplified Aid for EVA Rescue test flight (STS-64 [1994]) using the virtual reality flight trainer (left) and on orbit (right).

Simplified Aid for EVA Rescue (SAFER) on STS-64 (1994).

The output of a dynamic simulation of the SAFER backpack control system and its flying characteristics, using zero-gravity as a parameter, drove the head-mounted display visual graphics. Inputs to the simulation were made using a flight-equivalent engineering unit hand controller. The EVA crew member practiced and refined the flight test maneuvers to be flown during on-orbit tests of the rescue unit. The crew member could see the on-orbit configuration of the shuttle payload bay, the robotic arm, and the Earth/horizon through the virtual reality head-mounted display at the orbital altitude planned for the mission. The EVA crew member was also able to interact with the robotic arm operator as well as see the motions of the arm, which was an integral part of the on-orbit tests. The robotic arm operator was also able to view the EVA crew member's motions in the simulated shuttle payload bay camera views made available to the operator as part of the dynamic man-in-the-loop robotic arm simulation.

As a result of the engineering flights of the SAFER unit on STS-64, NASA was able to validate the virtual reality simulation and it became the ground-based SAFER training simulator used by all EVA crew members assigned to space station assembly missions.

Each EVA crew member was required to have at least four 2-hour training classes prior to a flight to practice flying rescue scenarios with the unit in the event he or she became separated from the space vehicle during an EVA.

NASA also developed a trainer that was flown on board the space station laptop computers. The trainer used the same simulation and display software as the ground-based simulator, but it incorporated a flat-screen display instead of a head-mounted display. It also used the same graphic model database as the ground-based simulators. ISS crew members used the on-board trainer to maintain SAFER hand controller proficiency throughout their time on the ISS.

Handling Large Objects During Extravehicular Activity

Learning to handle large objects in the weightlessness of space also posed a unique problem for EVA crew members training in ground-based facilities. In the microgravity environment of space, objects may be weightless but they still have mass and inertia as well as a mass distribution around a center of gravity.

NASA engineers developed a tendon-driven robot and a set of dynamic control software to simulate the feel and motion of large objects being handled by an EVA crew member within the zero-gravity parameter. The basic concept was to mount a reel of cable and an electric drive motor at each of the eight corners of a structure that measured approximately 3 m (10 ft) on a side. Each cable was then attached to one of the eight corners of an approximately 0.6-m (2-ft) cube. In this configuration, the position and orientation of the smaller cube within the large structure could be controlled by reeling in and out the cables. Load cells were mounted to the smaller cube

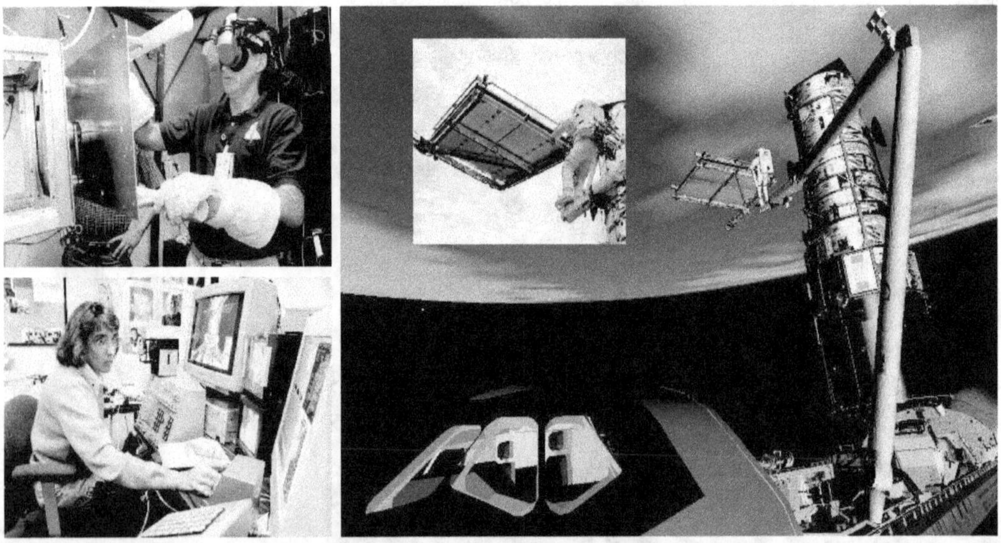

Astronauts Richard Linnehan (above left) and Nancy Currie (below) use the zero-gravity mass handling simulation and the Shuttle Robotic Arm simulation to practice combined operations prior to flight. The large image on the right is a rendering of the simulation. The inset is an actual photo of Astronaut Richard Linnehan (STS-109 [2002]) unfolding a solar array while anchored to the end of the robotic arm.

International Space Station Expedition 10 crew members Leroy Chiao (left) and Salizhan Sharipov train in virtual reality to photograph an approaching Orbiter through the space station windows. The lower pictures show what each sees through his respective camera view finder.

while handrails or other handling devices were attached to the load cells. As a crew member applied force to the handling device, the load cells measured the force and fed those values to a dynamic simulation that had the mass characteristics of the object being handled as though it were in weightlessness. Output from the computer program then drove the eight motors to move the smaller cube accordingly. Once these elements were integrated into graphics in the head-mounted display, the crew member not only felt the resulting six-degree-of-freedom motion of the simulated object, he or she also saw a three-dimensional (3-D) graphical representation of the real-world object in its actual surrounding environment.

The mass handling simulation—called kinesthetic application of mechanical force reflection—was qualitatively validated over a number of shuttle flights starting with STS-63 (1995). On that flight, EVA crew members were scheduled to handle objects that weighed from 318 to 1,361 kg (700 to 3,000 pounds) during an EVA. After their flight, they evaluated the ability of the application to simulate the handling conditions experienced in microgravity.

Kinesthetic application of mechanical force reflection was deemed able to faithfully produce an accurate simulation of the feel of large objects being handled by EVA crew members following a number of postflight evaluations.

Kinesthetic application of mechanical force reflection was also integrated with the Shuttle Robotic Arm simulation, which allowed the EVA crew member riding on the end of the arm to actually feel the arm-induced motion in a large payload that he or she would be holding during a construction or repair operation around the ISS or Hubble.

NASA built two kinesthetic application of mechanical force reflections so that two EVA crew members could train to handle the same large object from two different vantage points. The forces and motion input by one crew member were felt and seen by the other crew member. This capability allowed crew members to evaluate mass handling techniques preflight. It also allowed them to work out not only the command protocol they planned to use, but also which crew member would be controlling the object and which would be stabilizing the object during the EVA.

Virtual Reality Simulates On-orbit Conditions

Following the Columbia accident in 2003, as a shuttle approached the space station, space station crew members photographed its Thermal Protection System from a distance of 183 m (600 ft) using digital cameras with 400mm and 800mm telephoto lenses.

As in previous scenarios, there was no place on Earth where crew members could practice photographing a Space Shuttle doing a 360-degree pitch maneuver at a distance of 183 m (600 ft). Virtual reality was again used to realistically simulate the on-orbit conditions and provide ground-based training to all space station crew members prior to their extended stay in space.

Engineers placed a cathode ray tube display from a head-mounted display inside a mocked-up telephoto lens. The same 3-D graphic simulation that was used to support the previous applications drove the display in the telephoto lens to show a shuttle doing the pitch maneuver at a range of 183 m (600 ft). With a real camera body attached to the mocked-up lens, each crew member could practice photographing the shuttle during its approach maneuver.

Summary

NASA took advantage of the benefits that virtual reality had to offer. Beginning in 1992, the space agency used the technology at JSC to support integrated EVA/robotics training for all subsequent EVA flights, including SAFER engineering flights, Hubble repair/servicing missions, and the assembly and maintenance of the ISS. Each EVA crew member spent from 80 to 120 hours using virtual reality to train for work in space.

Integrated Solutions for Space Shuttle Management… and Future Endeavors

Kennedy Space Center (KSC) developed an integrated, wireless, and paperless computer-based system for management of the Space Shuttle and future space program products and processes. This capability was called Collaborative Integrated Processing Solutions. It used commercial off-the-shelf software products to provide an end-to-end integrated solution for requirements management, configuration management, supply chain planning, asset life cycle management, process engineering/process execution, and integrated data management. This system was accessible from stationary workstations and tablet computers using wireless networks.

Collaborative Integrated Processing Solutions leveraged the successful implementation of Solumina® (iBASEt, Foothill Ranch, California)—a manufacturing execution system that provided work instruction authorization, electronic approval, and paperless work execution. Solumina® provided real-time status updates to all users working on the same document. The system provided for electronic buy off of work instructions, electronic data collection, and embedded links to reference materials. The application included electronic change tracking and configuration management of work instructions. Automated controls provided constraints management, data validation, configuration, and reporting of consumption of parts and materials.

In addition, KSC developed an interactive decision analysis and refinement software system known as Systems Maintenance Automated Repair Tasks. This system used evaluation criteria for discrepant conditions to automatically populate a document/procedure with predefined steps for safe, effective, and efficient repair. It stored tacit (corporate) knowledge, merging hardware specification requirements with actual "how-to" repair methods, sequences, and required equipment. Although the system was developed for Space Shuttle applications, its interface is easily adaptable to any hardware that can be broken down by component, subcomponent, discrepancy, and repair.

Systems Maintenance Automated Repair Tasks Solution Philosophy—Variables

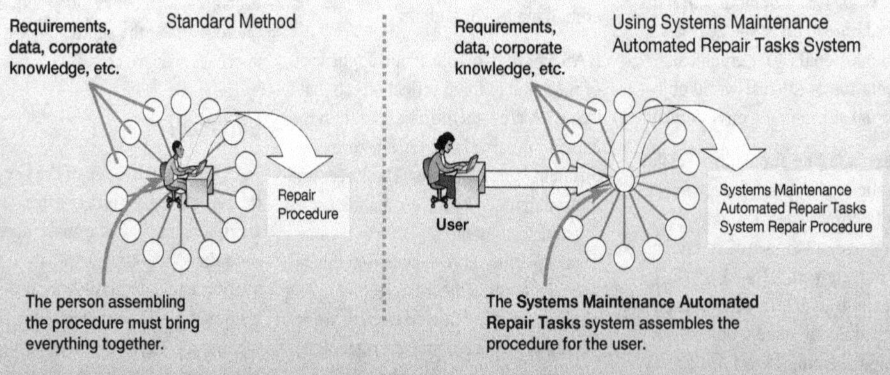

The Systems Maintenance Automated Repair Tasks allowed corporate knowledge to be kept in-house while increasing efficiency and lowering cost.

Three-Dimensional Graphics Provide Extraordinary Vantage Points

Astronauts' accomplishments in space seem effortless, yet they spent many hours on the ground training and preparing for missions.

Some of the earliest engineering concept development and training took place in the Johnson Space Center Virtual Reality Laboratory and involved the Dynamic Onboard Ubiquitous Graphics (DOUG) software package. NASA developed this three-dimensional (3-D) graphics-rendering package to support integrated training among the Shuttle Robotic Arm operators, the International Space Station (ISS) Robotic Arm operators, and the extravehicular activity (EVA) crew members. The package provided complete software and model database commonality among ground-based crew training simulators, ground-based EVA planning tools, on-board robotic situational awareness tools, on-board training simulations, and on-board EVA/robotic operations review tools for both Space Shuttle and ISS crews.

Level-of-detail Capability

Originally, the software was written as an application programming interface—an interface that enables the software to interact with other software—around the graphics-rendering package developed to support the virtual reality

Additional Extravehicular Activity Support

The International Space Station (ISS) has more than 2,300 handrails located on its exterior. These handrails provide translation paths for extravehicular activity (EVA) crew members. Pull-down menus in the Dynamic Onboard Ubiquitous Graphics (DOUG) software allow the user to highlight and locate each handrail. Entire translation paths can be highlighted and displayed for review by crew members prior to performing an EVA.

More than 620 work interface sockets are located on the external structure of the ISS, and nine articulating portable foot restraints can be relocated to any of the work interface sockets. Each articulating portable foot restraint has three articulating joints and a rotating base that produce 33,264 different orientations for an EVA crew member standing in that particular foot restraint. Each work interface socket can be located in the software package, and each articulating portable foot restraint can be configured to show all potential worksites and worksite configurations to support EVA planning.

The DOUG software package also contains and can highlight the locations of externally mounted orbital replacement units on the ISS, thruster and antenna keep-out zones that affect EVA crew member positioning, and articulating antennas, radiators, and solar arrays—all of which are configurable.

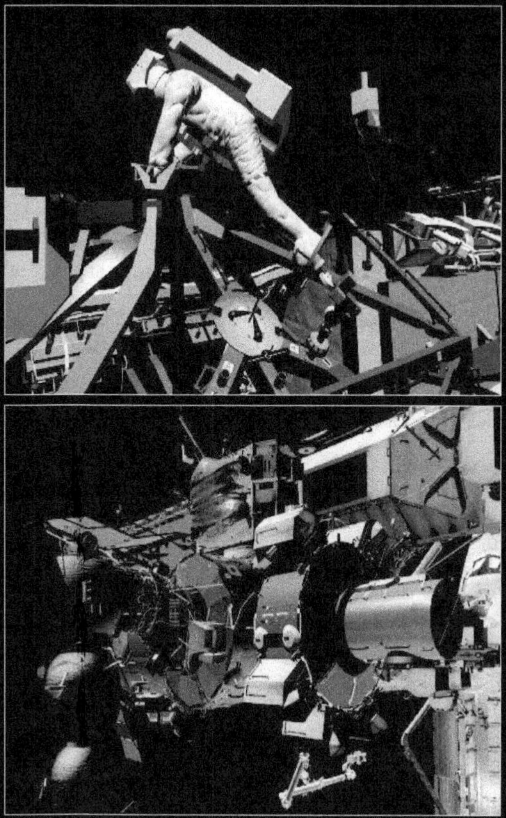

Articulated portable foot restraints configuration (top) and highlighted translation path (bottom).

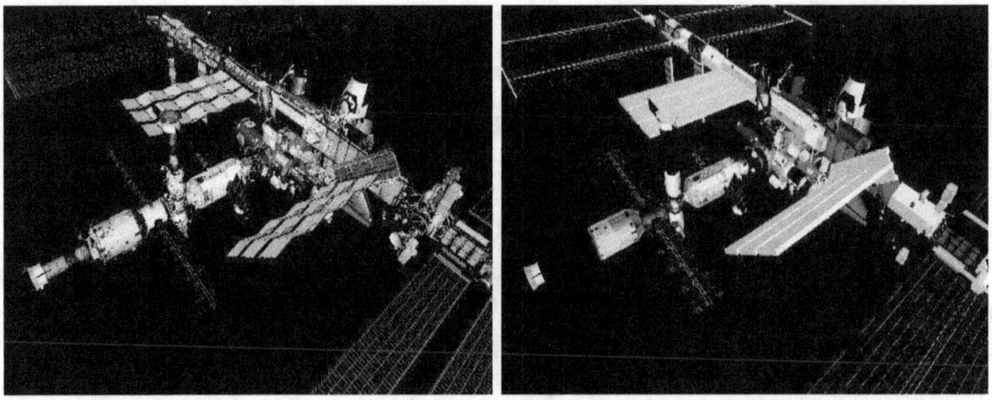

These two views show the effect of level-of-detail control. The left view is a high-resolution image compared to the low-resolution image on the right.

training simulation. The Simplified Aid for EVA Rescue (SAFER) on-board trainer required software that would run on the original IBM 760 laptop computers on board the ISS and thus required the UNIX-based code to be ported to a Windows-based operating system. The limited graphics capability of those computers also required additional model database artifacts that provided level-of-detail manipulation to make the simulation adequate for its intended purpose. This additional level-of-detail capability allowed the same high-fidelity model database developed for EVA training in the virtual reality facility to be used on the laptop computers on the ISS.

To obtain adequate graphics performance and screen update rates for simulating SAFER flying, crew members could select a low level-of-detail scene, which still displayed enough detail for the recognition of station landmarks and motion cues.

The DOUG software package, when not in use as a trainer, also provided a highly detailed, interactive 3-D model of the ISS that was viewable from any vantage point via keyboard inputs. The software first flew on board both shuttle and station in March 2001, and during Space Transportation System (STS)-102, and was on all subsequent shuttle and station flights with the exception of STS-107 (2003). That flight did not carry a robotic arm, had no planned EVAs, and did not dock with the ISS.

Benefits for Robotic Arm Operations

The DOUG software package supported SAFER training. The software was also capable of providing the situational awareness function during Space Station Robotic Arm operations by connecting to the on-board payload general support computer and using the telemetry from the arm to update the graphic representation in the program display.

The same software was compatible with laptop computers flown on the shuttle, and the graphical Shuttle Robotic Arm could be similarly driven with shuttle arm telemetry. Different viewpoints

Dynamic Onboard Ubiquitous Graphics displays multiple simulated camera and synthetic eye-point views on the same screen. The simulated camera views show the Japanese Experiment Module and the Columbus Laboratory in the top left image, the Mini Research Module-1 in the top right image, and the International Space Station in the bottom image.

could be defined in the software to represent the locations of various television cameras located around station and shuttle. The various camera parameters were defined in the software to display the actual field of view, based on the pan and tilt capabilities as well as the zoom characteristics of each camera.

The second ISS crew (2001) used these initial capabilities to practice for upcoming station assembly tasks with the Space Station Robotic Arm prior to the actual components arriving on a shuttle flight. The crew accomplished this by operating the real robotic arm using the real hand controllers and configuring a "DOUG laptop" to receive remote manipulator joint angle telemetry.

The graphics contained the station configuration with the shuttle docked and the station airlock component located in the shuttle's payload bay. The arm operator could see synthetic end-effector camera views produced in the program. These views showed the airlock with its grapple fixture in the payload bay of the Orbiter even though no Orbiter actually existed. The operator practiced maneuvering the real arm end-effector onto an imaginary grapple fixture and then maneuvering the real arm with the imaginary airlock attached, through the prescribed trajectory to berth the imaginary airlock onto the real common berthing mechanism on the ISS Unity Node.

Through DOUG the arm operator also had access to synthetic views from all the shuttle cameras, as well as the Space Station Robotic Arm cameras that would be used during the actual assembly operations. This made training much more effective than simply driving the robotic arm around in open space.

The colors displayed in Dynamic Onboard Ubiquitous Graphics indicate direction of approach of the robotic arm booms with respect to the closest object: green = opening; yellow = closing; and red = envelope violation.

Proximity Detection

As the ISS grew in complexity, NASA added capabilities to the DOUG software. Following a near collision between the Space Station Robotic Arm and one of the antennas located on the laboratory module of the ISS, the space agency added the ability to detect objects close to one another—i.e., proximity detection. The software calculated and displayed the point of closest approach for the main robotic arm booms and the elbow joint to any station or shuttle component displayed in the model database.

A vector was drawn between each of the three robotic arm components and the nearest structure. When DOUG received robotic arm telemetry data and was being used for situational awareness during robotic arm operations, the color of these vectors indicated whether measured distance was increasing or decreasing. It also indicated whether the relative distance was within a user-defined, keep-out envelope around the robotic arm. Both audible and graphical warnings were selectable to indicate when a keep-out envelope was breached.

Thermal Protection System Evaluation

During the preparation for Return to Flight following the Columbia accident in 2003, NASA incorporated the entire shuttle Thermal Protection System database and a "painting" feature into the DOUG software package. The database consisted of all 25,000+ tiles, thermal blankets, reinforced carbon-carbon wing leading edge panels, and nose cap.

The software was used preflight to develop the trajectories of the Shuttle Robotic Arm and Orbiter Boom Sensor System used to perform in-flight Orbiter inspections. The software allowed engineers to "paint" the areas that were within the specifications

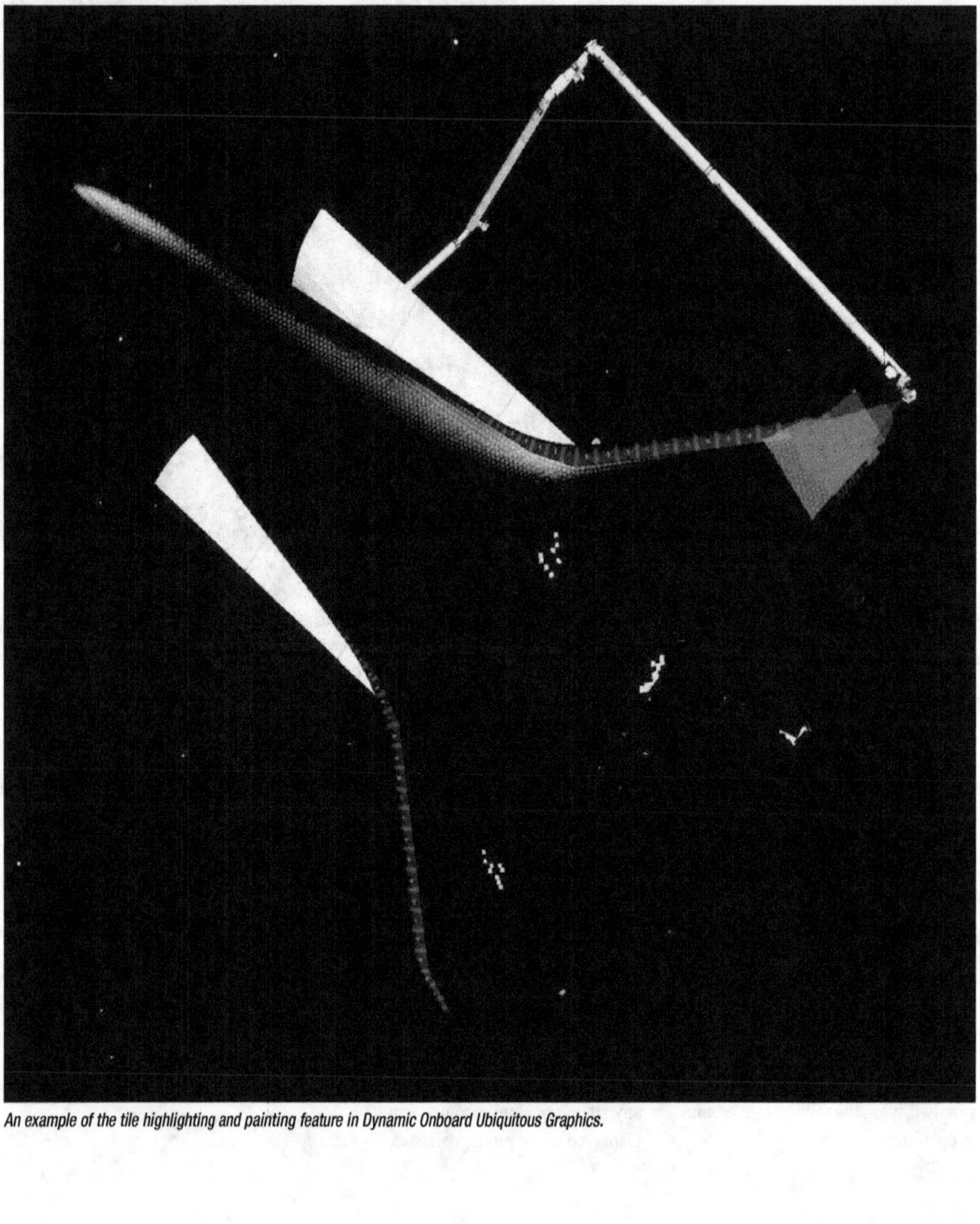

An example of the tile highlighting and painting feature in Dynamic Onboard Ubiquitous Graphics.

of various sensors on the Orbiter Boom Sensor System (e.g., range, field of view, incidence angle) to make sure the Thermal Protection System was completely covered during on-orbit surveys.

The same configuration models and tile database used on the ground were also loaded on the on-board laptop computers. This allowed the areas of interest found during the survey data analysis to be highlighted and uplinked to the shuttle and station crews for further review using the DOUG program.

Inspection of the STS-114 (2005) survey data showed protruding gap fillers between tiles on the Orbiter. These protrusions were of concern for re-entry into Earth's atmosphere. Ground controllers were able to highlight the surrounding tiles in the database, develop a Space Station Robotic Arm configuration with an EVA crew member in a foot restraint on the end, and uplink that configuration file to the station laptop computers. The crew members were then able to use the software to view the area of concern, understand how they would need to be positioned underneath the Orbiter, get a feel for the types of clearances they had with the structure around the robotic arm, and evaluate camera views that would be available during the operation.

Having the 3-D, interactive viewing capability allowed crew members to become comfortable with their understanding of the procedure in much less time than would have been required with just "words" from ground control. A key aspect to the success of this scenario was the software and configuration database commonality that DOUG provided to all participants—station and shuttle crews, ground analysis groups, procedure developers, mission controllers, and simulation facilities.

DOUG was loaded on more than 1,500 machines following the Columbia accident and was used as a tool to support preflight planning and procedures development as well as on-orbit reviews of all robotic and EVA operations. In addition to its basic capabilities, the software possessed many other features that made it a powerful planning and visualization tool.

Expansion of Capabilities

DOUG has also been repackaged into a more user-friendly application referred to as Engineering DOUG Graphics for Exploration (EDGE). This application is a collection of utilities, documentation, development tools, and visualization tools wrapped around the original renderer. DOUG is basically the kernel of the repackaged version, which includes the addition of various plug-ins, models, scripts, simulation interface code, graphical user interface add-ons, overlays, and development interfaces to create a visualization package. The project allows groups to quickly visualize their simulations in 3-D and provides common visuals for future program cockpits and training facilities.
It also allows customers to expand the capabilities of the original software package while being able to leverage off the development and commonality achieved by that software in the Space Shuttle and ISS Programs.

EDGE is now publicly available. To request a copy, call or email:

Technology Transfer and
Commercialization Office
NASA Johnson Space Center
Phone: 281-483-3809
Email: jsc-techtran@mail.nasa.gov

Summary

The graphics-rendering software developed by NASA to support astronaut training and engineering simulation visualization during the shuttle era provided the cornerstone for commonality among ground-based training facilities for both the Space Shuttle and the ISS. The software has evolved over the years to take advantage of ever-advancing computer graphics technology to keep NASA training simulators state of the art and to provide a valuable resource for future programs and missions.

Structural Design

Introduction
Gail Chapline

Orbiter Structural Design
Thomas Moser
Glenn Miller

Shuttle Wing Loads—Testing and Modification Led to Greater Capacity
Tom Modlin

Innovative Concept for Jackscrews Prevented Catastrophic Failures
John Fraley
Richard Ring
Charles Stevenson
Ivan Velez

Orbiter Structure Qualification
Thomas Moser
Glenn Miller

Space Shuttle Pogo—NASA Eliminates "Bad Vibrations"
Tom Modlin

Pressure Vessel Experience
Scott Forth
Glenn Ecord
Willard Castner

Nozzle Flexible Bearing—Steering the Reusable Solid Rocket Motor
Coy Jordan

Fracture Control Technology Innovations—From the Space Shuttle Program to Worldwide Use
Joachim Beek
Royce Forman
Glenn Ecord
Willard Castner
Gwyn Faile

Space Shuttle Main Engine Fracture Control
Gregory Swanson
Katherine Van Hooser

The Space Shuttle—a mostly reusable, human-rated launch vehicle, spacecraft, space habitat, laboratory, re-entry vehicle, and aircraft—was an unprecedented structural engineering challenge. The design had to meet several demands, which resulted in innovative solutions. The vehicle needed to be highly reliable for environments that could not be simulated on Earth or fully modeled analytically for combined mechanical and thermal loads. It had to accommodate payloads that were not defined or characterized. It needed to be weight efficient by employing a greater use of advanced composite materials, and it had to rely on fracture mechanics for design with acceptable life requirements. It also had to be certified to meet strength and life requirements by innovative methods. During the Space Shuttle Program, many such structural design innovations were developed and extended to vehicle processing from flight to flight.

Orbiter Structural Design

NASA faced several challenges in the structural design of the Orbiter. These challenges were greater than those of any previous aircraft, launch vehicle, or spacecraft, and the Orbiter was all three. Yet, the space agency proceeded with tenacity and confidence, and ultimately reached its goals. In fact, 30 years of successful shuttle flights validated the agency's unique and innovative approaches, processes, and decisions regarding characteristics of design.

A few of the more significant challenges NASA faced in Orbiter structural design included the evolution of design loads. The Orbiter structure was designed to an early set of loads and conditions and certified to a later set. The shuttle achieved first-flight readiness through a series of localized structural modifications and operational flight constraints. During the early design phase, computer analyses using complex calculations like finite-element models and techniques for combined thermal and mechanical loads were not possible. Later advances in analytical methods, coupled with test data, allowed significant reductions in both scope and cost of Orbiter structural certification. The space agency had to face other challenges. Structural efficiency had to be compromised to assure versatile payload attachment and payload bay door operations. Skin buckling had to be avoided to assure compatibility with the low-strength Thermal Protection System tiles. Composite materials beyond the state of the art were needed. The crew compartment had to be placed into the airframe such that the pressurized volume would effectively "float." And it was impractical to test the full airframe under combined mechanical and thermal loads.

Thousands of analytical design loads and conditions were proven acceptable with flight data with one exception: the ascent wing loads were greater than predicted because of the effect the rocket exhaust plume had on the aerodynamic pressure distribution. As a result, early flights were flown within limited flight regimes to assure that the structural capability of the wings was not exceeded. The wings were later "strengthened" with minor changes in the design and weight.

Shuttle Wing Loads—Testing and Modification Led to Greater Capacity

Orbiter wing loads demonstrated the importance of anchoring the prediction or grounding the analysis with flight data in assuring a successful flight. The right wing of Columbia was instrumented with strain gauges for the test flights and was load-calibrated to verify the in-flight air load distribution. The wing was also instrumented with pressure gauges; however, the number was limited due to on-board recorder space limitations. This resulted in the need to obtain additional pressure data.

Space Transportation System (STS)-1 (1981) data indicated higher shear in the aft spar web than was predicted. NASA conducted analyses to determine the location and magnitude of forces causing this condition. The results indicated an additional load along the outboard wing leading edge (elevon hinge line). Data obtained on STS-2 (1981) through STS-4 (1982) substantiated these results. This caused concern for the operational wing limits that were to be imposed after the flight test period.

The additional load caused higher bending and torsion on the wing structure, exceeding design limits. The flight limits, in terms of angle of attack and sideslip, would have to be restricted with an attendant reduction in performance.

The recovery plan resulted in modification to the wing leading edge fittings. The major impact was to the structure between the upper and lower wing skins, which were graphite-epoxy. These required angle stiffeners on each flat to increase the buckling stress. The weight of the modifications resulted in a loss of performance. The resulting flight envelope was slightly larger than the original when accounting for the negative angle-of-attack region of the flight regime.

Payload Access and Structural Attachments—Mid-Fuselage and Payload Bay Doors

NASA designed the mid-fuselage of the Orbiter to be "flexible" so as to accommodate the closing of payload bay doors in space. The design also had to accommodate a wide range of payload sizes, weights, and number.

The payload bay doors were an integral part of the fuselage structure. The classical structural design would have the doors provide strength when the fuselage encountered loads from bending, twisting, shear, internal pressure, and thermal gradients. The doors also had to open in space to provide access to the payload and enable the radiators to radiate heat to space. Equally important, the doors had to close prior to re-entry into Earth's atmosphere to provide aerodynamic shape and thermal protection.

To balance the functional and strength requirements, engineers designed the doors to be flexible. The flexibility and zipper-like closing ensured that the doors would close in orbit even if distorted thermally or by changes in the gravity environment (from Earth gravity to microgravity). If the latches did not fully engage, the doors could not be relied on to provide strength during re-entry for fuselage bending, torsion, and aerodynamic pressure. Thus, the classical design approach for ascent was not possible for re-entry. The bulkheads at each end of the payload section and the longerons on each side required additional strength. To reduce weight and thermal distortion, engineers designed the doors using graphite epoxy. This was the largest composite structure on any aircraft or spacecraft at the time.

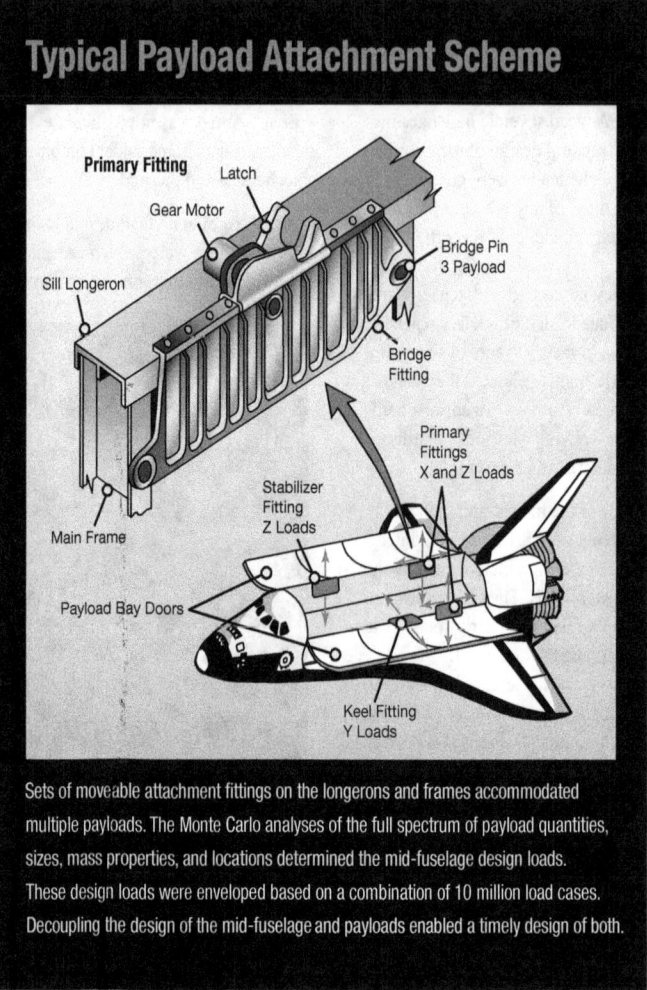

Typical Payload Attachment Scheme

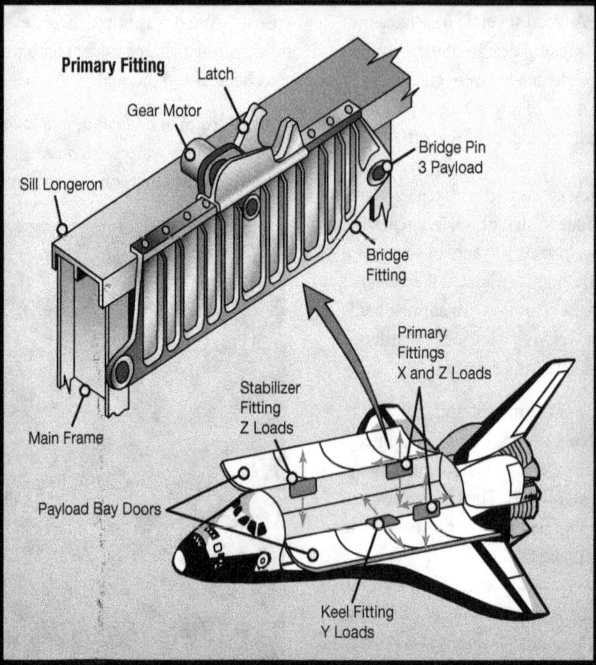

Sets of moveable attachment fittings on the longerons and frames accommodated multiple payloads. The Monte Carlo analyses of the full spectrum of payload quantities, sizes, mass properties, and locations determined the mid-fuselage design loads. These design loads were enveloped based on a combination of 10 million load cases. Decoupling the design of the mid-fuselage and payloads enabled a timely design of both.

The mid-fuselage had to accommodate the quantity, size, weight, location, stiffness, and limitations of known and unknown payloads. An innovative design approach needed to provide a statically determinant attachment system between the payloads and mid-fuselage. This would decouple the bending, twisting, and shear loads between the two structures, thus enabling engineers to design both without knowing the stiffness characteristic of each.

Designing to Minimize Local Deflections

The Orbiter skin was covered with more than 30,000 silica tiles to withstand the heat of re-entry. These tiles had a limited capacity to accommodate structural deflections from thermal gradients. The European supersonic Concorde passenger aircraft (first flown in 1969 and in service from 1976 to 2003) and the SR-71 US military aircraft encountered significant thermal gradients during flight. The design approach in each was to reduce stresses induced by the thermal gradients by enabling expansion of selected regions of the structure; e.g., corrugated wing skins for the SR-71 and "slots" in the Concorde fuselage. After consulting with the designers of both aircraft, NASA concluded that the Orbiter design should account for thermally induced stresses but resist large expansions and associated skin buckling. This brute-force approach protected the attached silica tile as well as simplified the design and manufacture of the Orbiter airframe.

NASA developed these design criteria so that if the thermal stresses reduced the mechanical stresses, the reductions would not be considered in the combined stress calculations.

To determine the thermally induced stresses, NASA established deterministic temperatures for eight initial temperature conditions on the Orbiter at the time of re-entry as well as at several times during re-entry. Engineers generated 120 thermal math models for specific regions of the Orbiter. Temperatures were extrapolated and interpolated to nodes within these thermal math models.

Orbiter Thermal Stress Analysis Modeling

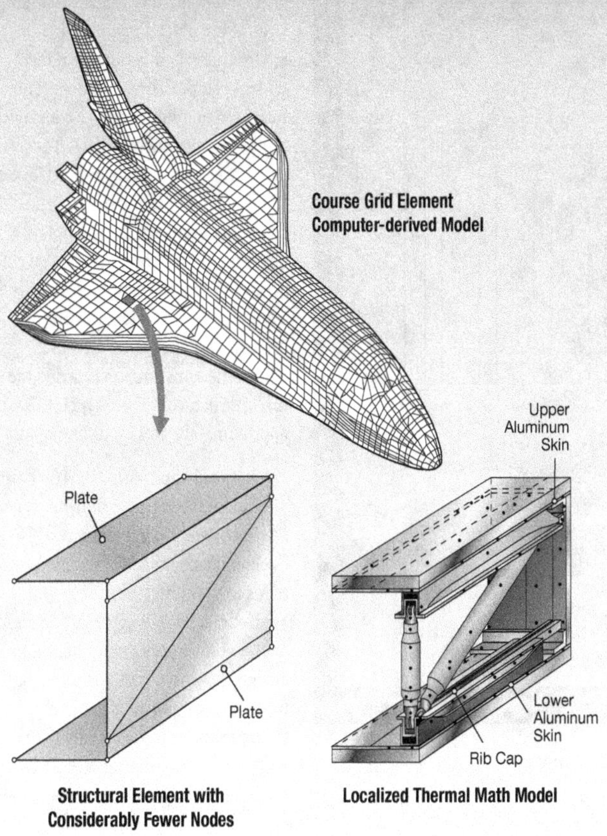

Use of Unique Advanced Materials

Even though the Orbiter was a unique aircraft and spacecraft, NASA selected a conventional aircraft skin/stringer/frame design approach. The space agency also used conventional aircraft material (i.e., aluminum) for the primary structure, with exceptions in selected regions where the use of advanced state-of-the-art composites increased efficiency due to their lower density, minimum thermal expansion, or higher modulus of elasticity.

Other exceptions to the highly reliable conventional structures were the graphite-epoxy Orbital Maneuvering System skins, which were part of a honeycomb sandwich structure. These graphite honeycomb structures had a vented core to relieve pressure differentials across the face sheets during flight. They also required a humidity-controlled environment while on the ground to prevent moisture buildup in the core. Such a buildup could become a source of steam during the higher temperature regimes of flight. Finally, during the weight-savings program instituted on Discovery, Atlantis, and Endeavour, engineers replaced the aluminum spar webs in the wing with a graphite/epoxy laminate.

Large doors, located on the bottom of the Orbiter, were made out of beryllium. These doors closed over the External Tank umbilical cavity once the vehicle

Early Trade Studies Showed Cost Benefits That Guided Materials Selection

Titanium offered advantages for the primary structure because of higher temperature capability—315°C vs. 177°C (600°F vs. 350°F). When engineers considered the combined mass of the structure and Thermal Protection System, however, they noted a less than 10% difference. The titanium design cost was 2.5 times greater. The schedule risk was also greater. NASA considered other combinations of materials for the primary structure and Thermal Protection System and conducted a unit cost comparison. This study helped guide the final selections and areas for future development.

Orbiter Structure/Thermal Protection System First Unit Cost Comparison

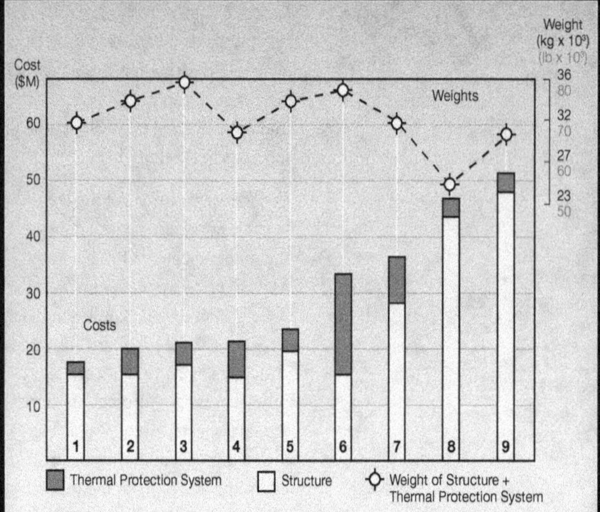

1—Aluminum Alloy 7075-T6 Structure, Ablator Thermal Protection System
2—Aluminum Alloy 7075-T6 Structure, Reusable Thermal Protection System LI-1500 (Lockheed-produced tiles)
3—Aluminum Alloy 2024-T81 Structure, Reusable Thermal Protection System LI-1500
4—Aluminum Alloy 7075-T6, Reusable Thermal Protection System on Beryllium Panels
5—Magnesium Alloy HM21A-T8 Structure, Reusable Thermal Protection System LI-1500
6—Aluminum Alloy 7075-T6, Metallic Inconel® Thermal Protection System
7—Combination Aluminum and Titanium Alloys Structure, Reusable Thermal Protection System LI-1500
8—Beryllium and Titanium Alloys Structure, Reusable Thermal Protection System LI-1500
9—Titanium Alloy 6Al-4V Structure, Reusable Thermal Protection System LI-1500

was on orbit. These approximately 1.3-m (50-in.) square doors maintained the out-of-plane deflection to less than 20 mm (0.8 in.) to avoid contact with adjacent tiles. They also had the ability to withstand a 260°C (500°F) environment generated by ascent heating. The beryllium material allowed the doors to be relatively lightweight and very stiff, and to perform well at elevated temperatures. The superior thermal performance allowed the door, which measured 25.4 mm (1 in.) in thickness, to fly without internal insulation during launch. Since beryllium can be extremely toxic, special procedures applied to those working in its vicinity.

The truss structure that supported the three Space Shuttle Main Engines was stiff and capable of reacting to over a million pounds of thrust. The 28 members that made up the thrust structure were machined from diffusion-bonded titanium. Titanium strips were placed in an inert environment and bonded together under heat, pressure, and time. This fused the titanium strips into a single, hollow, homogeneous mass. To increase the stiffness, engineers bonded layers of boron/epoxy to the outer surface of the titanium beams. The titanium construction was reinforced in select areas with boron/epoxy tubular struts to minimize weight and add stiffness. Overall, the integrated metallic composite construction reduced the thrust structure weight by 21%, or approximately 409 kg (900 pounds).

NASA used approximately 168 boron aluminum tubes in the mid-fuselage frames as stabilizing elements. Technicians bonded these composite tubes to titanium end fittings and saved approximately 139 kg (305 pounds) over a conventional aluminum tube design. During ground operations, however, composite tubes in high traffic areas were repeatedly damaged and were eventually replaced with an aluminum design to increase robustness during vehicle turnaround.

Orbiter Structure—Structural Arrangement and Location of Composite Materials

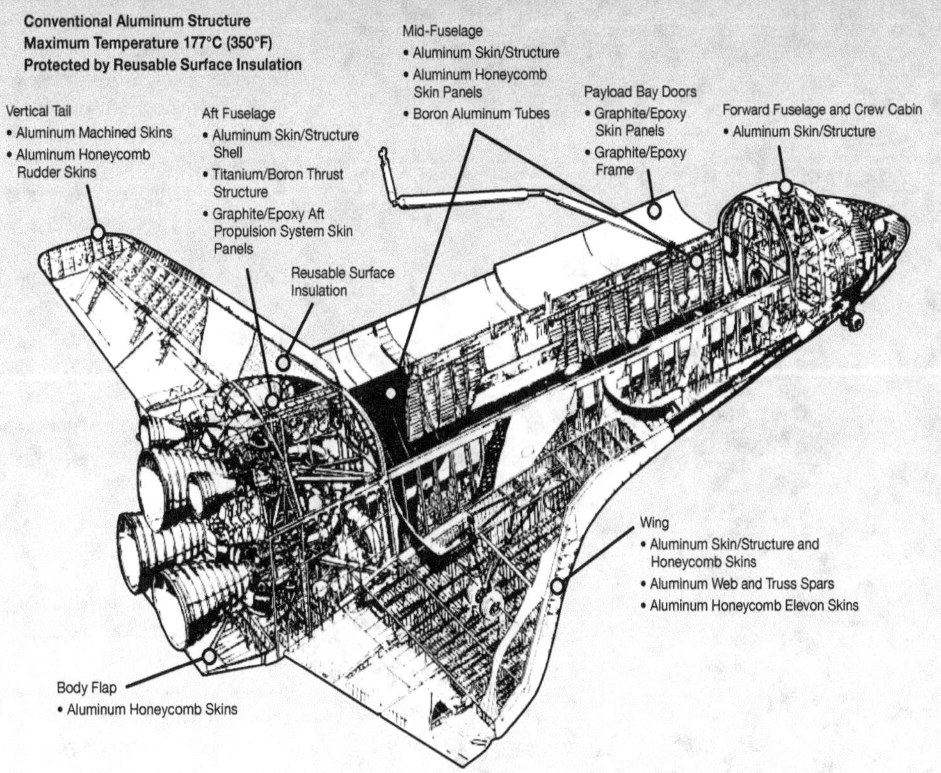

After the initial design of Challenger and Columbia, NASA initiated a weight-savings program for the follow-on vehicles—Discovery, Atlantis, and Endeavour. The space agency achieved weight savings through optimization of aluminum structures and replaced the aluminum spar webs in the wing with a graphite/epoxy laminate.

"Floating" Crew Compartment

The crew compartment structure "floated" inside the forward fuselage. The crew compartment was attached to the forward fuselage at four discrete points, thus enabling a simpler design (for pressure and inertia loads only) and greater thermal isolation. The crew compartment was essentially a pressure vessel and the only pressurized compartment in the Orbiter. To help assure pressure integrity, the aluminum design withstood a large noncritical crack while maintaining cabin pressure. The "floating" crew compartment reduced weight over an integrated forward fuselage design and simplified manufacturing.

The crew cabin being installed in the forward fuselage.

Innovative Concept for Jackscrews Prevented Catastrophic Failures

More than 4,000 jackscrews were in use around Kennedy Space Center (KSC) during the Space Shuttle era. NASA used some of these jackscrews on critical hardware. Thus, a fail-safe, continue-to-operate design was needed to mitigate the possibility of a catastrophic event in case of failure.

A conventional jackscrew contained only one nut made of a material softer than that of the threaded shaft. With prolonged use, the threads in the nut would wear away. If not inspected and replaced after excessive wear, the nut eventually failed. KSC's fail-safe concept for machine jackscrews incorporated a redundant follower nut that would begin to bear the axial jack load on the failure of the primary nut.

Unlike the case of a conventional jackscrew, it was not necessary to relieve the load to measure axial play or disassemble the nut from the threaded shaft to inspect the nut for wear. Instead, wear could be determined by measuring the axial gap between the primary nut and the follower nut.

Additionally, electronic and mechanical wear indicators were used to monitor the gap during operation or assist during inspection. These devices would be designed to generate a warning when the thread was worn to a predetermined thickness. The fail-safe, continue-to-operate design concept offered an alternative for preventing catastrophic failures in jackscrews, which were used widely in aeronautical, aerospace, and industrial applications.

Orbiter Structure Qualification

The conventional strength and life certification approach for a commercial or military aircraft is to demonstrate the ultimate strength and fatigue (life) capacities with a dedicated airframe for each. Similarly, NASA planned two full-scale test articles at the outset of the Orbiter design, development, test, and evaluation program. Ultimately, the Orbiter structure was certified with an airframe that became a flight vehicle and a series of smaller component test articles that comprised about 30% of the flight hardware. The space agency did not take additional risks, and the program costs for ground tests were reduced by several hundred million dollars.

Ultimate Strength Integrity

Virtually all of the Orbiter's primary structure had significant thermal stress components. Therefore, thermal stress had to be accounted for when certifying the design for ultimate strength. Yet, it was impractical—if not impossible—to simulate the correct combination of temperatures and mechanical loads for the numerous conditions associated with ascent, spaceflight, and re-entry into Earth's atmosphere, especially for transient cases of interest. NASA reached this conclusion after consulting with the Concorde aircraft structural experts who conducted multiyear, expensive combined environment tests.

Orbiter strength integrity would be certified in a bold and unconventional approach that used the Challenger (Orbiter) as the structural test article. Rather than testing the ultimate load (140% of maximum expected loads), NASA would test to 120% of limit mechanical load, use the test data to verify the analytical stress models, and analytically prove that the structure could withstand 140% of the combined mechanical and thermal stresses.

The structural test article was mounted in a horizontal position at the External Tank reaction points and subjected to a ground test program at the Lockheed test facility in Palmdale, California. The 390,900-kg (430-ton) test rig contained 256 hydraulic jacks that distributed loads across 836 application points to simulate various stress levels. Initial influence coefficient tests involved the application of approximately 150 load conditions as point loads on the vehicle. These unit load cases exercised the structure at the main engine gimbal and actuator attachments, payload fittings, and interfaces on the wing, tail, body flap, and Orbital Maneuvering System pods. Engineers measured load vs. strain at numerous locations and then used those measurements for math model correlation. They also used deflection measurements to substantiate analytical stiffness matrices.

The Orbiter airframe was subjected to a series of static test conditions carried to limit plus load levels (approximately 120% of limit). These conditions consisted of a matrix of 30 test cases representative of critical phases (boost, re-entry, terminal area energy management, and landing) to simulate design mechanical loads plus six thrust vector-only conditions. These tests verified analytically predicted internal load distributions. In conjunction with analysis, the tests also confirmed the structural integrity of the Orbiter airframe for critical design limit loads. Engineers used these data to support evaluation of the ultimate factor of safety by analysis. Finally, they used the test series to evaluate strains from the developmental flight instrumentation.

Space Shuttle Pogo—NASA Eliminates "Bad Vibrations"

Launch vehicles powered by liquid-fueled, pump-fed rocket engines frequently experience a dynamic instability that caused structural vibrations along the vehicle's longitudinal axis. These vibrations are referred to as "Pogo."

As Astronaut Michael Collins stated, "The first stage of Titan II vibrated longitudinally so that someone riding on it would be bounced up and down as if on a pogo stick."

In technical terms, Pogo is a coupled structure/propulsion system instability caused by oscillations in the propellant flow rate that feeds the engines. The propellant flow rate oscillations can result in oscillations in engine thrust. If a frequency band of the thrust oscillations is in phase with the natural frequency of engine structure and is of sufficient magnitude to overcome structural damping, the amplitude of the propellant flow rate oscillation will increase. Subsequently, this event will increase the amplitude of the thrust oscillation. This sequence can lead to Pogo instability, with the possible result in an unprogrammed engine shutdown and/or structural failure—both of which would result in loss of mission.

Most NASA launch vehicles experienced Pogo problems. Unfortunately, the problem manifested itself in flight and resulted in additional testing and analytical work late in the development program. The solution was to put an accumulator in the propellant feedline to reduce propellant oscillations.

The Space Shuttle Program took a proactive approach with a "Pogo Prevention Plan" drafted in the early 1970s. The plan called for comprehensive stability analysis and testing programs. Testing consisted of modal tests to verify the structural dynamic characteristics, hydroelastic tests of External Tank and propellant lines, and pulse testing of the Space Shuttle Main Engines. The plan baselined a Pogo suppression system—the first NASA launch vehicle to have such

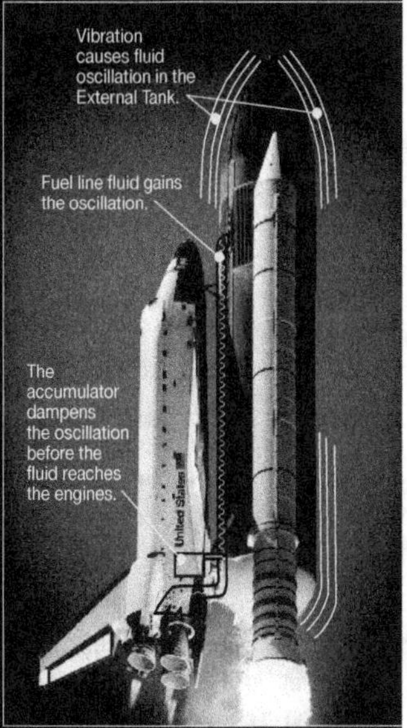

Vibration causes fluid oscillation in the External Tank.
Fuel line fluid gains the oscillation.
The accumulator dampens the oscillation before the fluid reaches the engines.
United States

a feature. The space agency selected and included an accumulator in the design of the main engines. This approach proved successful. Flight data demonstrated that the Space Shuttle was free of Pogo.

Test rig surrounds the Orbiter structural test article, Challenger, at the Lockheed Test Facility in Palmdale, California.

After the limit plus tests, the forward fuselage of the structural test article was subjected to a thermal environment gradient test. This testing entailed selective heating of the external skin regions with 25 zones. Gaseous nitrogen provided cooling. NASA used the data to assess the effects of thermal gradients and assist in the certification of thermal stresses by analysis techniques. Finally, the aft fuselage of the structural test article was subjected to internal/external pressures to provide strain and deflection data to verify the structural adequacy of the aft bulkhead and engine heat shield structures.

The structural test article subjected the Orbiter airframe to approximately 120% of limit load. To address ultimate load (140%) in critical areas, NASA conducted a series of supplemental tests on two major interfaces and 34 component specimens. The agency chose these specimens based on criticality of failure, uncertainty in analysis, and minimum fatigue margin. Designated specimens were subjected to fatigue testing and analysis to verify the 100-mission life requirement. Finally, NASA tested all components to ultimate load and gathered data to compare predictions.

This unprecedented approach was challenged by NASA Headquarters and reviewed by an outside committee of experts from the "wide body" commercial aircraft industry. The experts concurred with the approach.

Acoustic Fatigue Integrity

Commercial and military aircraft commonly have a design life of 20,000 hours of flight composed of thousands of take offs and landings. As a result, the fatigue life is a design factor. The Orbiter, on the other hand, had a design life of 100 missions and a few hundred hours of flight in the atmosphere, but the acoustic environment during ascent was very high. Certification of acoustic fatigue life had to be accomplished.

The challenge was to certify this large, complex structure for a substantial number of combined acoustic, mechanical, and thermal conditions. No existing test facilities could accommodate a test article the size of the Orbiter or simulate all of the loads and environments.

The acoustic fatigue certification program was as innovative as that of the ultimate strength certification. The approach was to test a representative structure of various forms, materials, and types of construction in representative acoustic environments until the structure failed. This

Orbiter Acoustic Fatigue Test Articles

These acoustic fatigue test articles (shaded in blue) are representative of structure and environment.

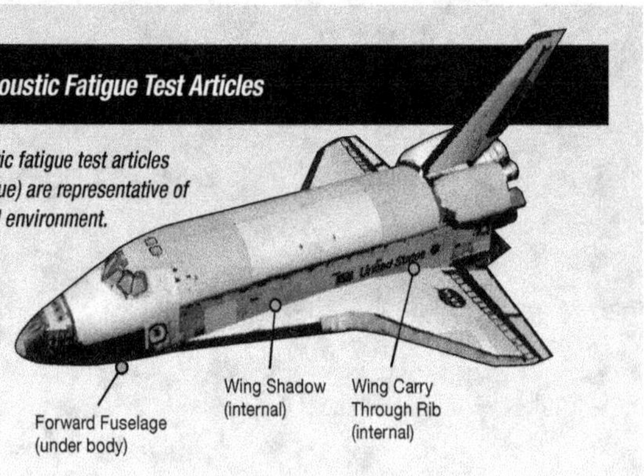

Nozzle Flexible Bearing—Steering the Reusable Solid Rocket Motor

At Space Shuttle liftoff, initial steering was controlled in large part by the reusable solid rocket motors' movable nozzles. Large hydraulic actuators were attached to each nozzle. On command, these actuators mechanically vectored the nozzle, thereby redirecting the supersonic flow of hot gases from the motor.

A flexible bearing allowed the nozzle to be vectored. At about 2.5 m (8 ft) in diameter and 3,200 kg (7,000 pounds), this bearing was the largest flexible bearing in existence. The component had to vector up to 8 degrees while maintaining a pressure-tight seal against the combustive gases within the rocket, withstand high loads imparted at splashdown, and fit within the constraints of the solid rocket motor case segments. It also had to be reusable up to nine times.

The structure consisted of alternating layers of natural rubber (for flexibility) and steel shims (for strength and stiffness). The layers were spherically shaped,

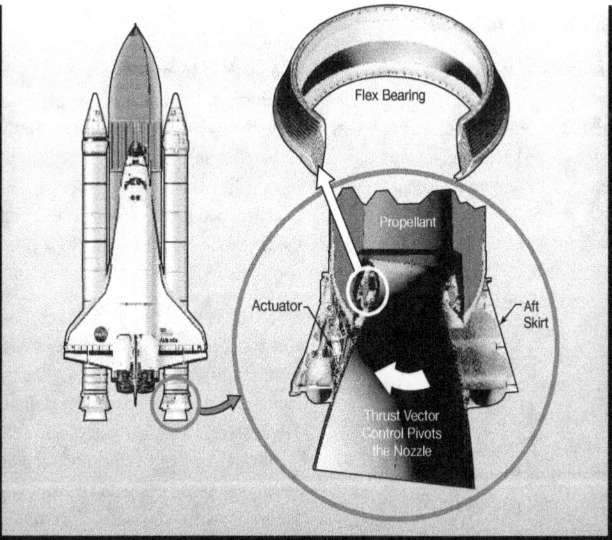

During the first minutes of flight, a Thrust Vector Control System housed at the base of each solid rocket motor provided a majority of the steering capability for the shuttle. A flexible bearing enabled nozzle movement. Two hydraulic actuators generated the mechanical force needed to move the nozzle.

allowing the nozzle to pivot in any direction. Forces from the actuators induced a torque load on the bearing that strained the rubber layers in shear, with each layer rotating a proportional part of the total vector angle. This resulted in a change in nozzle angular direction relative to the rocket motor centerline.

The most significant manufacturing challenge was producing a vulcanization bond between the rubber and the shims. Fabrication involved laying up the natural rubber by hand between the spherically shaped shims. Vulcanization was accomplished by applying pressure while controlling an elevated temperature gradient through the flexible bearing core. This process cured the rubber and vulcanized it to the shims in one step. The completed bearing underwent rigorous stretching and vectoring tests, including testing after each flight, as part of the refurbishment process.

established the level of damage that would be allowed for each type of structure. NASA selected 14 areas of the Orbiter to represent the various structural configurations.

The allowable damage was reduced analytically to account for the damage induced by the flight loads and temperature cycles for all regions of the vehicle.

Because of the high fatigue durability of the graphite-epoxy construction of the payload bay doors and Orbital Maneuvering System pods, these structures were not tested to failure. Instead, the strains measured during the acoustic tests were correlated with mathematical models and adequate fatigue life was demonstrated analytically. These test articles were subsequently used as flight hardware.

Summary

The unique approaches taken during the Space Shuttle Program in validating the structural integrity of the Orbiter airframe set a precedent in the NASA programs that followed. Even as more accurate analysis software and faster computers are developed, the need for anchoring predictions in the reality of testing remains a cornerstone in the safe flight of all space vehicles.

Pressure Vessel Experience

In the 1970s, NASA made an important decision—one based on previous experience and emerging technology—that would result in significant weight savings for shuttle. The agency implemented the Composite Overwrapped Pressure Vessels Program over the use of all-metal designs for storing high-pressure gases, 2,068 – 3,361 N/cm² (3,000 – 4,875 psi) oxygen, nitrogen, and helium. The agency used 22 such vessels in the Environmental Control and Life Support System, Reaction Control System, Main Propulsion System, and Orbital Maneuvering System. The basic new design consisted of a gas or liquid impermeable, thin-walled metal liner wrapped with a composite overwrap for primary pressure containment strength.

Safety—Always a Factor

The Space Shuttle Program built on the lessons learned from the Apollo Program. The pressure vessels were constructed of titanium and designed such that the burst pressure was only 1.5 times the operating pressure (safety factor). This safety factor was unprecedented at the time. To assure the safety of tanks with such a low margin of safety, NASA developed a robust qualification and acceptance program. The technical knowledge gained during the Apollo Program was leveraged by the shuttle, with the added introduction of a new type of pressure vessel to further reduce mass.

The Brunswick Corporation, Lake Forest, Illinois, developed, for the shuttle, a composite overwrapped pressure vessel for high-pressure oxygen, nitrogen, and helium storage. The metallic liners were made of titanium (Inconel® for the oxygen systems) overwrapped with DuPont™ Kevlar® in an epoxy matrix. Switching from solid titanium tanks to composite overwrapped pressure vessels reduced the Space Shuttle tank mass by approximately 209 kg (460 pounds).

Since the shuttle was reusable and composite overwrapped pressure vessels were a new technology, the baseline factor of safety was 2.0. As development progressed, NASA introduced and instituted a formal fracture control plan based on lessons learned in the Apollo Program. As the composite overwrapped pressure vessels were fracture-critical items—e.g., their failure would lead to loss of vehicle and crew—fracture control required extensive lifetime testing of the vessels to quantify all failure modes. The failure mechanisms of the composite were just beginning to be understood. Kevlar® is very durable, so minor damage to the overwrap was not critical. NASA, however, discovered that the composite could fail when under a sustained stress, less than its ultimate capability, and could fail without indication. This failure mode of the composite was called "stress rupture," and could lead to a catastrophic burst of the pressure vessel since the metallic liner could not carry the pressure stress alone.

In the late 1970s, engineers observed unexpectedly poor stress rupture performance in the testing of Kevlar® strands at the Lawrence Livermore Nationale Laboratory in Livermore, California. As a result, NASA contracted with that laboratory to study the failure modes of the Kevlar® fiber for application in the shuttle tanks. Technicians conducted hundreds of tests on individual Kevlar® fibers, fiber/epoxy strands, and subscale vessels.

The development program to characterize all the failure modes of the composite overwrapped pressure vessels set the standard for all spaceflight programs. Therefore, as tank development proceeded, NASA used the fracture control test program to

Composite Overwrapped Pressure Vessels

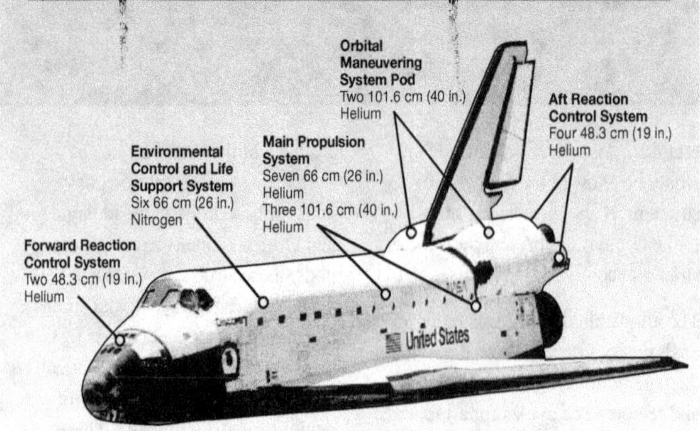

NASA Puts Vessels to the "Stress Test"

In 1978, NASA developed and implemented a "fleet leader" test program to provide Orbiter subscale vessel stress rupture data for comparison to existing strand and subscale vessel data. Vessels in the test program were subscale in size and used aluminum liners instead of titanium, yet they were built by the same company manufacturing the Orbiter composite overwrapped pressure vessels using the same materials, equipment, and processes/procedures. These vessels were put to test at Johnson Space Center in Houston, Texas.

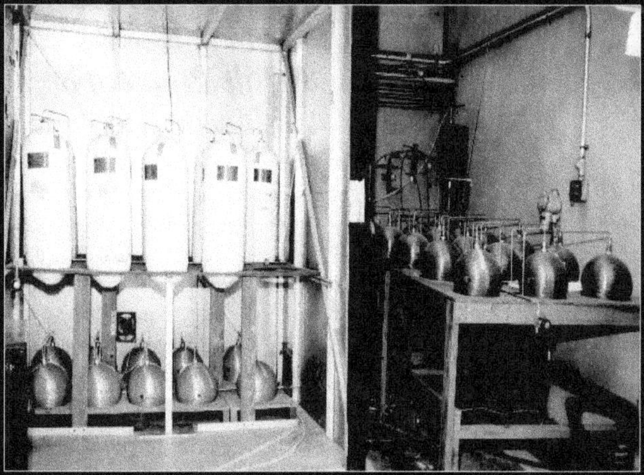

The test program consisted of two groups of vessels—15 vessels tested at ambient temperature conditions and an approximate stress level of 50% of ultimate strength; and 10 vessels tested at approximately 50% of average strength and an elevated temperature in an attempt to accelerate stress rupture failure. For the elevated temperature testing, 79°C (175°F) was chosen as the test temperature for both groups. Engineers performed periodic depressurizations/repressurizations to simulate Orbiter usage and any potential effects.

The ambient temperature vessels were pressurized for nearly 25 years without failure before NASA stopped testing. The flight vessels only accumulated a week or two worth of pressure per mission, so the ground tests led the fleet by a significant margin.

For the accelerated 79°C (175°F) temperature testing, the first failure occurred after approximately 12 years and the second at 15 years of pressure. These stress rupture failures indicated that the original stress rupture life predictions for composite overwrapped pressure vessels were conservative.

justify a safe reduction in the factor of safety on burst from 2.0 to 1.5, resulting in an additional 546 kg (1,203 pounds) of mass saved from the Orbiter.

Even with all of the development testing, two non-stress rupture composite overwrapped pressure vessels failures occurred on shuttle. The complexity of the welding process on certain materials contributed to these failures. To build a spherical pressure vessel, two titanium hemispheres had to be welded together to form the liner. Welding titanium is difficult and unintentional voids are sometimes created. Voids in the welds of two Main Propulsion System vessels had been missed during the acceptance inspection. In May 1991, a Main Propulsion System helium pressurization vessel started leaking on the Atlantis prior to the launch of Space Transportation System (STS)-43. NASA removed these vessels from the Orbiter.

The subsequent failure investigation found that, during manufacture, 89 pores formed in the weld whereas the typical number for other Orbiter vessels was 15. Radiographic inspection of the welds showed that the pores had initiated fatigue cracks that eventually broke through the liner, thereby causing the leak. While this inspection was ongoing, the other Main Propulsion System vessel on Atlantis started leaking helium—once again due to weld porosity. NASA reviewed all other vessels in service, but none had weld porosity levels comparable to the two vessels that had leaked.

Space Shuttle Experiences Influence Future Endeavors

NASA's Orbiter Project pushed the technology envelope for pressure vessel design. Lessons learned from development, qualification, and in-service failures prompted the International Space Station (ISS) and future space and science missions to develop more robust requirements and verification programs. The ISS Program instituted structure controls based on the shuttle investigation of pressure vessels. No other leaks in pressure vessel tanks occurred through 2010—STS-132. For instance, the factor of safety on burst pressure was 1.5; damage tolerance of the composite and metallic liner was clearly addressed through qualification testing and operational damage control plans; radiographic inspection of liner welds was mandatory with acceptable levels of porosity defined; and material controls were in place to mitigate failure from corrosion, propellant spills, and stress rupture. These industry standard design requirements for composite overwrapped pressure vessels are directly attributable to the shuttle experience as well as its positive influence on future spaceflight.

Fracture Control Technology Innovations—From the Space Shuttle Program to Worldwide Use

A fundamental assumption in structural engineering is that all components have small flaws or crack-like defects that are introduced during manufacturing or service. Growth of such cracks during service can lead to reduced service life and even catastrophic structural failure. Fracture control methodology and fracture mechanics tools are important means for preventing or mitigating the adverse effects of such cracks. This is important for industries where structural integrity is of paramount importance.

Prior to the Space Shuttle, NASA did not develop or implement many fracture mechanics and fracture control applications during the design and build phases of space vehicles. The prevailing design philosophy at the time was that safety factors on static strength provided a margin against fracture and that simple proof tests of tanks (pressure vessels) were sufficient to demonstrate the margin of safety. In practice, however, the Apollo Program experienced a number of premature test failures of pressure vessels that resulted in NASA implementing a version of fracture control referred to as "proof test logic." It was not until the early 1960s that proof tests were sufficiently understood from a fracture mechanics point of view—that proof tests could actually be used, in some cases, to ensure the absence of initial flaws of a size that could cause failure within a pressure vessel's operating conditions.

The application of proof test logic required the determination of environmental crack growth thresholds for all environments to which the pressure vessels were exposed while pressurized as well as development of fracture toughness values and cyclic crack growth rates for materials used in the pressure vessels. The thresholds resulted in pressurization restrictions and environmental control of all Apollo pressure vessels. In effect, proof test logic formed the first implementation of a rigorous fracture control program in NASA.

Fracture Control Comes of Age

The legacy of the Apollo pressure vessel failure experience was that NASA, through the Space Shuttle Program, became an industry leader in the development and application of fracture mechanics technology and fracture control methodology. Although proof test logic worked successfully for the Apollo pressure vessels, the Space Shuttle Program brought with it a wide variety of safety-critical, structurally complex components (not just pressure vessels), materials with a wide range of fracture properties, and an aircraft-like fatigue environment—all conditions for which proof test logic methodology could not be used for flaw screening purposes. The shuttle's reusable structure demanded a more comprehensive fracture control methodology. In 1973, the Orbiter Project released its fracture control plan that set the requirements for and helped guide the Orbiter hardware through the design and build phases of the project.

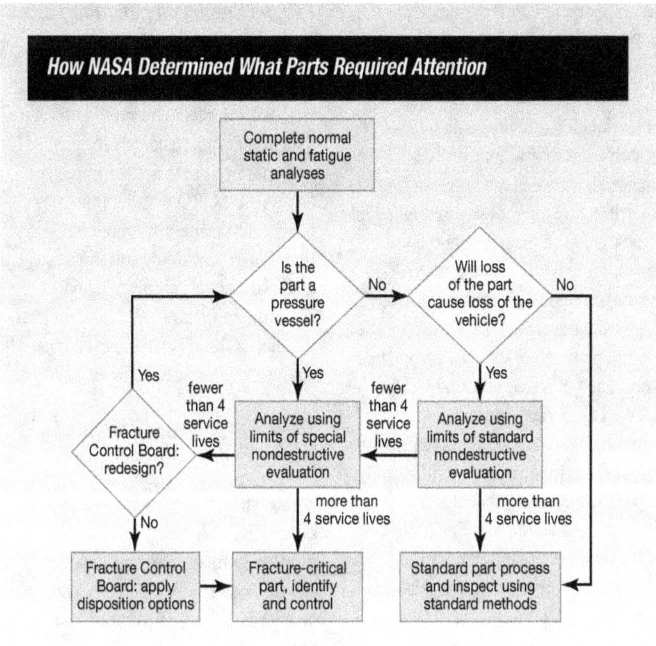

Early Shuttle Fracture Control

Fracture control, as practiced early in the Space Shuttle Program, was a three-step process: select the candidate fracture critical components, perform fracture mechanics analyses of the candidates, and disposition the components that had insufficient life.

Design and stress engineers selected the candidate fracture critical components. The selection was based on whether failure of the component from crack propagation could lead to a loss of life or vehicle. Certain components, such as pressure vessels, were automatically considered fracture critical. Performing a fracture mechanics analysis of the candidates started with an assumed initial crack located in the most unfavorable location in the component. The size of the assumed crack was typically based on the nondestructive inspection that was performed on the component. The fracture mechanics analysis required knowledge of the applied stress, load spectrum, environment, assumed initial crack size, materials fracture toughness, and materials fatigue and environmental crack growth properties. Fracture analysis was required to show a service life of four times the shuttle's 100-mission design life.

There were a number of options for dispositioning components that had insufficient life. These options included the following:

- Redesigning the component when weight and cost permitted
- Conducting nondestructive inspection with a more sensitive technique where special nondestructive evaluation procedures allowed a smaller assumed crack size
- Limiting the life of the component
- Considering multiple element load paths
- Demonstrating life by fracture mechanics testing of the component
- Refining the loading based on actual measurements from the full-scale structural test articles

In addition to being a fundamental part of the structural design process, fracture mechanics became a useful tool in failure analysis throughout the Space Shuttle Program.

Fracture Control Evolves with Payloads

The shuttle payload community further refined the Orbiter fracture control requirements to ensure that a structural failure in a payload would not compromise the Space Shuttle or its Orbiter. NASA classified payloads by the nature of their safety criticality. Typically, a standard fracture criticality classification process started by removing all exempt parts that were nonstructural items—i.e., items not susceptible to crack propagation such as insulation blankets or certain common small parts with well-developed quality-control programs and use history.

All remaining parts were then assessed as to whether they could be classified as non-fracture critical. This category included the following classifications:

- Low-released mass—parts with a mass low enough that, if released during a launch or landing, would cause no damage to other components
- Contained—a failed part confined in a container or otherwise restrained from free release
- Fail-safe—structurally redundant designs where remaining components could adequately and safely sustain the loading that the failed member would have carried or failure would not result in a catastrophic event
- Low risk—parts with large structural margins or other conditions making crack propagation extremely unlikely

- Nonhazardous leak-before-burst—pressure vessels that did not contain a hazardous fluid where loss of fluid would not cause a catastrophic hazard such as loss of vehicle and crew, and where the critical crack size was much greater than the vessel wall thickness

NASA processed non-fracture critical components under conventional aerospace industry verification and quality assurance procedures.

All parts that could not be classified as exempt or non-fracture critical were classified as fracture critical. Fracture critical components had to have their damage tolerance demonstrated by testing or by analysis. To assure conservative results, such tests or analyses assumed that a flaw was located in the most unfavorable location and was subjected to the most unfavorable loads. The size of the assumed flaw was based on the nondestructive inspections that were used to inspect the hardware. The tests or analyses had to demonstrate that such an assumed crack would not propagate to failure within four service lifetimes.

Fracture Control Software Development

Few analytical tools were available for fracture mechanics analysis at the start of the Space Shuttle Program. The number of available analytical solutions was limited to a few idealized crack and loading configurations, and information on material dependency was scarce. Certainly, computing power and availability provided no comparison to what eventually became available to engineers. Improved tools to effect the expanded application of fracture mechanics and fracture control were deemed necessary for safe operation of the shuttle.

With Space Shuttle Program support, Johnson Space Center (JSC) initiated a concerted effort in the mid 1970s to create a comprehensive database of materials fracture properties. This involved testing virtually all metallic materials in use in the program for their fracture toughness, environmental crack growth thresholds, and fatigue crack growth rate properties. NASA manufactured and tested specimens in the environments that Space Shuttle components experienced—cryogenic, room, and elevated temperatures as well as in vacuum, low- and high-humidity air, and selected gaseous or fluid environments. Simultaneously, a parallel program created a comprehensive library of analytical solutions. This involved compiling the small number of known solutions from various sources as well as the arduous task of deriving new ones applicable to shuttle configurations.

Fatigue Crack Computer Program

By the early 1980s, JSC engineers developed a computer program—NASA/FLAGRO—to provide fracture data and fracture analysis for crewed and uncrewed spacecraft components. NASA/FLAGRO was the first known program to contain comprehensive libraries of crack case solutions, material fracture properties, and crack propagation models. It provided the means for efficient and accurate analysis of fracture problems.

NASGRO® Becomes a Worldwide Standard in Fracture Analysis

Although NASA/FLAGRO was essentially a shuttle project, NASA eventually formed an agencywide fracture control methodology panel to standardize fracture methods and requirements across the agency and to guide the development of

Crack Models and Material Properties Required for Fracture Analyses

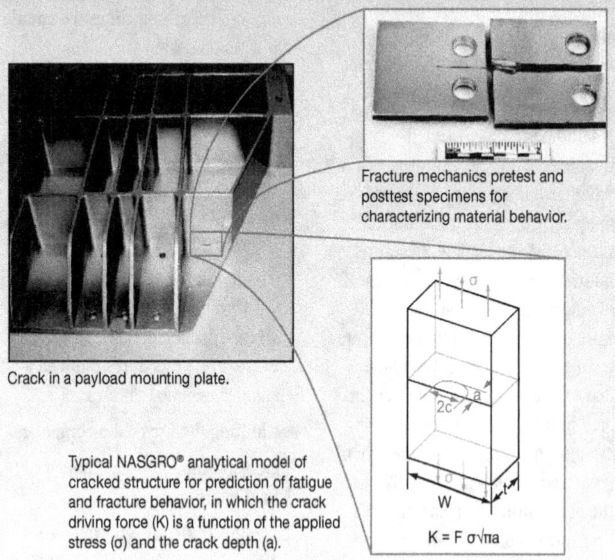

Fracture mechanics pretest and posttest specimens for characterizing material behavior.

Crack in a payload mounting plate.

Typical NASGRO® analytical model of cracked structure for prediction of fatigue and fracture behavior, in which the crack driving force (K) is a function of the applied stress (σ) and the crack depth (a).

Space Shuttle Main Engine Fracture Control

The early Space Shuttle Main Engine (SSME) criteria for selecting fracture critical parts included Inconel® 718 parts that were exposed to gaseous hydrogen. These specific parts were selected because of their potential for hydrogen embrittlement and increased crack growth caused by such exposure. Other parts such as turbine disks and blades were included for their potential to produce shrapnel. Titanium parts were identified as fracture critical because of susceptibility to stress corrosion cracking. Using these early criteria, approximately 59 SSME parts involving some 290 welds were identified as being fracture critical.

By the time the alternate turbopumps were introduced into the shuttle fleet in the mid 1990s, fracture control processes had been well defined. Parts were identified as fracture critical if their failure due to cracking would result in a catastrophic event. The fracture critical parts were inspected for preexisting cracks, a fracture mechanics assessment was performed, and materials traceability, and part-specific life limits were imposed as necessary. This combination of inspection, analysis, and life limits ensured SSME fracture critical parts were flown with confidence.

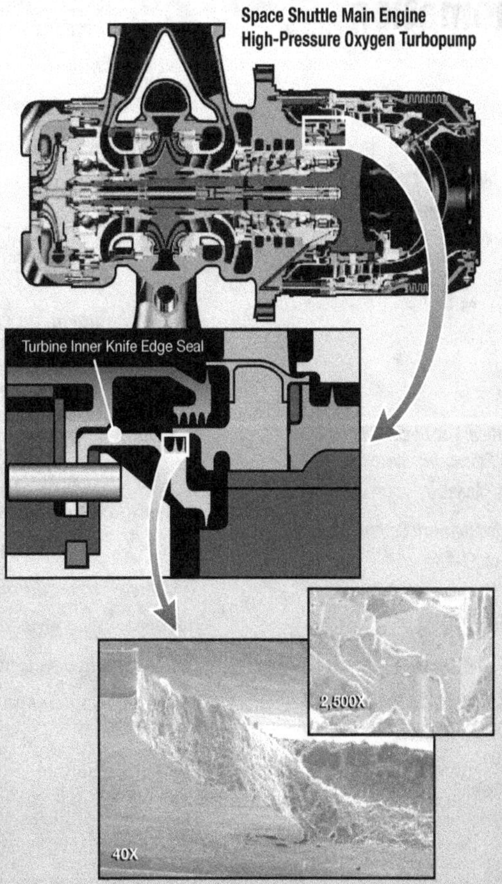

These two photographs show the fracture surface indicative of Stage I crystallographic fatigue growth.

NASA/FLAGRO, renamed NASGRO®, for partnership with industry.
While other commercial computer programs existed by the end of the Space Shuttle Program, none had approached NASGRO® in its breadth of technical capabilities, the size of its fracture solution library, and the size of its materials database. In addition to gaining several prestigious engineering awards, NASGRO® is in use by organizations and companies around the world.

Summary

Fracture mechanics is a technical discipline first used in the Apollo Program, yet it really came of age in the Space Shuttle Program. Although there is still much to be learned, NASA made great strides in the intervening 4 decades of the shuttle era in understanding the physics of fracture and the methodology of fracture control. It was this agency's need to analyze shuttle and payload fracture critical structural hardware that led to the development of fracture mechanics as a tool in fracture control and ultimately to the development of NASGRO®—the internationally recognized fracture mechanics analysis software tool.
The shuttle was not only a principal benefactor of the development of fracture control, it was also the principal sponsor of its development.

Robotics and Automation

Introduction
Gail Chapline
Steven Sullivan

Shuttle Robotic Arm
Henry Kaupp
Elizabeth Bains
Rose Flores
Glenn Jorgensen
Y.M. Kuo
Harold White

Automation: The Space Shuttle Launch Processing System
Timothy McKelvey

Integrated Network Control System
Wayne McClellan
Robert Brown

Orbiter Window Inspection
Bradley Burns

Robotics System Sprayed Thermal Protection on Solid Rocket Booster
Terry Huss
Jack Scarpa

Although shuttle astronauts made their work in space look like an everyday event, it was in fact a hazardous operation. Using robotics or human-assisted robotics and automation eliminated the risk to the crew while still performing the tasks needed to meet the mission objectives. The Shuttle Robotic Arm, commonly referred to as "the arm," was designed for functions that were better performed by a robotic system in space.

Automation also played an important role in ground processing, inspection and checkout, cost reduction, and hazardous operations. For each launch, an enormous amount of data from verification testing, monitoring, and command procedures were compiled and processed, often simultaneously. These procedures could not be done manually, so ground automation systems were used to achieve accurate and precise results. Automated real-time communication systems between the pad and the vehicle also played a critical role during launch attempts. In addition, to protect employees, automated systems were used to load hazardous commodities, such as fuel, during tanking procedures. Throughout the Space Shuttle Program, NASA led the development and use of the most impressive innovations in robotics and automation.

Shuttle Robotic Arm—Now That You Have the "TRUCK," How Do You Make the Delivery?

Early in the development of the Space Shuttle, it became clear that NASA needed a method of deploying and retrieving cargo from the shuttle payload bay. Preliminary studies indicated the need for some type of robotic arm to provide both capabilities. This prompted the inclusion of a Shuttle Robotic Arm that could handle payloads of up to 29,478 kg (65,000 pounds).

In December 1969, Dr. Thomas Paine, then administrator of NASA, visited Canada and extended an offer for Canadian participation with a focus on the Space Shuttle. This was a result of interest by NASA and the US government in foreign participation in post-Apollo human space programs. In 1972, the Canadian government indicated interest in developing the Shuttle Robotic Arm. In 1975, Canada entered into an agreement with the US government in which Canada would build the robotic arm that would be operated by NASA.

The Shuttle Robotic Arm was a three-joint, six-degrees-of-freedom, two-segment manipulator arm to be operated only in the microgravity environment. From a technical perspective, it combined teleoperator technology and composite material technology to produce a lightweight system useable for space applications. In fact, the arm could not support its own weight on Earth. The need for a means of grappling the payload for deployment and retrieval became apparent. This led to an end effector—a unique electromechanical device made to capture payloads.

Unique development and challenges of hardware, software, and extensive modeling and analysis went into the Shuttle Robotic Arm's use as a tool for delivery and return of payloads to and from orbit. Its role continued in the deployment and repair of the Hubble

Backdropped by the blackness of space and Earth's horizon, Atlantis' Orbiter Docking System (foreground) and the Canadarm—the Shuttle Robotic Arm developed by Canada—in the payload bay are featured in this image photographed by an STS-122 (2008) crew member during Flight Day 2 activities.

Space Telescope, its use in the building of the space station and, finally, in Return to Flight as an inspection and repair tool for the Orbiter Thermal Protection System.

Evolution of the Shuttle Robotic Arm

The initial job of the Shuttle Robotic Arm was to deploy and retrieve payloads to and from space. To accomplish this mission, the system that was developed consisted of an anthropomorphic manipulator arm located in the shuttle cargo bay, cabin equipment to provide an interface to the main shuttle computer, and a human interface to allow an astronaut to control arm operations remotely.

The manipulator arm consisted of three joints, two arm booms, an end effector, a Thermal Protection System, and a closed-circuit television system. Arm joints included a shoulder joint with two degrees of freedom (yaw and pitch), an elbow joint with one degree of freedom (pitch), and a wrist joint with three degrees of freedom (pitch, yaw, and roll). Each joint degree of freedom consisted of a motor module driving a gear box to effect joint movement and appropriate local processing to interpret drive commands originating from the cabin electronics.

The cabin electronics consisted of a displays and controls subsystem that provided the human-machine interface to allow a crew member to command the arm and display appropriate information, including arm position and velocity, end effector status, temperature, and caution and warning information. Additionally, in the displays and controls subsystem, two hand controllers allowed man-in-the-loop control of the end point of the arm. The main robotic arm processor—also part of the cabin electronics—handled all data transfer among the arm, the displays and controls panel, and the main shuttle computer. The main shuttle computer processed commands from the operator via the displays and controls panel; received arm data to determine real-time position, orientation, and velocity; and then generated rate and current limit commands that were sent to the arm-based electronics.

The arm was thermally protected with specially designed blankets to reduce the susceptibility of the hardware to thermal extremes experienced during spaceflight and had an active thermostatically controlled and redundant heater system.

Shuttle Robotic Arm System

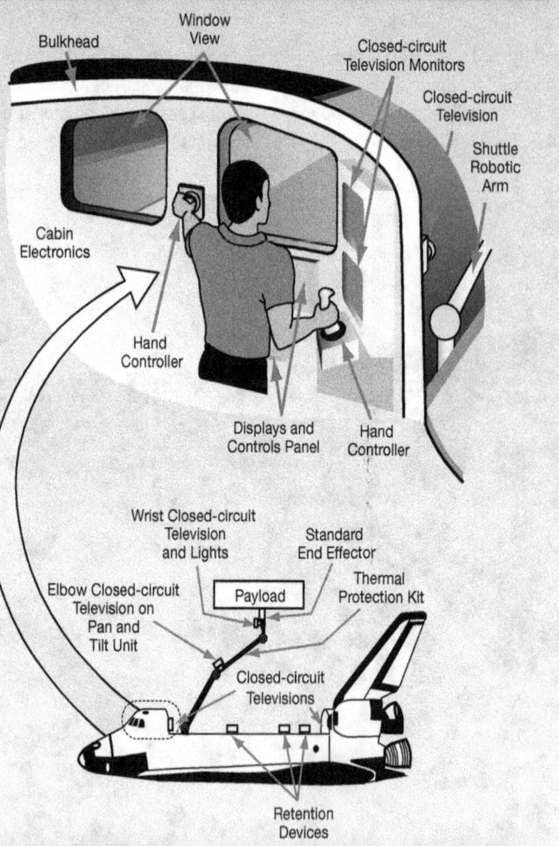

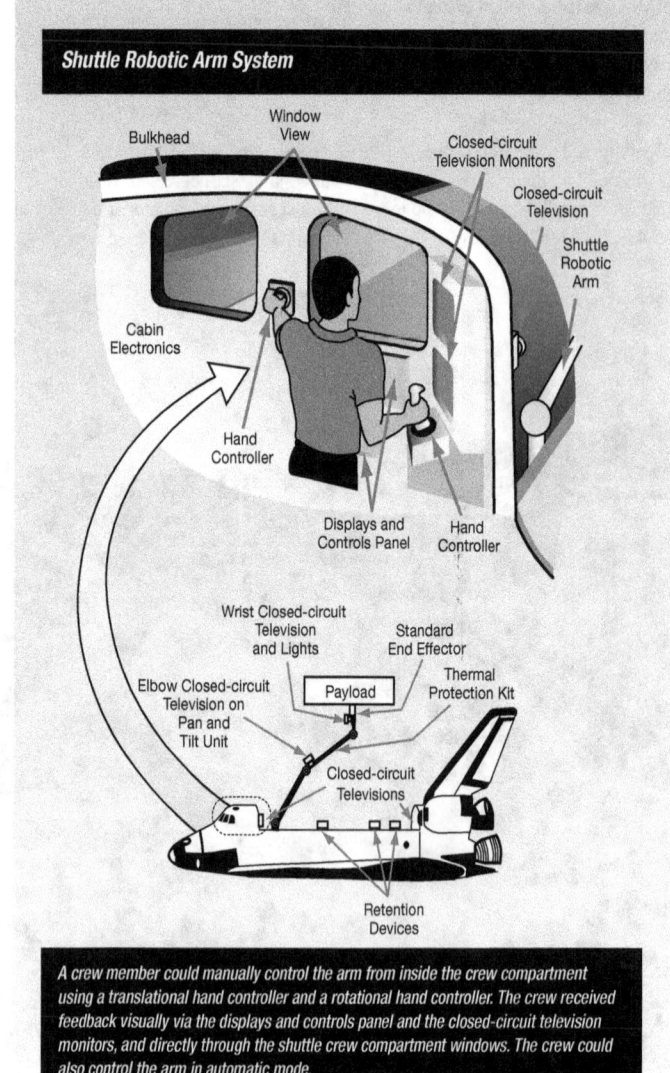

A crew member could manually control the arm from inside the crew compartment using a translational hand controller and a rotational hand controller. The crew received feedback visually via the displays and controls panel and the closed-circuit television monitors, and directly through the shuttle crew compartment windows. The crew could also control the arm in automatic mode.

Components of the Shuttle Robotic Arm

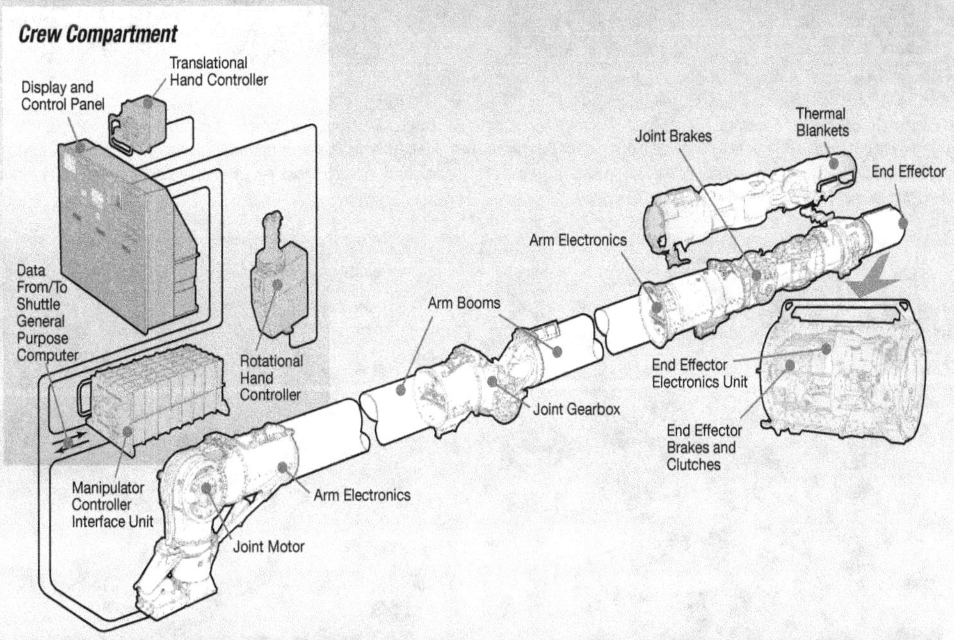

With a total length of 15.24 m (50 ft), the Shuttle Robotic Arm consisted of two lightweight high-strength tubes, each 0.381 m (1.25 ft) in diameter and 6.71 m (22 ft) in length, with an elbow joint between them. From a shoulder joint at the base of the arm providing yaw and pitch movement, the upper boom extended outward to the elbow joint providing pitch movement from which the lower arm boom stretched to a wrist joint providing pitch, yaw, and roll movement. The end effector was used to grapple the payload.

The closed-circuit television system consisted of a color camera on a pan/tilt unit near the elbow joint and a second camera in a fixed location on the wrist joint, which was primarily used to view a grapple fixture target when the arm was capturing a payload.

Self checks existed throughout all the Shuttle Robotic Arm electronics to assess arm performance and apply appropriate commands to stop the arm, should a failure occur. Caution and warning displays provided the operator with insight into the cause of the failure and remaining capability to facilitate the development of a workaround plan.

The interfacing end of the Shuttle Robotic Arm was equipped with a fairly complicated electromechanical construction referred to as the end effector. This device, the analog to a human hand, was used to grab, or grapple, a payload by means of a tailored interface known as a grapple fixture.

The end effector was equipped with a camera and light used to view the grapple fixture target on the payload being captured. The robotic arm provided video to the crew at the aft flight deck, and the camera view helped the crew properly position the end

Close-up View of End Effector and Grapple Fixture

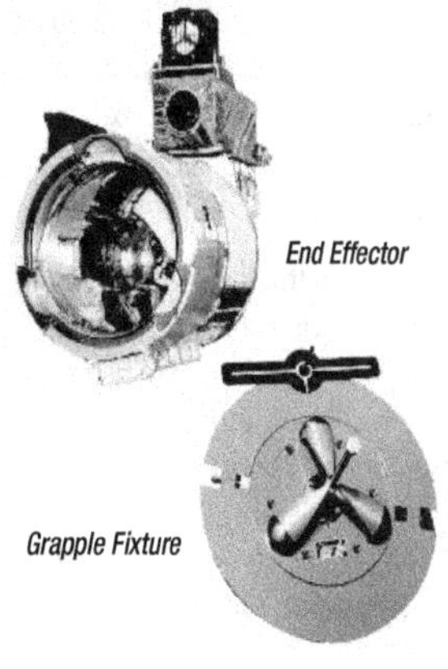

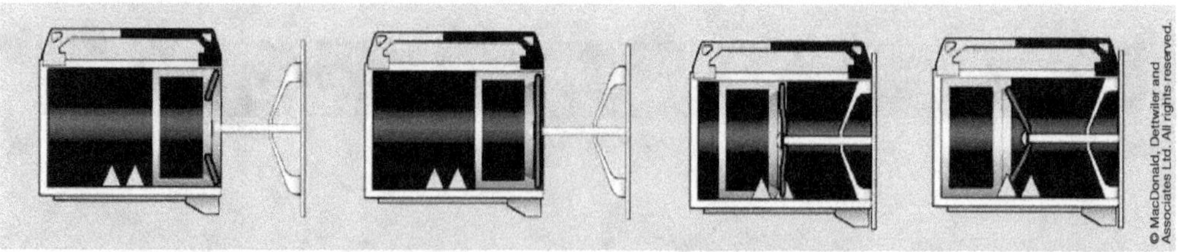

End Effector Capture/Rigidize Sequence: The left frame illustrates the snares in the open configuration, and the second frame shows the snares closed around the grapple shaft and under the grapple cam at the tip of the grapple shaft. The next frame illustrates the snares pulling the grapple shaft inside the end effector so the three lobes are nested into the mating slots in the end effector, and the final frame shows the snare cables being pulled taut to ensure a snug interface that could transfer all of the loads.

Flat floor testing of the Shuttle Robotic Arm.

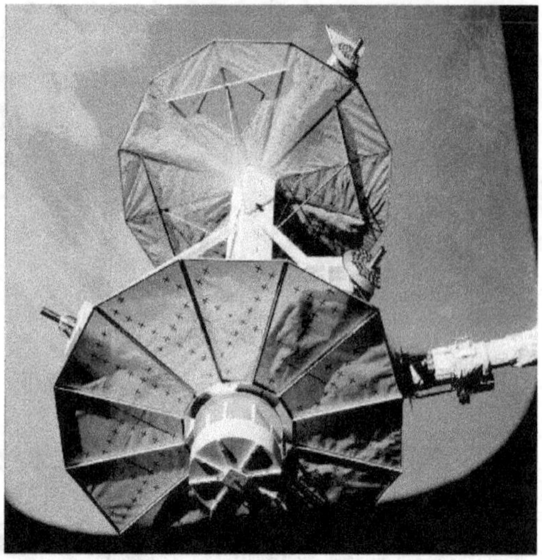

Challenger's (STS-8 [1983]) payload flight test article is lifted from the payload bay and held over clouds and water on Earth.

effector relative to the grapple fixture prior to capturing a payload. When satisfied with the relative position of the end effector to the payload grapple fixture using the grapple fixture target, the crew executed a command to capture and secure the payload.

Since the Shuttle Robotic Arm could not lift its own weight on Earth, all proposed operations had to be tested with simulations. In fact, terrestrial certification was a significant engineering challenge. Developing the complex equations describing the six-degrees-of-freedom arm was one technical challenge, but solving equations combining 0.2268-kg (0.5-pound) motor shafts and 29,478-kg (65,000-pound) payloads also challenged computers at the time. Canada—the provider of the Shuttle Robotic Arm—and the United States both developed simulation models. The simulation responses were tested against each other as well as data from component tests (e.g., motors, gearboxes) and flat floor tests. Final verification could be completed only on orbit. During four early shuttle flights, strain gauges were added to the Shuttle Robotic Arm to measure loads during test operations that started with an unloaded arm and then tested the arm handling progressively heavier payloads up to one emulating the inertia of a 7,256-kg (16,000-pound) payload—the payload flight test article. These data were used to verify the Shuttle Robotic Arm models.

Future on-orbit operations were tested preflight in ground-based simulations both with and without an operator controlling the Shuttle Robotic Arm. Simulations with an operator in the loop used mock-ups of the shuttle cockpit and required calculation of arm response between the time the operator commanded arm motion with hand controllers or computer display entries and the time the arm would respond to commands on orbit. This was a significant challenge to then-current computers and required careful simplification of the arm dynamics equations. During the late 1970s and early 1980s, this necessitated banks of computers to process dynamic equations and specialized computers to generate the scenes. The first electronic scene generator was developed for simulations of shuttle operations, and payload handling simulations drove improvements to this technology until it became attractive to other industries. Simulations that did not require an operator in the loop were performed with higher complexity equations. This allowed computation of loads within the Shuttle Robotic Arm and detailed evaluation of performance of components such as motors.

Since the Shuttle Robotic Arm's job was to deploy and retrieve payloads to and from space, NASA determined two cameras on the elbow and wrist would be invaluable for mission support viewing since the arm could be maneuvered to many places the fixed payload bay cameras could not capture. As missions and additional hardware developed, unique uses of the arm emerged. These included "cherry picking" in space using a mobile foot restraint that allowed a member of the crew to have a movable platform from which tasks could be accomplished; "ice busting" to remove a large icicle that formed on the shuttle's waste nozzle; and "fly swatting" to engage a switch lever on a satellite that had been incorrectly positioned.

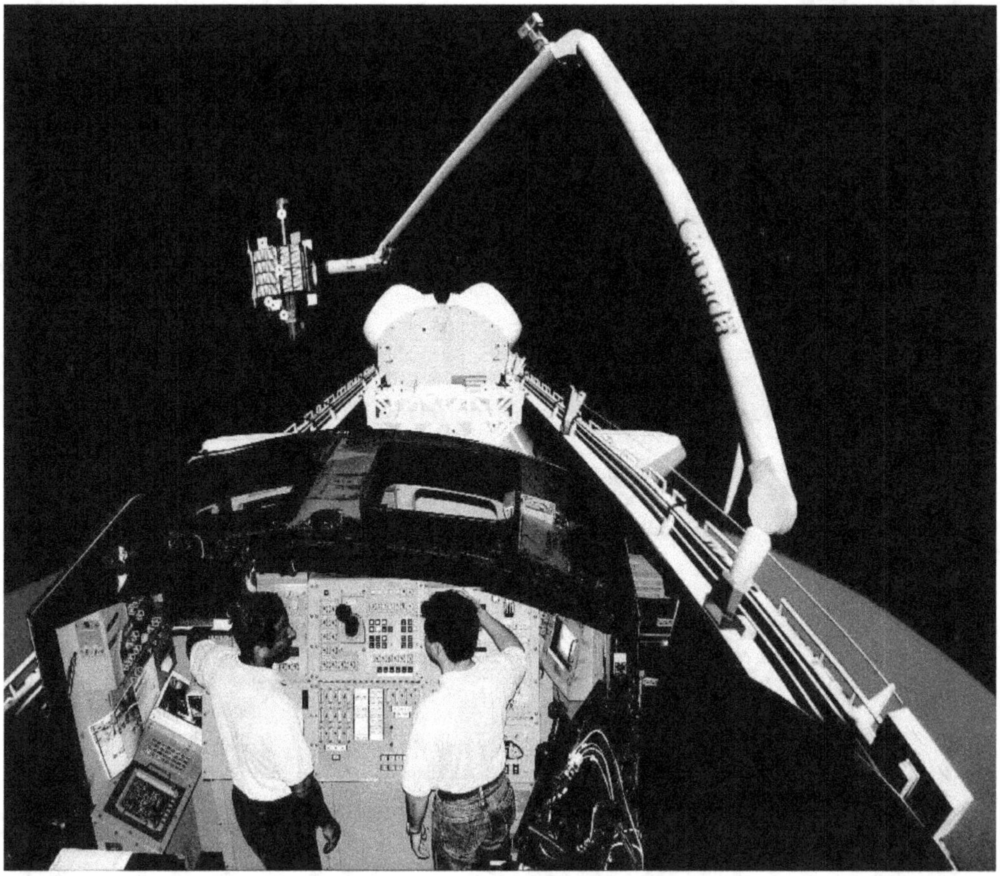

Astronauts Joseph Acaba and Akihiko Hoshide in the functional shuttle aft cockpit in the Systems Engineering Simulator showing views seen out of the windows. The Systems Engineering Simulator is located at NASA Johnson Space Center, Houston, Texas.

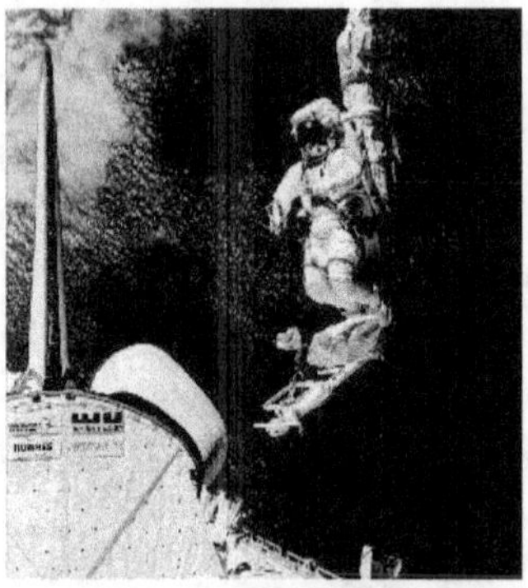

Cherry picking—On STS-41B (1984), Astronaut Bruce McCandless tests a mobile foot restraint attached to the Shuttle Robotic Arm. This device, which allowed a crew member to have a movable platform in space from which tasks could be accomplished, was used by shuttle crews throughout the program.

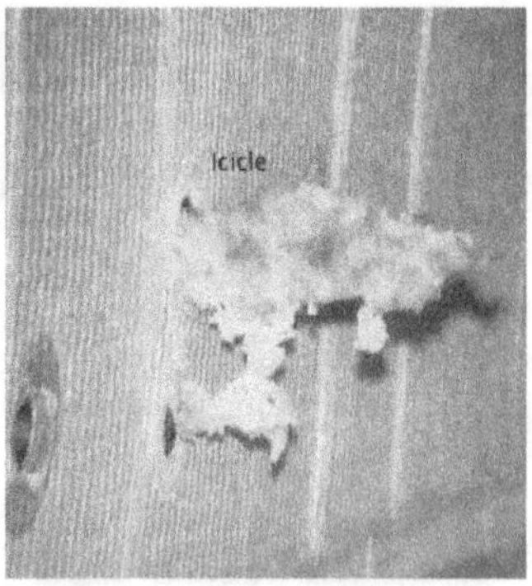

Ice busting—On STS-41D (1984), a large icicle formed on the shuttle's waste nozzle. NASA decided that the icicle needed to be removed prior to re-entry into Earth's atmosphere. The Shuttle Robotic Arm, controlled by Commander Henry Hartsfield, removed the icicle.

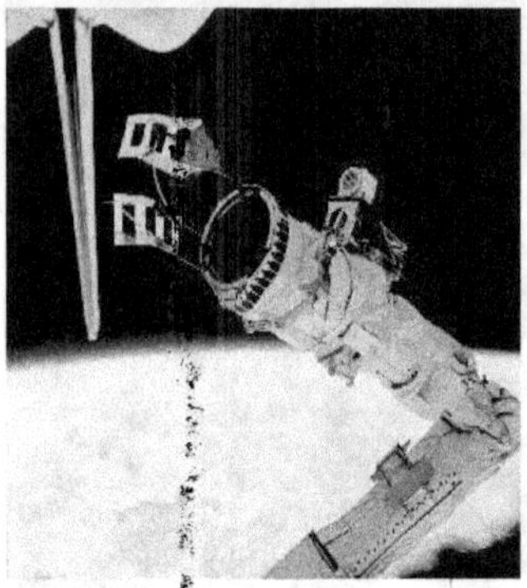

Fly swatting—On STS-51D (1985), the spacecraft sequencer on the Leasat-3 satellite failed to initiate antenna deployment, spin-up, and ignition of the perigee kick motor. The mission was extended 2 days to make the proper adjustments. Astronauts David Griggs and Jeffrey Hoffman performed a spacewalk to attach "fly swatter" devices to the robotic arm. Rhea Seddon engaged the satellite's lever using the arm and the attached "fly swatter" devices.

The Hubble Missions

The Hubble Space Telescope, deployed on Space Transportation System (STS)-31 (1990), gave the world a new perspective on our understanding of the cosmos. An initial problem with the telescope led to the first servicing mission and the desire to keep studying the cosmos. The replacement and enhancement of the instrumentation led to a number of other servicing missions: STS-61 (1993), STS-82 (1997), STS-103 (1999), STS-109 (2002), and STS-125 (2009). From a Shuttle Robotic Arm perspective, the Hubble servicing missions showcased the system's ability to capture, berth, and release a relatively large payload as well as support numerous spacewalks to complete repair and refurbishment activities.

In the case of Hubble, the crew captured and mated the telescope to a berthing mechanism mounted in the payload bay to facilitate the repair and refurbishment activities. In this scenario, a keel target mounted to the bottom of Hubble was viewed with a keel camera and the crew used the Shuttle Robotic Arm to position the Hubble properly relative to its berthing interface to capture and latch it.

The Era of Space Station Construction

With STS-88 (1998)—the attachment of the Russian Zarya module to the space station node—the attention of the shuttle and, therefore, the Shuttle Robotic Arm was directed to the construction of the space station. Early space station flights can be divided broadly into two categories: logistics flights and construction flights. With the advent of the three Italian-built Multi-Purpose Logistic Modules, the Shuttle Robotic Arm was needed to berth the modules to the station. The construction flights meant attaching a new piece of hardware to the existing station. Berthings were used to install new elements: the nodes; the modules, such as the US Laboratory Module and the Space Station Airlock; the truss segments, many of which contained solar panels for power to the station; and the Space Station Robotic Arm. These activities required some modifications to the Shuttle Robotic Arm as well as the addition of systems to enhance alignment and berthing operations.

During preliminary planning, studies evaluated the adequacy of the Shuttle Robotic Arm to handle the anticipated payload operations envisioned for the space station construction. These studies determined that arm controllability would not be satisfactory for the massive payloads the arm would need to manipulate.

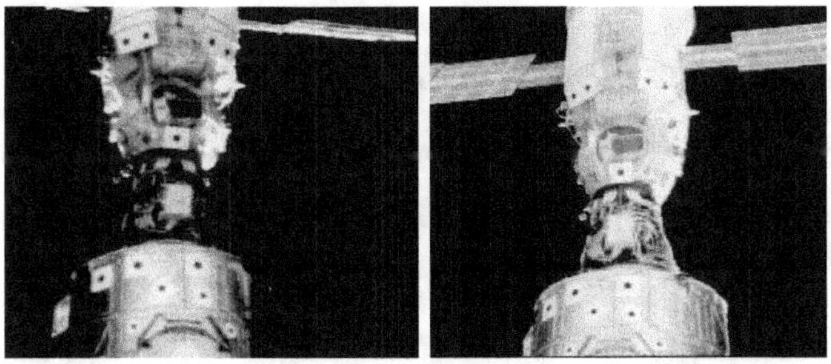

A robotic vision system known as the Space Vision System was used for the first space station assembly flight (STS-88 [1988]) that attached Node 1 to the Russian module Zarya. This Space Vision System used a robotic vision algorithm to interpret relative positions of target arrays on each module to calculate the relative position between the two berthing interfaces. The crew used these data to enhance placement to ensure a proper berthing. The two panes above show the camera views from the shuttle payload bay that the robotic vision system analyzed to provide a relative pose to the crew.

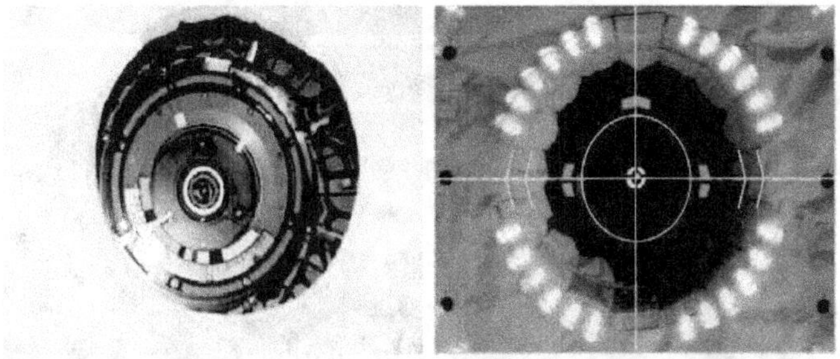

Centerline Berthing Camera System: A Centerline Berthing Camera System was later adopted to facilitate ease of use and to enhance the ability of the crew to determine relative placement between payload elements. The left pane shows the centerline berthing camera mounted in a hatch window with its light-emitting diodes illuminated. The right pane shows the display the crew used to determine relative placement of the payload to the berthing interface. The outer ring of light-emitting diode reflections come from the window pane that the camera was mounted against. However, these reflections never moved and were ignored. The small ring at the center of the crosshairs is the reflection of the Centerline Berthing Camera System light-emitting diodes in the approaching payload window being maneuvered by the Shuttle Robotic Arm system. This was used to determine the angular misalignment (pitch and yaw) of the payload. The red chevrons to the left and right were used to determine vertical misalignment and roll while the top red chevron was used to determine horizontal misalignment. The green chevrons in the overlay were used to determine the range of the payload. This system was first used during STS-98 (2001) to berth the US Laboratory Module (Destiny) to Node 1.

Redesigning the arm-based electronics in each joint provided the necessary controllability. The addition of increased self checks also assured better control of hardware failures that could cause hazardous on-orbit conditions.

During the process of assembling the space station, enhanced berthing cue systems were necessary to mate complicated interfaces that would need to transmit loads and maintain a pressurized interior. The complexity and close tolerance of mating parts led to the development of several berthing cue systems, such as the Space Vision System and the Centerline Berthing Camera System, to enhance the crew's ability to determine relative position between mating modules.

Return to Flight After Columbia Accident

During the launch of STS-107 (2003), a piece of debris hit the shuttle, causing a rupture in the Thermal Protection System that is necessary for re-entry into Earth's atmosphere, thereby leading to the Columbia accident. The ramifications of this breach in the shuttle's Thermal Protection System changed the role of the robotic arm substantially for all post-Columbia-accident missions. Development of the robotically compatible 15.24-m (50-ft) Orbiter Boom Sensor System provided a shuttle inspection and repair capability that addressed the Thermal Protection System inspection requirement for post-Columbia Return to Flight missions. Modification of the robotic arm wiring provided power and data capabilities to support inspection cameras and lasers at the tip of the inspection boom.

Two shuttle repair capabilities were provided in support of the Return to Flight effort. The first repair scenario required the Shuttle Robotic Arm, grappled to the space station, to position the shuttle and the space station in a configuration that would enable a crew member on the Space Station Robotic Arm to perform a repair. This was referred to as the Orbiter repair maneuver. The second repair scenario involved the Shuttle Robotic Arm holding the boom with the astronaut at the tip.

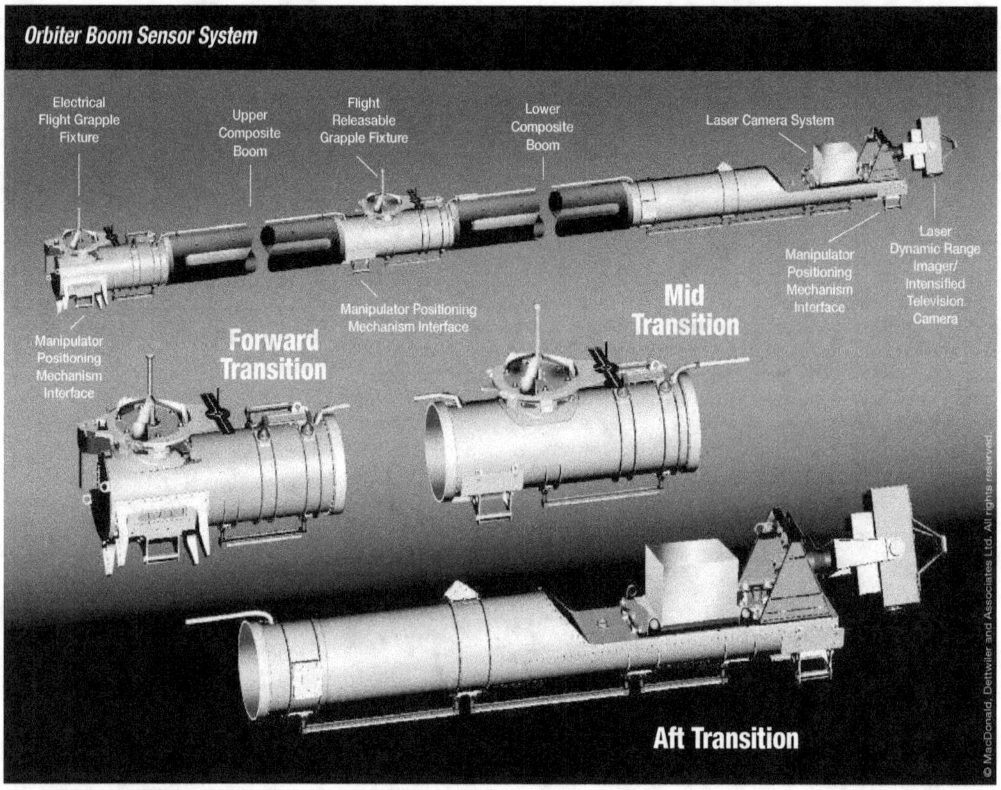

The operational scenario was that, post ascent and pre re-entry into Earth's atmosphere, the robotic arm would reach over to the starboard side and grapple the Orbiter Boom Sensor System at the forward grapple fixture and unberth it. The robotic arm and boom would then be used to pose the inspection sensors at predetermined locations for a complete inspection of all critical Thermal Protection System surfaces. This task was broken up into phases: inspect the starboard side, the nose, the crew cabin, and the port side. When the scan was complete, the robotic arm would berth the Orbiter Boom Sensor System back on the starboard sill of the shuttle and continue with mission objectives.

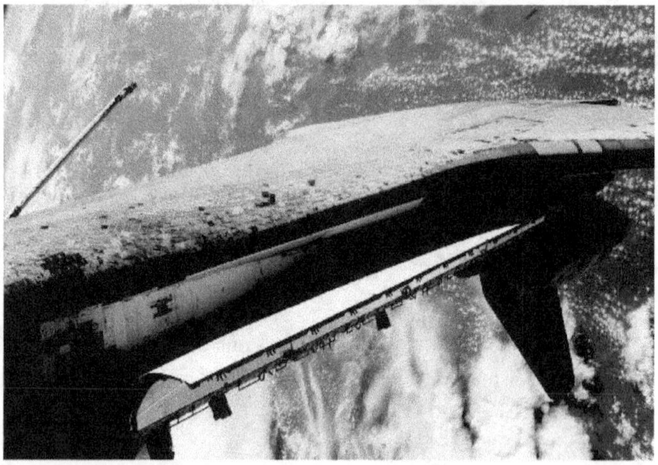

Image from STS-114 (2005) of the Orbiter Boom Sensor System scanning the Orbiter.

All post-Columbia-accident missions employed the Shuttle Robotic Arm and Orbiter Boom Sensor System combination to survey the shuttle for damage. The robotic arm and boom were used to inspect all critical Thermal Protection System surfaces. After the imagery data were processed, focused inspections occasionally followed to obtain additional images of areas deemed questionable from the inspection. A detailed test objective on STS-121 (2006) demonstrated the feasibility of having a crew member

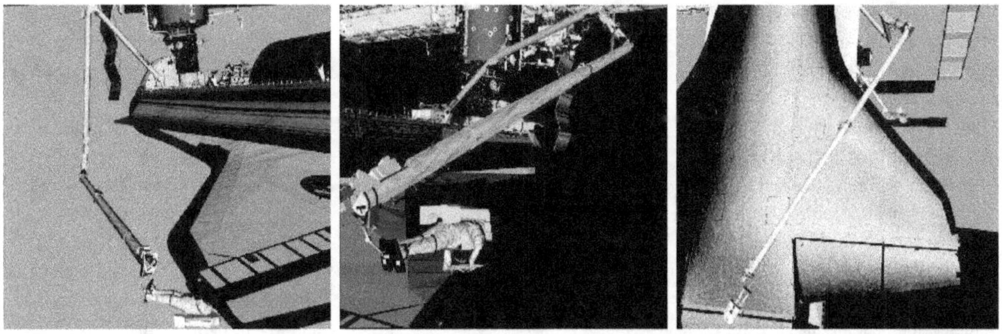

Graphic simulation of Shuttle Robotic Arm/Orbiter Boom Sensor System-based repair scenario for port wing tip, starboard wing, and Orbiter aft locations.

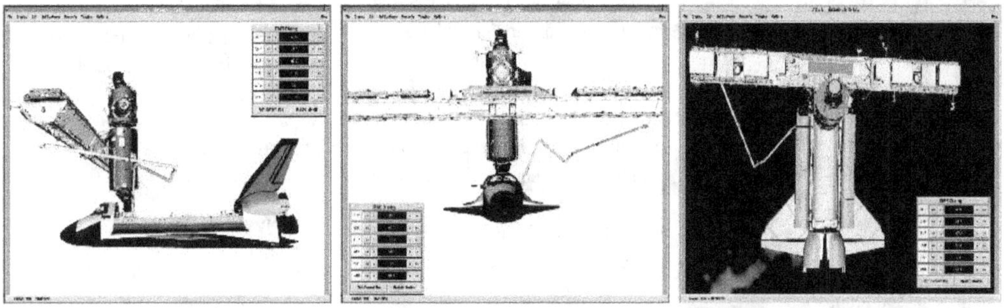

Graphic simulation of the configuration of the Shuttle Robotic Arm/Orbiter Boom Sensor System for STS-121 (2006) flight test.

on the end of the combined system performing actions similar to those necessary for Thermal Protection System repair. Test results showed that the integrated system could be used as a repair platform and the system was controllable with the correct control parameters, good crew training, and proper extravehicular activity procedures development.

In support of shuttle repair capability and rescue of the crew, simulation tools were updated to facilitate the handling of both the space station and another shuttle as "payloads." The space station as a payload was discussed earlier as a Return to Flight capability, known as the Orbiter repair maneuver. The shuttle as a payload came about due to the potential for a

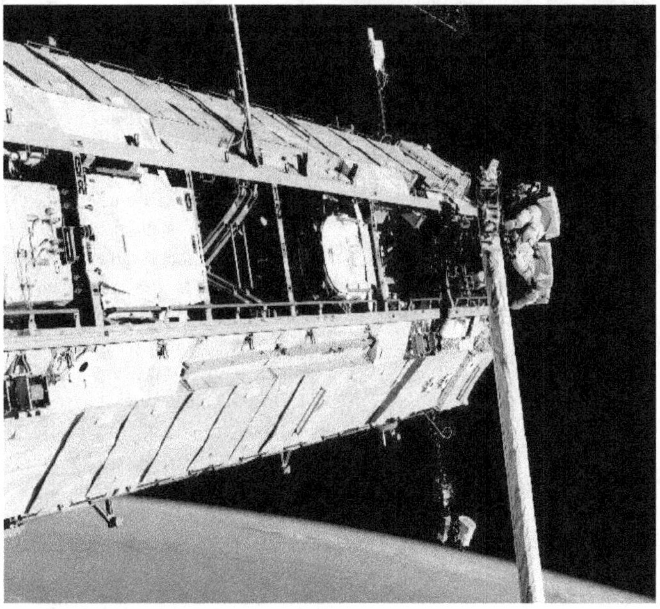

In addition to performing inspections, the Orbiter Boom Sensor System's role was expanded to include the ability to hold a crew in position for a repair to the Thermal Protection System. Considering that this was a 30.48-m (100-ft) robotic system, there was concern over the dynamic behavior of this integrated system. The agency decided to perform a test to evaluate the stability and strength of the system during STS-121 (2006).

Hubble rescue mission. Given that the space station would not be available for crew rescue for the final Hubble servicing mission, another shuttle would be "ready to go" on another launch pad in the event the first shuttle became disabled. For the crew from the disabled shuttle to get to the rescue shuttle, the Shuttle Robotic Arm would act as an emergency pole between the two vehicles, thus making the payload for the Shuttle Robotic Arm another shuttle. Neither of these repair/rescue capabilities—Orbiter repair maneuver or Hubble rescue—ever had to be used.

Summary

The evolution of the Shuttle Robotic Arm represents one of the great legacies of the shuttle, and it provided the impetus and foundation for the Space Station Robotic Arm. From the early days of payload deployment and retrieval, to the development of berthing aids and techniques, to the ability to inspect the shuttle for damage and perform any necessary repairs, the journey has been remarkable and will serve as a blueprint for space robotics in the future.

Automation: The Space Shuttle Launch Processing System

The Launch Processing System supported the Space Shuttle Program for over 30 years evolving and adapting to changing requirements and technology and overcoming obsolescence challenges.

Designed and developed in the early 1970s, the Launch Processing System began operations in September 1977 with a focused emphasis on safety, operational resiliency, modularity, and flexibility. Over the years, the system expanded to include several firing rooms and smaller, specialized satellite sets to meet the processing needs of multiple Space Shuttles—from landing to launch.

Architecture and Innovations

The architecture of the system and innovations included in the original design were major reasons for the Launch Processing System's outstanding success. The system design required that numerous computers had the capability to share real-time measurement and status data with each other about the shuttle, ground support equipment, and the health and status of the Launch Processing System itself. There were no commercially available products to support the large-scale distributed computer network required for the system. The solution to this problem was to network the Space Shuttle firing room computers using a centralized hub of memory called a common data buffer—designed by NASA at Kennedy Space Center (KSC) specifically for computer-to-computer communication. The buffer was a high-speed memory device that provided shared memory used by all command and control computers supporting a test. Each computer using the buffer was assigned a unique area of memory where only that computer could write data; however, every computer on the buffer could read those data. The buffer could support as many as 64 computers simultaneously and was designed with multiple layers of internal redundancy, including error-correcting software. The common data buffer's capability to provide fast and reliable intercomputer communication made it the foundation of the command and control capability of the firing room.

The System Console

Other outstanding features of the Launch Processing System resided in the human-to-machine interface known as the console. System engineers used the console to control and monitor the particular system for which they were responsible. Each firing room contained 18 consoles—each connected to the common data buffer, and each supporting three separate command and control workstations. One of the key features of the console was its ability to execute up to six application software programs, simultaneously. Each console had six "concurrencies"—or areas in console memory—that could independently support an application program. This capability foreshadowed the personal computer with its ability to multitask using different windows. With six concurrencies available to execute as many as six application programs, the console operator could monitor

Launch Processing System

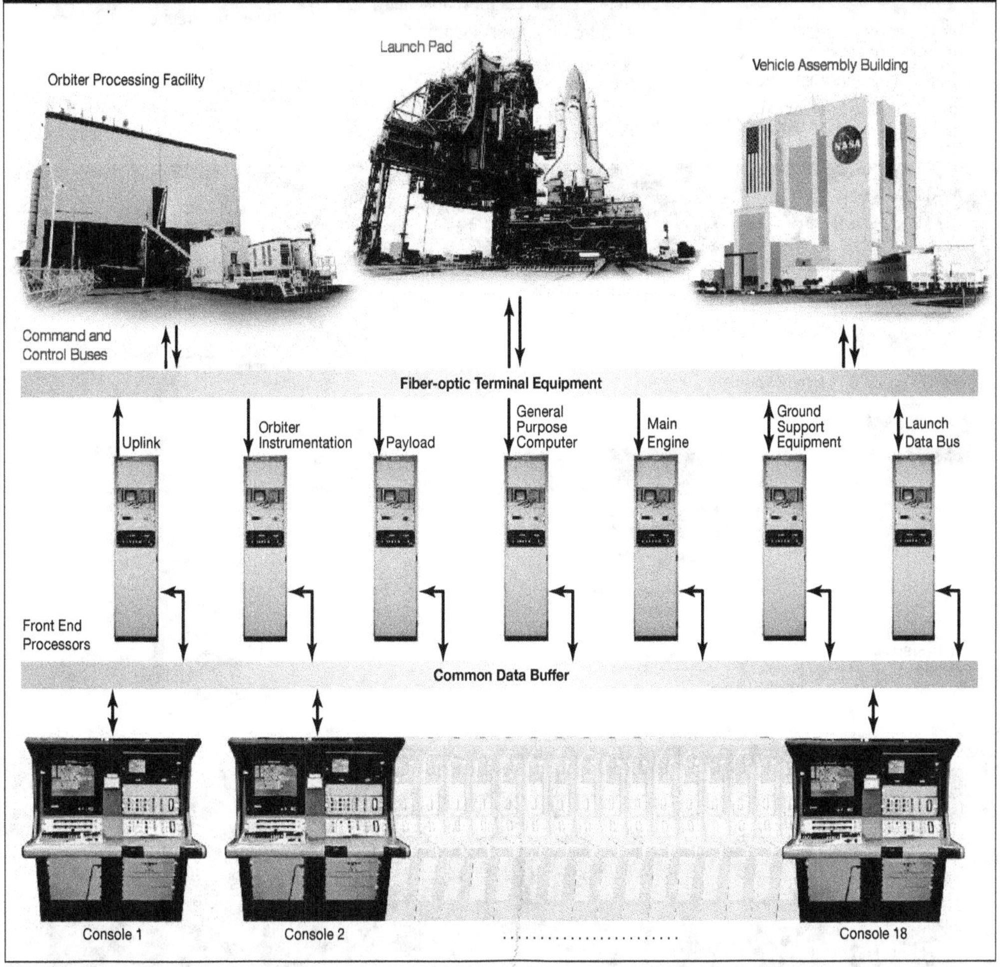

The Launch Processing System provides command and control of the flight vehicle elements and ground support equipment during operations at Kennedy Space Center.

thousands of pieces of information within his or her area of responsibility from a single location. Each console in the firing room was functionally identical, and each was capable of executing any set of application software programs. This meant any console could be assigned to support any system, defined simply by what software was loaded. This flexibility allowed for several on-demand spare consoles for critical or hazardous tests such as launch countdown. The console also featured full color displays, programmable function keys, a programmable function panel, full cursor control, and a print screen capability. Upgrades included a mouse, which was added to the console, and modernized cursor control and selection.

System Integrity

Fault tolerance, or the ability to both automatically and manually recover from a hardware or software failure, was designed and built into the Launch Processing System. An equivalent analogy for distributed computer systems would be the clustering of servers for redundancy. Most critical computers within the system were operated in an active/standby configuration. A very high degree of system reliability was achieved through automated redundancy of critical components.

A software program called System Integrity, which constantly monitored the health and status of all computers using the common data buffer,

Integrated Network Control System

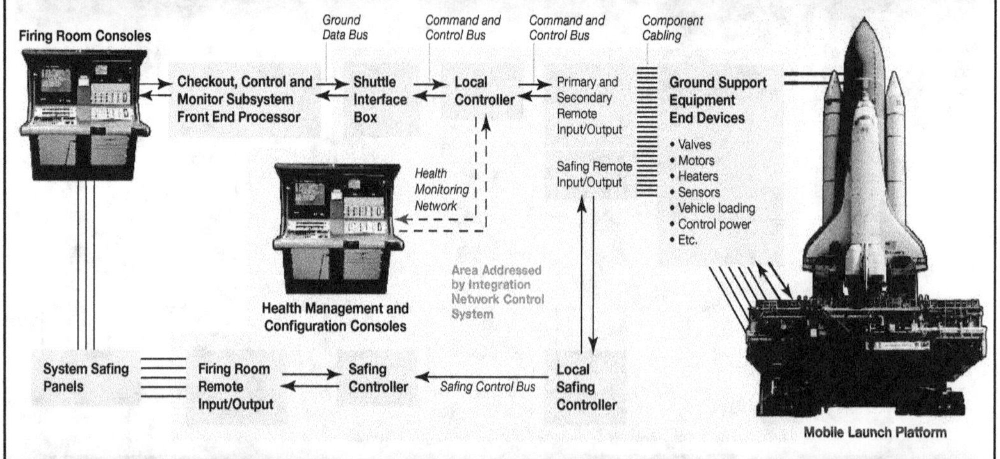

The Integrated Network Control System was a reliable, automated network system that sent data and commands between the shuttle Launch Control Center and hardware end items. It bridged industry automation technologies with customized aerospace industry communication protocols and associated legacy end item equipment. The design met several challenges, including connectivity with 40,000 end items located within 28 separate ground systems, all dispersed to 10 facilities. It provided data reliability, integrity, and emergency safing systems to ensure safe, successful launch operations.

Ground control and instrumentation systems for the Space Shuttle Launch Processing System used custom digital-to-analog hardware and software connected to an analog wire-based distribution system. Loss of a data path during critical operations would compromise safety. To improve safety, data integrity, and network connectivity, the Integrated Network Control System design used three independent networks.

The network topology used a quad-redundant, fiber-optic, fault-tolerant ring for long-distance distribution over the Launch Control Center, mobile launcher platforms, Orbiter processing facilities, and two launch pads. Shorter distances were accommodated with redundant media over coaxial cable for distribution over system and subsystem levels. This network reduced cable and wiring for ground processing over the Launch Complex 39 area by approximately 80% and cable interconnects by 75%. It also reduced maintenance and troubleshooting. This system was the first large-scale network control and health management system for the Space Shuttle Program and one of the largest, fully integrated control networks in the world.

governed the automatic recovery of failed critical computers in the firing room. In the event of a critical computer failure, System Integrity commanded a redundant switch, thereby shutting down the unhealthy computer and commanding the standby computer to take its place. Launch Processing System operators could then bring another standby computer on line from a pool of ready spares to reestablish the active/standby configuration.

Most critical portions of the Launch Processing System had redundancy and/or on-demand spare capabilities. Critical data communication buses between the Launch Control Center and the different areas where the shuttles were processed used both primary and backup buses. Critical ground support equipment measurements were provided with a level of redundancy, with a backup measurement residing on a fully independent circuit and processed by different firing room computers than the primary measurement. Electrical power to the firing room was supplied by dual uninterruptible power sources, enabling all critical systems to take advantage of two sources of uninterruptible power.

Critical software programs, such as those executed during launch countdown, were often part of the software load of two different consoles in the event of a console failure. The System Integrity program was executed simultaneously on two different firing room consoles. The fault tolerance designed into the Launch Processing System spanned from the individual measurement up through subsystem hardware and software, providing the Space Shuttle test team with outstanding operational resiliency in almost any failure scenario.

Orbiter Window Inspection

Bradley Burns, lead engineer in the development of the window inspection tool, monitors its progress as it scans an Orbiter window.

As the Orbiter moved through low-Earth orbit, micrometeors collided with it and produced hypervelocity impact craters that could produce weak points in its windows and cause the windows to fail during extreme conditions. Consequently, locating and evaluating these craters, as well as other damage, was critically important. Significant effort went into the development and use of ground window inspection techniques.

The window inspection tool could be directly attached to any of the six forward windows on any Orbiter. The tool consisted of a dual-camera system—a folded microscope and a direct stress imaging camera that was scanned over the entire area of the window. The stress imaging camera "saw" stress by launching polarized light at the window from an angle such that it bounced off the back of the window, then through the area being monitored, and finally into the camera where the polarization state was measured. Defects caused stress in the window. The stress changed the polarization of the light passing through it. The camera provided direct imaging of stress regions and, when coupled with the microscope, ensured the detection of significant defects.

The Portable Handheld Optical Window Inspection Device is vacuum attached to a window such that the small camera and optical sensor (black tube) were aimed at a defect.

The portable defect inspection device used an optical sensor. A three-dimensional topographic map of the defect could be obtained through scanning. Once a defect was found, the launch commit criteria was based on measuring the depth of that defect. If a window had a single defect deeper than a critical value, the window had to be replaced.

Robotics System Sprayed Thermal Protection on Solid Rocket Booster

Many Solid Rocket Booster (SRB) components were covered with a spray-on thermal protection material that shielded components from aerodynamic heating during ascent. The application process took place at the SRB Assembly and Refurbishment Facility at Kennedy Space Center. The process resulted in overspray and accounted for 27% of hazardous air emissions.

An SRB aft skirt receives a robotically controlled layer of Marshall Convergent Coating-1 Thermal Protection System material.

To address this drawback, NASA developed Marshall Convergent Coating-I, which consisted of improved mixing and robotic spray processes. The coating's ingredients were mixed (or *converged*) only during spraying. Hazardous waste was virtually eliminated after implementation of the system in the mid 1990s.

After each flight, the boosters were refurbished. This process began at NASA's Hangar AF Booster Recovery Facility at Cape Canaveral Air Force Station. There, a robotic high-pressure water jet, or "hydrolase," stripped the components of their Thermal Protection System materials.

NASA installed the hydrolase system in 1998. Each booster structure was numerically modeled. These models were used to program the robot to follow the contour of each component.

The Hangar AF wash facilities used a specially designed water filtration and circulation system to recycle and reuse the waste water.

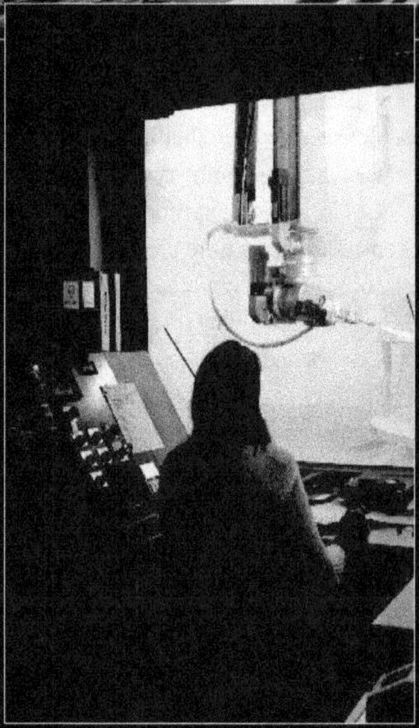

A technician in a control booth monitors the robotic high-pressure hydrolase as it strips Thermal Protection System material from an SRB forward skirt.

Exception Monitoring

Another key concept designed into the Launch Processing System software was the capability to recognize and automatically react to out-of-bounds measurements. This capability was called exception monitoring, and it monitored for specific measurements exceeding a predefined set of limits. When a Launch Processing System computer detected a measurement exception—for example, the pressure in a fuel tank exceeded its upper limit—the computer immediately notified the console responsible for that fuel tank. A software program at the console promptly reacted to the exception and automatically sent a command or series of commands to resolve the problem. Similar software could also prevent inadvertent damage by verifying required parameters prior to command issuance, such as confirming that pressures were appropriate prior to commanding a valve opening. Commands could also be manually sent by the console operator.

Survivability

Although the Launch Processing System's flexible architecture and distribution of hardware functionality allowed it to support the program consistently over 30 years, that support would not have been possible without a comprehensive and proactive sustaining engineering, maintenance, and upgrade approach. This is true for any large-scale computer system where an extended operational lifetime is desired.

The approach that kept the Launch Processing System operationally viable for over 3 decades was called the Survivability Program. Survivability was initiated to mitigate risk associated with the natural obsolescence of commercial off-the-shelf hardware products and the physical wear and tear on the electrical and mechanical subsystems within the Launch Processing System.

One of the main tenets of survivability was the desire to perform each upgrade with an absolutely minimal impact to system software. Hardware was upgraded to duplicate the existing hardware in form, fit, and function. The emphasis on minimizing software impacts was a distinct strength in survivability due to the resultant reduction of risk. Survivability projects were selected through careful analysis of maintenance failure data and constant surveillance of electronic manufacturers and suppliers by logistics to identify integrated circuits and other key components that were going to be unavailable in the near future. Through this process, NASA purchased a "lifetime" buy of some electronic components and integrated circuits to ensure the Launch Processing System had ample spares for repair until the end of the program. It could also redesign a circuit board using available parts or replace an entire subsystem if a commercial off-the-shelf or in-house design solution offered the most benefit.

NASA eventually upgraded or replaced about 70% of the original Launch Processing System hardware under the survivability effort. The proactive application of the Survivability Program mitigated obsolescence and continued successful operational support.

Summary

These innovations and the distributed architecture of the Launch Processing System allowed upgrades to be performed over the years to ensure the system would survive through the life of the program. This success demonstrated that, with appropriate attention paid to architecture and system design and with proactive sustaining engineering and maintenance efforts, a large, modular, integrated system of computers could withstand the inevitable requirements change and obsolescence issues. It also demonstrated that it could successfully serve a program much longer than originally envisioned.

The Launch Processing System was vital to the success of KSC fulfilling its primary mission of flying out the Space Shuttle Program in a safe and reliable manner, thus contributing to the shuttle's overall legacy.

Systems Engineering for Life Cycle of Complex Systems

Introduction
Gail Chapline
Steven Sullivan

Systems Engineering During Development of the Shuttle
Gail Chapline

Intercommunication Comes of Age—The Digital Age
John Hirko

Restoring Integration and Systems Thinking in a Complex Midlife Program
John Muratore

Electromagnetic Compatibility for the Space Shuttle
Robert Scully

Process Control
Glen Curtis
Steven Sullivan
David Wood

Alliant Techsystems, Inc. and United Space Alliance
Holly Lamb
Dennis Moore
David Wood

Michoud Assembly Facility
Jeffery Pilet
Kenneth Welzyn
Patrick Whipps

Pratt & Whitney Rocketdyne Manufacturing
Eric Gardze

Rockwell International and The Boeing Company
Bob Kahl
Larry Kauffman

NASA and the Environment—Compatibility, Safety, and Efficiency
Samantha Manning

Protecting Birds and the Shuttle
Stephen Payne

All complex systems require systems engineering that integrates across the subsystems to meet mission requirements. This interdisciplinary field of engineering traditionally focuses on the development and organization of complex systems. However, NASA applied systems engineering throughout the life cycle of the Space Shuttle Program—from concept development, to production, to operation and retirement. It may be surprising to many that systems engineering is not only the technical integration of complex space systems; it also includes ground support and environmental considerations. Engineers require the aid of many tools to collect information, store data, and interpret interactions between shuttle systems. One of the shuttle's legacies was the success of its systems engineering. Not only did the shuttle do what it was supposed to do, it went well beyond meeting basic requirements.

This section is about systems engineering innovations, testing, approaches, and tools that NASA implemented for the shuttle. Companies that developed, built, and maintained major shuttle components are highlighted. As manufacturers, contractors, NASA, and industry employees and management came and went, the shuttle stayed the same during its lifetime, primarily because of its well-honed process controls. All of these systems engineering advances are a legacy for the International Space Station and for future space vehicles.

Systems Engineering During Development of the Shuttle

Systems engineering is a complex, multilevel process that involves deconstructing a customers' overall needs into functions that the system must satisfy. But even in ordinary situations, that's just the beginning. Functional requirements are then allocated to specific components in the system. Allocated functions are translated into performance requirements and combined with design constraints to form requirements that a design team must satisfy. Requirements are then synthesized by a team of engineers into one or more concepts, which are traded off against each other. These design concepts are expanded into preliminary and detailed designs interspersed with reviews. Specialists from many disciplines work as a team to obtain a solution that meets the needs and requirements. Selected designs are translated into manufacturing, planning, procurement, operations, and program completion documents and artifacts.

Systems engineering for the Space Shuttle presented an extraordinary situation. The shuttle was the most complex space vehicle for its time and, therefore, required the evolution of systems engineering with significantly advanced new tools and modeling techniques. Not only was the vehicle sophisticated, it required the expertise of many people. Four prime contractors and thousands of subcontractors and suppliers, spread across the United States, designed and built the major elements of the shuttle. The complexity of the element interfaces meant the integration of elements would present a major systems engineering challenge. One prime contractor was in charge of building the main engines, which were mounted inside the Orbiter. A different prime contractor built the Orbiter. A third prime contractor built the External Tanks, which contained the fuel for the main engines. And, a fourth prime contractor built the Solid Rocket Boosters. As problems occurred, they involved multiple NASA engineering organizations, industry partners, subject matter experts, universities, and other government agencies. NASA's ability to bring together a wide group of technical experts to focus on problems was extremely important. Thus, one legacy of the Space Shuttle was the success of its systems engineering. Not only did the shuttle do what it was supposed to do, it went well beyond meeting basic requirements.

A discussion of all the systems engineering models and new tools developed during the lifetime of the Space Shuttle Program would require volumes. All elements of the Space Shuttle Program had successes and failures. A few of the most notable successes and failures in systems engineering are discussed here.

Change and Uncertainty

Space Shuttle Main Engines

NASA recognized that advancements were needed in rocket engine technology to meet the design performance requirements of the shuttle. Thus, its main engine was the first contract awarded.

A high chamber pressure combined with the amplification effect of the staged combustion cycle made this engine a quantum leap in rocket engine technology for its time. The engine also had to meet the multiple interface requirements to the vehicle, extensive operation requirements, and several design criteria. A major challenge for systems engineering was

Intercommunication Comes of Age—The Digital Age

As the shuttle progressed, it became evident that the existing communication system could not meet the multi-flow and parallel processing requirements of the shuttle. A new system based on digital technology was proposed and Operational Intercommunication System-Digital was born, and is now in its third generation. This system provided unlimited conferencing on 512 communication channels and support for thousands of end users. The system used commercially available off-the-shelf components and custom-designed circuit boards.

Digital communication systems included, among other things, the voice communication system at Kennedy Space Center (KSC). The voice communication system needed to perform flawlessly 24/7, 365 days a year. This need was met by Operational Intercommunication System-Digital—a one-of-a-kind communication system conceived, designed, built, and operated by NASA engineers and a team of support contractors. The system was installed in every major processing facility, office building, and various labs around KSC. This widespread distribution allowed personnel working on specific tasks to communicate with one another, even in separate facilities.

Dominic Antonelli
Commander, US Navy.
Pilot on STS-119 (2009) and STS-132 (2010).

"At the end of the day, people comprise the system that ultimately propelled the Space Shuttle Program to its stellar place in history. The future of space travel will forever be indebted to the dedication, hard work, and ingenuity of the men and women, in centers across the country, who transformed the dream into a tangible reality and established a foundation that will inspire generations to come."

that all of these requirements and design criteria were interrelated.

In most complex systems, verification testing is performed at various stages of the buildup and design. NASA followed this practice on previous vehicles. In component-level tests, engineers find problems and solve them before moving to the next higher assembly level of testing. The main engine components, however, were very large. Test facilities that could facilitate and perform the component and higher assembly level tests did not exist. The valves alone required a relatively large specialized test facility. Plans to build such facilities had been developed, but there was not enough time to complete their construction and maintain the schedule. Therefore, the completed main engine became the test bed.

A concurrent engineering development philosophy associated with the shuttle forced the engine to be its own test bed. The engine test stands at Stennis Space Center in Mississippi were already in place, so NASA decided to assemble the engines and use them as the breadboard or facility to test the components. This was a risky scenario. The engine proved to be unforgiving. NASA lost 13 engines from catastrophic failures on the test stand before first flight. Each of these failures was a rich learning experience that significantly enabled the engineers to improve the engine's design. Still, at times it seemed the technical challenges were insurmountable.

Another philosophy that prevailed in the development of the main engines was "test, test, and test some more." Testing was key to the success of this shuttle component. Technicians conducted tests with cracked blades, rough bearings, and seals with built-in flaws to understand the limitations. By late 1979, as noted in a paper written by Robert Thompson, Space Shuttle manager at the time: "We have conducted 473 single engine tests and seven multiple engine tests with a cumulative total running time of 98 times mission duration and with 54 times mission duration at the engine rated power level. Significant engine test activities still remain and must be completed successfully before the first flight, but the maturity of this vital system is steadily improving."

The test, test, and test some more philosophy reduced risk, built robustness, and added system redundancy. Testing also allowed engineers to understand interactions of failures with other systems during the 30 years of the program. In all, the main engines were upgraded three times. These upgrades improved the engines' performance and reliability, reduced turnaround costs, and were well-planned system engineering efforts.

Throughout the life of the Space Shuttle Program—and through many technical challenges and requirement changes—the main engine not only performed, but was also a technological leap for spacecraft rocket engines.

Where Was Systems Engineering When the Shuttle Needed It Most?

Thermal Protection System

Early development problems with the Orbiter's Thermal Protection System probably could have been avoided had a systems engineering approach been implemented earlier and more effectively.

The Thermal Protection System of the Orbiter was supposed to provide for the thermal protection of the structure while maintaining structural integrity. The engineers did a magnificent job in designing tiles that accepted, stored, and dissipated the heat. They also created a system that maintained the aerodynamic configuration. However, early in the process, these engineers neglected to design a system that could accept the loads and retain the strength of the tiles. Furthermore, it was not until late in the Thermal Protection System development process that NASA discovered a major problem with the attachment of tiles to the Orbiter's aluminum skin surfaces.

In 1979, when Columbia—the first flight Orbiter—was being ferried from Dryden Flight Research Center in California to Kennedy Space Center in Florida on the back of the 747 Shuttle Carrier Aircraft, several tiles fell off. This incident focused NASA's attention on the tile attachment problem. The solution ultimately delayed the maiden flight of Columbia (Space Transportation System [STS]-1) by nearly 1½ years. The problem resided in the bond strength of the tiles, which was even lower than the overall low strength of the tile material. Tile load analyses kept showing increasing loads and lower margins on tile strength. This low bond strength was related to stress concentrations at the bondline interface between the tile and the strain isolation pad. Attachment of the tiles to the Orbiter's aluminum skin required that the strains from structural deflections be isolated from the tiles. In other words, the tiles could not be bonded directly to the Orbiter structure.

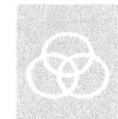

Strain isolation was accomplished with Nomex® felt pads bonded to the structure. In turn, the tiles bonded to the pads. Needling of the Nomex® pads through the thickness to control thickness resulted in straight through fibers ("stiff spots") that induced point loads in the bottom of the tiles. These point loads caused early localized failure of the tile material at the bondline. This did not meet design requirements.

After more than 1 year of intense, around-the-clock proof testing, bonding, removing, and re-bonding of tiles on the vehicle at Kennedy Space Center, tile densification proved to be the solution. Stress concentrations from the strain isolation pad were smoothed out and the full tile strength was regained by infusing the bottom of the tiles, prior to bonding, with a silica-based solution that filled the pores between tile fibers for a short distance into the bottom of the tile. This example demonstrates that a systems approach to the tile design, taking into consideration not only the thermal performance of the tile but also the structural integrity, would have allowed the tile attachment problem to be solved earlier in the design process.

The Importance of Organizational Structure

The structure of the Space Shuttle Program Systems Integration Office was a key element in the successful execution of systems engineering. It brought together all shuttle interfaces and technical issues. Design and performance issues were brought forward there. The office, which integrated all technical disciplines, also had a technical panel structure that worked the technical details from day to day.

The panels were composed of engineers from multiple NASA centers, prime contractors, and subcontractors. NASA also brought in technical experts when needed.

These panels varied in size. The frequency of discussions depended on the technical areas of responsibility and the difficulty of the problems encountered. The panels operated in an environment of healthy tension, allowing for needed technical interchange, questioning, and probing of technical issues. The technical panel structure has been recognized as a significant and an effective means to manage complex systems.

Initially, there were 44 formalized panels, subpanels, and working groups in the Space Shuttle Program Office.

Space Shuttle Systems Integration Program Structure

It takes a lot of people to integrate.

Systems Integration

Representatives

Systems Engineering

Prime	Support
Flight Performance Integration	Ancillary Hardware Requirements
Loads and Structural Dynamics	Commonality
Guidance, Navigation, and Control Integration	Quality Assurance
Integrated Avionics	Change Assessment
Integrated Prop. and Fluids	
Mechanical Systems	
Ascent Flight System Integration	
Thermal Design Integration	

Technical Integration

Prime	Support
Performance and Design Spec	Configuration Management
Flight Test Requirements	Change Integration
System Interfaces	Operational Requirements
Mass Properties	System Reviews
Systems/Ops Data Books	Major Ground Test Integration
Integrated Schematics	Network Interfaces
Materials and Processes	Element Reviews
Computer Systems Integration	Rockwell-Space Division Work Breakdown Structure
Integrated Systems Verification	

Test and Ground Operations

Prime	Support
Ground Systems Integration	Reliability
Maintainability	System Interfaces
Integrated Logistics	Safety
Integrated Test	Flight Test Requirements
Ground Support Equipment Requirements and Analysis	Systems Analysis and Design
Payload Integration for Design, Development, Test, and Evaluation	System Requirements

- Simulation Planning Panel
- Crew Safety Panel
- Configuration Management Panel
- Ground Interface Control Board
- Crew Procedures Control Board
- Information Management Systems Panel
- Payloads Interface Panel
- Program Information Coordination and Review Service Working Group
- Systems Integration Reviews
- Management Information Center Integration Panel
- Performance Management Panel
- Integrated Entry Performance Panel
- Crew Related Government Furnished Equipment Configuration Control Board
- Flight Test Program Panel
- Electromagnetic Effects Panel
- Ascent Performance Panel
- Abort Performance Panel
- Separation Performance Panel
- Aerothermodynamics Performance Panel
- Aerodynamic Performance Panel
- Main Propulsion Systems Panel
- Pogo Integration Panel
- Loads and Structural Dynamics Panel
- Ground Vibration Test Panel
- Spacecraft Mechanisms Panel
- Shuttle Vehicle Attachment and Separation Subpanel
- Payloads Docking, Retention, and Deployment Subpanel
- Landing Systems and Facilities Subpanel
- Shuttle Training Aircraft Review Board
- Communications and Data Systems Integration Panel
- Functional Requirements Subpanel
- Vehicle Communications Interface Subpanel
- Ground-Based Data Systems Subpanel
- Science and Engineering Data Processing Subpanel
- Flight Operations Panel
- Operations Integration Reviews
- Computer Systems Hardware/Software Integration Reviews
- Training Simulator Control Panel
- Ascent Flight Control/Structural Integration Panel
- On-Orbit Guidance and Control Panel
- Entry Guidance and Control Panel
- Approach and Landing Test Guidance and Control Panel
- Guidance and Navigation Systems Panel
- Safety, Reliability, and Quality Assurance Management Panel

The structure of the Space Shuttle Program was instrumental to its success. The panels listed on the right debated technical issues and reached technical decisions. These panels influenced multiple subsystems and were integrated by the Systems Integration Office.

However, because of the complexity, by 1977 the number had grown to 53 panels, subpanels, and working groups. These critical reviews provided guidance to maintain effective and productive technical decisions during the shuttle development phase. Also during this phase of the program, NASA established the definition and verification of the interfaces and associated documentation, including hazard analysis and configuration control.

Biggest Asset—People Working Together

Owen Morris, manager of the Systems Integration Office from 1974 to 1980, was an effective and a respected manager. When asked to describe the biggest challenge of that position, Owen answered, "People. Of course, all the people involved had their own responsibilities for their part of the program, and trying to get the overall program put together in the most efficient manner involved people frequently giving up part of their capability, part of their prerogative, to help a different part of the program, solve a problem, and do it in a manner that was better for everyone except them. And, that's a little difficult to convince people to do that. So, working with people, working with organizations, and getting them to work together in a harmonious manner was probably the most difficult part of that."

The challenge of getting people to work together successfully has been an enduring one. NASA stepped up to multiple challenges, including that of having various people and organizations working together toward a common goal. By working together, the space agency engineered many successes that will benefit future generations.

Restoring Integration and Systems Thinking in a Midlife Program

Aviation lore says that, during World War II, a heavily overworked crew chief confronted an aircraft full of battle damage and complained, "That's not an airplane, that's a bunch of parts flying in loose formation."

One of the greatest challenges during system development is transforming parts into a fully integrated vehicle. Glenn Bugos' book titled *Engineering the F-4 Phantom II* is subtitled *Parts into Systems* in recognition of this challenge. NASA also long realized this. In the standard NASA cost model for space systems, the agency planned that 25% of a program's development effort would go into systems engineering and integration. Efforts made during the initial development of the shuttle to ensure its integrated performance led to a successful and an enduring design.

NASA Learns an Expensive Lesson

NASA's experience in human spacecraft prior to the shuttle was with relatively short-lived systems. The agency developed four generations of human spacecraft—Mercury, Gemini, Apollo, and Skylab—in fewer than 15 years. Designers and project managers intuitively anticipated rapid replacement of human space systems because, at the time of shuttle development, they had no experience to the contrary. The initial design parameters for the Orbiter included 100 missions per Orbiter in 10 years. During the design phase, NASA did not plan for the 30-year operational life the shuttle actually flew.

The space agency, therefore, had no experience regarding the role of systems engineering and integration during the extended operational part of a system life cycle. Given the cost of a strong systems engineering and integration function, this was a topic of significant debate within NASA, particularly as budgets were reduced. As late as 1990—9 years after the shuttle's first flight—the systems engineering and integration effort was approximately $160 million per year, or approximately 6.4% of the $2.5 billion shuttle annual budget. Starting in 1992, to meet reduced operating budgets, this level of resource came under scrutiny. It was argued that, given major development of the shuttle system was complete, all system changes were under tight configuration control and all elements understood their interfaces to other elements, the same level of systems engineering and integration was no longer required. The effort was reduced to 2.2% of the shuttle annual budget in 1992. Occurrences of in-flight anomalies were decreasing during this period, thereby lending to the belief that the proper amount of integration was taking place.

This seemed to be a highly efficient approach to the problem until the loss of Columbia in 2003. In retrospect, the Columbia Accident Investigation Board determined there were clear indicators that the program was slowly losing the necessary degree of systems engineering and integration prior to the loss of Columbia. Critical integration documentation no longer reflected the vehicle configuration being flown. Furthermore, the occurrence of integrated anomalies was increasing over the years.

Crucial Role of Systems Engineering

Known Changes

Change was constantly occurring in the shuttle systems. Changes with known effects required a large and expensive integrated engineering effort but were usually the easiest to deal with. For example, when NASA upgraded the Space Shuttle Main Engines to a more-powerful configuration, a number of changes occurred in terms of avionics, electrical, and thrust performance. These changes had to be accommodated by the other parts of the system.

Known changes with unknown effects were more difficult to deal with. For example, as a cost-reduction effort, NASA decided not to replace the connectors on the Orbiter umbilicals after every flight. At the time, NASA did not know that the Solid Rocket Booster exhaust and salt-spray environment of the pad created corrosion on the connectors. This corrosion would eventually interrupt safety-critical circuits. On Space Transportation System (STS)-112 (2002), half the critical pyrotechnic systems, which release the shuttle from the launch pad, did not work. Because the systems had redundancy, the flight launched successfully.

Unknown Changes—Manufacturing Specification

There were many sources of unknown change during the Space Shuttle Program. First, the external environment was continually changing. For example, the electromagnetic environment changed as radio-frequency sources appeared and disappeared in terrain over which the shuttle flew. These sources could influence the performance of shuttle systems.

Second, the characteristics of new production runs of materials such as adhesives, metals, and electronic components changed over time. It was impossible to fully specify all characteristics of all materials on a large system. Changes in assembly tooling or operators could have resulted in a product with slightly different characteristics. For instance, major problems with fuel quality circuits caused launch delays for flights after the Columbia accident. The circuits were intended to identify a low fuel level and initiate engine shutdown, thus preventing a probable engine catastrophe. These circuit failures were random. While these anomalies remained unexplained, the circuit failures seemed to stop after improvements were made to the engine cutoff sensors. However, following another failure on STS-122 (2008), the problem was isolated to an electrical connector on the hydrogen tank and was determined to be an open circuit at the electrical connector's pin-to-socket interface. The increased failure rate was likely caused by a subtle change to the socket design by the vendor, combined with material aging within the connector assembly. The connector was redesigned, requiring soldering the sockets directly to the pins.

Solution—Systems Engineering

The only way to deal with known and unknown change was to have a significant effort in systems engineering and integration that monitored integrated flight performance and was attuned to the issues that could impact a system. One of the best approaches for maintaining this vigilance was comparing in-flight anomalies to established analyses of hazards to the integrated system. These integrated hazard analyses were produced at the start of the program but had not been updated at the time of the Columbia accident to reflect the present vehicle configuration. Further, the in-flight anomaly process was not tied to these analyses. In the period before Return to Flight, the systems engineering and integration organization tried to fix these analyses but determined the analyses were so badly out of date that they had to be completely redone. Thus, systems engineering and integration replaced 42 integrated hazards with 35 new analyses that used fault-tree techniques to determine potential causes of hazards to the integrated system. These analyses were also tied into a revamped in-flight anomaly process. Any problem occurring in flight that could cause a hazard to the integrated system required resolution prior to the next flight.

Preparing for Return to Flight After the Columbia Accident

When internal NASA evaluations and the Columbia Accident Investigation Board determined that shuttle systems engineering and integration would need to be rebuilt, NASA immediately recognized that systems engineering and integration could not be rebuilt to 1992 levels. There were simply not enough available, qualified systems engineers who were familiar with the shuttle configuration. Further, it was unlikely that NASA could afford to maintain the necessary level of staffing. NASA accomplished a modest increase of about 300 engineers by selective hiring. Also, NASA worked with the Aerospace Corporation (California), along with establishing agreements with other NASA centers, such as integration personnel at Marshall Space Flight Center and Kennedy Space Center. This returned systems engineering and integration activities to 1995 levels. More impressive was the way in which these resources were deployed.

The most immediate job for systems engineering and integration during this

Left photo: Ames Research Center wind tunnel test.
Right photo: Aerothermal test at Calspan-University of Buffalo Research Center.

period was determining design environments for all redesigns mandated by the Columbia Accident Investigation Board. The standard techniques for establishing design environments prior to this effort involved constructing environment changes to the basic environments by making conservative calculations based on the nature of the change.

A large number of configuration changes over the years resulted in an accumulation of conservative design environments. However, this cumulative approach was the only basis for estimating the environments. A new baseline effort would have required extensive calculations and ground tests. For the Return to Flight effort, systems engineering and integration decided to re-baseline the critical design environments to eliminate non-credible results. Fortunately, technology had advanced significantly since the original baseline environments were constructed in the 1970s. These advances enabled greater accuracy in less time.

The shuttle aerodynamics model was refurbished to the latest configuration for aerodynamics and aerodynamic loads. Shuttle wind tunnel tests were completed at Ames Research Center in California and the Arnold Engineering Development Center in Tennessee. Engineers employed new techniques, such as pressure-sensitive paint and laser velocimetry in addition to more advanced pressure and force instrumentation. The purpose of these tests was to validate computational fluid dynamics models because design modifications were evolving as the design environments were being generated. Thus, continued wind tunnel tests could not generate the final design environments. Validated computational fluid dynamics models were necessary to generate such environments for the remainder of the Space Shuttle Program to avoid the accumulation of conservative environments.

Engineers performed similar tests using the aerothermal model at the Calspan-University of Buffalo Research Center (New York) shock tunnel. Engineers used a combination of computational fluid dynamics and other engineering methods to generate an updated thermal database.

Another major task for systems engineering and integration was to understand the debris transport problem. A 0.76-kg (1.67-pound) piece of foam debris was liberated from the External Tank. This foam debris was responsible for the damage that caused the Columbia accident. Systems engineering and integration enabled engineers to identify the transport paths of debris to the shuttle to determine the hazard level of each debris item as well as determine the impact velocities that the structure would have to withstand. When analysis or testing revealed the elements could not withstand impact, systems engineering and integration worked with the debris-generating element to better understand the mechanisms, refine the estimated impact conditions, and determine whether debris-reduction redesign activities were sufficient to eliminate or reduce the risk. To understand debris transport, NASA modeled the flow fields with computational fluid dynamics and flight simulation models. Fortunately, NASA had entered into an agreement, post-Columbia, to create the world's largest supercomputer at Ames Research Center. This 10,240-element supercomputer came on line in time to perform extensive computational fluid dynamics and simulation analysis of debris transport.

Debris Transport During Launch Remained a Potential Hazard

NASA cataloged both the size and the shape of the debris population as well as the debris aerodynamics over a wide speed range. A large part of this

NASA validated computational fluid dynamics and flight simulation models of the foam debris in flight tests using the Dryden Flight Research Center (California) F-15B Research test bed aircraft. In these tests, debris fell from foam panels at simulated shuttle flight conditions. High-speed video cameras captured the initial flight of the foam divots.

effort involved modeling the flight characteristics of foam divots that came off of the tank. NASA first addressed this problem by firing small plastic models of foam divot shapes at the NASA Ames Research Center, California, ballistic range. When these results correlated well with computational fluid dynamics, the agency conducted more extensive tests. Engineers tested flight characteristics of foam debris in the Calspan-University of Buffalo Research Center tunnel and Dryden Flight Research Center, California. Results showed that foam would stay intact at speeds up to Mach 4 and, therefore, remain a potential hazard.

Other Return to Flight Activities

Two other major tasks were part of the systems engineering and integration Return to Flight effort. The first task involved integrated test planning to ensure that the system design was recertified for flight. The second task was to install additional instrumentation and imagery acquisition equipment to validate the performance of system design changes.

The diversity of integrated system testing was remarkable. Integrated tests included the first-ever electromagnetic interference tests run on the shuttle system. NASA ran a test to determine the effects of the crawler transporter on the vibration/fatigue of shuttle structures. This effort required construction of improved integrated structural models. First performed on a limited scale during the Return to Flight period, this effort expanded under Marshall Space Flight Center leadership. The integrated test effort also included two full-up tanking tests of the shuttle system. In addition to validating the performance of the new foam system on the tank, these tanking tests discovered two major problems in the shuttle: failures of the propellant pressurization system and problems with the engine cutoff sensors.

The instrumentation added to the shuttle system as part of the systems engineering and integration effort was also diverse. NASA added instrumentation to the External Tank to understand the vibration and loads on major components attached to the skin. These data proved vital after Return to Flight assessment because a loss of foam associated with these components required additional modification. This instrumentation gave the program the confidence to make these modifications. NASA also added instrumentation to help them understand over-pressure effects on the shuttle due to ignition transients of the Space Shuttle Main Engine and motion of the Orbiter-ground system umbilicals. The agency added ground-based radar and video imaging equipment to provide greater visibility into the debris environment and validate design modifications.

Integration Becomes the Standard

NASA learned some difficult yet valuable lessons about the importance of systems engineering and integration over the course of the Space Shuttle Program—especially in the years following the loss of Columbia. The lack of systems engineering and integration was a contributing cause to the accident. The shuttle had become "a collection of parts flying in loose formation." It took a major engineering effort over a 2-year period to reestablish the proper amount of integration. This effort significantly improved the shuttle system and laid the groundwork and understanding necessary for the successful flights that followed.

Electromagnetic Compatibility for the Space Shuttle

Electromagnetic compatibility is extremely complex and far reaching. It affects all major vehicle engineering disciplines involving multiple systems and subsystems and the interactions between them. By definition, electromagnetic compatibility is the capability of electrical and electronic systems, equipment, and devices to operate in their intended electromagnetic environment within a defined margin of safety, and at design levels of performance. But, that is just the beginning. This must be accomplished without causing unacceptable degradation as a result of any conducted or radiated electromagnetic energy that interrupts, obstructs, or otherwise limits the effective performance of telecommunications or other electrical and electronic equipment.

Design and Verification Requirements— A Learning Process

In 1973—when NASA was first defining the shuttle systems—military models offered the best available means of providing control of the system design leading to acceptable levels of electromagnetic compatibility. Previous requirements for Mercury, Gemini, and Apollo were cut from the same cloth, but none of those programs had a vehicle that could compare to the shuttle in terms of size and complexity.

Admittedly, these comprehensive requirements addressed a multiplicity of concerns. These included: subsystem criticality; degradation criteria; interference and susceptibility control;

wiring and cable design and installation; electrical power; electrical bonding and grounding; control of static electricity and its effects; electromagnetic hazards to personnel, explosives, and ordnance; and definition of, and design for, the external electromagnetic environment.

Detailed design and verification requirements for protection from the damaging effects of lightning were also included and developed independently by NASA. These shuttle lightning requirements became the foundation for a plethora of military and commercial aerospace requirements, culminating in a detailed series of Society of Automotive Engineers documents universally employed on an international basis.

A Custom Fit Was Needed

Unfortunately, without a solid basis for the tailoring of requirements, shuttle electromagnetic compatibility engineers chose to levy the baseline requirements with virtually no change from previous Apollo efforts. Although this was a prudent and conservative approach, it led to misinterpretation and misapplication of many requirements to the shuttle. As a result, NASA granted an unacceptably large number of waivers for failure to comply with the requirements. The problem continued to grow until 2000, at which time NASA made a major effort to completely review and revise the electromagnetic compatibility requirements and compliance approach. This effort eliminated or tailored requirements so that the content was directly and unequivocally applicable to the shuttle. This effort also allowed for a systematic and detailed revisitation of previously granted waivers against the backdrop of the new requirements' definitions.

Making Necessary Adjustments...and Succeeding

Original requirements and new requirements were tabulated together to facilitate direct comparison. For each set of requirements, NASA needed to examine several characteristics, including frequency range, measurement circuit configuration, test equipment application, and the measured parameter limits. As an example, certain conducted emissions requirements in the original set of requirements measured noise currents flowing on power lines whereas the equivalent new requirements measured noise voltages on the same power lines. To compare limits, it was necessary to convert the current limits to voltage limits using the linear relationship between voltage, current, and circuit impedance. In other cases, frequency bandwidths used for testing were different, so NASA had to adjust the limits to account for the bandwidth differences.

In all, NASA engineers were able to work through the complexity of electromagnetic compatibility—to follow all of the threads inherent in the vehicle's multiple systems and subsystems—and find a way to tailor the requirements to accommodate the shuttle.

Process Control

The design and fabrication of the Space Shuttle's main components took place in the early 1970s while Richard Nixon was president. The Space Shuttle was assembled from more than 2.5 million parts that had to perform per design with very little margin of error. NASA constantly analyzed and refurbished flight systems and their components to ensure performance. The success of the Space Shuttle Program was due in great part to diligent process control efforts by manufacturing teams, contractors, and civil service engineers who carefully maintained flight hardware.

Five Key Elements Ensure Successful Process Control

Process control consists of the systems and tools used to ensure that processes are well-defined, perform correctly, and are maintained such that the completed product conforms to requirements. Process control managed risk to ensure safety and reliability in a complex system. Strict process control practices helped prevent deviations that could have caused or contributed to incidents, accidents, mishaps, nonconformances, and in-flight

Electromagnetically Shielded Zones Within the Orbiter Structure

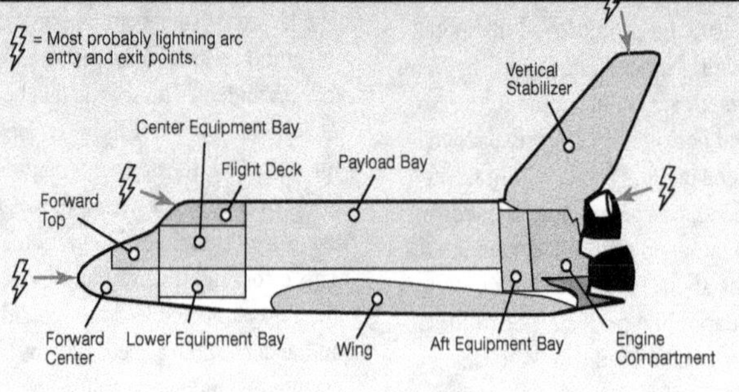

anomalies. As defined by NASA, the five key elements of a process are: people, methods/instructions, materials, equipment, and environment. It has been long understood that qualified, conscientious people are the heart of any successful operation. High-quality process control efforts require skilled, detail-oriented individuals who understand and respect the importance of process and change control. The methods or instructions of a process, often called "specifications" or "requirements," are those documented techniques used to define and perform a specific process. The term "equipment" refers to the tools, fixtures, and facilities required to make products that meet specifications and requirements while "materials" refers to both product and process materials used to manufacture and test products. Finally, the environmental conditions required to properly manufacture and test products must also be maintained to established standards to ensure safety and reliability.

Solid Engineering Design—A Fundamental Requirement

A clear understanding of the engineering design is fundamental when changes occur later in a program's life. Thousands of configuration changes occurred within the Space Shuttle Program. These changes could not have been made safely without proper process controls that included a formal configuration control system. This

Alliant Techsystems, Inc. and United Space Alliance

The signature twin reusable solid rocket motors of the Space Shuttle carried the fingerprints of thousands of people who designed, manufactured, tested, and evaluated the performance of these workhorse motors since 1982. The manufacturing facility in Promotory, Utah, is now owned and operated by Alliant Techsystems, Inc. (ATK). Originally developed to manufacture and test large-scale rocket motors for intercontinental ballistic missiles, the site provided 72% of the liftoff thrust to loft each shuttle beyond Earth's bounds.

Technicians process the solid rocket motor case segments at the ATK case lining facility in Utah.

The Assembly Refurbishment Facility complex—managed and operated by United Space Alliance (USA), headquartered in Houston, Texas—is located at Kennedy Space Center, Florida. The complex began operations in 1986 and was the primary integration and checkout facility for boosters. Refurbished and new hardware were assembled and submitted to rigorous testing to assure the assemblies were ready for human-rated flight. The facility was equipped to handle assembly, testing, and troubleshooting of thrust vector control systems, avionics, and recovery systems for the Space Shuttle Program.

Solid Rocket Booster case preparation.

Propellant mixing.

Solid Rocket Booster aft skirt processing at the Assembly and Refurbishment Facility at Kennedy Space Center.

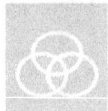

Michoud Assembly Facility

By the end of the Space Shuttle Program, NASA's Michoud Assembly Facility—located near New Orleans, Louisiana, and managed by Marshall Space Flight Center in Huntsville, Alabama—delivered 134 External Tanks (ETs) for flight. Two additional tanks were built but not scheduled to fly, and three assemblies were delivered for major tests, resulting in a total of 139 tanks. As one of the world's largest manufacturing plants, Michoud's main production building measured 17 hectares (43 acres) under one roof, including a 61-m (200-ft) vertical assembly building, and a port that permitted transportation of ETs via oceangoing barges and towing vessels to Kennedy Space Center in Florida.

ETs were produced at Michoud by prime contractor Lockheed Martin (headquartered in Bethesda, Maryland) over a 37-year period. The contractor procured parts and materials from hundreds of subcontractors across the country. In full production, 12 tanks were in various phases of production across the facility—each tank requiring approximately 3 years to complete. Each ET included over 0.8 km (0.5 miles) of welds, thousands of rivets and bolts, redundant inspections within each process, and sophisticated pressure and electrical testing.

Throughout the history of the program, Michoud continually improved the processing, materials, and components of ETs. Improvements included the introduction of a stronger, lighter aluminum-lithium alloy—which saved over 2.7 metric tons (3 tons) of weight—and transitioning to virtually defect-free friction stir welding. Additionally, Michoud developed thermal protection foam spray systems and process controls that reduced weight and minimized foam loss during the extreme environments of flight.

Liquid oxygen tank.

Liquid oxygen tank and intertank in a checkout cell.

Liquid hydrogen tank showing slosh and vortex baffle inside.

External Tank processing.

system involved the use of review boards, material review analyses, and tool controls.

A Team Effort

Hardware for the Space Shuttle Program was manufactured by a broad supplier base using a variety of processes. If these processes were not controlled, a deterioration of the end product could have occurred, thereby increasing risk. In essence, NASA depended on the process controls at over 3,000 flight hardware suppliers' facilities across the United States. Any subtle changes or deviations from any established processes could have negatively affected the outcome.

Think of the thousands of vendors and processes that might have affected manufacturing—from material pedigree to the material of gloves worn by a technician. All of these nuances affected the outcome of the product. Coordination and communication between NASA and its manufacturers were critical in this complicated web of hardware suppliers. The Space Shuttle was only as strong as its weakest link.

Strong process controls resulted in highly predictable processes. Built-in tests were critical because many flight components/systems could not be tested prior to their actual use in flight. For example, Thermal Protection Systems, pyrotechnics, and solid rocket motors could only be tested at the manufacturer's facilities before they were installed aboard the shuttle. This fact demonstrated once again that NASA was highly dependent on the integrity of its hardware suppliers to follow the tried and true "recipe" of requirements, materials, people, and processes to yield predictable and reliable components.

Processes Continue Well Beyond Flight

Because shuttles were reusable vehicles, process control was also vital to refurbishment and postflight evaluation efforts. After each flight, NASA closely monitored the entire vehicle to evaluate factors such as heat exposure, aging effects, flight loads, shock loads, saltwater intrusion, and other similar environmental impacts. For example, did you know that each heat tile that protected the underbelly of the vehicle from the extreme heat of re-entry into Earth's atmosphere was numbered and checked following each flight? Tiles that did not pass inspection were either repaired or replaced. This effort was a major undertaking since there were 23,000 thermal protection tiles.

Postflight recovery and inspections were an important part of process control. For example, NASA recovered the Solid Rocket Boosters, which separated from the vehicle during launch and splashdown in the Atlantic Ocean, and brought them back to Kennedy Space Center in Florida where they were examined and inspected. These standardized forensic inspections provided valuable data that determined whether the booster system operated within its requirements and specifications. Data collected by the manufacturer represented the single most important feedback process since this system had to function as intended every time without the ability to pretest.

Best Practices Are Standard Practice

Each of NASA's manufacturers and suppliers had unique systems for process control that guaranteed the integrity of the shuttle's hardware.

Pratt & Whitney Rocketdyne Manufacturing

The Space Shuttle Main Engine required manufacturing and maintenance across the entire United States. Pratt & Whitney Rocketdyne (Canoga Park, California), under contract to NASA, developed the main engine, which successfully met the challenges of reusability, high performance, and human-rated reliability. With every launch, the team continued to make improvements to render it safer and more reliable.

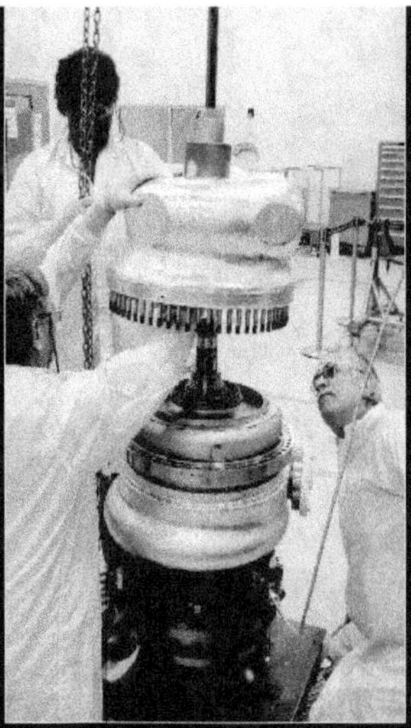

High-pressure fuel turbopump recycling.

The Pratt & Whitney Rocketdyne facility at the West Palm Beach, Florida, campus designed and assembled the critical high-pressure turbomachinery for the shuttle. The high pressures generated by these components allowed the main engine to attain its extremely high efficiency. At the main facility in Canoga Park, California, the company fabricated and assembled the remaining major components. The factory included special plating tanks for making the main combustion chamber (the key components to attain high thrust with the associated high heat transfer requirements), powerhead (the complex structural heart of the engine), and nozzle (another key complex component able to withstand temperatures of 3,300°C [6,000°F] degrees during operation). In addition, the company employed personnel in Huntsville, Alabama, and Stennis Space Center in Mississippi. The Huntsville team created and tested critical software. The Stennis team performed testing and checkout of engines and engine components before delivery to the launch site. Finally, at Kennedy Space Center in Florida, Pratt & Whitney Rocketdyne personnel performed all the hands-on work required to support launch, landing, and turnaround activities.

Space Shuttle Main Engine assembly.

Rockwell International and The Boeing Company

Rockwell of Downey, California (now Boeing) executed the Orbiter design, development, test, and evaluation contract, the production contract, and the system integration contract for the mated shuttle vehicle. Engineers were the primary producers of specifications, vehicle loads/environments, analysis, drawing release, certification/qualification testing, and certification documentation. Engineers performed key system-level integration and testing for many Orbiter subsystems including software, avionics hardware, flight controls/hydraulics, and thermal protection. At this same location, technicians manufactured the crew module, forward fuselage, and aft fuselage, which were integrated into the Orbiter at the Boeing facility in Palmdale, California.

Boeing engineers, technicians, and support personnel assembled and tested all six Space Shuttle Orbiter vehicles. The first shuttle vehicle, Enterprise, was delivered in January 1977. Being a non-orbital vehicle, it was used for fit checks, support equipment procedures, and the Approach and Landing Test Program conducted at Dryden Flight Research Center on the Edwards Air Force Base runway in California beginning in 1977. Columbia, the first space-rated Orbiter, was delivered in the spring of 1979 and later flew the Space Shuttle Program's maiden voyage in April 1981. Challenger was rolled out in 1982, followed by Discovery in 1983 and Atlantis in 1985. The newest shuttle, Endeavour, was authorized following the loss of Challenger in 1986 and was delivered in April 1991. From 1985 to 2001, engineers performed eight major modifications on the Orbiter fleet.

Orbiter assembly.

The communication and establishment of specific best practices as standards helped the program improve safety and reliability over the years. The following standards were the minimum process control requirements for all contractors within the Space Shuttle Program:

- Detect and eliminate process variability and uncoordinated changes.
- Eliminate creep—or changes that occur over time—through process controls and audits.
- Understand and reduce process risks.
- Identify key design and manufacturing characteristics and share lessons learned that relate to the processes.
- Be personally accountable and perform to written procedures.
- Promote process control awareness.
- Identify and evaluate changes to equipment and environment.
- Capture and maintain process knowledge and skills.

NASA witnessed a significant evolution in their overall process control measures during the shuttle period. This lengthy evolution of process control, a continuous effort on the part of both NASA and its contractors, included multiple initiatives such as:

- establishing reliable processes
- monitoring processes
- reinforcing the process-control philosophy or "culture"
- maintaining healthy systems

Establishing reliable processes included open communications (during and after the design process) among numerous review boards and change boards whose decisions dictated process-control measures. Monitoring processes involved postflight inspections, safety management systems, chemical fingerprinting, witness panels, and other monitoring procedures. Process control also referred to relatively new programs like the "Stamp and Signature Warranty" Program where annual audits were performed to verify the integrity of products/components for the shuttle era. Finally, maintaining healthy systems focused on sustaining engineering where design or operating changes were made or corrective actions were taken to enhance the overall "health" of the program.

An Enduring Success

Although NASA's process control measures have always been rigorous, additional enhancements for improved communication and information-sharing between shuttle prime contractors and suppliers created highly restrictive, world-class standards for process control across the program. Many of these communication enhancements were attainable simply because of advances in technology. The computer, for example, with its increased power and capabilities, provided faster and better documentation, communication, data tracking, archiving, lot number tracking, configuration control, and data storage. As manufacturers, contractors, and other businesses came and went—and as employees, managers, and directors came and went—the program stayed the same over its lifetime and continued to operate successfully primarily because of its well-honed process-control measures.

The Case of the Chloride Sponges

Let's look at "The Case of the Chloride Sponges" to further demonstrate the importance of process control and the complexities of maintaining the Space Shuttle fleet. Postflight maintenance requirements included applying a corrosion inhibitor (sodium molybdate) to the Space Shuttle Main Engine nozzles. Following the STS-127 (2009) flight, engineers observed increased nozzle corrosion instances in spite of the application of the corrosion inhibiter. A root-cause investigation found that the sponges used to apply the corrosion inhibitor contained high levels of chlorides. Apparently, the sponges being used to apply the corrosion inhibitor were themselves causing more corrosion.

It was determined that the commercial vendor for the sponges had changed their sponge fabrication process. They began adding magnesium chloride for mold prevention during their packaging process and since NASA did not have a specification requirement for the chloride level in the sponges, the sponge fabrication change initially went unnoticed. To solve this problem, NASA added a requirement that only chloride-free sponges could be used. The agency also added a specification for alternate applicator/wipes. Case closed!

NASA and the Environment—Compatibility, Safety, and Efficiency

As conscientious stewards of US taxpayers dollars, NASA has done its part to mitigate any negative impacts on the wildlife and environment that the agency's processes may impart. For NASA, it is not about technical issues; in this case, it is about the coexistence of technology, wildlife, and the environment.

Compatibility

The 56,700 hectares (140,000 acres) controlled by Kennedy Space Center (KSC) symbolize a mixture of technology and nature. Merritt Island National Wildlife Refuge was established in 1963 as an overlay of the center. The refuge consists of various habitats: coastal dunes; saltwater estuaries and marshes; freshwater impoundments; scrub, pine flatwoods; and hardwood hammocks. These areas provide habitat for more than 1,500 species of plants and animals. Hundreds of species of birds reside there year-round, with large flocks of migratory waterfowl arriving from the North and staying for the winter. Many endangered wildlife species are native to the area. Part of KSC's coastal area was classified as a national seashore by agreement between the NASA and the Department of the Interior.

Most of the terrain is covered with extensive marshes and scrub vegetation, such as saw palmettos, cabbage palm, slash pine, and oaks. Citrus groves are in abundance, framed by long rows of protective Australian pine. More than 607 hectares (1,500 acres) of citrus groves are leased to individuals who tend to the trees and harvest their fruit. Beekeepers maintain the health of the trees by collecting honey from—and maintaining—the hives of bees essential to the pollination of the citrus trees. Merritt Island National Wildlife Refuge manages the leases. Other NASA centers such as White Sands Test Facility and Wallops Flight Facility are also close to National Wildlife Refuges.

Safety

There is a limit as to what NASA can do to actually protect itself from the wildlife. During launch countdown of Space Transportation System (STS)-70 on Memorial Day 1995, the launch team discovered a pair of northern flicker woodpeckers trying to burrow a nesting hole in the spray-on foam insulation of the shuttle External Tank on Pad B. Spray-on foam insulation was comparable to the birds' usual nesting places, which include the soft wood of palm trees or dead trees. However, on reaching the aluminum skin of the tank beneath the spray-on foam insulation layer, the woodpeckers would move to a different spot on the tank and try again. In the end, there were at least 71 holes on the nose of the tank that couldn't be repaired at the pad. As a result, the stack was rolled back to the Vehicle Assembly Building for repairs to the damaged insulation.

The problem of keeping the woodpeckers from returning and continuing to do damage to the tank's spray-on foam insulation proved to be complex. The northern flicker is a protected species so the birds could not be harmed. In NASA fashion, shuttle management formed the Bird Investigation Review and Deterrent (BIRD) team to research the flicker problem and formulate a plan for keeping the birds away from the pads. After studying flicker behavior and consulting ornithologists and wildlife experts, the team devised a three-phase plan. Phase 1 of the plan consisted of an aggressive habitat management program to make the pads more unattractive to flickers and disperse the resident population of these birds. NASA removed palm trees, old telephone poles, and dead trees from the area around the pads. The agency allowed the grass around the pad to grow long to hide ants and other insects—the flickers' favorite food. Phase 2 implemented scare and deterrent tactics at the pads. NASA used plastic owls, water sprays, and "scary eye" balloons to make the area inhospitable to the birds and frighten them away without injuring them. Phase 3 involved the implementation of bird sighting response procedures. With the BIRD team plans in place and the flickers successfully relocated, STS-70 was able to launch approximately 6 weeks later.

Woodpeckers are not the only form of wildlife attracted to the External Tank. On STS-119 (2009), a bat was found clinging to Discovery's external fuel tank during countdown. Based on images and video, a wildlife expert said the small creature was a free tail bat that likely had a broken left wing and some problem with its right shoulder or wrist. Nevertheless, the bat stayed in place and was seen changing positions from time to time. The temperature never dropped below 15.6°C (60°F) at that part of the tank, and infrared cameras showed that the bat was 21°C (70°F) through launch. Analysts concluded that the bat remained with the spacecraft as it cleared the tower. This was not the first bat to land on a shuttle during a countdown. Previously, one landed on the tank during the countdown of STS-90 (1998).

Another species that NASA dealt with over the life of the Space Shuttle Program was a type of wasp called a mud dauber. Although the mud daubers aren't very aggressive and don't pose an immediate threat to people, the nests they build can pose a problem. Mud daubers tend to build nests in small openings and tubes such as test ports. This can be an annoyance in some cases, or much more serious if the nests are built in the openings for the pitot-static system (i.e., a system of pressure-sensitive instruments) of an aircraft. Nests built in these openings can affect functionality of the altimeter and airspeed indicator.

Efficiency

In keeping with imparting minimal negative impact on the environment, NASA also took proactive steps to reduce energy usage and become more "green." At KSC, NASA contracted several multimillion-dollar energy projects with Florida Power & Light Company that were third-party-financed projects. There was no out-of-pocket expense to NASA. The utility was repaid through energy savings each month. The projects included lighting retrofits; chilled water modifications for increased heating, ventilation, and air-conditioning efficiency; and controls upgrades. As an example, NASA installed a half-sized chiller in the utility annex—the facility that supplies chilled water to the Launch Complex 39 area—so as to better match generation capacity with the demand and reduce losses. The agency also retrofitted lighting and lighting controls with the latest in fluorescent lamp and ballast technology. In total, these multimillion-dollar projects saved tens of millions of kilowatt-hours and the associated greenhouse emissions.

Protecting Birds and the Shuttle

During the July 2005 launch of Discovery, a vulture impacted the shuttle's External Tank. With a vulture's average weight ranging from 1.4 to 2.3 kg (3 to 5 pounds), a strike at a critical area on the shuttle could have caused catastrophic damage to the vehicle. To address this issue, NASA formed the avian abatement team. The overall goal was to increase mission safety while dispersing the vulture population at Kennedy Space Center (KSC).

Through its research, the team attributed the large vulture population to an abundant food source—carrion (road kill). A large educational awareness effort was put into place for the KSC workforce and local visitors. This effort included determining wildlife crossing hot spots, ensuring the placement of appropriate signage on the roadways to increase traveler awareness, and timely disposal of the carrion.

NASA added new radar and video imaging systems to electronically monitor and track birds at the pads. Already proven effective, the avian radar—known as Aircraft Birdstrike Avoidance Radar—provided horizontal and vertical scanning and could monitor either launch pad for the movement of vultures. If data relayed from the avian radar indicated large birds were dangerously close to the vehicle, controllers could hold the countdown.

Endeavour, STS-100 (2001), roars into space, startling a flock of birds.

In addition to the energy-saving benefits of the projects, NASA was also able to modernize KSC infrastructure and improve facility capability. As an example, when the Vertical Assembly Building transfer aisle lighting was redesigned, better local control and energy saving fixtures were provided. At the same time, this increased light levels and color rendering capability. As another example, although KSC had a 10-megawatt emergency generator plant capable of servicing critical loads in a power outage, this same plant could not start the chillers needed for cooling these systems. As such, the backup plant was unable to sustain these loads for more than a few minutes before overheating conditions began. Soft start drives were installed on two of the five chiller motors, thus allowing the motors to be started from the generator plant and providing a true backup capability for the Launch Complex 39 area.

In yet another partnership with Florida Power & Light Company, KSC opened a 10-megawatt solar power plant on 24 hectares (60 acres) of old citrus groves. This plant could generate enough electricity for more than 1,000 homes and reduce annual carbon dioxide emissions by more than 227,000 tons. Florida Power & Light Company estimated that the 35,000 highly efficient photovoltaic panels were 50% more efficient than conventional solar panels. This solar power plant, in addition to the 1-megawatt plant, has been supplying KSC with electricity since 2009. The opening of the 10-megawatt solar field made Florida the second-largest solar-power-producing state in the country.

Summary

Throughout the shuttle era, NASA was a conscientious steward of not only the taxpayer's dollars but also of nature and the environment. Not only was the space agency aware of the dangers that wildlife could pose to the shuttle, it was also aware of the dangers that humans pose to the environment and all its inhabitants. As NASA moves forward, the agency continues to take proactive steps to assure a safe and efficient coexistence.

Major Scientific Discoveries

The Space Shuttle and Great Observatories

Carol Christian
Kamlesh Lulla

Space Shuttle Bestows On Hubble the Gift of "Perpetual Youth"
David Leckrone

NASA's "Great Observatories" are symbolic of our urge to explore what lies at the edge of the universe, as we know it. Humans had stared at the stars for centuries before the advent of simple telescopes brought them a little closer to the amazing formations in our solar system. Telescopes became larger, technologies were developed to include invisible wavelengths from the shortest to longest, and locations of instruments were carefully chosen to gain better sights and insights into our vast universe. Then, the Space Age dawned and we sent humans to the moon. The desire to explore our universe became even more intense. NASA probes and rovers landed on destinations in our solar system—destinations we once thought remote and beyond reach. These initiatives forever changed our perception of the solar system and galaxies.

Scientists have long desired space-based observation platforms that would provide a better view of our universe. NASA's Great Observatories (satellites) are four large and powerful space-based telescopes that have made outstanding contributions to astronomy. The satellites are:

- *Hubble Space Telescope*
- *Compton Gamma Ray Observatory*
- *Chandra X-ray Observatory*
- *Spitzer Space Telescope*

Of these, only the Spitzer Space Telescope was not launched by the Space Shuttle. In June 2000, the Compton Gamma Ray Observatory was deorbited and parts splashed into the Pacific Ocean.

While Hubble has become the people's telescope due its public and media impact, all the Great Observatories made enormous science contributions including: new wave bands; high-resolution, high-sensitivity observations; and a sharper, deeper look into distant galaxies.

Space Shuttle Bestows On Hubble the Gift of "Perpetual Youth"

The Space Shuttle and Hubble Space Telescope were conceived and advocated as new NASA programs in the same era—roughly the late 1960s and the decade of the 1970s. It was recognized early on that a partnership between the Hubble program and the Space Shuttle Program would be mutually beneficial at a time when both were being advocated to Congress and the Executive Branch.

A telescope designed to be periodically serviced by astronauts could be viewed as a "permanent" astronomical observatory in space, modeled after the observatories on Earth's surface. At Hubble's core would be a large, high-quality optical telescope that, with its surrounding spacecraft infrastructure periodically serviced by shuttle crews, could have an operating lifetime measured in decades. Its heart would be scientific instruments that could be regularly replaced to take advantage of major advances in technology. Thus, the shuttle brought to Hubble the prospect of a long life and, at the same time, the promise of "perpetual youth" in terms of its technological prowess.

Hubble provided a splendid example of the value of the shuttle in allowing regular access to low-Earth orbit for a large crew and heavy payloads. The shuttle enabled modes of working in space that were not otherwise possible, and the Hubble program was both the proof of concept and the immediate beneficiary. The two programs represented the nexus of human spaceflight and robotic exploration of the universe.

Hubble Space Telescope after capture by STS-61 (1993).

Hubble's design was optimized with its relationship to the shuttle in mind. The optical telescope and surrounding structure needed to be small enough to fit into a shuttle payload bay. On the other hand, the scientific value of the telescope hinged on making its aperture as large as possible. The final aperture size, 2.4 m (7.9 ft), was large enough to allow one of the observatory's most important science objectives—precisely measuring the distance scale and age of the universe—yet could be packaged to fit inside the shuttle payload bay.

Many of the Hubble spacecraft's subsystems were designed in modular form, and were removable and replaceable with relative ease by astronauts in spacesuits. However, this was not the case for every subsystem. One of the most telling demonstrations of the value of human beings working in space comes from the creativity and ingenuity of the astronauts and their engineering colleagues on the ground in devising methods for replacing or repairing components that were not designed to be worked on easily in orbit.

It is well known that Hubble, launched in 1990, was seriously defective. Its 2.4-m (7.9-ft) primary mirror—a beautifully ground and polished optic—was accidentally ground to the wrong prescription. The result, which became apparent when the newly launched telescope first turned its gaze to starlight, was a blurry image that could not be corrected with any adjustments in the telescope's focus mechanism or with the 24 actuated pressure pads placed under the primary mirror to adjust its shape, as needed. The erroneous curvature of the mirror produced a common form of optical distortion called "spherical aberration." In addition, Hubble's two flexible solar arrays shuddered significantly due to thermal stresses introduced every time the spacecraft passed from darkness to daylight or vice versa. This phenomenon introduced jitter into the pointing of the telescope, further smearing out its images.

In the years immediately after Hubble's deployment, the observatory did produce some interesting astronomical science but nothing at all like what had been expected throughout its design and development. It quickly became a national embarrassment and the butt of jokes on late-night talk shows.

It is interesting to consider how the history of Hubble and NASA might have transpired if the spacecraft had not been designed as an integral part of the world of human spaceflight but, rather, had been launched with an expendable rocket and not been serviceable in space. The scandal and embarrassment would likely have persisted for a while longer and then faded as the less-than-memorable science being produced also faded from public interest. One wonders if the champions of the Hubble mission could have stimulated the political and public will to try again—to develop and launch a second Hubble. Certainly, any such project would have taken a decade or longer and required new expenditures of public funds, probably $2 billion or more. In any event, the original Hubble Space Telescope would have long ago failed and today would be orbiting Earth as a large and expensive piece of space junk.

Hubble's history has played out in an entirely different and much more satisfying manner precisely because it was built to be cared for by human beings in low-Earth orbit. Scientists and engineers quickly identified the nature of Hubble's optical flaw and created optical countermeasures to correct the telescope's eyesight. The European Space Agency devised a new thermal design to mitigate the jitter-inducing flexure of the European solar arrays. The time required to design, fabricate, test, and fly these fixes to Hubble on the first servicing mission was approximately 3.5 years. In late 1993, public scorn turned into adulation, both because of the exquisite imagery that a properly performing Hubble returned to the ground and the heroism of the astronauts and the dedication of a team of NASA employees and contractors who refused to give up on the original dream of what Hubble could accomplish. The public image of NASA as a "can-do" agency certainly received a major boost. The techniques of working with precision on large structures in space surely contributed to the acceptance of the feasibility of constructing the International Space Station.

The possibility of periodically servicing Hubble added a degree of flexibility, timeliness, and creativity that was not possible in the world of robotic science missions, which must be planned and executed over periods of many years or even decades. Hubble's scientific capabilities have never grown out of date because it was regularly updated by shuttle servicing. It is the most in-demand, scientifically successful, and important astronomical observatory in human history, after Galileo's original telescope. Arguably, it is one of the most important scientific instruments of any kind. There is simply no way this level of achievement could have been possible without the Space Shuttle.

Hubble—A Work of Ingenuity

On September 9, 2009, NASA declared Hubble to be in full working order following the tremendously successful fifth shuttle mission to service the telescope. As a result of coordination across the extensive Hubble team, the crew of Space Shuttle Atlantis (Space Transportation System [STS]-125) left behind an essentially new telescope with six working instruments. Two superb instruments—the Wide Field Camera 3 and the Cosmic Origins Spectrograph—replaced older devices. Two instruments that had suffered electronic failure in flight were restored to working order through repair activities that, to date, were the most ambitious ever attempted in space. Specifically, the Advanced Camera for Surveys and the Space Telescope Imaging Spectrograph were returned to service to make Hubble the most powerful optical telescope in the world. The STS-125 spacewalks were long and arduous, presenting unforeseen challenges over and above the demanding activities scheduled on the manifest; however, the payoff, seen in the first data, was the reward.

At the launch of STS-125, hopeful astronomers were already planning more research programs using the advanced capabilities of the new telescope. They were confident in the knowledge that over Hubble's 19-year track record, the telescope had greatly surpassed expectations and would continue to do so. Hubble is not the facility that eminent scientist Lyman Spitzer envisioned in the 1940s; it has dramatically exceeded the imagination of all who contributed to the dream of such a capable observatory.

John Mather, PhD
Nobel Prize in Physics (2006).
Senior project scientist for the James Webb Space Telescope, Goddard Space Flight Center.

"The Space Shuttle is a 'brilliant engineering' accomplishment but it was a poor decision on the part of senior leadership as it 'swallowed' other expendable launch vehicles. This decision was not well received by members of the science community.

"The Hubble Space Telescope was conceived and designed to be repaired by the shuttle. In that shuttle had brilliant success. What would have been a 'black eye' forever for American science, the shuttle made the capabilities of Hubble 10 times better, over and over with each servicing mission. In addition to the significant repairs, the shuttle greatly expanded the capabilities of the Hubble by upgrading several key components. The upgrades have allowed superlative science to be accomplished from the Hubble. What has been learned from the Hubble is being used in assembling the James Webb Space Telescope. In my view, there would be no Hubble telescope without the Space Shuttle and no James Webb Space Telescope without the Hubble.

"In many ways, the most important scientific contribution of the Space Shuttle was that it kept the agency (NASA) alive after the Apollo Program. Thus it kept science alive at NASA indirectly.

"Human spaceflight captures people's imagination at gut level. The Space Shuttle was also a product of the Cold War environment of the nation. Humans go into space for more than just science."

The fidelity of Hubble's wide-field imagery is superb because the telescope's exquisite optical quality is not limited by the jitter and distortion caused by the shifting atmosphere that affect images obtained from the ground. Additionally, as instruments with new technologies—primarily more sensitive light detectors—were placed on board, additional wavelengths of light blocked by the Earth's atmosphere could be detected in the ultraviolet region of the spectrum where the Earth's atmosphere is opaque, and for some regions of the infrared that suffer from absorption due to water vapor and other molecules.

Hubble has been a crown jewel in the Space Shuttle Program, providing scientific return and unparalleled public acknowledgment over its lifetime. Launched by STS-31 in 1990, Hubble has contributed to every aspect of astrophysics, achieved its original design goals, and opened new areas of investigation not envisioned in the original proposals for its construction.

Shuttle-enabled refurbishments of Hubble have allowed astronomers to:

- Determine the expansion rate of the universe to 5% accuracy (10% was the goal)
- Discover the existence of dark energy (unexpected) and thus resolve the age of the universe to be 13.7 billion years old
- Identify the host objects for powerful gamma-ray bursts
- Observe some of the deepest images of the cosmos
- Discover protoplanetary disks
- Observe chemical constituents of the atmospheres of planets orbiting other stars
- Characterize the nature of black holes, from supermassive objects in galaxies to stellar-sized objects in star clusters
- Explore numerous views of solar system objects revealing planetary weather and distant dwarf planets still bound to the sun

There were early times in the Hubble program, however, when such amazing accomplishments seemed unachievable.

Astronomical Terms

Astronomical unit: A unit of length used for measuring astronomical distances within the solar system equal to the mean distance from Earth to the sun, approximately 150 million km (93 million miles).

Black hole: Formed when the core of a very massive star collapses from its own gravity. A black hole has such a strong pull of gravity that not even light can escape from it.

Dark energy: Dark energy is inferred from observations of gravitational interactions between astronomical objects and is not directly observed. It permeates space and exerts a negative pressure.

Dark matter: Physicists infer the existence of dark matter from gravitational effects on visible matter, such as stars and galaxies. It is a form of matter particle that does not reflect or emit electromagnetic radiation.

Galaxy: A collection of stars, gas, and dust bound together by gravity. The largest galaxies have thousands of billions of stars.

Light-year: The distance that light travels in a vacuum in 1 year, approximately 9.46 trillion km (5.88 trillion miles).

Nebula: A diffuse mass of interstellar dust or gas or both, visible as luminous patches or areas of darkness depending on the way the mass absorbs or reflects incident radiation.

Planetary nebulae: A nebula, such as the Ring Nebula, consisting of a hot, blue-white central star surrounded by an envelope of expanding gas.

Quasars: Celestial objects that emit extremely high levels of electromagnetic radiation (including light). The amount of energy emitted by a quasar is higher than even the brightest stars. The closest known quasar is 780 million light-years away.

Supermassive black hole: A gigantic black hole, with a mass ranging from millions up to billions of times the mass of our sun, residing at the core of almost every galaxy.

Supernova: The explosive death of a massive star whose energy output causes its expanding gases to glow brightly for weeks or months. A supernova remnant is the glowing, expanding gaseous remains of a supernova explosion.

The Launch of Hubble—First Results

On April 24, 1990, Hubble was launched into orbit with Space Shuttle Discovery (STS-31). The shuttle carried five instruments: the Wide Field Planetary Camera; the Goddard High Resolution Spectrograph; the Faint Object Camera; the Faint Object Spectrograph; and the High Speed Photometer.

During the years of advocacy for the telescope and the subsequent detailed design period, astronomers described some of the amazing results that would be forthcoming from Hubble; however, the much-anticipated first images showed, quite clearly, that something was amiss with the telescope.

Despite their disenchantment, astronomers worked hard to understand and model the Hubble images, and interesting research was accomplished nonetheless. In the first year, the campaign to characterize the nature of black holes in the universe was initiated with the confirmation that a

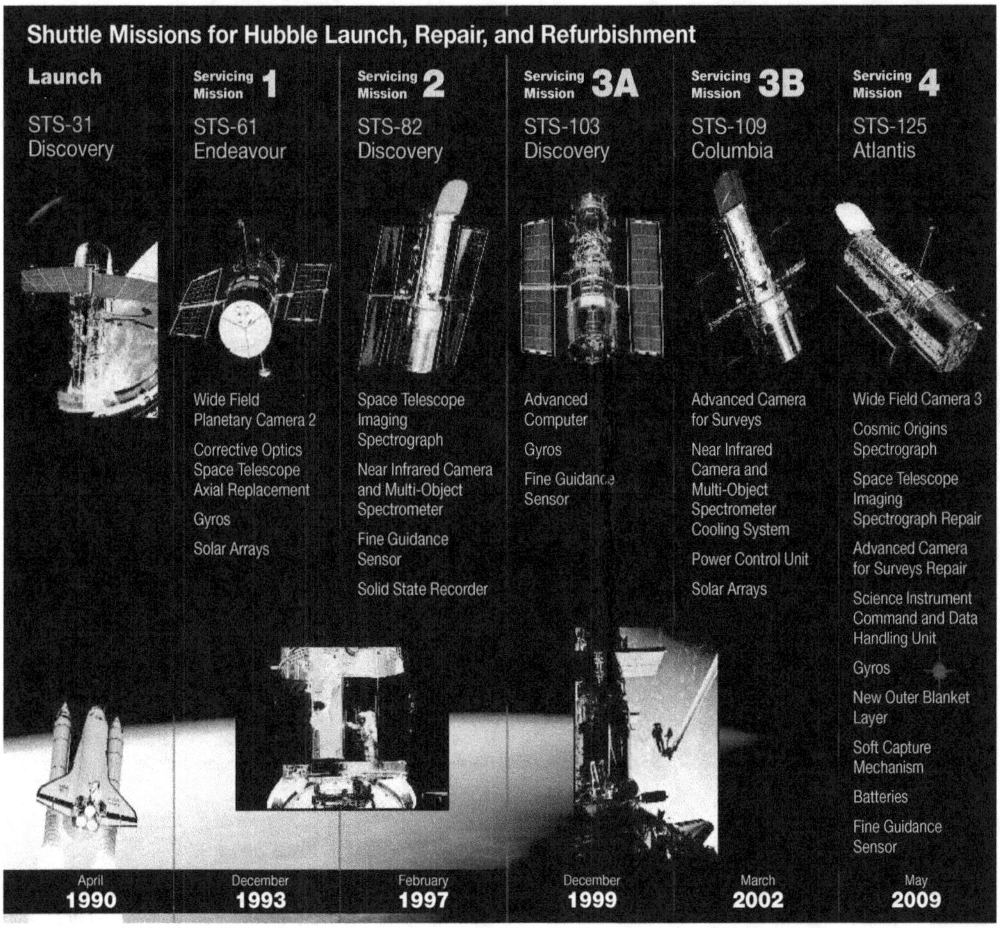

supermassive black hole with mass about 2.6 billion times the mass of the sun resides in the center of the giant elliptical galaxy M87. This result was based on Wide Field Planetary Camera and Faint Object Camera imagery and Faint Object Spectrograph spectroscopy. In addition to that scientific result, optical counterparts of radio jets in galaxies were resolved, spectroscopic observations helped to disentangle the nature of intergalactic clouds absorbing light from near and far galactic systems, and the monitoring of surface features of solar system planets was initiated.

Servicing Mission 1

To correct for the telescope's optical flaw, Hubble scientists and engineers designed and fabricated a new instrument, the Wide Field Planetary Camera 2, and another device called Corrective Optics for Space Telescope Axial Replacement, the latter intended to correct the instruments already on board. The first Hubble servicing mission (STS-61 [1993]) was the ambitious shuttle flight to install the corrective optics and resolve other spacecraft problems. It was a critical mission for NASA. The future of the Hubble program depended on the astronauts' success, and the Space Shuttle Program hung in the balance as well as the future of the agency. The struggle to keep the first repair mission funded was a day-by-day battle that served to cement the cooperation between NASA and the university research community.

As the first images came into focus, overjoyed researchers and engineers began to gain confidence that the promise of Hubble could now be realized.

New Results After Servicing Mission 1

Immediately, NASA obtained impressive results. For example, Wide Field Planetary Camera 2 images of the Orion Nebula region resolved tiny areas of compact dust around newly formed stars. These protoplanetary disks, sometimes called *proplyds*, were the first hint that Hubble would contribute in a significant way to the studies of the formation of extrasolar planetary systems. In another observation, Hubble detected a faint galaxy around a luminous quasar (short for quasi-stellar object), suggesting that luminous quasars and galaxies were fundamentally linked. In our own galaxy, the core of an extremely dense, ancient cluster of stars—the globular cluster 47 Tucanae—was resolved, demonstrating definitively to the skeptical scientific community that individual stars in crowded fields could be distinguished with the superb imaging power of Hubble.

Gas pillars in the Eagle Nebula: Pillars of Creation in star-forming region captured by the Wide Field Planetary Camera 2 in 1995. The region is trillions and trillions of miles away in the constellation Serpens. The tallest pillar is 4 light-years long and the colors show emissions from different atoms.

Shoemaker-Levy

Early Hubble observations of solar system objects included the spectacular crash of Comet Shoemaker-Levy 9 into Jupiter in 1994. This event was witnessed from start to finish, from the first fragment impact to the aftermath on the Jovian atmosphere. Images were also taken in visible blue light and ultraviolet light to determine the depth of the impacts and the nature of Jupiter's atmospheric composition.

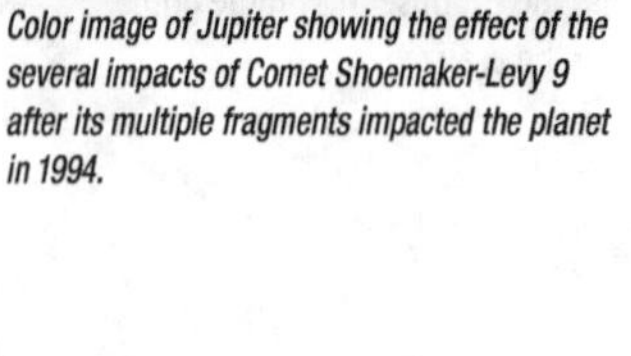

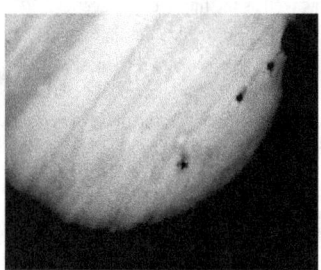

Color image of Jupiter showing the effect of the several impacts of Comet Shoemaker-Levy 9 after its multiple fragments impacted the planet in 1994.

Pillars of Creation

The famous "Pillars of Creation" image of the Eagle Nebula captured the public imagination and contributed to the understanding of star-formation processes. The images captured in 1995 with Wide Field Planetary Camera 2 showed narrow features protruding from columns of cold gas and dust. Inside the gaseous "towers," interstellar material collapsed to form young stars. These new hot stars then heated and ionized the gas and blew it away from the formation sites. The dramatic scene, published in newspapers far and wide, began to redeem the public reputation of Hubble.

Existence of Supermassive Black Holes

From ground-based data, scientists knew that galaxies exhibit jets and powerful radio emission that extends well beyond their optical periphery. Huge x-ray emissions and spectroscopic observations of galaxies suggested that some of these objects might contain a large amount of mass near their centers. Even Wide Field Planetary Camera 2 observations of the innards of several galaxies suggested that black holes might be hidden there. However, it was the observation of the giant elliptical galaxy M87 with the Faint Object Spectrograph that conclusively demonstrated that supermassive black holes exist in large galaxies. This was the turning point in black hole studies, with spectroscopy being the powerful diagnostic tool astronomers could use to begin the Hubble census of these exotic objects.

Building Blocks of Early Galaxies

One of the planned goals for Hubble research was to understand the nature of the universe and look back in time to the earliest forming galaxies. In December 1995, 2 years after the first servicing mission, Hubble's Wide Field Planetary Camera 2 was pointed at a field in Ursa Major for 10 days, accumulating 342 exposures. The final image—the Hubble Deep Field—was, at the time, the deepest astronomical image ever acquired. The field probes deep into the universe and contains over 1,500 galaxies at various distances.

After the Hubble Deep Field data were produced, telescopes were pointed at the same part of the sky to obtain data in every conceivable way. Besides bolstering the idea that galaxies form from building blocks of smaller components that are irregularly shaped and that the rate of star and galaxy formation was much higher in the past, analysis of the data pushed the observable universe back to approximately 12 billion years. Papers written on Hubble Deep Field data alone number in the hundreds and document a new understanding of cosmological and astrophysical phenomena.

The immediate release of Hubble Deep Field data represented a watershed in astronomical research as well. A new method was born for concentrating astronomical facilities and the collective brainpower of the scientific community on a specific research problem. Thus, the Hubble Deep Field represents not only a leap forward in scientific understanding of the universe, but a significant alteration in the way astronomy was conducted.

Edward Weiler, PhD
Chief scientist for the Hubble Space Telescope (1979-1998).
NASA associate administrator, Science Mission Directorate.

"It's fair to say that Hubble, today, would be a piece of orbiting space debris if it hadn't been for the Space Shuttle Program. If Hubble had been launched on an expendable launch vehicle, we would have discovered the optical problem yet been unable to fix it. Hubble would have been known as one of the great American scientific disasters of our time. Hubble's redemption is due to the Space Shuttle Program and, most importantly, to the astronauts who flew the shuttle and did things (in repairing Hubble) that we never thought could be done in space. Hubble became a symbol of excellence in technology and science, and the shuttle made that happen.

"I've spent 34 years on Hubble in one way or another. I was on top of Mount Everest at the launch, with all of us astronomers who had never done an interview. I was on the Today Show *and* Nightline *on the same day. I experienced the ecstasy in April 1990, to the bottom of the Dead Sea 2 months later when a spherical aberration was detected in the Hubble. In our hearts, we knew we could fix it. We promised the press we would fix it by December 1993, and nobody believed us. Then, on December 20, 1993, we saw the first image come back. It was spectacular. It was fixed. And the rest is history. We went from the bottom of the Dead Sea back to the top of Mount Everest and beyond…we were elated!"*

Subsequent Servicing Missions

Servicing Mission 2

By the end of 1996, Hubble was a productive scientific tool with instruments for optical and ultraviolet astronomy. During the second servicing mission in February 1997, the STS-82 crew installed two new scientific instruments: the Near Infrared Camera and Multi-Object Spectrometer, extending Hubble's capabilities to the infrared, and the Space Telescope Imaging Spectrograph, offering ultraviolet spectroscopic capability. Astronomers now expanded their research to probe astrophysical phenomena using the excellent imaging performance of Hubble coupled with new capability over a larger range of wavelengths.

Servicing Missions 3A and 3B

The third servicing mission was intended to replace aging critical telescope and control parts to retain Hubble's superb pointing ability and to install new computer equipment and a new instrument; however, when a third (out of six) gyroscope on Hubble failed—three gyros are needed for target acquisition—NASA elected to split these missions into two parts. To add to the drama, a fourth gyroscope failed on November 13, 1999. Hubble was safe, but it could not produce scientific observations. Another bit of tension was created by concern about the transition to the year 2000 and the hidden computer problems that might occur. Just in time, on December 19, 1999, Space Shuttle Discovery (STS-103) delivered new gyroscopes, one fine guidance sensor, a central computer, and other equipment, restoring Hubble to reliable operation and making it better than ever.

The next servicing mission occurred during March 2002 when Space Shuttle Columbia (STS-109) was launched to further upgrade the telescope. The new science instrument, the Advanced Camera for Surveys, was installed with a wide field of view, sharp image quality, and enhanced sensitivity. The Advanced Camera for Surveys field was twice that of the Wide Field Planetary Camera 2 and collected data 10 times faster. The astronauts also installed new solar array panels, a power control unit, and a new cooler for the Near Infrared Camera and Multi-Object Spectrometer to extend its life. They also installed a refurbished Fine Guidance Sensor and a reaction wheel to ensure telescope steering and fine pointing.

Hubble Deep Field and Hubble Ultra Deep Field

With the new infrared capability installed during the second servicing mission, astronomers turned the Near Infrared Camera and Multi-Object Spectrometer to view part of the original Hubble Deep Field for a series of long exposures. Extremely distant objects were revealed, objects that had been undetected in the optical Hubble Deep Field because their light was red-shifted due to the expansion of the universe. The original Hubble Deep Field is located in the northern celestial hemisphere. In 1998, NASA added a second field, the Hubble Deep Field-South, to the collection. The second field represented another "core sample" of the universe compared and contrasted to the northern observation to verify that Hubble Deep Field-North is representative of the universe in general. Researchers took advantage of Space Telescope Imaging Spectrograph and Near Infrared Camera and Multi-Object Spectrometer cameras to obtain deep adjacent fields as additional samples of the universe in the ultraviolet and infrared.

The Hubble Ultra Deep Field showing thousands of galaxies reaching back to the epoch when the first galaxies formed.

After astronauts installed Advanced Camera for Surveys during the third servicing mission, astronomers pushed the limits of observation even further in an additional field called the Hubble Ultra Deep Field. Deep Advanced Camera for Surveys and Near Infrared Camera and Multi-Object Spectrometer data revealed thousands of galaxies, some of which existed a mere 800 million years after the Big Bang. The optical detections reached 31 to 32 magnitudes, at least seven times deeper than ever before, and there were hints from the new Near Infrared Camera and Multi-Object Spectrometer data that galaxies as young as a few million years after the creation of the universe were detected. Observations with NASA's Spitzer Space Telescope produced deep images of the Hubble Ultra Deep Field in the infrared. These data were analyzed

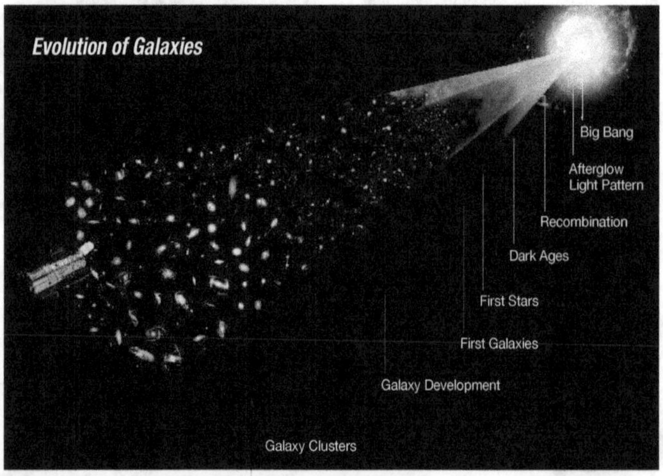

Schematic of Hubble sampling galaxies through space and time.

along with the Hubble data to provide a more complete catalog of very distant galaxies with the result that at least one surprisingly massive galaxy was identified in the field where only small "precursor" galaxies were expected.

Astronomers were quick to test that result using Wide Field Camera 3, deployed during a servicing mission. The faintest galaxies found are blue and should be deficient in heavy elements, meaning they are from a population that formed extremely early when the universe was only 600 million years old. More data from Wide Field Camera 3 may reach even 100 million years earlier. Beyond that, astronomers anticipate continuing to push earlier in the universe with the launching of the James Webb Telescope.

Age of the Universe

The cornerstone investigation to be carried out by Hubble was the determination of the age of the universe. Previous work provided a wide range for this age: from 10 to 20 billion years old—a factor of two. Hubble research was to address one of the most basic questions about the cosmos, and further refinement was to be based on more accurate measurement of the cosmological expansion rate; i.e., the Hubble constant. From this expansion rate, the age of the universe can be determined by tracing the expansion back to the origin of cosmos. In fact, this key project was used as prime justification for fabrication of the telescope.

In particular, it was well known that data for variable stars called Cepheids were critical to answer this fundamental question. Cepheid variables were discovered in the early 1900s when Henrietta Leavitt studied photographic plate material while working at the Harvard College Observatory. She

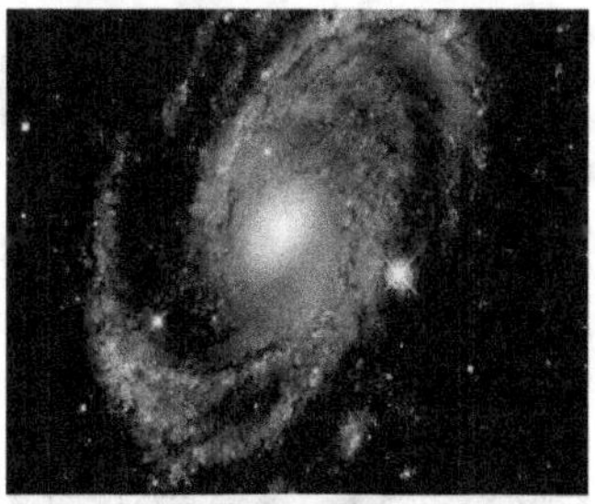

The spiral galaxy NGC 4603 is the most distant galaxy used to study the pulsating Cepheid variables for the Hubble constant study. This galaxy is associated with the Centaurus cluster, one of the most massive groupings of galaxies in the closer universe. This image was assembled from Hubble Wide Field Planetary Camera 2 data obtained in 1996 and 1997.

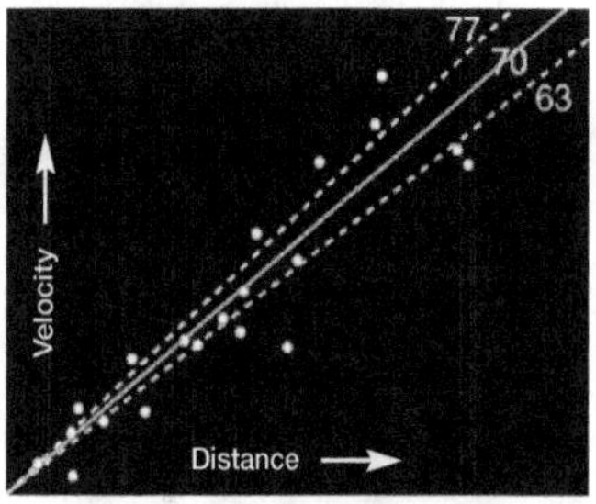

This Hubble Diagram for Cepheids shows a plot of galaxy distance (determined from the Cepheid variables in it) vs. the velocity that the galaxy appears to be receding from Earth (determined from spectroscopy). The graph shows that the value of Hubble constant is the best fit from the key project observations.

carefully compiled a list of stars that changed brightness regularly in the nearby Large and Small Magellanic Clouds, companion galaxies to our own Milky Way. While classifying the subset of variable stars that were Cepheids, she noticed that objects with longer periods of variation were brighter. Her "period-luminosity" relation is the basis for the use of Cepheids as a standard to be used for distance measurements. Before Hubble observations were taken, distances to nearby galaxies had been determined from Cepheids using ground-based telescopes to map the local structure, motions, and expansion.

Since the results from many previous studies of the nearby universe produced such disparity, a goal of the Hubble observational program was to push the measurements out farther to more distant, fainter objects and determine Hubble constant with greater accuracy. It also was understood at the time of the launch of Hubble that the oldest objects known, the globular clusters, had ages of about 15 billion years, and this result served as an independent measure of the age of the universe (the universe has to be at least as old the objects in it).

The key project team measured superb resolution Hubble Wide Field Planetary Camera 2 images over many years. The team identified nearly 800 Cepheids in 18 galaxies out to 65 million light-years. Data from 13 other galaxies were combined for a total of 31 galaxies with measured distances. The recession velocity of each galaxy was plotted against each galaxy's distance as measured from the Cepheids for a self-consistent measurement. This plot indicated the expansion rate exhibited by the benchmark galaxies was within 10% of Hubble constant. The results, published in 2001, also compared favorably with the Hubble flow calibrated with several secondary distance indicators that could also be used in more remote objects. Type Ia supernova is a category of cataclysmic stars that formed as the violent explosion of a white dwarf star. It produces consistent peak luminosity and is used as standard candles to measure the distance to their host galaxies. The brightnesses of Type Ia supernovae, being much brighter than Cepheids, are critical for measuring Hubble constant at even larger distances, and those measurements could be combined with the Cepheid values. At that point, one of Hubble's major objectives was achieved.

While the measurements of Hubble constant were converging to a consistent value, the simplest cosmological model in favor (the Einstein-de-Sitter model), used to convert the expansion rate into an age for the universe, resulted in a value of about 9 billion years. The situation was clearly impossible. The ever-refined globular cluster ages dropped slightly with better understanding of stellar astrophysics, but the big question in cosmology remained: 13 billion or 9 billion? The quandary was finally resolved for the most part with the discovery that the expansion rate is changing over time and the universe is actually accelerating, so the age derived from the simple model is not correct. The new model, which accommodates this circumstance, has resolved the discrepancy, resulting in an age of the universe of 13.7 billion years that is consistent with the independent globular cluster ages.

The story is not complete, however. A study reported in 2009, using Near Infrared Camera and Multi-Object Spectrometer data, produced a value of Hubble constant to within 5% uncertainty. This measurement represents a factor of two in improvement and is in general agreement with the key project report. The acceleration and age of the universe will continue to be investigated and refined. Thus, the determination of Hubble constant and the detailed nature of the expansion of the universe will be important research topics for future Hubble studies.

Interacting Galaxies

Galaxies occur in a variety of environments: small groups, such as those surrounding our own Milky Way; medium-sized and large clusters; and tight formations of interacting objects.

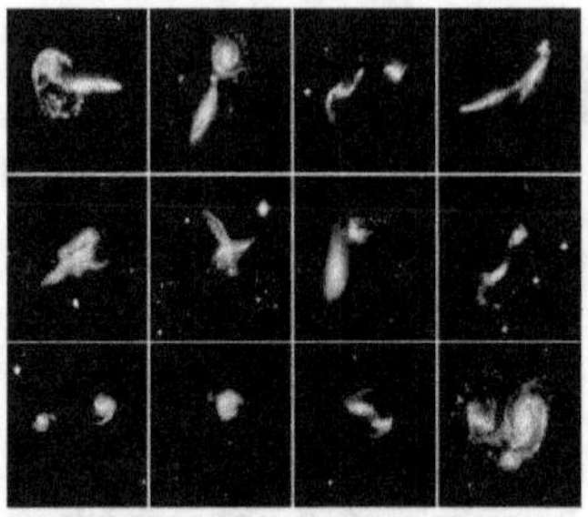

A small selection of the hundreds of interacting galaxies observed by Hubble.

The study of interacting or colliding galaxies yields information about how galaxies may have formed and merged in the early universe and how star formation is triggered across the span of a spiral galaxy's disk. From the first days of Hubble observations to years later, magnificent images of pairs, groups, and small clusters of galaxies have been obtained for this research.

Gamma-ray Bursts

Knowledge of the existence of energetic bursts of emission in gamma rays from all across the sky was traced to the 1960s with the serendipitous detection of gamma-ray bursts by the US Vela satellites designed to detect gamma rays from nuclear weapon tests. The nature of the bursts was enigmatic and posed a problem for astrophysics once it was understood that the energy originated from somewhere in the sky. Data from the Burst and Transient Source Experiment instrument of the Compton Gamma Ray Observatory, launched in 1991, represented a watershed in understanding by demonstrating that gamma-ray bursts come from *everywhere* in the sky. The search was on until 1997 when another gamma-ray satellite, BeppoSAX, with an Italian/Dutch instrument, detected a gamma-ray burst called GRB 970228 associated with a fading x-ray emitter. The breakthrough in understanding came as scientists identified the optical counterpart in Hubble images and realized that the source resided in a distant galaxy. Hubble monitored the object and traced its rate of fading over time. The observations demonstrated that although the source was in a distant galaxy, it was not near its center, suggesting that the bursts were associated with a single object but not the galaxy's nucleus.

Hubble research identified a number of gamma-ray bursts over time, and all were attributed to objects in distant galaxies. For example, a staggeringly bright object in a host galaxy was identified with Wide Field Planetary Camera 2 after detections by the BeppoSAX and Compton satellites in 1997. In general, Hubble data are used to monitor the fading of the object months after the initial burst, when the emission is no longer observable by other facilities. An accumulation of such observations of over 40 objects with Space Telescope Imaging Spectrograph, Wide Field Planetary Camera 2, and, later, Advanced Camera for Surveys clarified that "long-duration" gamma-ray bursts reside in the brightest regions of small, irregular galaxies. The analysis suggests that the progenitors are massive stars, roughly 20 or more times the mass of the sun, in regions with a dearth of heavy chemical elements. Overall, gamma-ray bursts appear associated with some sort of stellar collapse sometimes involving magnetic fields and the creation of stellar black holes, often associated with supernovae explosions.

Hubble Wide Field Planetary Camera 2 recorded the brightest supernova gamma-ray burst that could be seen with the naked eye halfway across the universe. The explosion was so far away, it took its light 7.5 billion years to reach Earth. In fact, the explosion

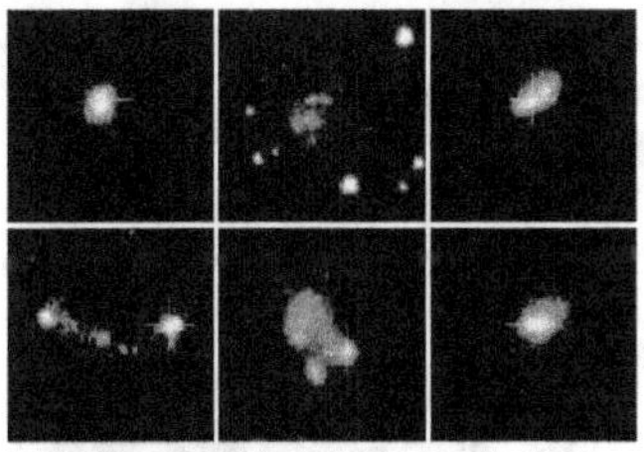

Hubble images probing the environments of gamma-ray burst detections found that long-duration events are located in small, faint, misshapen, (irregular) galaxies, which are usually deficient in heavier chemical elements. Only one of the bursts was spotted in a spiral galaxy like our Milky Way. The burst sources are concentrated in the brightest regions of the host galaxies, suggesting they may come from some of the most massive stars, for example those that are 20 times the mass of the sun.

took place so long ago that Earth had not yet come into existence. This object may be a star more than 50 times the mass of the sun that had exploded much more violently than the "usual" supernovae. These objects, called *hypernovae*, fade more slowly than other gamma-ray bursts.

Black Hole Census

Astronomers had avidly searched for the existence of black holes in galaxies with a variety of instrumentation and telescopes, and it was spectroscopic observations of large galaxies that revealed that supermassive black holes might be quite common. After servicing mission 2, astronomers were able to employ a full suite of Hubble instruments to continue the ongoing inventory of black holes in galaxies. Researchers eventually inferred that the smaller black holes exist in smaller galaxies, so that a correlation between galaxy size and black hole mass was uncovered. Near Infrared Camera and Multi-Object Spectrometer and Wide Field Planetary Camera 2 data uncovered evidence for black holes in a growing list of objects. The detailed profiles of black holes were traced with spectroscopic data from Space Telescope Imaging Spectrograph. Astronomers observed material surrounding the cores of numerous galaxies. This material exhibited features particular to material spiraling into black holes. In addition, jets, bubbles, and dense star clusters were detected. A black hole also was discovered in our own galaxy's nearby companion, M31. The exotic nature of star clusters close to the black hole in the center of the Milky Way was characterized through infrared observations with Hubble. The picture that emerged is that black holes are pervasive in the center of galaxies rather than a rarity. Giant elliptical galaxies and spiral galaxies with enormous bulge components seem to be the hosts of supermassive black holes, whereas galaxies such as the Milky Way, with smaller bulges, have smaller black holes. Another link between galaxies and black holes is that it now appears that very active nuclei, called active galactic nuclei, and luminous quasars are linked to black hole and galaxy formation.

Image of the entire galaxy M87 taken with Advanced Camera for Surveys.

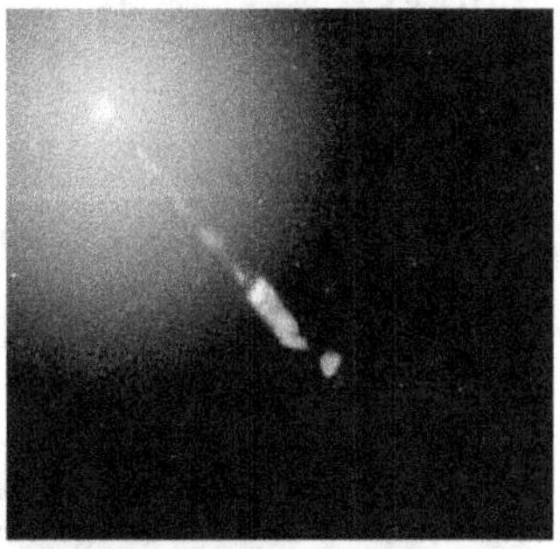

Detail of the jet and tight core containing an enormous black hole in the giant elliptical galaxy M87 reobserved with Hubble's Wide Field Planetary Camera 2.

The black hole in the center of the giant elliptical galaxy M87 is the best studied with Hubble. Since Hubble was first launched, the instruments on board have been used to image the detail of the galaxy's core, the structure of its jet, and, more recently, the flare-up of the jet as observed with Advanced Cameras for Surveys and Space Telescope Imaging Spectrograph. The mysterious brightening and fading is likely due to activity around the black hole.

Astronomers also have pushed Hubble to observe smaller-sized black holes; for example, mapping the chaotic fluctuations in the ultraviolet light exhibited by Cygnus XR-1, one of the first stellar black holes known. The observations verified the existence of material sliding through the event horizon of the black hole. Apparently, medium-sized stellar black holes do exist as well, as determined from Wide Field Planetary Camera 2 images and Space Telescope Imaging Spectrograph spectroscopic observations of the globular cluster M15. Since these star clusters contain the oldest stars in the universe, they probably contained black holes when they originally formed. An intermediate-mass black hole was similarly discovered in the giant cluster "G1" in M31. With improvements in instrumentation coupled with excellent pointing stability, the multiyear Hubble black hole campaign has provided insights into the black holes in the violent cores of galaxies and possible linkages to stellar-mass black holes formed in the early universe.

Star Formation

Luminous nebulae comprised of ionized hydrogen with numerous and sometimes hundreds of young stars can be seen in our own galaxy, in nearby galaxies, and in distant galaxies. These star-forming regions are sites of clusters of stars containing some massive objects that are synthesizing many of the heavy chemical elements, later to be spewed out in stellar explosions. Studies of these objects with Hubble allowed the details of the nebulae to be mapped along with the interaction of the hot stars emitting intense ultraviolet radiation causing the nebular material to be ionized and glow. The first images of such regions included the Orion star-forming region and the Eagle Nebula.

One such huge complex is 30 Doradus—the largest in the local group of galaxies. It is located 170,000 light-years from Earth in the Large Magellanic Cloud, a companion galaxy to the Milky Way. It has been called an astronomical "Rosetta Stone" because detailed examination of the object gives a clue to the nature of star-forming regions that are seen, but unresolved, in distant galaxies across the universe.

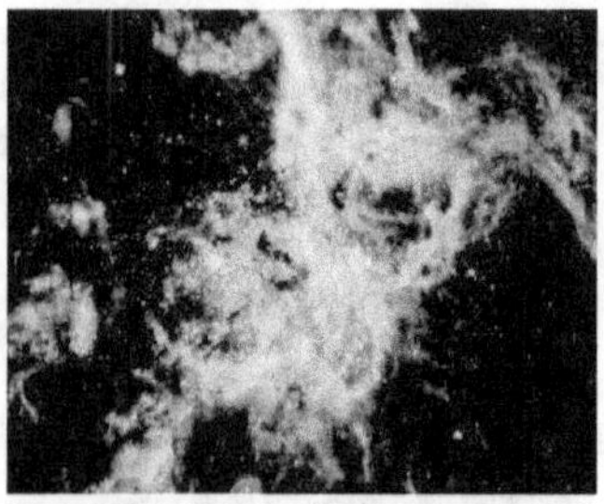

The Wide Field Planetary Camera 2 image of 30 Doradus in the Large Magellanic Cloud contains one of the most spectacular clusters of massive stars, called R136, in our cosmic neighborhood of about 25 galaxies.

The Orion Nebula is a star-forming region in our own galaxy and close enough to be seen in small telescopes as it is 1,500 light-years away. Because this region is so vast, the large mosaic image was created after the Advanced Camera for Surveys was installed. Detailed examination of parts of the image shows stars, gas, and dust as well as several regions revealing clusters of stars forming. The nebula itself is being disrupted by radiation from those stars, leaving loops, bubbles, and rings of material, all of which can be distinguished with the high-resolution Advanced Camera for Surveys composite.

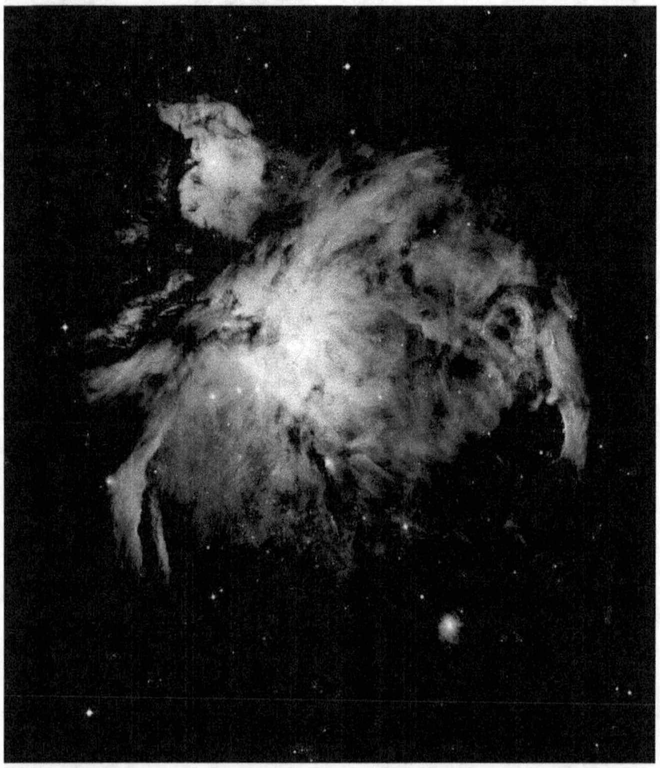

Image mosaic of the Orion Nebula exhibiting clumps of stars forming and the detailed sculpting of the nebula by radiation from the bright young stars formed there. (Hubble Advanced Camera for Surveys and European Southern Observatory at La Silla [Chile] 2.2-m [7.2-ft] telescope.)

With the installation of the infrared Wide Field Camera 3 during servicing mission 4, Hubble observers can peer into the dust of these tumultuous regions.

Stellar Death Throes

Planetary Nebulae

Some of the most photogenic nebulae are remnants of the last stages of stellar life, called planetary nebulae. These nebulae are formed from stars with mass similar to the sun while stars larger than eight times the solar mass end as supernovae. In small telescopes, these nebulae appear as roundish, smooth objects but, in fact, no two planetary nebulae are alike. With Hubble observations, it has become clear that planetary nebulae formation is very complex. Material is often ejected in rings and loops, and the nebulae chaotic structures suggest that these stars shed mass in several

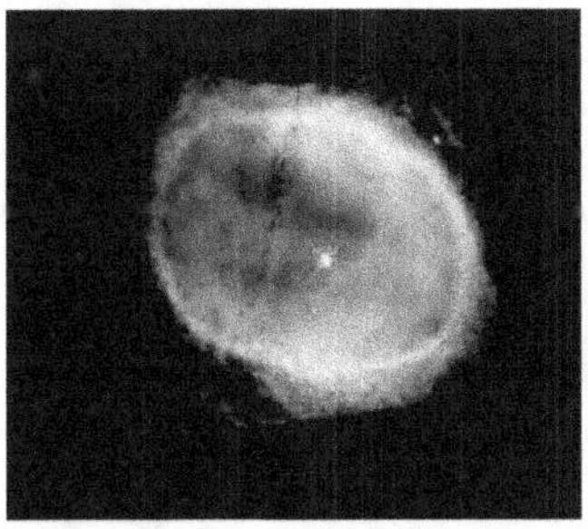

The diversity of planetary nebulae is shown in this image. The nearly symmetric appearance of NGC 3132 in this Wide Field Planetary Camera 2 image shows the more "classic" morphology of a planetary nebula.

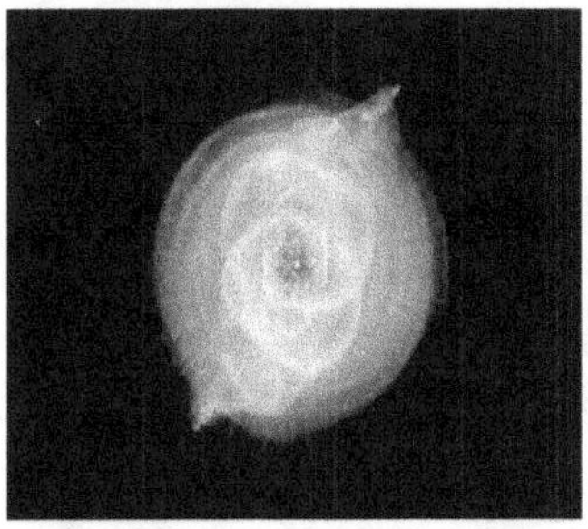

This nebula, called NGC 6543 or the Cat's Eye Nebula, was one of the first planetary nebulae to be discovered. The Advanced Camera for Surveys image shows how complex these objects can be. Planetary nebulae exhibit a huge range of diverse morphologies due to their formation process.

episodes. Some of the nebulae exhibit irregular streamers and nodules as well. It is likely that the interplay of stellar winds and radiation emitted by the star causes the structures, but the exact manner in which this occurs is still poorly understood.

Supernovae and Supernova 1987A

Stars larger than about eight times the mass of the sun end their lives in a different, spectacular way as their nuclear fuel is exhausted. The violent explosion, a supernova, blows off a significant fraction of the star's mass into a nebula or remnant, emitting radiation from the x-ray to the radio. The resulting nebula is a complex twist of material and magnetic fields, giving these objects complicated shapes. The detailed, exceptional imagery from Hubble has allowed researchers to examine the morphologies of these objects.

There are several classifications of supernova reflecting different features and formation mechanisms. The supernovae called type Ia are sometimes formed by binary stars. The importance of these types of supernova is that they appear to have a signature luminosity increase and a particular relationship between the various energies emitted. Because they have unique characteristics, they are considered "standard candles"; i.e., they have a known intrinsic brightness so that, when they are discovered in distant galaxies, the distance to them can be fairly accurately determined. These objects are lynchpins in the study of the expansion of the universe and the discovery of dark energy.

One well-known supernova in our own galaxy, the Crab Nebula, has been imaged by Hubble over several years. In addition to the intricate appearance of the nebula, the actual explosive event was witnessed by Japanese and Chinese astronomers in 1054 and most likely was also seen by Native Americans.

Many supernovae remnants in the galaxy are so large they cannot be imaged easily with a few exposures of Hubble. The supernova remnant called N132D is one of several such objects imaged by Hubble. It is located in

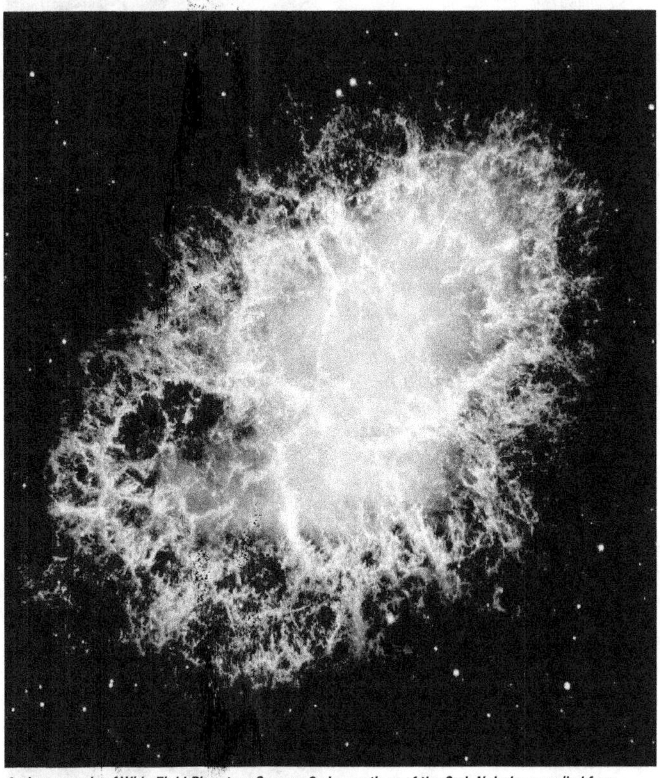

A giant mosaic of Wide Field Planetary Camera 2 observations of the Crab Nebula, compiled from observations accumulated in 1999 and 2000.

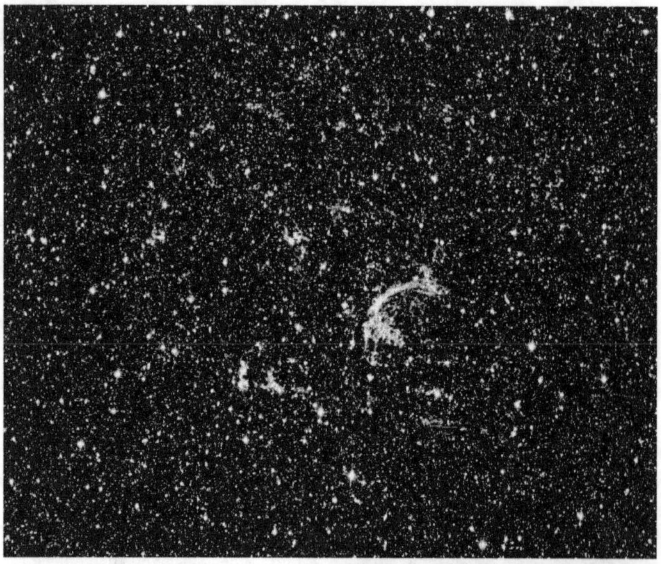

This Advanced Camera for Surveys and Wide Field Camera 3 image is of supernova remnant N132D in the Large Magellanic Cloud.

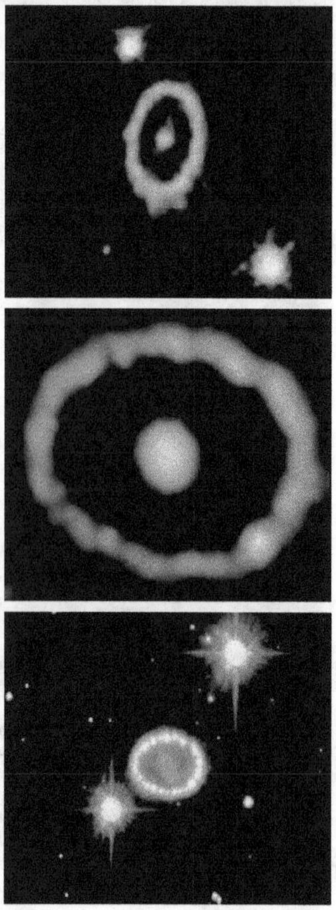

Light from the explosion illuminated the remnant material around supernova 1987A. The Wide Field Planetary Camera 2 images revealed an hourglass structure evidenced in the two overlapping rings. The central ring is apparently in a plane parallel to the other rings, which are in front of and behind the central ring as seen from this vantage point. The upper image was obtained in 1994. The next image, also obtained in 1994, shows the brightening of the inner ring caused by the explosion shockwave impacting the ring. Twenty years after the explosion, Hubble was able to resolve multiple sites that were illuminated due to the shockwave continuing to expand outward into the remnant material from prior events. The lower image, from Advanced Camera for Surveys, shows the fully illuminated ring and the outer ring structure from 2006.

the Large Magellanic Cloud, close enough for detailed examination, but sufficiently far away to allow the whole structure of the nebula to be examined. The observation of N132D is actually a composite of the newly restored Advanced Camera for Surveys, repaired during servicing mission 4, and the new Wide Field Camera 3. A spectrum of this object was also obtained with the new Cosmic Origins Spectrograph instrument to analyze the chemical composition of the nebula.

The most famous and scientifically important supernova is supernova 1987A, an object that exploded in the Large Magellanic Cloud in February 1987. The light from the explosion expanded outward and illuminated material far from the progenitor star, suggesting prior outflows and explosions may have occurred. Astronomers have used nearly every Hubble camera to monitor changes in supernova 1987A. Merged with observations from other observatories, the Hubble images have contributed to the understanding of this particular object. This information also has helped with understanding of type Ia supernovae in general.

Dark Energy

At its inception, Hubble was designed to determine the age of the universe through measurements of cosmological expansion—the value of Hubble constant. Every improvement in instrumentation, computing systems, and telescope capability has led to greater knowledge and sometimes extraordinary results about the cosmos. As details of the universe's expansion unfolded, astronomers derived an unexpected nuance of the expansion. It appears from Hubble observations that the universe is not expanding at a constant rate or slowing down under the tug of gravity as astronomers expected. Instead its expansion is speeding up and has been for the past 4 to 5 billion years.

A key to this discovery is the understanding that, like Cepheid variables, supernovae can be used as distant light posts or standard candles, but supernovae are about a million times brighter. One type of supernova explosion, a *Type Ia Supernova* (abbreviated SN Ia), is thought to explode as a result of binary stars exchanging matter. The explosive output, 1-2 x 10^{44} joules or about 3.5 x 10^{28} megatons of TNT, has a specific profile: a fast rise in a few hours or days and a decline over about a month or so. These objects also achieve a more or less typical intrinsic brightness—the characteristic that makes SN Ia a valuable standard for measuring the distances to its very remote host galaxies in which the supernova is imbedded.

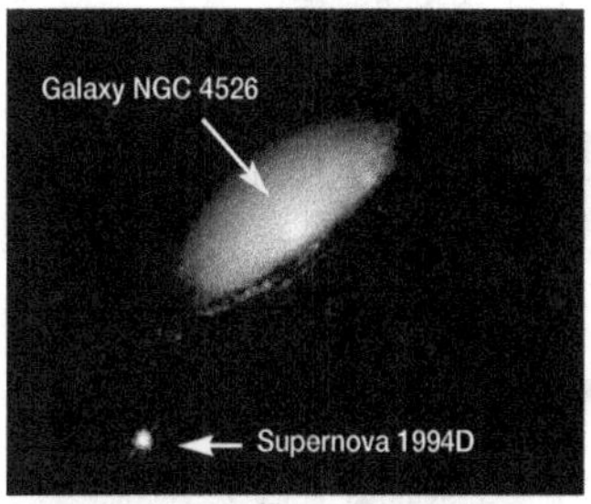

Supernova 1994D in galaxy NGC 4526 can be seen as the bright star in the lower left corner of this Hubble Wide Field Planetary Camera 2 image.

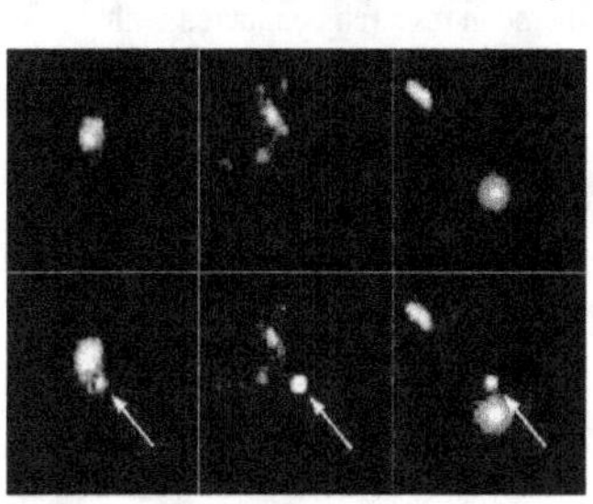

Images of three of the most-distant supernovae known, discovered using the Advanced Camera for Surveys.

Steven Hawley, PhD
Astronaut on STS-41D (1984), STS-61C (1986), STS-31 (1990), STS-82 (1997), and STS-93 (1999).

"I have been very fortunate to be among a very small group of individuals to have seen the Hubble Space Telescope in space—twice. A memory that I will cherish forever is seeing the Hubble Space Telescope as we approached on STS-82, 7 years after I released it from Discovery in April 1990. To see Hubble Space Telescope once in a career is special, but to see it twice is truly a privilege. I remember when we were able to see the back side of Hubble Space Telescope for the first time on the 1997 mission. Hubble Space Telescope keeps one side preferentially pointed at the sun and that side is opposite the side to which you approach in the shuttle to grapple the telescope. When we saw the far side, we were able to see that the thermal insulation resembling aluminum foil looked brittle and had peeled away from the telescope in some locations. Prior to the last extravehicular activity for that mission our crew was asked to fabricate some temporary patches from material that we had on board and to install them over some of the worst damaged sites. Before we did that we all signed the foil patches, so for a while my signature was on the Hubble Space Telescope."

Hubble was employed along with several powerful ground-based telescopes to seek out and measure SN Ia across the universe. Hubble Wide Field Planetary Camera 2 was first used to map SN Ia and then deep Advanced Camera for Surveys observations probed the most distant supernova. The amassed observations helped refine Hubble constant. Since the measurements extended to some of the farthest reaches of the universe, it was possible to use all the SN Ia observations pieced together to measure another important cosmological parameter: the cosmological constant. The cosmological constant was proposed by Einstein in his General Relativity as a kind of "repulsive gravity," a means of keeping the universe static so that it would not collapse under its own gravity. When he learned from Edwin Hubble that the universe is not static but is in fact expanding, Einstein removed the cosmological constant from his equations (and referred to it as "my greatest blunder"). The observations by the Hubble Space Telescope and its partner ground-based telescopes that the expansion of the universe is in fact accelerating under the influence of some completely baffling force, a kind of repulsive gravity, strongly suggests that the cosmological constant may not have been a "blunder" after all.

This result is problematic as we currently do not have a succinct theory to explain why this situation exists. For example, we know the Big Bang that originated the universe causes objects to recede from each other when measured over cosmological distances. We also know that gravity is the retarding force that slows the expansion due to mutual attraction between all matter in the universe. Therefore, either the universe would keep expanding because there is not enough matter (gravity) to slow it or its expansion would slow (decelerate) because there is enough gravitational force to retard that expansion. Acceleration of the expansion does not fit into this picture. The unexplained cause of the acceleration, called dark energy, is the focus of additional observations and theoretical work. The existence of the acceleration has been confirmed by detailed analysis of the Wilkinson Microwave Anisotropy Probe observations designed to measure the cosmic microwave background, the remnant radiation from the Big Bang. Other observations of x-ray emission, further observations of supernovae, and other results have contributed to the confirmation of this puzzle.

Needless to say, the discovery of evidence for dark energy was not predicted for Hubble or for any other observatory constructed to date. This significant problem in physics and astrophysics is expected to be a driving part of the design for new telescopes to be commissioned in the next decade.

This is an image of a gravitational lens obtained after servicing mission 4 with the newly repaired Advanced Camera for Surveys camera. Abell 370 is one of the very first galaxy clusters where astronomers observed the phenomenon of gravitational lensing.

Dark Matter

An interesting phenomena produced by gravitational fields is *gravitational lensing*. A warping of space by a large mass such as a cluster of galaxies can distort light from more distant objects. The distortions appear as shreds of images, stretched into arcs and streaks. Gravitational lenses are of interest for two main reasons: first, the very distant objects can be analyzed since the lens also enhances the brightness of the far galaxy or luminous quasar; and second, the total mass of the lensing cluster can be determined. The total mass is a composite of luminous mass (the galaxies detected by Hubble) plus dark, unseen matter. Reconstruction of the mass distribution gives clues to the nature of dark matter that cannot be seen through telescopes. Such observations also were combined and used to create a three-dimensional map of dark matter in the universe, although the true nature of the material is still unknown.

Extrasolar Planets

A planet outside the solar system is commonly categorized as an extrasolar planet. Scientists have made confirmed detections of 473 such planets. The vast majority were detected through velocity calculations observations and other indirect methods rather than actual imaging. The search for planets forming around other stars has been a consistent theme in research conducted with Hubble. Besides probing star-formation regions, Hubble is used to detect planetary disks around stars where planets are likely to be forming. While it was not expected that Hubble would contribute significantly to the detection and characterization of extrasolar planets, the opposite has been true.

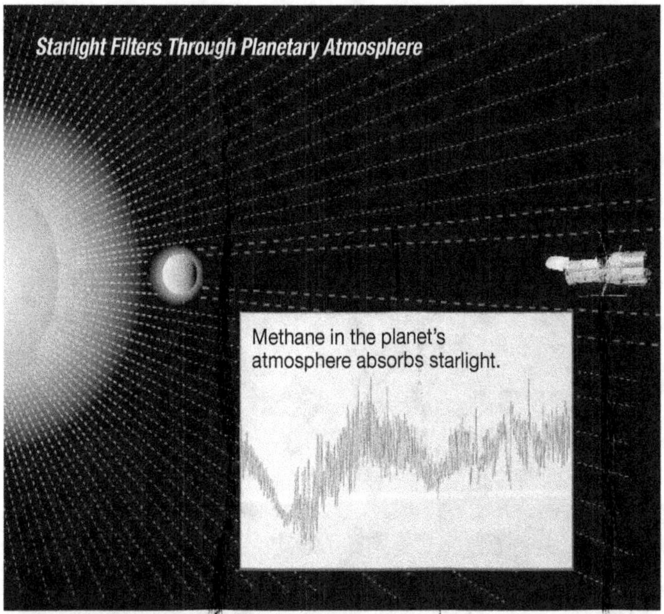

An illustration of the spectrum obtained from an extrasolar planet and the configuration of the parent star, the planet, and Hubble to obtain the observation.

In 2001, Hubble observed the first transit of an extrasolar planet across the disk of its parent star. The yellow dwarf star HD 209458 has a Jupiter-sized planet in a tight, 3½-day orbit around it. The extremely close orbit causes the planet to lose its atmosphere; i.e., the atmosphere is blowing off its surface into space. It is the planet plus the atmospheric material that caused a slight dip in the brightness of the star that could be observed with precise observations.

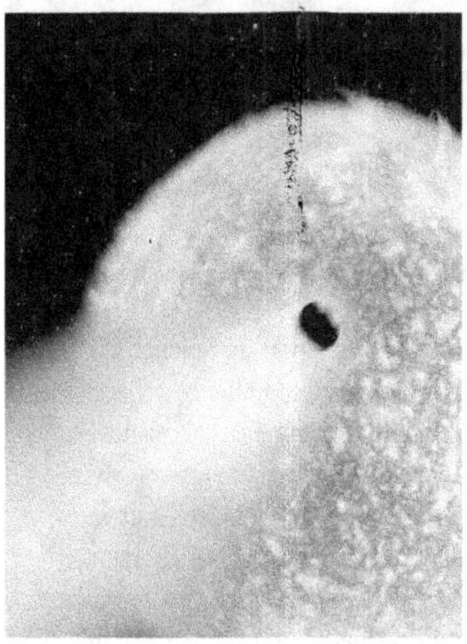

Scientists reported the first-ever optical detection of an extrasolar planet, which passed in front of a huge star in the constellation Pegasus. This transit dimmed the light of the star by a measurable 1.7%. This shows the capability of Hubble to detect extrasolar planets.

In 2007, Hubble actually detected the atmosphere of an extrasolar planet, a new achievement in planetary research. The light from the star passed through the atmosphere of the planet and was detected by Hubble's Near Infrared Camera and Multi-Object Spectrometer. The atmosphere contains methane, carbon dioxide, carbon monoxide, and water molecules. This exciting observation was an important achievement because it demonstrated that prebiotic materials are present in the atmosphere of at least one extrasolar planet, and that as such measurements are possible with Hubble they bode a bright future for such research with the James Webb Telescope.

Solar System

Hubble has not been idle in contributing to the understanding of our solar system objects. The first spectacular solar system observation was that of the 1994 crash of Comet Shoemaker-Levy 9 into Jupiter. Subsequently, Mars is and has been actively researched with Hubble. Wide Field Planetary Camera 2, Near Infrared Camera and Multi-Object Spectrometer, Space Telescope Imaging Spectrograph, and Advanced Camera for Surveys have all monitored weather conditions, observed seasonal changes, mapped the polar caps, watched dust storms, and conducted remote "site surveys" of landing spots for Martian probes. In the Advanced Camera for Surveys image of the sharpest Earth-based image ever taken of Mars, small craters and other surface markings only about a few tens of kilometers (a dozen

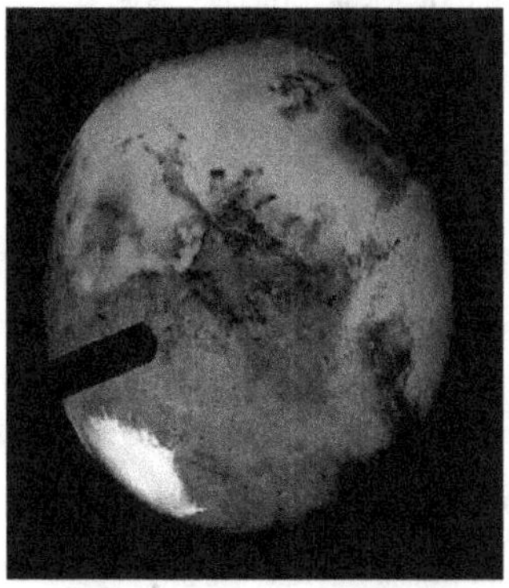

An Advanced Camera for Surveys image of Mars exhibiting the sharpest view ever taken from Earth. In view are numerous craters, several large volcanoes of the great Tharsis plateau along the upper left limb, and a large multi-ring impact basin, called Argyre, near image center. There is a reddish tinge over the southern ice cap suggesting dust contamination in the clouds or the ground ice.

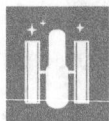

miles) across can be seen. Hubble continues to support the NASA Mars mission and probe activities.

Other phenomena observed include the changing atmosphere of Jupiter, spectacular views of Jupiter's moons, the rings of Saturn in various phases, an aurora on Uranus, clouds on Neptune, and the first map of the surface features of Pluto. Hubble observations contributed to the characterization of asteroids and support of NASA probes landing on such objects, discovery of outer solar system Kuiper belt objects, and measurements of Quaoar and the dwarf planet Eris. The latter observations, in concert with data from the W.M. Keck Observatory in Hawaii, helped lead to the reclassification of Pluto as a "dwarf planet."

Most Popular Results

In addition to extensive research results obtained through the use of Hubble observations, public enthusiasm for NASA's endeavors—and Hubble in particular—is a consequence of the open and active press release system for Hubble.

Public understanding of astronomy and somewhat of science in general comes from the free availability of Hubble results. Particular images become popular by nature of their image quality, such as nebulae and galaxies. Other images are fascinating due to the astrophysical processes they depict, such as extrasolar planets, the distant universe, and Mars. Many images are also used in education to improve science literacy. All Hubble press release material can be found at: *http://hubblesite.org/newscenter/archive/releases/YEAR/PR.*

Hubble Scorecard

The initial primary driver for Hubble was cosmological studies; specifically, the determination of the age of the universe. Other important research areas involved the nature of galaxies and black holes, and the details of the intervening material permeating the universe. Below are a few examples of the anticipated and unanticipated science results. The qualities of Hubble, such as diffraction limited, high-sensitivity imagery, excellent spectroscopic capability, and high-contrast imaging from the ultraviolet through the visible to the infrared has provided for the exemplary science achieved.

Anticipated science:

- Measurement of the expansion rate of the universe since the Big Bang
- Confirmation of the existence of massive black holes in galaxies and a census of less-massive black holes in smaller galaxies and black holes in binary star systems
- Observation of emission revealing the physical nature of energetic active galactic nuclei
- Discovery of the host galaxies associated with enigmatic quasi-stellar objects (quasars)
- Detection of the intergalactic medium and the interstellar medium through absorption of light from distant quasars

Unanticipated science:

- Characterization of conditions for galaxy formation in the early universe through mergers and black hole formation
- Detection of the acceleration of the universe corresponding to the discovery of dark energy, the cosmic mechanism that counteracts the slowdown of the universe caused by gravity
- Unveiling the nature of gamma-ray bursts through identification of the host galaxies
- Observations of planetary disk formation
- Detection of extrasolar planets and several atmospheres of planets orbiting other stars

Other Science and Technology

The development of Hubble and its relationship to the shuttle, as well as other NASA programs, yielded advances in science and technology beyond discoveries about the universe. The advancement of optical and infrared detectors for use in space and the evolution of various sensors, circuitry, and navigation systems are all part of the contribution toward technologies needed to support the science and instrumentation. Other benefits of the program include the manufacture of robust electronic chips, hard drives, computation systems, and software. The science and technology required for human and robotic space exploration transformed due to the partnership between the Hubble science endeavor and the Space Shuttle Program.

Compton Gamma Ray Observatory

Hubble was the first Great Observatory, while Compton Gamma Ray Observatory was the second. Its launch on Space Shuttle Atlantis (Space Transportation System [STS]-37) in 1991 represented a benchmark in shuttle lift capability since it was the heaviest astrophysical payload flown to date. As planned, Compton spent almost a decade enabling insight into the nature and origin of enigmatic gamma-ray sources and was safely deorbited and reentered the Earth's atmosphere on June 4, 2000. The observatory was named in honor of Nobel Prize winner Dr. Arthur Compton for his physics research on scattering of high-energy photons by electrons, a critical process in the detection of gamma rays.

Compton Gamma Ray Observatory being prepared for deployment from Atlantis, STS-37 (1991).

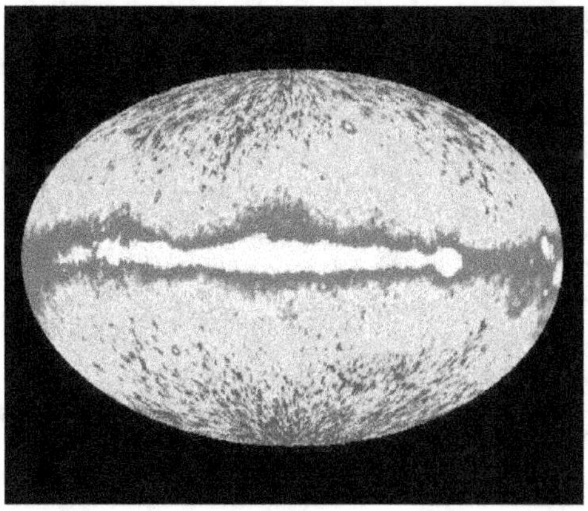

Can you imagine "seeing" gamma rays? This computer-processed image allows you to "see" the entire sky at photon energies above 100 million electron volts. These gamma-ray photons are 10,000 times more energetic than visible-light photons and are blocked from reaching Earth's surface by the atmosphere. A diffuse gamma-ray glow from the plane of our Milky Way is seen across the middle belt in this image.

Instrumentation

Compton was designed to detect high-energy gamma-ray emissions caused by diverse astrophysical phenomena including solar flares, pulsars, nova and supernova explosions, black holes accreting material, quasars, and the bombardment of the interstellar medium by cosmic rays. Four scientific instruments—Burst and Transient Source Experiment, Oriented Scintillation Spectrometer Experiment, Imaging Compton Telescope, and Energetic Gamma Ray Experiment Telescope—were intended to cover the high end of the electromagnetic spectrum.

While previous gamma-ray missions sampled astrophysical sources (after the original chance detection of gamma rays by the Vela military satellite in the 1960s), Compton pushed to a factor of 10 sensitivity improvement in each instrument. Based on the spectacular results, specifications emerged for new gamma-ray satellites.

Compton Science Results

All-sky surveys are an important tool for uniformly mapping the sky and understanding the overall relationship of various components of the nearby neighborhood as well as the universe. The Energetic Gamma Ray Experiment Telescope instrument provided a high-energy map that demonstrated the interaction between the interstellar gas that pervades the disk of our galaxy with cosmic rays. The telescope also sampled variable extragalactic sources such as quasars that emit in high-energy "blazers."

All-sky maps also were obtained with the Imaging Compton Telescope and Oriented Scintillation Spectrometer Experiment. The Imaging Compton Telescope surveyed a narrow energy band of gamma rays. It also detected neutrons from a solar flare early on in the program. The Oriented Scintillation Spectrometer Experiment survey mapped the center of our galaxy and was also sensitive to solar flares caused by accelerating particles colliding with the sun's surface.

The workhorse of the Compton observatory was the Burst and Transient Source Experiment, designed to detect gamma-ray bursts. The first result was to confirm that the bursts came from all over the sky, suggesting a cosmic origin rather than a local solar neighborhood cause or some phenomena restricted to our galaxy. The brief flashes were eventually traced to chaotic events, some associated with the collapse of stars in distant galaxies. The instrument also detected gamma-ray burst repeaters and a few sources that were identified by monitoring x-ray sources and watching them wink out as the Earth occulted the object. These discoveries began to narrow in on the types of phenomena that could produce gamma rays.

Compton ended its impressive science career in 1999 with a gyro failure. A safe re-entry into Earth's atmosphere was successfully executed in 2000.

The Chandra X-ray Observatory

NASA named its x-ray observatory to honor the scientific achievements of American Astrophysicist Dr. Subrahmanyan Chandrasekhar who was awarded the Nobel Prize in Physics (1983) for his theoretical studies of the physical processes of importance to the structure and evolution of the stars.

X-rays are emitted by a plethora of objects including galaxies, exploding stars, black holes, and the sun. The Chandra X-ray Observatory was designed to probe x-ray emitters across the universe. When Chandra was deployed from Space Shuttle Columbia—Space Transportation System (STS)-93 (1999)—it was the longest satellite and provided a new heaviest-science-payload benchmark. Chandra is the third Great Observatory launched by NASA.

Chandra X-ray Observatory.

Scientific Research with Chandra

Chandra detected many types of sources, but the nature of black holes definitely caught the attention of both the scientific community and the public. Even in our own locale, the black hole at the inner 10 light-years of our galaxy was mapped. This source emits x-rays due to the extremely hot temperature (millions of degrees) of the material that has been gravitationally captured by the black hole and is spiraling into it. Chandra detected a "cool" black hole at the center of the Andromeda Galaxy, and more black holes were found that were confirmed as "supermassive" black holes in other galaxies.

Chandra data on individual stars have shown that binary star systems in collapse can produce x-rays, and normal stars in formation can produce x-rays through their stellar winds. Chandra showed that nearly all normal stars on the main sequence emit x-rays.

Chandra also provided a gallery of observations of supernova remnants. Research allowed scientists to understand how some supernovae are produced by binary stars, and how remnant neutron stars and pulsars interact with their surroundings. The dynamic of the shock wave, interactions with the interstellar medium, and the origin of cosmic rays are all in evidence in the x-ray emissions. The detailed compositions and distribution of the ejecta are traced in the x-rays.

Chandra also provided insight into the "hard x-ray" background—energies in the 2-10 keV range was a mystery for several decades. Some of these sources appear to be quasars as expected, and others are associated with nuclei of active galaxies that are fainter and possibly obscured by surrounding dusty material.

Observations of the "deep fields"—the Hubble Deep Fields and also the fields selected to survey deep x-ray emission—bolster the idea that some sources are quasars and active galaxies. The supermassive black holes in these objects cause intense x-rays to be emitted. Other distinct sources are galaxies with modest x-ray luminosity.

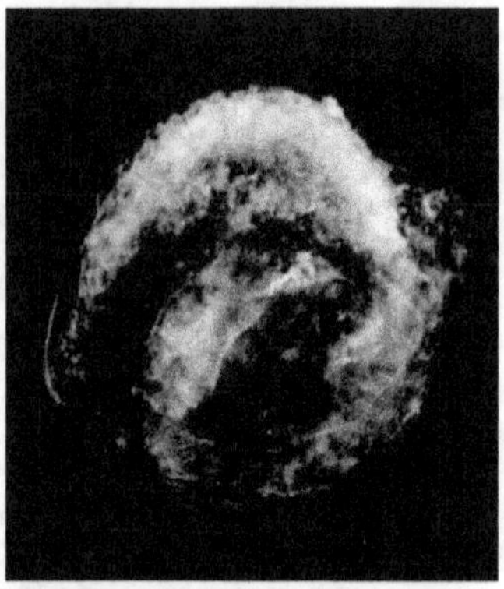

The Chandra image of one of the youngest supernova remnants in the galaxy, Kepler's supernova. This remnant may have been produced by the collapse of a single star relatively early in its lifetime.

Gamma-ray bursts were mysterious sources. Once the gamma ray is detected, rapid scheduling of telescopes allows the observation of the afterglow, including in the x-ray. Chandra data can assist in the determination of the elements present near the object.

The combined observations of optical, infrared, and x-ray emission from clusters of galaxies led to the identification of dark matter. It is suspected that most of the universe is filled with dark matter and the luminous material represents a few percent of the universe's contents. Observations of several clusters of galaxies showed that the collision of these massive clusters left a clump of dark matter behind. This implies that dark matter is not exactly the same as the luminous material seen in optical images of the galaxies in the clusters. The material left behind also produces impressive gravitational lensing of more distant objects. What dark matter is exactly remains a mystery.

Eileen Collins
Colonel, US Air Force (retired).
NASA's first woman Space Shuttle pilot and commander.
Pilot on STS-63 (1995) and STS-84 (1997).
Commander on STS-93 (1999) and STS-114 (2005).

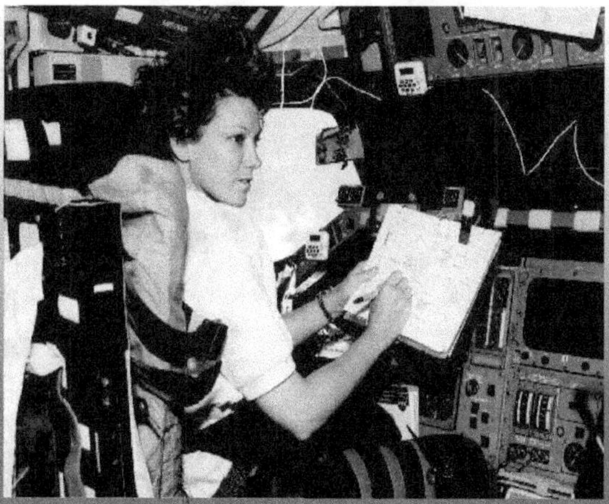

The Chandra X-ray Observatory: One of the shuttle's many success stories

"On July 23, 1999, I had the incredible privilege of commanding the Space Shuttle Columbia, which took the Chandra X-ray Observatory into space.

"Some fun facts about Chandra: the observatory can focus so well it could read a newspaper at half a mile. If the surface of the Earth was as smooth as Chandra's mirrors, the highest mountain would be no greater than 1.8 m (6 ft) tall.

"STS-93 was a dream mission for me. Not only did I have an opportunity to command a shuttle mission, I could marry it with a longtime hobby: astronomy. When I was a child in Upstate New York, I would look to the stars at night and feel inspired and excited. I wanted to travel to each one of those points of light, know what was there, what were they made of. Were there people there?

"I moved to Oklahoma for US Air Force pilot training. The wide open, dark, clear skies encouraged me to buy my first telescope. I bought books and magazines on astronomy and spent most of my spare time reading! Many shuttle astronauts came to Vance Air Force Base for training. This combination of exposure to the night skies and the emerging Space Shuttle Program inspired me to plan my career around my eventual application to the astronaut program!

"After over a year of training for STS-93 and several unexpected launch delays, my crew headed to the launch pad on July 20, 1999, which coincided with the 30th anniversary of Apollo 11. Our launch was manually halted at T minus 8 seconds by a sharp engineer who saw the 'hydrogen spike' in the aft compartment. A sensor had failed, and we were subsequently cleared to launch again in 2 days. After a single weather scrub, we rescheduled for the 23rd and lit up the sky shortly after midnight. Well, this was no ordinary launch! Five seconds after liftoff, we saw a 'Fuel Cell pH' message, received a call from Houston about an electrical short, which took out two main engine controllers! Unbeknownst to us, there was a second problem: at start-up, a pin had popped loose from a main engine injector plate. It hit several cooling tubes, causing us to leak hydrogen. Due to the shuttle redundancy and robustness of the main engines, they did not fail. The shuttle fleet was grounded to conduct thorough wiring inspections, resulting in many lessons learned for aging spacecraft.

"Despite the launch issues, I believe it was the right decision to launch Chandra on the shuttle vs. an expendable launch vehicle. The mission reaped the benefits of a human presence. True, a shuttle launch is more costly, but it is similar to buying insurance for missions with irreplaceable payloads.

"Today, the Chandra X-ray Observatory is increasing our understanding of the origin, evolution, and destiny of the universe. It is an incredible product of human ingenuity. The data will be around for generations of worldwide scientists to digest as we discover our place in the universe. I see Chandra as an expression of our curiosity as humans. As we search to discover what makes up this wondrous universe we live in, creations like Chandra will be far and away worth the investment we put into them. Chandra is one of the successful, productive, and mighty success stories of the Space Shuttle Program!"

Other Space Science Missions

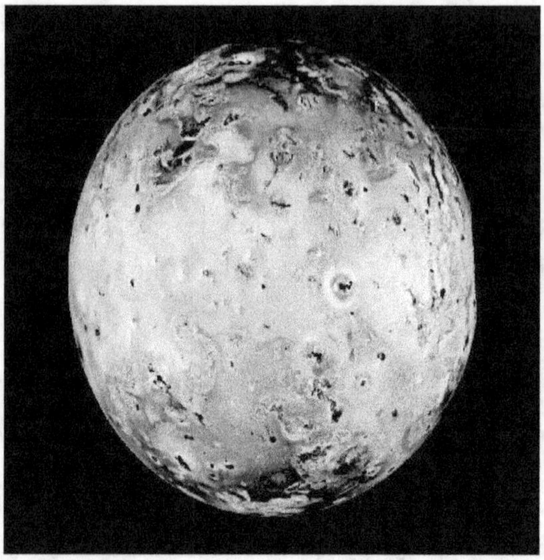

Image of Jupiter's moon Io obtained with the Galileo spacecraft. Io has the most volcanic activity in the solar system, giving it this mottled appearance. Features down to 2.5 km (1.6 miles) in size are seen along with mountains, volcanic craters, and impact craters from asteroids and comets.

Ultraviolet Programs

NASA devoted two shuttle flights to instrument packages designed to study the ultraviolet universe. A pallet of telescopes called the "Astro Observatory" were mounted together to fly several times. Astro-1 comprised three ultraviolet telescopes and an x-ray telescope while Astro-2 concentrated on the ultraviolet. Astro-1 flew on Columbia—Space Transportation System (STS)-35 (1990)—and Astro-2 flew on Endeavour—STS-67 (1995). The missions were designed to probe objects in the solar system, our galaxy, and beyond. Data on supernovae such as the Crab Nebula, planetary nebula, globular clusters, and young stellar disks were obtained.

Space Shuttle Columbia (STS-35) carries Astro-1 for observations of the ultraviolet universe in December 1990.

Exploring Stellar Surfaces: Hot and Cold Stars

The Orbiting and Retrievable Far and Extreme Ultraviolet Spectrometer Shuttle Pallet Satellite missions were designed to be free-flying missions supported by the shuttle. Space Shuttle Discovery (STS-51) deployed this satellite in 1993, the first of a series of missions. Ultraviolet spectra of hot stars, the coronae of cool stars, and the interstellar medium were observed. The second mission observed nearly 150 astronomical targets including the moon, nearby and more distant stars in the Milky Way, other galaxies, a few active galaxies, and the energetic quasar 3C273.

Chasing Jupiter and Its Moons

NASA's Galileo Mission was designed to study Jupiter and its system of moons. The spacecraft was launched by Space Shuttle Atlantis (STS-34) in 1989. Galileo was fitted with a solid-fuel upper stage that accelerated the spacecraft out of Earth orbit toward Venus. Galileo arrived at Jupiter and entered orbit in December 1995.

The spacecraft orbited through the Jovian system, measuring the moons as well as the planet Jupiter. Galileo sent a probe into Jupiter's atmosphere, finding the planet's composition to differ from that of the sun—important for understanding how the solar system formed. It provided the first close-up views of the large moons—Io, Europa, Ganymede, and Callisto—showing the dynamic Io volcanic activity and evidence that Europa may have a frozen surface with liquid underneath. Discoveries of many new moons around Jupiter, flybys of asteroids, and an interaction with a comet are part of Galileo's accomplishments. The spacecraft also was fortuitously in position to image the full sequence of more than 20 fragments of Comet Shoemaker-Levy impacting Jupiter in 1994.

The Galileo mission ended on September 21, 2003, when the spacecraft plummeted into Jupiter's atmosphere. From launch to impact, Galileo traversed trillions of kilometers (miles) on a single tank of gas, not counting the fuel for the shuttle. The total amount of data returned during its 14-year lifetime was 30 gigabytes, including 14,000 memorable pictures.

Studying the Anatomy of the Sun

On February 14, 1980, NASA launched the Solar Maximum Satellite (SolarMax) aimed at studying the maximum part of the sun's cycle. During this intense period, the sun's surface activity is characterized by massive ejections of high-energy particles extending into the solar system. SolarMax's life was almost cut short by a malfunction, but it fortunately was extended due to servicing by Space Shuttle Challenger (STS-41C) in 1984. Astronauts performed maintenance and repairs by replacing the attitude control system and one of the main electronics boxes, demonstrating that satellites could be repaired successfully and given extended life when serviced by the shuttle. SolarMax's career ended with re-entry on December 2, 1989.

The SolarMax instruments were mainly designed to study the x-ray and gamma-ray emissions from the sun. Two of the instruments also were capable of observing celestial sources outside the solar system. Observations showed that due to the bright faculae in the vicinity of dark sunspots are so intense that they increase the overall brightness of the sun. Therefore, the sun not only emits many charged particles but is also more intense during sunspot maximum.

The Magellan Mission: Mapping Venus

The Magellan spacecraft was launched on May 4, 1989, by Space Shuttle Atlantis from Kennedy Space Center, Florida, arrived at Venus on August 10, 1990, and was inserted into a near-polar elliptical orbit. Radio contact with Magellan was lost on October 12, 1994. At the completion of radar mapping, 98% of the surface of Venus was imaged at resolutions better than 100 m (328 ft), and many areas were imaged multiple times. The Magellan mission scientific objectives were to study land forms and tectonics, impact processes, erosion, deposition, and chemical processes and to model the interior of Venus. Magellan showed us an Earth-sized planet with no evidence of Earth-like plate tectonics.

Our Amazing Star: The Ulysses Mission

To fully understand our amazing star, it was necessary to study the sun at near maximum conditions. During the solar maximum, Ulysses reached the maximum Southern latitude of our sun on November 27, 2000, and traveled through the High Northern latitude September through December 2001.

After more than 12 years in flight, Ulysses had returned a wealth of data that led to a much broader understanding of the global structure of the sun's environment—the heliosphere.

Summary

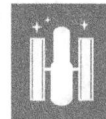

Many hundreds of years ago, our ancestors came out of their caves, gazed at the stars in the sky, and wondered, "How did we get here?" and "Are we alone?" They likely asked themselves, "Is there more out there?" and "How did this world begin?" They tried to comprehend their place in this complex puzzle between the Earth and the skies. We live in an age that has seen an explosion of science and technology and the beginnings of space exploration. We are still asking the same questions.

The Space Shuttle played a significant role in leading us toward some of the answers. Space science missions discussed here are opening a new window on our universe and providing a glimpse of galaxies far beyond.

Clearly, the partnership between the Space Shuttle Program and the Hubble Space Telescope, as well as other missions, contributed to the science productivity and outstanding reputation of NASA as a science-enabling agency. The obstacles that faced NASA throughout the journey were actually stepping-stones that led to a higher level of understanding not only of the universe, but of our own capabilities as a space agency and as individuals.

"...and measure every wand'ring planet's course,

Still climbing after knowledge infinite..."

– Christopher Marlowe

Atmospheric Observations and Earth Imaging

Introduction
Jack Kaye
Kamlesh Lulla

The Space Shuttle as a Laboratory for Instrumentation and Calibration
Ozone Calibration Experiments
Ernest Hilsenrath
Richard McPeters
Understanding the Chemistry of the Air
Jack Kaye
Aerosols in the Atmosphere
Zev Levin

The Space Shuttle as an Engineering Test Bed
Lidar In-space Technology Experiment
Patrick McCormick

A National Treasure—Space Shuttle-based Earth Imagery
Kamlesh Lulla

Earth is a dynamic, living oasis in the desolation of space. The land, oceans, and air interact in complex ways to give our planet a unique set of life-supporting environmental resources not yet found in any other part of our solar system. By understanding our planet, we can protect vital aspects, especially those that protect life and affect weather patterns. The shuttle played an integral role in this process. In the mid 1980s, NASA developed a systems-based approach to studying the Earth and called it "Earth System Science" to advance the knowledge of Earth as a planet. Space-based observations, measurements, monitoring, and modeling were major focuses for this approach. The Space Shuttle was an important part of this agency-wide effort and made many unique contributions.

The shuttle provided a platform for the measurement of solar irradiance. By flying well above the atmosphere, its instruments could make observations without atmospheric interference. Scientists' ability to calibrate instruments before flight, make measurements during missions, and return instruments to the laboratory after flight meant that measurements could be used to help calibrate solar-measuring instruments aboard free-flying satellites, which degrade over their time in space. The Atmospheric Laboratory for Applications and Science payload, which flew three times on the shuttle in the early 1990s, had four such instruments—two measuring total solar irradiance and two measuring solar spectral irradiance. The Shuttle Solar Backscatter Ultraviolet Instrument, which flew numerous times, also made solar spectral irradiance measurements as part of its ozone measurements.

The shuttle's low-light-level payload bay video imaging led to the discovery of upper-atmosphere phenomena of transient luminous events of electrical storms called "Elves." NASA pointed the first laser to the Earth's atmosphere from the shuttle for the purpose of probing the particulate composition of our air. The agency used the shuttle's many capabilities to image Earth's surface and chronicle the rapidly changing land uses and their impact on our ecosystems.

"Every shuttle mission is a mission to planet Earth" was a commonly heard sentiment from scientists involved in Earth imaging. In addition to working with many Earth observing payloads during the course of the Space Shuttle Program, "Earth-Smart" astronauts conducted scientific observations of the Earth systems. Thus, the shuttle provided an extraordinary opportunity to look back at our own habitat from low-Earth orbit and discover our own home, one mission at a time.

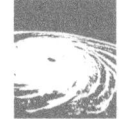

The Space Shuttle as a Laboratory for Instrumentation and Calibration

Global environmental issues such as ozone depletion were well known in the 1970s and 1980s. The ability of human by-products to reach the stratosphere and catalytically destroy ozone posed a serious threat to the environment and life on Earth. NASA and the National Oceanic and Atmospheric Administration (NOAA) assumed responsibility for monitoring the stratospheric ozone. A national program was put into place to carefully monitor ground levels of chlorofluorocarbon and stratospheric ozone, and the shuttle experiments became part of the overall space program to monitor ozone on a global scale. The NASA team successfully developed and demonstrated ozone-measuring methods. NOAA later took responsibility for routinely measuring ozone profiles using the Solar Backscatter Ultraviolet 2 instrument, while NASA continued to map ozone with a series of Total Ozone Mapping Satellite instruments.

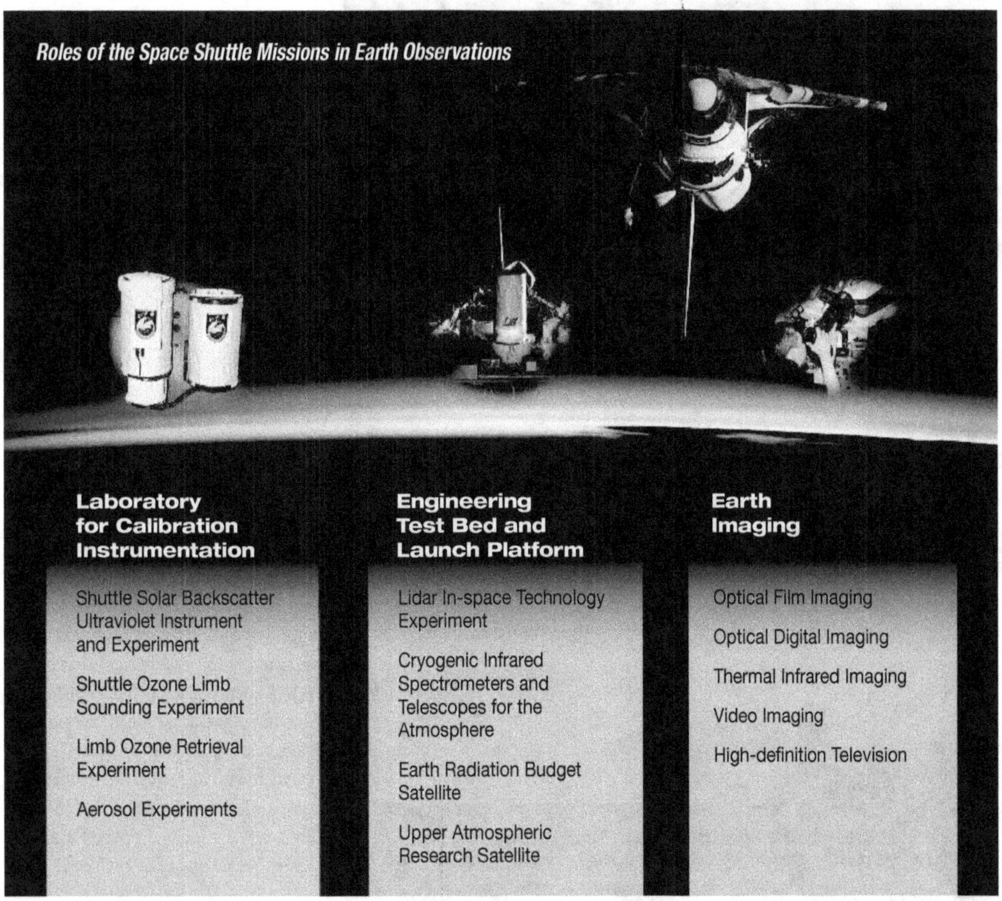
Roles of the Space Shuttle Missions in Earth Observations
Laboratory for Calibration Instrumentation
Shuttle Solar Backscatter Ultraviolet Instrument and Experiment
Shuttle Ozone Limb Sounding Experiment
Limb Ozone Retrieval Experiment
Aerosol Experiments
Engineering Test Bed and Launch Platform
Lidar In-space Technology Experiment
Cryogenic Infrared Spectrometers and Telescopes for the Atmosphere
Earth Radiation Budget Satellite
Upper Atmospheric Research Satellite
Earth Imaging
Optical Film Imaging
Optical Digital Imaging
Thermal Infrared Imaging
Video Imaging
High-definition Television

Some examples of multiple roles of the Space Shuttle: orbiting laboratory, engineering test bed, Earth imaging, and launch platform for several major Earth-observing systems.

Ozone Depletion and Its Impact—Why Research Is Important

The Earth's ozone layer provides protection from the sun's harmful radiation. The atmosphere's lower region, called the troposphere (about 20 km [12 miles]), is the sphere of almost all human activities. The next layer is the stratosphere (20 to 50 km [12 to 31 miles]), where ozone is found. The occurrence of ozone is very rare, but it plays an important role in absorbing the ultraviolet portion of the sun's radiation. Ultraviolet radiation is harmful to all forms of life. Thus, depletion in the ozone layer is a global environmental issue. Space-based measurements of ozone are crucial in understanding and mitigating this problem.

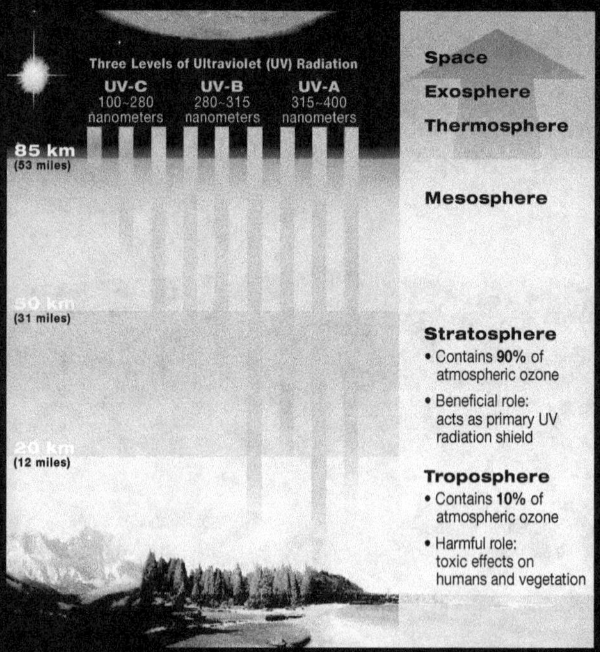

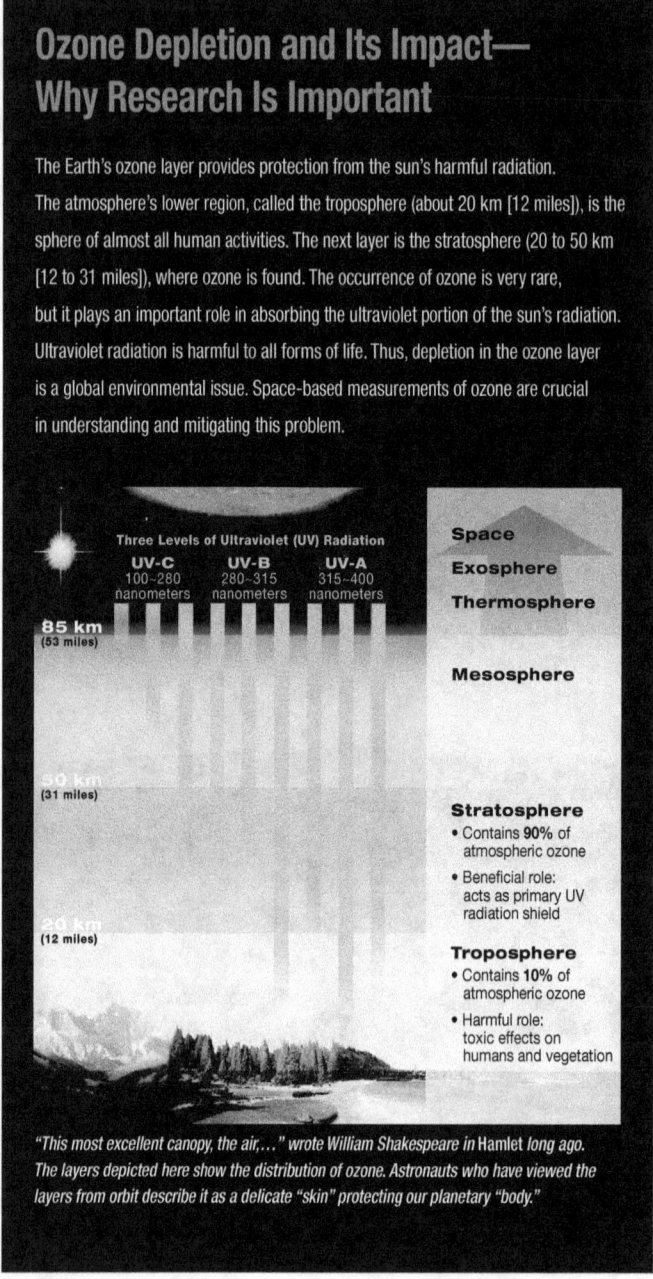

"This most excellent canopy, the air,..." wrote William Shakespeare in Hamlet *long ago. The layers depicted here show the distribution of ozone. Astronauts who have viewed the layers from orbit describe it as a delicate "skin" protecting our planetary "body."*

A Unique "Frequent Flyer" for Ozone Measurements

The Shuttle Solar Backscatter Ultraviolet experiment was dubbed "NASA's frequent flyer" since it flew eight times over a 7-year period (1989 to 1996)—an unprecedented opportunity for a shuttle science. Its primary mission was to provide a calibration or benchmark for concurrent ozone-monitoring instruments (Solar Backscatter Ultraviolet 2) flying on the NOAA operational polar orbiting crewless weather satellite. The NOAA satellite monitored stratospheric ozone and provided data for weather forecasts. Other satellites, such as NASA's Upper Atmosphere Research Satellite, Aura satellite, and the series of Stratospheric Aerosol and Gas Experiment and Total Ozone Mapping Spectrometer missions, measured ozone as well. Comparison of these ozone data was a high priority to achieve the most accurate ozone record needed for determining the success of internationally agreed-upon regulatory policy.

How Did Shuttle Solar Backscatter Ultraviolet Work?

Repeated shuttle flights provided the opportunity to check the calibration of NOAA instruments with those of the Shuttle Solar Backscatter Ultraviolet instrument by comparing their observations. The shuttle instrument was carefully calibrated in the laboratory at Goddard Space Flight Center before and after each of flight.

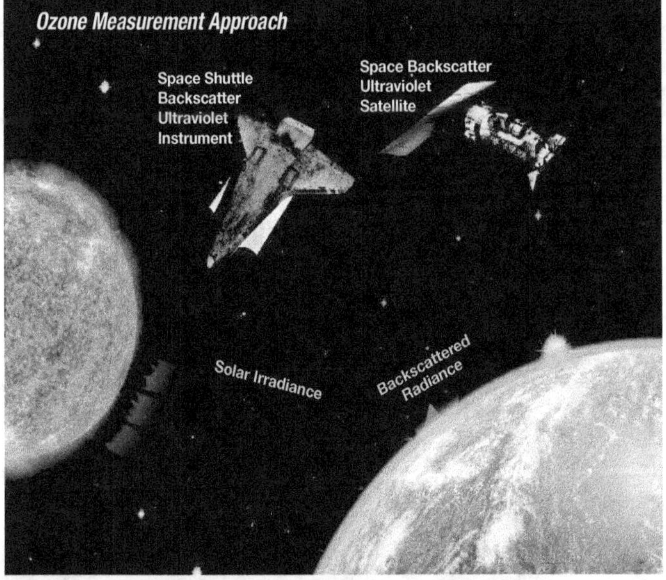

The sun is the source of radiation reaching the atmosphere. A spacecraft carrying an ozone-measuring instrument receives the backscattered radiation. Ozone is derived from the ratio of the observed backscattered radiance to the solar irradiance in the ultraviolet region.

The National Oceanic and Atmospheric Administration (NOAA) satellite carries an ozone instrument similar to the one that flew in the shuttle payload bay. The shuttle-based instrument was carefully calibrated at Goddard Space Flight Center. The shuttle's orbital path and satellite flight pass overlapped over the same Earth location within a 1-hour window during which the measurements took place and were later analyzed by scientists.

The sun's output in the ultraviolet varies much more than the total solar irradiance, which undergoes cycles of about 11 years. Changes in ozone had to be attributed accurately from solar changes and human sources. The Shuttle Solar Backscatter Ultraviolet instrument flew along with other solar irradiance monitors manifested on Space Transportation System (STS)-45 (1992), STS-56 (1993), and STS-66 (1994). Measurements from these three Atmospheric Laboratory for Applications and Science missions were intercompared and reprocessed, resulting in an accurate ultraviolet solar spectrum that became the standard for contemporary chemistry/climate models. This spectrum was also used to correct the continuous solar measurements taken by Solar Backscatter Ultraviolet 2 on the NOAA satellite.

Ozone Instrument Calibrations—Success Stories

- Comparisons with NOAA-11 satellite measurements over a period of about 5 years were within 3%—a remarkable result. The key to Shuttle Solar Backscatter Ultraviolet success was the careful calibration techniques, based on National Institute of Standards developed by the NASA team at Goddard Space Flight Center. These techniques were also applied to the NOAA instruments. The shuttle was the only space platform that could provide this opportunity.

- Although the instrument flew intermittently, it independently helped confirm ozone depletion at 45 km (28 miles), where chlorine chemistry is most active. Measurements made in October 1989 were compared with the satellite Nimbus-7 Solar Backscatter Ultraviolet measurements made in October 1980, an instrument that was also known to have an accurate calibration. Detected ozone loss of about 7% was close to predictions of the best photochemical models at that time.
- Calibration techniques were applied to all international satellites flying similar instruments—from the European Space Agency, European Meteorological Satellite, and the Chinese National Satellite Meteorological Center—thus providing a common baseline for ozone observations from space.

More Good News

An international environmental treaty designed to protect the ozone layer by phasing out the production of a number of chemicals linked to ozone depletion was ratified in 1989 by 196 countries and became known as the Montreal Protocol. This protocol and its amendments banned the production and use of chlorofluorocarbons. Once the ban was in place, chlorofluorocarbons at ground level and their by-products in the stratosphere began going down. The latest observations from satellites and ground-based measurements indicate ozone depletion has likely ended, with good signs that ozone levels are recovering.

Ellen Ochoa, PhD
Astronaut on STS-56 (1993), STS-66 (1994), STS-96 (1999), and STS-110 (2002).

Atmospheric Observations and Ozone Assessments

"The three Atmospheric Laboratory for Applications and Science missions in the early 1990s illustrated the collaborative role that the shuttle could play with unmanned science satellites. While the satellites had the advantage of staying in orbit for years at a time, providing a long-term set of measurements of ozone and chemicals related to the creation and destruction of ozone, their optics degraded over time due to interaction with ultraviolet light. The Space Shuttle carried up freshly calibrated instruments of the same design and took simultaneous measurements over a period of 9 or 10 days; the resulting data comparison provided correction factors that improved the accuracy of the satellite data and greatly increased their scientific value.

"One of the fortunate requirements of the mission was to videotape each sunrise and sunset for use by the principal investigator of the Fourier transform spectrometer, an instrument that used the sunlight peeking through the atmosphere as a light source in collecting chemical information. Thus, one of the crew members needed to be on the flight deck to start and stop the recordings, a job we loved as it gave us the opportunity to view the incredible change from night to day and back again. I would usually pick up our pair of gyro-stabilized binoculars and watch, fascinated, as the layers of the atmosphere changed in number and color in an incredible spectacle that repeated itself every 45 minutes as we orbited the Earth at 28,200 km per hour (17,500 miles per hour)."

Advancing a New Ozone Measurement Approach

From the calibration experiments conducted on five flights from 1989 to 1994, NASA expanded research on ozone elements.

The Total Ozone Mapping Spectrometer (satellite) and Solar Backscatter Ultraviolet instruments measured ozone using nadir viewing spectrometers. This approach was good for determining the spatial distribution (i.e., mapping the ozone depletion) but did a poor job of determining the vertical distribution of ozone. A spectrometer that measures light scattered from the limb of the Earth could be used for measuring how ozone varies with altitude; however, a test was needed to show that this approach would work.

While early models predicted that the largest changes in ozone as a result of the introduction of chlorofluorocarbons into the atmosphere would be observed in the upper stratosphere—in the 40- to 45-km (25- to 28-mile) region—the discovery of the ozone hole demonstrated that large changes were occurring in the lower stratosphere as a result of heterogeneous chemistry. The Solar Backscatter Ultraviolet instruments flown by NASA and NOAA were well designed to measure ozone change in the upper stratosphere.

For changes occurring below 25 km (16 miles), Solar Backscatter Ultraviolet offered little information about the altitude at which the change was occurring. Occultation instruments, such as the Stratospheric Aerosol and Gas Experiment, were capable of retrieving ozone profiles from the troposphere to nearly 60 km (37 miles) with approximately 1-km (0.6-mile) vertical resolution, but they could measure only at sunrise and sunset. Thus, the sampling limitations of occultation instruments limited the accuracy of the ozone trends derived for the lower stratosphere while the poor vertical resolution of the Solar Backscatter Ultraviolet instruments severely limited their ability to determine the altitude dependence of these trends. An instrument was needed with vertical resolution comparable to that of an occultation instrument but with coverage similar to that of a backscatter ultraviolet instrument.

The measurement of limb scattered sunlight offered the possibility of combining the best features of these two measurement approaches. The Shuttle Ozone Limb Sounding Experiment was a test of this concept.

How Did the Shuttle Ozone Limb Sounding Experiment and the Limb Ozone Retrieval Experiment Work?

To measure ozone in the upper stratosphere, scientists needed the large ozone cross sections available in the ultraviolet. To measure ozone at lower altitudes, scientists needed to use wavelengths near 600 nanometers (nm). The Shuttle Ozone Limb Sounding Experiment mission addressed these needs through the use of two

Shuttle Ozone Limb Sounding Experiment Concept

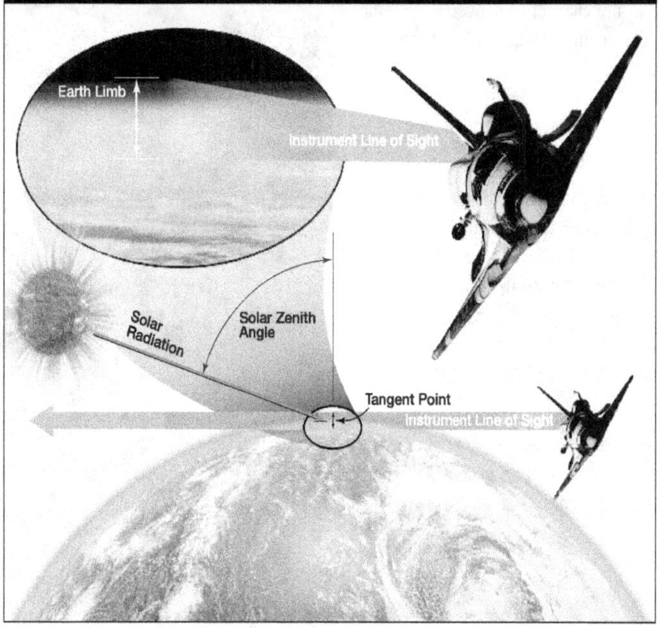

Light scattered from the limb of the Earth is measured to determine how ozone varies with the altitude.

instruments—the Shuttle Ozone Limb Sounding Experiment and the Limb Ozone Retrieval Experiment—flown as a single payload on STS-87 (1997).

The Shuttle Ozone Limb Sounding Experiment instrument measured ozone in the 30- to 50-km (19- to 31-mile) region. This ultraviolet imaging spectrometer produced a high-quality image of the limb of the Earth while minimizing internal scattered light.

The Limb Ozone Retrieval Experiment measured ozone in the 15- to 35-km (9- to 22-mile) region. This multi-filter imaging photometer featured bands in the visible and near infrared, and included a linear diode array detector. The 600-nm channel was the ozone-sensitive channel, the 525- and 675-nm channels were used for background aerosol subtraction, a 1,000-nm channel was used to detect aerosols, and a 345-nm channel gave overlap with the instrument and was used to determine the pointing.

New Ozone Measurement Approach Proven Successful

Comparisons with other satellite data showed that the calibration of Shuttle Ozone Limb Sounding Experiment instrument was consistent to within 10%, demonstrating the potential of limb scattering for ozone monitoring.

This approach compared the limb ozone measurements with data from ground observations and showed that this new approach indeed worked.

Space Shuttle Columbia's Final Contributions—Ozone Experiments

The loss of Columbia on re-entry was a heartbreaking event for NASA and for the nation. It was a small consolation that at least some data were spared. The ozone experiments were re-flown on STS-107 (2003) to obtain limb scatter data over a wider range of latitudes and solar zenith angles with different wavelengths. For this mission, Shuttle Ozone Limb Sounding Experiment was configured to cover the wavelength range from 535 to 874 nm. Seventy percent of the data was sent to the ground during the mission. In 2003, NASA identified an excellent coincidence between Columbia (STS-107) ozone measurement and data from an uncrewed satellite.

Summary of Ozone Calibration Research

In all, the Space Shuttle experiments showed that limb scattering is a viable technique for monitoring the vertical distribution of ozone. On the basis of these experiments, a

An ozone limb scatter instrument designed on the basis of successful Shuttle Ozone Limb Sounding Experiment measurements will be included in the uncrewed National Polar-orbiting Operational Environmental Satellite System. This interagency satellite system will monitor global environmental conditions and collect and disseminate data related to weather, atmosphere, oceans, land, and near-space environment.

limb scatter instrument on a newly designed, uncrewed National Polar-orbiting Operational Environmental Satellite System has been included. This is an outstanding example of the successful legacy of these shuttle science flights.

"The Space Shuttle is the only space platform that could provide an opportunity to calibrate the ozone monitoring instruments on orbiting satellites in order to measure ozone depletion in stratosphere. This role of Space Shuttle in ozone research has been invaluable."

– NASA Ozone Processing Team

Understanding the Chemistry of the Air

Atmospheric Trace Molecule Spectroscopy Experiments

The Atmospheric Trace Molecule Spectroscopy experiments investigated the chemistry and composition of the middle atmosphere using a modified interferometer. The interferometer obtained high-resolution infrared solar spectra every 2 seconds during orbital sunsets and sunrises, making use of the solar occultation technique in which the instrument looks through the atmosphere at the setting or rising sun. The availability of a bright source (i.e., the sun), a long atmospheric path length, the self-calibrating nature of the observation, and the high spectral and temporal resolution all combined to make the Atmospheric Trace Molecule Spectroscope one of the most sensitive atmospheric chemistry instruments to ever fly in space.

The instrument was first flown on the Spacelab 3 (STS-51B) mission in April 1985 and then re-flown as part of the Atmospheric Laboratory for Applications and Science (ATLAS) series of payloads. The solar occultation nature of the observations provided limited latitude ranges for each mission, but the combination of shuttle orbit characteristics (e.g., launch time) and the occultation viewing geometry provided unique opportunities. For example, the flight in 1993 (STS-56) made sunrise observations at high Northern latitudes to best observe the atmospheric concentrations of "reservoir species" relevant to polar ozone depletion. The flight in 1994 (STS-66) provided the first opportunity to acquire comprehensive space-based atmospheric composition measurements on the state of large-scale, persistent polar cyclonic conditions. These allowed comparisons of photochemical conditions inside and outside the region of maximum ozone loss.

The results of these observations included several first detections of critical atmospheric species in addition to the 30 or more constituents for which profiles were derived at altitudes between 10 and 150 km (6 and 93 miles). These measurements, widely used to test the photochemical models of the stratosphere, have been important in addressing the vertical distribution of halogen- and nitrogen-containing molecules in the troposphere and stratosphere as well as in characterizing the isotopic composition of atmospheric water vapor. Atmospheric Trace Molecule Spectroscopy observations served as important validation information for instruments that flew aboard NASA's Upper Atmosphere Research Satellite on STS-48 (1991). Through its high-resolution infrared observations, the spectroscope also left an important legacy leading to observations aboard the Earth Observing System's Aura satellite, launched in 2004. Aura's instruments studied the atmosphere's chemistry and dynamics and enabled scientists to investigate questions about ozone trends and air quality changes and their linkage to climate change.

The measurements also provided accurate data for predictive models and useful information for local and national agency decision support systems. Shuttle's efforts provided the impetus for the Canadian Atmospheric Chemistry Experiment satellite, launched in 2003.

Aerosols in the Atmosphere—Tiny Particles, Big Influence

Aerosols play an important role in our planet's dynamic atmosphere and globally impacted our climate. For example, aerosols interact with clouds and influence their rain production, which could affect the health of oceanic life and coral reefs as they carry minerals. Scientists have documented that Africa's Saharan dust particles (aerosols) travel all the way to South America to nourish the Amazonian rain forest. The Space Shuttle was well suited to facilitate research on these tiny particles that exert such a big influence on our atmosphere.

The vantage point of space has proven essential for understanding the global distribution of atmospheric aerosols, including horizontal and vertical distribution, chemical

Aerosols—A Mystery Revealed

Have you ever wondered why sunsets appear redder on some days? Or why the Earth becomes cooler after a volcanic eruption? The reason is aerosols.

Aerosols are minute particles suspended in the atmosphere (e.g., dust, sea salt, viruses, and smog). When these particles are sufficiently large, their presence is noticeable as they scatter and absorb sunlight. Their scattering of sunlight can reduce visibility (haze) and redden sunrises and sunsets. Aerosols affect our daily weather and have implications for transportation, among other impacts.

Aerosols interact both directly and indirectly with the Earth's climate. As a direct effect, aerosols scatter sunlight directly back into space. As an indirect effect, aerosols in the lower atmosphere can modify the size of cloud particles, changing how the clouds reflect and absorb sunlight, thereby affecting the Earth's energy budget and climatic patterns.

Sizes of Different Aerosols

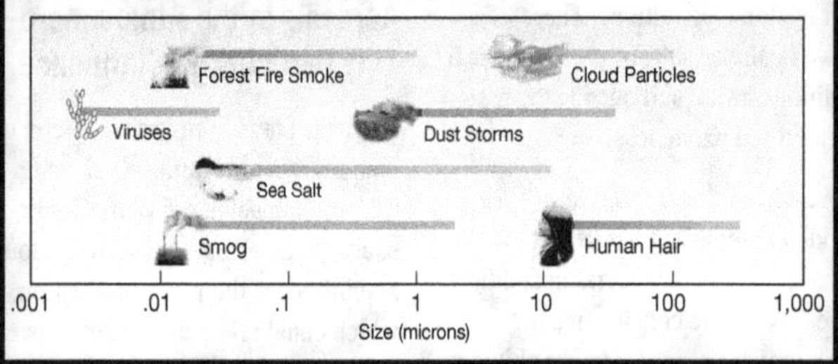

and optical properties, and interaction with the atmospheric environment. The diversity of aerosol characteristics makes it important to use a variety of remote sensing approaches. Satellite instruments have added dramatically to our body of knowledge. The Mediterranean Israeli Dust Experiment that flew on board STS-107 in 2003 complemented these observations due to its viewing geometry (the inclined orbit of the shuttle provided data at a range of local times, unlike the other instruments in polar sun-synchronous orbits that only provided data at specific times of the day) and its range of wavelengths (from ultraviolet through visible into near-infrared).

The Space Shuttle Columbia and Israeli Dust Experiment

Space Shuttle Columbia's final flight carried the Mediterranean Israeli Dust Experiment by Tel Aviv University and the Israeli Space Agency.

The primary objective of this experiment was the investigation of desert aerosol physical properties and transportation, and its effect on the energy balance and chemistry of the ambient atmosphere with possible applications to weather prediction and climate change. The main region of interest for the experiment was the Mediterranean Sea and its immediate surroundings.

How Do We Know the Distribution of Dust Particles?

The experiment included instruments for remote as well as in-situ measurements of light scattering by desert aerosol particles in six light wavelengths starting from the ultraviolet region to the near-infrared. The supporting ground-based and airborne measurements included optical observations as well as direct sampling. Airborne measurements were conducted above dust storms under the shuttle orbit ground-track during the passage of the shuttle over the target area. The collocation and simultaneity of shuttle, aircraft, and ground-based correlated data were aimed to help validate the remote spaceborne observations from Columbia and other space platforms.

Since most data from this experiment were transmitted to the ground for backup, the experiment's data were saved almost entirely and, after years of analysis, yielded a wealth of scientific data.

Insights From the Mediterranean Dust Experiment

Over 30% of the dust particles that pass over the Mediterranean Sea are coated with sulfate or sea salt. These particles play a crucial role in the development of clouds and precipitation as they often act as giant cloud condensation nuclei and enhance the development of rain. On January 28, 2003, a dust storm that interacted with a cold front, which produced heavy rain and flooding, was studied during this experiment. This is an example of how dust aerosols influence the local climate.

Observing Transient Luminous Events

In addition to measuring the dust particle distribution, the other major objective of the Israeli Dust Experiment was to use the same instruments at night to study electrical phenomena in the atmosphere. Scientists have known that large thunderstorms produce these electrical phenomena called "transient luminous events."

These events occur in upper atmospheric regions of the stratosphere, mesosphere, or ionosphere. The most common events include Sprites and Elves. It is interesting to note that Elves were discovered in 1992 by video camera in the payload bay of the Space Shuttle.

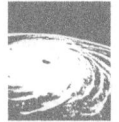

Sprites and Elves— Phenomenal Flashes of Light

So what are transient luminous events? They can best be defined as short-lived electrical phenomena generated as a result of enormous thunderstorms, and are categorized into Sprites and Elves.

Sprites are jellyfish-shaped, red, large, weak flashes of light reaching up to 80 km (50 miles) above the cloud tops. They last only a few tens of microseconds. Seen at night, Sprites can be imaged by cameras and only rarely seen by human eyes.

Elves are disk-shaped regions of glowing light that can expand rapidly to large distances up to 483 km (300 miles) across. They last fewer than thousandths of a second. Space Shuttle low-light video cameras were the first to record the occurrence of Elves.

Record-setting Measurements from Columbia (STS-107 [2003])

The experiment succeeded in a spectacular fashion as almost all data on Sprites and Elves were saved, thereby yielding the first calibrated measurements of their spectral luminosity, first detection of Sprite emission in the near-infrared, and clear indication for the generation of Elves by intra-cloud lightning flashes. The global observations of transient luminous events enabled calculation of their global occurrence rate. These shuttle-based results are considered a benchmark for satellite observations.

Elves over the South Pacific ***Sprites over Southeast Australia***

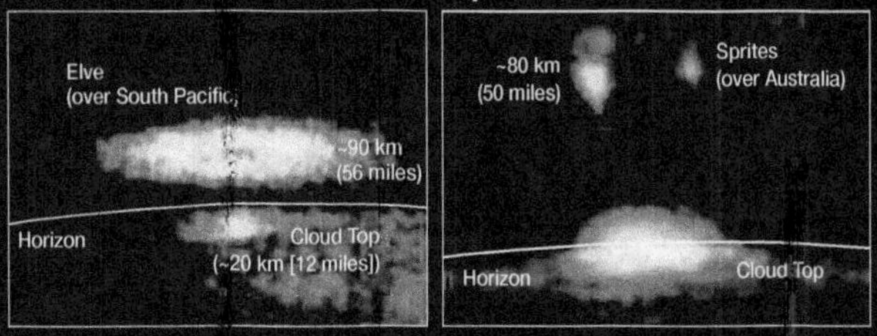

Short-lived electrical phenomena in upper atmosphere in disk-shaped regions (termed Elves) were imaged over the South Pacific. This was the first calibrated measurement of their spectral luminosity from space.

The Space Shuttle as an Engineering Test Bed

The Lidar In-space Technology Experiment

Scientists need the inventory of clouds and aerosols to understand how much energy is transmitted and lost in the atmosphere and how much escapes to space. To gain insight into these important questions, NASA explored the potential of lidar technology using the Space Shuttle as a test bed. Why lidar? Lidar's ability to locate and measure aerosols, water droplets, and ice particles in clouds gave scientists a useful tool for scientific insights.

Why Use the Space Shuttle as a Test Bed for Earth-observing Payloads?

The Space Shuttle could carry a large payload into low-Earth orbit, thereby allowing Earth-observing payloads an opportunity for orbital flight. Similarly, science goals might have required a suite of instruments to provide its measurements and, taken together, the instruments would have exceeded the possible spacecraft resources. Further, the shuttle provided a platform for showing a proof of concept when the technology was not mature enough for a long-duration, uncrewed mission. All of the above applied to the Lidar In-space Technology Experiment.

Laser technology was not at a point where the laser efficiencies and lifetime requirements for a long-duration mission were feasible; however, the shuttle could fly the experiment with its over 1,800-kg (4,000-pound), 4-kilowatt requirements.

The Lidar In-space Technology Experiment, which was the primary payload on Space Transportation System (STS)-64 (1994), orbited the Earth for 11 days and ushered in a new era of remote sensing from space. It was the first time a laser-based remote sensing atmospheric experiment had been flown in low-Earth orbit.

Of Lasers and Lidar: What is Laser? What is Lidar?

You have heard about use of lasers in eye surgery or laser printer for your computer or laser bar code readers in stores. So, what is a laser? Laser is short for Light Amplification by Stimulated Emission of Radiation. Unlike ordinary light composed of different wavelengths, laser light is one wavelength. All of its energy is focused in one narrow beam that can produce a small point of intense energy. Lasers are used in "radar-like" applications and are known as Lidars.

What is a lidar? It stands for Light Detection and Ranging and is an optical technology that uses pulsed lasers. It measures properties of scattered light to find range and/or other information of a distant target. As with similar radar technology, which uses radio waves, with a lidar the range to an object is determined by measuring the time delay between transmission of a pulse and detection of the reflected signal. Lidar technology has application in many Earth Science disciplines.

How Did Lidar Work in Space?

A spaceborne lidar can produce vertical profile measurements of clouds and aerosols in the Earth's atmosphere by accurately measuring the range and amount of laser light backscattered to the telescope. Using more than one laser color or wavelength produces information on the type of particle and/or cloud that is scattering the laser light from each altitude below.

The Lidar In-space Technology Experiment employed a three-wavelength laser transmitter. The lidar return signals were amplified, digitized, stored on tape on board the shuttle, and simultaneously telemetered to the ground for most of the mission using a high-speed data link.

The Lidar In-space Technology Experiment took data during ten 4½-hour data-taking sequences and five 15-minute "snapshots" over specific target sites. The experiment made measurements of desert dust layers, biomass burning, pollution outflow off continents, stratospheric

Lidar Maps Saharan Dust Transport

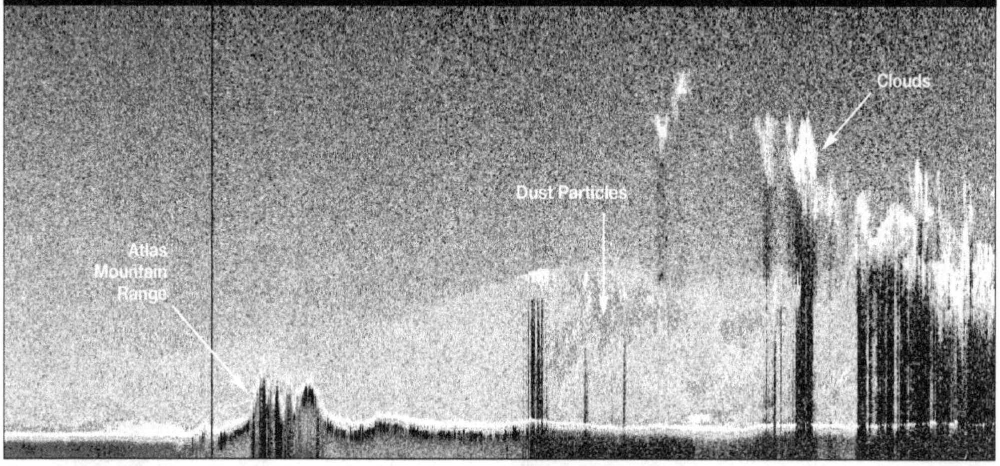

Lidar data during STS-64 (1994) depict widespread transport of dust aerosols over the African Sahara. The Atlas Mountain range appears to separate a more optically thick aerosol air mass to the Southeast from a relatively cleaner air mass to the Northwest. Over the desert interior, the aerosol plume extends in altitude to about 5 km (3 miles) with complex aerosol structures embedded within the mixed layer.

volcanic aerosols, and storm systems. It observed complex cloud structures over the intertropical convergence zone, with lasers penetrating the uppermost layer to four and five layers below.

Six aircraft, carrying a number of up- and down-looking lidars, performed validation measurements by flying along the shuttle footprint. NASA also coordinated ground-based lidar and other validation measurements—e.g., balloon-borne dustsondes—with the experiment's overflights. Photography took place from the shuttle during daylight portions of the orbits. A camera, fixed and bore sighted to the Lidar In-space Technology Experiment, took pictures as did the astronauts using two Hasselblad cameras and one camcorder to support the experiment's measurements.

Space Shuttle Legacy of Lidar In-space Technology Experiment

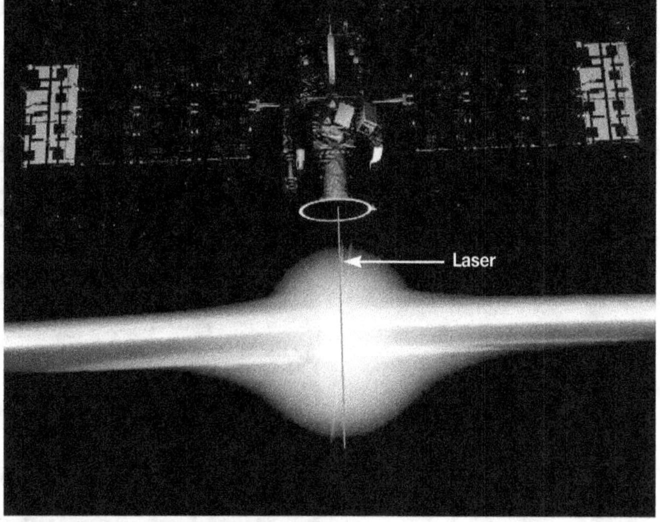

The Cloud-Aerosol Lidar and Infrared Pathfinder Satellite Observations satellite was launched in 2006 on a delta rocket to provide new information about the effects of clouds and aerosols on changes in the Earth's climate. The major instrument is a three-channel lidar.

Lidars in Space—A New Tool for Earth Observations

The Lidar In-space Technology Experiment mission proved exceedingly successful. It worked flawlessly during its 11-day mission. Data were used to show the efficacy of measuring multiple-layered cloud systems, desert dust, volcanic aerosols, pollution episodes, gravity waves, hurricane characterization, forest fires, agricultural burning, and retrieving winds near the ocean's surface. All measurements were done near-globally with a vertical resolution of 15 m (49 ft), which was unheard of using previous remote sensors from space. The Lidar In-space Technology Experiment even showed its utility in measuring land and water surface reflectivity as well as surface topography.

- It showed that space lidars could penetrate to altitudes of within 2 km (1.2 miles) of the surface 80% of the time and reach the surface 60% of the time, regardless of cloud cover. It appeared that clouds with optical depths as high as 5 to 10 km (3 to 6 miles) could be studied with lidars. The comparison of shuttle lidar data and lidar data acquired on board the aircraft was remarkable, with each showing nearly identical cloud layering and lower tropospheric aerosol distributions.
- The mission introduced a new technology capable of a global data set critical for understanding many atmospheric phenomena, such as global warming and predicting future climates.
- It provided a benefit in developing long-duration lidars for uncrewed satellite missions. Simulations using the experiment's characteristics and data have been carried out by groups all over the world in developing the feasibility of various lidar concepts for space application. This effort manifested itself, for example, in the Cloud-Aerosol Lidar and Infrared Pathfinder Satellite Observations experiment—a joint US/French mission flying a lidar as its centerpiece experiment.

A National Treasure—Space Shuttle-based Earth Imagery

Have you ever imagined gazing through the Space Shuttle windows at our own magnificent planet? Have you wondered what an ultimate field trip experience that could be?

Space Shuttle astronauts have experienced this and captured their observations using a wide variety of cameras. To these astronauts, each

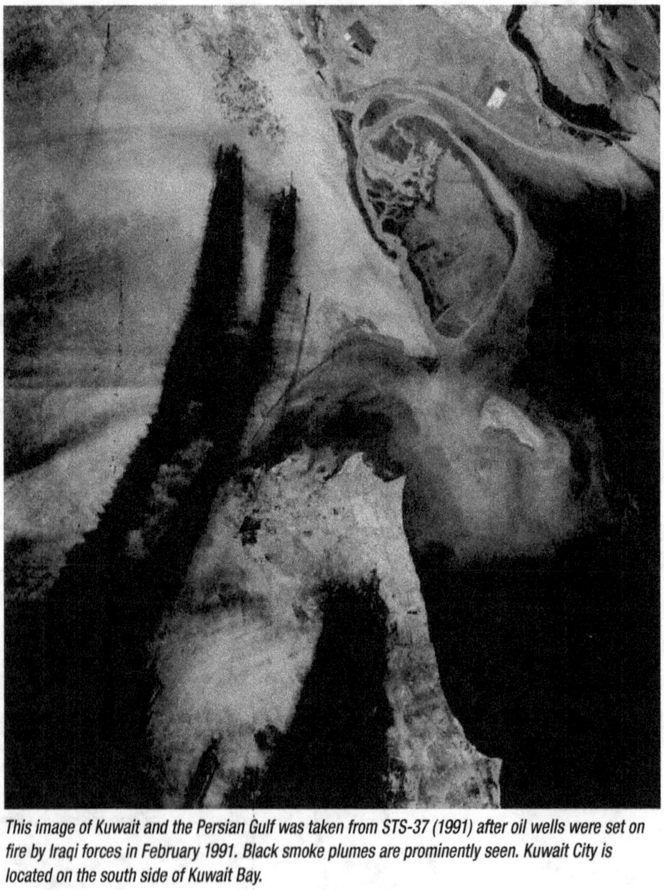

This image of Kuwait and the Persian Gulf was taken from STS-37 (1991) after oil wells were set on fire by Iraqi forces in February 1991. Black smoke plumes are prominently seen. Kuwait City is located on the south side of Kuwait Bay.

Shuttle Imagery Captures Earth's Dynamic Processes

Lake Chad 1968* 1982* 1992* 2000*

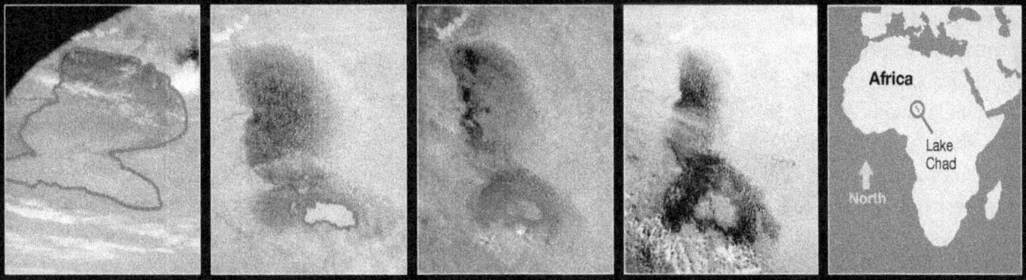

Astronauts photographed many sites of ecological importance from their missions over the 30 years of the Space Shuttle Program. These images yielded unprecedented insights into the changes occurring on Earth's surface.

One such site repeatedly imaged by shuttle crews was Lake Chad. This vast, shallow, freshwater lake in Central Africa straddles the borders of Chad, Niger, Nigeria, and Cameroon. Once the size of Lake Erie in the United States some 40 years ago, the shrinking of this lake was recorded on shuttle Earth imagery. First photographed by Apollo 7 astronauts in 1968—when the lake was at its peak—the decline in water levels is clearly seen from a small sampling of time series from shuttle flights in 1982, 1992, and 2000. While estimates of decline vary due to seasonal fluctuations, experts confirm that less than 25% of the water remains in the southern basin.

What has caused the shrinking of this life-supporting source of water for millions of people in Central Africa? Researchers point to a combination of factors—natural climatic changes ushering in drier climate, deforestation, aquatic weed proliferation, overgrazing in the region, and water use for agriculture and other irrigation projects.

Images not rectified to scale

shuttle mission offered a window to planet Earth in addition to whatever else the mission involved.

Astronauts have used handheld cameras to photograph the Earth since the dawn of human spaceflight programs. Beginning with the Mercury missions in the early 1960s, astronauts have taken more than 800,000 photographs of Earth. During the Space Shuttle Program, astronauts captured over 400,000 images using handheld cameras alone.

Making Astronauts "Earth Smart"

Shuttle astronauts were trained in scientific observation of geological, oceanographic, environmental, and meteorological phenomena as well as in the use of photographic equipment and techniques. Scientists on the ground selected and periodically updated a series of areas to be photographed as part of the crew Earth observations. Flight notes were routinely sent to the shuttle crew members, listing the best opportunities for photographing target site areas. The sites included major deltas in South and East Asia, coral reefs, major cities, smog over industrial regions, areas that typically experience floods or droughts triggered by El Niño cycles, alpine glaciers, long-term ecological research sites, tectonic structures, and features on Earth—such as impact craters—that are analogous to structures on Mars.

Scientific and Educational Uses of Astronaut Earth Imagery

Shuttle Earth imagery filled a niche between aerial photography and imagery from satellite sensors and complemented these two formats with additional information. Near real-time information exchange between the crew and scientists expedited the recording of dynamic events of scientific importance.

Critical environmental monitoring sites are photographed repeatedly over time; some have photographic records dating back to the Gemini and Skylab missions. Images are used to develop change-detection maps. Earth limb pictures taken at sunrise and sunset document changes in the Earth's atmospheric layering and record such phenomena as auroras and noctilucent clouds. Shuttle photographs of hurricanes, thunderstorms, squall lines, island cloud wakes, and the jet stream supplement satellite images. Other observations of Earth made by flight crews are used not only as scientific data but also to educate students and the general public about the Earth's ever-changing and dynamic systems. Over 3,000,000 images are downloaded, globally, each month by the public (*http://eol.jsc.nasa.gov/*). Educators, museums, science centers, and universities routinely use the imagery in their educational pursuits.

This imagery, archived at NASA, is a national treasure that captures the unique views of our own habitat acquired by human observers on orbit.

A mighty volcanic eruption of Mount St. Helens in 1980 and a large earthquake altered the landscape of this serene region in a blink of an eye. Landslides and rivers of rocks rushed downhill, causing havoc. Volcanic ash traveled more than 322 km (200 miles). This shuttle image from STS-64 (1994) captures the impact of these dynamic events in the US Pacific Northwest.

Jack Kaye, PhD
Associate director for research at NASA Headquarters Earth Science Division.

The Role of the Space Shuttle in Earth System Science

"The Space Shuttle played a significant role in the advancement of Earth System Science. It launched major satellites that helped revolutionize our study of the Earth. Its on-board experiments provided discoveries and new climatologies never before available, such as the tropospheric carbon monoxide distributions measured by the Measurement of Air Pollution from Satellite experiment, the stratospheric vertical profiles of many halogen-containing species important in ozone depletion measured by the Atmospheric Trace Molecule Spectroscopy instrument, and the high-resolution surface topography measurements made by the Shuttle Radar Topography Mission. It provided for multiple flight opportunities for highly calibrated instruments used to help verify results from operational and research satellites, most notably the eight flights of the Shuttle Solar Backscatter Ultraviolet instrument. Shuttle flights provided for on-orbit demonstration of techniques that helped pave the way for subsequent instruments and satellites. For example, the Lidar In-space Technology Experiment, with its demonstration of space-based lidar to study aerosols and clouds, paved the way for the US-French Cloud-Aerosol Lidar and Infrared Pathfinder Satellite Observation satellite. Similarly, the Shuttle Ozone Limb Sounding Experiment and Limb Ozone Retrieval Experiment provided demonstrations of the experimental technique to be used by the Ozone Mapping and Profiling Suite's limb sensor aboard the National Polar-orbiting Operational Environmental Satellite System Preparatory Project. The shuttle enabled international cooperation, including the multinational Atmospheric Laboratory for Applications and Science payload that included instruments with principal investigators from Germany, France, and Belgium among its six instruments, as well as deployment of the German Cryogenic Infrared Spectrometers & Telescopes for the Atmosphere-Shuttle Palette Satellite. The shuttle provided launch capability for Earth Science-related experiments to the International Space Station, such as the launch of the French Solar Spectrum Measurement instrument. Finally, the shuttle provided outstanding education and outreach opportunities."

Summary

The Space Shuttle played a significant role in NASA's missions to study, understand, and monitor Earth system processes. The shuttle was an integral component of the agency's missions for understanding and protecting our home planet. In the end, Space Shuttle missions for Earth observations were not only about science or instruments or images—these missions were also about humanity's journey into space to get a glimpse of our planet from a new perspective and rediscover our own home.

Mapping the Earth: Radars and Topography

Michael Kobrick
Kamlesh Lulla

One of the Space Shuttle's enduring science legacies is the near-global topographic mapping of the Earth with innovative radar remote sensing technologies. The shuttle also served as an important engineering test bed for developing the radar-based mapping technologies that have ushered in a quiet revolution in mapping sciences. The Shuttle Radar Topography Mission data set, in particular, has had an enormous impact on countless scientific endeavors and continues to find new applications that impact lives. This mission helped create the first-ever global high-resolution data for Earth topography—a data set for the ages. On average, one Shuttle Radar Topography Mission-derived topographic data set is downloaded from the US Geological Survey's servers every second of every day—a truly impressive record. Experts believe the mission achieved what conventional human mapmaking was unable to accomplish—the ability to generate uniform resolution, uniform accuracy elevation information for most of the Earth's surface.

In all, the development of imaging radars using the shuttle demonstrated, in dramatic fashion, the synergy possible between human and robotic space operations. Radar remote sensing technology advanced by leaps and bounds, thanks to the five shuttle flights, while producing spectacular science results.

"Seeing" Through the Clouds

What Is Imaging Radar?

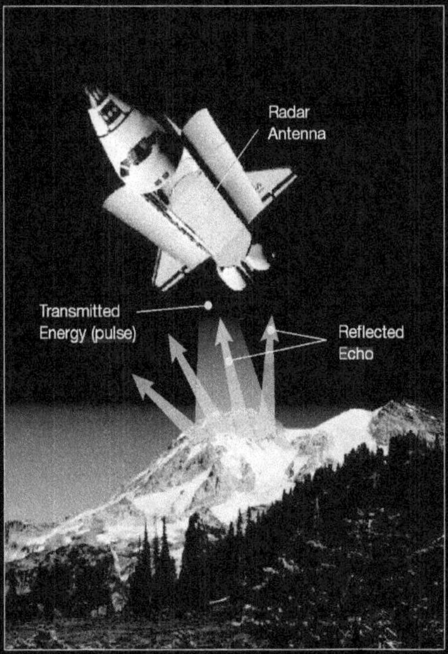

How does radar work? A radar transmits a pulse, then measures the reflected echo.

The term radar stands for Radio Detection and Ranging. You have seen radar images of weather patterns on television. Typical radar works like a flash camera, so it can operate day or night. But, instead of a lens and a film, radar uses an antenna to send out energy ("illumination") and computer tapes to record the reflected "echoes" of pulses of "light" that comprise its image. Radar wavelengths are much longer than those of visible light so it can "see" through clouds, dust, haze, etc. Radar antenna alternately transmits and receives pulses at a particular microwave wavelength (range of 1 cm [0.4 in.] to several meters [feet]). Typical imaging radar systems transmit around 1,500 high-power pulses per second toward the area or surface to be imaged.

What Is Synthetic Aperture Radar?

When a radar is moving along a track, it is possible to combine the echoes received at various positions to create a sort of "radar hologram" that can be further processed into an image. The improved resolution that results would normally require a much larger antenna, or aperture, thus a "synthetic aperture" is created.

Why Do We Need Accurate Topographic Maps of the Earth?

If you have ever used a global positioning system for navigation, you know the value of accurate maps. But, have you ever wondered how accurate the height of Mount Everest is on a map or how its height was determined? One of the foundations of many science disciplines and their applications to societal issues is accurate knowledge of the Earth's surface, including its topography. Accurate elevation maps have numerous common and easily understood civil and military applications, like locating sites for communications towers and ground collision avoidance systems for aircraft. They are also helpful in planning for floods, volcanic eruptions, and other natural disasters, and even predicting the viewscape for a planned scenic highway or trail.

It is hard to imagine that the global topographic data sets through the end of the 20th century were quite limited. Many countries created and maintained national mapping databases, but these databases varied in quality, resolution, and accuracy. Most did not even use a common elevation reference so they could not be easily combined into a more global map. Space Shuttle radar missions significantly advanced the science of Earth mapping.

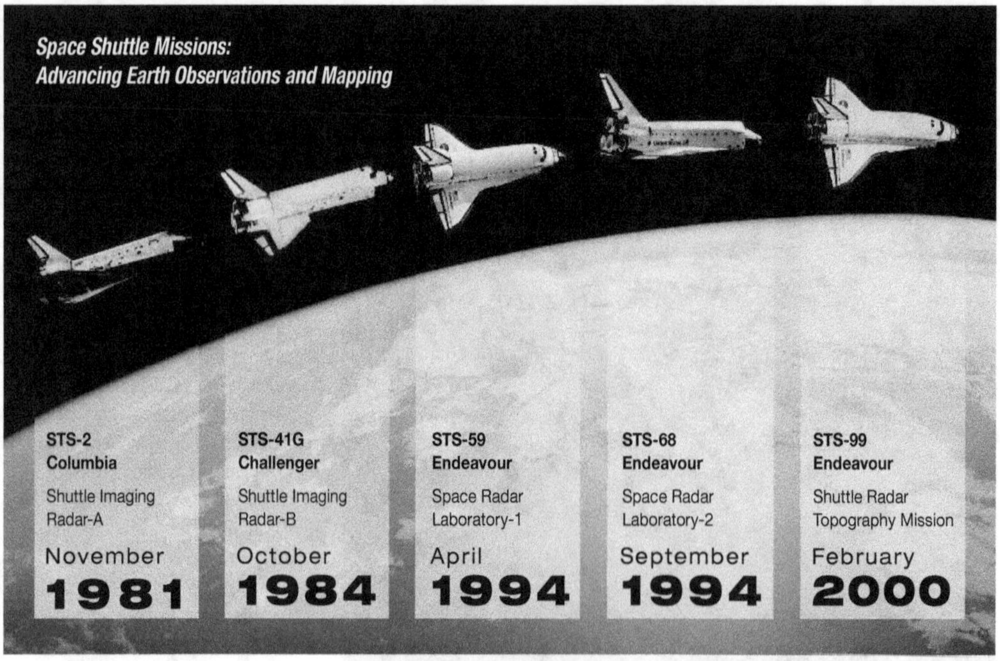

Shuttle and Imaging Radars—A Quiet Revolution in Earth Mapping

The First Mission

The Shuttle Imaging Radar-A flew on Space Shuttle Columbia (Space Transportation System [STS]-2) in November 1981. This radar was comprised of a single-frequency, single-polarization (L-band wavelength, approximately 24-cm [9-in.]) system with an antenna capable of acquiring imagery at a fixed angle and a data recorder that used optical film. Shuttle Imaging Radar-A worked perfectly, and the radar acquired images covering approximately 10 million km^2 (4 million miles^2) from regions with surface covers ranging from tropical

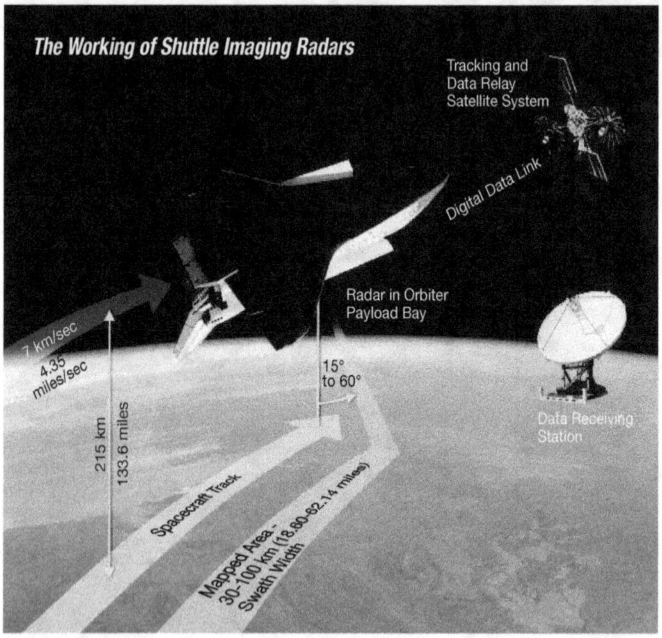

Space Shuttle's track at the altitude of 215 km (134 miles) with changing radar antenna look angle allowed the mapping of swaths up to 100 km (62 miles) wide.

forests in the Amazon and Indonesia to the completely arid deserts of North Africa and Saudi Arabia. Analysts found the data to be particularly useful in geologic structure mapping, revealing features like lineaments, faults, fractures, domes, layered rocks, and outcrops. There were even land-use applications since radar is sensitive to changes in small-scale roughness, surface vegetation, and human-made structures. Urban regions backscatter strongly, either because the walls of buildings form corner reflectors with the surface or because of the abundance of metallic structures—or both.

The Shuttle Imaging Radar-A's most important discovery, however, resulted from a malfunction. STS-2 was planned as a 5-day excursion and the payload operators generated an imaging schedule to optimize use of the radar's 8-hour supply of film. But early on, one of the three Orbiter fuel cells failed, which by mission rules dictated a minimum-duration flight—in this case, a bit over 2 days. So, the operators quickly retooled the plan to use the film in that time frame and ended up running the system whenever the Orbiter was over land. The result was a number of additional unplanned image passes over Northern Africa, including the hyper-arid regions of the Eastern Sahara.

Left: *Optical view of the Sahara region (Africa) showing the vast, featureless expanse of sand. The white lines depict the radar flight path.*

Right: *Radar imagery over the same region, taken during STS-2 (1981), reveals the network of channels and dried-up rivers (radar rivers) beneath the sand sheets, thereby illustrating the power of radar for archeological mapping.*

"Radar Rivers" Uncovered

This Sahara region, particularly the Selima Sand Sheet straddling the Egypt/Sudan border, is one of the driest places on our planet. Photographs from orbit show nothing but vast, featureless expanses of sand, and with good reason. The area gets rain no more than two or three times per century, and rates a 200 on the geological aridity index. For comparison, California's Death Valley—the driest place in the United States—rates no more than a 7 on the geological aridity index.

But when scientists got their first look at the Shuttle Imaging Radar-A images, they said "Hey, where's the sand sheet?" Instead of the expected dark, featureless plain, they saw what looked like a network of rivers and channels that covered virtually all the imaged area and might extend for thousands of kilometers (miles). To everyone's surprise, the radar waves had penetrated 5 or more meters (16 or more feet) of loose, porous sand to reveal the denser rock, gravel, and alluvium marking riverbeds that had dried up and been covered over tens of thousands of years ago.

Scientists knew the Sahara had not always been dry because some 50 million years ago, large mammals roamed its lush savannahs, swamps, and grasslands. Since then, the region has fluctuated between wet and dry, with periods during which rivers carved a complex drainage pattern across the entire Northern part of the continent. The existence of wadis (dry valleys) carved in Egypt's nearby Gilf Kebir Plateau, as well as other geologic evidence, supports this idea.

Subsequent field expeditions and excavations verified the existence of what came to be called the "radar rivers" and even found evidence of human habitation in the somewhat wetter Neolithic period, about 10,000 years ago. This discovery of an evolving environment was a harbinger of current concerns about global climate change, evoking historian Will Durant's statement, "Civilization exists by geological consent, subject to change without notice."

The Second Mission

The Shuttle Imaging Radar-B mission launched October 5, 1984, aboard the Space Shuttle Challenger (STS-41G) for an 8-day mission. This radar, again L-band, was a significant improvement, allowing multi-angle imaging—a capability achieved by using an antenna that could be mechanically tilted. It was also designed as a digital system, recording echo data to a tape recorder on the flight deck for subsequent downlink to the ground but with Shuttle Imaging Radar-A's optical recorder included as a backup. The results, deemed successful, included the cartography and stereo mapping effort that produced early digital-elevation data.

Next Generation of Space Radar Laboratory Missions

The Shuttle Imaging Radar instrument expanded to include both L-band (24 cm [9 in.]) and C-band (6 cm [2 in.]) and, with the inclusion of the German/Italian X-band (3 cm [1 in.]), radar. For the first time, an orbiting radar system not only included three wavelengths, the instrument was also fully polarimetric, capable of acquiring data at both horizontal and vertical polarizations or anything in between. It also used the first "phased-array" antenna, which meant it could be electronically steered to point at any spot on the ground without any motion of the antenna or platform. The resulting multiparameter images could be combined and enhanced to produce some of the most spectacular and information-rich radar images ever seen.

The Space Radar Laboratory missions (1 and 2) in 1994 were an international collaboration among NASA, the Jet Propulsion Laboratory, the German Space Agency, and the Italian Space Agency and constituted a real quantum leap in radar design, capability, and performance.

Tom Jones, PhD
Astronaut on STS-59 (1994), STS-68 (1994), STS-80 (1996), and STS-98 (2001).

"Space Radar Laboratory-1 and -2 orbited a state-of-the-art multifrequency radar observatory to examine the changing state of Earth's surface. Our STS-59 and STS-68 crews were integral members of the science team. Both missions returned, in total, more than 100 terabytes of digital imagery and about 25,000 frames of detailed Earth photography targeted on more than 400 science sites around the globe.

"For our crews, the missions provided a glorious view of Earth from a low-altitude, high-inclination orbit. Earth spun slowly by our flight deck windows, and we took advantage of the panorama with our 14 still and TV cameras. On September 30, 1994, on Space Radar Laboratory-2, we were treated to the awesome sight of Kliuchevskoi volcano in full eruption, sending a jet-like plume of ash and steam 18,288 m (60,000 ft) over Kamchatka. Raging wildfires in Australia, calving glaciers in Patagonia, plankton blooms in the Caribbean, and biomass burning in Brazil showed us yet other faces of our dynamic Earth. These two missions integrated our crews into the science team as orbital observers, providing 'ground truth' from our superb vantage point. Flight plan duties notwithstanding, I found it hard to tear myself away from the windows and that breathtaking view.

"Both missions set records for numbers of individual Orbiter maneuvers (~470 each) to point the radars, and required careful management of power resources and space-to-ground payload communications. The demonstration of precise orbit adjustment burns, enabling repeat-pass interferometry with the radar, led to successful global terrain mapping by the Shuttle Radar Topography Mission (STS-99) in 2000."

The Shuttle Radar Topography Mission—A Quantum Leap in Earth Mapping

The Shuttle Radar Topography Mission was major a breakthrough in the science of Earth mapping and remote sensing—a unique event. NASA, the Jet Propulsion Laboratory, the National Geospatial-Intelligence Agency (formerly the Defense Mapping Agency Department of Defense), and the German and Italian Space Agencies all collaborated to accomplish the goals of this mission. The 11-day flight of the Space Shuttle Endeavour for the Shuttle Radar Topography Mission acquired a high-resolution topographic map of the Earth's landmass (between 60°N and 56°S) and tested new technologies for deployment of large, rigid structures and measurement of their distortions to extremely high precision.

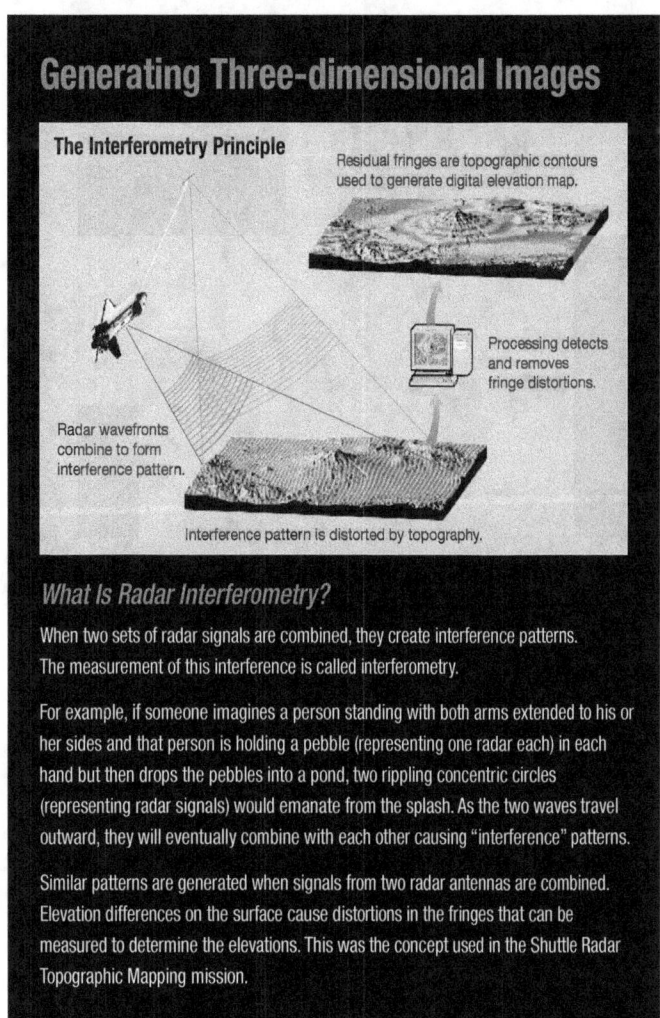

Generating Three-dimensional Images

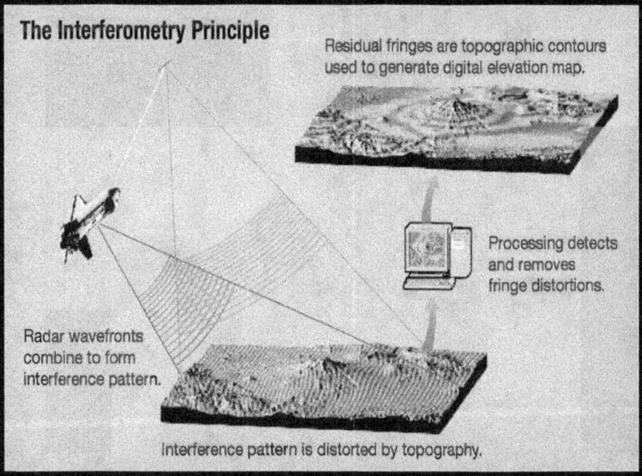
The Interferometry Principle
Residual fringes are topographic contours used to generate digital elevation map.
Processing detects and removes fringe distortions.
Radar wavefronts combine to form interference pattern.
Interference pattern is distorted by topography.

What Is Radar Interferometry?

When two sets of radar signals are combined, they create interference patterns. The measurement of this interference is called interferometry.

For example, if someone imagines a person standing with both arms extended to his or her sides and that person is holding a pebble (representing one radar each) in each hand but then drops the pebbles into a pond, two rippling concentric circles (representing radar signals) would emanate from the splash. As the two waves travel outward, they will eventually combine with each other causing "interference" patterns.

Similar patterns are generated when signals from two radar antennas are combined. Elevation differences on the surface cause distortions in the fringes that can be measured to determine the elevations. This was the concept used in the Shuttle Radar Topographic Mapping mission.

How Did the Shuttle Radar Topography Mission Work?

The heart of this mission was the deployable mast—a real engineering marvel. At launch, it was folded up inside a canister about 3 m (10 ft) long. The mast had 76 bays made of plastic struts reinforced with carbon fiber, with stainless-steel joints at the corners and titanium wires held taut by 227 kg (500 pounds) of tension. The strict requirements of interferometry dictated that the mast be incredibly rigid and not flex by more than a few centimeters (inches) in response to the firing of the Orbiter's attitude control vernier jets. It didn't. Once in orbit, a helical screw mechanism pulled the mast open and unfurled it one "bay" at a time to the mast's full length of 60 m (197 ft).

A crucial aspect of the mapping technique was determination of the interferometric baseline. The Shuttle Radar Topography Mission was designed to produce elevations such that 90% of the measured points had absolute errors smaller than 16 m (52 ft), consistent with National Mapping Accuracy Standards, and to do so without using ground truth—information collected "on location." Almost all conventional mapping techniques fit the results to ground truth, consisting of arrays of points

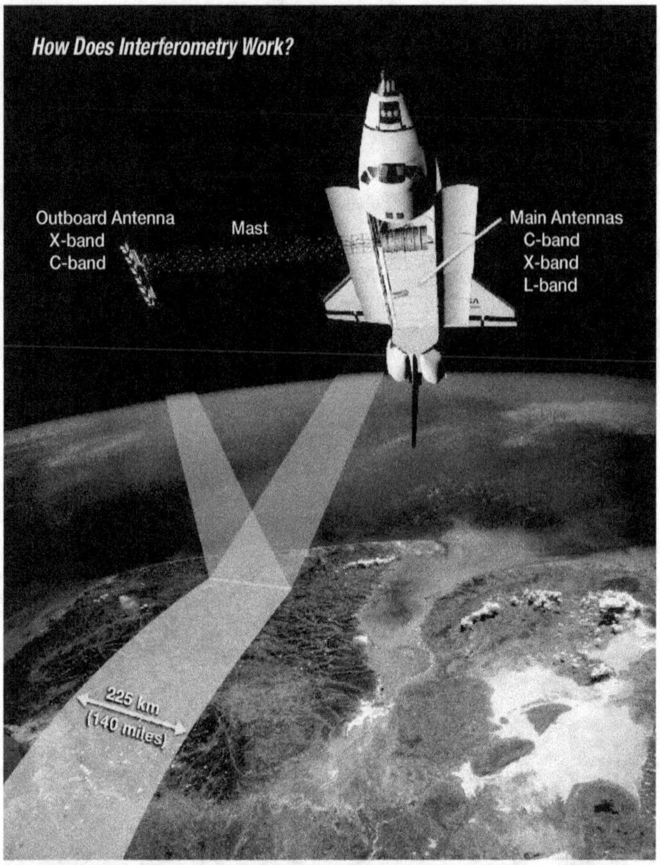

This interferometry concept was used in the Shuttle Radar Topography Mapping mission. Radars on the mast (not to scale) and in the shuttle payload bay were used to map a swath of 225 km (140 miles), thus covering over 80% of the Earth's landmass.

with known locations and elevations, to remove any residual inaccuracies. But, because the Shuttle Radar Topography Mission would be mapping large regions with no such known points, the system had to be designed to achieve that accuracy using only internal measurements.

This was a major challenge since analysis showed that a mere 1-arc-second error in our knowledge of the absolute orientation of the mast would result in a 1-m (3-ft) error in the elevation measurements. A 1-arc-second angle over the 60-m (197-ft) baseline is only 0.3 mm (0.1 in.)—less than the thickness of a penny.

This problem was solved by determining the Orbiter's attitude with an inertial reference unit borrowed from another astronomy payload, augmented with a new star tracker. To measure any possible bending of the mast, the borrowed star tracker was mounted on the main antenna to stare at a small array of light-emitting diodes mounted on the other antenna at the end of the mast. By tracking the diodes as if they were stars, all mast flexures could be measured and their effects removed during the data processing.

The mast-Orbiter combination measured 72 m (236 ft) from wingtip to the end of the mast, making it the largest solid object ever flown in space at that time. This size created one interesting problem: The Orbiter had to perform a small orbit maintenance burn using the Reaction Control System about once per day to maintain the proper altitude, and analysis showed that the resulting impulse would generate oscillations in the mast that would take hours to die out and be too large for the Shuttle Radar Topography Mission to operate.

By collaborating, Johnson Space Center flight controllers and Jet Propulsion Laboratory mechanical engineers arrived at a firing sequence involving a series of pulses that promised to stop the mast dead at the end of the burn. They called it the "flycast maneuver" since it mimicked the way a fisherman controls a fly rod while casting. The maneuver involved some tricky flying by the pilots and required much practice in the simulators, but it worked as planned. It also gave the crew an excuse to wear fishing gear in orbit—complete with hats adorned with lures—and produced some amusing photos.

NASA developed the original flight plan to maximize the map accuracy by imaging the entire landscape at least twice while operating on both ascending and descending orbits,

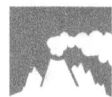

Turkey: Mount Ararat was mapped with a Shuttle Radar Topographic Mapping elevation model and draped with a color satellite image. This view has been vertically exaggerated 1.25 times to enhance topographic expression. This peak is a well-known site for searches for the remains of Noah's Ark. The tallest peak rises to 5,165 m (16,945 ft).

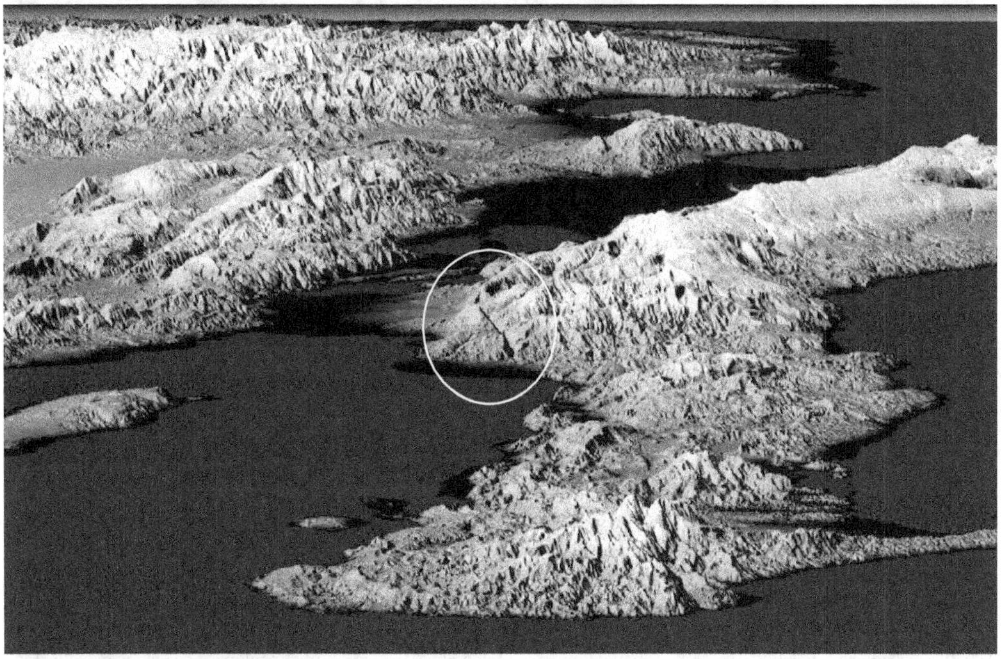

Haiti: This pre-earthquake image clearly shows the Enriquillo fault that probably was responsible for the 7.0-magnitude earthquake on January 12, 2010. The fault is visible as a prominent linear landform that forms a sharp diagonal line at the center of the image. The city of Port-au-Prince is immediately to the left (North) at the mountain front and shoreline.

but it turned out that a limited region was covered only once because the mapping had to be terminated a few orbits early when the propellant ran low. This had a minor impact, however, because even a single image could meet the accuracy specifications. In addition, the affected regions were mostly within the already well-mapped US terrain near the northern and southern limits of the orbits where the swaths converged were covered as much as 15 to 20 times. In all, the instrument covered 99.96% of the targeted landmass.

Converting Data Sets Into Real Topographic Maps

NASA assembled a highly effective computerized production system to produce topographic maps for users. Successful completion of radar data collection from Endeavour's flight was a major step, but it was only the first step. Teams from several technical areas of microwave imaging, orbital mechanics, signal processing, computer image processing, and networking worked together to generate the products that could be used by the public and other end users. Major steps included: rectifying the radar data to map coordinates, generating mosaics for each continent, performing quality checks at each stage, and assessing accuracy.

Africa: Tanzania's Crater Highlands along the East African Rift Valley are depicted here as mapped with the Shuttle Radar Topographic Mapping elevation model with vertical exaggeration of two times to enhance topographic variations. Lake Eyasi (top of the image, in blue) and a smaller crater lake are easily seen in this volcanic region. Mount Loolmalasin (center) is 3,648 m (11,968 ft).

Results of Shuttle Radar Topography Mission

The mission collected 12 terabytes of raw data—about the same volume of information contained in the US Library of Congress. Processing those data into digital elevation maps took several years, even while using the latest supercomputers. Yet, the Shuttle Radar Topography Mission eventually produced almost 15,000 data files, each covering 1° by 1° of latitude and longitude and covering Earth's entire landmass from the tip of South America to the southern tip of Greenland. The data were delivered to both the National Geospatial Agency and the Land Processes Distributed Active Archive Center at the US Geological Survey's EROS (Earth Resources Observation and Science) Data Center in Sioux Falls, South Dakota, for distribution to the public. The maps can be downloaded from their Web site (*http://srtm.usgs.gov/*) at no charge, and they are consistently the most popular data set in their archive.

Elevation accuracy was determined by comparing the mission's map to other higher-resolution elevation maps. Results confirmed the findings of the National Geospatial-Intelligence Agency and the US Geological Survey that Shuttle Radar Topography Mission data exceeded their 16-m (52-ft) height accuracy specification by at least a factor of 3.

In all, the Shuttle Radar Topography Mission successfully imaged 80% of Earth's landmass and produced topographic maps 30 times as precise as the best maps available at that time.

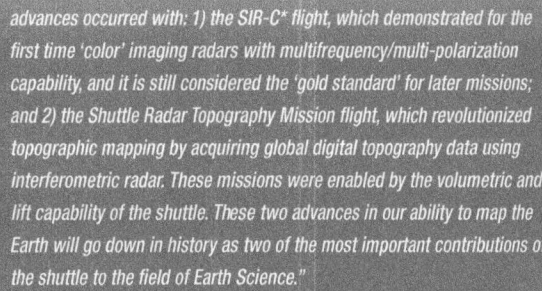

Charles Elachi, PhD
Director of Jet Propulsion Laboratory
California Institute of Technology.

"The Space Shuttle played a key role, as the orbiting platform, in advancing the field of radar observation of the Earth. Five flights were conducted between 1981 and 2004, each one with successively more capability. Probably the two most dramatic advances occurred with: 1) the SIR-C flight, which demonstrated for the first time 'color' imaging radars with multifrequency/multi-polarization capability, and it is still considered the 'gold standard' for later missions; and 2) the Shuttle Radar Topography Mission flight, which revolutionized topographic mapping by acquiring global digital topography data using interferometric radar. These missions were enabled by the volumetric and lift capability of the shuttle. These two advances in our ability to map the Earth will go down in history as two of the most important contributions of the shuttle to the field of Earth Science."*

** Shuttle Imaging Radar-C*

US: California's San Andreas fault (1,200 km [800 miles]) is one of the longest faults in North America. This view of a section of it was generated using a Shuttle Radar Topographic Mapping elevation model and draped with a color satellite image. The view shows the fault as it cuts along the base of Temblor Range near Bakersfield, California.

Summary

The successful shuttle radar missions demonstrated the capabilities of Earth mapping and paved the way for the Shuttle Radar Topography Mission. This mission was bold and innovative, and resulted in vast improvement by acquiring a new topographic data set for global mapping. It was an excellent example of a mission that brought together the best engineering and the best science minds to provide uniform accuracy elevation information for users worldwide. This success has been enshrined at the Smithsonian Air and Space Museum's Udvar-Hazy Center in Virginia, where the radar mast and outboard systems are displayed.

Astronaut Health and Performance

Introduction

Helen Lane

Laurence Young

How Humans Adapt to Spaceflight: Physiological Changes

Vision, Orientation, and Balance

William Paloski

Sleep

Laura Barger

Charles Czeisler

Muscle and Exercise

John Charles

Steven Platts

Daniel Feeback

Kenneth Baldwin

Judith Hayes

Cardiovascular

John Charles

Steven Platts

Nutritional Needs in Space

Helen Lane

Clarence Alfrey

Scott Smith

Immunology and Infectious Diseases

Brian Crucian

Satish Mehta

Mark Ott

Duane Pierson

Clarence Sams

Habitability and Environmental Health

Habitability

Janis Connolly

David Fitts

Dane Russo

Vickie Kloeris

Environmental Health

John James

Thomas Limero

Mark Ott

Chau Pham

Duane Pierson

Astronaut Health Care

Philip Stepaniak

Human travel to Mars and beyond is no longer science fiction. Through shuttle research we know how the body changes, what we need to do to fix some of the problems or—better yet—prevent them, the importance of monitoring health, and how to determine the human body's performance through the various sequences of launch, spaceflight, and landing. Basically, we understand how astronauts keep their performance high so they can be explorers, scientists, and operators.

Astronauts change physically during spaceflight, from their brain, heart, blood vessels, eyes, and ears and on down to their cells. Many types of research studies validated these changes and demonstrated how best to prevent health problems and care for the astronauts before, during, and after spaceflight.

*During a shuttle flight, astronauts experienced a multitude of gravitational forces. Earth is 1 gravitational force (1*g*); however, during launch, the forces varied from 1 to 3*g*. During a shuttle's return to Earth, the forces varied from nearly zero to 1.6*g*, over approximately 33 minutes, during the maneuvers to return. In all, the shuttle provided rather low gravitational forces compared with other rocket-type launches and landings.*

The most pervasive physiological human factor in all spaceflight, however, is microgravity. An astronaut perceives weightlessness and floats along with any object, large or small. The microgravity physiological changes affect the human body, the functions within the space vehicle, and all the fluids, foods, water, and contaminants.

We learned how to perform well in this environment through the Space Shuttle Program. This information led to improvements in astronauts' health care not only during shuttle flights but also for the International Space Station (ISS) and future missions beyond low-Earth orbit. Shuttle research and medical care led directly to improved countermeasures used by ISS crew members. No shuttle mission was terminated due to health concerns.

How Humans Adapt to Spaceflight: Physiological Changes

Vision, Orientation, and Balance Change in Microgravity

Gravity is critical to our existence. As Earthlings, we have come to rely on Earth's gravity as a fundamental reference that tells us which way is down. Our very survival depends on our ability to discern down so that we can walk, run, jump, and otherwise move about without falling. To accomplish this, we evolved specialized motion-sensing receptors in our inner ears—receptors that act like biological guidance systems. Among other things, these receptors sense how well our heads are aligned with gravity. Our brains combine these data with visual information from our eyes, pressure information from the soles of our feet (and the seats of our pants), and position and loading information from our joints and muscles to continuously track the orientation of our bodies relative to gravity. Knowing this, our brains can work out the best strategies for adjusting our muscles to move our limbs and bodies about without losing our balance. And, we don't even have to think about it.

At the end of launch phase, astronauts find themselves suddenly thrust into the microgravity environment. Gravity, the fundamental up/down reference these astronauts relied on throughout their lives for orientation and movement, suddenly disappears. As you might expect, there are a number of immediate consequences. Disorientation, perceptual illusions, motion sickness, poor eye-head/eye-hand coordination, and whole-body movements are issues each astronaut has to deal with to some degree.

Laurence Young, ScD
Principal investigator or coinvestigator on seven space missions, starting with STS-9 (1983).
Alternate payload specialist on STS-58 (1993).
Founding director of the National Space Biomedical Research Institute.
Apollo Program professor of astronautics and professor of health sciences and technology at Massachusetts Institute of Technology.

"The Space Shuttle Program provided a golden era for life sciences research. The difference between science capabilities on spacecraft before and after the Space Shuttle is enormous: it was like doing science in a telephone booth in the Gemini-Apollo era while shuttle could accommodate a school-bus-size laboratory. This significantly added to the kind of research that could be done in space. We had enormous success in life sciences, especially with the Spacelabs, for quality of instrumentation, their size, and opportunity for repeated measurements on the astronauts on different days of flight and over many different flights including Space Life Sciences flights 1 and 2 and ending with Neurolab.

"Our research led to a much more complete understanding of the neurovestibular changes in spaceflight and allowed us to know what issues require countermeasures or treatment, such as space motion sickness, as well as what research needed to continue in Earth laboratories, such as the role of short radius centrifuges for intermittent artificial gravity to support a Mars exploration mission."

One thing we learned during the shuttle era, though, is that astronauts' nervous systems adapt very quickly. By the third day of flight, most crew members overcame the loss of gravitational stimulation. Beyond that, most exhibited few functionally significant side effects. The downside to this rapid adaptation was that, by the time a shuttle mission ended and the astronauts returned to Earth, they had forgotten how to use gravity for orientation and movement. So, for the first few days after return, they suffered again from a multitude of side effects similar to those experienced at the beginning of spaceflight. During the Earth-readaptation period, these postflight affects limited some types of physical activities, such as running, jumping, climbing ladders, driving automobiles, and flying planes.

The Space Shuttle—particularly when carrying one of its Spacelab or Spacehab modules and during the human-health-focused, extended-duration Orbiter medical missions (1989 through 1995)—provided unique capabilities to study neurological adaptation to space. By taking advantage of the shuttle's ability to remove and then reintroduce the fundamental spatial orientation reference provided by gravity, many researchers sought to understand the brain mechanisms responsible for tracking and responding to this stimulus. Other researchers used these stimuli to investigate fundamental and functional aspects of neural adaptation, while others focused on the operational impacts of these adaptive responses with an eye toward reducing risks to space travelers and enabling future missions of longer duration.

Space Motion Sickness

What Is Space Motion Sickness?

Many people experience motion sickness while riding in vehicles ranging from automobiles to airplanes to boats to carnival rides. Its symptoms include headache, pallor, fatigue, nausea, and vomiting. What causes motion sickness is unknown, but it is clearly related to the nervous system and almost always involves the specialized motion-sensing receptors of the inner ear, known as the vestibular system.

The most popular explanation for motion sickness is the sensory-conflict theory. This theory follows from observations that in addition to planning the best strategies for movement control, the brain also anticipates and tracks the outcome of the movement commands it issues to the muscles. When the tracked outcome is consistent with the anticipated outcome, everything proceeds normally; however, when the tracked outcome is inconsistent, the brain must take action to investigate what has gone wrong. Sensory conflict occurs when some of the sensory information is consistent with the brain's anticipated outcome and some information is inconsistent. This might occur in space, for example, when the brain commands the neck muscles to tilt the head. The visual and neck joint receptors would provide immediate feedback indicating that the head has tilted, but because gravity has been reduced, some of the anticipated signals from the inner ear would not arrive. Initially, this would cause confusion, disorientation, and motion sickness symptoms. Over time, however, the brain would learn not to anticipate this inner-ear information during head tilts and the symptoms would abate.

How Often Do Astronauts Have Space Motion Sickness?

Many astronauts report motion sickness symptoms just after arrival in space and again just after return to Earth. For example, of the 400 crew members who flew on the shuttle between 1981 and 1998, 309 reported at least some motion sickness symptoms, such as stomach awareness, headache, drowsiness, pallor, sweating, dizziness, and, of course, nausea and vomiting. For most astronauts, this was a short-term problem triggered by the loss of gravity stimuli during ascent to orbit and, again, by the return of gravity stimuli during descent back to Earth. It usually lasted only through the few days coinciding with neural adaptations to these gravity transitions. While the symptoms of space motion sickness were quite similar to other types of motion sickness, its incidence was not predicted by susceptibility to terrestrial forms, such as car sickness, sea sickness, air sickness, or sickness caused by carnival rides. To complicate our understanding of the mechanisms of space motion sickness further, landing-day motion sickness was not even predicted by the incidence or severity of early in-flight motion sickness. The only predictable aspect was that repeat flyers usually had fewer and less severe symptoms with each subsequent flight.

How Do Astronauts Deal With Space Motion Sickness?

Crew members can limit head movements during the first few days of microgravity and during return to Earth to minimize the symptoms of space motion sickness. For some astronauts, drugs are used to reduce the symptoms. Promethazine-containing drugs emerged as the best choice during the early 1990s, and were frequently used throughout the remaining shuttle flights. Scientists also investigated preflight adaptation training in devices that simulate some aspects of the sensory conflicts during spaceflight, but more work is necessary before astronauts can use this approach.

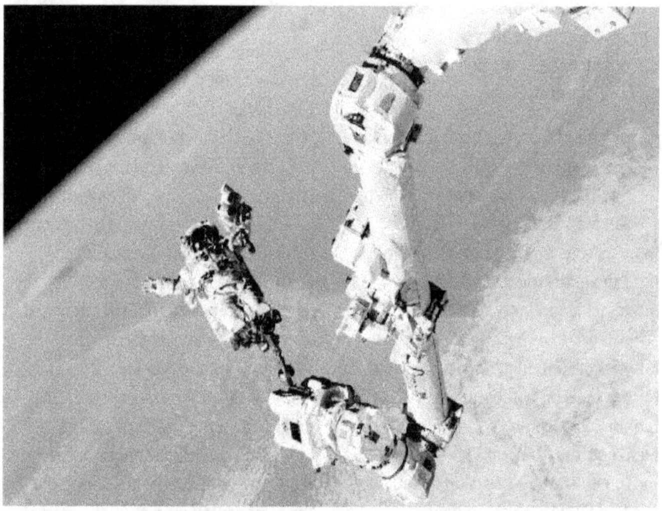

Some crew members experience height vertigo or acrophobia during extravehicular activities. Astronaut Stephen Robinson is anchored by a foot restraint on the International Space Station Robotic Arm during STS-114 (2005).

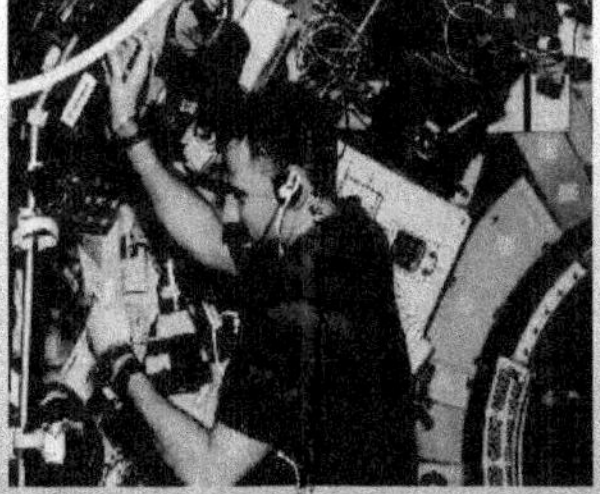

Dafydd (Dave) Williams, MD
Canadian astronaut on STS-90 (1998) and STS-118 (2007).

"Humans adapt remarkably well to the physiologic challenges associated with leaving the Earth's gravitational environment. For me, these started at main engine cutoff. After 7 minutes of the 8½-minute ride, G forces pushed me like an elephant sitting on my chest. The crushing pressure resolved as I was thrown forward against my harness when the main engines shut down. This created a sense of tumbling, head over heels, identical to performing somersaults as a child. I pulled myself down in the shuttle seat to re-create the gravitational sense of sitting in a chair and the tumbling stopped. I had experienced my first illusion of spaceflight!

"On the first day, many changes took place. My face felt puffy. I had a mild headache. Over the first few days, I experienced mild low back pain. Floating freely inside the shuttle with fingertip forces gently propelling us on a somewhat graceful path reminded me of swimming underwater—with the notable absence of any resistance.

"During re-entry into the Earth's atmosphere, I felt the forces of gravity gradually building. Standing on the middeck after landing, I felt gravitationally challenged. As I walked onto the crew transfer vehicle I felt as though my arms weighed twice what they normally do. Moving my head created an instant sense of vertigo.

"On my second spaceflight, when I arrived in space it seemed like I had never left and as I floated gracefully, looking back at Earth, it reminded me that I will always remain a spacefarer at heart."

Spatial Disorientation: Which Way Is Down?

Astronauts entering the microgravity environment of orbital spaceflight for the first time report many unusual sensations. Some experience a sense of sustained tumbling or inversion (that is, a feeling of being upside down). Others have difficulty accepting down as being the direction one's feet are pointing, preferring instead to consider down in terms of the module's orientation during preflight training on the ground. Almost all have difficulty figuring out how much push-off force is necessary to move about in the vehicle. While spacewalking (i.e., performing extravehicular activities [EVAs]), many astronauts report height vertigo—a sense of dizziness or spinning—that is often experienced by individuals on Earth when looking down from great heights. Some astronauts also experienced transient acrophobia—an overwhelming fear of falling toward Earth—which can be terrifying.

After flight, crew members also experience unusual sensations. For example, to many crew members everyday objects (e.g., apples, cameras) feel surprisingly heavy. Also, when walking up stairs, many experience the sensation that they are pushing the stairs down rather than pushing their bodies up. Some feel an overwhelming sense of translation (sliding to the side) when rounding corners in a vehicle. Many also have difficulty turning corners while walking, and some experience difficulty while bending over to pick up objects. Early after return to Earth, most are unable to land from a jump; many report a sensation that the ground is coming up rapidly to meet them. For the most part, all of these sensations abate within a few days; however, there have been some reports of "flashbacks" occurring, sometimes even weeks after a shuttle mission.

Eye-Hand Coordination: Changes in Visual Acuity and Manual Control

Manual control of vehicles and other complex systems depends on accurate eye-hand coordination, accurate perception of spatial orientation, and the ability to anticipate the dynamic response of the vehicle or system to manual inputs. This function was extremely important during shuttle flights for operating the Shuttle Robotic Arm, which required high-level coordination through direct visual, camera views, and control feedback. It was also of critical importance to piloting the vehicle during rendezvous, docking, re-entry, and landing.

Clear vision begins with static visual acuity (that is, how well one can see an image when both the person and the image are stationary). In most of our daily activities, however, either we are

Eye-Hand Coordination

Catching a ball is easy for most people on Earth. Yet, we don't usually realize how much work our brains do to predict when and where the ball will come down, get our hand to that exact place at the right time, and be sure our fingers grab the ball when it arrives. Because of the downward acceleration caused by gravity, the speed of a falling object increases on Earth. Scientists think that the brain must anticipate this to be able to catch a ball. Objects don't fall in space, however. So, scientists wondered how well people could catch objects without gravity. To find out, astronauts were asked to catch balls launched from a spring-loaded canon that "dropped" them at a constant speed rather than a constant acceleration as on Earth. In flight, the astronauts always caught the balls, but their timing was a little bit off. They reacted as if they expected the balls to move faster than they did, suggesting that their brains were still anticipating the effects of gravity. The astronauts eventually adapted, but some of the effects were still evident after 15 days in space. After flight, the astronauts were initially surprised by how fast the balls fell, but they readapted very quickly. This work showed that, over time in microgravity, astronauts could make changes in their eye-hand coordination, but that it took time after a gravity transition for the brain to accurately anticipate mechanical actions in the new environment.

Payload specialist James Pawelczyk, STS-90 (1998).

moving or the object we wish to see is moving. Under these dynamic visual conditions, even people with 20/20 vision will see poorly if they can't keep the image of interest stabilized on their retinas. To do this while walking, running, turning, or bending over, we have evolved complex neural control systems that use information from the vestibular sensors of the inner ear to automatically generate eye movements that are equal and opposite to any head movements. On Earth, this maintains a stable image on the retina whenever the head is moving.

Since part of this function depends on how the inner ear senses gravity, scientists were interested in how it changes in space. Many experiments performed during and just after shuttle missions examined the effects of spaceflight on visual acuity. Static visual acuity changed mildly, mainly because the headward fluid shifts during flight cause the shape of the eyes to change. Dynamic visual acuity, on the other hand, was substantially disrupted early in flight and just after return to Earth. Even for simple dynamic vision tasks, such as pursuing a moving target without moving the head, eye movements were degraded. But the disruption was found to be greatest when the head was moving, especially in the pitch plane (the plane your head moves in when you nod it to indicate "yes"). Scientists found that whether pursuing a target, switching vision to a new target of interest (the source of a sudden noise, for instance), or tracking a stationary target while moving (either voluntarily or as a result of vehicle motion), eye movement control was inaccurate whenever the head was moving.

Vision (eye movements) and orientation perceptions are disrupted during spaceflight. Scientists found that some kinds of anticipatory actions are inaccurate during flight. The impact of these changes on shuttle operations was difficult to assess. For example, while it appears that some shuttle landings were not as accurate as preflight landings in the Shuttle Training Aircraft, many confounding factors (such as crosswinds and engineering anomalies) precluded rigorous scientific evaluation. It appears that the highly repetitive training crew members received just before a shuttle mission might have helped offset some of the physiological changes during the flight. Whether the positive effects of this training will persist through longer-duration flights is unknown. At this point, training is the only physiological countermeasure to offset these potential problems.

Postflight Balance and Walking

When sailors return to port following a long sea voyage, it takes them some time to get back their "land legs." When astronauts returned to Earth following a shuttle mission, it took them some time to get back their "ground legs." On landing day, most crew members had a wide-based gait, had trouble turning corners, and could not land from a jump. They didn't like bending over or turning their heads independent of their torsos. Recovery usually took about 3 days; but the more time the crew member spent in microgravity, the longer it took for his or her balance and coordination to return to normal. Previous experience helped, though; for most astronauts, each subsequent shuttle flight resulted in fewer postflight effects and a quicker recovery.

Balance: Eye, Ear, and Brain Working in Concert

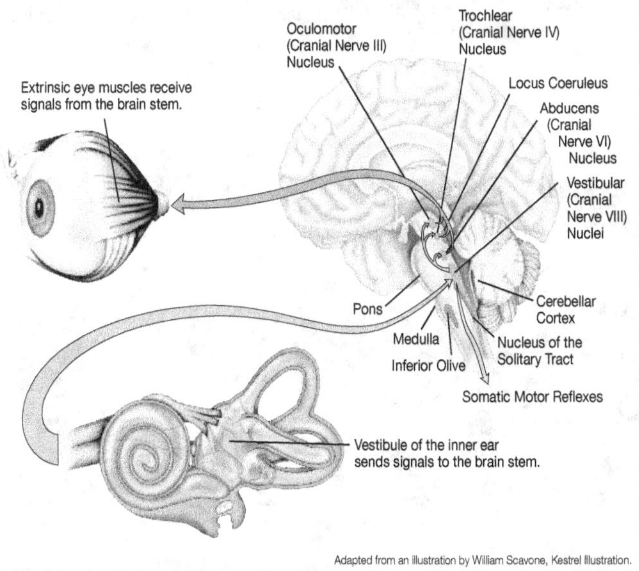

Adapted from an illustration by William Scavone, Kestrel Illustration.

For us to see clearly, the image of interest must be focused precisely on a small region of the retina called the fovea. This is particularly challenging when our heads are moving (think about how hard it is to make a clear photograph if your camera is in motion). Fortunately, our nervous systems have evolved very effective control loops to stabilize the visual scene in these instances. Using information sensed by the vestibular systems located in our inner ears, our brains quickly detect head motion and send signals to the eye muscles that cause compensatory eye movements. Since the vestibular system senses gravity as well as head motion, investigators performed many experiments aboard the shuttle to determine the role of gravity in the control of eye movements essential for balance. They learned that the eye movements used to compensate for certain head motions were improperly calibrated early in flight, but they eventually adapted to the new environment. Of course, after return to Earth, this process had to be reversed through a readaptation process.

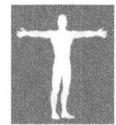

Scientists performed many experiments before and after shuttle missions to understand the characteristics of these transient postflight balance and gait disorders. By using creative experimental approaches, they showed that the changes in balance control were due to changes in the way the brain uses inner-ear information during spaceflight. As a result, the crew members relied more on visual information and body sense information from their ankle joints and the bottoms of their feet just after flight. Indeed, when faced with a dark environment (simulated by closing their eyes), the crew members easily lost their balance on an unstable surface (like beach sand, deep grass, or a slippery shower floor), particularly if they made any head movements. As a result, crew members were restricted from certain activities for a few days after shuttle flights to help them avoid injuring themselves. These activities included the return to flying aircraft.

In summary, experiments aboard the Space Shuttle taught us many things about how the nervous system uses gravity, how quickly the nervous system can respond to changes in gravity levels, and what consequences flight-related gravity changes might have on the abilities of crew members to perform operational activities. We know much more now than we did when the Space Shuttle Program started. But, we still have a lot to learn about the impacts of long-duration microgravity exposures, the effects of partial gravity environments, such as the moon and Mars, and how to develop effective physiological countermeasures to help offset some of the undesirable consequences of spaceflight on the nervous system. These will need to be tackled for space exploration.

Sleep Quality and Quantity on Space Shuttle Missions

Many people have trouble sleeping when they are away from home or in unusual environments. This is also true of astronauts. When on a shuttle mission, however, astronauts had to perform complicated tasks requiring optimal physical and cognitive abilities under sometimes stressful conditions.

Astronauts have had difficulty sleeping from the beginning of human spaceflight. Nearly all Apollo crews reported being tired on launch day and many gave accounts of sleep disruption throughout the missions, including some reporting continuous sleep periods lasting no more than 3 hours. Obtaining adequate sleep was also a serious challenge for many crew members aboard shuttle missions.

Environmental Factors

Several factors negatively affect sleep: unusual light-dark cycles, noise, and unfavorable temperatures. All of these factors were present during shuttle flights and made sleep difficult for crew members. Additionally, some crew members reported that work stress further diminished sleep.

When astronauts completed a daily questionnaire about their sleep, almost 60% of the questionnaires indicated that sleep was disturbed during the previous night. Noise was listed as the reason for the sleep disturbance approximately 20% of the time. High levels of noise negatively affect both slow-wave (i.e., deep sleep important for physical restoration) and REM (Rapid Eye Movement) sleep (i.e., stage at which most dreams occur and important for mental restoration), diminishing subsequent alertness, cognition, and performance. A comfortable ambient temperature is also important for promoting sleep. On the daily questionnaire, approximately 15% of the disturbances were attributed to the environment being too hot and approximately 15% of the disturbances were attributed to it being too cold. Thus, the shuttle environment was not optimal for sleep.

Circadian Rhythms

Appropriately timed circadian rhythms are important for sleep, alertness, performance, and general good health. Light is the most important time cue to the body's circadian clock, which has a natural period of about 24.2 hours. Normally, individuals sleep when it is dark and are awake when it is light.

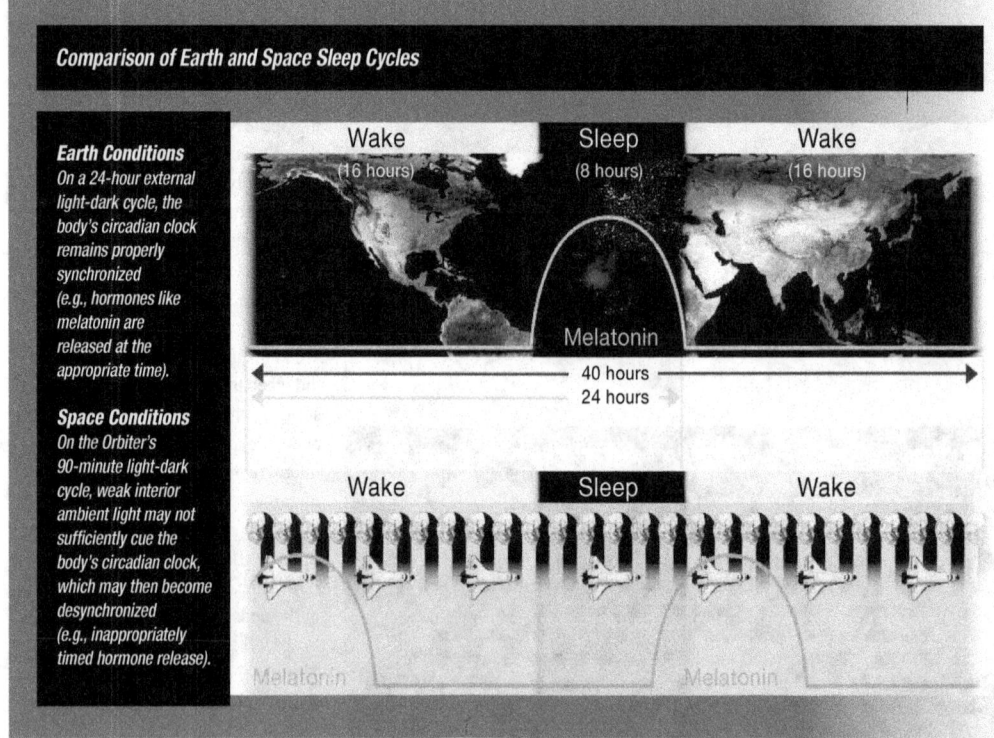

Comparison of Earth and Space Sleep Cycles

Earth Conditions
On a 24-hour external light-dark cycle, the body's circadian clock remains properly synchronized (e.g., hormones like melatonin are released at the appropriate time).

Space Conditions
On the Orbiter's 90-minute light-dark cycle, weak interior ambient light may not sufficiently cue the body's circadian clock, which may then become desynchronized (e.g., inappropriately timed hormone release).

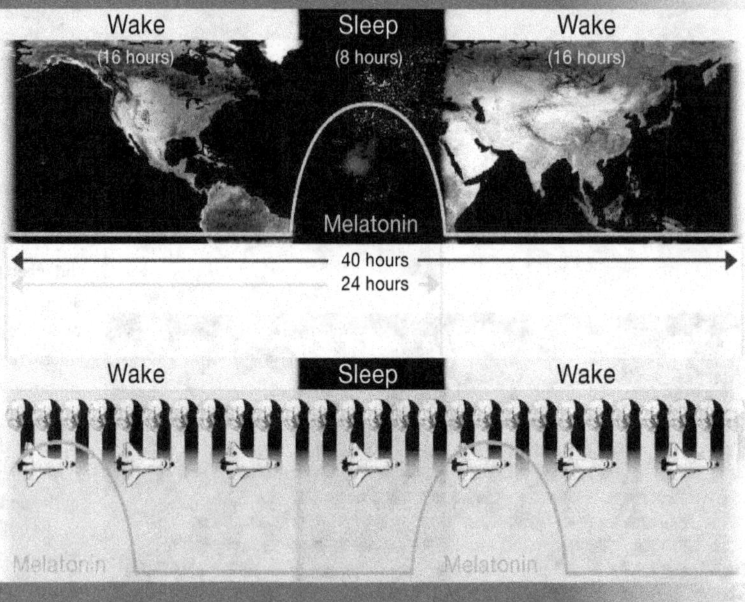

This 24-hour pattern resets the body's clock each day and keeps all of the body's functions synchronized, maximizing alertness during the day and consolidating sleep at night. Unlike the 24-hour light-dark cycle that we experience on Earth, shuttle crew members experience 90-minute light-dark cycles as they orbited the Earth.

Not only is the timing of light unsuitable, but the low intensity of the light aboard the shuttle may have contributed to circadian misalignment. Light levels were measured in the various compartments of the shuttle during Space Transportation System (STS)-90 Neurolab (1998) and STS-95 (1998) missions. In the Spacelab, light levels were constant and low (approximately 10 to 100 lux) during the working day. In the middeck, where the crew worked, ate, and slept, the light levels recorded were relatively constant and very dim (1 to 10 lux). Laboratory data showed that these light levels are insufficient to entrain the human circadian pacemaker to non-24-hour sleep-wake schedules. Normal room lighting (200 to 300 lux) would be required to keep the circadian system aligned under 24-hour light-dark cycles.

Crew members also were often scheduled to work on 23.5-hour days or had to shift their sleep-wake schedule several hours during flight. Moreover, deviations from the official schedule were frequently required by operational demands typical of space exploration. Therefore, the crew members' circadian rhythms often became misaligned, resulting in them having to sleep during a time when their circadian clock was promoting alertness, much as a shift worker on Earth.

Actually, difficulties with sleep began even before the shuttle launched. Often in the week prior to launch crew members had to shift their sleep-wake schedule, sometimes up to 12 hours. This physiological challenge, associated with sleep disruption, created "fatigue pre-load" before the mission even began.

All US crew members participated in the Crew Health Stabilization Program where they were housed together for 7 days prior to launch to separate them from potential infectious disease from people and food. During this quarantine period, scientists at Harvard Medical School, in association with NASA, implemented a bright-light treatment program for crew members of STS-35 (1990), the first Space Shuttle mission requiring both dual shifts and a night launch. Scheduled exposure to bright light (about 10,000 lux—approximately the brightness at sunrise), at appropriate times throughout the prelaunch period at Johnson Space Center and Kennedy Space Center, was used to prepare shuttle crew members of the Red Team of STS-35 for both their night launch and their subsequent night-duty shift schedule in space. A study confirmed that the prescribed light exposure during the prelaunch quarantine period successfully induced circadian realignment in this crew. Bright lights were installed at both centers' crew quarters in 1991 for use when shuttle flights required greater than a 3-hour shift in the prelaunch sleep-wake cycle.

Studies of Sleep in Space

NASA studied sleep quality and quantity and investigated the underlying physiological mechanisms associated with sleep loss as well as countermeasures to improve sleep and ultimately enhance alertness and performance in space. Scientists conducted a comprehensive sleep study on STS-90 and STS-95 missions using full polysomnography, which monitors brain waves, tension in face muscles, and eye movements, and is the "gold standard" for evaluating sleep. Scientists also made simultaneous recordings of multiple circadian variables such as body temperature and cortisol, a salivary marker of circadian rhythms. This extensive study included performance assessments and the first placebo-controlled, double-blind clinical trial of a pharmaceutical (melatonin) during spaceflight. Crew members on these flights experienced circadian rhythm disturbances, sleep loss, and decrements in neurobehavioral performance.

For another experiment, crew members wore a watch-like device, called an actigraph, on their wrists to monitor sleep. The actigraph contained an accelerometer that measured wrist motion. From that recorded motion scientists were able to use software algorithms to estimate sleep duration. Fifty-six astronauts (approximately 60% of the Astronaut Corps between 2001 and 2010) participated in this study. Average nightly sleep duration across multiple shuttle missions was approximately 6 hours. This level of sleep disruption has been associated with cognitive performance deficits in numerous ground-based laboratory and field studies.

Pharmaceuticals were the most widespread countermeasure for sleep disruption during shuttle flights. Indeed, more than three-quarters of astronauts reported taking sleep medications during missions. Astronauts took sleep medications during flight half the time. Wake-promoting therapeutics gained in popularity as well, improving alertness after sleep-disrupted nights.

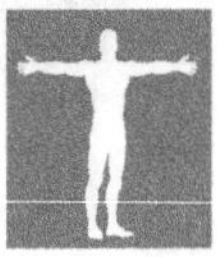

Richard Searfoss
Colonel, US Air Force (retired).
Pilot for STS-58 (1993) and STS-76 (1996).
Commander for STS-90 (1998).

Perspectives on Neurolab

"I was privileged to command STS-90 Neurolab, focusing on the effects of weightlessness on the brain and nervous system. Although my technical background is in engineering and flight test, it was still incredibly rewarding to join a dedicated team that included not just NASA but the National Institutes of Health and top researchers in the world to strive with disciplined scientific rigor to really understand some of the profound changes to living organisms that take place in the unique microgravity environment. I viewed my primary role as science enabler, calling on my operational experience to build the team, lead the crew, and partner with the science community to accomplish the real 'mission that mattered.'

"Even though at the time STS-90 flew on Columbia humans had been flying to space nearly 40 years, much of our understanding of the physiological effects was still a mystery. Neurolab was extremely productive in unveiling many of those mysteries. The compilation of peer-reviewed scientific papers from this mission produced a 300-page book, the only such product from any Space Shuttle mission. I'll leave it to the scientists to testify to the import, fundamental scientific value, and potential for Earth-based applications from Neurolab. It's enough for me to realize that my crew played an important role in advancing science in a unique way.

"With STS-90 as the last of 25 Spacelab missions, NASA reached a pinnacle of overall capability to meld complex, leading-edge science investigations with the inherent challenges of operating in space. Building on previous Spacelab flights, Neurolab finished up the Spacelab program spectacularly, with scientific results second to none. What a joy to be part of that effort! It was unquestionably the honor of my professional life to be a member of the Neurolab team in my role as commander."

Although sleep-promoting medication use was widespread in shuttle crew members, investigations need to continue to determine the most acceptable, feasible, and effective methods to promote sleep in future missions. Sleep monitoring is ongoing in crew members on the International Space Station (ISS) where frequent shifts in the scheduled sleep-wake times disrupt sleep and circadian alignment. Sleep most certainly will also be an issue when space travel continues beyond low-Earth orbit. Private sleep quarters will probably not be available due to space and mass issues. Consequently, ground-based studies continue to search for the most effective, least invasive, and least time-consuming countermeasures to improve sleep and enhance alertness during spaceflight. Currently, scientists are trying to pinpoint the most effective wavelength of light to use to ensure alignment of the circadian system and improve alertness during critical tasks.

Spaceflight Changes Muscle

Within the microgravity environment of space, astronauts' muscles are said to be "unweighted" or "unloaded" because their muscles are not required to support their body weight. The unloading of skeletal muscle during spaceflight, in what is known as "muscle atrophy," results in remodeling of muscle (atrophic response) as an adaptation to the spaceflight. These decrements, however, increase the risk of astronauts being unable to adequately perform physically demanding tasks during EVAs or after abrupt transitions to environments of increased gravity (such as return to Earth at the end of a mission).

A similar condition, termed "disuse muscle atrophy," occurs any time muscles are immobilized or not used as the result of a variety of medical conditions, such as wearing a cast or being on bed rest for a long time. Space muscle research may provide a better understanding of the mechanisms underlying disuse muscle atrophy, which may enable better management of these patients. In the US human space program, the only tested in-flight preventive treatment for muscle atrophy has been physical exercise. In-flight exercise hardware and protocols varied from mission to mission, somewhat dependent on mission duration as well as on the internal volume of the spacecraft. Collective knowledge gained from these shuttle missions aided in the evolution of exercise hardware and protocols to prevent spaceflight-induced muscle atrophy and the concomitant deficits in skeletal muscle function.

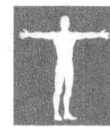

How Was Muscle Atrophy Measured, and What Were the Results?

Leg and Back Muscle Size Decreases

Loss of muscle and strength in the lower extremities of astronauts was initially found in the Gemini (1962-1966) and Apollo missions (1967-1972) and was further documented in the first US space station missions (Skylab, 1973-1974) of 28, 59, and 84 days' duration. NASA calculated crude muscle volumes by measuring the circumference of the lower and upper legs and arms at multiple sites.

For shuttle astronauts, more sophisticated, accurate, and precise measures of muscle volume were made by magnetic resonance imaging (MRI). MRI is a common diagnostic medical procedure used to image patient's internal organs that was adapted to provide volume measurements of a crew member's lower leg, thigh, and back muscles before and after flight. The leg muscle volume was evaluated in eight astronauts (seven males and one female, age range 31 to 39 years) who flew on either one of two 9-day missions. Scientists obtained MRI scans of multiple leg cross sections prior to flight and compared them to scans obtained at 2 to 7 days after flight. The volumes of various leg muscles were reduced by about 4% to 6% after spaceflight. In another study of longer missions (9 to 16 days' duration—two males and one female, mean age 41 years), the losses were reported to be greater, ranging from 5.5% to 15.9% for specific leg muscles. This study found that daily volume losses of leg muscles normalized for duration of flight were from 0.6% to 1.04% per mission day.

Muscle Strength Decreases

Decreases in muscle strength persisted throughout the shuttle period in spite of various exercise prescriptions. Measurements of muscle strength, mass, and performance helped NASA determine the degree of muscle function loss and assess the efficacy of exercise equipment and determine whether exercise protocols were working as predicted.

Muscle strength, measured with a dynamometry (an instrument that measures muscle-generated forces, movement velocity, and work) before launch and after landing consistently showed loss of strength in muscles that extend the knee (quadriceps muscles) by up to 12% and losses in trunk flexor strength of as much as 23%. The majority of strength and endurance losses occurred in the trunk and leg muscles (the muscle groups that are active in normal maintenance of posture and for walking and running) with little loss noted in upper body and arm muscle strength measurements. In contrast, four STS-78 (1996) astronauts had almost no decrease in calf muscle strength when they participated voluntarily in high-volume exercise in combination with the in-flight, experiment-specific muscle strength performance measurements. This preliminary research suggested that such exercises may prevent loss of muscle function leading to implementation of routine combined aerobic and resistive exercise for ISS astronauts.

Muscle Fiber Changes in Size and Shape

An "average" healthy person has roughly equal numbers of the two major muscle fiber types ("slow" and "fast" fibers). Slow fibers contract (shorten) slowly and have high endurance (resistance to fatigue) levels. Fast fibers contract quickly and fatigue readily. Individual variation in muscle fiber type composition is genetically (inherited) determined. The compositional range of slow fibers in the muscles on the front of the thigh (quadriceps muscles) in humans can vary between 20% and 95%, a percentage found in many marathon runners. On the other hand, a world-class sprinter or weight lifter would have higher proportions of fast fibers and, through his or her training, these fibers would be quite large (higher cross-sectional diameter or area). Changing the relative proportions of the fiber types in muscles is possible, but it requires powerful stimulus such as a stringent exercise program or the chronic unloading profile that occurs in microgravity. NASA was interested in determining whether there were any changes in the sizes or proportions of fiber types in astronauts during spaceflight.

In the only biopsy study of US astronauts to date, needle muscle biopsies from the middle of the vastus lateralis muscle (a muscle on the side of the thigh) of eight shuttle crew members were obtained before launch (3 to 16 weeks) and after landing (within 3 hours) for missions ranging in duration from 5 to 11 days. Three of the eight crew members (five males and three females, age range 33 to 47 years) flew 5-day missions while the other five crew members completed 11-day flights. Five of the eight crew members did not participate in other medical studies that might affect muscle fiber size and type. NASA made a variety of measurements in the biopsy samples, including relative proportions of the two major muscle fiber types, muscle fiber cross-sectional area by muscle fiber type, and muscle capillary (small blood vessel) density. Slow fiber-type cross-sectional area decreased by 15% as compared to a 22% decrease for fast fiber muscle fibers. Biopsy samples from astronauts who flew on the 11-day mission showed there were relatively more fast fiber types and fewer slow fiber types, and the density of muscle capillaries was reduced when the samples taken after landing were compared to those taken before launch. NASA research suggests that fiber types can change in microgravity due to the reduced loads. This has implications for the type and volume of prescriptive on-orbit exercise.

Research conducted during the shuttle flights provided valuable insight into how astronauts' muscles responded to the unloading experienced while living and working in space. Exercise equipment and specific exercise therapies developed and improved on during the program are currently in use on the ISS to promote the safety and health of NASA crew members.

The "Why" and "How" of Exercise on the Space Shuttle

Why Exercise in Space?

Just as exercise is an important component to maintain health here on Earth, exercise plays an important role in maintaining astronaut health and fitness while in space. While living in space requires very little effort to maneuver around, the lack of gravity can decondition the human body.

Knowledge gained during the early years of human spaceflight indicated an adaptation to the new environment. While the empirical evidence was limited, the biomedical data indicated that microgravity alters the musculoskeletal, cardiovascular, and neurosensory systems. In addition, the responses to spaceflight varied from person to person. Space adaptation was highly individualized, and some human systems adjusted at different rates. Overall, these changes were considered to have potential implications on astronaut occupational performance as well as possible impacts to crew health and safety. There was concern that space-related deconditioning could negatively influence critical space mission tasks, such as construction of the space station, repair of the orbiting Hubble Space Telescope, piloting and landing operations, and the ability to egress in an emergency.

Historically, NASA worked on programs to develop a variety of strategies to prevent space deconditioning, thus migrating toward the use of exercise during spaceflight to assure crew member health and fitness. In general, exercise offered a well-understood approach to fitness on Earth, had few side effects, and provided a holistic approach for addressing health and well-being, both physically and psychologically.

NASA scientists conducted experiments in the 1970s to characterize the effects of exercise during missions lasting 28, 56, and 84 days on America's first orbiting space station—Skylab. This was the first opportunity for NASA to study the use of exercise in space. These early observations demonstrated that exercise modalities and intensity could improve the fitness outcomes of astronauts, even as missions grew in length. Armed with information from Skylab, NASA decided to provide exercise on future shuttle missions to minimize consequences that might be associated with spaceflight deconditioning to guarantee in-flight astronaut performance and optimize postflight recovery.

Benefits of Exercise

Space Shuttle experience demonstrated that for the short-duration shuttle flights, the cardiovascular adaptations did not cause widespread significant problems except for the feelings of light-headedness—and possibly fainting—in about one-fifth of the astronauts and a heightened concern over irregular heartbeats during spacewalks. During the Space Shuttle Program, however, it became clear from these short-duration missions that exercise countermeasures would be required to keep astronauts fit during long-duration spaceflights. Although exercise was difficult in the shuttle, simple exercise devices were the stationary bike, a rowing machine, and a treadmill. Astronauts, like those from Skylab, found it difficult to raise their heart rate high enough for adequate exercise. NASA demonstrated that in-flight exercise could be performed and helped maintain some aerobic fitness, but much research remained to be done. This finding led to providing the ISS with a bicycle ergometer, a treadmill, and a resistive exercise device to ensure astronaut fitness.

Deconditioning due to a lack of aerobic exercise is a concern in the area of EVAs, as it could keep the astronaut from performing spacewalks and other strenuous activities. Without enough in-flight aerobic exercise, astronauts experienced elevated heart rates and systolic blood pressures.

Ken Baldwin, PhD
Principal investigator on three Spacelab missions—STS-40 (1991), STS-58 (1993), and STS-90 (1998)—and a Physiological Anatomical Rodent Experiment. Muscle Team leader, 2001-2009, for the National Space Biomedical Research Institute.

"The space life sciences missions (STS-40, STS-58, and STS-90) provided a state-of-the-art laboratory away from home that enabled scientists to customize their research studies in ways that were unheard of prior to the Space Shuttle Program. In using such a laboratory, my research generated unique insights concerning the remodeling of muscle structure and function to smaller, weaker, fatigue-prone muscles with a contractile phenotype that was poorly suitable to apposing gravity. These unique findings became the cornerstone of recommendations that I spearheaded to redesign the priority of exercise during spaceflight from one of an aerobic exercise focus (treadmill and cycling exercise) to a greater priority of exercise paradigms favoring heavy-resistance exercise in order to prevent muscle atrophy in microgravity. Additionally, our group also made an important discovery in ground-based research supported by NASA's National Space Biomedical Research Institute showing that it is not necessarily the contraction mode that the muscles must be subjected to, but rather it is the amount and volume of mechanical force that the muscle must generate within a given contraction mode in order to maintain normal muscle mass. Thus, the early findings aboard the Space Shuttle have served as a monument for guiding future research to expand humankind's success in living productively on other planets under harsh conditions."

The deconditioned cardiovascular system must work harder to do the same or even less work (exercise) than the well-conditioned system.

Exercise capacity was measured preflight on a standard upright bike. Exercise was stepped up every 3 minutes with an increase in workload. Maximal exercise was determined preflight by each astronaut's maximum volume of oxygen uptake. A conditioned astronaut may have little increase in heart rate above sitting when he or she is walking slowly. The heart rate and systolic blood pressure (the highest blood pressure in the arteries, just after the heartbeats during each cardiac cycle) increase as the astronaut walks fast or runs until the heart rate cannot increase any more.

In-flight exercise testing showed that crew members could perform at 70% of the preflight maximum exercise level with no significant issues. This allowed mission planners to schedule EVAs and other strenuous activities that did not overtax the astronauts' capabilities.

How Astronauts Exercised on the Space Shuttle

Because of the myriad restrictions about what can be launched within a space vehicle, tremendous challenges exist related to space exercise equipment. Systems need to be portable and lightweight, use minimal electrical power, and take up limited space during use and stowage. In addition, operation of exercise equipment in microgravity is inherently different than it is on Earth. Refining the human-to-machine interfaces for exercise in space was a challenging task tested throughout the shuttle missions. Providing exercise concepts with the appropriate physical training stimulus to maintain astronaut performance that operates effectively in microgravity proved to be a complex issue.

Exercise systems developed for shuttle included: treadmill, cycle ergometer, and rower. The devices offered exercise conditioning that simulated ambulation, cycling, and rowing activities. All exercise systems were designed for operations on the shuttle middeck; however, the cycle could also be used on the flight deck so that astronauts could gaze out the overhead windows during their exercise sessions.

Each of the three systems had its own challenges for making Earth-like exercise feasible while in space within the limits of the shuttle vehicle. Most traditional exercise equipment has the benefit of gravity during use, while spaceflight systems require unique approaches to exercise for the astronaut users. While each system had its unique issues for effective space operations, the exercise restraints were some of the biggest challenges during the program. These restraints included techniques for securing an astronaut to the exercise device itself to allow for effective exercise stimuli.

"Shuttle left a legacy, albeit incomplete, of the theory and practice for exercise countermeasures in space."

William Thornton, MD, astronaut, principal investigator and original inventor of the shuttle treadmill.

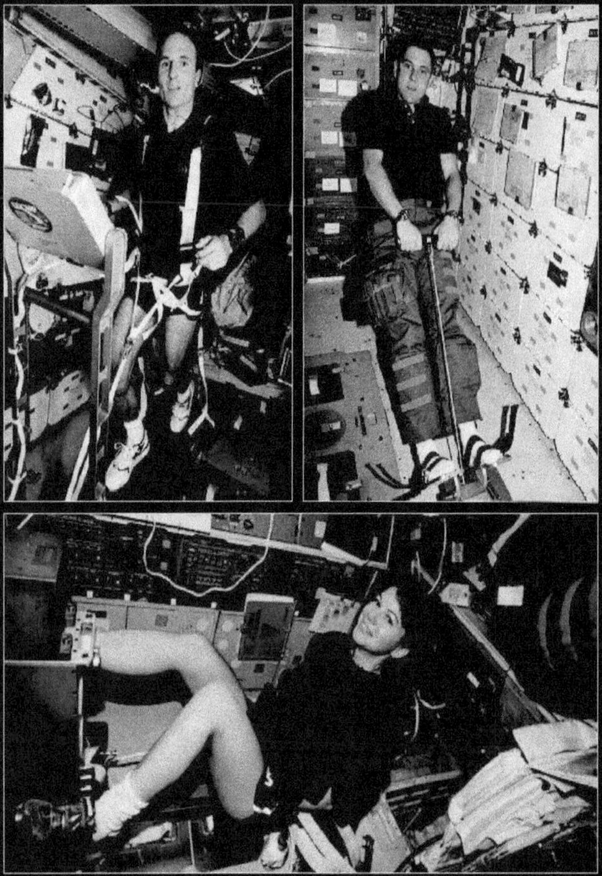

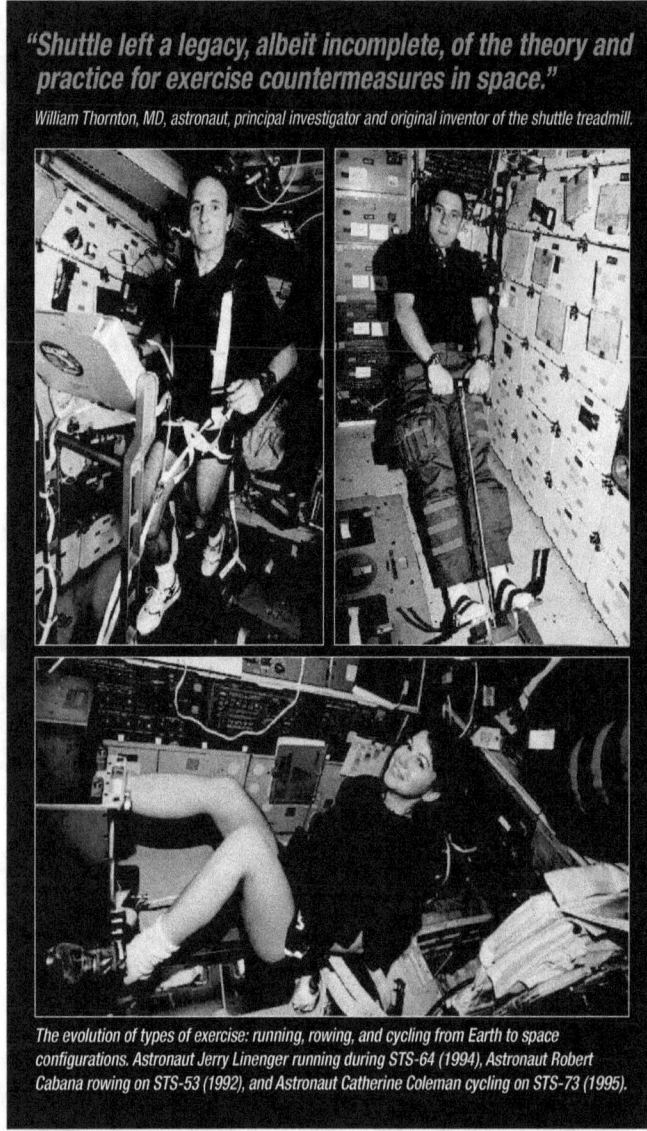

The evolution of types of exercise: running, rowing, and cycling from Earth to space configurations. Astronaut Jerry Linenger running during STS-64 (1994), Astronaut Robert Cabana rowing on STS-53 (1992), and Astronaut Catherine Coleman cycling on STS-73 (1995).

In-flight exercise quality and quantity were measured on all modalities using a commercial heart rate monitor for tracking work intensity and exercise duration. This allowed for a common measure across devices. Heart rate is a quality indicator of exercise intensity and duration (time) is a gauge of exercise quantity—common considerations used for generating exercise prescriptions. Research showed that target heart rates could be achieved using each of the three types of exercise during spaceflight.

Treadmill

Running and walking on a treadmill in the gym can be computer controlled with exercise profiles that alter speed and grade. The shuttle treadmill had limits to its tread length and speed and had no means for altering grade. Treadmill ambulation required the astronaut to wear a complex over-the-shoulder bungee harness system that connected to the treadmill and held the runner in place during use. Otherwise, the runner would propel off the tread with the first step. While exercise target heart rates were achieved, the treadmill length restricted gait length and the harness system proved quite uncomfortable. This information was captured as a major lesson learned for the development of future treadmill systems for use in space.

Cycle Ergometer

The shuttle cycle ergometer (similar to bicycling) operated much like the equipment in a gym. It used a conventional flywheel with a braking band to control resistance via a small motor with a panel that displayed the user's speed (up to 120 rpm) and workload (up to 350 watts). The restraint system used commercial pedal-to-shoe bindings, or toe clips, that held the user to the cycle while leaning on a back pad in a recumbent position. The cycle had no seat, however, and used a simple lap belt to stabilize the astronaut during aerobic exercise. While the cycle offered great aerobic exercise, it was also used for prebreathe operations in preparation for EVAs. The prebreathe exercise protocol allowed for improved nitrogen release from the body tissues to minimize the risk of tissue bubbling during the EVA that could result in decompression sickness or "the bends." Exercise accelerated, "washout" nitrogen that may bubble in the tissues during EVA, causing decompression sickness and, thereby, terminating the EVA and risking crew health.

Rower

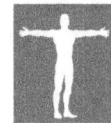

The rower offered total body aerobic exercise, similar to gym rowers. It also had limited capability for resistance exercise. Similar to the cycle, it was seatless since the body floats. The astronaut's feet were secured with a Velcro® strap onto a footplate that allowed for positioning. The rower used a magnetic brake to generate resistance.

Summary

In summary, exercise during Space Shuttle flights had physical and psychological benefits for astronauts. In general, it showed that astronauts could reduce the deconditioning effects that may alter performance of critical mission tasks using exercise in space, even on the relatively short shuttle missions. As a result, a "Flight Rule" was developed that mandated astronauts exercise on missions longer than 11 days to maintain crew health, safety, and performance.

Each device had the challenge of providing an appropriate exercise stimulus without the benefit of gravity and had a unique approach for on-orbit operations. Engineers and exercise physiologists worked closely together to develop Earth-like equipment for the shuttle environment that kept astronauts healthy and strong.

Cardiovascular: Changes in the Heart and Blood Vessels That Affect Astronaut Health and Performance

The cardiovascular system, including the heart, lungs, veins, arteries, and capillaries, provides the cells of the body with oxygen and nutrients and allows metabolic waste products to be eliminated through the kidneys (as urine) and the gastrointestinal tract. All of this depends on a strong heart to generate blood pressure and a healthy vascular system to regulate the pressure and distribute the blood, as needed, throughout the body via the blood vessels.

For our purposes, the human body is essentially a column of fluid; the hydrostatic forces that act on this column, due to our upright posture and bipedal locomotion, led to a complex system of controls to maintain—at a minimum—adequate blood flow to the brain.

On Earth, with its normal gravity, all changes in posture—such as when lying down, sitting, or standing as well as changes in activity levels such as through exercising—require the heart and vascular system to regulate blood pressure and distribution by adjusting the heart rate (beats per minute), amount of blood ejected by the heart (or stroke volume), and constriction or dilation of the distributing arteries. These adjustments assure continued consciousness by providing oxygen to the brain or continued ability to work, with oxygen going to the working muscles.

Removing the effects of gravity during spaceflight and restoring gravity after a period of adjustment to weightlessness present significant challenges to the cardiovascular control system. The cardiovascular system is stressed very differently in spaceflight, where body fluids are shifted into the head and upper body and changes in posture do not require significant responses because blood does not drain and pool in the lower body. Although the cardiovascular system is profoundly affected by spaceflight, the basic mechanisms involved are still not well understood.

During the shuttle era, flight-related cardiovascular research focused on topics that could benefit the safety and well-being of crew members while also revealing the mechanisms underlying the systemic adjustments to spaceflight. NASA researchers studied the immediate responses to the effects of weightlessness during Space Shuttle flights and the well-developed systemic adjustments that followed days and weeks of exposure. Most such research related to the loss of orthostatic tolerance after even brief flights and to the development of potentially detrimental disturbances in cardiac rhythm during longer flights.

Scientists also evaluated the usefulness of several interventions such as exercise, fluid ingestion, and landing-day gravity suits (g-suits) in protecting the astronauts' capacities for piloting the Orbiter—an unpowered, 100-ton glider—safely to a pinpoint landing, and especially for making an unaided evacuation from the Orbiter if it landed at an alternate site in an emergency.

Orthostatic Intolerance: Feeling Light-headed and Fainting on Standing Upright

One of the most important changes negatively impacting flight operations and crew safety is landing day orthostatic intolerance. Astronauts who have orthostatic intolerance (literally, the inability to remain standing upright) cannot maintain adequate arterial blood pressure and have decreased brain blood levels when upright, and they experience light-headedness and perhaps even fainting. This may impair their ability to stand up and egress the vehicle after landing, and even to pilot the vehicle while seated upright as apparent gravity increases from weightlessness to 1.6*g* during atmospheric re-entry.

The orthostatic intolerance condition is complicated and multifactorial. Its hallmarks are increased heart rate, decreased systolic blood pressure,

Blood Volume Changes During Spaceflight

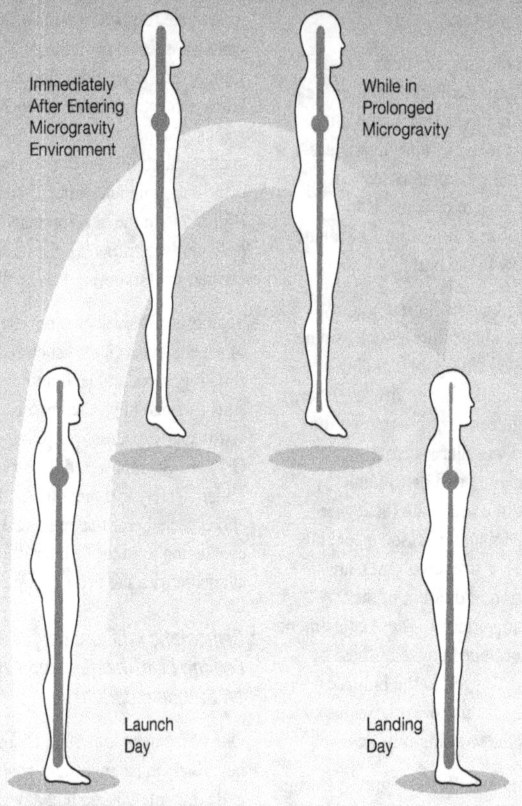

The distribution of blood changes in microgravity more in the upper torso and less in the legs. At landing, the astronaut is light-headed because of less blood and the pooling of the blood in the feet.

and decreased stroke volume during 5 minutes of standing shortly after landing. The decrease in blood volume frequently observed is an important initiating event in the etiology of orthostatic intolerance, but it is the subsequent effects and the physiological responses (or lack thereof) to those effects that may result in orthostatic intolerance after shuttle flights. This is highlighted by the fact that while all shuttle crew members who were tested had low blood volume on landing day, only one-quarter of them developed orthostatic intolerance during standing or head-up tilting.

The group of astronauts that developed orthostatic intolerance lost comparable amounts of plasma (the watery portion of the blood, which the body can adjust quickly) to the group that did not develop orthostatic intolerance. But, the group that was not susceptible had a more pronounced increase in the functioning of the sympathetic nervous system, which is important in responding to orthostatic stress after returning to Earth. Thus, it is not the plasma volume loss alone that causes light-headedness but the lack of compensatory activation of the sympathetic system.

Another possible mechanism for post-spaceflight orthostatic hypotension (low blood pressure that causes fainting) is cardiac atrophy and the resulting decrease in stroke volume (the amount of blood pushed out of the heart at each contraction). Orthostatic hypotension occurs if the fall in stroke volume overwhelms normal compensatory mechanisms such as an increase in heart rate or constriction in the peripheral blood vessels in the arms, legs, and abdomen.

The vast majority of astronauts have been male. Consequently, any conclusions drawn regarding the physiological responses to spaceflight are male biased. NASA recognized significant differences in how men and women respond to spaceflight, including the effects of spaceflight on cardiovascular responses to orthostatic stress. More than 80% of female crew members tested became light-headed during postflight standing as compared to about 20% of men tested, confirming a well-established difference in the non-astronaut population. This is an important consideration for prevention, as treatment methods may not be equally effective for both genders.

How Can This Risk be Changed?

While orthostatic intolerance is perhaps the most comprehensively studied cardiovascular effect of spaceflight, the mechanisms are not well understood. Enough is known to allow for the implementation of some countermeasures, yet none of these countermeasures have been completely successful at eliminating spaceflight-induced orthostatic intolerance following spaceflight.

In 1985, ingestion of fluid and salt (or "fluid loading") prior to landing became a medical requirement through a Flight Rule given the demonstrated benefits and logic that any problem caused—at least in part—by a loss in plasma volume should be resolved—at least in part—by fluid restoration. Starting about 2 hours before landing, astronauts ingest about 1 liter (0.58 oz) of water along with salt tablets. Subsequent refinements to enhance palatability and tolerance include the addition of sweeteners and substitution of bouillon solutions. Of course, any data on plasma volume acquired after 1985 do not reflect the unaltered landing day deficit. But, in spite of the fluid loading, astronauts still returned from shuttle missions with plasma volume deficits ranging from 5% to 19% as well as with orthostatic intolerance.

Shuttle astronauts returned home wearing a lower-body counterpressure garment called the anti-g suit. These suits have inflatable bladders at the calves, thighs, and lower abdomen that resist blood pooling in those areas and force the blood toward the head. The bladders can be pressurized from 25 mmHg (0.5 psi) to 130 mmHg (2.5 psi). In addition, ISS crew members landing on the shuttle used recumbent seats (as opposed to the upright seats of the shorter-duration shuttle crews) and only inflated their suit minimally to 25 mmHg (0.5 psi). All astronauts deflated their anti-g suit slowly after the shuttle wheeled to a stop to allow their own cardiovascular systems time to readjust to the pooling effects of Earth's gravity.

Other treatments for orthostatic intolerance were also evaluated during the program. A technique called "lower body negative pressure," which used slight decompression of an airtight chamber around the abdomen and legs to pool blood there and thus recondition the cardiovascular system, showed promise in ground studies but was judged too cumbersome and time consuming for routine shuttle use. A much simpler approach used a medication known as fludrocortisone, a synthetic corticosteroid known to increase fluid retention in patients on Earth. It proved unsuccessful, however, when it was not well-tolerated by crew members and did not produce any differences in plasma volume or orthostatic tolerance.

Thus, the countermeasures tested were not successful in preventing postflight orthostatic intolerance, at least not in an operationally compatible manner. The knowledge gained about spaceflight-induced cardiovascular

How Red Blood Cells Are Lost in Spaceflight

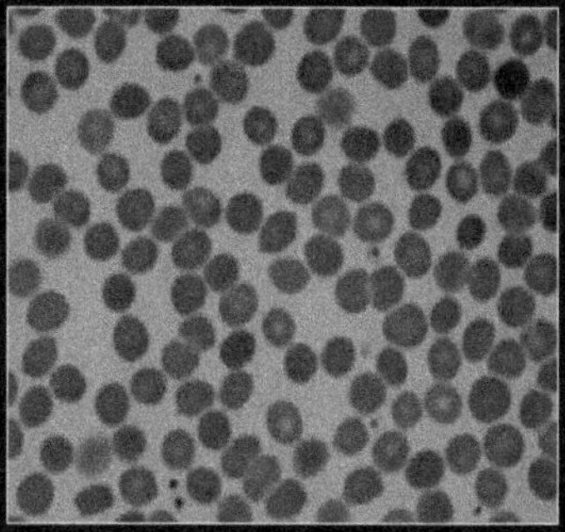

What do astronauts, people traveling from high altitudes to sea level, and renal (kidney) failure patients have in common? All experience changes in red blood cell numbers due to changes in the hormone erythropoietin, synthesized in the kidneys.

Red blood cells bring oxygen to tissues. When astronauts enter microgravity or high-altitude residents travel to sea level, the body senses excess red blood cells. High-attitude residents produce an increased number because of decreased ambient oxygen levels but, at sea level, excess cells are not needed. Astronauts experience a 15% decrease in plasma volume as the body senses an increase in red blood cells per volume of blood. In these situations, erythropoietin secretion from the kidneys ceases. Prior to our research, we knew that when erythropoietin secretion stops, the bone marrow stops production of pre red blood cells and an increase in programmed destruction of these cells occurs.

Another function was found in the absence of erythropoietin, the loss of the newly secreted blood cells from the bone marrow—a process called neocytolysis. Since patients with renal failure are unable to synthesize erythropoietin, it is administered at the time of renal dialysis (a process that replaces the lost kidney functions); however, blood levels of erythropoietin fell rapidly between dialysis sessions, and neocytolysis occurs. Thus, the development of long-lasting erythropoietin now prevents neocytolysis in these patients. Erythropoietin is, therefore, important for human health—in space and on Earth—and artificial erythropoietin is essential for renal failure patients.

changes and differences between orthostatic tolerance groups, however, provided a base for development of future pharmacological and mechanical countermeasures, which will be especially beneficial for astronauts on long-duration missions on space stations and to other planets.

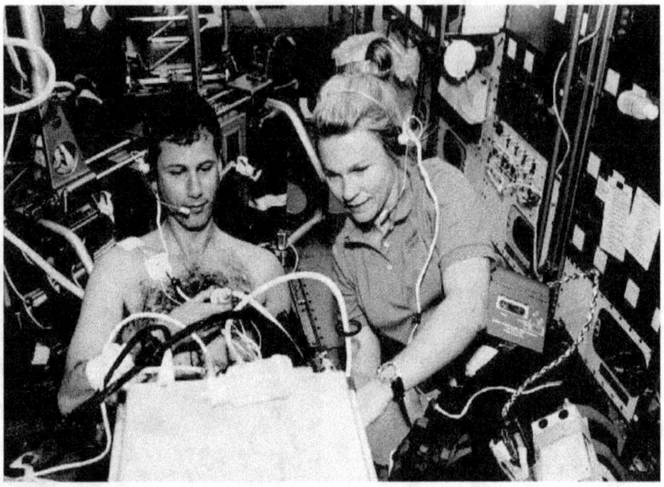

In the Spacelab (laboratory in Orbiter payload bay) Astronaut Rhea Seddon, MD, measures cardiac function on Martin Fettman during Columbia life sciences mission STS-58 (1993).

Cardiovascular Changes During Spaceflight

Headward fluid shift was inferred from reports containing astronaut observations of puffy faces and skinny legs, and was long believed to be the initiating event for subsequent cardiovascular responses to spaceflight. The documentation of this shift was an early goal of Space Shuttle-era investigators, who used several techniques to do so. Direct measurement of peripheral venous blood pressure in an arm vein (assumed to reflect central venous pressure in the heart, an indication of headward fluid shift) was done in 1983 during in-flight blood collections. Actual measurement of central venous pressure was done on a small number of astronauts on dedicated space life sciences Spacelab missions starting in 1991. These studies, and particularly the direct central venous pressure measurements, demonstrated that central venous pressure was elevated in recumbent crew members even before launch, and that it increased acutely during launch with acceleration loads of up to three times Earth's surface gravity. This increased the weight of the column of blood in the legs "above" the heart and the central venous pressure decreased to below baseline values immediately on reaching orbit. Investigators realized that the dynamics of central blood volume changes were more complex than originally hypothesized.

By measuring and recording arterial blood pressures, heart rate, and rhythm, two-dimensional echocardiography demonstrated the variety of changes in the cardiovascular system in flight. In-flight heart rate and systolic and diastolic blood pressure decreased when compared to the preflight values. During re-entry into Earth's atmosphere, these values increased past their preflight baseline, reaching maximal values at peak deceleration loading. When crew members stood upright for the first time after landing, both systolic and diastolic pressures significantly decreased from their seated values and the decrease in diastolic pressure was greater in crew members who did not fully inflate their g-suits. Systolic pressure and heart rate returned to preflight values within an hour of landing, whereas all other spaceflight-induced cardiovascular changes were reversed within a week after landing. Furthermore, stress hormones such as adrenaline (involved in the primal "fight or flight response") were increased postflight, whether the astronauts were resting supine or standing.

So, What Does This Mean?

During weightlessness, there is reduced postural stress on the heart. As expected, the cardiovascular response is muted: blood pressure and heart rate are lower in the resting astronaut than before flight. The volume of blood ejected from the heart with each beat initially increases because of the headward fluid shift, but it becomes lower than preflight levels after that due to the decreased blood volume.

Cardiac Rhythm Disturbances

Contrary to popular opinion, shuttle astronauts were not monitored extensively throughout their flights. Electrocardiograms were recorded and transmitted for crew health assurance only on up to two crew members (out of crews numbering up to seven) and only during launch and landing through the 14th shuttle mission, STS-41G (1984). Subsequently, given the established confidence that healthy astronauts could tolerate spaceflight without difficulty, the requirement for even such minimal medical monitoring was eliminated. Later, a purpose-built system for on-board recording of electrocardiograms and blood pressure was used on select volunteer astronauts between 1989 and 1994.

At present, there is little evidence to indicate that cardiovascular changes observed in spaceflight increase susceptibility to life-threatening disturbances in cardiac rhythms. Certain findings, however, suggest that significant cardiac electrical changes occurred during short and long flights.

NASA systematically studied cardiac rhythm disturbances during some shuttle missions in response to medical reports of abnormal rhythms in nine of 14 spacewalking astronauts between 1983 and 1985. In subsequent studies on 12 astronauts on six shuttle flights, investigators acquired 24-hour continuous Holter recordings of the electrocardiograms during and after altitude chamber training, then again 30 days before launch, during and after each EVA, and after return to Earth. These investigators observed no change in the number of premature contractions per hour during flight compared to preflight or postflight. Given the fact that these data disagreed with other previous reports on astronauts, the investigators recommended that further study was required.

Summary

The Space Shuttle provided many opportunities to study the cardiovascular system due to the high number of flights and crew members, along with an emphasis on life sciences research. This research provided a better understanding of the changes in spaceflight and provided focus for the ISS research program.

Nutritional Needs in Space

Do Astronauts Have Special Nutritional Needs?

If elite athletes like Olympians have special nutritional needs, do astronauts too? During the shuttle flights, nutrition research indicated that, in general, the answer is no. Research, however, provided the groundwork for long-duration missions, such as for the ISS and beyond. Additionally, as the expression goes, while good nutrition will not make you an Olympic-quality athlete, inadequate nutrition can ruin an Olympic-quality athlete.

Nutritional needs drive the types and amounts of food available on orbit. Since shuttle flights were short (1 to 2 weeks), nutritional needs were more like those required for a long camping trip. Accordingly, NASA's research focused on the most important nutrients that related to the physiological changes that microgravity induced for such short missions. The nutrients studied were water, energy (calories), sodium, potassium, protein, calcium, vitamin D, and iron.

Many astronauts eat and drink less in flight, probably due to a combination of reduced appetite and thirst, high stress, altered food taste, and busy schedules. Because the success of a flight is based on the primary mission, taking time for eating may be a low priority. Astronauts are healthy adults, so NASA generally uses Earth-based dietary nutrient recommendations; however, researchers commonly found inadequate food intake and corresponding loss of body weight in astronauts. This observation led to research designed to estimate body water and energy needed during spaceflight.

How Much Water Should an Astronaut Consume?

Water intake is important to prevent dehydration. About 75% of our bodies is water, located mostly in muscles. The fluid in the blood is composed of a noncellular component (plasma) and a cellular component (red blood cells).

NASA measured the various body water compartments using dilution techniques: total body water; extracellular volume (all water not in cells), plasma volume, and blood volume. Because of the lack of strong gravitational force, a shift of fluid from the lower body to the upper body occurs. This begins on the launch pad, when crew members may lie on their backs for 2 to 3 hours for many flights. Scientists hypothesize that the brain senses this extra upper-level body water and adapts through reduced thirst and, sometimes, increased losses through the kidney—urine. An initial reduction of about 15% water (0.5 kg [1.15 pounds]) occurred in the plasma in flight, thus producing a concentrated blood that is corrected by reducing the levels of red blood cells through a mechanism that reduces new blood cells. Soon after entering space, these two compartments (plasma and red blood cells) return to the same balance as before flight but with about 10% to 15% less total volume in the circulation than before flight. Through unknown mechanisms, extracellular fluid is less and total body water does not change or may decrease slightly, 2% to 3% (maximum loss of 1.8 kg [4 pounds]). From this NASA scientists inferred that the amount of intracellular fluid is increased, although this has not been measured. These major fluid shifts affect thirst and, potentially, water requirements as well as other physiological functions. Water turnover decreases due to a lower amount of water consumed and decreased urine volume—both occur in many astronauts during spaceflight. Since total body water does not change much, recommended water intakes are around 2,000 ml/d (68 oz, or 8.5 cups). Astronauts may consume this as a combination of beverages, food, and water.

Because of potentially reduced thirst and appetite, astronauts must make an effort to consume adequate food and water. Water availability on the shuttle was never an issue, as the potable water was a by-product of the fuel cells. With flights to the Russian space station Mir and the ISS, the ability to transfer water to these vehicles provided a tremendous help as the space agencies no longer needed to launch water, which is very heavy.

A much-improved understanding of water loss during EVAs occurred during the shuttle period. This information led to the ISS EVA standards. Dehydration may increase body heat, causing dangerously high temperatures. Therefore, adequate water intake is essential during EVAs. NASA determined how much water was needed for long EVAs (6 hours outside the vehicle, with up to 12 hours in the EVA suit). Due to the concern for dehydration, water supplies were 710 to 946 ml (24 to 32 oz, or 3 to 4 cups) in the in-suit drink bag (the only nutrition support available during EVA).

How Spaceflight Affects Kidney Function

Does the headward fluid shift decrease kidney function? The kidneys depend on blood flow, as it is through plasma that the renal system removes just the right amount of excess water, sodium, metabolic end products like urea and creatinine, as well as other metabolic products from foods and contaminates. So, what is the affect of reduced heart rates and lower blood volumes? Astronauts on several Spacelab flights participated in research to determine any changes in renal function and the hormones that regulate this function. When the body needs to conserve water, such as when sweating or not hydrating enough, a hormone called antidiuretic hormone prevents water loss. Similarly, when the body has too little sodium, primarily due to diet and sweating, aldosterone keeps sodium loss down. All the experiments showed that these mechanisms worked fine in spaceflight. We learned not to worry about the basic functions of the kidney.

Renal Stones

As stated, the kidney controls excess water. But, what happens if a crew member is dehydrated due to sweating or not consuming enough water? During spaceflight, urine becomes very concentrated with low levels of body water. This concentrated urine is doubly changed by immediately entering microgravity, and the bone starts losing calcium salts. Although these losses were not significant during the short shuttle flights, this urinary increase had the potential to form calcium oxalate renal stones. Furthermore, during spaceflight, protein breakdown increases due to muscle atrophy and some of the end products could also promote renal stones. Due to the potential problem of renal stones, crew members were strongly encouraged to consume more water than their thirst dictated. This work led to the development of countermeasures for ISS crew members.

Sodium and Potassium: Electrolytes Important for Health

The electrolytes sodium (Na) and potassium (K) are essential components of healthy fluid balance; Na is a primarily extracellular ion while K is a primarily intracellular ion. They are essential for osmotic balance, cell function, and many body chemical reactions. K is required for normal muscle function, including the heart. With changes in fluid balance, what happens to these electrotypes, especially in their relationship to kidney and cardiovascular function?

Total body water levels change with changes in body weight. With weight loss, liver glycogen (polymers of glucose) stores that contain significant associate water are lost, followed by tissue water—fat 14% and lean body mass 75% water. Antidiuretic hormone conserves body water. Aldosterone increases the volume of fluid in the body and drives blood pressure up, while atrial natriuretic peptide controls body water, Na, K, and fat (adiposity), thereby reducing blood pressure. In the first few days of spaceflight, antidiuretic hormone is high but it then readjusts to controlling body water. Aldosterone and atrial natriuretic peptide reflect Na and water intakes to prevent high blood pressure.

Research from several Spacelab missions demonstrated that in microgravity, astronauts' bodies are able to adjust to the changes induced by microgravity, high Na intakes, and the stress of spaceflight. During spaceflight, Na intakes are generally high while K intakes are low as compared to needs. The astronauts adjust to microgravity within a few days. Although astronauts have less body water and a headward shift of water, these regulatory hormones primarily reflect dietary intakes.

The implications of these data for long-duration flights, such as the ISS, remain unknown. While on Earth, high Na intakes are most often associated with increasing blood pressure. Such intakes also may exacerbate bone loss, which is a problem for astronauts on long-duration spaceflights.

How Many Calories Do Astronauts Need in Spaceflight?

Because astronauts eat less, research determined the energy level (calories) needed during spaceflight. For selected missions, astronauts completed food records with a bar code reader to obtain good information about dietary intake during spaceflights. These studies showed that most astronauts ate less than their calculated energy needs—on average, about 25% less.

Scientists completed two types of research for measuring astronauts' body energy use. Energy can be determined from the products of energy metabolism: carbon energy sources like carbohydrates, protein, or fat + oxygen (O_2) = heat + carbon dioxide (CO_2). We used two methods for shuttle flights. For most flights, all the expired CO_2 was removed by chemical reaction with lithium hydroxide (LiOH) so the amount of CO_2 produced during a flight could be determined. CO_2 that was absorbed into the LiOH could be measured at the end of the flight to determine the energy use by the crew over the entire mission. The second method was to determine the amount of CO_2 and water loss over 3 to 5 days of time per astronaut. Astronauts consumed two stable isotopes (not radioactive), deuterium and ^{18}O, and the levels of these isotopes in urine were measured over a period of several days. The O_2 occurs in the CO_2 and water, but deuterium is only in the water; thus the method allowed for the determination of the CO_2 produced by an astronaut. Surprisingly for both methods, the levels of energy used were the same in flight as on Earth. As a result of this research, NASA dietitians use gender and weight, along with allowing for moderate activity values, to calculate astronauts' energy needs for spaceflight. This method has worked for many years to ensure adequate provision of space foods.

One of the major contributions of EVA research is the increased ability to predict energy expenditure during spacewalks. EVAs were routinely conducted from the shuttle. Energy expenditure was important for both suit design and dietary intakes before and after a spacewalk. After conducting thousands of EVA hours, NASA knows that the energy expenditure was not high for a short period of time, similar to walking 4 to 6.4 kph (2.5 to 4.0 mph). Nearly all EVAs lasted around 6 hours, however, and thus energy expenditure added up to a fairly high level. The lower energy levels occurred when crew members were within the payload bay, primarily doing less-demanding work for short periods. With the construction of the ISS, EVA activity increased along with duration to about 4 to 8 kJ/hr (250 to 500 kcal/hr). For an 8-hour EVA, this was significant. Of course, as previously described, increased energy expenditure increased water needs.

Protein and Amino Acids: Essential for Maintenance of Muscle Function

Protein and its components (amino acids) are essential for all body chemical reactions, structure, and muscles. In spaceflight, total body protein turnover increases as measured by the loss of the orally ingested stable isotope ^{15}N-glycine, which was measured in body tissues such as saliva and blood. Glycine is an amino acid that occurs abundantly in proteins, so changes in blood levels indicate the amount of glycine moved to the tissues for protein syntheses. Some of the increased turnover may be due to the catabolic state of weight loss found with many astronauts due to lower-than-needed energy intakes. There is evidence, even with short-term shuttle flights, that skeletal muscle function decreases. The mild stress of spaceflight found with hard-working astronauts may increase protein breakdown. Increased stress was determined by increased levels of blood and urinary cortisol. Dietary protein levels are already high in spaceflight. Protein recommendations are the same as ground-based dietary guidelines.

Bones Need Calcium and Vitamin D

Studies with Skylab astronauts in the 1970s and shuttle crew members found calcium (Ca) losses increased during flight, probably through removal from bone. NASA confirmed this initial observation of bone loss in the 1990s by using the latest biological markers technology. In fact, research showed that as soon as the astronauts arrived in space, they started losing bone.

Vitamin D is essential for the body to absorb the dietary Ca that is used for bone and other tissue functions. Vitamin D syntheses occur in the skin during exposure to sunlight. In spacecraft, however, sunlight is not tolerated: the rays are too strong because flights take place above the protective atmosphere. Studies completed during the Shuttle-Mir and European Space Agency research programs showed low vitamin D levels could be a problem for Ca absorption and good bone health. A vitamin D supplement is provided for ISS long-duration spaceflights.

Too Much Iron May Be Toxic

Changes in astronaut's red blood cells and iron (Fe) levels are similar to those of a person who lives at a high altitude (e.g., 3,658 m [12,000 ft]) coming to sea level. Both have too much available Fe (i.e., not bound up in red blood cells).

Fe is an important part of red blood cells that brings oxygen from the lungs to the tissues. Low levels of red blood cells cause fatigue. The initial decrease in plasma volume produces an increased concentration of red blood cells. The body may then perceive too many red blood cells and make adjustments accordingly. A 12% to 14% decrease in the number of red blood cells occurs within a couple of weeks of spaceflight. To maintain the correct percent of red blood cells (about 37% to 51% of the blood), newly formed red blood cells are destroyed until a new equilibrium is achieved. The red blood cell Fe is released back into the blood and tissues, and no mechanism except bleeding can reduce the level of body Fe. Excess Fe could potentially have toxic effects, including tissue oxidation and cardiovascular diseases. Shuttle research showed that the dietary Fe need is below that needed on Earth because of the reduced need for red blood cell production.

Summary of Nutritional Needs Found for Space Shuttle Astronauts

Nutrient	Level
Energy men 70 kg (~154 pounds)	12.147 MJ/d (2,874 kcal/d)
Energy women 60 kg (~132 pounds)	9.120 MJ/d (2,160 kcal/d)
Protein	12% to 15% of energy intake < 85 g/d
Water	2,000 ml/d
Na	1,500 to 3,500 mg/d
K	3,500 mg/d
Fe	10 mg/d
Vitamin D	10 ug/d
Calcium	800 to 1,200 mg

Changes in Immunity and Risk of Infectious Disease During Spaceflight

Humans are healthy most of the time, despite being surrounded by potentially infectious bacteria, fungi, viruses, and parasites. How can that be? The answer is the immune system. This highly complex and evolved system is our guardian against infectious diseases and many cancers. It is essential that astronauts have a robust, fully functional immune system just as it is for us on Earth. Astronauts are very healthy, exquisitely conditioned, and well nourished—all factors promoting healthy immunity. In addition, exposures to potential microbial pathogens are limited by a series of controls. All shuttle consumables (e.g., drinking water and food) and environment (breathing air and surfaces) are carefully examined to ensure the health and safety of the astronauts. Preflight restrictions are in place to limit exposure of astronauts to ill individuals. This system works very well to keep astronauts healthy before, during, and after spaceflight. Since spaceflight is thought to adversely affect the immune system and increase disease potential of microorganisms, the shuttle served as a platform to study immunity and microbes' ability to cause disease.

The Immune System

Your immune system quietly works for you, a silent army within your body protecting you from microorganisms that can make you sick. If it is working well, you never know it. But, when it's not working well, you will probably feel it.

The human immune system consists of many distinct types of white blood cells residing in the blood, lymph nodes, and various body tissues. The white blood cells of the immune system function in a coordinated fashion to protect the host from invading pathogens (bacteria, fungi, viruses, and parasites).

There are various elements of immunity. Innate immunity is the first line of defense, providing nonspecific killing of microbes. The initial inflammation associated with a skin infection at a wound site is an example of innate immunity, which is primarily mediated by neutrophils, monocytes, and macrophages. Cell-mediated immunity provides a specific response to a particular pathogen, resulting in immunologic "memory" after which immunity to that unique pathogen is conferred. This is the part of the immune system that forms the basis of how vaccines work. T cells are part of cell-mediated immunity, while B cells provide the humoral immune response. Humoral immunity is mediated by soluble antibodies—highly specific antimicrobial proteins that help eliminate certain types of pathogens and persist in the blood to guard against future infections. Upon initial exposure to a unique pathogen such as a herpes virus, the number of specific types of T and B cells expands in an attempt to eliminate the infection. Afterward, smaller numbers of memory cells continue to patrol the body, ever vigilant for another challenge by that particular pathogen. An immune response can be too strong at times, leading to self-caused illness without a pathogen. Examples of this are allergies and autoimmune diseases. At other times an immune response is not strong enough to fight an infection (immunodeficiency). Acquired Immunodeficiency Syndrome (AIDS) and cancer chemotherapy are both examples of immunodeficiency conditions caused by the loss of one or more types of immune cells.

Spaceflight-associated Changes in Immune Regulation

Changes in regulation of the immune system are found with both short- and long-duration spaceflight. Studies demonstrated that reduced cell mediated immunity and increased reactivation of latent herpes viruses occur during flight. In contrast, humoral (antibody) immunity was found to be normal when astronauts were immunized during spaceflight. Other shuttle studies showed reduced numbers of T cells and natural killer cells (a type of white blood cell important for fighting cancer and virally infected cells), altered distribution of the circulating leukocyte (white blood cell) subsets, altered stress hormone levels, and altered cytokine levels. Reduced antimicrobial functions of monocytes, neutrophils, and natural killer cells also occur when measured soon after spaceflight. Cytokines are small proteins produced by immune cells; they serve as molecular messengers that control the functions of specialized immune cells. Cytokines are released during infection and serve to shape the immune response. There are many cytokines, and they can be grouped in several ways. Th1 cytokines are produced by specialized T cells to promote cell-mediated immunity, whereas Th2 cytokines promote humoral immunity. One hypothesis to explain immune dysregulation during spaceflight is a shift in the release of cytokines from Th1 toward Th2 cytokines. Data gained from the shuttle research support this theory.

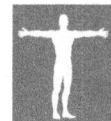

Selected Space Shuttle Immune Studies

Hypersensitivity

Hypersensitivity occurs when the immune response to a common antigen is much stronger than normal. Usually, this manifests itself as a rash and is commonly measured via skin testing. Briefly, seven common antigens, bacteria, Proteus (common in urinary track infections), Streptococcus, tuberculin and Trichophyton (skin diseases), and yeast, Candida (known to increase in the immune compromised), are injected into the forearm skin. For most normal individuals, the cell-mediated arm of the immune system reacts to these antigens within 2 days, resulting in a visible red, raised area at the site of the injections. These reactions are expected and represent a healthy immune response. The red, raised circular area for each antigen can be quantified. To test astronauts, antigens were injected 46 hours before landing, and the evaluation of the reaction took place 2 hours after landing. Data showed that, as compared to preflight baseline testing, the cell-mediated immunity was significantly reduced during flight. Both the number of reactions and the individual reaction size were reduced during flight. These data indicated for the first time that immunity was reduced during short-duration spaceflight. Any associated clinical risks were unknown at the time. The possibility that this phenomenon would persist for long-duration flight was also unknown. Similar reductions in cell-mediated immunity were reported in Russian cosmonauts during longer missions.

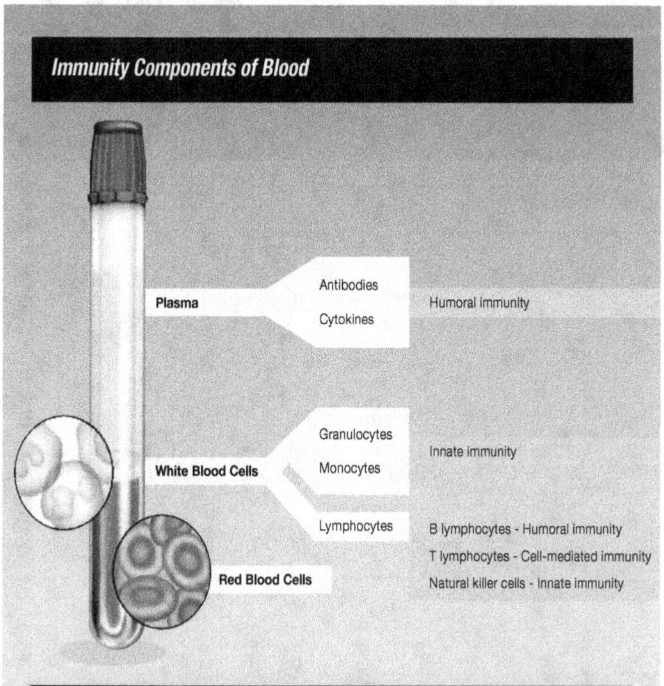

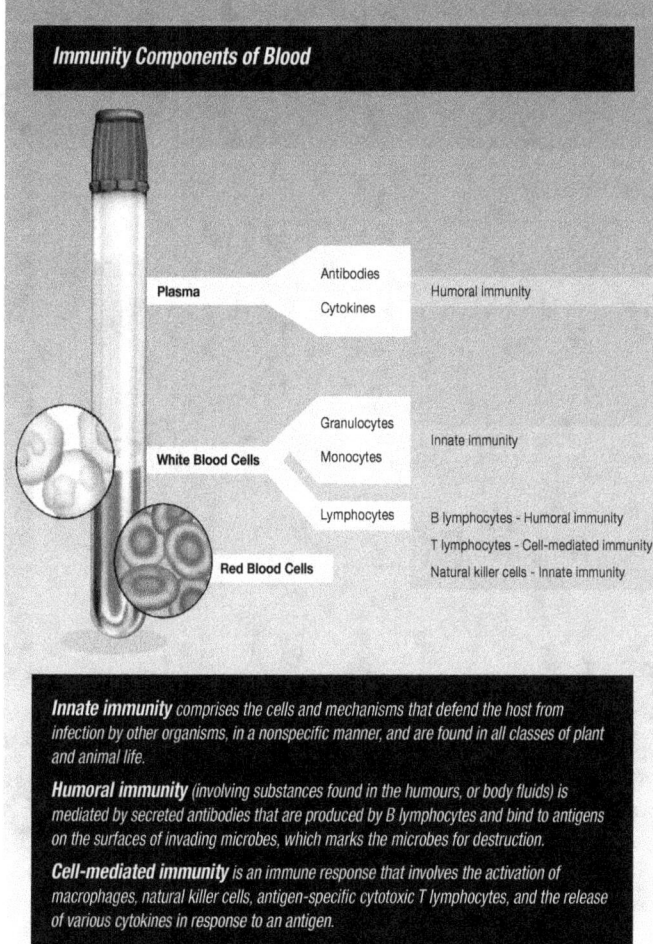

Innate immunity *comprises the cells and mechanisms that defend the host from infection by other organisms, in a nonspecific manner, and are found in all classes of plant and animal life.*

Humoral immunity *(involving substances found in the humours, or body fluids) is mediated by secreted antibodies that are produced by B lymphocytes and bind to antigens on the surfaces of invading microbes, which marks the microbes for destruction.*

Cell-mediated immunity *is an immune response that involves the activation of macrophages, natural killer cells, antigen-specific cytotoxic T lymphocytes, and the release of various cytokines in response to an antigen.*

Studies of the Peripheral Mononuclear Cells

Peripheral mononuclear cells are blood immune cells. Their numbers are a measure of the current immune status of a subject. During the latter stages of the 11-day STS-71 (1995) shuttle mission, the shuttle astronauts and the returning long-duration astronauts (from Mir space station) stained samples of their peripheral

Herpes Viruses Become Active During Spaceflight

Herpes viruses, the most commonly recognized latent viruses in humans, cause specific primary diseases (e.g., chicken pox), but may remain inactive in nervous tissue for decades. When immune response is diminished by stress or aging, latent viruses reactivate and cause disease (e.g., shingles).

Epstein-Barr virus reactivated and appeared in astronauts' saliva in large numbers during spaceflight. Saliva collected during the flight phase contained tenfold more virus than saliva collected before or after flight. This finding correlated with decreased immunity in astronauts during flight. The causes of reduced immunity are unknown, but stress associated with spaceflight appears to play a prominent role, as the levels of stress hormones increase during spaceflight. The resulting decreased immunity allows the viruses to multiply and appear in saliva. The mechanism for Epstein-Barr virus reactivation seems to be a reduction in the number of virus-specific T cells leading to decreased ability to keep Epstein-Barr virus inactive.

Cytomegalovirus, another latent virus, also reactivated and appeared in astronaut urine in response to spaceflight. Healthy individuals rarely shed cytomegalovirus in urine, but the virus is commonly found in those with compromised immunity.

Scientists also studied Varicella-Zoster virus, the causative agent of chicken pox and shingles. These astronaut studies were the first reports of the presence of this infectious virus in saliva of asymptomatic individuals. A rapid, sensitive test for use in doctors' offices to diagnose shingles and facilitate early antiviral therapy resulting in reductions in nerve damage was a product of this study.

Role of Varicella-Zoster Virus in Chicken Pox and Shingles

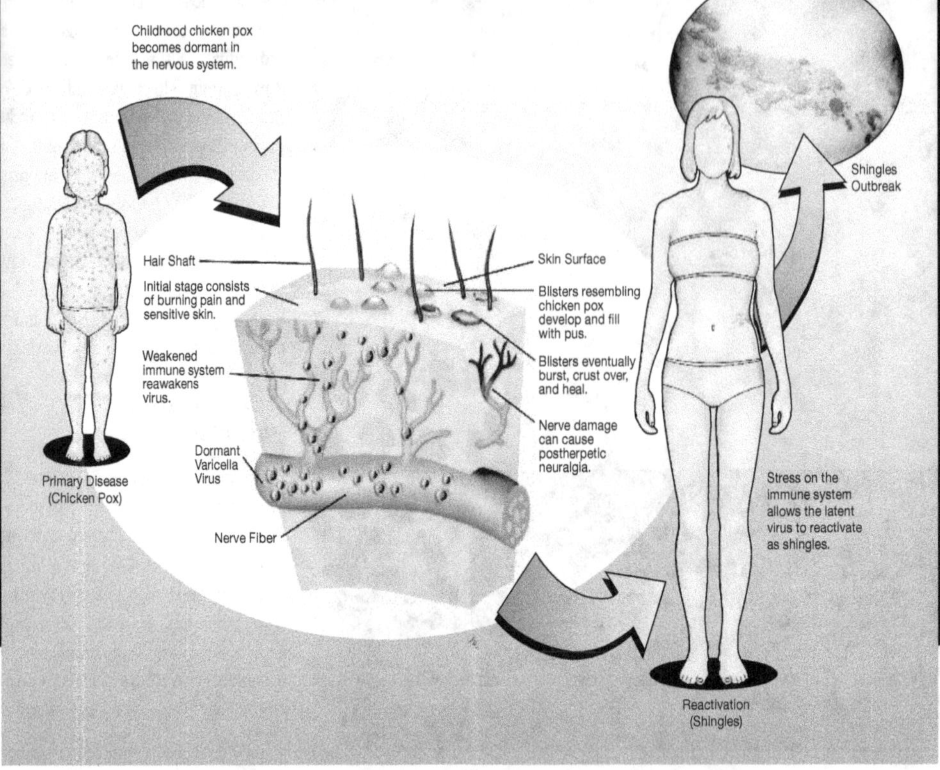

blood immune cells with various dyes using unique and patented equipment developed at Johnson Space Center. These data showed that the major "bulk" levels of peripheral blood immune cells did not appear to be altered during flight.

Summary

The laboratory capabilities of the Space Shuttle allowed our first systematic assessment of the effects of space travel on the human immune system. Most indicators of immunity were altered during short-duration spaceflight, which is a uniquely stressful environment. These stressors were likely major contributors to the observed changes in immunity and the increased viral reactivation. Latent viruses were shown to be sensitive indicators of immune status. Bacterial pathogens were also shown to be more virulent during spaceflight. It is unknown whether these are transient effects or whether they will persist for long-duration missions. These important data will allow flight surgeons to determine the clinical risk for exploration-class space missions (moon, Mars) related to immunology, and to further the development of countermeasures for those risks. These studies and the hardware developed to support them serve as the platform from which new studies on board the ISS were initiated. It is expected that the ISS studies will allow a comprehensive assessment of immunity, stress, latent viral reactivation, and bacterial virulence during long-duration spaceflight.

Habitability and Environmental Health

Habitability

The shuttle contributed significantly to advances in technologies and processes to improve the habitability of space vehicles and enable humans to live and work productively in space. These shuttle-sponsored advances played a key role in our coming to view living and working in space as not only possible but also achievable on a long-term basis.

Habitability can be defined as the degree to which an environment meets an individual's basic physiological and psychological needs. It is affected by multiple factors, including the size of the environment relative to the number of people living and working there and the activities to be undertaken. Other habitability factors include air, water, and food quality as well as how well the environment is designed and equipped to facilitate the work that is to be done.

Resource limitations conspire to severely limit the habitability of space vehicles. Spacecraft usually provide minimal volume in which crew members can live and work due to the high cost of launching mass into space. The spacecraft's environmental control system is usually closed to some degree, meaning that spacecraft air and water are recycled and their quality must be carefully maintained and monitored. It may be several months between when food is prepared and when it is consumed by a space crew. There is normally a limited fresh resupply of foods. Care must be taken to assure the quality of the food before it is consumed.

The following sections illustrate some of the technologies and processes that contributed to the habitability of the shuttle and provided a legacy that will help make it possible for humans to live safely and work productively in space.

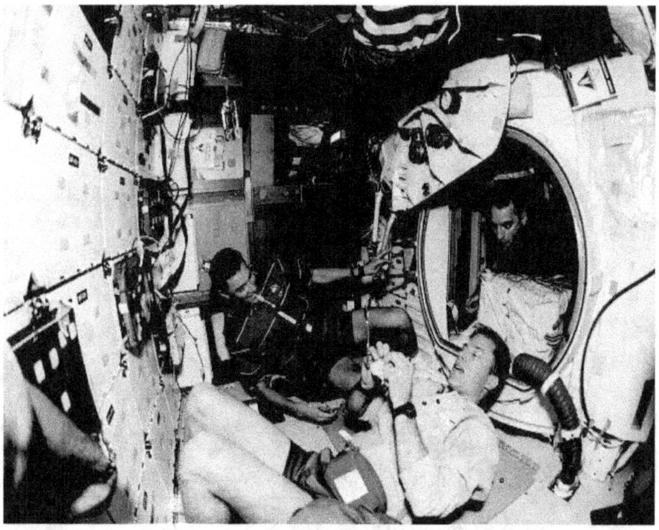

On STS-90 (1998), three Space Shuttle Columbia crew members—Astronauts James Pawelczyk, Richard Searfoss, and Richard Linnehan—meet on the middeck, where the crew ate, slept, performed science, prepared for extravehicular activities (spacewalks), exercised, took care of personal hygiene needs, and relaxed.

Innovations Improve Habilitability

Restraints and Mobility Aids

One of the most successful aids developed through the program, and one that will be used on future spacecraft, to support crew member physical stability in microgravity is foot restraints. It is nearly impossible to accomplish tasks in microgravity without stabilizing one's feet. NASA scientists developed several designs to make use of the body's natural position while in space. One design has foot loops and two-point leg/foot restraints used while a crew member works at a glove box. These restraints stabilize a crew member. The effectiveness of a restraint system relates to the simplicity of design, comfort, ease of use, adjustability, stability, durability, and flexibility for the range of the task. Other restraint systems developed include handrails, bungee cords, Velcro®, and flexible brackets. Furthermore, foot restraints aid in meeting other challenges such as limited visibility and access to the activity area. The latter difficulties can lead to prolonged periods of unnatural postures that may potentially harm muscles or exacerbate neurological difficulties.

Cursor Control Devices

The shuttle spacecraft environment included factors such as complex lighting scenarios, limited habitable volume, and microgravity that could render Earth-based interface designs less than optimal for space applications. Research in space human factors included investigating ways to optimize interfaces between crew members and spacecraft hardware, and the shuttle proved to be an excellent test bed for evaluating those interfaces.

For example, while computer use is quite commonplace today, little was known about how, or if, typical cursor control devices used on Earth would work in space. NASA researchers conducted a series of experiments to gather information about the desirable and undesirable characteristics of cursor control devices using high-fidelity environments. Experiments began in ground laboratories and then moved to the KC-135 aircraft for evaluation in a short-duration microgravity environment during parabolic flight. The experiments culminated with flight experiments on board Space Transportation System (STS)-29 (1989), STS-41 (1990), and STS-43 (1991). These evaluations and experiments used on-board crew members to take the devices through the prescribed series of tasks.

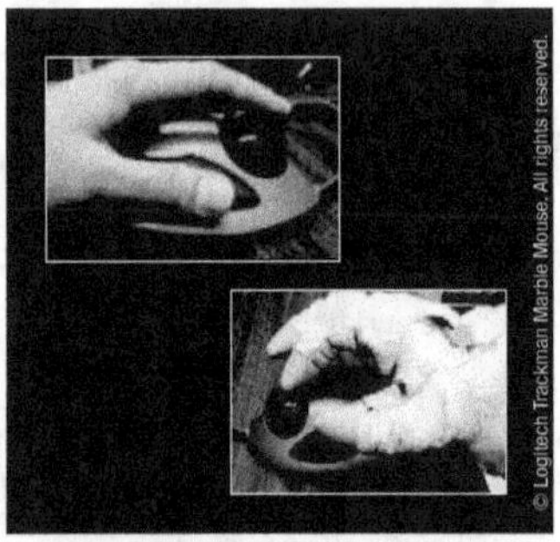

Example of a cursor control device with a trackball as used with ungloved and gloved hands.

Anchoring Improves Performance

Without Constraints

On STS-73 (1995) Astronaut Kathryn Thornton works at the Drop Physics Module on board the Spacelab science module located in the cargo bay of the Earth-orbiting shuttle. Notice that Dr. Thornton is anchoring her body by using a handrail for her feet and right hand. This leaves only one free hand to accomplish her tasks at that workstation and would be an uncomfortable position to hold for a long period of time.

With Constraints

Also on STS-73, Astronaut Catherine Coleman uses the advanced lower body extremities restraint at the Spacelab glove box. With Dr. Coleman's feet and knees anchored for body stability, she has both hands free to work for longer periods, providing her stability and comfort.

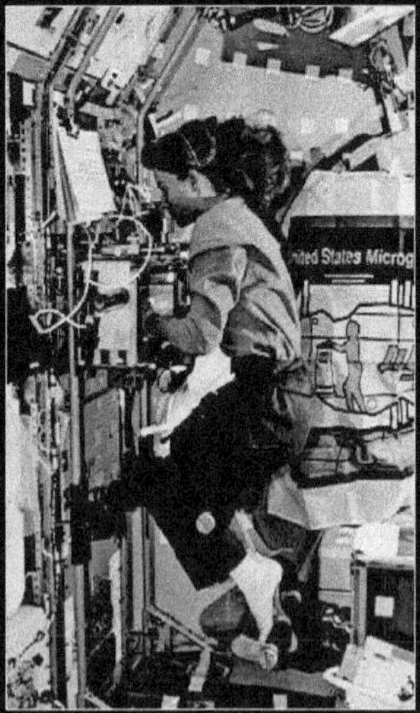

It cannot be assumed that computer equipment, like cursor control devices (e.g., a trackball, an optical mouse), used on Earth will behave the same way in space. Not only does microgravity make items "float," in general the equipment might be used while a crew member is wearing gloves—and the gloves could be pressurized at the time. For example, a trackball has a certain amount of movement allowed within its casing. In space, the ball will float, making it much more difficult to use the trackball and be accurate. During STS-43, the shuttle crew worked with a trackball that was modified to reduce the "play," and they reported that the mechanism worked well. This modification resulted in the fastest and most accurate responses.

Those tests in the flight environment paved the way for the types of equipment chosen for the International Space Station (ISS). The goal was to provide the best equipment to ensure quick and precise execution of tasks by crew members. As computer technology advances, NASA will continue investigations involving computer hardware as spacecraft and habitats are developed.

Shuttle Food System Legacy

Does NASA have a grocery store in space? The answer is no. One significant change NASA made to the space food system during the Space Shuttle Program, however, was the addition of a unique bar code on each food package to facilitate on-orbit science.

When crew members began participating in experiments on orbit that required them to track their food consumption, a method was needed that would promote accurate data collection while minimizing crew time; thus, the

White Light-emitting Diode Illuminators

As the shuttle orbited Earth, the crew experienced a sunrise and sunset every 45 minutes on average. This produced dramatic changes in lighting conditions, making artificial light sources very important for working in space.

Because of power and packaging constraints during the Space Shuttle Program, most artificial lights were restricted to fixed locations. With the assembly of the International Space Station and the maintenance of the Hubble Space Telescope, NASA felt it would be a great improvement to have lights mounted on all of the shuttle cameras. These light sources had to be durable, lightweight, and low in power requirements—the characteristics of light-emitting diodes (LEDs).

In 1995, NASA began using white LED lights for general illumination in camera systems several years in advance of industry. These early lights were designed as rings mounted around the lens of each camera. The four payload bay cameras were equipped with four LED light systems capable of being pointed with the pan-and-tilt unit of each camera. NASA also outfitted the two robotic arm cameras with LED rings. In June 1998, the first white 40 LED illumination system was flown. In May 1999, white 180 LED illuminators were flown. These lighting systems remained in use on all shuttle flights.

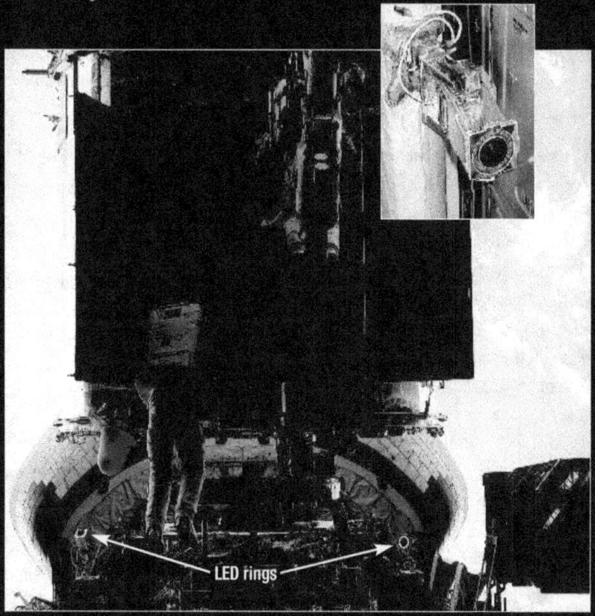

Light-emitting diode (LED) rings mounted on the two shuttle cameras in the aft payload bay of shuttle.

bar code. Crew members simply used a handheld scanning device to scan empty food packages after meals. The device automatically recorded meal composition and time of consumption. Not only did bar codes facilitate science, they also had the additional benefit of supporting the Hazard Analysis and Critical Control Point program for space food.

Hazard Analysis and Critical Control Point is a food safety program developed for NASA's early space food system. Having a unique bar code on each food package made it easy to scan the food packages as they were stowed into the food containers prior to launch. The unique bar code could be traced to a specific lot of food. This served as a critical control point in the event of a problem with a food product. If a problem had arisen, the bar code data collected during the scanning could have been used to locate every package of food from that same lot, making traceability much easier and more reliable. This system of bar coding food items carried over into the ISS food system.

Food preparation equipment also evolved during the shuttle era. The earliest shuttles flew with a portable water dispenser and a suitcase-sized food warmer. The first version of the portable water dispenser did not measure, heat, or chill water, but it did allow the crew to inject water into foods and beverages that required it. This dispenser was eventually replaced by a galley that, in addition to measuring and injecting water, chilled and heated it as well. The shuttle galley also included an oven for warming foods to serving temperature. Ironically, the food preparation system in use on the ISS does not include chilled water and, once again, involves the use of the suitcase-sized food warmer for heating US food products.

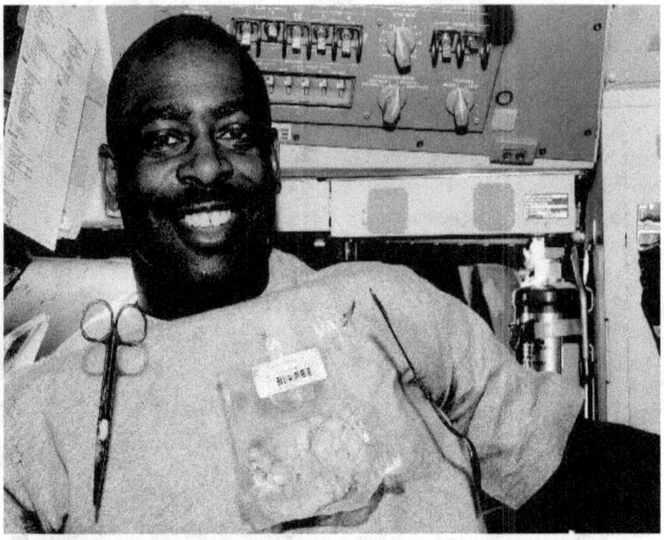

On STS-122 (2007), Astronaut Leland Melvin enjoys his dessert of rehydrated peach ambrosia. Also shown is the pair of scissors that is needed to open the pouch. On the pouch is a bar code that is used to track the food. The blue Velcro® allows the food to be attached to the walls.

Food packaging for shuttle foods also changed during the course of the program. The original rigid, rectangular plastic containers for rehydratable foods and beverages were replaced by flexible packages that took up less room in storage and in the trash. The increase in crew size and mission duration that occurred during the program necessitated this change. These improvements continue to benefit the ISS food system.

Environmental

Environmental Conditions

Maintaining a Healthy Environment During Spaceflight

The shuttle crew compartment felt like an air-conditioned room to astronauts living and working in space, and the Environmental Control and Life Support System created that habitable environment. In fact, this system consisted of a network of systems that interacted to create such an environment, in addition to cooling or heating various Orbiter systems or components. The network included air revitalization, water coolant loop, active thermal control, atmosphere revitalization pressure control, management of supply and wastewater, and waste collection.

The Air Revitalization System assured the safety of the air supply by using lithium hydroxide to maintain carbon dioxide (CO_2) and carbon monoxide at nontoxic levels. It also removed odors and trace contaminants through active charcoal, provided ventilation in the crew compartment via a network of fans and ducting, controlled the cabin's relative humidity (30% to 75%) and temperature (18°C [65°F] to 27°C [80°F]) through cabin heat exchangers for additional comfort, and supplied air cooling to various flight deck and middeck electronic avionics as well as the crew compartment.

The water coolant loop system collected heat from the crew compartment cabin heat exchanger and from some electronic units within the crew compartment. The system transferred the excess heat to the water coolant/Freon®-21 coolant loop heat exchanger of the Active Thermal Control System, which then moved excess heat from the various Orbiter systems to the system heat sinks using Freon®-21 as a coolant.

During ground operations, the ground support equipment heat exchanger in the Orbiter's Freon®-21 coolant loops rejected excess heat from the Orbiter through ground systems cooling. Shortly after liftoff, the flash evaporator (vaporization under reduced pressure) was activated and provided Orbiter heat rejection of the Freon®-21 coolant loops through water boiling. When the Orbiter was on orbit and the payload bay doors were opened, radiator panels on the underside of the doors were exposed to space and provided heat rejection. If combinations of heat loads and the Orbiter attitude exceeded the capacity of the radiator panels during on-orbit operations, the flash evaporator was activated to meet the heat rejection requirements. At the end of orbital operations, through deorbit and re-entry, the flash evaporator was again brought into operation until atmospheric pressure, about 30,480 m (100,000 ft) and below, no longer permitted the flash evaporation process to provide adequate cooling. At that point, the ammonia boilers rejected heat from the Freon®-21 coolant loops by evaporating ammonia through the remainder of re-entry, landing, and postlanding until ground cooling was connected to the ground support equipment heat exchanger.

Atmosphere revitalization pressure control kept cabin pressure around sea-level pressure, with an average mixture of 80% nitrogen and 20% oxygen. Oxygen partial pressure was maintained between 20.3 kPa (2.95 pounds per square inch, absolute [psia]) and 23.8 kPa (3.45 psia), with sufficient nitrogen pressure of 79.3 kPa (11.5 psia) added to achieve the cabin total pressure of 101.3 kPa (14.7 psia) +/-1.38 kPa (0.2 psia). The Pressure Control System received oxygen from two power reactant storage and distribution cryogenic oxygen systems in the mid-fuselage of the Orbiter. Nitrogen tanks, located in the mid-fuselage of the Orbiter, supplied gaseous nitrogen—a system that was also used to pressurize the potable and wastewater tanks located below the crew compartment middeck floor.

Three fuel-cell power plants produced the astronauts' potable water, to which iodine was added to prevent bacterial growth, that was stored in water tanks. Iodine functions like the chlorine that is added to municipal water supplies, but it is less volatile and more stable than chlorine. Condensate water and human wastewater were collected into a wastewater tank, while solid waste remained in the Waste Collection System until the Orbiter was serviced during ground turnaround operations.

Space Shuttle Environmental Standards

We live on a planet plagued with air and water pollution problems because of the widespread use of chemicals for energy production, manufacturing, agriculture, and transportation. To protect human health and perhaps the entire planet, governmental agencies set standards to control the amount of potentially harmful chemicals that can be released into air and water and then monitor the results to show compliance with standards. Likewise, on the shuttle, overheated electronics, systems leaks, propellants, payload chemicals, and chemical leaching posed a risk to air and water quality. Standards were necessary to define safe air and water, along with monitoring systems to demonstrate a safe environment.

Air

Both standards and methods as well as instruments to measure air quality were needed to ensure air quality. For the shuttle, NASA had a formalized process for setting spacecraft maximum allowable concentrations. Environmental standards for astronauts must consider the physiological effects of spaceflight, the continuous nature of airborne exposures, the aversion to drinking water with poor aesthetic properties, and the reality that astronauts could not easily leave a vehicle if it were to become dangerously polluted.

On Earth, plants remove CO_2—a gas exhaled in large quantities as a result of human metabolism—from the atmosphere. By contrast, CO_2 is one of the most difficult compounds to deal with in spaceflight. For example, accumulation of CO_2 was a critical problem during the ill-fated Apollo 13 return flight. As the disabled spacecraft returned to Earth, the crew had to implement unanticipated procedures to manage CO_2. This involved duct-taping filters and tubing together to maintain CO_2 at tolerable levels. Such extreme measures were not necessary aboard shuttle; however, if the crew forgot to change out filters, the CO_2 levels could have exceeded exposure standards within a few hours.

Although older limits for CO_2 were set at 1%, during NASA's new standard-setting process with the National Research Council it became

Combustion Product Analyzer Ensured Crew Breathed Clean Air After Small Fire in Russian Space Station

The combustion product analyzer flew on every Space Shuttle flight from 1990 through 1999 and proved its value during the Shuttle-Mir Program (1995-1998). On the seventh joint mission in 1998, no harm seemed to have occurred during an inadvertent valve switch on an air-purifying scrubber. In fact, during this time, the crew—including American Andrew Thomas—participated in a video presentation transmitted back to Earth; however, shortly after the valve switch, the crew experienced headaches. As on Earth, when occupants of a house or building experience headaches simultaneously, it can indicate that the air has been severely degraded. The crew followed procedures and activated the combustion product analyzer, designed to detect carbon monoxide (CO), hydrogen cyanide, hydrogen chloride, and hydrogen fluoride. The air contained over 500 parts per million of CO, significantly above acceptable concentrations. This high concentration was produced by hot air flowing through a paper filter and charcoal bed and then into the cabin when the valve was mistakenly switched on. The combustion product analyzer was used to follow the cleanup of the CO. Archival samples confirmed the accuracy of the analyzer's results. The success of this analyzer and its successor—the compound specific analyzer-combustion products—led to the inclusion of four units (compound specific analyzer-combustion products) on the International Space Station and a combustion products analyzer on future crew exploration vehicles.

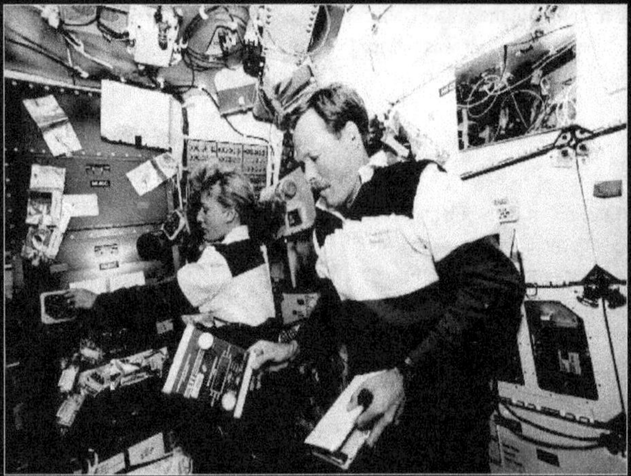

Commander Robert Gibson and Astronaut Jan Davis check the combustion product analyzer during STS-47 (1992).

clear that 1% was too high and, therefore, the spacecraft maximum allowable concentration was reduced to 0.7 %. Even this lower value proved to be marginal under some conditions. For example, the shuttle vehicle did not have the capability to measure local pockets of CO_2, and those pockets could contain somewhat higher levels than were found in the general air. That was especially true in the absence of gravity where convection was not available to carry warm, exhaled air upward from the astronaut's breathing zone. Use of a light-blocking curtain during a flight caused the crew to experience headaches on awakening, and this was attributed to accumulation of CO_2 because the crew slept in a confined space and the curtain obstructed normal airflow.

Setting air quality standards for astronaut exposures to toxic compounds is not a precise science and is complicated. NASA partnered with the National Research Council Committee on Toxicology in 1989 to set and rigorously document air quality standards for astronauts during shuttle spaceflight.

The spaceflight environment is like Earth in that exposure standards can control activities when environmental monitors suggest the need for control. For example, youth outdoor sports activities are curtailed when ozone levels exceed certain standards on Earth. Likewise, spacecraft maximum allowable concentrations for carbon monoxide, a toxic product of combustion, were used to determine criteria for the use of protective masks in the event of an electrical burn. The shuttle Flight Rules provided the criteria. Ranges for environmental monitoring instruments were also based on spacecraft maximum allowable

Measuring Airborne Volatile Organic Compounds

Volatile organic compounds are airborne contaminants that pose a problem in semi-closed systems such as office buildings with contributions from carpets, furniture, and paper products as well as in closed systems such as airplanes and spacecraft. These contaminates cause headaches, eye and skin irritation, dizziness, and even cancer.

NASA needed to be able to measure such compounds for the International Space Station (ISS), a long-term closed living situation. Therefore, in the latter 1990s, the shuttle was used as a test bed for instruments considered for use on the ISS.

Shuttle flights provided the opportunity to assess the performance of a volatile organic analyzer-risk mitigation experiment in microgravity on STS-81 (1997) and STS-89 (1998). Results confirmed component function and improved the instrument built for ISS air monitoring.

Contaminants: Elektron Incident on the Russian Space Station Mir

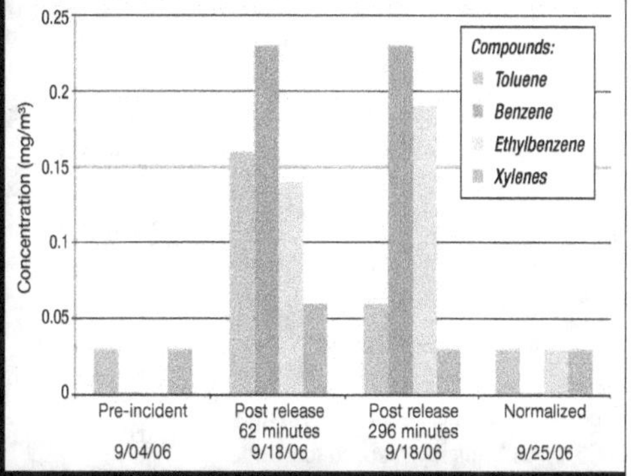

This chart plots the course of the Elektron incident showing the concentrations of toluene, benzene, ethylbenzene, and xylenes—all serious toxins—released into the air. In 2004, the levels of the four contaminates were very low, as measured by the volatile organic analyzer and grab samples returned to Earth for analysis. During the incident, the analyzer measured increases in the four compounds. Grab samples confirmed the higher levels for these compounds and verified that the analyzer had worked. The next available data showed the contaminants had returned to very low levels.

The volatile organic analyzer operated episodically on ISS since 2001 and provided timely and valuable information during the Elektron (Russian oxygen generation system) incident in September 2006 when the crew tried to restart the Elektron and saw what appeared to be smoke emanating from the device. The volatile organic analyzer collected and analyzed samples prior to the event and during cleanup. Data showed that the event had started before the crew noticed the smoke, but the concentrations of the contaminants released were not a health hazard.

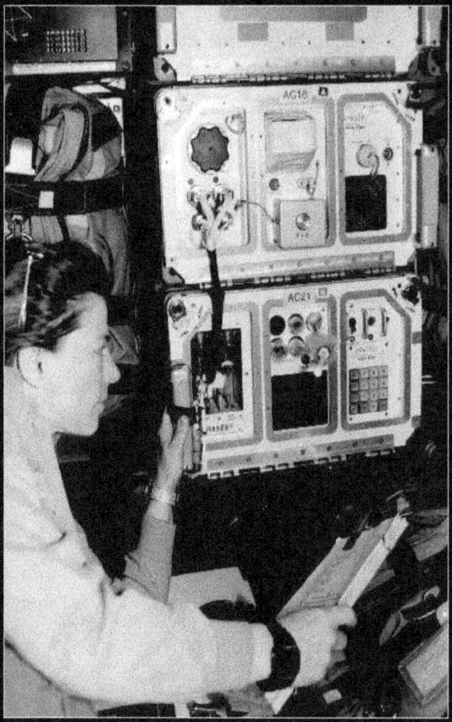

During the STS-89 shuttle dock with Russian space station Mir, Astronaut Bonnie Dunbar goes through her checklist to start the volatile organic analyzer sample acquisition sequence.

concentrations. For example, the monitoring requirements for hydrogen cyanide, another toxic combustion product, were based on spacecraft maximum allowable concentrations to determine how sensitive the monitor must be. By analogy with Earth-based environmental monitoring, spaceflight monitors needed the ability to indicate when safe conditions had returned so that normal operations could resume.

Water

NASA recognized the need for unique water-quality standards. Although the effort to set specific water-quality standards, called spacecraft water exposure guidelines, did not begin until 2000, NASA quickly realized the value of these new limits. One of the first spacecraft water-exposure guidelines set was for nickel, a slightly toxic metal often found in water that has been held in metal containers for some time. The primary toxic effect of concern was nickel's adverse effect on the immune system. High nickel levels had been observed from time to time in the shuttle water system based on the existing requirements in NASA documents. This sometimes caused expensive and schedule-breaking activity at Kennedy Space Center to deal with these events. When National Research Council experts accepted a new, higher standard, the old standard was no longer applied to shuttle water and the nickel "problem" became history.

Toxicants From Combustion

Fire is always a concern in any environment, and a flame is sometimes difficult to detect. First responders must have instruments to quickly assess the contaminants in the air on arriving at the scene of a chemical spill, fire, or building where occupants have been overcome by noxious fumes. Additionally, these instruments must be capable of determining when the cleanup efforts have made it safe for unprotected people to return. When a spill, thermodegradation, or unusual odor occurs on a spacecraft, crew members are the first responders. They need the tools to assess the situation and track the progress of the cleanup. As a result of shuttle experiments, NASA was able to provide crews with novel instruments to manage degradations in air quality caused by unexpected events.

The combustion products analyzer addressed spacecraft thermodegradations events, which can range from overheated wiring to a full-fledged fire. Fire in a sealed, remote capsule is a frightening event. A small event—overheated wire (odor produced)—occurred on STS-6 (1983), but it wasn't until 1988, when technology advances improved the reliability and shrank the size of monitors, that a search for a combustion products analyzer was initiated. Before the final development of the analyzer, however, a more significant event occurred on STS-28 (1989) that hastened the completion of the instrument. On STS-28, a small portion of teleprinter cable pyrolyzed and the released contaminants could have imperiled the crew if more of the cable had burned. The combustion products analyzer requirements were to measure key contaminants in the air following thermodegradation incidents, track the effectiveness of cleanup efforts, and determine when it was safe to remove protective gear.

Toxic containments may be released from burning materials depending on the type of materials and level of oxygen. For spaceflight, NASA identified five marker compounds: carbon monoxide (odorless and colorless gas) released from most thermodegradation events; hydrogen chloride released from polyvinyl chloride; hydrogen fluoride and carbonyl fluoride associated with Teflon®; and hydrogen cyanide released from Kapton®-coated wire and polyurethane foam. The concentration range monitored for each marker compound was based on the established spacecraft maximum allowable concentrations at the low end and, at the other end of the range, an estimated highest concentration that might be released in a fire.

An upgraded combustion product analyzer is now used on the ISS, demonstrating that the technology and research on fire produced methods that detect toxic materials. The results indicate when it is safe for the astronauts to remove their protective gear.

Safeguarding the Astronauts From Microorganisms—Prevention of Viral, Bacterial, and Fungal Diseases

Certain bacteria, fungi, and viruses cause acute diseases such as upper respiratory problems, lung diseases, and gastrointestinal disease as well as chronic problems such as some cancers and serious liver problems. In space, astronauts are exposed to microorganisms and their by-products from the food, water (both used for food and beverage rehydration, and for personal hygiene), air, interior surfaces, and scientific investigations that include animals and microorganisms. The largest threat to the crew members, however, is contact with their crewmates.

The shuttle provided an opportunity to better understand the changes in microbiological contamination because, unlike previous US spacecraft for human exploration, the shuttle was designed to be used over many years with limited refurbishment between missions. Risks associated with the long-term accumulation of microorganisms in a crewed compartment were unknown at the start of the shuttle flights; however, many years of studying these microorganisms produced changes that would prevent problems for the ISS and the next generation of crewed vehicles. With assistance from industry and government standards (e.g., Environmental Protection Agency) and expert panels, NASA established acceptability limits for bacteria and fungi in the environment (air and surfaces) and consumables (food and water). Preflight monitoring for spaceflight was thorough and included the crew, spaceflight food, potable water, and vehicle air and surfaces to ensure compliance with these acceptability standards. NASA reviewed all flight payloads for biohazardous materials. Space Shuttle acceptability limits evolved with time and were later used to develop contamination limits for the ISS and the next generation of crewed vehicles.

Microbial growth in the closed environment of spacecraft can lead to a wide variety of adverse effects including infections as well as the release of volatile organics, allergens, and toxins. Biodegradation of critical materials, life support system fouling, and bio-corrosion represent other potential microbial-induced problems. Shuttle crew members sometimes reported dust in the air and occasional eye irritation. In-flight monitoring showed increased bacterial levels in the shuttle air as the number of days in space increased. Dust, microbes, and even water droplets from a simple

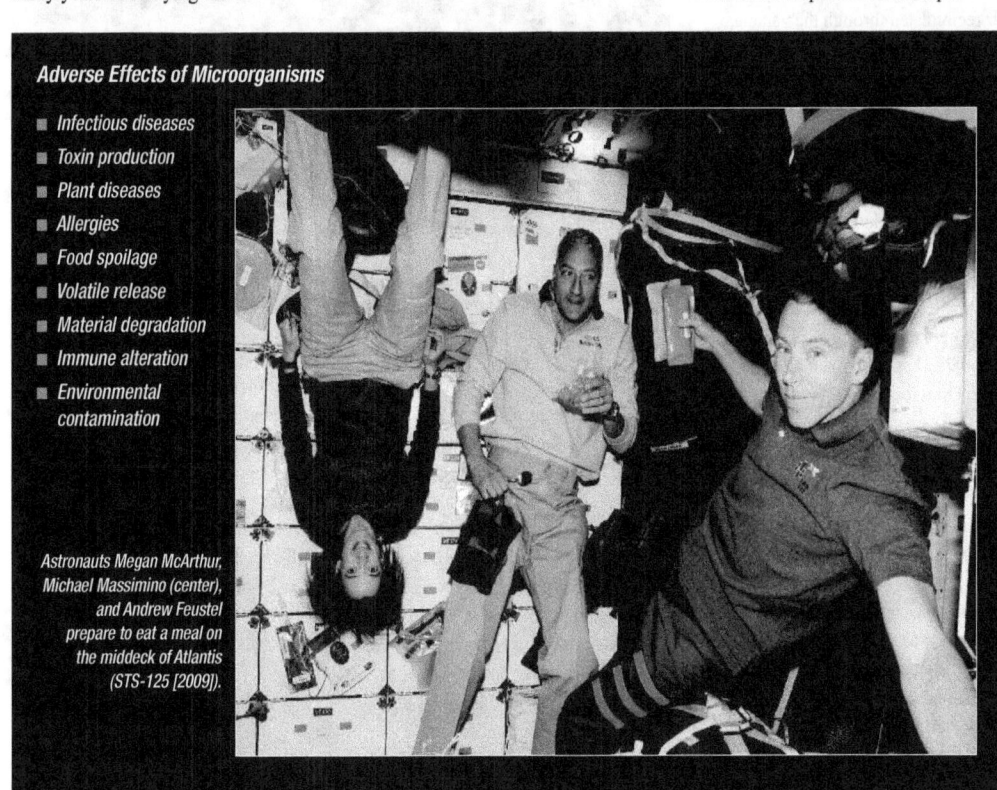

Adverse Effects of Microorganisms

- *Infectious diseases*
- *Toxin production*
- *Plant diseases*
- *Allergies*
- *Food spoilage*
- *Volatile release*
- *Material degradation*
- *Immune alteration*
- *Environmental contamination*

Astronauts Megan McArthur, Michael Massimino (center), and Andrew Feustel prepare to eat a meal on the middeck of Atlantis (STS-125 [2009]).

sneeze settle out on Earth. The human body alone sheds about 1 billion skin cells every week. Particles remain suspended in space and carry microorganisms and allergens that pose a health risk to the crew.

The shuttle's air filters were designed to remove particles greater than 70 micrometers. The filters removed most skin cells (approximately 100 micrometers) and larger airborne contaminants (e.g., lint); however, they did not quickly remove smaller contaminants such as bacteria, viruses, and particulates. When the shuttle was modified for longer flights of up to 2 weeks, an auxiliary cabin air cleaner provided filtration that removed particles over 1 micrometer. As the air recirculated through the vehicle, the filter captured skin cells, lint, microorganisms, and other debris. This resulted in much-improved air quality. These high-efficiency particulate air (HEPA) filters (99.97% efficient at removing particles >0.3 micrometers) provide dust- and microbe-free air. This led to the inclusion of HEPA filters in the Air Revitalization System on the ISS where monitoring has shown that air quality has been maintained below stringent microbial requirements. HEPA filters are also planned for other crewed vehicles.

Microbial growth can result in volatile chemicals that can produce objectionable odors or irritants. For example, during the STS-55 (1993) mission, the crew reported a noxious odor that was later found by extensive ground studies to be a mixture of three dimethyl sulfides resulting from the bacterial metabolism of urine in a waste storage container.

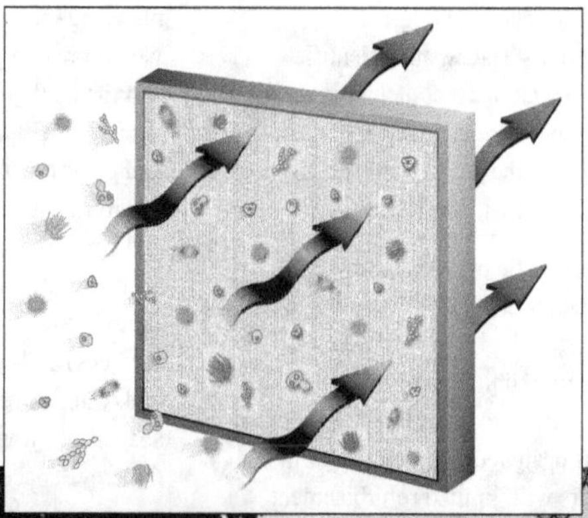

As illustrated, a high-efficiency particulate air (HEPA) filter removes particles from recirculated air, resulting in improved air quality. The HEPA filter in the air-purification system on the International Space Station (ISS), as pictured below, is of a higher quality than purification systems used in offices and homes.

These challenges provided opportunities for improvements that served as "lessons learned," which were applied to all future missions. Lessons learned from the shuttle experiences led to NASA's current approach of prevention first and mitigation second. Many microbiological risks associated with living in space can be prevented or mitigated to acceptable levels through engineering approaches. Prevention strategy begins with the design phase and includes steps that discourage excessive microbial growth. Use of antimicrobial materials, maintaining relative humidity below 70%, avoiding condensation buildup, implementing rigorous housekeeping, maintaining air and water filtration, and judicially using disinfectants are effective steps limiting the adverse effects of microorganisms. In all, the microbiological lessons learned from the Space Shuttle era resulted in improved safety for all future spacecraft.

Astronaut Health Care

Astronaut health care includes all issues that involve flight safety, physiological health, and psychological health. During the Space Shuttle Program, space medicine was at the "heart" of each issue.

Space medicine evolved during the shuttle's many transitional phases, from the experimental operational test vehicle to pre-Challenger (1986) accident, post-Challenger accident, unique missions such as Department of Defense and Hubble, Spacelab/Spacehab, Extended Duration Orbiter Project, Shuttle–Mir, Shuttle-International Space Station (ISS), post-Columbia (2003) accident and, finally, the ISS assembly completion. All of these evolutionary phases required changes in the selection of crews for spaceflight, preparation for spaceflight, on-orbit health care, and postflight care of the astronauts.

Astronauts maintained their flight status, requiring both ambulatory and preventive medical care of their active and inactive medical conditions. Preflight, on-orbit, and postflight medical care and operational space medicine training occurred for all flights. The medical team worked with mission planners to ensure that all facets of coordinating the basic tenets of personnel, equipment, procedures, and communications were included in mission support. During the shuttle era, the Mission Control Center was upgraded, significantly improving communications among the shuttle flight crew, medical team, and other flight controllers with the flight director for the mission. Additionally, the longitudinal study of astronaut health began with all medical data collected during active astronaut careers. NASA used post-retirement exams, conducted annually, to study the long-term effect of short-duration spaceflight on crews.

Space Adaptation Syndrome

The first thing an astronaut noticed was a fluid shift from his or her lower extremities to his or her torso and upper bodies, resulting in a facial fullness. Ultimately, this fluid shift caused a stretch on the baroreceptors in the arch of the aorta and carotid arteries and the astronaut would lose up to 1.5 to 2 L (1.6 to 2.1 qt) of fluid.

Secondly, over 80% of crew members experienced motion sickness, from loss of appetite to nausea and vomiting. Basic prevention included attempting to maintain an Earth-like orientation to the vehicle. Also, refraining from exaggerated movements helped. If symptoms persisted despite preventive measures, medications in an oral, suppository, or injectable form were flown to treat the condition.

The next thing crew members noticed was a change in their musculoskeletal system. In space, the human body experiences a lengthening and stretching of tendons and ligaments that hold bones, joints, and muscles together. Also, there was an unloading of the extensor muscles that included the back of the neck and torso, buttocks, and back of the thighs and calves. Preventive measures and treatment included on-orbit exercise, together with pain medications.

Additional changes were a mild decrease in immune function, smaller blood cell volume, and calcium loss. Other problems included headache, changes in visual acuity, sinus congestion, ear blocks, nose bleeds, sore throats, changes in taste and smell, constipation, urinary infections and difficulty in urination, fatigue, changes in sleep patterns with retinal flashes during sleep, minor behavioral health adjustment reactions, adverse reactions to medications, and minor injuries.

Astronaut Selection and Medical Standards

Due to increasing levels of flight experience and changes in medical delivery, medical standards for astronaut selection evolved over the shuttle's 30 years, as it was important that the selected individuals met certain medical criterion to be considered as having the "right stuff." The space agency initially adopted these standards from a combination of US Air Force, US Navy, and Federal Aviation Administration as well as previous standards from the other US space programs. The shuttle medical standards were designed to support short-duration spaceflights of as many as 30 days. NASA medical teams, along with experts in aerospace medicine and systems specialties, met at least every 2 years to review and update standards according to a combination of medical issues related to flights and the best evidence-based medicine at that time. These standards were very strict for selection, requiring optimum health, and they eventually led to the ISS medical requirements for long-duration spaceflight.

Preventive medicine was the key to success. Astronauts had an annual spaceflight certification physical exam to ensure they remained healthy for spaceflight, if assigned. Also, if a potential medical condition or problem was diagnosed, it was treated appropriately and the astronaut was retained for spaceflight. Medical exams were completed 10 days prior to launch and again at 2 days prior to launch to ensure that the astronaut was healthy and met the Flight Readiness Review requirements for launch. Preventive health successfully kept almost 99% of the astronauts retained for spaceflight duties during their careers with NASA.

Crew Preparation for Flight

Approximately 9 months prior to each shuttle flight, the medical team and flight crew worked together to resolve any medical issues. The flight medical team provided additional medical supplies and equipment for the crew's active and inactive medical problems.

Spaceflight inspired some exceptional types of medical care. Noise was a hazard and, therefore, hearing needed to be monitored and better hearing protection was included. Due to the presence of radiation, optometry was important for eye health and for understanding the impact of radiation exposure on cataract development. Also, in space visual changes occurred with elongation of the eye, thus requiring special glasses prescribed for flight. All dental problems needed to be rectified prior to flight as well. Behavioral health counseling was also available for the crews and their families, if required. This program, along with on-orbit support, provided the advantage of improved procedures and processes such as a family/astronaut private communication that allowed the astronaut another avenue to express concerns.

Over the course of the Space Shuttle Program, NASA provided improved physical conditioning and rehabilitation medicine throughout the year to keep crews in top physical shape. Before and during all shuttle flights, the agency provided predictions on solar activity and accumulation of the radiation astronauts received during their careers to help them limit their exposure.

Prior to a shuttle mission, NASA trained all astronauts on the effects of microgravity and spaceflight on their bodies to prepare them for what to expect in the environment and during the physiological responses to microgravity. The most common medical concerns were the space adaptation syndrome that included space motion sickness and the cardiovascular, musculoskeletal, and neurovestibular changes on orbit. Other effects such as head congestion, headaches, backaches, gastrointestinal, genitourinary, crew sleep, rest, fatigue, and handling of injuries were also discussed. The most common environmental issues were radiation, the biothermal considerations of heat and cold stress, decompression sickness from an extravehicular activity (EVA), potable water contamination, carbon dioxide (CO_2), and other toxic exposures. Re-entry-day (return to Earth) issues were important because the crew transitioned quickly from microgravity into a hypergravity, then into a normal Earth environment. Countermeasures needed to be developed to overcome this rapid response by the human body. These countermeasures included the control of cabin temperature, use of the g-suit, and entry fluid loading, which helped restore fluid in the plasma volume that was lost on orbit during physiological changes to the cardiovascular system. It was also important to maximize the health and readaptation of the crew on return to Earth in case emergency bailout, egress, and escape procedures needed to be performed.

The addition of two NASA-trained crew medical officers further improved on-orbit medical care. Training included contents of the medical kits with an understanding of the diagnostic and therapeutic procedures contained within the medical checklist. These classes were commonly referred to as "4 years of medical school in three 2-hour sessions." Crew medical officers learned basic emergency and nonemergency procedures common to spaceflight. This training included how to remove foreign bodies from the eye; treat ear blocks and nose bleeds; and start IVs and give medications that included IV, intramuscular, and subcutaneous injections and taught the use of oral and suppository intake. Emergency procedures included training in cardiopulmonary resuscitation, airway management and protection, wound care with Steri-Strip™ and suture repair, bladder catheterization, and needle thoracentesis. NASA taught special classes on how to mitigate the possibility of decompression sickness from an EVA. This incorporated the use of various EVA prebreathe protocols developed for shuttle only or shuttle-ISS docking missions. Crews were taught to recognize decompression sickness and how to medically manage this event by treating and making a disposition of the crew member if decompression sickness occurred during an EVA.

Environmental exposure specialty classes included the recognition and management of increased CO_2 exposure, protection and monitoring in case of radiation exposure from either artificial or solar particle events, and the

Shuttle Medical Kit

The Shuttle Orbiter Medical System had generic and accessory items and provided basic emergency and nonemergency medical care common to spaceflight. The contents focused on preventing illnesses and infection as well as providing pain control. It also provided basic life support to handle certain life-threatening emergencies, but it did not have advanced cardiac life support capabilities. Initially, it included two small kits of emergency equipment, medications, and bandages; however, this evolved into a larger array of sub packs as operational demands required during the various phases of the program. The generic equipment remained the same for every flight, but accessory kits included those mission-specific items tailored for the crew's needs. Overall, the Shuttle Orbiter Medical System included: a medical checklist that helped the on-board crew medical officers diagnose and treat on-orbit medical problems; an airway sub pack; a drug sub pack; an eye, ear, nose, throat, and dental sub pack; an intravenous sub pack; saline supply bags; a trauma sub pack; a sharps container; a contamination cleanup kit; patient and rescue restraints; and an electrocardiogram kit.

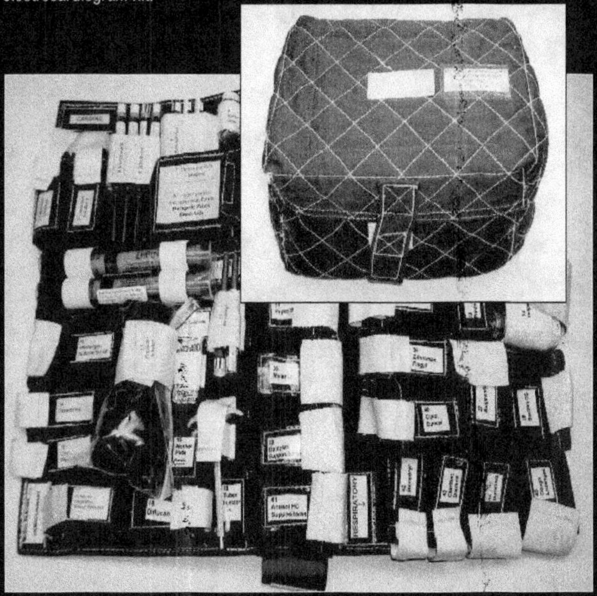

biothermal consideration of heat stress in case the Orbiter lost its ability to maintain cooling. Toxicology exposure specialty classes focused on generic toxic compounds unique to the Orbiter and included hypergolic exposure to hydrazines and nitrogen tetroxide, ammonia, and halogenated hydrocarbons such as halon and Freon®. Certain mission-specific toxic compounds were identified and antidotes were flown in case of crew exposure to those compounds. NASA trained crew members on how to use the toxicology database that enabled them to readily identify the exposed material and then provide protection to themselves during cleanup of toxic compounds using a specialty contamination cleanup kit. Astronauts were also trained on fire and smoke procedures such as the rapid quick-don mask for protection while putting out the fire and scrubbing the cabin atmosphere. In such an incident, the atmosphere was monitored for carbon monoxide, hydrogen cyanide, and hydrogen chloride. When those levels were reduced to nontoxic levels, the masks were removed.

The potable water on the shuttle was monitored 15 and 2 days preflight to ensure quality checks for iodine levels, microbes, and pH. Crews were instructed in limiting their iodine (bacteriostatic agent added to stored shuttle water tank) intake by installing/reinstalling a galley iodine-reduction assembly device each day that limited their intake of iodine from the cold water. The crews also learned how to manage the potable water tank in case it became contaminated on orbit.

Over the course of the program, NASA developed Flight Rules that covered launch through recovery after landing and included risky procedures such as EVAs. These rules helped prevent medical conditions and were approved through a series of review boards that included NASA missions managers, flight directors, medical personnel, and outside safety experts. The Flight Rules determined the preplanned decision on how to prevent or what to do in case something went wrong with the shuttle systems. Other controlled activities were rules and constraints that protected and maintained the proper workload, rest, and sleep prior to flight and for on-orbit operations during the presleep, work, and post-sleep periods. The flight-specific sleep and work schedule was dependent on the launch time and included the use of bright and dim lights, naps, medications, and shifts in sleep and work patterns. NASA developed crew schedules to prevent crew fatigue—an important constraint for safety and piloted return.

Although implemented in the Apollo Program, preflight crew quarantine proved to be essential during the Space Shuttle Program to prevent infectious disease exposure prior to launch. The quarantine started 7 days prior to launch. At that point, all crew contacts were monitored and all contact personnel received special training in the importance of recognizing the signs and symptoms of infectious disease, thus limiting their contact with the flight crew if they became sick. This program helped eliminate the exposure of an infectious disease that would delay launch and was successful in that only one flight had to be delayed because of a respiratory illness.

Readiness for Launch and On-orbit Health Care

Launch day is considered the most risky aspect of spaceflight. As such, medical teams were positioned to work directly with mission managers as well as the shuttle crew during this critical stage. On launch day, one crew medical doctor was stationed in the Launch Control Center at Kennedy Space Center (KSC) with KSC medical emergency care providers. They had direct communication with Johnson Space Center Mission Control, Patrick Air Force Base located close to KSC, alternate landing sites at Dryden Flight Research Center/Edwards Air Force Base, White Sands Space Harbor, and transoceanic abort landing medical teams. Another crew medical doctor was pre-staged near a triage site with the KSC rescue forces and trauma teams at a site determined by wind direction. Other forces, including military doctors and US Air Force pararescuers in helicopters, stood on "ready alert" for any type of launch contingency.

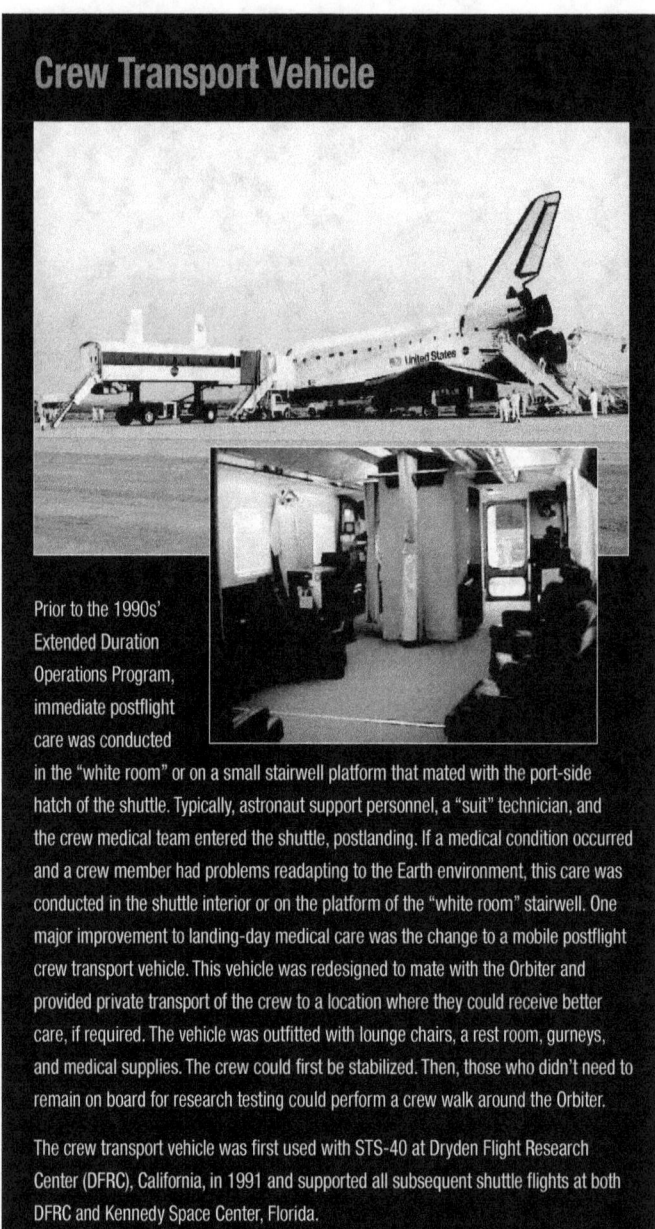

Once launch occurred and the crew reached orbit in just over 8 minutes, physiologic changes began. Every crew member was unique and responded to these changes differently on a various scale.

Crew Transport Vehicle

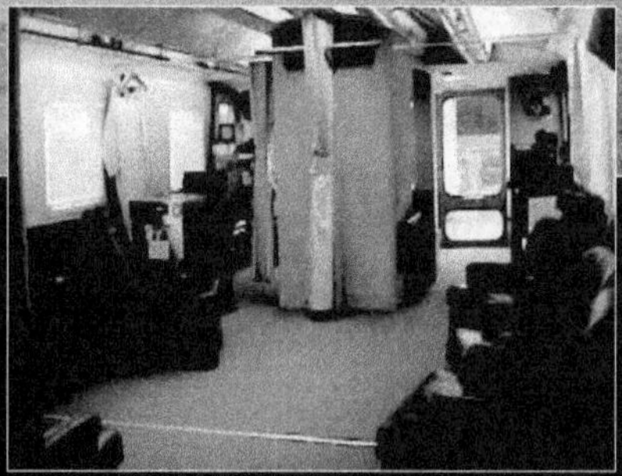

Prior to the 1990s' Extended Duration Operations Program, immediate postflight care was conducted in the "white room" or on a small stairwell platform that mated with the port-side hatch of the shuttle. Typically, astronaut support personnel, a "suit" technician, and the crew medical team entered the shuttle, postlanding. If a medical condition occurred and a crew member had problems readapting to the Earth environment, this care was conducted in the shuttle interior or on the platform of the "white room" stairwell. One major improvement to landing-day medical care was the change to a mobile postflight crew transport vehicle. This vehicle was redesigned to mate with the Orbiter and provided private transport of the crew to a location where they could receive better care, if required. The vehicle was outfitted with lounge chairs, a rest room, gurneys, and medical supplies. The crew could first be stabilized. Then, those who didn't need to remain on board for research testing could perform a crew walk around the Orbiter.

The crew transport vehicle was first used with STS-40 at Dryden Flight Research Center (DFRC), California, in 1991 and supported all subsequent shuttle flights at both DFRC and Kennedy Space Center, Florida.

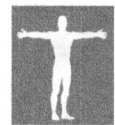

All medical conditions were discussed during a private medical communication with the crew every flight day. The results at the end of a discussion were one of the following: no mission impact (the majority); possible mission impact; or mission impact. With possible mission impacts, further private discussion with the crew and flight director, other crew members, and other medical care specialists occurred. Fortunately for the program, all possible mission impacts were resolved with adjustments to the timeline and duties performed by the crew so the mission could continue to meet its objectives. If a mission impact were to occur, changes would be made public but not the specifics of those changes. Due to the Medical Privacy Act of 1974, details of these private medical conferences could not be discussed publicly.

Private family communication was another important aspect, psychologically, of on-orbit health care. Early in the program, this was not performed but, rather, was implemented at the start of the Extended Duration Orbiter Medical Project (1989-1996) and involved flights of 11 days or longer.

The second riskiest time of spaceflight was returning to Earth. To overcome hypotension or low blood pressure during re-entry, the crew employed certain countermeasures. The crew would fluid load to restore the lost plasma volume by ingesting 237 ml (8 oz) of water with two salt tablets every 15 minutes, starting 1 hour prior to the time of deorbit ignition and to finish this protocol by entry interface (i.e., the period right before the final return stage) for a total fluid loading time of 90 minutes. Body weight determined the total amount ingested. After the Challenger accident, NASA developed a launch and re-entry suit that transitioned from the standard Nomex® flight suit, to a partial pressure suit, then on to a full pressure suit called the advanced crew escape suit. An incorporated g-suit could be used to compress lower extremities and the abdomen, which prevented fluid from accumulating in those areas. Another post-Challenger accident lesson learned was to cool the cabin and incorporate the liquid cooling within the launch and re-entry suit to prevent heat loads that could possibly compromise the landing performance of the vehicle by the commander and pilot (second in command). Finally, each crew member used slow, steady motions of his or her head and body to overcome the neurovestibular changes that occurred while transitioning from a microgravity to an Earth environment. All items were important that assisted the crew in landing the vehicle on its single opportunity in a safe manner.

Postflight Care

Once the landed shuttle was secured from any potential hazards, the medical team worked directly with returning crew members. Therefore, medical teams were stationed at all potential landing sites—KSC in Florida, Dryden Flight Research Center in California, and White Sands in New Mexico.

When the crew returned to crew quarters, they reunited with their families and then completed a postflight exam and mini debrief. Crew members were advised not to drive a vehicle for at least 1 day and were restricted from aircraft flying duties due to disequilibrium—problems with spatial and visual orientation. NASA performed another postflight exam and a more extensive debrief at return plus 3 days and, if passed, the crew member was returned to aircraft flight duties. Mission lessons learned from debriefs were shared with the other crew medical teams, space medicine researchers, special project engineers, and the flight directors. All of these lessons learned over time, especially during the transitional phases of the program, continued to refine astronaut health and medical care.

Accidents and Emergency Return to Earth

Main engine or booster failures could have caused emergency returns to KSC or transoceanic abort landing sites. NASA changed its handling of post-accident care after the two shuttle accidents. Procedures specific for the medical team were sessions on emergency medical services with the US Department of Defense Manned Spaceflight Support Office and included search and rescue and medical evacuation. This support and training evolved tremendously after the Challenger and Columbia accidents, incorporating lessons learned. It mainly included upgrades in training on crew equipment that supported the scenarios of bailout, egress, and escape.

The Future of Space Medicine

NASA's medical mission continues to require providing for astronaut health and medical care. Whatever the future milestones are for the US space program, the basic tenets of selecting healthy astronaut candidates by having strict medical selection standards and then retaining them through excellent preventive medical care are of utmost importance. Combining these with the operational aspects of coordinating all tenets of understanding the personnel, equipment, procedures, and communications within the training to prepare crews for flight will enhance the success of any mission.

At the closing stages of the Space Shuttle Program, no shuttle mission was terminated or aborted because of a medical condition, and this was a major accomplishment.

The Space Shuttle: A Platform That Expanded the Frontiers of Biology

Kenneth Souza

The Space Shuttle brought a new dimension to the study of biology in space. Prior to the shuttle, scientists relied primarily on uncrewed robotic spacecraft to investigate the risks associated with venturing into the space environment. Various biological species were flown because they were accepted as models with which to study human disease and evaluate human hazards. The results from the pioneering biological experiments aboard uncrewed robotic spacecraft not only provided confidence that humans could indeed endure the rigors of spaceflight, they also formed the foundation on which to develop risk mitigation procedures; i.e., countermeasures to the maladaptive physiological changes the human body makes to reduced gravity levels. For example, the musculoskeletal system reacts by losing mass. This may pose no hazard in space; however, on returning to Earth after long spaceflights, such a reaction could result in an increased risk of bone fractures and serious muscle atrophy.

Unfortunately, most biological research in uncrewed spacecrafts was limited to data that could only be acquired before and/or after spaceflight. With crew support of the experiments aboard the Space Shuttle and Spacelab, and with adequate animal housing and lab support equipment, scientists could train the crew to obtain multiple biospecimens during a flight, thus providing windows into the adaptation to microgravity and, for comparison, to samples obtained during readaptation to normal terrestrial conditions postflight.

With the Space Shuttle and its crews, earthbound scientists had surrogates in orbit—surrogates who could be their eyes and hands within a unique laboratory. The addition of Spacelab and Spacehab, pressurized laboratory modules located in the shuttle payload bay, brought crews and specialized laboratory equipment together, thus enabling complex interactive biological research during spaceflight. Crew members conducted state-of-the-art experiments with a variety of species and, in the case of human research, served as test subjects to provide in-flight measurements and physiological samples.

In addition to the use of biological species to evaluate human spaceflight risks, research aboard the shuttle afforded biologists an opportunity to examine the fundamental role and influence of gravity on living systems. The results of such research added new chapters to biology textbooks. Life on Earth originated and evolved in the presence of a virtually constant gravitational field, but leaving our planet of origin creates new challenges that living systems must cope with to maintain the appropriate internal environment necessary for health, performance, and survival.

How Does Gravity Affect Plants and Animals?

Throughout the course of evolution, gravity has greatly influenced the morphology, physiology, and behavior of life. For example, a support structure—i.e., the musculoskeletal system—evolved to support body mass as aquatic creatures transitioned to land. To orient and ambulate, organisms developed ways to sense the gravity vector and translate this information into a controlled response; hence, the sensory-motor system evolved. To maintain an appropriate blood supply and pressure in the various organs of the mammalian body, a robust cardiovascular system developed. Understanding how physiological systems sense, adapt, and respond to very low gravity cannot be fully achieved on the ground; it requires the use of spaceflight as a tool. Just as we need to examine the entire light spectrum to determine how the visual organs of living systems work, so too we must use the complete gravity spectrum, from hypogravity to hypergravity, to understand how gravity influences life both on and off the Earth.

Space biologists have identified and clarified the effects of spaceflight on a few representative living systems, from the cellular, tissue, and system level to the whole organism. NASA achieved many "firsts" as well as other major results that advanced our understanding of life in space and on Earth. The agency also achieved many technological advances that provided life support for the study of the various species flown.

Baruch Blumberg, MD
Nobel Prize winner in medicine, 1976.
Professor of Medicine
Fox Chase Cancer Center.
Former director of Astrobiology
Ames Research Center, California.

"The United States and other countries are committed to space travel and to furthering the human need to explore and discover. Since April 12, 1981, the shuttle has been the major portal to space for humans; its crews have built the International Space Station (ISS), a major element in the continuum that will allow humans to live and work indefinitely beyond their planet of origin. The shuttle has provided the high platform that allows observations in regions that were previously very difficult to access. This facilitates unique discoveries and reveals new mysteries that drive human curiosity.

"In the final paragraph of Origin of Species *Darwin wrote:*

> There is grandeur in this view of life, with its several powers, having been originally breathed by the Creator into a few forms or into one; and that, whilst this planet has gone circling on according to the fixed law of gravity, from so simple a beginning endless forms most beautiful and most wonderful have been, and are being evolved.

"The shuttle and the ISS now provide a means to study life and its changes without the constraints of gravity. What will be the effect of this stress never before experienced by our genome and its predecessors (unless earlier forms of our genes came to Earth through space from elsewhere) on physiology, the cell, and molecular biology? Expression of many genes is altered in the near-zero gravity; how does this conform to the understanding of the physics of gravity at molecular and atomic dimensions?

"In time, gravity at different levels, at near-zero on the ISS, at intermediate levels on the moon and Mars, and at one on Earth, can provide the venues to study biology at different scales and enlarge our understanding of the nature of life itself."

Gravity-sensing Systems—How Do Plants and Animals Know Which Way Is Down or Up?

As living systems evolved from simple unicellular microbes to complex multicellular plants and animals, they developed a variety of sensory organs that enabled them to use gravity for orientation. For example, plants developed a system of intracellular particles called statoliths that, upon seed germination, enabled them to sense the gravity vector and orient their roots down into the soil and their shoots up toward the sun. Similarly, animals developed a variety of sensory systems (e.g., the vestibular system of the mammalian inner ear) that enabled them to orient with respect to gravity, sense the body's movements, and transduce and transmit the signal to the brain where it could be used together with visual and proprioceptive inputs to inform the animal how to negotiate its environment.

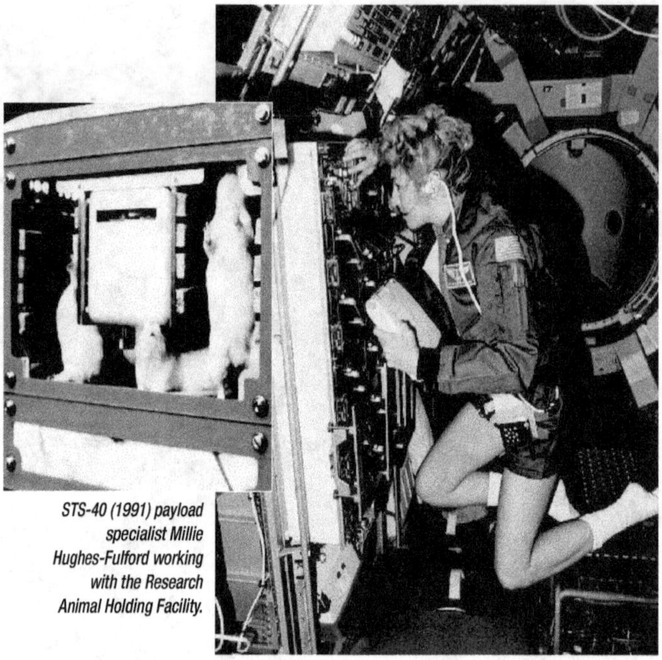

STS-40 (1991) payload specialist Millie Hughes-Fulford working with the Research Animal Holding Facility.

Why Do Astronauts Get Motion Sickness in Spaceflight?

One consequence of having gravity-sensing systems is that while living in microgravity, the normal output from the vestibular system is altered, leading to a confusing set of signals of the organism's position and movement. Such confusion is believed to result in symptoms not too different from the typical motion sickness experienced by seafarers on Earth. This affliction, commonly termed "space motion sickness," affects more than 80% of astronauts and cosmonauts during their first few days in orbit. Interestingly, one of the two monkeys flown in a crewed spacecraft, the Space Transportation System (STS)-51B (1985) Spacelab-3 mission, displayed symptoms resembling space motion sickness during the first few days of spaceflight.

The basic process of space motion sickness became one of the main themes of the first two dedicated space life sciences missions: STS-40 (1991) and STS-58 (1993). Scientists gained insights into space motion sickness by probing the structural changes that occur in the vestibular system of the mammalian balance organs. Using rodents, space biologists learned for the first time that the neural hair cells of the vestibular organ could change relatively rapidly to altered gravity. Such neuroplasticity was evident in the increased number of synapses (specialized junctions through which neurons signal to each other) between these hair cells and the vestibular nerve that occurred as the gravity signal decreased. In effect, the body tried to turn up the gain to receive the weaker gravitational signal in space. This knowledge enabled medical doctors and crew members to have a better understanding of why space motion sickness occurs.

Is Gravity Needed for Successful Reproduction?

Amphibian Development

Studies of the entire life span of living systems can provide insights into the processes involved in early development and aging. The Frog Embryology Experiment flown on STS-47 (1992) demonstrated for the first time that gravity is not required for a vertebrate species, an amphibian, to ovulate, fertilize, achieve a normal body pattern, and mature to a free-swimming tadpole stage. This experiment put to rest the "gravity requirement" question that had been debated by embryologists since the late 19th century.

In Earth gravity, frog eggs, when shed, have a bipolar appearance; i.e., the spherical egg has a darkly pigmented hemisphere containing the nucleus and much of the cell machinery needed for development while the opposite, lightly pigmented hemisphere is rich in yolk that provides the energy to drive the cell machinery during early development. Shortly after being shed, the eggs can be fertilized by sperm released by an adjoining male frog. Once fertilized, a membrane lifts off the egg surface and the egg responds to gravity by orienting the dense yolk-rich hemisphere down and the darkly pigmented hemisphere up with respect to the gravity vector. This geotropic response was what spurred early embryologists to interfere with egg rotation and thereby tried to determine whether the response to gravity was required for normal development. Unfortunately, research on the ground yielded ambiguous results due primarily to the trauma imparted to the eggs by the scientists' attempts to interfere with rotation.

During the STS-47 flight, adult female frogs (*Xenopus laevis*) were injected with hormone to induce the shedding of eggs, followed by the addition of a sperm suspension. Half of a group of fertilized eggs were placed in special water-filled chambers and on a rotating centrifuge to provide an acceleration environment equivalent to terrestrial gravity. The other half were placed in the same type of water-filled chambers, but in a temperature-controlled incubator and were kept in a microgravity environment. Samples of developing embryos were taken during the flight to capture important developmental stages for examination postflight. Some were returned to Earth as free-swimming tadpoles. The results of the experiment ended the centuries-old debate as to whether gravity is needed for successful reproduction, and demonstrated for the first time that a vertebrate species could be fertilized and develop normally to a free-swimming stage in the virtual absence of gravity. It remains to be seen, however, whether metamorphosis, maturation, and a complete life cycle of an amphibian or other complex organisms can occur in the absence of gravity.

Fertilized frog eggs (above) and free-swimming tadpoles (below).

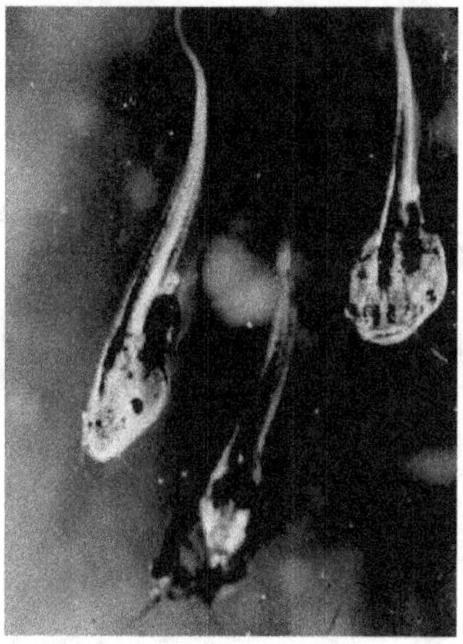

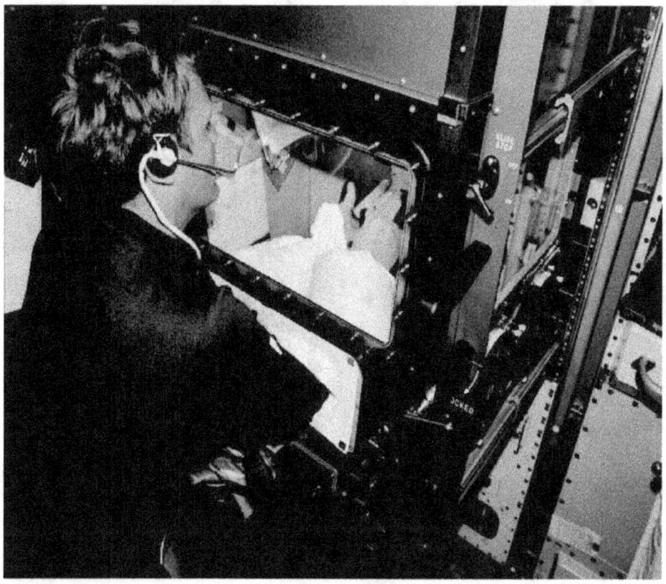

Astronaut Mark Lee working on the Frog Embryology Experiment in the General Purpose Work Station during the STS-47 (1992) mission.

In summary, for the first time, a vertebrate species was fertilized and developed through to a free-swimming stage in the virtual absence of gravity.

William Thornton, MD
Principal investigator for the first in-flight studies of space motion sickness on shuttle. Astronaut on STS-8 (1983) and STS-51B (1985).

Bring 'em Back Alive

The First Human Flight in Space with an Animal

"My training for the Spacelab 3 animal payload began as a toddler in North Carolina, surrounded by and growing up with a great variety of domestic and wild animals. Their humane treatment was my first lesson.

"After additional years of formal and informal education in medicine and biophysics, I used my training for research on space motion sickness. For some 18 months during the first shuttle flights, we completed human studies, which produced an array of first-time procedures in the US space program, including evoked potentials, coordination, complex reaction time, gastrointestinal activity and pressure, ambulatory blood pressure, and electrocardiograms, etc. These experiments answered some urgent operational questions and provided points of departure for the more formal studies that followed.

"Like so much of medical science, elemental knowledge of our nervous system comes from animal studies on Earth. On my first flight (STS-8 in 1983), 24 rats were flown in a research animal holding facility. But, to fly animals for study in the small, enclosed environment of the shuttle is a complex challenge that required years of preparation.

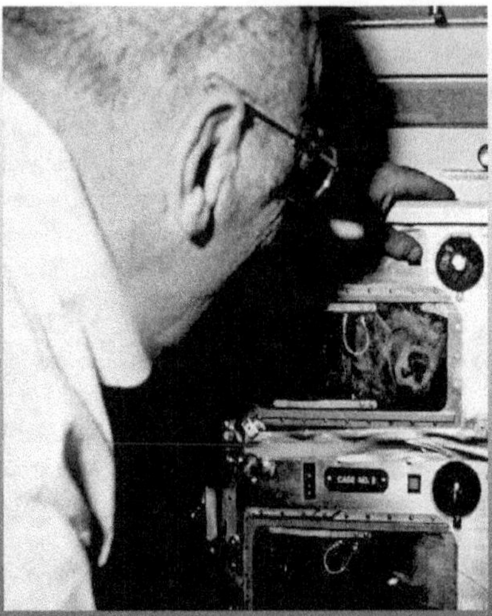

Dr. Thornton is taking care of one of the two squirrel monkeys on STS-51B.

"Finally flying on Challenger, we were able to open the cage inspection ports. All was well except for the monkey who had been a laboratory favorite (this is the animal in the photo) but who was now in deep withdrawal. He didn't eat or drink for 2 days and by the third day, dehydration was real. I used some tricks learned while feeding wild pets and he took a banana pellet and another—and more and more, then cage food.

"We returned with all animals alive and well and a great deal of experience subsequently incorporated into the shuttle legacy of astronauts and animals in space. Now, those of us who work with humans and space motion sickness have such remarkable aid as the molecular and ultra-microscopic studies from animals in Neurolab, another shuttle legacy."

Animal Development

Studies with rodents aboard the Space Shuttle identified stages of early mammalian development that are sensitive to altered gravity. They also provided insights into what might happen if humans experience abnormal gravity levels during early development. Pregnant rats on STS-66 (1994) and STS-70 (1995) showed that spaceflight resulted in striking changes in the structure of the fetal balance organ—the vestibular system. On STS-90 (1998), rat pups were launched at 8 or 14 days *postpartum*. After 16 days in microgravity, their sensorimotor functions were tested within several hours of landing; e.g., walking, and righting (rolling over). Postflight, the righting response of postnatal pups was profoundly deficient compared to ground control animals, suggesting that removal of gravitational cues during early postnatal development can significantly alter inherent patterns of behavior. In addition, neonatal animals exposed to microgravity during this Neurolab mission failed to undergo normal skeletal muscle growth and differentiation, suggesting that gravity stimuli are essential for generating the structure needed to perform basic ambulatory and righting movements when subjected, postflight, to terrestrial gravity.

The Biomass Production System installed on STS-111 (2002) carrying plants grown in the International Space Station (ISS) for return to Earth. ISS Flight Engineer Dan Bursch (pictured) conducted all of the plant experiments.

Plant Biology

Germination

The importance of gravity in the germination and development of plants has been observed and studied for centuries; however, it wasn't possible to unravel how a plant detects and responds to gravity until access to space was achieved. NASA had to develop specialized equipment to grow plants and study their response to gravity. The agency developed a plant growth unit to fit within a shuttle middeck locker. This unit provided light, water, and an appropriate substrate to support plant growth. On the STS-51F (1985) mission, seedlings were grown in enclosed chambers within the plant growth unit; i.e., mung beans (*Vigna radiata*), oats (*Avena sativa*), and pine (*Pinus elliotti*). Mung beans and oat seeds were planted 16 hours before launch and germination occurred in space. Pine seedlings were 4 or 10 days post-germination at launch. Although the mung bean and oat seeds germinated in orbit, root growth was somewhat disoriented and oats grew more slowly during spaceflight. In addition, the amount of lignin, a biochemical component of a plant's "skeletal" system, was significantly reduced in all three species, indicating that gravity is an important factor in lignification necessary for plant structure.

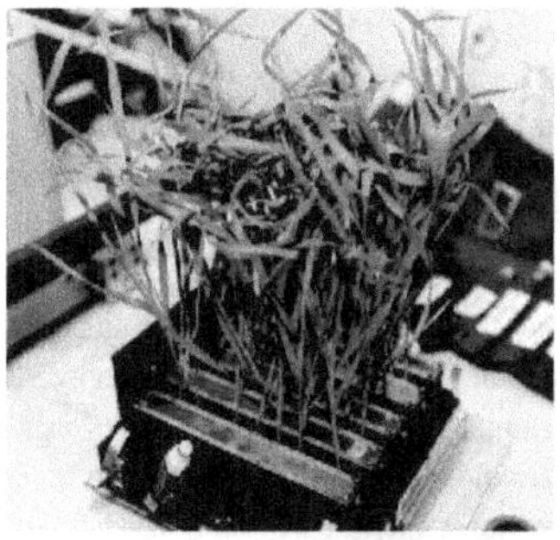

Multiple generations of plants grew in spaceflight for the first time. Examples include Apogee Wheat (top) and Brassica rapa *(bottom).*

Plant Growth

Another pioneering experiment in the study of plant responses to gravity was the Gravitational Threshold Experiment flown on the STS-42 (1992) mission. It tested plant sensitivity to altered gravitational fields during spaceflight. The Gravitational Plant Physiology Facility was built to support plant growth and stimulate plants with different levels of gravity using four centrifuge rotors contained within the facility. Two centrifuge rotors (culture rotors) were used to grow small seedlings in a 1 gravitational force (1*g*) environment (normal terrestrial level). The other two rotors provided gravity stimulations of varying strength and duration (test rotors). After stimulation on the test rotors, images of the seedling responses were captured on video recorders. This research identified for the first time the threshold stimulus for a biological response to gravity. Oat seedlings were used in the experiment and, when the seedlings reached the proper stage of growth on the 1*g* centrifuge rotor, an astronaut transferred them to the test centrifuge to expose them to a *g*-stimulus for different durations and intensities. The threshold was found to be very low—about 15 or 20 *g*-seconds; i.e., it took a force of 1*g* applied for 15 to 20 seconds to generate a plant response.

Following the pioneering plant experiments, NASA and others developed equipment with a greater range of capabilities, thus enabling more complex and sophisticated scientific experiments. This equipment included the European Space Agency's Biorack flown on Spacelabs; the Russian Svet and Lada systems flown on Mir and the International Space Station (ISS), respectively; NASA's Biomass Production System; and the European Modular Cultivation System flown on the ISS. This latter device enabled more in-depth studies of plant geotropisms than had been possible in any of the previous flight experiments with plants.

Arabidopsis ***plant. This small plant is related to cabbge and mustard and is widely used as a model for plant biology research.***

Arabidopsis seedlings were subjected to 1*g* in space on a Biorack centrifuge while a separate group was held under microgravity conditions. The experiments provided evidence that intracellular starch grains (statoliths) sediment in the presence of a gravity stimulus and influence how plants are oriented with respect to the gravity vector. Experiments within the Biomass Production System revealed much about growing plants within spaceflight hardware, particularly about plant metabolism in the absence of normal terrestrial gravity. Biomass Production System investigators concluded that plant photosynthesis and transpiration processes did not differ dramatically from those on the ground.

Multiple Generations of Growth—Fresh Foods

The early shuttle experiments with plants focused on basic questions about gravity-plant interactions. The scientific results as well as the knowledge gained in the design and fabrication of plant growth habitats greatly contributed to the development of the next generation of growth chambers. Russian investigators from the Institute of Biomedical Problems, Moscow, with support of US scientists and engineers, provided the equipment necessary to achieve multiple generations of plants in space. Multiple generations of wheat and mustard species were obtained during spaceflight on Mir and the ISS. In addition, a variety of edible vegetables were grown during spaceflight, demonstrating that plants can be used to provide fresh food supplements for future long-duration space exploration missions.

Space Biology Payloads

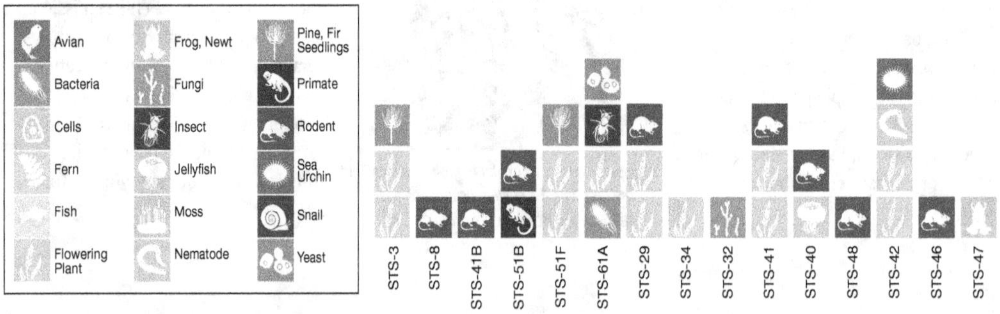

Bacteria More Dangerous in Space Environment

As reported by Cheryl Nickerson, the interplay between the human immune system and the invading microorganism determines whether infection and disease occur. Factors that diminish immune capability or increase the virulence of the microorganism will greatly increase the likelihood of disease.

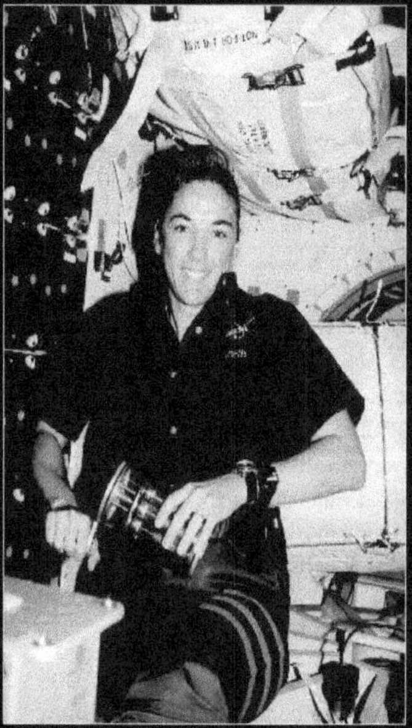

Astronaut Heidemarie Stefanyshyn-Piper, in the middeck of the Space Shuttle Atlantis, activates the MICROBE experiment, which investigated changes to Salmonella *virulence after growth in space.*

To gain insight to this issue, investigators compared responses of the food-borne bacterial pathogen *Salmonella typhimurium*, grown in the microgravity of spaceflight, to otherwise identical ground-based control cultures. Interestingly, they found that the spaceflight environment profoundly changed the gene expression and virulence characteristics (disease-causing potential) of the pathogen in novel ways that are not observed when growing the cells with traditional culture methods. This work also identified a "master molecular switch" that appears to regulate many of the central responses of *Salmonella* to the spaceflight environment.

On both the STS-115 (2006) and the STS-123 (2008) shuttle missions, scientists investigated the spaceflight response of *Salmonella* grown in various growth media containing different concentrations of five critical ions. The effects of media ion composition on the disease-causing potential of *Salmonella* were dramatic. Flight cultures grown in media containing lower levels of the ions displayed a significant increase in virulence as compared to ground control cultures, whereas flight cultures grown in higher ion levels did not show an increase in virulence. The wealth of knowledge gained from these *Salmonella* gene expression and virulence studies provides unique insight into both the prevention of infectious disease during a spaceflight mission and the development of vaccines and therapeutics against infectious agents on Earth.

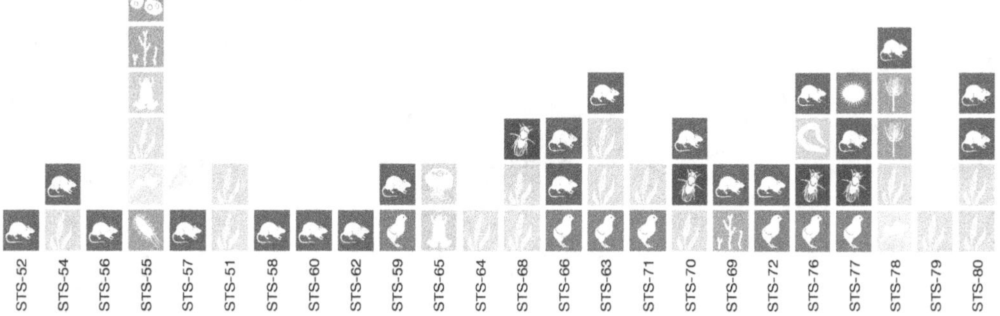

Why Do Astronauts Get Weak Muscles and Bones?

Muscles

The Space Transportation System (STS)-58 (1993) mission opened a new window on how weightlessness affects muscle structure and function. Previously, scientists knew that skeletal unloading (lack of gravity) resulted in the atrophy of muscle fibers. Until this flight, all of the skeletal muscles studied were obtained from humans or rats postflight, several hours after readapting to terrestrial gravity. Consequently, distinguishing the structural and biochemical changes made in response to microgravity from changes readapting to Earth postflight was very difficult. During the STS-58 mission, crew members obtained tissue samples from animals and processed these samples for detailed structural and biochemical analyses postflight, thus avoiding the effects of re-entry and readapting to Earth's gravity. Danny Riley of Wisconsin Medical College summed up how sampling in flight changed his understanding. "When we looked at muscle samples that we obtained from previous missions, we saw muscle atrophy and muscle lesions, small tears. Samples taken from rats during and following the STS-58 flight enabled us to determine that the atrophy was clearly a response to microgravity while the muscle lesions were a result of re-entry and readaptation stresses," Riley said.

For the STS-58 mission, muscle physiologists examined the contractile properties of rat muscles and demonstrated large changes that correlated well with the biochemical and morphological changes they had previously observed. As Ken Baldwin of the University of California at Irvine stated, "The uniqueness of performing spaceflight studies aboard STS-58 using the rodent model was that we discovered marked remodeling of both structure and function of skeletal muscle occurring after such a short duration in space. The results enabled scientists to better predict what could happen to humans if countermeasures (i.e., exercise) were not instituted early on in long-duration space missions."

The fundamental research with animals aboard the shuttle Spacelabs contributed markedly to the current understanding of the effects of spaceflight on skeletal muscle. The results laid the foundation for defining optimal countermeasures that minimize the atrophy that occurs in the human response to microgravity.

Bones

Skeletal bone, much like skeletal muscle, atrophies when unloaded. Bone mass loss as a consequence of skeletal unloading during spaceflight is a well-established risk for long-term human space exploration. A great deal of the insight into "why" and "how" bone mass loss occurs in flight resulted from research with rodents both on board the US Space Shuttle and the Russian Bion biosatellites. Such research revealed that bone formation becomes uncoupled from resorption (process of minerals leaving the bone) and normal bone mineral homeostasis is compromised. Consistent with several previous studies, results from the Physiological Systems Experiments series of payloads (STS-41 [1990], STS-52 [1992], STS-57 [1993], and STS-62 [1994]) showed that bone formation in the weight-bearing bones of male rats was inhibited by short-term spaceflight. Radial bone growth in the humerus (long bone in the arm or forelimb that runs from the shoulder to the elbow) was also decreased, though no changes in longitudinal bone growth in the tibia (shin bone in leg) were detected. These effects were associated with a decrease in the number and activity of bone-forming cells (osteoblasts). Results of experiments on board STS-58 and STS-78 (1996) provided further

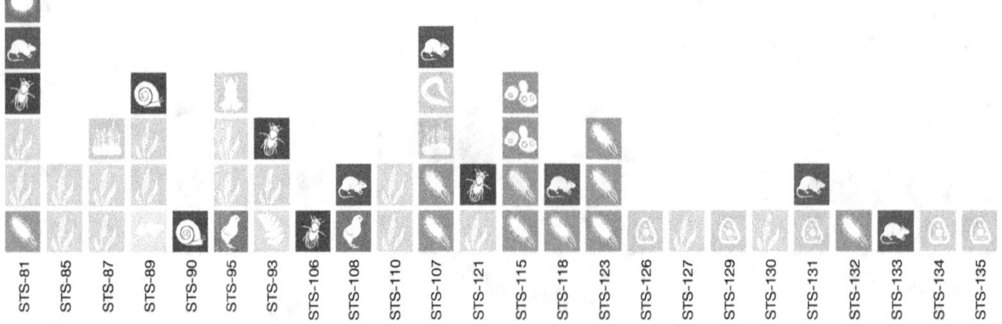

evidence of changes at both the structural and the gene expression levels associated with spaceflight-induced bone loss. Alterations also occurred in bone mineral distribution, ultrastructure and geometry, and mechanical properties as well as in site and gene-specific decreases in expression of bone matrix proteins (structural proteins with minerals attached). Taken together, these results suggest that significant tissue-specific alterations at the structural and molecular levels accompany bone loss in microgravity.

At the cellular level, spaceflight was also shown to affect bone, cartilage, and tendons, resulting in reduced matrix production or altered matrix composition.

How do bone cells sense and respond to changes in gravity? Some scientists suggest that certain cell types, when exposed to microgravity, reduce their activity or metabolism as well as the amount of new protein normally produced and enter a "resting" phase. This microgravity effect could be due to a direct effect on the mature differentiated cell (final cell type for a specific organ like bone; e.g., osteoblast) so that the cell generates some "signal" during spaceflight, thus driving the cell to enter a resting phase. Another possibility is that the cell division cycle is delayed so that cells simply develop into their differentiated state more slowly than normal.
A series of experiments was flown on STS-118 (2007) and STS-126 (2008) that studied bone marrow cell (the progenitors of bone cells) population changes in microgravity

New Technology for Three-dimensional Imaging

Rodent inner-ear hair cells are almost identical to human inner-ear hair cells. These cells are important for the vestibular system. Space biological research contributed novel technologies for diagnostic medicine on Earth. NASA developed three-dimensional (3-D) imaging software to facilitate and expedite the microscopic analysis of thin sections of the body's balance organ—the vestibular system of the inner ear. The software enabled reconstruction of the innervation pattern of the rodent's inner ear much faster than traditional manual methods. Not only did the technology greatly accelerate the analyses of electron microscopic images, it also was adapted to construct 3-D images from computerized axial tomography (CAT) and magnetic resonance imaging (MRI) scans of humans, providing surgeons with 3-D dynamic simulations for reconstructive breast cancer surgery, dental reconstruction, plastic surgery, brain surgery, and other delicate surgeries. Such simulations enable doctors to visualize and practice procedures prior to surgery, resulting in a much shorter time for the patient to be under anesthesia and a lower risk surgery.

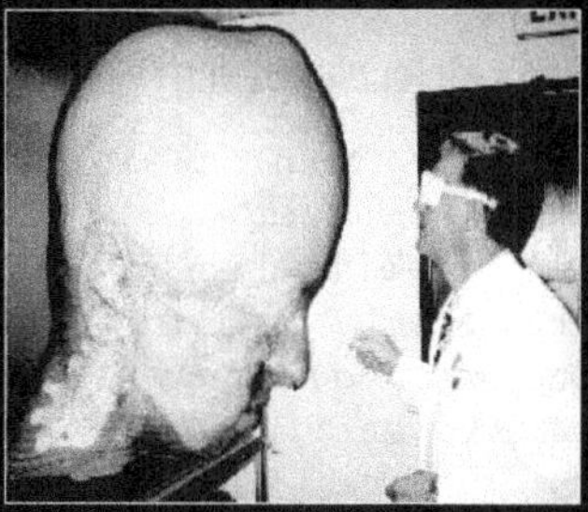

Surgical planning using 3-D virtual imaging software. Dr. Stephen Schendel, Stanford University.

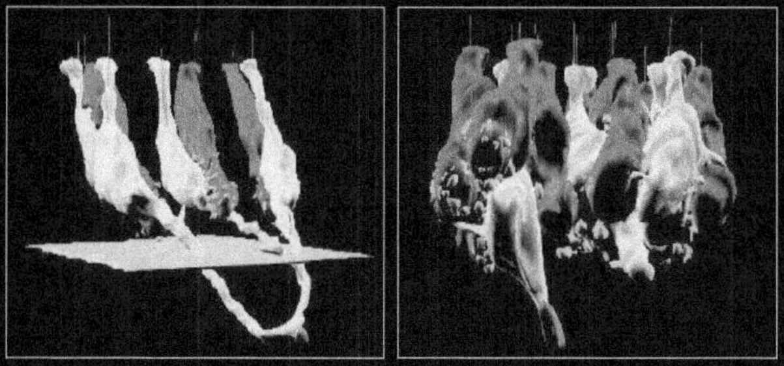

3-D reconstructions of rodent inner-ear hair cells using Ross software.

using mouse primary white blood cell (macrophage) cultures, respectively. The mouse study identified phenotypic (any observable characteristic or trait of an organism, such as its structure or function) shifts in the bone marrow cell subpopulations, including a subpopulation of macrophages.

On STS-95 (1998) scientists placed bone cartilage cells into cartridges carried in a special cell culture device built by the Walter Reed Army Institute for Research, Washington, D.C. Samples of these cells were collected on Flight Days 2, 4, 7, and 9. The media in which the cells grew were also collected on the same days, and the conversion of glucose to lactate in the media—a sign of metabolic activity—was determined postflight. Following flight, these cells were analyzed for their state of differentiation and parameters showing the cell cycle activity. The results strongly indicated that these cells were affected by flight. Flight cells were metabolically less active and produced fewer matrix components (necessary for structure) than the cells grown on the ground. In contrast to this, the flight cells showed a greater content of cyclins (proteins related to different stages of the cell cycle), suggesting that these cells were undergoing more proliferation (producing more cells) than their ground control counterparts. Exposure to spaceflight also resulted in cartilage cells undergoing more cell division, less cell differentiation (maturation), and less metabolic activity compared to ground controls. This is the first time that cell cultures flown in space were shown to exhibit alterations in their normal cell cycles.

Do Cells Grow Differently in Spaceflight and Affect Crew Health?

Cell and Molecular Biology

A large number of experiments with microorganisms were flown. Nearly all revealed that higher populations of cells are obtained from cultures grown under microgravity conditions than are obtained in cultures grown statically on the ground, possibly due to a more homogeneous distribution of cells. Recent studies of microbial cultures grown in space resulted in a substantial increase in virulence in the space-grown cultures when used postflight to infect mice.

The response of terrestrial life to microgravity at the molecular level is reflected in the response of many of an organism's genes when gravity is significantly reduced in the environment. Human renal (kidney) cell cultures flown on the Space Transportation System (STS)-90 (1998) mission exhibited a genetic response to microgravity that exceeded all expectations. More than 1,600 of the 10,000 genes examined in the renal cells showed a change in expression (i.e., increased or decreased production of the protein products of the genes) as a result of spaceflight. Armed with these results, investigators are now focusing on specific groups of genes and their functions to try and unravel why certain genes and metabolic pathways may be amplified or reduced due to a change in gravity.

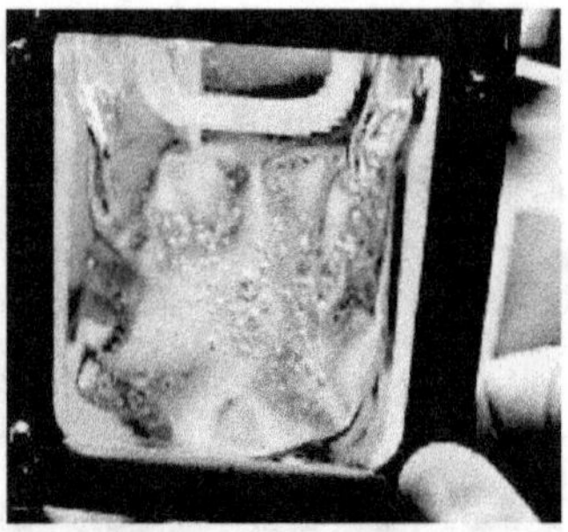

Human renal cortical cell culture grown on STS-90 (1998).

Summary

The Space Shuttle's unique capabilities, coupled with the unbounded curiosity, energy, and creativity of scientists and engineers, enabled huge leaps in our knowledge of how biological species, including humans, react and adapt to the near weightlessness of orbital spaceflight. Over the past 3 decades, space biologists demonstrated that gravity, and the lack thereof, affects life at cellular and molecular levels. They determined how amazingly durable and plastic biological systems can be when confronted with a strange new environment like space. Even in the Columbia Space Transportation System (STS)-107 (2003) tragedy, the survival of the small soil nematode worms and the mosses on board was an extremely stunning example of plant and microbial responses and resiliency to severe stress.

Over the past 4 decades of space biology research, our textbooks were rewritten, whole new areas of study were created, new technologies were developed for the benefit of science and society, and thousands of new

Microgravity—A Tool to Provide New Targets for Vaccine Design

The use of spaceflight as a tool for new discoveries has piqued the interest of scientists and engineers for decades. Relatively recently, spaceflight also gained the attention of commercial entities that seek to use the unique environment of space to provide opportunities for new product design and development.

For example, Astrogenetix, Inc. was formed by Astrotech Corporation, Austin, Texas, to commercialize biotechnology products processed in the unique environment of microgravity. Astrogenetix developed capabilities to offer a turnkey platform for preflight sample preparation, flight hardware, mission planning and operations, crew training, and certification processes needed within the highly regulated and complex environment of human spaceflight.

Astrogenetix's primary research mission is to discover therapeutically relevant and commercially viable biomarkers—substances used as indicators of biologic states—in the microgravity environment of space. By applying a biotechnology model to this unique discovery process, the company finds novel biomarkers that may not be identifiable via terrestrial experimentation. Through this method, Astrogenetix expects to shorten the drug development time frame and guide relevant therapeutics agents (or diagnostics) into the clinical trial process more quickly and cost-effectively.

Astronaut John Phillips activating a Fluid Processing Apparatus containing tubes of microorganisms on STS-119 (2009).

Specifically, Astrogenetix used assays of bacterial virulence in the microscopic worm *Caenorhabditis elegans*. Bacteria, worms, and growth media were launched separated in different chambers of the Fluid Processing Apparatus, which was developed by Bioserve Space Technologies, Boulder, Colorado. Astronauts hand-cranked the hardware twice, first to initiate the experiment by mixing bacteria, worms, and growth media and at a later scheduled time to add fixative to halt the process. This was the first direct assay of bacterial virulence in space without the effects of re-entry into Earth's atmosphere and delays due to offloading the experiment from the Space Shuttle. This experimental model identified single gene deletions of both *Salmonella sp* and *Methicillin-resistant Staphylococcus aureus* for potential acceleration of vaccine-based applications. The investigative team included Timothy Hammond, Patricia Allen, Jeanne Becker, and Louis Stodiek.

scientists and engineers were educated and trained. In the words of Nobel Prize-winner Baruch Blumberg, "Like the pioneering voyages of the early European explorers to the New World, Darwin's voyages of scientific discovery, and the historic lunar missions by the Apollo astronauts, the Space Shuttle expanded the boundaries of biology, providing insights into the role and influence of gravity on living systems."

Microgravity Research in the Space Shuttle Era

Bradley Carpenter

Cells in Space
Neal Pellis

Physical Sciences in Microgravity
Bradley Carpenter
Iwan Alexander

Commerical Ventures Take Flight
Charles Walker

The Space Shuttle cargo capability in the early 1980s stimulated a wave of imaginative research. Space-based microgravity research gave new insights into technologies critical to the space program, medical research, and industry.

NASA dedicated over 20 shuttle missions to microgravity research as a primary payload, and many more missions carried microgravity research experiments as secondary payloads. The space agency's microgravity research strived to increase understanding of the effects of gravity on biological, chemical, and physical systems. Living systems benefited as well. Cells, as they adapted to microgravity, revealed new applications in biotechnology.

Shuttle-era microgravity research was international in scope, with contributions from European, Japanese, and Russian investigators as well as commercial ventures. Several missions in which the Spacelab module was the primary payload were either officially sponsored by a partner agency, such as the Japanese or German space agency, or they carried a large number of research experiments developed by, or shared with, international partners. NASA and its partners established close working relationships through their experience of working together on these missions. These collaborations have carried over to operation of the International Space Station (ISS) and will provide the foundation for international cooperation in future missions to explore space.

Much of the Space Shuttle's legacy continues in research currently under way on the ISS—research that is building a foundation of engineering knowledge now being applied in the design of vehicle systems for NASA's next generation of exploration missions.

Cells in Space

Question: *Why fly cells in space?*

Answer: *It helps in space exploration and provides novel approaches to human health research on Earth.*

The NASA Biotechnology Program sponsored human and animal cell research, and many of the agency's spacecraft laboratory modules supported the cell research and development necessary for space exploration and Earth applications. The shuttle, in particular, hosted experiments in cell biology, microbiology, and plant biology.

The rationale for studying cells in space is the same as it is on Earth. Cells can be a model for investigating various tissues, tumors, and diseases. NASA's work with cells can reveal characteristics of how terrestrial life adapts to the space environment as well as give rise to technologies and treatments that mitigate some of the problems humans experience in space exploration scenarios. Embarking on cell biology experiments in space spawned an almost revolutionary approach to accommodate cells in a controlled culture environment. The design of equipment for propagation of cells in microgravity involved special considerations that the Earth-based cell biologist seldom accommodates.

Cell Growth in Microgravity: Going Without the Flow

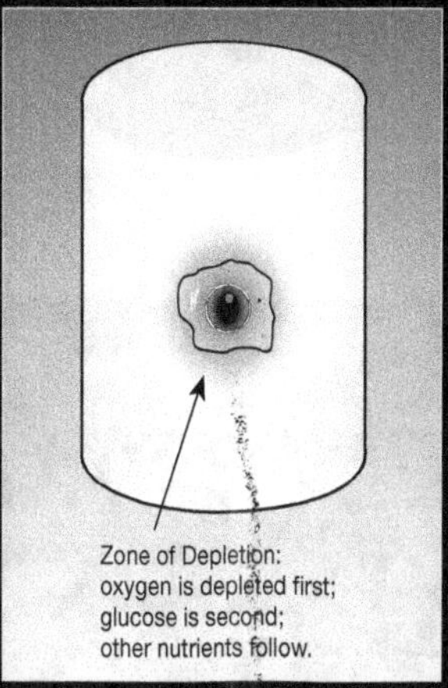
Zone of Depletion:
oxygen is depleted first;
glucose is second;
other nutrients follow.

In the early stages of planning for cell culture in space, scientists theorized that cells may not survive for long because of a potential inability to assimilate nutrients from the culture fluid. Although undisturbed fluid appears not to be moving, gravity-driven convection mixes the fluid. Gravity continually moves colder, more dense fluid to the bottom of the vessel, displacing the warmer fluid to the top. As the fluid on the bottom is heated, the process is repeated. In space, there is no gravity-driven convection to mix the medium and keep nutrients well distributed and available to cells. Therefore, theoretically, cells should experience a decrease in the availability of nutrients, thus slowing assimilation down to their intrinsic rate of diffusion—a rate potentially insufficient to support life. Oxygen should be the first essential to be depleted within a matter of minutes, followed by glucose. In reality, the cells do not die. Instead, they adapt to the lower rate of nutrient delivery and proceed to survive. Apparently, other more subtle convections (e.g., surface tension driven) may supply sufficient transport of nutrients. Understanding these concepts was essential to the design of cell culture systems for humans in space.

Unique Conditions Created by Microgravity

In microgravity, gravity-driven convection is practically nonexistent. Gravity-driven convection is familiar to us in a different context. For example, air conditioners deliver cool air through the vents above. Cooler air is more dense than warm air and gravity settles the more dense cool air closer to the floor, thereby displacing the warm air up to be reprocessed. These same convective flows feed cells on Earth-based cultures where the cooler fluid streams toward the bottom of the vessel, displacing warmer medium to the upper regions of the container. This process provides sufficient nutrient transport for the cells to thrive.

What Happens in Microgravity?

Scientists theorized that, in microgravity, cells would rapidly assimilate nutrients from the medium and, in the absence of gravity-driven convection, the cells would consume all the nutrients around them. Nutrient transport and the mechanical sensing mechanism operate differently in the absence of gravity. NASA conducted research on the Space Shuttle over the last 2 decades of the program to elucidate the nature of cell response to microgravity and showed that, while most cell cultures can survive in microgravity, substantial adaptation is required. The outcome of this cellular research is the emergence of space cell biology as a new scientific discipline.

Suite of Equipment

To meet the various requirements for a full complement of cell biology experiments, NASA developed a suite of equipment that spans from relatively simple passive cell cultures to complicated space bioreactors with automated support systems. The experiments that were supported included space cellular and molecular biology, tissue engineering, disease modeling, and biotechnology. Space cell biology includes understanding the adaptive response to microgravity in the context of metabolism, morphology, and gene expression, and how cells relate to each other and to their environment.

Analog and Flight Research

The cell culture in space, and to a certain extent in microgravity analogs, is an environment where mammalian cells will associate with each other spontaneously, in contrast to Earth culture where cells sediment to the lower surface of the container and grows as a sheet that is one cell layer thick. In space and in an analog culture, the association results in the assembly of small tissue constructs. A construct may be made up of a single type of cell, or it may be designed to contain several types of cells. As the assembly proceeds, cells divide and undergo a process of differentiation where they specialize into functions characteristic of their tissue of origin. For example, as liver cells go through this process, they produce constructs that look and function akin to a native liver specimen. In other instances, colon cancer cells mixed with normal cells will produce assemblies that look and act like a fresh tumor biopsy.

Transition of Cells

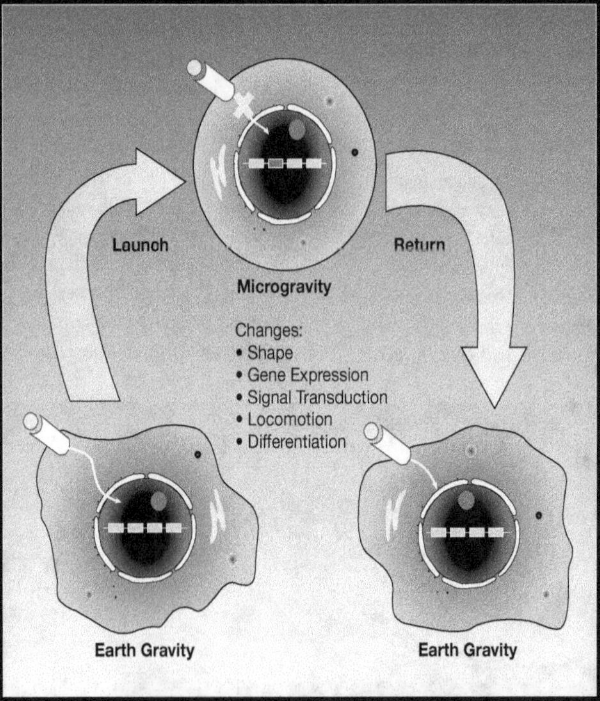

As cells transition to space, changes occur that provide new insights into life systems and offer the prospect of understanding the role of gravity in life as it developed on this planet. A stylized cell with its nucleus (red) containing genetic material (blocks), an example of a cell surface receptor and its communication linkage to the nucleus, and the external simulating factor (yellow ball) are displayed above in three phases: 1) on Earth at unit gravity; 2) following launch into microgravity; and 3) return to Earth. Within a few seconds after arriving in microgravity, the cell becomes round and, thereafter, a cascade of changes follows over the next few days and weeks. As the cell adapts to the new environment, it turns on some genes and turns off others. The ability to respond to certain external stimuli is diminished. This is due to a disruption (indicated by the "✖") of some cell surface receptor signal transduction pathways. In addition, cells locomote (move) very poorly in microgravity. The ability to mature and develop into functional tissues and systems seems to be favored. These observations provide a basis for robust investigation of microgravity cell biology as a means to understand terrestrial life in space and to use the space environment to foster goals in biomedical research on Earth.

These microgravity-inspired technologies are now used in cell culture and some tissue engineering studies. Scientists and physicians can produce tissues to be used as research models (e.g., cardiac tissues; cancers of the kidney, liver, colon, prostate, breast, and brain). Microgravity cultures are used in biotechnology to produce cell by-products that can be used to treat diseases and produce vaccines to prevent diseases.

Mary Ellen Weber, PhD
Astronaut on STS-70 (1995) and STS-101 (2000).

Colon Cancer Cells' unique response in microgravity: reassembly and reconstruction of their tissue origin.

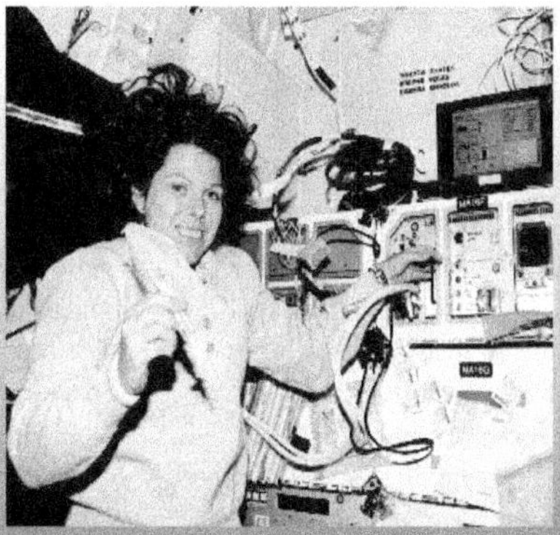

Astronaut Mary Ellen Weber with the space bioreactor on STS-70.

"One of my fondest memories of my shuttle missions was working preflight with the bioreactor team on its first experiment in space. I can still vividly remember my awe in watching colon cancer cells growing into cancer tissue, and the satisfaction in seeing it all come together. The experiment held so much promise early on that it was manifested on the mission well before all its details were worked out, and this gave me, its assigned crew member, the opportunity to work far more closely with these dedicated scientists than usual in getting it ready to go as well as the opportunity to learn far more about the science. Most researchers get to see their hard work come to fruition first hand, and as I watched the bioreactor successfully working in space, I was really struck—unexpectedly so—by the fact that they could not be there to witness it with me. It gave me a great sense of responsibility to do right by them, and it made me all the more proud to be a part of it."

NASA Develops Special Equipment to Grow Cells—Space Bioreactor

The use of microgravity cell culture to engineer tissues from individual cells began in systems where cells were grown in a tubular vessel containing a bundle of hollow fibers that carried nutrients to the cells in the tube. As concepts for space bioreactors matured, the cylindrical rotating systems emerged because of several advantages: greater volume; a format that supported both analog culture on Earth and space cell culture; and a natural association of cells with each other rather than with the plastic or glass vessel. The system could be rendered compatible with Earth or space by setting the rotation regime to the gravitational conditions. NASA performed a validation of the first rotating bioreactor system on Space Transportation System (STS)-44 (1991). No cells were used for the validation test. Instead, scientists used small beads made of inert polymer as surrogate cells. This enabled observation of the media delivery system and movement of "cells" along flow streams in the culture fluid. Results of the experiment showed characteristics consistent with maintaining live cells and set the stage for the first rotating bioreactor experiments in space.

The first investigation on the shuttle (STS-70 [1995]) used colon cancer cells as the test population to determine whether the new bioreactor system was compatible with cell assembly, growth, and maturation. The bioreactor was composed of a cylindrical culture vessel, culture medium reservoir, waste reservoir, pump (functions as a heart), and gas exchange module that delivered oxygen and removed carbon dioxide (essentially acting as a lung). The results showed that microgravity afforded continuous suspension of the cells, spontaneous association, cell propagation, and formation of a tissue construct.

The space bioreactor facilitated rapid assembly, substantially larger constructs, and metabolically active cells. The experiment confirmed the hypothesis that microgravity facilitates tissue morphogenesis (formation) and set the stage for use of the space environment to identify the essential stages in tissue engineering that are novel to microgravity. The ability to engineer tissue from individual cells provided tissue for research, drug testing, disease modeling and, eventually, transplantation into afflicted individuals. Subsequent colon cancer experiments on STS-85 (1997) identified some of the novel metabolic properties and demonstrated the mechanism used by the cancer to spread to other organs.

Interest in space cell culture opened the new vista of space cell biology. Mammalian cells are enclosed by a pliable lipid membrane. On Earth, those cells have a characteristic shape; however, when in microgravity, most mammalian cells become more spherical. Following this shape change, a cascade of adaptive changes occurs. Some genes are turned on while others are turned off, some receptors on the surfaces of cells cease to transduce signals to the inside, many cells cease locomotion (movement), and other cells will mature and change function spontaneously.

Microgravity-induced Changes at the Cell Level

Cells Adapt to Microgravity

On STS-62 (1994), NASA demonstrated that cells could grow in microgravity culture without succumbing to the lack of convective mixing of the medium. This demonstration occurred in a static culture system wherein rapidly dividing colon cancer cells and slowly dividing cartilage cells were placed in small culture vessels held at 4°C (39°F) (refrigeration temperature) until arriving in microgravity and reaching

Colon Cancer Cell Cultures

The first experiment using living tissue in the space bioreactor developed at Johnson Space Center used human colon cancer cells to determine whether there are specific advantages to propagation of cells in space. NASA conducted this experiment on STS-70 (1995) and again on STS-85 (1997). The right panel shows the large tissue assemblies that readily formed within a few days in microgravity when compared with the ground-based bioreactor analog, where the assemblies were much smaller and less well developed. For reference, the left panel shows the same cells in standard culture on Earth, where the cells grew and attached to the petri dish in a single layer with little evidence of tissue formation. This experiment set the stage for using space cell culture to produce tissues with a greater parity to the actual tumor in situ in the patient. Furthermore, unlike the standard culture, it demonstrated the signature biochemicals associated with the disease.

orbit where the temperature was raised to 37°C (98.6°F) (body temperature) to initiate growth. Results showed that colon cancer cells rapidly assimilated nutrients from the medium while cartilage took more than twice as long to deplete nutrients. Neither cell population succumbed to the depletion but, rather, changed their metabolic profile to adapt to more stringent conditions. Thus, bioreactors to support these cells for long-term experiments needed to accommodate re-feeding and waste disposal to ensure health of the tissue. The results of this experiment set the requirements for final design of the space bioreactors to grow bulk culture in microgravity.

Immune Cells Have Diminished Locomotion in Microgravity

The immune cells known as lymphocytes locomote and traverse many environments within the body to engage invading microbes and effect their destruction or inactivation. Experiments conducted on STS-54 (1993) and STS-56 (1993) were the first

Cell Locomotion

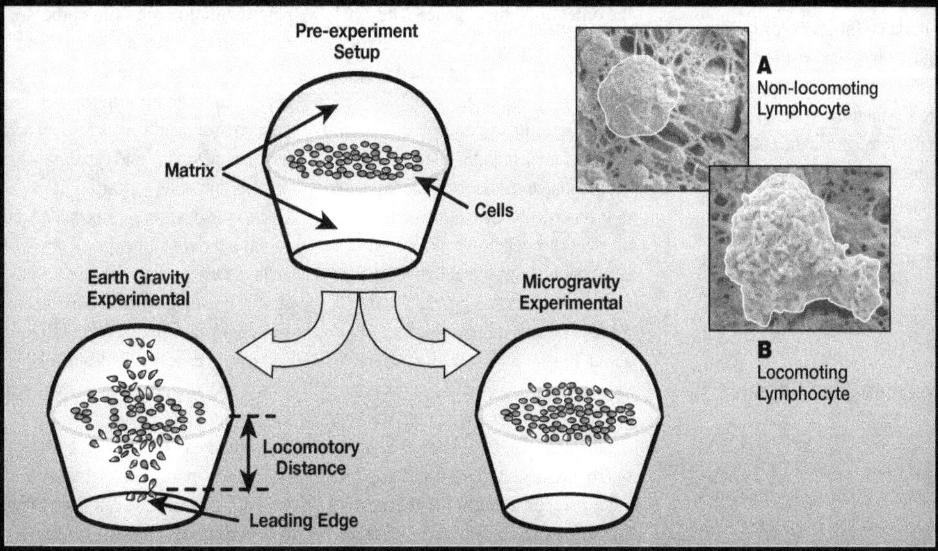

Human immune cells (lymphocytes) locomote through tissue matrix (intercellular cement) as part of their normal function in mediating immunity. Experiments performed in the analog culture system indicated a profound loss of the ability to locomote through matrix. This experiment described above was performed on STS-54 (1993) and STS-56 (1993). The matrix material is gelled collagen cast in two separate upper and lower phases, and the interface is loaded with human lymphocytes. Some were incubated as ground controls and others were transported to the shuttle. Locomotion remained arrested throughout the preparation and transport to space by maintaining them at 4°C (39°F). Upon arrival in microgravity, the temperature was raised to 37°C (99°F) in the experimental and control specimens. The lower left control shows how the lymphocytes locomote symmetrically up and down. Distance of locomotion to the leading edge can be measured using a microscope. In space, the experimental specimens evidenced very little locomotion. Non-locomoting lymphocytes are round and incapable of deforming (photo A), whereas locomoting lymphocytes deform and extend the process toward the direction of movement (photo B). The loss of locomotion in space indicates a potential defect in immunity in space. Loss of locomotion for extended periods of time can profoundly impact immunity. Locomotion is essential to this trafficking of lymphocytes through lymphoid organs and to sites of infection or invasion by cancer cells.

to show that these important immune cells have diminished locomotion in microgravity. Lymphocytes from a total of six donors were introduced into natural matrix (collagen) and kept at 4°C (39°F) until achieving orbit, where the temperature was raised to 37°C (98.6°F). Results showed that locomotion was inhibited by more than 80% in all specimens. Locomotion is a critical function in the immune system. Cessation does not have immediate effects; however, if sustained, it can contribute to a decline in immune function in space. Preparation for long-duration (in excess of 1 year) excursions in space will require extensive research and preparation to ensure the immune system functions normally throughout the entire mission. From strictly a cell biology perspective, the experiment was a milestone demonstration that locomotion can be modulated by a physical factor (gravity) rather than a biochemical factor.

Gene Expression Changes

Gene expression—defined as which genes are turn on and/or off in response to changing conditions—changes with almost every stimulus, stress, or alteration offered by our environment and activities. Most of these responses at the gene level occur in suites of genes that have been refined through evolution. This is why life systems can adapt to various environmental stimuli to survive and even thrive. Since all Earth organisms evolved in Earth gravity, the effect of microgravity on these genetic suites was unknown. Understanding the response at the genetic level to microgravity will give new insights to the changes necessary for adaptation.

New technology allows for the investigation of changes in more than 10,000 genes in a single experiment. The first genetic signatures for cells in microgravity were conducted on STS-106 (2000) using human kidney cells as a test model. The results provided a provocative revelation. Out of 10,000 genes tested, more than 1,600 were significantly changed in expression. Normally, a suite of genes refined through evolution is on the order of 20 to 40 genes. The enormous response to microgravity suggests there is not a refined suite, and the response is made up of genes that are essential to adaption—some are incidental and unrelated to adaptation, and some are consequential to the incidental activation of unnecessary genes. Analysis of gene expression showed that hypergravity (centrifugations at 3 gravitational force [*g*]) has a more refined set of about 70 genes. This is likely due to terrestrial life experiencing hypergravity during accelerations (running, starting, or stopping). On the other hand, analog microgravity culture on Earth also had a large response suite of 800 genes. Of those genes, only about 200 were shared with the microgravity suite.

The significance of these results is multifold. For short-duration missions, we will want to manage any untoward effects brought about by the response. For long-duration missions in space and permanent habitation on planetary surfaces, we will want to know whether there is a refinement in the gene suite and whether, in conjunction with the new environment, it poses the possibility for permanent changes.

STS-105 (2001) hosted an experiment on human ovarian carcinoma, asking whether space cell culture gave a gene expression profile more like the actual tumor in the patient or like that observed in standard cell culture on Earth. Results showed tissue-like assemblies that expressed genes much in the same profile as in the tumor. This is significant because these results give scientists a more robust tool to identify specific targets for chemotherapy as well as other treatments.

Space cell culture offers a unique opportunity to observe life processes that otherwise may not be apparent. Forcing terrestrial life to muster its adaptive mechanisms to survive the new environment makes evident some new characteristic and capabilities of cells and other terrestrial life. One of the observations is the induction of differentiation (the process by which cells mature and specialize). The shuttle hosted numerous experiments that confirmed unique differentiation patterns in cancer cells from colon, ovary, and adrenals as well as human kidney cells and mouse cells that differentiate into red blood cells. All but the mouse cells were on STS-105. The mouse cell experiment was performed on STS-108 (2001).

In summary, these experiments opened a new understanding of the differentiation process and products of cells. The processes revealed aspects useful in proposing new approaches to treatment of disease and tissue engineering and to understanding complex developmental pathways. On the product side, materials were produced that may lead to new biopharmaceuticals, dietary supplements, and research tools.

Gene Expression Differs at Three Gravity Levels

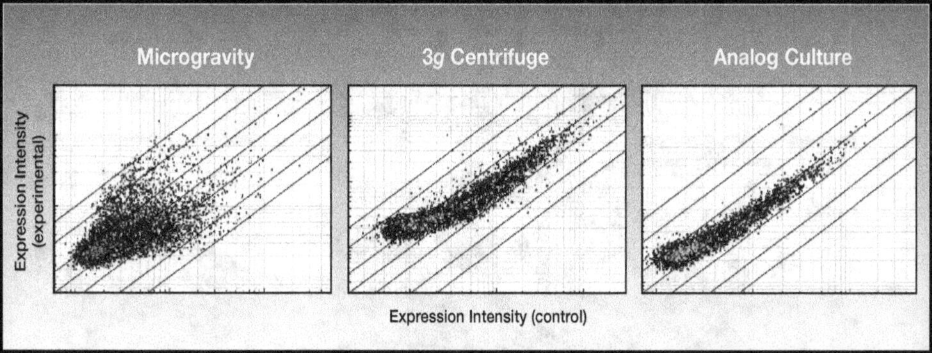

NASA performed experiments using human kidney cell cultures on STS-105 (2001) and STS-106 (2000) to investigate the gene expression response to microgravity and compare it to hypergravity and to an analog culture system on Earth. In a sample set (10,000 genes), the genes turned on and off compared with the control in normal culture on Earth. If the expression is identical in control and experimental conditions, the dots line up on the diagonal line passing through the origin. Genes that are turned on are above and beyond the first parallel diagonal line. Genes below and beyond the first parallel diagonal are decreased in expression compared with the control. In microgravity, more than 1,600 of the 10,000 genes are up-regulated or down-regulated compared with the control, meaning that it is unlikely that terrestrial life has a preformed, inherited set of genes used to adapt to microgravity. The cells were then subjected to 3 gravitational force (*g*) using a centrifuge. The array is more compacted. Fewer than 70 genes are affected, suggesting that terrestrial life has a history of responding to hypergravity. The last panel shows the same cells in response to microgravity analog cell culture. More than 700 genes modified in response to the analog system that rotates the cell culture, such that the cells are falling continuously. Analysis indicated that it shared about 200 genes with that observed in microgravity.

Observations from early experiments strongly suggested that the space environment may promote conditions that favor engineering of normal tissues for research and transplantation. Experiments in ground-based analog culture suggested that microgravity can facilitate engineering of functional cartilage starting from individual cells. (Cartilage is the tissue that forms the joints between bones.) Cartilage tissue was chosen because of its low metabolic demand on the culture system, durability, and conveniently observed characteristics of maturity and functionality. STS-79 (1996) flew a bioreactor containing beef cartilage cells to the Russian space station Mir. The culture set a landmark for 137 consecutive days of culture in microgravity. Results from this experiment and subsequent ground-based research: 1) confirmed the utility of microgravity in tissue engineering; 2) showed that generation of cartilage in microgravity produces a very pliable product when contrasted to native cartilage; and 3) showed that on transplantation the less mature, more pliable space cartilage remodels into the recipient site much better than mature cartilage. The study suggests that microgravity and space technology are useful in developing strategies for engineering tissues from a small number of cells.

Human Prostate Cancer Cells

On STS-107 (2003), NASA performed an experiment to investigate a model of metastatic prostate cancer. Prostate cancer is more manageable as a local disease, which is why there is such emphasis on preventive measures. Management of the disease becomes difficult when the tumor metastasizes to bone. Therein, the tumor establishes a relationship to that which contributes to its intractable state. Space cell culture offers an environment consistent with culturing two different kinds of cells harmoniously and also favors reassembly of cells into ordered tissue arrays. The upper cylinder shows the rapid assembly of the cells into tissue constructs that are much larger than those in the lower cylinder (controlled on Earth). The assemblies propagated in space achieved diameters approaching 2 cm (0.79 in.), while those on the ground were about 3 mm (0.19 in.). The result demonstrated the value of space cell culture in providing robust models for investigating human disease. These specimens were not analyzed, since they were not recovered from the ill-fated Columbia mission.

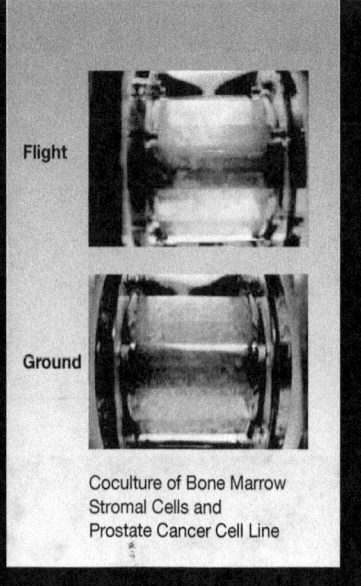

Coculture of Bone Marrow Stromal Cells and Prostate Cancer Cell Line

Human Prostate Cancer Cells Grew Well in Microgravity

In pursuit of using space to understand disease processes, NASA conducted experiments on STS-107 (2003) to understand the special relationship between prostate cancer and bone marrow cells. Prostate cancer, like breast cancer, is a glandular tumor that is a manageable disease when treated at its origin. In contrast, when tumors spread to other areas of the body, the disease becomes intractable. The experiment on STS-107 modeled the metastatic site in the bone for prostate cancer. Results showed the largest tissue constructs grown in space and demonstrated the outcome of the cohabitation of these two cell types. It also showed that we could produce these models for research and provide a platform for demonstrating the contribution of the normal cell environment to the establishment and maintenance of the tumor at a new site. With such a model, we may identify new targets for therapy that help prevent establishment of metastases.

The Future of Space Cell Biology

Research in cell science plays a significant role in space exploration. Cells, from bacteria to humans, are the basic unit of all life. As is true for Earth-based biomedical research in cells, the observations must be

Important Questions:

Does the cell respond directly to the change in gravitation force, or is it responding to conditions created by microgravity?

What does terrestrial life do to adapt and thrive in space?

Does microgravity influence how life might evolve after many generations in space?

What is the effect of microgravity on cells from major organs and the immune and digestive systems?

How much gravity is necessary to have normal function?

What Is the Relationship Between Gravity and Biological Responses?

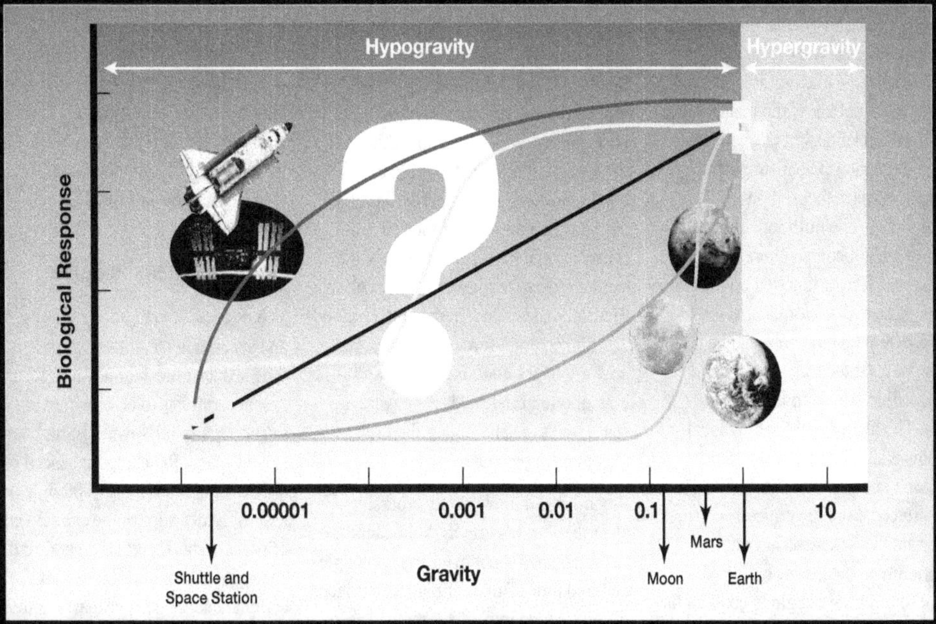

The future of space cell biology includes a critical question regarding the relationship of gravity to various biological responses within the systems of the human body as well as in microbes, plants, animals, and bioprocessing systems. The possible relationships are depicted as lines on the graph, where values are known for the shuttle, space station, and Earth. The knowledge of the actual relationship will enable better understanding of human adaptability on the moon (1/6 gravitational force [*g*]) and Mars (3/8*g*). Furthermore, it will assist in the design of artificial *g* technologies. Knowledge of biologic responses on Earth reveals that the response relationships to stimuli are sigmoid, as in the yellow and green curve, and that the range of the response is usually within one tenfold increment of the normal physiologic state (Earth). Thus, the green relationship may be the most likely one. With this probability, research on moon and Mars gravity becomes more important in exploration planning. Depending where on the "*g*" scale the s-shaped part of the curve flexes, that is the amount of *g* that will begin to restore normal function.

consistent at the tissue, organ, and whole-organism level to be useful in developing treatments. Because we cannot perform experiments that may be difficult or even unethical in humans, biomedical researchers rely on cell-based research to investigate fundamental life process, diseases, and the effects of drugs and environment on life. Thus, part of our understanding of microgravity, hypogravity (such as the level found on Mars or the moon), radiation, and environmental factors will come from cell studies conducted in space and in analog culture systems.

The answer to the last question may have the most impact on risk reduction for humans exploring space. The answer will not only reveal the gravity force necessary to have acceptable physiologic function (bone health, muscle conditioning, gastrointestinal performance, etc.), it also may set requirements for the design of vehicles, habitats, exercise systems, and other countermeasures. The pervasive question is: How much gravity do you need? We do not know the mathematical basis of the relationship of gravity to biologic function. The history of research in space focused on microgravity (one millionth of Earth gravity) and, of course, there is a wealth of data on biologic function on Earth. Given these two sets of data, at least four different relationships can be envisioned. Of the four, the sigmoid (s-shaped) relationship is the most likely. The likely level for biological systems will be around 1/10*g*. Since the moon and Mars are 1/6 and 3/8*g*, respectively, it will be critically important that scientists have an opportunity to determine biological response levels and begin to conduct the mathematical relationship between *g* and biological function.

As NASA proceeds toward a phase of intensified use of the International Space Station (ISS) for research, it is important to have a robust plan that will continue the foundational research conducted on the shuttle and procure the answers that will reduce health risks to future spacefarers. When the United States enacted the national laboratory status of the ISS, it set the stage for all federal agencies to use the microgravity environment for their research. Increasing the science content of orbiting facilities will bring answers that will enable reduction of risks to explorers and help ensure mission success.

Physical Sciences in Microgravity

What is Gravity?

Gravity is a difficult thing to escape. It also turns out to be a difficult thing to explain. We all know enough to say that things fall because of gravity, but we don't have easy answers for how gravity works; i.e., how the mass of one object attracts the mass of another, or why the property that gives matter a gravitational attraction (gravitational mass) is apparently the same property that gives it momentum (inertial mass) when in motion. Gravity is a fundamental force in physics, but how gravity is bound to matter and how gravitational fields propagate in space and time are among the biggest questions in physics.

Regardless of how gravity works, it's clear that Earth's gravity field cannot be easily escaped—not even from a couple hundred miles from our planet's surface. If you stepped into a hypothetical space elevator and traveled to the 100,000th floor, you would weigh almost as much as you do on Earth's surface. That's because the force that the Earth exerts on your body decreases at a rate inversely proportional to twice your distance from the center of the Earth. In an orbit around the Earth, the force exerted by our planet's mass on a spacecraft and its contents keeps them continually falling toward the Earth with an acceleration inversely proportional to the square of the distance from the center of the planet. That's Newton's law of gravitation.

Gravity certainly works on and in airplanes. When you are traveling in an airplane during a steady flight, gravity keeps you firmly in your seat. The lift created by air flowing around the wings keeps an airplane and your seat aloft under you—and that's a good thing. Now imagine being in an airplane that has somehow turned off its lift. In this scenario, you would fall as fast as the airplane was falling. With your seat falling out from under you at the same rate, the seat would no longer feel your weight. No force would be holding you in it. In fact, you would be approximately weightless for a short period of time.

Weightlessness in Space

The essence of conquering gravity and sustaining weightlessness for longer than a few seconds is velocity. A spacecraft has to be moving very fast to continually fall toward Earth but still stay in space. Reaching that speed of a little over 27,500 km (17,000 miles) per hour provides a lot of the excitement of spaceflight. It takes a great deal of energy to put an object into Earth orbit, and that energy goes primarily into attaining orbital velocity. An astronaut in Earth orbit has kinetic energy equivalent to the explosion of around 454 kg (1,000 pounds) of TNT. Once an astronaut reaches orbital velocity, he or she is a long way toward the velocity needed to escape Earth's gravity, which is 1.4 times orbital velocity.

When you're in a vehicle moving fast enough to fall continually toward the Earth, it doesn't look or feel like you're falling. At least, not the kind of falling that people are accustomed to—the kind that ends in a painful collision with the ground. You have the feeling of being light, and the things around you are light, too. In fact, everything floats if not fastened to something. Items in the spacecraft are falling with

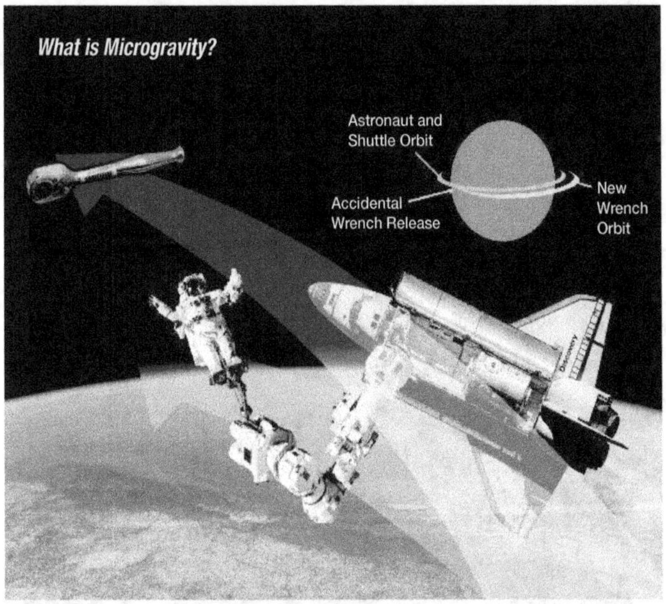

Imagine an astronaut tethered to the outside of the shuttle. The astronaut and the shuttle are in orbit together. If the astronaut releases a tool, the tool generally goes into a slightly different orbit because it has to maintain a different speed to achieve the same orbit as the shuttle. The astronaut, shuttle, and tool are in orbit with their outward acceleration from the Earth, balanced by Earth's gravity. The slight differences in orbit make it seem, to the astronaut, that a small acceleration is pushing the wrench away. This is microgravity.

you. With everything accelerating toward Earth at precisely the same rate within this falling frame of reference, Earth's gravity is not apparent. To an outside observer, gravity is still obvious—it's the reason you're in an orbit and not flying away from Earth in a line to space.

Early Low-gravity Technology

The consequences of being weightless were merely hypothetical until the dawn of space travel, with one small exception: One hundred years prior to the launch of the first rocket beyond Earth's atmosphere, spherical lead shot was manufactured by allowing molten lead to solidify in free fall inside a shot tower. As long as the shot wasn't falling fast enough for air resistance to deform it, the absence of gravitationally created hydrostatic pressure in the falling lead drop that allowed it to assume a spherical shape as the liquid was driven by thermodynamics into a volume of minimum exposed surface. The falling shot quickly hardened as it cooled, and it collected in a water bath at the bottom of the tower. The shot-manufacturing industry relied on this early low-gravity technology until the first decade of the 20th century.

Physics Environment in Space

Spaceflight provides a good place to conduct experiments in physics—experiments that would not be possible on Earth. Wernher von Braun (center director at Marshall Space Flight Center from 1960 to 1970) had more practical applications, such as making ball bearings in space. Several simple experiments were flown on Apollo 14 and performed by the crew on the return from the moon. More experiments were conducted on the three Skylab missions—an early space station built in the 1970s—with promising results reported in areas such as semiconductor crystal growth. By the time of Skylab, however, the next era of space exploration was on the horizon with the approval of the Space Shuttle Program in 1972.

Fundamental Physics

One of the great questions of physics is the origin of long-range order in systems of many interacting particles. The concept of order among particles is a broad one—from simple measures of order, such as the density of a collection of molecules or the net magnetization of the atomic nuclei in an iron bar, to complex patterns formed by solidifying alloys, turbulent fluids, or even people milling about on an urban sidewalk. In each of these systems, the "particles" involved interact nearly exclusively with only their near neighbors; however, it's a common observation in nature that systems composed of many interacting elements display ordering or coherent structures over length scales much larger than the lengths describing the particles or the forces that act between them. The term for the distinctive large-scale behavior that results from cooperatively interacting constituent particles is "emergent phenomena." Emergent phenomena are of interest to science because they appear to be present at virtually every scale of the natural world—from the microscopic to the galactic—and they suggest that common principles underlie many different complex natural phenomena.

Phase transitions at a critical point provide physicists with a well-controlled model of an emergent phenomenon. Pure materials, as determined by thermodynamics, exist in a particular state (a "phase") that is a function only of temperature and pressure. At a point called a "critical point," simple single-phase behavior breaks down and collective fluctuations sweep through the system at all length scales—at least in theory. The leading theory that has been developed to describe emergent phenomena, such as critical point fluctuations, is called "renormalization group theory." It provides a model that explains how the behavior of a system near a critical point is similar over a large range of scales because the physical details of many interacting molecules appear to average out over those scales as a result of

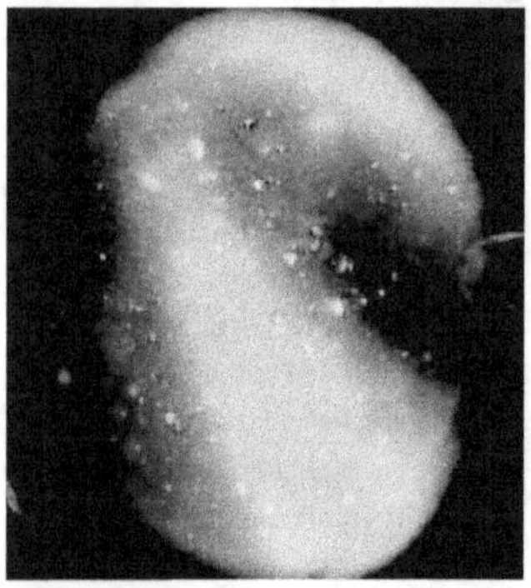

Small particles in a colloidal solution assemble to form an ordered crystalline structure, such as the opalescent crystalline particles shown in this image taken on STS-73 (1995). Building an understanding of emergent phenomena remains one of the great challenges of physics. Explaining the origins of long-range order and structures in complex systems is key to advancing potential breakthroughs, and the experiments in fundamental physics aboard the shuttle played a significant role.

Critical Point Experiments Test Theories

The critical point of xenon is 289 K, 5.8 MPa—or 15.85°C (60.53°F), 57.2 atm. Note that the axis on the left is logarithmic. Research on STS-52 (1992) measured the phase boundary between normal liquid helium and superfluid helium. (Superfluids, such as supercooled helium-4, exhibit many unusual properties. The superfluid component has zero viscosity, zero entropy, and infinite thermal conductivity.) This shuttle research confirmed the renormalization group theory better than any Earth research. These types of research questions are now being studied on the International Space Station.

NASA tested the theory for gas-liquid critical phenomena on STS-97 (1997).

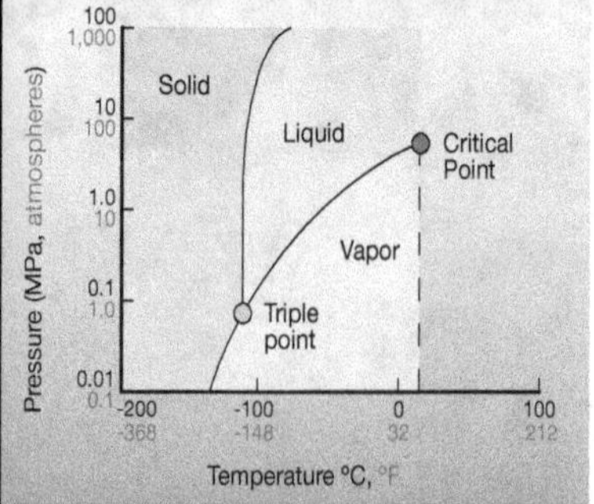

The Lambda Point Experiment cryostat assembly (identified by the JPL insignia) in the STS-52 (1992) payload bay.

cooperative behavior. Renormalization group theory is one of the great developments of physics during the 20th century. The most precise tests of this theory's predictions for critical point phenomena relied on experiments carried aboard the shuttle.

Careful critical point experiments required the ultimate in precise control of pressure and temperature to the extent that the difference in pressure, caused by gravity, between the top and the bottom of a small fluid sample in a laboratory on Earth by the mid 1970s became the limiting factor in experimental tests of renormalization group theory.

Research on Space Transportation System (STS)-52 (1992) measured the phase boundary between normal liquid helium and superfluid helium. Superfluids, such as supercooled helium-4, exhibit many unusual properties. The superfluid component has zero viscosity, zero entropy, and infinite thermal conductivity. This shuttle research confirmed the renormalization group theory better than any Earth research.

Protein Crystal Growth

A foundation for the explosion of knowledge in biological science over the past 50 years has been the understanding of the structure of molecules involved in biological functions. The most powerful tool for determining the structure of large biomolecules, such as proteins and DNA, is x-ray crystallography. In traditional x-ray crystallography, an x-ray beam is aimed at a crystal made of the molecule of interest. X-rays impacting the crystal are diffracted by the electron densities of each atom of each molecule arranged in a highly ordered crystal array. Because nearly each atom of each molecule is in a highly ordered and symmetrical crystal, the x-ray diffraction pattern with a good crystal is also highly ordered and contains information that can be used to determine the structure of the molecule. Obtaining high-quality protein crystals has been a critical step in determining a protein's three-dimensional structure since the time when Max Perutz first used x-ray crystallography to determine the structure of hemoglobin in 1959. A few proteins are easily crystallized. Most require laborious trial-and-error experimentation.

The first step in growing protein crystals is preparation of as pure a protein sample as can be obtained in quantity. This step was made easier for many molecules in recent years with the ability to increase the products of individual genes through gene amplification techniques; however, every purification step is still a tradeoff with loss of starting material and the likelihood that some of the molecules in solution will denature or permanently change their shape, effectively becoming contaminants to the native molecules. After biochemists have a reasonably pure sample in hand, they turn to crystal-growing recipes that vary many parameters and hunt for a combination that will produce suitable crystals. Although usable crystals can

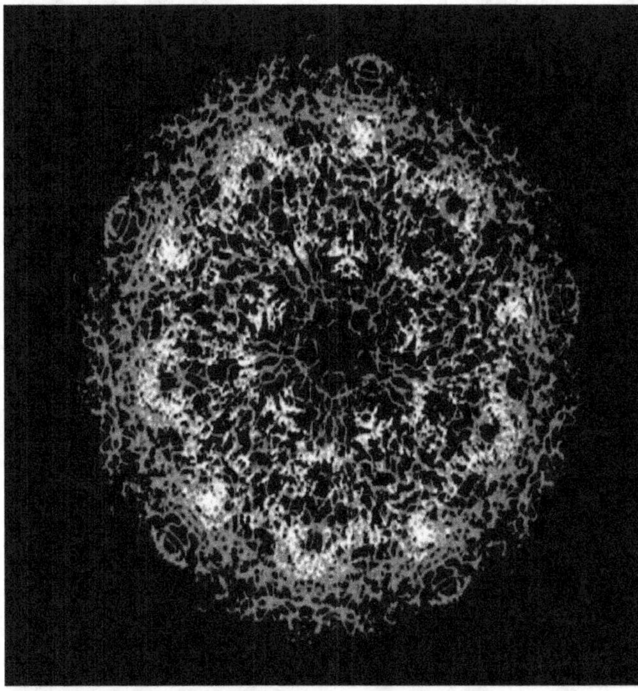

This molecular structure of the Satellite Tobacco Mosaic Virus was captured at 1.8-angstrom (0.18-nanometer) resolution from analysis of crystals obtained on experiments performed on the International Microgravity Laboratory-2 mission (STS-65) in 1994. The best of these crystals was 30 times larger and produced 237% more data than any previous Earth-grown crystals and yielded what was, at that time, the highest-resolution structure of a virus ever obtained.

be as small as 0.1 mm (0.004 in.) on a side, the crystals often take weeks or even months to grow, so biochemists will normally try many combinations simultaneously and in specially designed trays. It is not unusual to spend several years finding good growth conditions for a protein.

Effects of Gravity on Protein Crystal Growth

Gravity has two principal effects in protein crystal growth. The first is to cause crystals to sink to the bottom of the solution in which they are growing. As a result, the growing crystals can pile up on each other and fuse, thus becoming a single mass that can't be used for data collection. The second effect of gravity is to produce weak but detectible liquid flow near the surface of the growing crystals. Having contributed some of its dissolved protein to the growing crystal, liquid near the crystal surface is lighter than liquid farther away. Due to gravity, the lighter liquid will rise. The consequences of this flow for crystal quality are complex and even now not fully understood. At the beginning of the shuttle era, German chemist Walter Littke thought that liquid flow near the growth surface would interfere with the molecules on the surface finding their places in a crystal. Before the first launch of the shuttle, he conducted several promising short rocket-launched experiments in which several minutes of low gravity were achieved in a suborbital flight.

Protein Crystallization on the Shuttle

The first protein crystallization experiments on the shuttle were conducted in a simple handheld device carried aboard in an astronaut's kit. Encouraging results from Professor Littke's experiment aboard STS-61A, the D-1 Spacelab mission (1985), where he reported achieving crystal volumes as much as 1,000 times larger than comparable Earth-based controls, opened a huge level of interest including many international and commercial investigators. Professor Charlie Bugg of the University of Alabama, Birmingham, working with Professor Larry DeLucas, who went on to fly on the US Microgravity Laboratory-1 mission (1992) as a payload specialist, eventually developed a community of nearly 100 investigators interested in flying proteins.

Some investigators obtained crystals that gave spectacular results, including the highest resolutions ever attained at the time for the structure of a virus and, in several instances, the first crystals suitable for structural analysis. Other proteins, however, seemed to show no benefit from space crystallization. A major focus of NASA's research was to explain this wide range of results.

Eugene Trinh, PhD

Payload Specialist and NASA expert in microgravity sciences on STS-50 (1992) US Microgravity Laboratory-1 Spacelab mission.

"The Space Shuttle gave scientists, for the first time, an opportunity to use the space environment as an experimental tool to rigorously probe the details of physical processes influenced by gravity to gather better theoretical insight and more accurate experimental data. This precious new information could not have been otherwise obtained. It furthered our fundamental understanding of nature and refined our practical earthbound industrial processes."

Eugene Trinh, PhD, a payload specialist for this mission, is working at the Drop Physics Module using the glove box inside the first US Microgravity Laboratory science module on STS-50.

Modeling Protein Crystal Growth

Physicists and biochemists constructed models of protein crystal growth processes to understand why some proteins produced better crystals in microgravity while others did not, and why crystals sometimes started growing well but later stopped. Investigators applied techniques like atomic force microscopy to examine the events involved in the formation of crystalline arrays by large and rather floppy protein molecules. The role of impurities in crystal growth and crystal quality was first documented through the work of Professor Alexander McPherson (University of California, Irvine), Professor Peter Vekilov (University of Houston, Texas), and Professor Robert Thorne (Cornell University, Ithaca, New York), along with many others. A simplified picture of a popular model is that proteins that grow better crystals in microgravity have small levels of contaminants in solution that preferentially adhere to the growing surface and slow the growth of the molecule-high step layers that form the crystal.

Accelerated transport of contaminant species due to buoyant flow on Earth will increase the population of contaminant species on the surface, eventually inducing the formation of defects. Such proteins will produce better crystals in microgravity because strongly adhering contaminants are transported by slower molecular diffusion rather than convection, and their surface concentration on the crystal remains lower.

This research has given a detailed scientific foundation to the art and technology of protein crystallization, thus providing structural biologists with a mechanistic understanding of one of their principal tools.

Biotechnology and Electrophoresis

In the 1970s and early 1980s, the biotechnology industry identified a large number of biological molecules with potential medical and research value. The industry discovered, however, that the difficulty of separating molecules of interest from the thousands of other molecules inside cells was a barrier to the production of therapeutic materials.

Separation techniques for biological molecules rely on using small differences between molecules to spatially separate the components of a mixture. The mobility differences that separation methods use can result from the size of the molecule, substrates to which the molecule binds, or charge on the molecule in solution.

Separation methods relying on the interaction of biological molecules with an applied electric field, including zone electrophoresis and isoelectric focusing, use the charge on a molecule that is dependent on the solution properties (pH, ionic strength, etc.) around the molecule to separate mixtures of molecules. The throughput and resolution of these techniques are limited by the flow induced in the solution containing the molecules, heat generated by the electric current passing through the liquid, and sedimentation of the large molecules during the necessary long separations. It was recognized that electrophoresis, one of the earliest candidates for space experiments, would solve the problem of the disruptive heat-driven flows by minimizing the effect of gravity. Warmer, lighter liquid wouldn't rise in the electrophoresis cell, and device performance might be dramatically improved.

Professor Milan Bier (University of Arizona, Tucson)—a pioneer in biological separations whose discoveries did much to establish electrophoresis as a laboratory tool—conducted several important flight experiments with NASA. As Professor Bier's work on the Isoelectric Focusing Experiment proceeded and flew on several early shuttle missions, he came to understand the impact of gravitational effects on Earth-based electrophoresis. He developed designs for electrophoresis equipment that minimized the impact of gravity. Within a few years, these designs became the industry standard and a basic tool of the biotechnology industry. Commercial organizations became interested in the potential of space-based bioseparations. McDonnell Douglas Astronautics Company sponsored seven flights of a large electrophoresis device—the Continuous Flow Electrophoresis System. Several flights included a McDonnell Douglas Astronautics Company technical expert who traveled on board as a payload specialist.

Using this facility on the shuttle middeck, Robert Snyder of the Marshall Space Flight Center, along with his colleagues, discovered a new mode of fluid behavior—electrohydrodynamic instability—that would limit the performance of electrophoresis devices even after the distortion of gravity was eliminated. The discovery of this instability in space experiments and subsequent confirmation by mathematical analysis allowed electrophoresis practitioners on Earth to refine their formulations of electrophoresis liquids to minimize the consequences of electrohydrodynamic effects on their separations. This led to experiments, conducted in a French-built facility by French pharmaceutical company Roussel-Uclaf SA, Paris.

The opportunity to conduct sequential experiments of increasing complexity was one of the benefits of these shuttle microgravity missions. Interest shown by these commercial and international organizations initiated in early shuttle missions continues today on the International Space Station (ISS).

Materials Processing and Materials Science

The semiconductor industry grew up with the space program. The progression from commercial transistors appearing in the 1950s to the first integrated circuits in the 1960s and the first microprocessors in the 1970s was paralleled, enabled, and driven by the demanding requirements of space vehicles for lightweight, robust, efficient electronics.

Since the beginning of semiconductor technology, a critical issue has been the production of semiconductor crystals from which devices can be fabricated. As device technology advanced, more stringent device performance and manufacturing requirements on crystal size, homogeneity, and defect density demanded advances in crystal growth technology. In the production of semiconductor crystals, when molten semiconductor freezes to form a crystalline solid, variations in the temperature and composition of the liquid produce density variations that cause flows as less-dense fluid rises. These flows can cause poor distribution of the components of the molten material, leading to nonuniformities and crystal defects. Studying semiconductor crystal growth in low gravity, where buoyancy-driven flows would be extremely weak, would give insight into other factors at work in crystal growth. There was also hope that in microgravity, quiescent conditions could be attained in which crystallization would be "diffusion controlled" (i.e., controlled by stable, predictable mechanisms proportional to simple gradients of temperature and composition) and that, under these conditions, material of higher quality than was attainable on Earth would be produced.

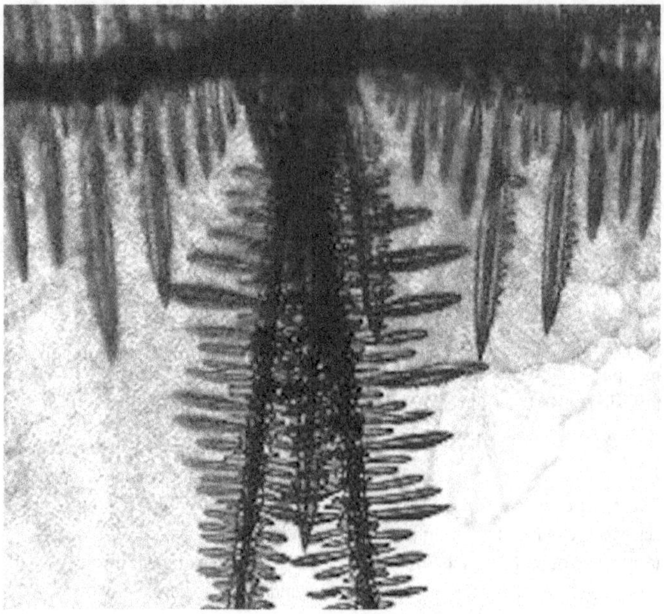
Solidification of a liquid is an unstable process under many conditions. An initially flat boundary will evolve into an elaborate web of branched dendrites. In metals, the properties of the resulting solid are highly dependent on the structure formed during solidification, making the understanding of interface evolution an important goal of materials science.

In the early 1970s, semiconductor crystal growth was one of the first concepts identified by the National Research Council for materials processing in space. Promising early results, especially on Skylab, spurred plans for semiconductor research on the shuttle. Materials processing and semiconductor crystal growth experiments were also a prominent part of Soviet microgravity research. Crystal growth in space was a challenge because of the power needed by the furnaces and the containment required to meet NASA safety standards. Eventually, however, furnaces were built and flown on the shuttle not only by NASA but also by the European Space Agency and the space agencies of Japan, Germany, and France. Large furnaces flew on pallets in the cargo bay and in Spacelab while small furnaces flew on the shuttle middeck. To quantify the role of gravity in semiconductor crystal growth, NASA supported a comprehensive program of experiments and mathematical modeling to build an understanding of the physical processes involved in semiconductor crystal growth.

The results of materials processing and materials science experiments strongly influenced scientific understanding in several technologically important areas:

- Control of homogeneity and structural defects in semiconductor crystals
- Control of conditions for production of industrial alloys in processes like sintering and precipitation hardening
- Measurement of accurate thermophysical properties, such as surface tension, viscosities, and diffusivities, required for accurate process modeling

Liquid phase sintering experiments performed in low gravity yielded the unexpected results that the shape distortion of samples processed in microgravity is considerably greater than that of terrestrially processed samples. Sintering is a method for making objects from powder by heating the material in a sintering furnace below the material's melting point (solid state sintering) until its particles adhere to each other. Sintering is traditionally used to manufacture ceramic objects and has also found uses in fields such as powder metallurgy. This result led to improved understanding of the underlying causes of the shape changes of powder compacts during liquid-phase sintering with significant impact on a $1.8 billion/year industry.

Space experiments on the prediction and control of microstructure in solidifying alloys advanced theories of dendritic (from *dendron*, the Greek word for tree) growth and yielded important contributions to the understanding of the evolution of solid-liquid interface morphologies and the consequences for internal structure of the solid material. Introductions to metallurgy traditionally begin with a triangle made of three interconnected concepts: process, structure, and properties. According to this triangle, the study of metallurgy concerns how processing determines structure for various metals and alloys and also determines properties. A solidifying metal develops a characteristic structure on several distinct interacting length scales. The microstructure (usually on the scale of tens of microns) is formed by the typically dendritic pattern of growth of the solid interface. The macroscale pattern of a whole casting is determined by, among other things, the distribution of solutes rejected from the solid, shrinkage of the solid during freezing, and thermal conditions applied to the metal. The formation of structures during the solidification of practical systems is further complicated by the multiplicity of liquid and solid phases that are possible in alloys of multiple elements.

Understanding the processes that control the growth of dendrites on a growing solid is a foundation for how processing conditions determine the internal structure of a metal. Gravity can have a visible influence on the growth of dendrites because of the disruptive effects of flow caused by temperature gradients near the dendrite. Therefore, removing the effects of gravity was essential to obtain benchmark data on the growth rates, shapes, and branching behavior. In the 1990s, a series of experiments designed by Professor Martin Glicksman, then at the Rensselaer Polytechnic Institute, Troy, New York, was conducted on shuttle missions using an instrument named the Isothermal Dendritic Growth Experiment. The experiments carefully measured the characteristics of single growing dendrites in an optically transparent liquid; accurately determined the relationship among temperature, growth rate, and tip shape; and established the importance of long-range interactions between dendrites. Data from those experiments are widely used by scientists who work to improve the physical understanding and mathematical models of pattern formation in solidification.

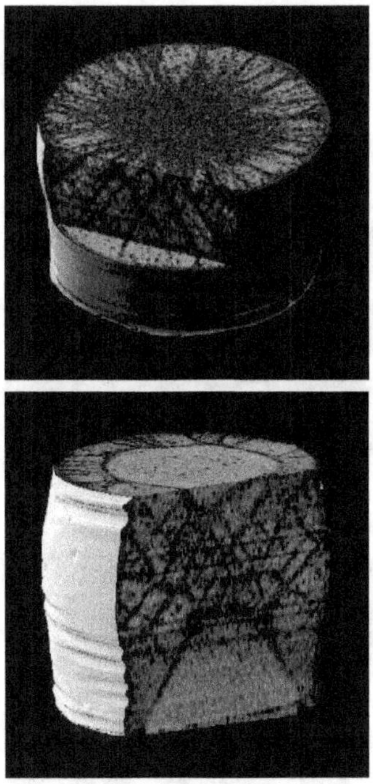

These two samples show fracture patterns in sand at two different low confining pressures. The confining pressure is an equal, all-sided pressure that is experienced, for example, by rock at some depth in the Earth. Very low confining pressures are not obtainable on Earth due to gravity.

We learned the underlying physics of freckle formation (a defect in the formation alloy that changes its physical characteristics) from early results of materials research. It was shown that convection was directly responsible for the formation of freckles, and that rotating the sample can suppress freckle formation.

The contributions of the materials effort led to many innovations in crystal growth and solidification technology, including the use of magnetic fields, rotating crucibles, and temperature-control techniques. In addition, the analytical tools developed to understand the results of space experiments were a major contribution to the use of computational modeling as a tool for growth process control in manufacturing.

Fluid Behavior Changes in Space

Many people connect the concept of liquids in space with the familiar image of an astronaut playing with a wiggly sphere of orange juice. And, yes, liquids in space are fun and surprising. But, because many space systems that use liquids—from propulsion and thermal management to life support—involve aspects of spaceflight where surprises are not a very good idea, understanding the behavior of liquids in space became a well-established branch of fluid engineering.

The design of space vehicles—fluid and thermal management systems, in particular—made low gravity a practical concern for engineers. Decades before the space program began, airplane designers had to create fuel systems that would perform even if the plane were upside down or in free fall. Rocket and satellite designers, however, had to create systems that would operate without the friendly hand of gravity to put liquids at the bottom of a tank, let bubbles rise to the top of a liquid, and cool hot electronic equipment with the natural flow of rising hot air.

Without gravity, liquid fuel distributes itself in a way that minimizes its total free energy. For most fuels, liquid at the surface of the tank has a lower energy than the liquid itself, which means the fuel spreads out to wet the solid surfaces inside the tank. When bubbles are created in a fluid in space, in the absence of other factors the bubbles will sit where they are. Buoyancy, which causes bubbles to rise in liquids or hot air to rise around a flame, is the result of gravity producing a force proportional to density within a fluid. Many aspects of a vehicle design, such as its mechanical structure, are driven primarily by the large forces experienced during launch. For fluid and thermal systems, low gravity becomes a design driver.

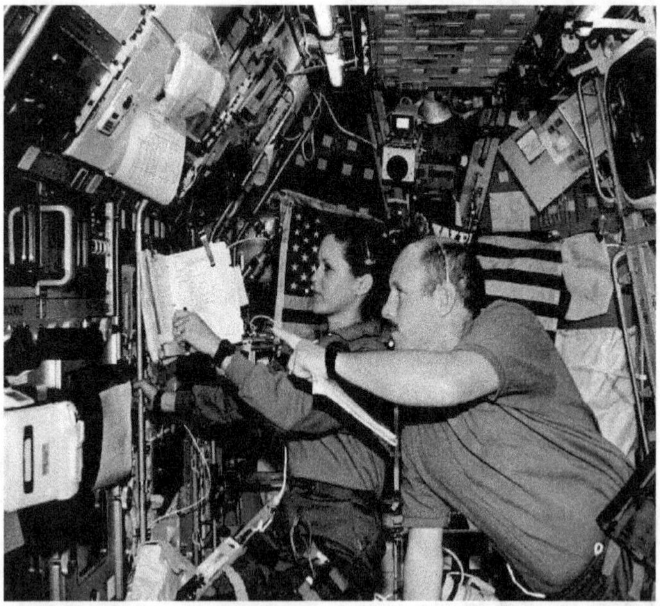

Astronauts Kathryn Thornton and Kenneth Bowersox observe a liquid drop's activity at the Drop Physics Module in the science module aboard the Earth-orbiting Space Shuttle Columbia (STS-73 [1995]). The two were joined by three other NASA astronauts and two guest researchers for almost 16 days of in-orbit research in support of the US Microgravity Laboratory mission.

A great deal of low gravity research performed in the 1960s focused on making liquid systems in space reliable. Low gravity experiments were performed by dropping the experiment from a tower or down a deep shaft or flying it in an aircraft on a parabolic trajectory that allowed the experiment to fall freely for about 20 seconds. The experiments possible in drop shafts and aircraft didn't allow enough time to test many technologies. As a result, engineers weren't sure how some familiar technologies would work in the space environment.

Low-gravity fluid engineering began with Apollo-era research focused on controlling liquid fuels; i.e., making sure liquid fuels didn't float around inside their tanks like an astronaut's orange juice. NASA performed most of this research in drop facilities, where experiments conducted in up to 5 seconds of free fall allowed basic ideas about fluid management to be investigated.

The arrival of the Space Shuttle opened the window for experiment duration from seconds to days and inspired the imaginations of scientists and engineers to explore new areas.

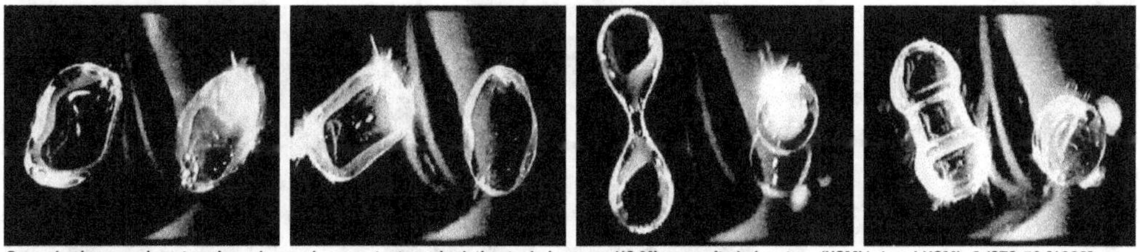

Drop physics experiments using advanced noncontact manipulating techniques on US Microgravity Laboratory (USML)-1 and USML-2 (STS-50 [1992] and STS-65 [1994], respectively) helped scientists understand the complex physical mechanisms underlying the seemingly simple processes of droplet shaping, splitting, and fusion.

The source of engineering problems with liquids in space is the partially filled container, or the gas-liquid interface. Without gravity, surface tension—the force that pulls a liquid drop into a sphere—together with the attraction of the liquid to the solid surfaces of the container determine the shape that a liquid will assume in a partially filled container.

To understand the unique behavior of liquids in space, researchers needed to look at the critical pieces of information in the liquid boundaries. Fluid physics experiments in the Spacelab Program, such as the Surface Tension-Driven Convection Experiment developed for Professor Simon Ostrach of Case Western Reserve University, Cleveland, Ohio, and the Drop Physics Module developed for Professors Robert Apfel of Yale University, New Haven, Connecticut, and Taylor Wang of Vanderbilt University, Nashville, Tennessee, led a wave of research into the properties of liquid interfaces and their roles in fluid motions. This research contributed to advances in other areas, such as microfluidics, in which the properties of liquid interfaces are important.

The shuttle enabled researchers to explore many new kinds of fluid behavior. Two examples out of many include: the Mechanics of Granular Materials experiment, and the Geophysical Fluid Flow Cell experiment. The Mechanics of Granular Materials experiment, developed by Professor Stein Sture at the University of Colorado, Boulder, examined the fluid-like behavior of loosely compressed soils and helped in understanding when and how, in situations like earthquakes, soils abruptly lose their load-bearing capability. Data from the experiment will also help engineers predict the performance of soils in future habitat foundations and roads on the moon, Mars, and other extraterrestrial

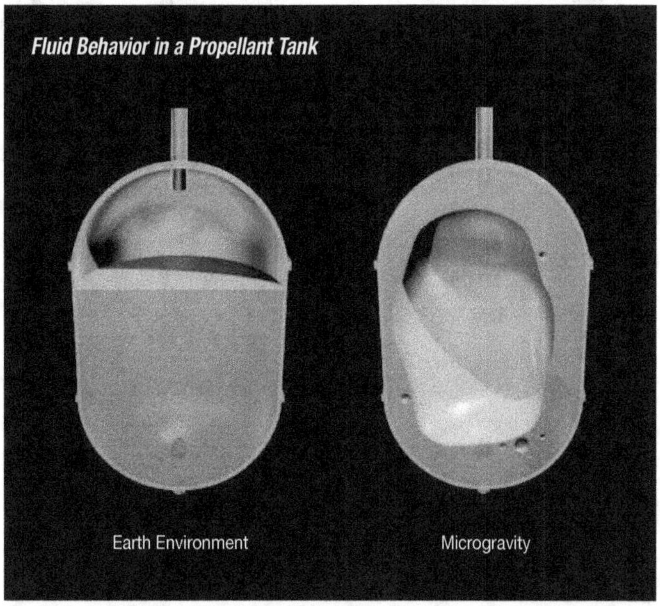

One of the earliest concerns about fluid behavior in microgravity was the management of propellants in spacecraft tanks as they orbited the Earth. On the ground, gravity pulls a fluid to a bottom of a tank (Earth environment, left). In orbit, fluid behavior depends on surface tension, viscosity, wetting effects with the container wall, and other factors. In some cases, a propellant can wet a tank and leave large gas bubbles in the center (microgravity, right). Similar problems can affect much smaller experiments using fluids in small spaces.

applications where the weight of the soil is much lower than on Earth.

The Geophysical Fluid Flow Cell experiment, developed by Professor John Hart at the University of Colorado, Boulder, used the microgravity environment to create a unique model of the internal motion in stars and gaseous planets, with a device that used an electric field to simulate gravity in a spherical geometry. The Geophysical Fluid Flow Cell flew on Spacelab 3 (1985), and again on US Microgravity Laboratory-2 (1995). Results from the experiment, which first appeared on the cover of *Science* magazine in 1986, provided many basic insights into the characteristics of gas flows in stars and gaseous planets. Hart and his colleagues were able to reproduce many of the flow patterns observed in gaseous planets under controlled and quantified conditions inside the Geophysical Fluid Flow Cell, thus providing a basis for analysis and physical interpretation of some of the distinctive dynamic features stars and gaseous planets.

Combustion in Microgravity

What Is Fire Like in Microgravity?

The crew of a spacecraft has few options in the event of a major fire. Fortunately, fires in spacecraft are rare; however, because both rescue and escape are uncertain possibilities at best, fire prevention, detection, and suppression continue to be an ongoing focus of NASA research even after more than 30 years of study.

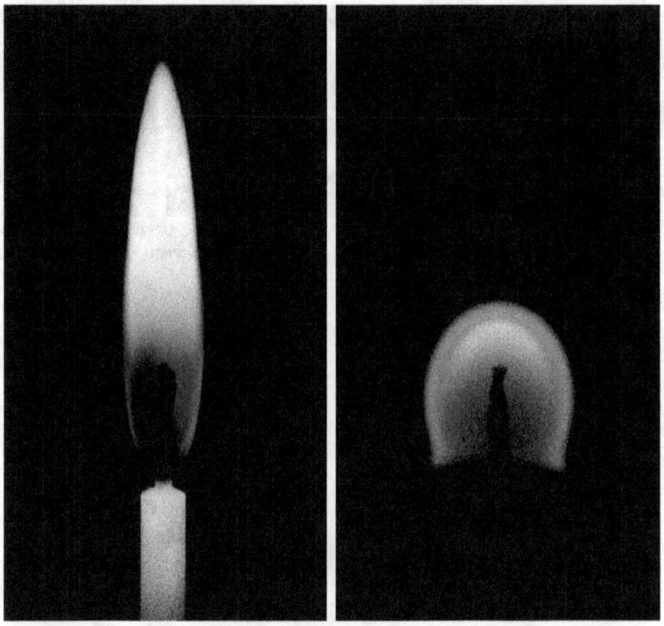

This demonstrates the difference between flames on Earth (left) and in microgravity (right). The flame in microgravity is different because there is no upward buoyant force causing air to rise, so flames in space produce no buoyant convective flow that carry them upward.

In the near-absence of gravity, fires ignite and spread differently than they do on Earth. Fires produce different combustion products, so experiments in space are essential to creating a science-based fire safety program. Research aboard the shuttle gave scientists an understanding of ignition, propagation, and suppression of fires in space. NASA is using the pioneering results of shuttle-era research to design a new generation of experiments for the ISS to help engineers design safer vehicles and better fire-suppression systems in the future.

The biggest difference between space- and Earth-based fires is that on Earth, the heat released by combustion will cause a vigorous motion of the neighboring atmosphere as warm gas, less dense than the gas around it, rises due to its buoyancy under gravity. The upward buoyant flow draws surrounding air into the fire, increasing reaction rates and usually increasing the intensity of the fire. In space, buoyancy is negligible. Fire safety specialists must take into account the effects of cooling and ventilating airflows, which can significantly accelerate fires. Under "typical" conditions, however, combustion in space is slower than on Earth and is less complete. Soot particles are larger in space because particles spend more time growing in the fuel-rich reaction zone. As a result,

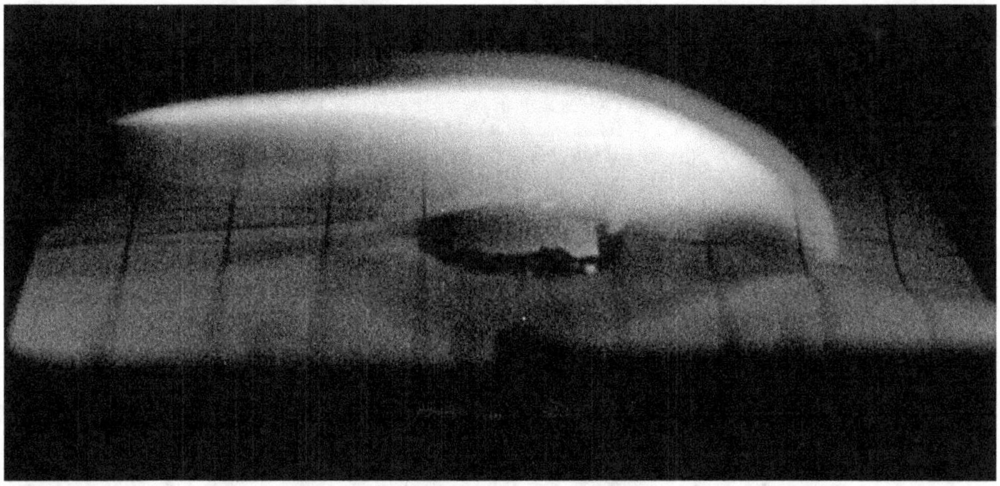

NASA fire safety experiments examined the effects of weak cabin airflows on fires. Here, a piece of paper burns in a flow like those used to cool avionics systems in space. NASA research showed that weak flows can have a strong influence on material flammability.

fire detectors in space need to be more sensitive to larger smoke particles than do fire detectors on Earth.

The experiments of David Urban of the NASA Glenn Research Center and his colleagues, included on the US Microgravity Payload-3 mission (1996), examined particulate-forming combustion in microgravity and observed that the larger particulates produced in microgravity were often not detected by the sensor technology employed in detectors deployed on the shuttle, even though the detectors worked reliably on Earth. An alternate technology more sensitive to large particulates provided superior detection. This technology, which uses scattering of a laser beam by particles in the airstream, is now deployed aboard the ISS.

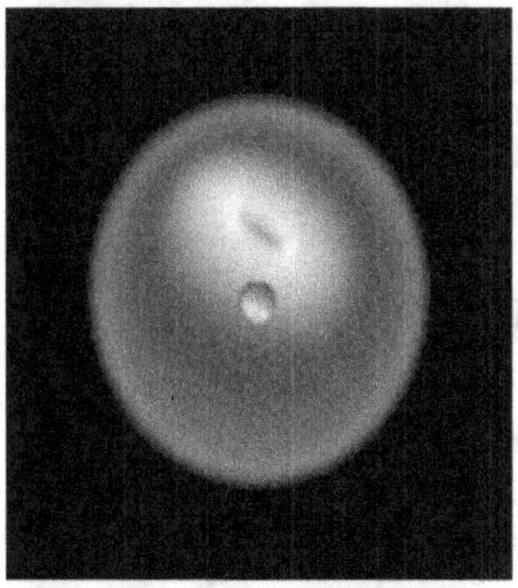

In nearly perfect weightlessness, an ethanol droplet on the Microgravity Science Laboratory-1 mission in 1997 burns with a spherical flame.

Combustion of Fuels for Power

Beyond its initial motivation, combustion research on the shuttle also helped scientists better understand the basic processes of burning hydrocarbon fuels that according to the US Department of Energy provide the US economy with 85% of its energy. Research by Forman Williams of the University of California, San Diego, and Fred Dryer of Princeton University, New Jersey, and their students on the burning of fuel drops has been used by both General Electric (Fairfield, Connecticut) and Pratt & Whitney (East Hartford, Connecticut) to improve the jet engines they manufacture. Droplet combustion experiments in space produced well-controlled data that allowed Williams and Dryer to validate a comprehensive model for liquid fuel combustion. This model was integrated into the simulations that engine manufacturers use to optimize designs. Another experiment, led by Paul Ronney of the University of Southern California, Los Angeles, used microgravity to study the weakest flames ever created—100 times weaker than a birthday candle. Data on how combustion reactions behave near the limits of flammability were used to help design efficient hydrogen-burning engines that may eventually meet the need for clean transportation technologies.

Commercial Ventures Take Flight

Industry Access to Space Shuttle-inspired Innovation

NASA's charter included "seek and encourage ... the fullest commercial use of space." Acting in that direction, NASA promoted the Space Shuttle during the 1970s as a platform for industry.

Private industry is in business to provide goods and services for a financial return. Innovation is important. Microgravity—a physical environment that was new to industry at the time—proved to be intriguing. High-efficiency processing and free-floating containerless manipulation and shaping of materials could become reality with an absence of convection, buoyancy, sedimentation, and density differentiation. Highly purified biological separations, new combinations and structures of materials with valuable properties, and contamination-free solidifications prepared in orbit and returned to Earth became industry objectives for prospective space processing research.

In 1985, NASA and the National Bureau of Standards were responsible for the first sale of a product created in space. Designated "Standard Research Material 1960," this product was highly uniform polystyrene latex microspheres (specifically, sizes of 10 and 30 micrometers mean diameter) used in the calibration of scientific and medical instruments. Dozens of companies purchased "space beads" for $350 per batch. This milestone came from an in-space investigation that produced both immediate science and an application.

Charles Walker

Payload specialist on STS-41D (1984), STS-51D (1985), and STS-61B (1985).

Charles Walker, payload specialist, works at the commercial Continuous Flow Electrophoresis System on STS-61B.

"As a corporate research engineer I had dreamed of building an industry in space. Business conducted in orbit for earthly benefit would be important. The Space Shuttle could begin that revolution.

"The first industry-government joint endeavor agreement, negotiated in 1979 between NASA and the McDonnell Douglas Astronautics Company, my employer, would facilitate space-enabled product research and development among different industrial sectors. It also presented an opportunity for me to realize that personal dream.

"NASA's astronauts had already successfully conducted limited company proprietary and public NASA research protocols during four flights with McDonnell Douglas' electrophoresis bioseparation equipment. Then NASA allowed one of our researchers to continue the work in person—exceedingly rare among researchers, and the first for industry.

"As the company's noncareer, non-NASA astronaut candidate, I had to pass the same medical and psychological screening as NASA's own. Training mixed in with my continuing laboratory work meant a frenzied year. Preparations for flight were exhilarating but they weren't free. McDonnell Douglas paid NASA for my flights as a payload specialist astronaut.

"Working with NASA and its contractor personnel was extraordinarily rewarding. I conducted successively more advanced applied commercial research and development as a crew member on board three shuttle missions over a 16-month period. It seemed the revolution had begun.

"I'm sorry to see these first-hand opportunities for applied research recede into history. Spaceflight is a unique, almost magical, laboratory environment. Disciplined research in microgravity can change human science and industry as surely as humanity's ancient experiences in the control of heat, pressure, and material composition."

For-profit businesses vary in their need for scientific research. Companies often prioritize the application (product) as more important than its scientific basis. For them, reliable, practical, and cost-effective process knowledge is sufficient to create marketable products. But, if convinced that research can add value, companies will seek it. Various industries looked at the shuttle as an applied science and technology laboratory and, perhaps, even a platform for space-based product production. Industry found that production was not especially feasible in small spacecraft such as the shuttle, but they were successful with scientific-technology advancements.

McDonnell Douglas' space-based research and development section was the first to fly on seven missions, and these missions took place from 1982 to 1985. The electrophoresis applications work was technically a success. It improved bio-separations over Earth gravitational force processing. For example, when a cell-cultured human hormone erythropoietin (an anemia therapy) was to be purified 100 times better than ground-based separations, a 223 times improvement was obtained. Protein product throughput per unit of time also improved 700 times. After the Challenger accident (Space Transportation System [STS]-51L) in 1986, access to space for commercial efforts was severely restricted, thus ending the business venture. The demonstration of possibilities, together with McDonnell Douglas' investments in ground-based cell culturing and assaying, made for the effort's enduring advances.

In 2009, Astrogenetix (Austin, Texas)—a subsidiary of Spacehab/Astrotech (Austin, Texas)—was organized to commercialize biotechnology products processed in microgravity. The company developed a proprietary means of assaying disease-related biomarkers through microgravity processing. This research objective was aimed at shortening and guiding drug development on Earth. From five rapid, shuttle-based flight opportunities (over a 15-month period), the company discovered a candidate for a salmonella vaccine. Even as Astrogenetix prepared to file an investigational new drug application with the US Food and Drug Administration, it was researching candidates for a methicillin-resistant Staphylococcus aureus vaccine. The company conducted this later work in microgravity on board the shuttle's final flights. Looking to the future, Astrogenetix is among the first commercial firms with an agreement from NASA for use of the International Space Station (ISS) national laboratory.

In the materials area, Paragon Vision Sciences (Mesa, Arizona) developed new contact lens polymers. During three flight experiments, the company looked into the effects of gravity-driven convection on long molecular chain formation, resulting in an improved ground-based process and Paragon's proprietary HDS® Technology materials product line.

Shuttle-based investigations amount to fewer than 6 months of laboratory time. Yet there have been significant outcomes across multiple disciplines. The national laboratory capability at the ISS seemingly offers a tremendous future of returns.

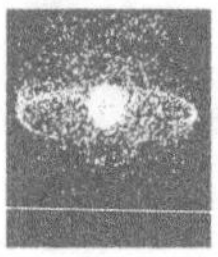

Space Environments

Introduction
Kamlesh Lulla

Orbital Debris
Eric Christiansen
Kamlesh Lulla

Space Radiation and Space Weather
Steve Johnson
Neal Zapp
Kamlesh Lulla

Where does "space" really begin?

The Earth's atmosphere begins to thin out as we ascend to higher altitudes. This thinning continues in the near-space environment. International aeronautics standards use the altitude of 100 km (62 miles) to mark the beginning of the space environment and the end of Earth's atmosphere. The Space Shuttle was flown at various altitudes from 185 to 593 km (100 to 320 nautical miles) during the Hubble Space Telescope missions, but it generally flew at an altitude of around 306 km (165 nautical miles) in what is commonly called low-Earth orbit.

What is environment like in space? Travel in space environment exposes vehicles and their occupants to: vacuum-like conditions, very low or zero gravity, high solar illumination levels, cosmic rays or radiation, natural micrometeoroid particles or fragments, and human-made debris—called "orbital debris"—from space missions. Thus, the space environment posed distinct challenges for both the shuttle flight crew and hardware.

You may be surprised to learn that, on average, one human-made object falls back to Earth from space each day. The good news is that most objects are small fragments that usually burn up as they reenter Earth's atmosphere. Those that survive re-entry likely land in water or in large, sparsely populated regions such as the Australian Outback or the Canadian Tundra. Of course, not all objects fall to Earth. Thousands remain in orbit for a considerable duration, giving rise to a population of "space junk" or "debris" that affected the shuttle and its operations.

Space radiation is also an inseparable component of the space environment. Radiation exposure is unavoidable and it affects space travelers, hardware, and operations. NASA conducted operations and experiments on the shuttle to characterize the radiation environment, document astronaut exposures, and find ways to minimize this exposure to protect both the humans and the hardware.

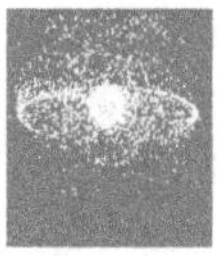

What Goes Up in Space May Not Always Return to Earth

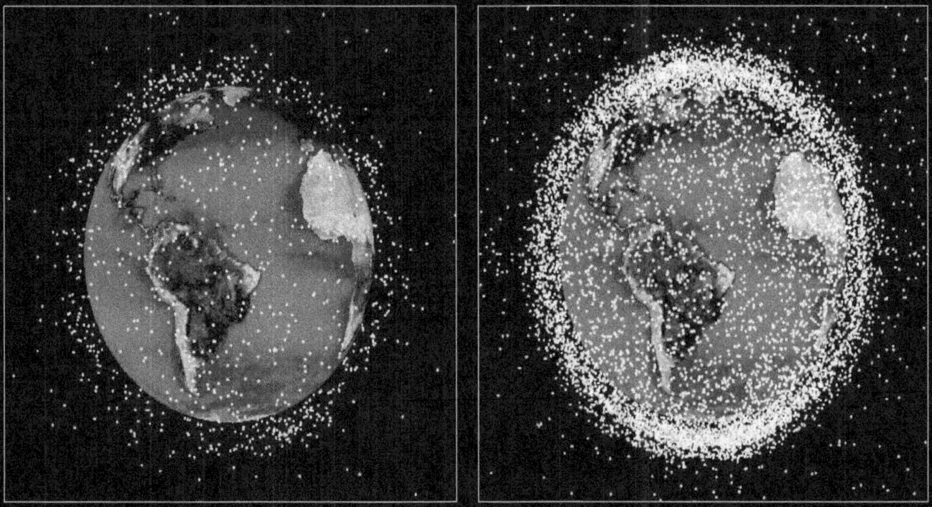

Growth of orbital debris: Each dot represents a debris object that is greater than 10 cm (4 in.) in diameter and has been cataloged. Comparison of 1970 (left) and 2010 maps shows clear evidence of rapid growth in debris population over the past 40 years.

What is orbital debris?

You have probably heard of human-made "space junk" or "space debris pollution." Since the dawn of space activities initiated with the launch of Sputnik in 1957, many nations have launched satellites, probes, and spacecraft into space. Some of these objects have come back to Earth and burned up in the atmosphere on re-entry. Many others remained in orbit and disintegrated into pieces that circle the Earth at around 27,000 kph (17,000 mph) in low-Earth orbit. This is orbital debris. It can be as small as a flake of paint from a spacecraft or as large as a school bus, and can impact operational spacecraft at very high impact speeds (up to 55,000 kph [34,000 mph). This space junk is of concern to all spacefaring nations.

What is a micrometeoroid?

Micrometeoroids are common, small pieces or fragments of rock or metal in orbit about the sun. These fragments have origins in the solar system and were generated from asteroids or comets, or left over from the birth of the solar system (i.e., they are natural debris). Micrometeoroids could pose a significant threat to space missions. They can impact at a higher velocity than orbital debris, and even the tiniest pieces can significantly damage spacecraft.

How much orbital debris is present, and how is it monitored?

Experts report more than 21,000 pieces of debris larger than 10 cm (4 in.) in diameter in orbit around Earth. The number of debris particles between 1 cm (0.4 in.) and 10 cm (4 in.) in diameter is estimated to be around 500,000. Experts think the number of particles smaller than 1 cm (0.4 in.) in size exceeds tens of millions.

The US Space Surveillance Network tracks large orbital debris (>10 cm [4 in.]) routinely. It uses ground-based radars to observe objects as small as 3 mm (0.12 in.) and provides a basis for a statistical estimate of its numbers. Orbital debris 1 mm (0.04 in.) in diameter and smaller is determined by examining impact features on the surfaces of returned spacecraft, such as the Orbiter.

How has the debris grown?

Debris population in space has grown as more and more space missions are launched. So, what are we doing about orbital debris?

In 1995, NASA became the world's first space agency to develop a comprehensive set of guidelines for mitigation of orbital debris. Since then, other countries have joined in the effort. NASA is part of the Inter-Agency Space Debris Coordination Committee consisting of 10 nations and the European Space Agency whose purpose includes identifying cooperative activities to mitigate orbital debris. This includes stimulation for engineering/research based on solutions.

Orbital Debris

You have probably seen video clips of US Airways Flight 1549 glide into the Hudson River for landing in 2009 after a flock of geese disabled its engines. This incident highlighted the dangers of the local aviation environment on Earth. In space, while no geese posed a threat, fast-traveling debris consisting of fragments of spacecrafts or tiny pieces of meteoroids posed potential dangers to the shuttle.

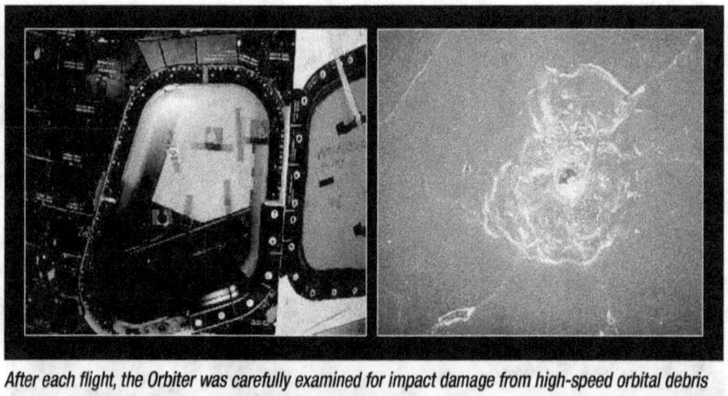

After each flight, the Orbiter was carefully examined for impact damage from high-speed orbital debris and meteoroids. Each of the shuttle windows were inspected with microscopes, which typically revealed several minor impacts (these images from STS-97, 2000). On average, one to two window panes were replaced after each mission due to these impacts or other contamination.

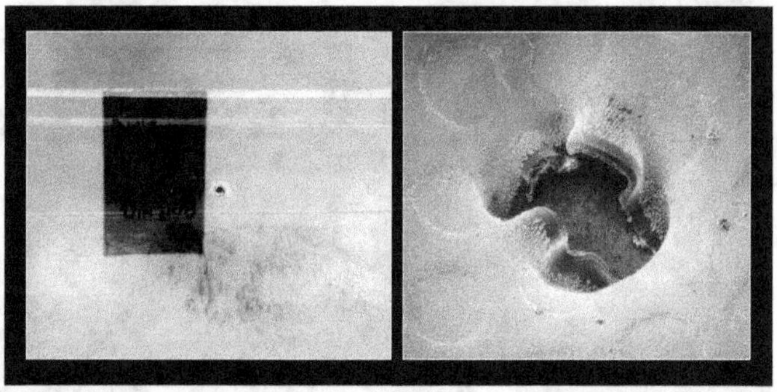

The large aluminum radiators attached to the inside of the cargo bay doors were examined for possible punctures (image on left from STS-115, 2006). Close-up inspections sometimes revealed complete penetrations of the radiator and debris from the impactor (magnified image on right from STS-90, 1998).

Have you ever wondered what a postflight inspection of the Orbiter might have revealed? During postflight assessments, NASA engineers found over 1,000 hits caused by micrometeoroids and orbital debris that had occurred over the course of several years.

Why is it important to be concerned about human-made debris or natural meteoroid particles? The damages caused by debris impacts required shuttle windows to be replaced, wing leading edge to be repaired, and payload bay radiator panels and connector lines to be refurbished. Thus, the mitigation of such impacts became a high priority at NASA in its efforts to safeguard the spacecraft and astronaut crews and conduct mission operations without a glitch.

Was the Space Shuttle Damaged by Debris?

The shuttle was damaged by micrometeoroid and orbital debris, but the extent of damages varied with each flight. Postflight inspections revealed numerous debris impact damages requiring repairs to the vehicle. For example, NASA scrapped and replaced more than 100 windows, repaired hundreds of small sites on the radiator, and refurbished pits from impacts on the wing leading edge.

Notable Damage

The Space Transportation System (STS)-50 mission in 1992 spent nearly 10 days in a payload-bay-forward attitude (to reduce exposure to debris) during a 16-day mission. Postflight inspections revealed a crater measuring 0.57 mm (0.02 in.) in depth with a diameter of 7.2 mm (0.28 in.) by 6.8 mm (0.27 in.) in the right-hand forward window. The crater was caused by a piece of titanium-rich orbital debris. Because of the damage, the window had to be removed and replaced. The STS-50 mission experienced a large increase in payload bay door radiator impacts when compared to previous missions. The largest radiator impact on STS-50 occurred on the left-hand forward panel, producing a hole measuring 3.8 mm (0.15 in.) in diameter in the thermal control tape, and a hole measuring 1.1 mm (0.04 in.) in diameter in the face sheet. This impact was due to a piece of paint.

The 16-day STS-73 mission in 1995 carried a US Microgravity Module Spacelab module and an Extended Duration Orbiter cryogenics pallet in

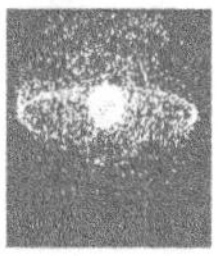

the payload bay. The vehicle was oriented with its port wing into the velocity vector for 13 days of the mission, and the port payload door was kept partially closed to protect the two payloads from debris impacts. Postflight inspections revealed a crater in the outside surface of the port payload bay door. The crater measured 17 mm (0.67 in.) in diameter and 6 mm (0.24 in.) deep. NASA found a 1.2-mm- (0.047 in.)-long fragment of a circuit board in the crater as well as many smaller pieces of circuit board and solder. Thus, a small piece of orbital debris (circuit board/solder) caused this particular impact damage.

After the STS-86 mission in 1997, NASA observed several significant debris impacts on the left-hand radiator interconnect lines. The aluminum tubes carried Freon® coolant between the Thermal Control System radiator panels. The largest impact, on the external line at a panel, penetrated just over halfway through the 0.9-mm- (0.035-in.)-thick coolant tube wall. A scanning electron microscope equipped with x-ray spectrometers examined samples of the damage. NASA decided the damage was likely due to impact by a small orbital debris particle composed of stainless steel. Additional inspections of the interior surface of the coolant tube wall determined that a small piece of the interior wall was removed directly opposite the impact crater on the exterior surface. This particular impact damage feature, called "detached spall," indicated that a complete penetration of the tube was about to happen. A tube leak would likely have resulted in a mission abort and possible loss of mission objectives.

After this mission, all external radiator lines on the Orbiter vehicles (flexible and hard lines) were toughened by installing a double-layer beta-cloth sleeve around the line. This sleeve was sewn together such that there was a gap between the two layers and a gap between the sleeve and coolant line that created a bumper-shield effect. Ground-based impact tests revealed that more effective protection from hypervelocity meteoroid and debris impacts could be obtained using several relatively thin layers (or "bumpers") that stood off from the item being protected.

Since the STS-86 mission, NASA has found more micrometeoroid and orbital debris impacts on the shuttle windows, radiators, and wing leading edge.

The Scientific Basis for Mitigating Orbital Debris Impact—How NASA Protected the Space Shuttle

NASA's active science and engineering program provided the agency with an understanding of orbital debris and its impact on the shuttle. Engineers implemented several techniques and changes to vehicle hardware design and operations to safeguard the shuttle from micrometeoroid and orbital debris impacts based on the scientific efforts discussed here.

NASA performed thousands of impact tests using high-velocity objects on representative samples of shuttle Thermal Protection System materials, extravehicular mobility unit materials, and other spacecraft components to determine impact parameters at the failure limits of the various subsystems. Engineers used test results to establish and improve "ballistic limit" equations that were programmed in the computer code tool used to calculate impact risks to specific Orbiter surfaces. NASA completed an integrated mission assessment with this code, including the effect of the different orientations the vehicle flew during a mission for varying amounts of time. This tool provided the basis for showing compliance of each shuttle mission to debris protection requirements.

Risk Assessment Using Mathematical Models

NASA, supported by these impact tests, used a computer code called BUMPER to assess micrometeoroid and orbital debris risk. The space agency used these risk assessments to evaluate methods to reduce risk, such as determining the best way to fly the shuttle to reduce debris damage and how much risk was reduced if areas of the shuttle were hardened or toughened from such impacts.

Design Modifications of Shuttle Components

NASA made several modifications to the shuttle to increase micrometeoroid and orbital debris protection, thereby improving crew safety and mission success.

The space agency improved the wing leading edge internal Thermal Protection System by adding Nextel™ insulation blankets that increased the thermal margins of the panel's structural attachment to the wing spar. This change allowed more damage to the wing leading edge panels before over-temperature conditions were reached on the critical structure behind those panels.

Another improvement involved toughening the radiator coolant flow tubes. This was accomplished by installing aluminum doublers over the coolant tubes in the payload bay door radiators. Additional protection to the flow loops was made in the form of adding a double-beta-cloth wrap that was attached via Velcro® around radiator panel-interconnect flexible and hard lines (0.63-cm [0.25-in.] gaps were sewn into the beta-cloth wraps to improve hypervelocity impact protection).

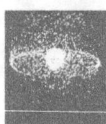

NASA added automatic isolation valves to each of the two thermal control flow loops on the vehicle to prevent excessive loss of coolant in the event of tube leak.

Operational Changes

Shuttle flight attitudes were identified (using BUMPER code) and flown whenever possible to reduce micrometeoroid and orbital debris risk. Impacts were quite directional. For the shuttle and the International Space Station (ISS), about 20 times more impacts would occur on the leading surfaces of the spacecraft (in the velocity direction) compared to the trailing surface and 200 times more impacts would occur on the leading surface compared to the Earth-facing surface (because the Earth provides shadowing). When the shuttle was docked to the ISS, the entire ISS-shuttle stack was yawed 180 degrees such that the ISS led and the shuttle trailed (i.e., the ISS was flying backward). This was done to protect sensitive surfaces on the belly of the shuttle from micrometeoroid and orbital debris impacts because the belly of the shuttle would be trailing when the ISS-shuttle stack completed the 180-degree yaw maneuver. The shuttle in free flight flew with tail forward and payload bay facing earthward whenever possible to again provide the greatest protection while conducting the mission.

An operational step to reduce micrometeoroid and orbital debris risk was made during the STS-73 mission, which flew predominately in a wing-forward, tail-to-Earth attitude. The Spacelab module, along with the Extended Duration Orbiter pallet containing high-pressure cryogenic oxygen and nitrogen, occupied the payload bay on this mission. To protect the payloads as well as reduce micrometeoroid and orbital debris risk to the radiators, the shuttle flew with the leading payload bay door nearly closed.

Another important step in reducing micrometeoroid and orbital debris risk for the shuttle was implemented with STS-114 (2005); this step included an inspection of vulnerable areas of the vehicle for damage. This inspection was performed late in the mission, just after undock from the ISS, using the Orbiter Boom Sensor System. The late inspection focused on the wing leading edge and nose cap of the Orbiter because those areas were relatively thin and sensitive to damage. If critical damage was found, the crew would perform a repair of the damage or would re-dock with the ISS and await a rescue mission to return to Earth.

On-orbit Damage Detection and Repair

With STS-114, NASA installed an on-orbit impact detection sensor system to detect impacts on the wing leading edge of the shuttle. The Wing Leading Edge Impact Detection System consisted of 132 single-axis accelerometers mounted along the length of the Orbiter's leading edge wing spars.

During launch, the accelerometers collected data at a rate of 20 kHz and stored these data on board for subsequent downlink to Mission Control. Within 6 to 8 hours of launch, summary files containing periodic subsamples of the data collected by each accelerometer were downlinked for analysis to find potential signatures of ascent damage. This analysis had to be completed within 24 to 48 hours of launch so the results could be used to schedule focused inspection using the Orbiter Boom Sensor System in orbit.

The Wing Leading Edge Impact Detection System was capable of detecting micrometeoroid and orbital debris impacts to the wing leading edge, although it was battery operated and did not continuously monitor for impacts. Rather, it was turned on during specific periods of the mission where the assessed risk was the highest.

Repair kits were developed to repair damages to the wing leading edge, nose cap, and Thermal Protection System tiles if damages didn't allow for safe return. Those repairs could be accomplished by the crew during an extravehicular activity.

Successfully Diminishing the Risk of Damage

Teams of NASA engineers and scientists worked diligently to enhance the safety of the Space Shuttle and the crew while in orbit by implementing threat mitigation techniques that included vehicle design change, on-orbit operational changes, and on-orbit detection and inspection. The design changes enhanced the survival ability of the wing leading edge and payload bay radiators.

Operational changes, such as flying low-risk flight attitudes, also improved crew safety and mission success. Inspection of high-risk areas

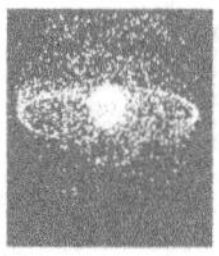

Kevin Chilton
General, US Air Force
United States Strategic Command/Joint Operations Command Center.
Pilot on STS-49 (1992) and STS-59 (1994).
Commander on STS-76 (1996).

The Need to Minimize Orbital Debris in Space

"Our Space Shuttle experiences gave us a deep appreciation and respect for the space environment—its vastness, its harshness, and its natural beauty. Hand in hand with this appreciation comes, in my view, a sense of stewardship for this domain we share, and will continue to share, with other countries and peoples. It's a realm over which no one has ownership, but for which all who traverse it are, in a sense, responsible.

"This imperative for responsibility became particularly poignant to me during one of my shuttle missions, when one day a crewmate noticed a disconcerting crack in the outer pane of the circular window on the side hatch. NASA scientists and engineers later determined the crack was caused by the high-speed impact of a miniscule piece of human-made debris. I'd prefer not to think what might have happened had it been something a bit larger. The event was a reminder to us that we were, in our fragile craft, mere travelers in a rather hazardous place of great velocities and hostile conditions. But, our collision with this other human-made object in space also made clear that we have a role in keeping the space environment as pristine as we can, and as we found it—if for nothing else, for the safety and freedom of space travels after ours.

"Later in my career, as Commander of U.S. Strategic Command, I saw this imperative for responsibility even more clearly in the aftermath of two significant debris-generating events: the January 2007 Chinese anti-satellite test, and the February 2009 collision between two satellites in low-Earth orbit. Both dramatically increased the debris count in low orbit and were wake-up calls for the imperative for more responsible behavior in the first case, and the need to better understand and to minimize—to the extent possible—the challenge of space debris in the latter. We've since taken steps to improve that understanding and to pursue debris mitigation, but there is still much more to be done.

"If we truly are to be good stewards of the space environment, we will need to make every reasonable effort to keep it habitable for both human and machine. This demands a deliberate effort to minimize orbital debris in the design, deployment, operation, and disposal of those spacecraft we send into orbit and beyond, as well as proactive efforts to mitigate the likelihood of spacecraft collisions with debris or other satellites in the future."

(e.g., wing leading edge and nose cap) along with repair were useful techniques pioneered by the Space Shuttle Program to further mitigate the risk of micrometeoroid and orbital debris impacts.

Summary

Experts estimate that, collectively, these implemented steps diminished the risk of damage from the orbital debris and micrometeoroids by a factor of 10 times or more.

Experience and knowledge gained from the shuttle orbital debris monitoring is valuable for current operations of the ISS and will have significant value as NASA develps future exploration concepts.

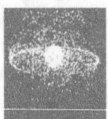

What Is Space Radiation?

Radiation may seem like a mystical, invisible force used in applications such as x-rays, nuclear power plants, and atomic bombs, and is the bread and butter of science fiction for creating mutant superheroes. The reality is that radiation is not so mysterious. Space radiation is composed of charged particles (90% protons) with high kinetic energies. Cellular damage results as a charged particle travels through the body, transferring its kinetic energy to the cellular molecules by stripping electrons and breaking molecular bonds.

Deoxyribonucleic acid (DNA) bonds may be broken if a charged particle travels through the cell nucleus. In fact, scientists can observe chromosomal damage in the white blood cells (lymphocytes) in astronauts by comparing postflight chromosome damage to the preflight chromosome condition. If the chromosomes do not correctly rejoin in the aftermath, stable abnormal DNA combinations can create long-term health implications for astronauts. Accumulated cellular damage may lead to cancer, cataracts, or other health effects that can develop at any time in life after exposure.

There are three sources of space radiation: galactic cosmic radiation, trapped radiation, and solar energetic particle events. Galactic cosmic radiation is composed of atomic nuclei, with no attached electrons, traveling with high velocity and therefore significant kinetic energy. In fact, the highest energy particles are traveling near the speed of light (relativistic). High energy galactic cosmic radiation is impossible to shield with any reasonable shield thickness. Most importantly, of the three sources, galactic cosmic radiation creates the biggest risk to astronaut health. Trapped radiation—Van Allen belts—is composed of protons and electrons trapped in the magnetic field. Trapped proton energy is much lower than galactic cosmic radiation energy and is easier to shield. Solar energetic particle events are composed primarily of large numbers of energetic protons emitted from the sun over the course of 1 to 2 days. Solar energetic particle energies generally reside between trapped proton and galactic cosmic radiation.

Radiation exposure in space is unavoidable and the potential for adverse health effects always remains. It is essential to understand the physics and biology of radiation interactions to measure and document astronaut exposures. It is equally important to conduct operations in such a way as to minimize crew exposures as much as practicable.

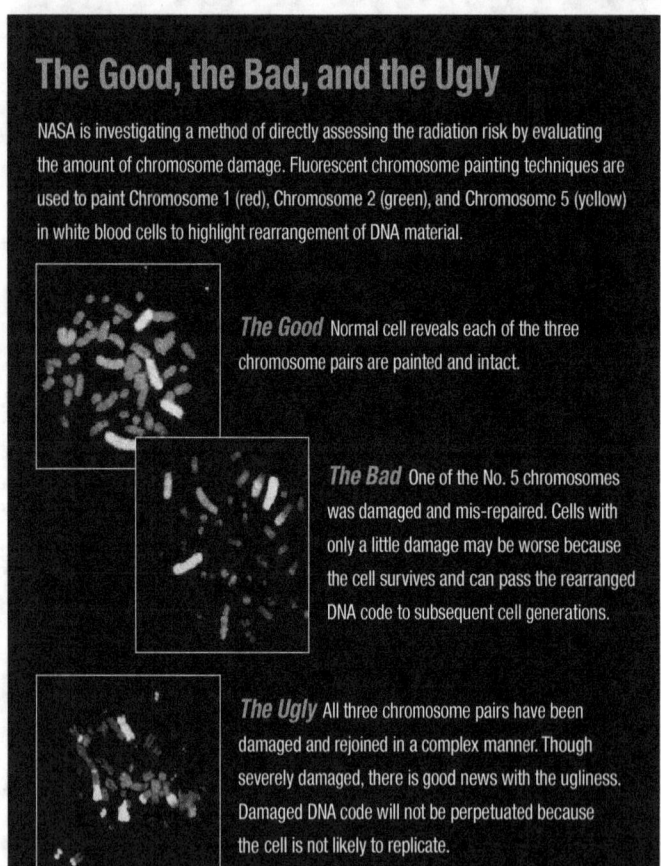

The Good, the Bad, and the Ugly

NASA is investigating a method of directly assessing the radiation risk by evaluating the amount of chromosome damage. Fluorescent chromosome painting techniques are used to paint Chromosome 1 (red), Chromosome 2 (green), and Chromosome 5 (yellow) in white blood cells to highlight rearrangement of DNA material.

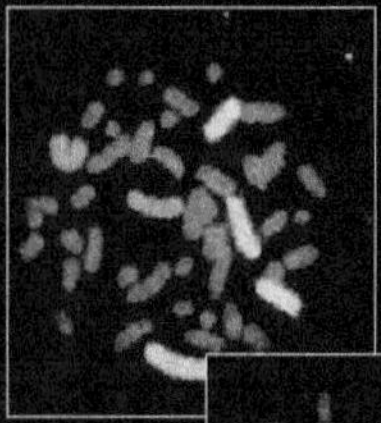

The Good Normal cell reveals each of the three chromosome pairs are painted and intact.

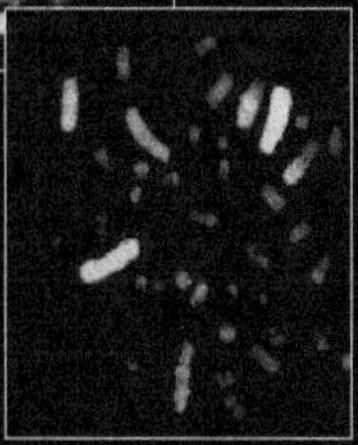

The Bad One of the No. 5 chromosomes was damaged and mis-repaired. Cells with only a little damage may be worse because the cell survives and can pass the rearranged DNA code to subsequent cell generations.

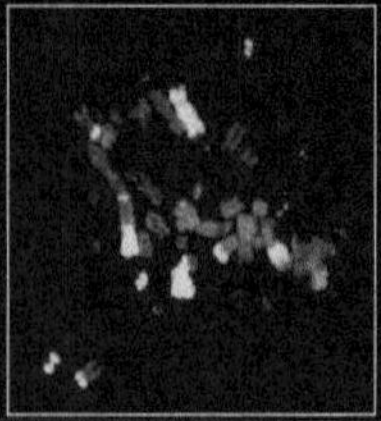

The Ugly All three chromosome pairs have been damaged and rejoined in a complex manner. Though severely damaged, there is good news with the ugliness. Damaged DNA code will not be perpetuated because the cell is not likely to replicate.

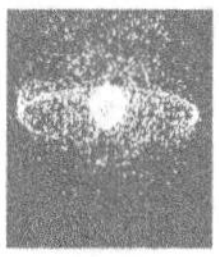

The Eyes Have It!

Could astronauts be more susceptible to developing cataracts from space radiation?

Researchers have recorded a higher-than-anticipated rate of cataracts in astronauts. Could the lens of the eye be more susceptible to developing cataracts from space radiation, especially as a result of exposure to biologically damaging heavy ion components of galactic cosmic radiation? Apollo astronauts were the first to report the effect known as "light flashes," which are generally attributed to heavy galactic cosmic radiation ions interacting within the eye. Astronauts on Skylab, shuttle, and the International Space Station have reported light flashes, but the reported frequency of flashes is greater during trajectories through higher latitudes in which radiation intensity is the highest.

Researchers used a pool of approximately 300 astronauts and divided them by their total mission doses. The "low-dose" group had exposures less than 800 mrem (8 mSv), and the "high-dose" group had greater exposures. The result: The high-dose group was more likely to develop cataracts than the low-dose group.

In addition, the astronauts were grouped by orbital inclination of their mission. The fraction of galactic cosmic radiation dose received by high-inclination missions (50 degrees) was greater than the galactic cosmic radiation dose fraction for low-inclination flights. This was due to the reduced magnetic shielding of radiation at higher latitudes encountered in trajectories of high-inclination flights; thus, these flights received more exposure to galactic cosmic radiation. This grouping allows for a comparison of astronauts with the same dose but with a different amount of exposure. As expected, the high-inclination group exhibited increased cataract incidence.

This research indicates that the risk of radiation-induced cataracts from heavy ion exposure is much higher than previously believed.

Radiation Intensity Inside the Shuttle

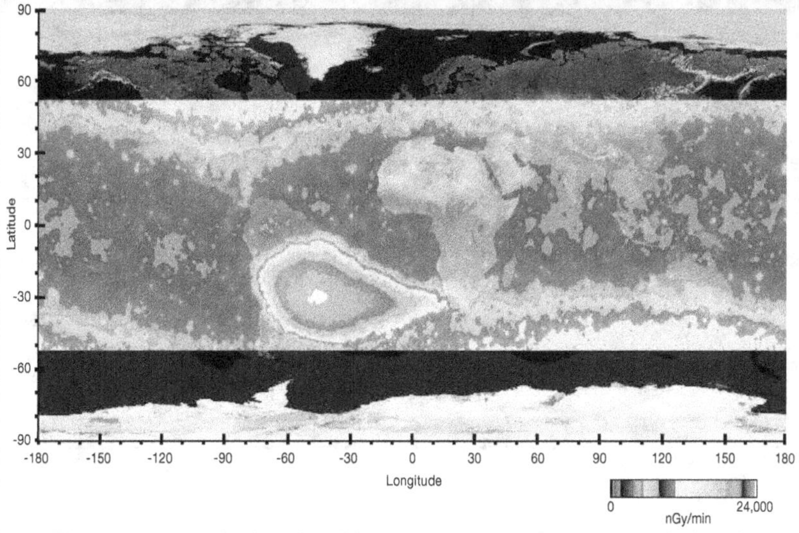

Radiation in low-Earth orbit is influenced by the magnetic field and follows a complex distribution pattern, as seen from measurements from STS-91 (1998). The prominent bull's-eye is a localized region of trapped radiation known as the South Atlantic Anomaly. The highest dose rates experienced by the shuttle occurred during transits through this region.

To manage the space radiation exposure risk to astronauts, NASA determined radiation exposure limits. Career exposure limits are established to limit the lifetime likelihood of adverse health effects from chronic exposure damage. Short-term exposure limits are established to ensure that astronauts do not receive acute exposures that might impair their ability to perform their duties.

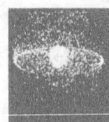

Using the Shuttle to Measure the Characteristics of Space Radiation

Scientists use two ways to measure radiation exposure to monitor astronaut health. The most frequent unit is the "dose" in units of rad or gray. Dose is solely a measure of the amount of energy deposited by the radiation. The second unit is "dose equivalent," which represents a level of biological effect of the radiation absorbed in the units of roentgen equivalents man (rem) or sievert (Sv). The amount of energy deposited by two different types of radiation may be the same, but the biological effect can differ vastly due to the damage density of different species of charged particles. A spectral weighting factor is used to adjust the dose into dose equivalent—the unit of interest when discussing astronaut exposures.

NASA developed an innovative instrument called the Tissue Equivalent Proportional Counter for experimentation on the shuttle to record the spectral distribution of measured radiation. Using the spectral information and the measured dose, an estimate of the dose equivalent could be made. Scientists used this instrument to conduct detailed assessments of the radiation environment surrounding the astronauts and their operational activities. Tissue Equivalent Proportional Counter measurements captured the dynamic changes in the radiation environment such as shift in locations and enhancements in trapped radiation. Far superior to the standard trapped radiation computer models, Tissue Equivalent Proportional Counter data became an effective tool for operational planning. Thus, mission planners were able to avoid additional exposure to the crew during extravehicular activities (EVAs).

Here is an example of why measurements are important: During a severe solar magnetic storm in March 1989, the electron population was enhanced by a factor of 50 relative to quiet conditions. Without these types of measurements, engineers would not have known about the belt enhancement and could not have considered this vital information in planning EVAs or evaluating astronaut radiation exposures.

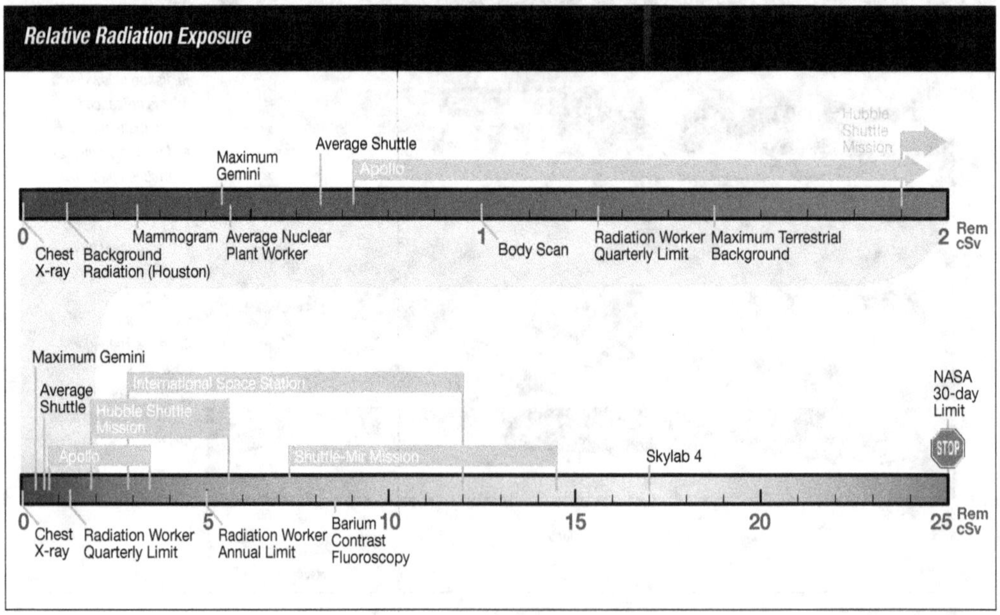

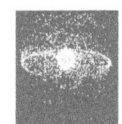

Space Shuttle Experiments Advance the Science of Radiation Shielding

How do the characteristics of radiation change as it travels through shielding or the body? What is the relative exposure to the internal organs compared to external exposure measurements? Answers to these questions assist in evaluating astronaut exposure risks.

Space Shuttle experiments, flown twice, used a set of multiple Tissue Equivalent Proportional Counters with detectors located at the center of polyethylene and aluminum spheres of different thicknesses to evaluate radiation source and transport/penetration models.

In polyethylene measurements, the galactic cosmic radiation dose equivalent was reduced by 40% with 12 cm (4.7 in.) of water. (Water is the international standard for shielding. Effectiveness of shielding is compared to this standard.) In contrast, aluminum shielding reduced the galactic cosmic radiation dose equivalent by a negligible amount using twice the polyethylene shield weight. The aluminum was significantly less effective and much heavier. Measurements of trapped radiation achieved a 70% reduction with 12 cm (4.7 in.) of polyethylene but required 50% more aluminum weight to achieve the same level of protection. Thus, polyethylene is a much better shield than aluminum for space radiation. These results contributed to improving radiation shielding on the International Space Station (ISS).

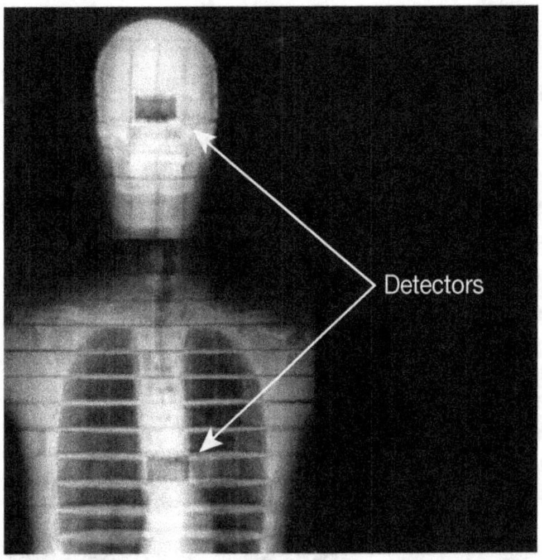

The phantom torso—a body phantom without arms or legs—was constructed out of skeletal bones and tissue-equivalent plastics to simulate internal organs. This x-ray image shows two locations of detectors as examples of multiple passive detectors.

Human Phantoms in Flight

The shuttle sphere shielding experiments were followed with an innovative way to measure radiation penetration. This innovation was called "body phantoms"—anthropomorphic density phantom (anatomical and tissue density) replicas of the human body. The first experiment used a head phantom; the second used a phantom torso along with the head phantom. The body phantom was constructed out of skeletal bones and tissue-equivalent plastics to simulate internal organs. The phantom torso was filled with 350 small holes, each containing multiple passive detectors. Five silicon detectors were placed at strategic organ sites.

Surprisingly, the phantom torso experiment revealed that the radiation penetration within the body did not decrease with depth as much as the models would indicate. Scientists found that the dose at blood-forming organs—some of the most radiosensitive sites—was 80% of the skin dose. The dose equivalent was nearly the same as the skin. The higher measured internal dose levels inferred more risk to internal organs for a given level of external radiation exposure.

The shuttle phantom torso experiment also provided an opportunity to make measurements of the neutron levels within the body. Neutrons are created as secondary products within the spacecraft. How does this happen? As an example, an energetic proton could hit the nucleus of an aluminum atom, causing the aluminum atom to break into several pieces that probably include neutrons. Neutrons have the potential to pose more biological risk to astronauts than do most charged particles. Also, neutrons are difficult to measure in space because charged particles interfere by producing many of the same interactions. The wide range of neutron energies increases the challenge because most neutron detectors only sample small energy ranges. Several experiments suggested that neutron-related risk is higher than anticipated.

Summary

The Space Shuttle experiments helped improve the characterization of the radiation environment that enabled scientists to better quantify the risk to astronaut health.

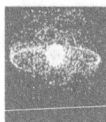

How did Space Weather Affect Astronauts and Shuttle Operations?

So what is space weather? The weather forecaster on the local television channel informs us of the trends and the degree of adverse weather to expect. Space weather is forecasting the trend and degree of changes in the space radiation environment. All dynamic changes in the radiation environment around Earth are driven by processes originating at the sun, such as flares and coronal mass ejections. Magnetic storms, shifts in the intensity and location of trapped radiation, and enhanced levels of solar protons—referred to as solar energetic particle events—are phenomena observed at Earth resulting from solar activity.

Astronaut health protection from space radiation during shuttle missions required an understanding of the structure, dynamics, and characteristics of the radiation environment. Radiation scientists who supported shuttle missions were as much "space weather forecasters" as they were radiation health physicists.

Space Shuttle Operations and Space Weather

During the course of the Space Shuttle Program, 20 flights (about 15%) were flown during enhanced solar proton conditions. In 1989, a period of maximum solar activity, all five flights encountered enhanced conditions from solar energetic particles; however, astronauts received little additional solar energetic particle dose due to a fortunate combination of orbital inclination, ground track timing, and event size. Almost all solar energetic particle dose exposures to any shuttle

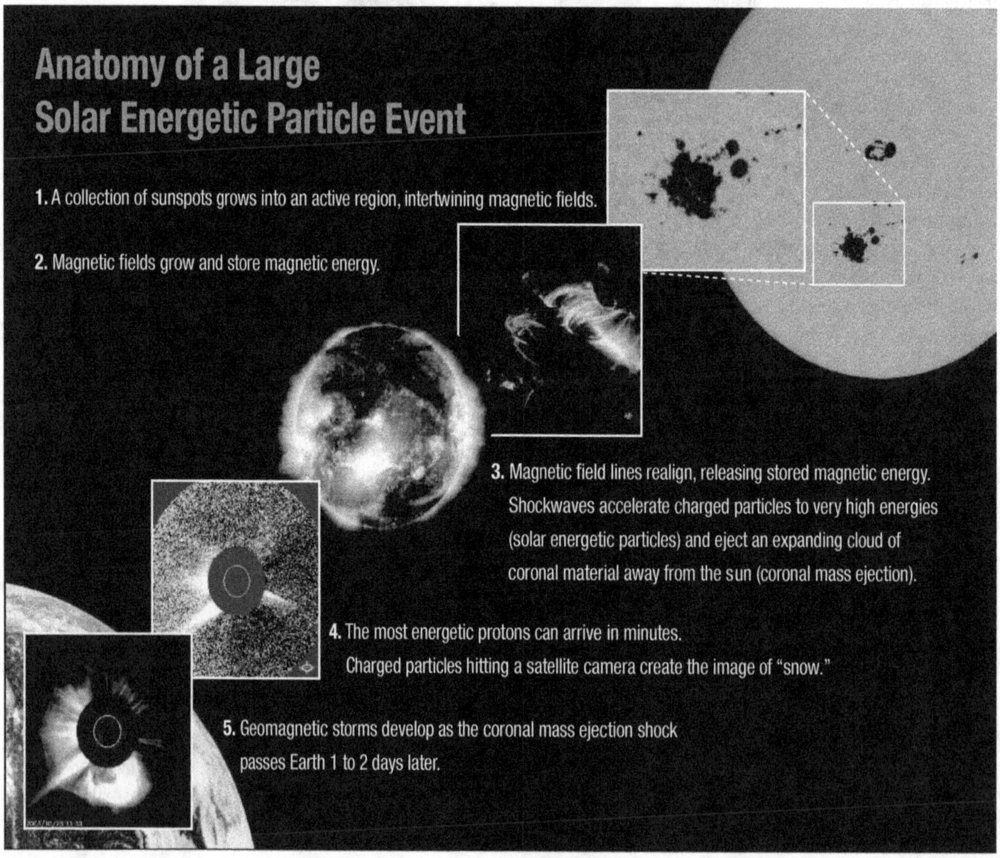
Anatomy of a Large Solar Energetic Particle Event
1. A collection of sunspots grows into an active region, intertwining magnetic fields.
2. Magnetic fields grow and store magnetic energy.
3. Magnetic field lines realign, releasing stored magnetic energy. Shockwaves accelerate charged particles to very high energies (solar energetic particles) and eject an expanding cloud of coronal material away from the sun (coronal mass ejection).
4. The most energetic protons can arrive in minutes. Charged particles hitting a satellite camera create the image of "snow."
5. Geomagnetic storms develop as the coronal mass ejection shock passes Earth 1 to 2 days later.

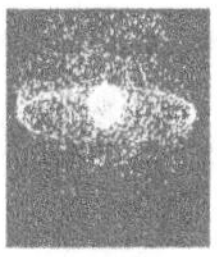

Space Radiation and the Shuttle Flying in Adverse Space Weather

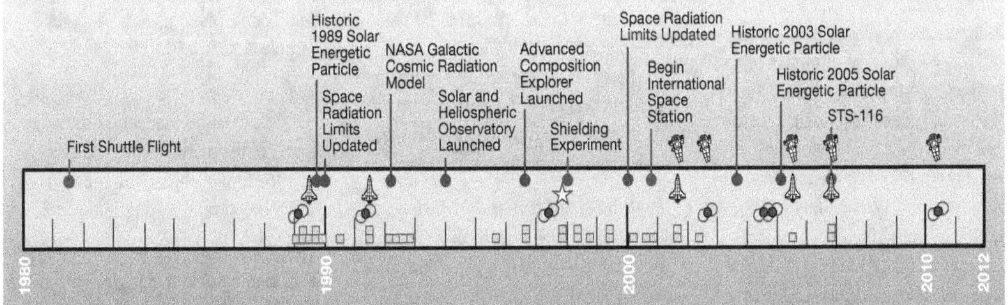

Several shuttle flights flew during solar energetic particle events but were not affected. Clusters of single event particles correspond to solar maximum (1980, 1990, 2001) periods of intense solar activity during the 11-year solar cycle.

- *Space Shuttle flight*
- *Radiation milestone*
- *Solar energetic particle event during a mission*
- *Two solar energetic particle events during a mission*
- *Temporary trapped radiation belt enhancement*
- *Internal solar energetic particle exposure during shuttle mission*
- *Extravehicular activity during solar energetic particle or belt enhancement*
- *Shuttle-Mir internal solar energetic particle exposure*

astronauts corresponded to less than an extra week of spaceflight daily exposure.

NASA conducted four EVAs supporting ISS construction during the course of solar energetic particle events. Astronauts received very little dose due to orbital timing and the magnitude of the events. The most interesting case occurred during Space Transportation System (STS)-116 in December 2006. NASA conducted this mission at a time when solar activity was at a minimum and solar energetic particle events were considered extremely unlikely. One event occurred just after the crew reentered the space station on the first EVA. A second event initiated while crew members were wrapping up the second EVA. Solar energetic particle exposures for both EVAs were negligible due to ground track timing;

Agencies Work Together to Assess Risks

The Space Weather Prediction Center at the National Oceanographic and Atmospheric Administration and the NASA Space Radiation Analysis Group worked together to support Space Shuttle flights. Space Weather Prediction Center forecasters reviewed available solar and environmental data to assess future environmental trends and provide a daily forecast. The NASA radiation operations group monitored environmental trends as well and reviewed the daily forecast with Space Weather Prediction Center personnel. The Space Radiation Analysis Group then interpreted the forecasted environmental trends and assessed potential impacts to the mission operations much in the way a local weather forecaster applies the National Weather Service forecast to the local area for the public to assess how the weather will impact its planned activities. During dynamic changes in the radiation environment, the radiation operations group tracked the progress of the event and advised the flight team when conditions warranted contingency procedures.

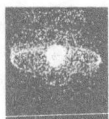

however, if the EVAs had been scheduled 3 hours later, the story would have been much different.

Inclination and ground track timing influence the degree of impact of a solar energetic particle. Flight inclination is the angle between the orbital plane and the equator. Inclination defined what ground track latitudes the orbit flew between. Low-inclination flights traveled between latitudes of 28.5 degrees to approximately 40 degrees. High-inclination flights flew between latitudes greater than 50 degrees. The geomagnetic field provided considerable protection to flight crews that flew low-inclination flights because the charged particles could not penetrate to the shuttle orbit. STS-34 flew in October 1989 during one of the historically largest solar energetic particle events but was unaffected by it because the geomagnetic field protected the low-inclination mission.

High-inclination missions, such as those to the ISS, flew through regions of virtually no geomagnetic protection. When the shuttle flew through those orbital regions during solar energetic particle events, the crew was exposed to solar energetic particle protons. During the remainder of the orbit, the crew was protected by the geomagnetic field and received no solar energetic particle dose.

Magnetic storms increase the size of the regions of no magnetic protection. A severe magnetic storm could have resulted in increased time spent in low protection, resulting in three times the exposure.

The good news is that high-risk time intervals of low geomagnetic protection can be accurately predicted, thus

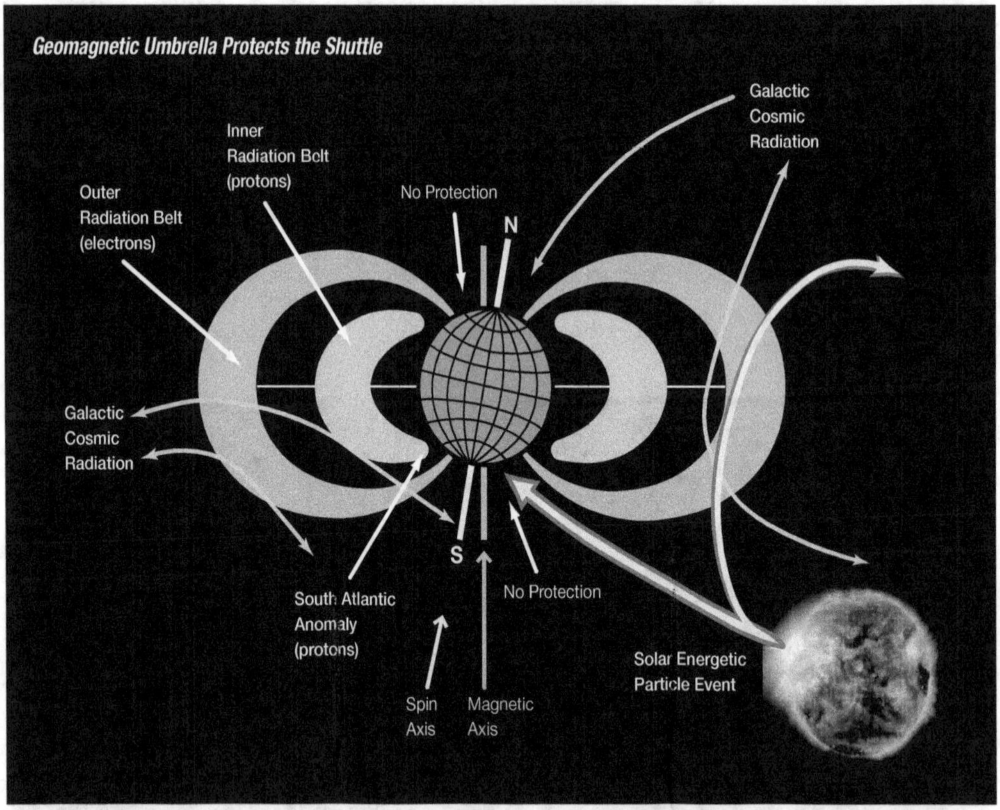

From strong protection at the equator to no protection at the poles, Earth's magnetic field provided considerable radiation protection to the shuttle by deflecting solar and galactic cosmic radiation. Usually, the shuttle was well protected; however, when the shuttle flew beyond 45 degrees latitude, there was usually little or no magnetic protection. The magnetic field also defined the regions of trapped radiation.

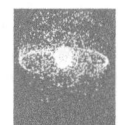

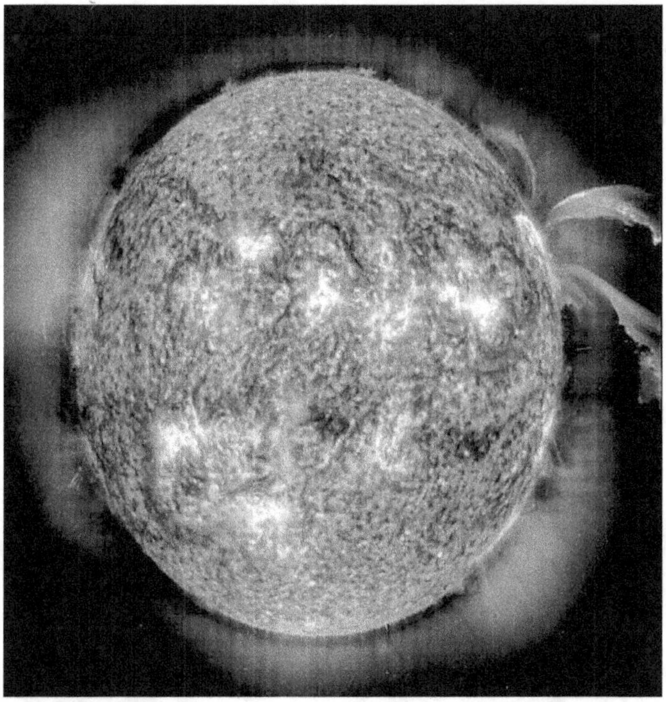

A pair of curving, erupting solar prominences on June 28, 2000. Prominences are huge clouds of relatively cool dense plasma suspended in the sun's hot, thin corona.

enabling operational response planning. Although the solar energetic particle magnitude cannot be predicted, the time intervals of when the crew will be subject to exposure can be quickly determined. If the particle is large and it is prudent for the crew to move to higher shielded areas of the station, shelter would be recommended.

Fortunately, the average exposure to shuttle crews—around 0.5 rem (5 mSv)—was far lower than the maximum exposure guideline of 25 rem/month (250 mSv/month) and also fell below the quarterly terrestrial exposure limits. During the course of the Space Shuttle Program, crew radiation exposures ranged from 0.008 rem (0.08 mSv) to 6 rem (60 mSv). The 10-day, high-altitude Hubble Space Telescope mission approached an exposure similar to an average 180-day mission to the ISS, which was 8 rem (80 mSv).

In all, operational tools and procedures to respond to space weather events matured during the course of the Space Shuttle Program and are being applied to space station operations.

Summary

During the Space Shuttle Program, great strides forward were gained in the operational effectiveness for managing radiation health protection for the astronauts. Knowledge gained via experiments vastly improved the characterization of the environment and illuminated factors that contribute to defining health risks from exposure to space radiation. These lessons will greatly benefit future generations of space travelers.

Social, Cultural, and Educational Legacies

NASA Reflects America's Changing Opportunities; NASA Impacts US Culture

Education: Inspiring Students as Only NASA Can

NASA Reflects America's Changing Opportunities; NASA Impacts US Culture

Jennifer Ross-Nazzal
Shannon Lucid
Helen Lane

The Space Shuttle, which began flying in 1981 and ushered in an entirely new human spaceflight program, was a watershed for cultural diversity within NASA and had substantial cultural impact outside the realm of spaceflight. In the 1950s and 1960s, opportunities for American women and minorities were limited as they were often segregated into pink collar and menial jobs. NASA's female and minority employees faced similar obstacles. The Space Shuttle Program opened up opportunities for these groups—opportunities that did not exist during Projects Mercury and Gemini or the Apollo and Skylab Programs. NASA's transformation was a direct consequence of a convergence of events that happened in the 1960s and 1970s and continued through the following 3 decades. These included: public policy changes instituted on the national level; the development of a spacecraft whose physical capabilities departed radically from the capsule concept; and an increase in the number of women and minorities holding degrees in the fields of science and engineering, making them attractive candidates for the space agency's workforce. Over the course of the program, the agency's demographics reflected this transformation: women and minorities were incorporated into the Astronaut Corps and other prominent technical and administrative positions.

The impact of NASA's longest-running program extends beyond these dramatic changes. Today, the shuttle—the crown jewel of NASA's spaceflight programs—symbolizes human spaceflight and is featured in advertisements, television programs, and movies. Its image exemplifies America's scientific and economic power and encourages dreamers.

Social Impact—NASA Reflects America's Changing Opportunities

Before the Space Shuttle was conceived, the aerospace industry, NASA employees, and university researchers worked furiously on early human spaceflight programs to achieve President John Kennedy's goal of landing a man on the moon by the end of the 1960s. Although these programs employed thousands of personnel across the United States, White men overwhelmingly composed the aerospace field at that time, and very few women and minorities worked as engineers or scientists on this project. When they did work at one of NASA's centers, women overwhelmingly served in clerical positions and minorities accepted low-paying, menial jobs. Few held management or professional positions, and none were in the Astronaut Corps, even though four women had applied for the 1965 astronaut class. By the end of the decade, NASA offered few positions to qualified minorities and women. Only eight Blacks at Marshall Space Flight Center in Alabama held professional-rated positions while the Manned Spacecraft Center (currently known as Johnson Space Center) in Texas had 21, and Kennedy Space Center in Florida had only five.

Signs of change appeared on the horizon as federal legislation addressed many of the inequalities faced by women and minorities in the workplace. During the Kennedy years, the president ordered the chairman of the US Civil Service Commission to ensure the federal government offered positions not on the basis of sex but, rather, on merit. Later, he signed into law the Equal Pay Act of 1963, making it illegal for employers to pay women lower wages than those paid to men for doing the same work. President Lyndon Johnson signed the Civil Rights Act of 1964, which prohibited employment discrimination (hiring, promoting, or firing) on the basis of race, sex, color, religion, or national origin. Title VII of the Act established the Equal Employment Opportunity Commission, which executed the law. The Equal Employment Opportunity Act of 1972 strengthened the commission and expanded its jurisdiction to local, state, and federal governments during President Richard Nixon's administration. The law also required federal agencies to implement affirmative action programs to address issues of inequality in hiring and promotion practices.

One year earlier, NASA appointed Ruth Bates Harris as director of Equal Employment Opportunity. In the fall

Changing Faces of the Astronauts From 1985 Through 2010

In 1985, STS-51F—Center: Story Musgrave, MD, mission specialist, medical doctor. To Musgrave's right, and going clockwise: Anthony England, PhD, mission specialist, geophysicist; Karl Henize, PhD, mission specialist, astronomer; Roy Bridges, pilot, US Air Force (USAF); Loren Acton, PhD, industry payload specialist; John-David Bartoe, PhD, Navy payload specialist; Gordon Fullerton, commander, USAF.

In 2010, STS-131 and International Space Station (ISS) Expedition 23—Clockwise from lower right: Stephanie Wilson, mission specialist, aerospace engineer; Tracy Caldwell Dyson, PhD, ISS Expedition 23 flight engineer, chemist; Dorothy Metcalf-Lindenburger, mission specialist, high school science teacher and coach; Naoko Yamazaki, Japanese astronaut, aerospace engineer.

Guion Bluford, PhD
Colonel, US Air Force (retired). Astronaut on STS-8 (1983), STS-61A (1985), STS-39 (1991), and STS-53 (1992).

Astronaut Guion Bluford conducting research on STS-53.

In 1983, Colonel Guion Bluford became the first African American to fly in space. He earned a Bachelor of Science in aerospace engineering from Pennsylvania State University, followed by flight school and military service as a jet pilot in Vietnam, which included missions over North Vietnam. He went on to earn a Master of Science and Doctor of Philosophy in aerospace engineering with a minor in laser physics from the Air Force Institute of Technology. He also earned a Master of Business Administration after joining NASA. Prior to joining NASA as a US Air Force astronaut, he completed research with several publications. Since leaving NASA, he has held many leadership positions.

As a NASA astronaut, he flew on four missions: two on Challenger (1983, 1985) and two on Discovery (1991, 1992).

Dr. Bluford has said, "I was very proud to have served in the astronaut program and to have participated on four very successful Space Shuttle flights. I also felt very privileged to have been a role model for many youngsters, including African American kids, who aspired to be scientists, engineers, and astronauts in this country. For me, being a NASA astronaut was a great experience that I will always cherish."

of 1973, Harris proclaimed NASA's equal employment opportunity program "a near-total failure." Among other things, the agency's record on recruiting and hiring women and minorities was inadequate. In October, NASA Administrator James Fletcher fired Harris and Congress held hearings to investigate the agency's affirmative action programs. Legislators concluded that NASA had a pattern of discriminating against women and minorities. Eventually, a resolution was reached, with Fletcher reinstating Harris as NASA's deputy assistant administrator for community and human relations. From 1974 through 1992, Dr. Harriett Jenkins, the new chief of affirmative action at NASA, began the process of slowly diversifying NASA's workforce and increasing the number of female and minority candidates.

Though few in number, women and minorities made important contributions to the Space Shuttle Program as NASA struggled with issues of race and sex. Dottie Lee, one of the few women engineers at Johnson Space Center and the subsystem manager for aerothermodynamics, encouraged engineers to use a French curve design for the spacecraft's nose, which is now affectionately called "Dottie's nose." NASA named Isaac Gillam as head of Shuttle Operations at the Dryden Flight Research Center, where he coordinated the Approach and Landing Tests. In 1978, he became the first African American to lead a NASA center. JoAnn Morgan of Kennedy Space Center served as the deputy project manager over the Space Shuttle Launch Processing Systems Central Data Subsystems used for Columbia's first launch in 1981.

Astronaut Corps

Forced to diversify its workforce in the 1970s, NASA encouraged women and minorities to apply for the first class of Space Shuttle astronauts in 1976. When NASA announced the names in January 1978, the list included six women, three African Americans, and one Japanese American, all of whom held advanced degrees. Two of the women were medical doctors, another held a PhD in engineering, and the others held PhDs in the sciences. Two of the three African Americans had earned doctorates, while the third, Frederick Gregory, held a master's degree. The only Asian member of their class, Ellison Onizuka, had completed a master's degree in aerospace engineering. This was the most diverse group of astronauts NASA had ever selected and it illustrated the sea change brought about within the Astronaut Corps by 1978. From then on, all astronaut classes that NASA selected included either women or minorities. In fact, the next class included both as well as the first naturalized citizen astronaut candidate, Dr. Franklin Chang-Diaz, a Costa Rican by birth.

Admitting women into the Astronaut Corps did require some change in the NASA culture, recalled Carolyn Huntoon, a member of the 1978 astronaut selection board and mentor to the first six female astronauts. "Attitude was the biggest thing we had to [work on]," she said.

Astronaut Richard Mullane, who was selected as an astronaut candidate in 1978, had never worked with professional women before coming to NASA. Looking back on those first few years, he remembered that "the women had to endure a lot because" so many of the astronauts came from military backgrounds and "had never worked with women and were kind of struggling to come to grips on working professionally with women."

When "everyone saw they could hold their own, they were technically good, they were physically fit, they would do the job, people sort of relaxed a little bit and started accepting them," explained Huntoon.

Sally Ride, one of the first six female astronauts selected, remembered the first few years a bit differently. The Gemini and Apollo-era astronauts in the office in 1978 were not used to working with women as peers. "But, they knew that this was coming," she said, "and they'd known it was coming for a couple of years." By 1978, the remaining astronauts "had adapted to the idea." As a sign of the changing culture within NASA, she could not recall any issues the women of her class encountered. This visible change signaled a dramatic shift within the agency's macho culture.

The 1978 group was unique in other ways. Several of the men and women came from the civilian world and their experiences differed greatly from those of their classmates who had come from the military. Previously, test pilots had comprised the majority of the office. Many of the PhDs were young, with less life experience, according to Mullane, than many of the military test pilots and flight test engineers who had completed tours in Vietnam.

The shuttle concept brought about other measurable changes. The versatility of the Space Shuttle, when compared with the first generation of spacecraft, provided greater opportunities for more participants. The shuttle was a much more flexible vehicle than the capsules of the past, when astronauts had to be 6 feet tall or under to fit into the spacecraft. (The Mercury astronauts could be no more than 5 feet 11 inches in height.) The capabilities of the shuttle were so unusual that astronauts of all sizes could participate; even James van Hoften—one of the tallest astronauts ever selected at 6 feet 4 inches—could fit inside the vehicle. Eventually, flight crews, which had previously consisted of one, two, or three American test pilots, expanded in size and the shuttle flew astronauts from across the globe, just as Nixon had hoped when he approved the shuttle in 1972. Indeed, the shuttle became the vehicle by which everyone, regardless of protected classes—sex, race, ethnicity, or national origin—could participate.

After the first four flights, the shuttle crews expanded to include mission specialists (a new category of astronauts that would perform research in space, deploy satellites in orbit, and conduct spacewalks). In addition to these scientists and engineers, the shuttle allowed room for a different category—the payload specialist. These individuals were not members of the Astronaut Corps. They were selected by companies or countries flying a payload on board the shuttle. Over the years, payload specialists from Saudi Arabia, Mexico, Canada, West Germany, France, Belgium, the Ukraine, Italy, Japan, the Netherlands, and Sweden flew on the shuttle as did two members of Congress: US Senator

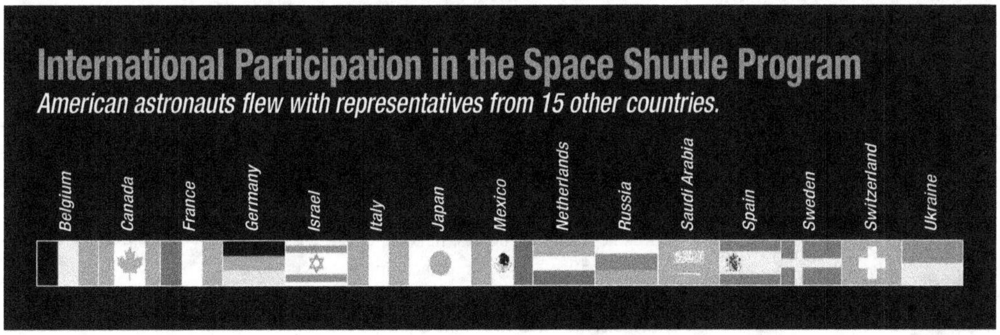

Diversity Succeeds

In 2005, NASA selected a new class of flight directors, one of the most diverse ever selected, which included the first African American (Kwatsi Alibaruho) and the first two Hispanics (Ginger Kerrick and Richard Jones). At the time of their selection, only 58 people had served in the position. All three began their careers with NASA as students and then rose through the ranks. Since their selection, Kerrick and Alibaruho have guided shifts in Russia and in the International Space Station flight control room, while Jones has supervised shuttle flights. In all, the class of 2005 dramatically changed the look of shuttle and station flight directors.

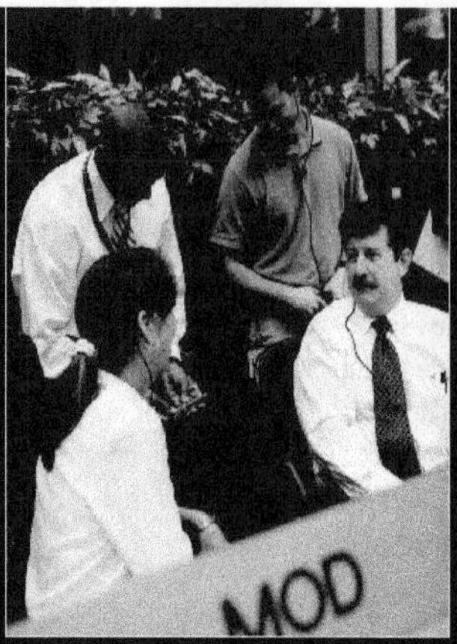

A diverse workforce.

Jake Garn of Utah and Congressman Bill Nelson of Florida. Industry also flew its own researchers, who managed their commercial payloads, with the first being McDonnell Douglas' Charles Walker. In 1972, NASA Deputy Administrator George Low remembered that this was one of the things Nixon liked about the program: "the fact that ordinary people," not just test pilots "would be able to fly in the shuttle, and that the only requirement for a flight would be that there is a mission to be performed."

Over the years, women and minorities also made their way into the pilot seat on board the shuttle and eventually went on to direct their own missions, with Eileen Collins serving as the first female pilot and commander. Space Transportation System (STS)-33 (1989) featured the first African American commander, Frederick Gregory, who later became NASA's deputy administrator. An example of NASA's diverse workforce, African American former Space Shuttle Commander Charles Bolden became NASA administrator in the summer of 2009. In all, 48 women flew on the shuttle over the course of the program between 1981 and 2010.

The female and minority shuttle astronauts quickly became heroes in the United States and abroad for breaking through barriers that had prevented their participation in the 1960s and 1970s. Millions celebrated the launches of Sally Ride, Guion Bluford, John Herrington, and Mae Jemison: first American woman, African American, Native American, and African American woman, respectively, in space.

When the crews of STS-51L (1986) and STS-107 (2003) perished, Americans grieved. Lost in two separate-but-tragic accidents, the astronauts immediately became America's heroes. In honor of their sacrifice, two separate memorials were erected at Arlington National Cemetery to the crews of the Challenger and Columbia accidents, and numerous other tributes (coins and songs, for instance) were made to the fallen astronauts. Naturally, national interest in the Return to Flight missions of STS-26 (1988) and STS-114 (2005) was high, with a great deal of attention showered on America's newest idols. Richard Covey, pilot of the STS-26 flight, recalled, "it was unprecedented, the attention that we got." The crews of the Return to Flight missions after the accidents also symbolized the changes within the Astronaut Corps. For Return to Flight after the Challenger accident, the crew members were all male. By 2005, the Return to Flight mission following the Columbia accident had a female commander.

Johnson Space Center, Texas, Changes

As the definition of the term "astronaut" became more fluid over time, America's idea of what constituted a flight director or flight controller also evolved. In NASA's heyday, all flight directors and nearly all flight controllers were men, with the exception of Frances Northcutt. She blazed the trail during the Apollo Program, becoming the first woman to work in the Mission Control Center. The number of women expanded over the years as the agency prepared for the orbital test flights. Opportunities to work in the cathedral of spaceflight (Mission Control) also expanded for other underrepresented groups, like African Americans. Angie Johnson, the first African American female flight controller in the control center in 1982, served as payloads officer for STS-2.

Over the years, the number of women working in mission operations increased dramatically. But, in general, NASA was slow to promote women into the coveted position of flight director, with the first selected in 1985—7 years after women were first named as astronaut candidates. Change came slowly, however. Eventually, flight teams became so open to women that they were nearly equally composed of men and women.

Kennedy Space Center, Florida, Changes

In the mid 1970s, women and minorities did not have a strong presence at Kennedy Space Center (KSC). In fact, many operational facilities at KSC did not even provide separate restroom facilities for women. Women had to work extra hard to gain acceptance within the KSC community. Nevertheless, a handful of talented and dedicated women and minorities broke through the cultural barriers that were in place. JoAnne Morgan became the first and, at the time, only female system engineer. By the mid 1980s, many men from the Apollo-era workforce began retiring from NASA, providing management opportunities for women and minorities. Ann Montgomery became the first female flow director for the shuttle and Ruth Harrison was one of the first system engineers within the External Tank Ground Support group. The first female senior executive—JoAnne Morgan—was soon joined by others. Ruth Harrison rose to the level of associate director of shuttle processing. By the 1990s, Arnold Postell, an African American engineer, and Hugo Delgado, a Hispanic American engineer, became branch chiefs for the shuttle Launch Processing System on their way to senior management. As of October 2010, all flow directors at KSC were women along with the lead test director and the directors for shuttle processing. The workforce culture at KSC clearly evolved into one of inclusion and equal opportunity.

Marshall Space Flight Center, Alabama, Changes

Alabama women broke the glass ceiling and accepted Space Shuttle management positions during the 1990s and the following years. From 1992 to 1996, Dewanna Edwards served as deputy manager of the Space Shuttle Main Engine Project Office. In 2002, Jody Singer was appointed manager of the Reusable Solid Rocket Booster Project, making her the first woman to lead a propulsion element office at NASA. She remained in that position until 2007, when she became deputy manager of the Shuttle Propulsion Office, which was responsible for the main engines, boosters, and External Tank. Management appointed Sandy Coleman project manager for the tank project in 2003—a position she held until 2006. From 2000 to 2004, Ann McNair managed the Ground Systems Department of Flight Projects. She was responsible for the Huntsville Operations Support Center and its key facilities, including the Payload Operations Integration Center that supported payload and science research for the International Space Station. During the same period, she led the development of the Chandra X-ray Observatory Operations Control Center. In 2004, McNair was appointed manager of the Mission Operations Laboratory in the Engineering Directorate. In 2007, she was named the center's director of operations.

Summary

Despite these advancements at NASA's shuttle field centers, women and minorities did not break into some key positions. As of 2010, not one minority or woman served as shuttle launch director or managed the Space Shuttle. NASA could, however, point to significant workforce diversification by the end of the program.

NASA Impacts US Culture

Since its inception, NASA has captivated the dreamers and adventurers, and its Apollo Program captured the public's interest and imagination. Similarly, the Space Shuttle broadly impacted art, popular music, film, television, and photos, as well as consumer culture. Over the years, the shuttle became a cultural icon—a symbol of America's technological prowess that inspired many people inside and outside of the agency.

Paintings and murals of the shuttle, payloads, and flight crews abound. Numerous pieces of art in a variety of mediums—fabric, watercolors, acrylic, oil, etching, triptych, and pencil—depict the launch and landing of the shuttle, simulations, spacewalks, and the launch facilities. Artist Henry Casselli used watercolors to depict Astronaut John Young as he suited up for the first shuttle flight (1981). Space artist Bob McCall painted several of the murals that adorn the walls of many of NASA's centers, including Johnson Space Center. "Opening the Space Frontier: The Next Giant Step"—the large mural in the now decommissioned visitor center—includes the shuttle and one of NASA's female astronauts. Coincidentally, at Young's urging, McCall designed the STS-1 patch.

Music

The shuttle, the crews, and the missions inspired many musicians, who composed songs about the shuttle and its flights. Canadian rockers Rush, who were present at the first launch, wrote their 1982 song "Countdown" about that event and dedicated that song to "Astronauts Young, [Robert] Crippen, and all the people of NASA for their inspiration and cooperation." When First Lady Hillary Rodham Clinton announced that a woman would command a mission for the first time in NASA's 40-year history, the NASA Arts Program asked Judy Collins to write a song to commemorate the occasion. She agreed and composed "Beyond the Sky" for that historic flight. The song describes the dream of a young girl to fly beyond the sky and heavens. The girl eventually achieves her goal and instills hope in those with similar aspirations. This is foreshadowed in the fifth verse.

She had led the way
beyond darkness

For other dreamers who
would dare the sky

She has led us to believe
in dreaming

Given us the hope that
we can try

Authored for NASA as part of the NASA Arts Program

Inspiration

The shuttle inspired so many people in such different ways. Much as the flag came to symbolize American pride, so too did the launch and landing of the shuttle. As an example, William Parsons, Kennedy Space Center's former director, witnessed his first launch at age 28 and recalled, "When I saw that shuttle take off at dusk, it was the most unbelievable experience. I got tears in my eyes; my heart pounded. I was proud to be an American, to see that we could do something that awesome."

Film and Television

IMAX® films built on the thrill of spaceflight by capturing the excitement and exhilaration of NASA's on-orbit operations. Shuttle astronauts were trained to use the camera and recorded some of the program's most notable events as the events unfolded in orbit, like the spacewalk of Kathryn Sullivan, America's first woman spacewalker. Marketed as "the next best thing to being there," the film *The Dream is Alive* documented living and working in space on board shuttle. *Destiny in Space* featured shots from the dramatic first Hubble Space Telescope servicing mission in 1993, which boasted a record-breaking five spacewalks. Other feature films like *Mission to Mir* took audiences to the Russian space station, where American astronauts and cosmonauts performed scientific research.

The excitement inspired by the Space Shuttle and the technological abilities—both real and imagined—did not escape screenwriters and Hollywood directors. In fact, the shuttle appeared as a "character" in numerous films, and several major motion pictures featured a few of NASA's properties. These films attracted audiences across the world and sold millions of dollars in tickets based on two basic themes: NASA's can-do spirit in the face of insurmountable challenges, and the flexibility of the shuttle. They include *Moonraker*, *Space Camp*, *Armageddon*, and *Space Cowboys*.

Television programs also could not escape the pull of the Space Shuttle. In 1994, the crew of Space Transportation System (STS)-61 (1993), the first Hubble servicing mission, appeared on ABC's *Home Improvement*. Six of the seven crew members flew to California for the taping, where they starred as guests of *Tool Time*—the fictional home improvement program—and showed off some of the tools they used to work on the telescope in space. Following this episode, astronauts from the US Microgravity Laboratory-2, STS-73 (1995), appeared on *Home Improvement*. Astronaut Kenneth Bowersox, who was pilot for one flight and commander of two flights, made three appearances on the show. Bowersox once brought Astronaut Steven Hawley, who also flew on STS-82 (1997).

The Space Shuttle and its space fliers were also the subject of the television drama *The Cape*. Based on the astronaut experience, the short-lived series captured the drama and excitement associated with training and flying shuttle missions. Set and filmed at Kennedy Space Center, the series ran for one season in the mid 1990s.

Consumer Culture

The enduring popularity of the Space Shuttle extended beyond film and television into consumer culture. During the shuttle era, millions of people purchased goods that bore images of shuttle mission insignias and the NASA logo—pins, patches, T-shirts, polos, mugs, pens, stuffed animals, toys, and other mementos. The shuttle, a cultural icon of the space program associated with America's progress in space, was also prominently featured on wares. Flight and launch and re-entry suits, worn by the astronauts, were particularly popular with younger children who had hopes of one day flying in space. People still bid on thousands of photos and posters signed by shuttle astronauts on Internet selling and trading sites.

Photos of the shuttle, its crews, astronaut portraits, and images of notable events in space are ubiquitous.

Chiaki Mukai, MD, PhD

Japanese astronaut.
Payload specialist on STS-65 (1994) and STS-95 (1998).
Deputy mission scientist for STS-107 (2003).

My Space Shuttle Memory

"From the mid 1980s to 2003, I worked for the space program as a Japanese astronaut. This was the golden time of Space Shuttle utilization for science. Spacelab missions, which supported diverse fields of research, were consecutively scheduled and conducted. The science communities were so busy and excited. I flew two times (STS-65/IML [International Microgravity Laboratory]-2 and STS-95) and worked as an alternate crew member for two other science missions (STS-47 and STS-90). On my last assignment, I was a deputy mission scientist for the STS-107 science mission on board the Space Shuttle Columbia.

I really enjoyed working with many motivated people for those missions. I treasure these memories. Among the many photographs taken during my time as an astronaut, I have one favorite sentimental picture. The picture was taken from the ground showing STS-65, Columbia, making its final approach to Kennedy Space Center. The classic line of the shuttle is clearly illuminated by the full moon softly glowing in the dawn's early light. When I see this photo, I cannot believe that I was actually on board the Columbia at that moment. It makes me feel like everything that happened to me was in a dream.

The Space Shuttle Program enabled me to leave the Earth and to expand my professional activities into space. My dream of 'Living and working in space' has been truly realized. Thanks to the enormous capacity of human and cargo transportation made by the Space Shuttles between Earth and space, people can now feel that 'Space is reachable and that it is ours.'

I want to thank the dedicated people responsible for making this successful program happen. The spirit of the Space Shuttle will surely live on, inspiring future generations to continue using the International Space Station and to go beyond."

They can be found in books, magazines, calendars, catalogs, on television news broadcasts, and on numerous non-NASA Web sites. They adorn the walls of offices and homes across the world. One of the most famous images captures the historic spacewalk of Astronaut Bruce McCandless in the Manned Maneuvering Unit set against the blackness of space. Another well-known photo, taken by the crew of STS-107 (2003), features the moon in a haze of blue.

Tourism

The Space Shuttle attracted vacationing travelers from the beginning of the program. Tourists from across the country and globe flocked to Florida to witness the launch and landing of the shuttle, and also drove to California, where the shuttle sometimes landed. Kennedy Space Center's Visitor Complex in Florida and the US Space and Rocket Center in Alabama welcome millions of sightseers each year—people who hope to learn more about the nation's human spaceflight program. Visitors at Kennedy Space Center have the unique opportunity to experience the thrill of a simulated launch on the Shuttle Launch Experience, with veteran shuttle Astronaut Bolden walking riders through the launch sequence. Others visit Space Center Houston in Texas and the Smithsonian's Udvar-Hazy Center in Virginia, the latter of which includes the Enterprise, the first Space Shuttle Orbiter rolled out in 1976.

One need only visit the areas surrounding the space centers to see the ties that bind NASA's longest-running program with their local and state communities. In the Clear Lake area (Texas), McDonald's restaurant attracted visitors by placing a larger-than-life astronaut model donned in a shuttle-era spacesuit on top of the roof. A mock Space Shuttle sits on the lawn of Cape Canaveral's city hall. Proud of its ties to the space program, Florida featured the shuttle on the state quarter released by the US Mint in 2004; Texas, by contrast, included the Space Shuttle on its state license plates.

Summary

For nearly 30 years, longer than the flights of Mercury, Gemini, Apollo, and Skylab combined, the Space Shuttle—the world's most complex spacecraft at the time—had a tremendous influence on all aspects of American culture. Television programs and motion pictures featured real-life and imaginary Space Shuttle astronauts; children, entertained by these programs and films, dreamed of a future at NASA. Twenty-five years after Sally Ride's first flight, thousands of girls—who were not even born at the time of her launch—joined Sally Ride's Science Club, inspired by her career as the first American woman in space.

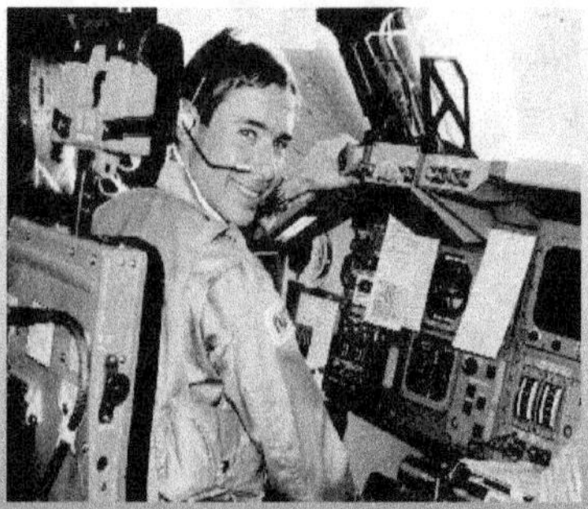

Brewster Shaw
Colonel, US Air Force (retired).
Pilot on STS-9 (1983).
Commander on STS-61B (1985) and STS-28 (1989).

Space Is For Everyone

"I was on STS-9 and we had waved off several revs before landing in California. My wife joined me after the postflight conference. I asked her what she thought. She replied that I said 'Space is for everyone.' I have reflected on that. I remember looking out the back window of the shuttle and looking at Earth as it passed by very quickly. I marveled at the fact the human brain has developed the capability to lift 250,000 pounds of mass into orbit and is flying around at the orbital velocity of 17,500 miles per hour—what an accomplishment of mankind! Looking at Earth from that vantage point made me realize that there are a lot of people on Earth who would give their arm and a leg to be where I am! Here I was a 30-something macho test pilot and I was humbled!

"Suddenly it occurred to me how privileged I was to be here in space! It was a revelation. I had no more right than any other human being to be here—I was just luckier than they were. There I realized that space is for everyone! I decided to dedicate my career to helping as many humans as possible experience what I was experiencing."

An Expansive Legacy

The Space Shuttle became an "icon" not only for the capabilities and technological beauty of the vehicles, but also for the positive changes NASA ultimately embraced and further championed. Through the efforts of those who recognized the need for diversity in the workplace, the Space Shuttle Program was ultimately weaved into the fabric of our nation—on both a social and a cultural level. The expansion of opportunities for women, minorities, industry, and international partners in the exploration of the universe not only benefitted those individuals who had the most to gain; the expansion also made the program an even greater success because of each individual's unique and highly qualified contributions. No longer regarded as a "manned" spaceflight in the most literal sense of the term, the shuttle ushered in a new era of "human" spaceflight that is here to stay.

View-Master
Space Shuttle
FLORIDA
CLINT EASTWOOD TOMMY LEE JONES
DONALD SUTHERLAND JAMES GARNER
SPACE
COWBOYS
Copyright Notice*
LITTLE MAMMOTH MEDIA
The Big
space
shuttle
THE BIG ADVENTURE SERIES!
SHUTTLE
CROSSING

Education: Inspiring Students as Only NASA Can

Introduction
Helen Lane
Kamlesh Lulla

The Challenger Center
June Scobee Rodgers

The Michael P. Anderson Engineering Outreach Project
Marilyn Lewis

Long-distance Calls from Space
Cynthia McArthur

Project Starshine
Gilbert Moore

Earth Knowledge Acquired by Middle School Students
Sally Ride
Kamlesh Lulla

Toys in Space
Carolyn Sumners
Helen Lane
Kamlesh Lulla

Flight Experiments
Dan Caron
John Vellinger

Spaceflight Science and the Classroom
Jeffery Cross

Teachers Learn About Human Spaceflight
Susan White

College Education
Undergraduate Engineering Education
Aaron Cohen
Graduate Student Science Education
Iwan Alexander

NASA's commitment to education is played out with the Space Shuttle, but why?

> "And to this end nothing inspires young would-be scientists and engineers like space and dinosaurs—and we are noticeably short of the latter."
>
> – Norman Augustine, former president and CEO of Lockheed Martin Corporation

Every Space Shuttle mission was an education mission as astronauts always took the time, while in orbit, to engage students in some kind of education activity. In fact, the shuttle served as a classroom in orbit on many missions.

Of the more than 130 flights, 59 included planned student activities. Students, usually as part of a classroom, participated in downlinks through ham radio (early in the program) to video links, and interacted with flight crews. Students asked lots of questions about living and working in space, and also about sleep and food, astronomy, Earth observations, planetary science, and beyond. Some insightful questions included: Do stars sparkle in space? Why do you exercise in space?

Through student involvement programs such as Get Away Specials, housed in the shuttle payload bay, individual students and classes proposed research. If selected, their research flew on the shuttle as a payload. Students also used the astronaut handheld and digital-camera photos for various research projects such as geology, weather, and environmental sciences in a program called KidSat (later renamed Earth Knowledge Acquired by Middle School Students [EarthKAM]). Teacher materials supported classroom EarthKAM projects. Concepts of physics were brought to life during Toys in Space payload flights. Playing with various common toys demonstrated basic physics concepts, and teacher materials for classroom activities were provided along with the video from spaceflight. Not all education projects were this specific, however. Starshine—a satellite partially built by middle school students and launched from the shuttle payload—provided data for scientific analysis completed by students from all over the world. In fact, most of the scientific missions contained student components. Students usually learned about research from the principal investigators, and some of the classrooms had parallel ground-based experiments. Teacher workshops provided instruction on how to use the space program for classrooms.

The Space Shuttle became a true focus for education when President Ronald Reagan announced the Teacher in Space Program in 1984. Of course, the pinnacle of NASA's educational involvement was the selection of Astronaut Christa McAuliffe, first teacher in space. Although her flight was cut short (Challenger accident in 1986), she inspired the nation's educators. Created as a legacy of the Challenger crew by June Scobee, Challenger Centers focus on scientific and engineering hands-on education to continue NASA's dedication to education. Barbara Morgan, the backup to Christa, flew 11 years later as the educator astronaut on Space Transportation System (STS)-118 (2007), and this program continues. From the Columbia accident (2003), the education legacy continued with the establishment of the Michael P. Anderson Engineering Outreach Project in Huntsville, Alabama, to promote education of minority students through hands-on science and engineering.

Educational activities were, indeed, an integral part of the Space Shuttle Program.

Donald Thomas, PhD
Astronaut on STS-65 (1994), STS-70 (1995), STS-83 (1997), and STS-94 (1997).

"The Space Shuttle has without a doubt demonstrated remarkable engineering and scientific achievement, but I believe an even more impressive accomplishment and enduring legacy will be its achievements in the field of education. The Space Shuttle was not just another space program that students were able to watch 'from the sidelines.' It was a program in which they could participate first-hand, speaking directly with the astronauts and performing their own original research in space with experiments like SEEDS, SAREX**, and many more. For the first time we made access to space available to the classroom, and many teachers and students from across the country and around the world were able to participate. Since its first flight in 1981, the Space Shuttle, its crews, and the NASA team have inspired a whole generation of students. By exciting them and motivating them to work hard in the STEM (Science, Technology, Engineering, and Mathematics) disciplines, the Space Shuttle Program has helped prepare this next generation of scientists and engineers to take over the torch of exploration as we move from the Space Shuttle to Orion*** and resume our exploration of the moon, Mars, and beyond."*

**SEEDS—Space Exposed Experiment Developed for Students*
***SAREX—Space Shuttle Amateur Radio Experiment*
****Crew Exploration Vehicle named Orion*

Sivaker Strithar, fifth-grade student at the Harry Eichler School, New York City Public School 56Q, compares the growth of seeds flown on the Space Shuttle with earthbound control seeds. NASA flew 10 million basil seeds on STS-118 (2007) to mark the flight of the first educator and mission specialist, Barbara Morgan. The seeds were distributed to students and educators throughout the country.

Kindergarten Through 12th Grade Education Programs

The Challenger Center

The Challenger Center for Space Science Education, created by the families of the Space Shuttle Challenger astronauts, is an outstanding example of how a tragic event can be transformed into a positive force for educational achievement across the nation.

Christa McAuliffe, payload specialist and first Teacher in Space, trains on shuttle treadmill for Challenger flight STS-51L. The Challenger accident occurred on January 28, 1986.

Education became the primary focus of the Challenger STS-51L (1986) mission as teacher Christa McAuliffe was to use the shuttle as a "classroom in space" to deliver lessons to children around the world. It was to be the ultimate field trip of discovery and exploration; however, the Space Shuttle Challenger and her crew perished shortly after liftoff, and the vision for education and exploration was not realized.

The goal of the Challenger Center and its international network of Challenger Learning Centers is to carry on the mission of Space Shuttle Challenger and continue "Inspiring, Exploring, Learning" for the next generation of space pioneers and teachers.

Since its inception in 1986, the Challenger Center has reached more than 8 million students and teachers through its 53 centers scattered across the globe. Using simulation in a Mission Control Center and space station environment, expert teachers foster learning in science, mathematics, engineering, and technology. In fact, each year, more than 500,000 students and 25,000 educators experience hands-on learning in those disciplines. The Challenger Center simulators provide cooperative learning, problem solving, decision making, and teamwork—all key ingredients of any successful mission. This experiential learning is structured to support the National Science Education Standards as well as national standards in mathematics, geography, technology, and language arts. Using "Mission to Planet Earth" as one of the themes, the center also inculcates, in young minds an awareness of global environmental issues.

The centers offer a wholesome, integrated, and engaging learning environment. It is truly an authentic science- and mathematics-based learning approach that grabs students' attention, engages them to develop problem-solving skills, and provides satisfaction of accomplishing a tough mission during a team effort that takes them to the moon, Mars, or even Jupiter.

Educators wholeheartedly support this learning environment. For example, the State Board of Education in Virginia considered the Challenger Center model to be highly effective, and the US Department of Education cited the center as significantly impacting science literacy in the country.

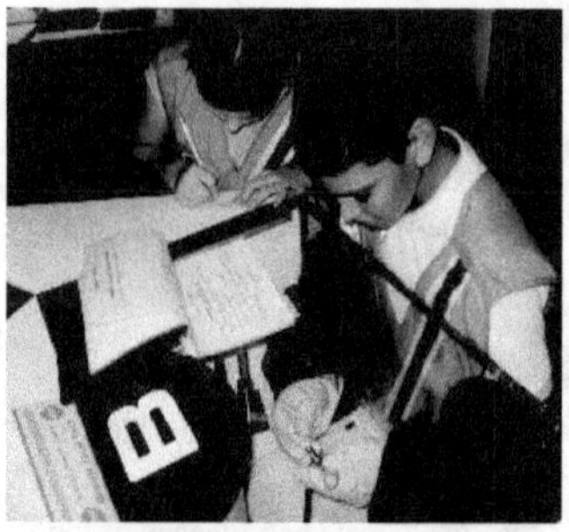

Students at a Challenger Center learning about shuttle science by working in a glove box.

A former governor of Kentucky requested three Challenger Learning Centers for his state to improve the science literacy of Kentucky's youth population. Police officials in Canada created a Challenger Center as a gift to the youth for nontraditional outreach uses. Other youth groups, such as the Girl Scout and Boy Scout organizations, also participated.

Tomorrow's aerospace and scientific workforce and the destiny of our nation's space exploration leadership are being shaped in Challenger Learning Centers across our nation. This is a powerful educational bridge that the Space Shuttle helped build for "teaching and touching the future."

The Michael P. Anderson Engineering Outreach Project

The Michael P. Anderson Engineering Outreach Project is part of the educational legacy of the Space Shuttle. Named for Columbia Astronaut Michael Anderson (who lost his life in the accident), the project seeks to engage underserved high school students in engineering design challenges in aerospace, civil, mechanical, and electrical engineering so these students become aware of engineering career options. Participating students learn about the life and accomplishments of Anderson, and they see him as a role model.

Astronaut Michael Anderson (Lieutenant Colonel, US Air Force) flew on STS-89 (1998) and then on the ill-fated Columbia (STS-107 [2003]).

The objectives are to inspire students to prepare for college by taking more advanced mathematics courses along with improved problem-solving skills, and by learning more about the field of engineering. Parents are involved in helping plan their child's academic career in science, mathematics, or engineering.

Students participate in a 3-week training program each summer. Alabama A&M School of Engineering faculty and NASA employees serve as students' leaders and mentors. At the end, the students present their engineering and mathematics projects. The curriculum and management design are disseminated from these activities to other minority-serving institutions.

Michael P. Anderson Project students Alecea Kendall, a tenth-grade New Century Technology student, and Hilton Crenshaw, a tenth-grade Lee High student, work as a team to assemble their LEGO NXT Mindstorm robot.

Long-distance Calls from Space

Students and teachers have friends in high places, and they often chat with them during shuttle missions. In November 1983, Astronaut Owen Garriott carried a handheld ham radio aboard Space Shuttle Columbia. The ham radio contacts evolved into the Space Shuttle Amateur Radio Experiment, which provided students with the opportunity to talk with shuttle astronauts while the astonauts orbited the Earth. Ham radio contacts moved from shuttle to the International Space Station, and this activity has transitioned to amateur radio on board the International Space Station. In addition to ham radio contacts, students and teachers participated in live in-flight education downlinks that included live video of the astronauts on orbit. The 20-minute downlinks provided a unique learning opportunity for students to exchange ideas with astronauts and watch demonstrations in a microgravity environment. Ham radio contacts and in-flight education downlinks allowed more than 6 million students to experience a personal connection with space exploration.

Astronauts Speak to Students Through Direct Downlink

The STS-118 (2007) crew answering a student's question.

Elementary school student asking the crew a question.

Student watching the downlink for STS-118.

Students participated in in-flight education downlinks that included live video of the astronauts on orbit. Students asked questions and exchanged ideas with astronauts.

Project Starshine

Project Starshine engaged approximately 120,000 students in more than 4,000 schools in 43 countries.

NASA deployed reflective spherical student satellites from two separate shuttle missions—STS-96 (1999) and STS-108 (2001). NASA had flown a third satellite on an expendable launch vehicle mission, and a fourth satellite was manifested on a shuttle mission but later cancelled following the Columbia accident (STS-107 [2003]). A coalition of volunteer organizations and individuals in the United States and Canada built the satellites. Each satellite was covered by approximately 1,000 small front-surface aluminum mirrors that were machined by technology students in Utah and polished by tens of thousands of students in schools and other participating organizations around the world. During the orbital lifetime of the satellites, faint sunlight flashes from their student-polished mirrors were visible to the naked eye during certain morning and evening twilight periods. The student observers measured the satellites' right ascension and declination by reference to known stars, and they recorded the precise timing of their observations through the use of stopwatches synchronized with Internet time signals. They used global positioning satellite receivers or US Geological Survey 7.5-minute quadrangle maps, or their equivalents in other countries, to measure the latitude, longitude, and altitude of their observing sites. They posted their observations and station locations on the Starshine Web site.

Students in the Young Astronauts/Astronomy Club at Weber Middle School in Port Washington, New York, proudly display a set of mirrors destined for Starshine.

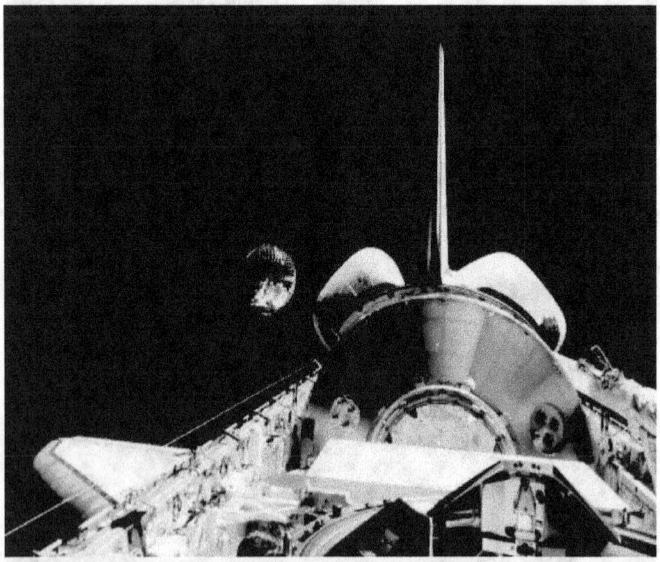

Launching Starshine satellite from Endeavour's payload bay during STS-108 (2001).

As an example of Project Starshine, children in the Young Astronauts/ Astronomy Club at Weber Middle School in Port Washington, New York, contributed to the project.

"The club members arrived at school at 7:30 a.m. every day to make sure the project would be completed on time. They worked diligently and followed instructions to the letter," said their science teacher, Cheryl Dodes.

Earth Knowledge Acquired by Middle School Students

How does one inspire school students to pursue science and engineering? Imagine creating an opportunity for students to participate in space operations during real Space Shuttle flights.

The brainchild of Dr. Sally Ride—first American woman in space—the **Earth K**nowledge **A**cquired by **M**iddle School Students (EarthKAM) education program, sponsored by NASA, gives students "hands-on" experience in space operations. During the Space Shuttle Program, NASA's EarthKAM was the next best thing to being on board for junior scientists.

The idea is as simple as it is elegant: by installing a NASA camera on board a spacecraft, middle school students across the United States and abroad had front-row seats on a space mission. They used images to study Earth science and other science disciplines by examining river deltas, deforestation, and agriculture. The hardware consisted of an electronic still camera and a laptop that was set up by an astronaut and then operated remotely from the ground with imaging requests coming directly from the students.

While this hands-on, science-immersive learning was cool for kids, the high-tech appeal was based on proper science

Students as Virtual Astronauts

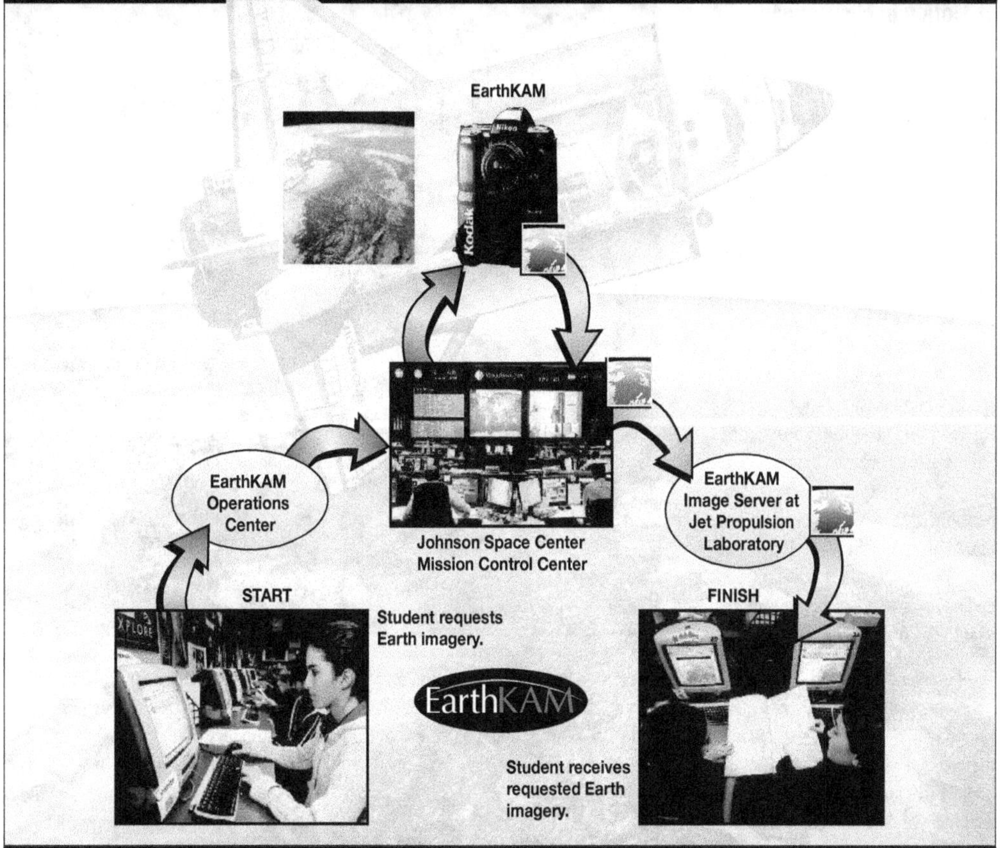

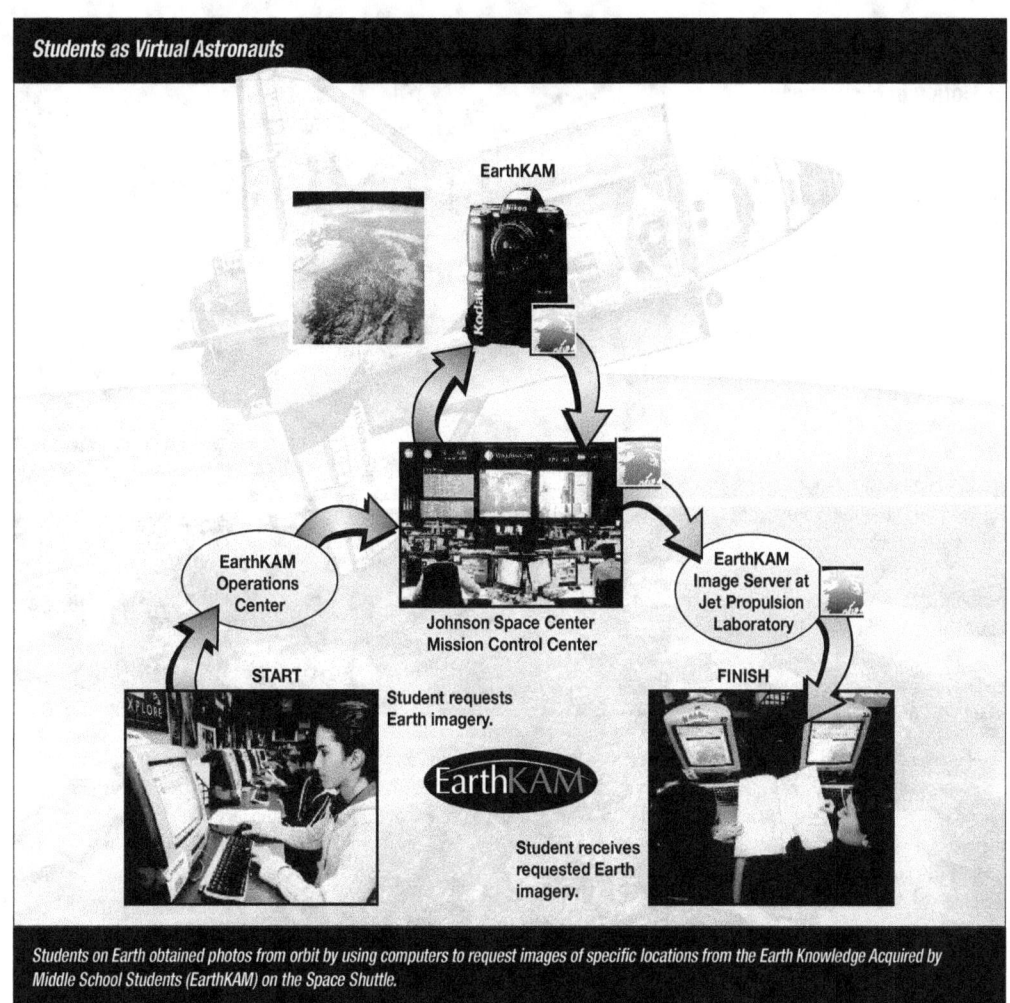

Students on Earth obtained photos from orbit by using computers to request images of specific locations from the Earth Knowledge Acquired by Middle School Students (EarthKAM) on the Space Shuttle.

methods. Students prepared a solid research proposal outlining the topic they wanted to study. The program was similar to a time-share facility. Schools were to take a certain number of photographs. During the Space Shuttle Program, students set up a 24-hour classroom Mission Control operation to track the shuttle's orbit. By calculating latitude and longitude, they followed the shuttle's route and monitored weather conditions. After choosing photo targets, students relayed those instructions over the Internet to University of California at San Diego operations unit. Undergraduate volunteers wrote the code that instructed the camera when to acquire imagery. The students received their photo images back through the Web site and began analyzing their data.

Since its first launch in 1996, EarthKAM flew on six shuttle missions and now continues operations on the International Space Station. To date, more than 73,000 students from 1,200 schools in 17 countries have participated in the program. This exciting adventure of Earth exploration from space is a great hit at schools all over the globe. While youngsters can learn latitude, longitude, and geography from a textbook, when their lesson comes first-hand from the Space Shuttle, they really pay attention. "In 20 years of teaching," says Sierra Vista Middle School (California) teacher Mark Sontag, "EarthKAM is by far the most valuable experience I've ever done with kids."

Toys in Space: Innovative Ways to Teach the Mechanics of Motion in Microgravity

Toys are the technology of childhood. They are tools designed to be engaging and fun, yet their behaviors on Earth and on orbit can illustrate science, engineering, and technology concepts for children of all ages. The STS-51D (1985) crew carried the first 11 toys into orbit. The STS-54 mission (1993) returned with some of those toys and added 29 more. The STS-77 (1996) mission crew returned with 10 of the STS-54 toys that had not been tested in space. For all these missions, crews also carried along the questions of curious children, teachers, and parents who had suggested toy experiments and predicted possible results. A few dozen toys and a few hours of the crew members' free time brought the experience of free fall and an understanding of gravity's pull to students of all ages.

Toys in Space on Discovery, STS-51D (1985)

Astronauts Jeffrey Hoffman and Rhea Seddon worked with a coiled spring. The spring demonstrated wave action in microgravity.

Astronaut Donald Williams plays with a paddleball. He could stick the ball at any angle because very little gravity pulled the ball.

Toys included acrobats (showing the positive and negative roles of gravity in earthbound gymnastics)—toy planes, helicopters, cars, and submarines (action-reaction in action), spinning tops, yo-yos, and boomerangs (all conserving angular momentum), magnetic marbles and coiled-spring jumpers (conserving energy), and the complex interplay of friction and Newton's Laws in sports, from basketball and soccer to horseshoes, darts, jacks, Lacrosse, and jump rope.

Toys are familiar, friendly, and fun—three adjectives rarely associated with physics lessons. Toys are also subject to gravity's downward pull, which often stops their most interesting behaviors. Crew members volunteered to perform toy experiments on orbit where gravity's tug would no longer affect toy activities. Toy behaviors on Earth and in space could then be compared to show how gravity shapes the motions of toys and of all other moving objects held to the Earth's surface.

The toys were housed at the Houston Museum of Natural Science after flights. A paper airplane toy used during the flight of US Senator Jake Garn (shuttle payload specialist) was displayed at the Smithsonian Air and Space Museum in Washington DC. McGraw-Hill published two books for teachers on using the Toys in Space Program in the classroom. NASA created a DVD on the International Toys in Space Program with the other Toys in Space videos included. The DVD also provided curriculum guides for all of the toys that traveled into space.

The Toys in Space Program integrated science, engineering, and technology. The National Science Education Standards recognized that scientists and engineers often work in teams on a project. With this program, students were technicians and engineers as they constructed and evaluated toys. They became scientists as they experimented with toys and predicted toy behaviors in space. Finally, they returned to an engineering perspective as they thought about modifying toys to work better in space or about designing new toys for space. Designing for space taught students that technical designs have constraints (such as the shuttle's packing requirements) and that perfect solutions are often not realistic. Space toys, like space tools, had to work in a new and unfamiliar environment. Ultimately, however, Toys in Space was about discovering how things work on Spaceship Earth.

Flight Experiments: Students Fly Research Projects in Payload Bay

The Space Shuttle provided the perfect vehicle for students and teachers to fly experiments in microgravity. Students, from elementary to college, participated in the Self-Contained Payload Program—popularly named Get Away Specials—and the Space Experiment Modules Program. These students experienced the wonders of space.

Get Away Specials

Get Away Specials were well suited to colleges and universities that wished for their students to work through the engineering process to design and build the hardware necessary to meet criteria and safety standards required to fly aboard the shuttle. Students, along with their schools, proposed research projects that met NASA-imposed standards, such as requiring that the experiment fit in the standard container, which could be no heavier than 91 kg (200 pounds), have scientific intent, and be safe. For biological experiments, only insects that could survive 60 to 90 days were allowed. The payload had to be self-contained, require no more than six crew operations, and be self-powered (not relying on the Orbiter's electricity). The payload bay was in the vacuum and thermal conditions of spaceflight, so meeting these goals was difficult.

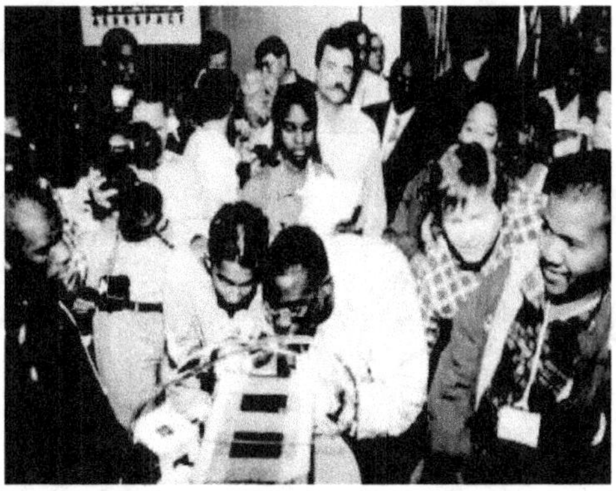

DuVal High School (Lanham, Maryland) students look inside a Get Away Special canister to see whether any of the roaches survived spaceflight.

DuVal High School in Lanham, Maryland, however, did experience success with their experiment—Get Away Special 238, which flew on STS-95 (1998). The National Capital Section of the American Institute of Aeronautics and Astronautics, a professional society, and the school district (through fund-raisers) financed this project.

From day one, the students wished to fly a biological experiment and debated whether to select termites or cockroaches since both could survive in a dark, damp environment. Once a decision was made, DuVal's project became known as the Roach MOTEL—an acronym for Microgravity Opportunity To Enhance Learning. The insects included three adults, three nymphs, and three egg cases sealed in separate compartments of a habitat inside a Get Away Special can that had sufficient life support systems for a journey into space and back—a journey lasting no longer than 6 months. The students expected the roaches to carry out all life functions (including reproduction) and return alive. The project stretched on for more than 7 years while students and teachers entered and left the program. The two factors that finally brought the project to completion were a team of administrators and teachers that was determined to see it through and NASA's relaxation of the dry nitrogen/dry air purge of the canister. The ability to seal the Get Away Special can with ambient air was the key to success for this experiment. Over the course of 7 years, 75 adults from 16 companies and organizations assisted with the project. Seventy-seven students were directly involved with engineering solutions to the many problems, while hundreds of other students were exposed to the project. Two roaches survived, and the egg cases never hatched.

Nelson Columbano, one of the students, described the experience as follows:

"I was involved with the Get Away Specials Program at DuVal High School in Lanham, Maryland, in 1996/97. Our project involved designing a habitat for insects (roaches) to survive in orbit for several days. I can't say the actual experiment is something I'm particularly proud of, but the indirect experiences and side projects associated with planning, designing, and building such a complex habitat were easily the most enriching part of my high school experience. The Get Away Specials Program introduced me to many aerospace industry consultants who volunteered to work with the class. It also presented me with real-world challenges like calling vendors for quotes, interviewing experts in person and over the phone, evaluating mechanical and electrical devices for the project and other activities that gave me a glimpse of what it's like to interface with industry professionals. At the end of the school year, some of the consultants came back to interview students for summer internships. I was lucky to receive an offer with Computer

Sciences Corporation, 11 years later becoming the proud IT Project Manager. I often think about how different my career path may have been without the Get Away Specials Program and all of the doors it opened for me."

The Get Away Specials Program was successful for both high school and university students. Over the years, it changed to the Space Experiment Module Program, which simplified the process for students and teachers.

Space Experiment Modules

To reduce costs to get more students involved, NASA developed the Space Experiment Module Program since much of the engineering to power and control experiments was done for the students. Space Experiment Module experiments, packaged 10 modules to a payload canister, varied from active (requiring power) to passive (no power). Since no cost was involved, students in kindergarten as well as college students proposed projects. During the mid 1990s, 50 teachers from the northeastern United States, participating in the NASA Educational Workshops at Goddard Space Flight Center and Wallops Flight Facility, designed Space Experiment Modules with activities for their students. During this 2-week workshop, teachers learned about the engineering design process and designed module hardware, completed the activities with their students, and submitted their experiment for consideration. One of the Get Away Special cans on STS-88 (1998) contained a number of Space Experiment Module experiments from NASA Educational Workshops participants. Students and teachers attended integration and de-integration activities as well as the launch.

Martin Crapnell, a retired technology education teacher who attended one of the NASA Educational Workshop sessions, explained.

"Experiencing the tours, briefings, and launch were once-in-a-lifetime experiences. I tried to convey that excitement to my students. The Space Experiment Modules and NASA Educational Workshops experience allowed me to share many things with my students, such as the physics of the thrust at launch and the 'twang' of the shuttle, long-term space travel and the need for food (Space Experiment Modules/Mars Lunchbox), spin-offs that became life-saving diagnostics and treatments (especially mine), job opportunities, and manufacturing and equipment that was similar to our Technology Lab.

"Even though delays in receiving all of the Space Experiment Modules materials affected the successful completion we desired, I believe I was able to share the experience and create more excitement and understanding among the students as a result of the attempt. The Space Experiment Modules and NASA Educational Workshops experiences allowed relevant transfer to lab and life experiences."

A Nutty Experiment of Interest

One of the many experiments conducted by students during the Space Shuttle Program was to determine the effects of microgravity and temperature extremes on various brands of peanut butter. Students microscopically examined the peanut butters, measured their viscosity, and conducted qualitative visual, spreadability, and aroma tests on the samples before and after flight. The students from Tuttle Middle School, South Burlington, Vermont, and The Gilbert School, Winsted, Connecticut, called this research "a nutty idea."

Students Go On to Careers in Engineering

John Vellinger, executive vice president and chief operating officer of Techshot, Inc. (Greenville, Indiana), is an example of how one participating student secured a career in engineering.

As an eighth-grade student in Lafayette, Indiana, Vellinger had an idea for a science project—to send chicken eggs into space to study the effects of microgravity on embryo development. Vellinger entered his project in a science competition called the Shuttle Student Involvement Program, sponsored by NASA and the National Science Teachers Association.

In 1985, after Vellinger's freshman year at Purdue University, NASA paired him with Techshot, Inc. co-founder Mark Deuser who was working as an engineer at Kentucky Fried Chicken (KFC). Through a grant from KFC, Deuser and Vellinger set out to develop a flight-ready egg incubator.

By early 1986, their completed "Chix in Space" hardware was launched aboard Space Shuttle Challenger on its ill-fated STS-51L (1986) mission. Regrouping after the tragic loss of the shuttle, its crew, and the Chix in Space incubator, Deuser and Vellinger continued to develop the payload for a subsequent flight. Together, the pair designed, fabricated, and integrated the flight hardware, coordinated the project with NASA, and assisted the scientific team.

More than 3 years after the Challenger accident, Chix in Space successfully reached orbit aboard Space Shuttle Discovery on mission STS-29 (1989). The results of the experiment were so significant that the project received worldwide interest from gravitational and space biologists, and it established a strong reputation for Techshot, Inc. as an innovative developer of new technologies.

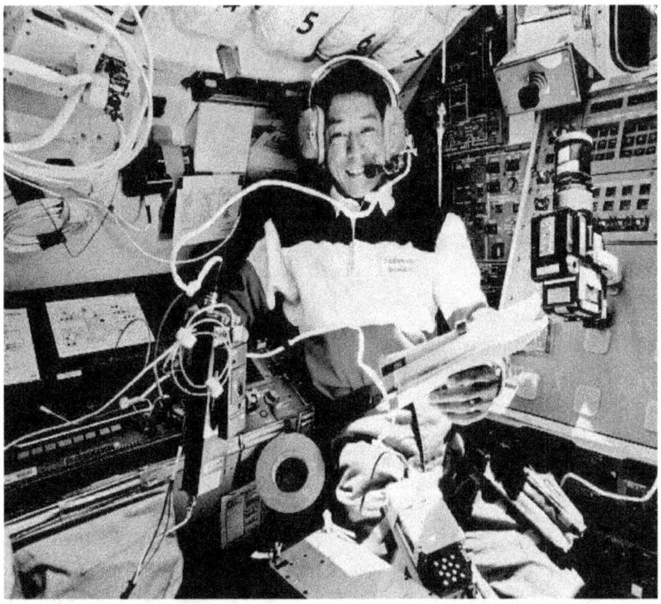

Japanese Astronaut Mamoru Mohri talks to Japanese students from the aft flight deck of the Space Shuttle Endeavour during the STS-47 (1992) Spacelab-J mission.

Spaceflight Science and the Classroom

Can students learn from Space Shuttle science? You bet they can. To prove this point, life sciences researchers took their space research to the classroom.

Bone Experiment

STS-58 (1993), a mission dedicated to life science research, had an experiment to evaluate the role of microgravity on calcium-essential element for health. With the assistance of Lead Scientist Dr. Emily Holton, three sixth-grade classes from the San Francisco Bay Area in California conducted parallel experiments to Holton's spaceflight experiment. Research staff members traveled to the schools 10 days prior to the launch date. They discussed the process of developing the experiment and assembling the flight hardware and reviewed what was needed to include the experiment on the shuttle flight. The students conducted experiments on cucumber, lettuce, and soybean plants using hydroponics—the growing of plants in nutrient solutions with or without an inert medium to provide mechanical support. Half the plants were fed a nutritionally complete food solution while the other half was fed a solution deficient in calcium. During the 2 weeks of the mission, students measured each plant's height and growth pattern and then recorded the data. Several of the students traveled to Edwards Air Force Base, California, to witness the landing of STS-58. The students analyzed their data and recorded their conclusions. The classes then visited NASA Ames Research Center, where they toured the life science labs and participated in a debriefing of their experiment with researchers and Astronaut Rhea Seddon.

Fruit Flies—How Does Their Immune System Change in Space?

Fruit flies have long been used for research by scientists worldwide because their genome has been completely mapped, their short life cycle enables multiple generations to be studied in a short amount of time, and they have many analogous processes to humans. The fruit fly experiment flew on STS-121 (2006). Its goal was to characterize the effects of space travel (including weightlessness and radiation exposure) on fruit flies' immune systems.

Middle school students (grades 5-8) were directed to a Web site to follow this experiment. The Web site provided information about current NASA space biology research, the scientific method, fruit flies, and the immune system.

Using documentation on the special site, teachers and their students conducted hands-on activities relating to this experiment. Students communicated with expert fly researchers, made predictions about the results, and asked questions of the scientists.

Frogs in Space—How Does the Tadpole Change?

In the United States and Japan's quest to learn how life responds to the rigors of the space environment, NASA launched STS-47 (1992)—a Japanese-sponsored life science mission. The question to be answered by this mission was: How would space affect the African clawed frog's life cycle? The life cycle of this particular frog fit nicely into this time period. Fertilized eggs were packaged in small grids, each housed in specially designed plastic cases. Some of these samples were allowed to experience microgravity during the mission, while others were placed in small centrifuges and kept at various simulated gravities between microgravity and Earth environment. The education portion of the experiment allowed student groups and teachers to learn about the frog embryology experiment by studying the adaptive development of frogs to the microgravity environment. NASA produced an education package and educational CD-ROM from this experiment.

Teachers Learn About Human Spaceflight

"Reach for your dreams, the sky is no limit," exclaimed Educator Astronaut Barbara Morgan while encouraging teachers to facilitate their students' discovery, learning, and sharing about human spaceflight.

The excitement of spectacular shuttle launches and on-orbit science enriched students' learning. For 30 years, the Space Shuttle Program provided teachers around the nation an unparalleled opportunity to participate in professional development workshops—promoting students to get hooked on science, technology, engineering, and mathematics careers. Historically, NASA has focused on teachers because of their profound impact on students. The main objective of NASA teacher programs was professional development while providing numerous classroom and curriculum resources.

Exciting educator workshops with themes such as "Blastoff into Learning" or "Ready, Set, and Launch" focused on the Space Shuttle as a classroom in space. Teachers responded enthusiastically to these initiatives.

Damien Simmons, an advanced placement physics teacher at an Illinois high school, said it best after attending a Network of Educator Astronaut Teachers workshop at the NASA Glenn Research Center in Cleveland, Ohio. "I'm taking home lessons and examples that you can't find in textbooks. When my students see the real-world applications of physics, I hope it will lead them to pursue careers in engineering."

Melanie Brink, another teacher honored by the Challenger Center, said, "Embracing the fundamentals of science has always been at the core of my curriculum. Preparing students to be successful young adults in the age of technology, math, and science is an exciting challenge."

NASA continues to provide teachers opportunities to use spaceflight in their classrooms to promote education.

City of Bellflower, California, luncheon "Reaching for the Stars/Growing Together" honored teacher Pam Leestma's second- and third-grade students for their spaceflight learning activities.
Back row (left to right): Kaylin Townsend, Jerron Raye, Brendan Mire, Payton Kooi, and Rylee Winters. Front row: Julianne Bassett and teacher Pam Leestma.

Barbara Morgan
Educator astronaut on STS-118 (2007).
Idaho teacher.

"Inspiring and educating future scientists and engineers are major accomplishments of the Space Shuttle Program. Much of this began with the Teacher in Space Program, despite the tragic 1986 loss of Space Shuttle Challenger and her crew.

"Before Challenger, American teachers were stinging from a report, titled 'A Nation at Risk,' that condemned the American education system and appeared to tar all teachers with the same broad brush. Even the noble call to teaching was dismissed, by many, with the saying, 'Those that can, do. Those who can't, teach.'

"But NASA was the first federal agency to start to turn that around, by making a school teacher the first 'citizen' spaceflight participant. NASA selected a stellar representative in New Hampshire social studies teacher Christa McAuliffe, who showed what great teachers all over the country do. I was fortunate to train as Christa's backup. Barely a day went by without NASA employees coming up to us to tell us about those teachers who had made a difference for them. We felt that Teacher in Space was more than just a national recognition of good teaching; it was also a display of gratitude by hundreds of NASA employees.

"Thousands of teachers gathered their students to watch Christa launch on board Challenger. The tragic accident shook all of us to the core. But for me, the pain was partly salved by what I saw in the reactions of many to the tragedy. Instead of defeatism and gloom, I heard many people say that they'd fly on the next Space Shuttle 'in a heartbeat.' Others told me how Challenger had inspired them to take bold risks in their own lives—to go back to college or to go into teaching. Also, 112 Teacher in Space finalists made lasting contributions to aerospace education in this country. And the families of the Challenger crew created the superlative Challenger Center for Space Science Education.

"After Challenger, NASA's education program grew in many ways, including establishing the Teaching From Space office within the Astronaut Office, and producing many astronaut-taught lessons from orbit to school children around the world. I returned to teaching in Idaho, and continued working with NASA, half-time, until I became an astronaut candidate in 1998. I am proud that NASA later selected three more teachers to be educator astronauts. It marked the first time since the scientist astronauts were selected for Apollo that NASA had made a major change in its astronaut selection criteria.

"So, certainly, the Space Shuttle Program has made a major impact on American education and on the way teachers are seen by the public. And this brings me back to that old comment of 'Those who can't, teach.' It reminds me of how, to pay tribute to those who went before, engineers and scientists are fond of quoting Sir Isaac Newton. He said, 'I stand on the shoulders of giants.' We teachers have a similar sense of tradition. We think of teachers who teach future teachers, who then teach their students, who go on to change the world. For example, Socrates taught Plato, who taught Aristotle, who taught Alexander the Great. So I'd like to end this little letter with a quote that far predates 'Those who can't, teach.' Two millennia ago, in about 350 BC, Aristotle wrote, 'Those who know, do. Those who understand, teach.' Aristotle understood.

"I want to thank the Space Shuttle Program for helping teachers teach. Explore, discover, learn, and share. It is what NASA and teachers do."

College Education

Undergraduate Engineering Education

A legacy of building the shuttle is strengthening the teaching of systems engineering to undergraduate students, especially in design courses. The shuttle could not have been designed without using specific principles. Understanding the principles of how systems engineering was used on the shuttle and then applying those principles to many other design projects greatly advanced engineering education.

Engineering science in all fields of engineering was advanced in designing the shuttle. In the fields of avionics, flight control, aerodynamics, structural analysis, materials, thermal control, and environmental control, many advances had to be made by engineers working on the Space Shuttle—advances that, in turn, were used in teaching engineering sciences and systems engineering in universities.

The basic philosophy underlying the teaching approach is that the design must be a system approach, and the entire project must be considered as a whole rather than the collection of components and subsystems. Furthermore, the life-cycle orientation addresses all phases of the system, encouraging innovative thinking from the beginning.

The use of large, complicated design projects rather than smaller, more easily completed ones forces students to think of the entire system and use advanced engineering science techniques. This was based on the fact that the shuttle itself had to use advanced techniques during the 1970s. The emphasis on hierarchical levels provides an appreciation for the relationship among the various functions of a system, numerous interface and integrating problems, and how the design options are essentially countless when one

Katie Gilbert

Inspired by NASA to become an aerospace engineer.

"In the school year of 2000, NASA released an educational project for elementary-aged students. Of course, this project reached the ears of my fun-seeking fourth-grade science teacher, Mrs. Maloney. For extra credit, we were to group ourselves up and answer the critical question: What product could be sent up to space on the shuttle to make our astronauts' lives easier?

"For weeks, our fourth-grade selves spent hours of time creating an experiment that would answer this question. My group tested cough drops; would they still have the same effectiveness after being in zero gravity for extended periods of time? We sent it in, and months later we received a letter. Four of our school's projects were to be sent up on the Space Shuttle Endeavour. Our projects were going to space!

"When the time finally came, we all flew down to Florida to watch Endeavour blast off with our experiments on board. This all gave me the opportunity to visit the Kennedy Space Center, see a real Space Shuttle, and talk to actual astronauts. The entire experience was one of the most memorable of my life. With all of the excitement and fascination of the world outside of ours, I knew right then that I wanted to be an astronaut and I made it my life goal to follow my cough drops into space.

"As it turns out, cough drops are not at all affected by zero gravity or extreme temperatures. The experiment itself didn't bring back alien life forms or magically transform our everyday home supplies into toxic space objects, but it wasn't a complete waste. The simple experiment opened my eyes to the outside world and the possibilities that exist within it. It captivated my interest and held it for over 8 years, and the life goals I made way back then were the leading factor in choosing Purdue University to study Aerospace Engineering."

considers all the alternatives for satisfying various functions and combinations of functions.

Also, learning to design a very complex system provides the skills to transfer this understanding to the design of any system, whereas designing a small project does not easily transfer to large systems. In addition, this approach provides traceability of the final system design as well as the individual components and subsystems back to the top-level need, and lowers the probability of overlooking an important element or elements of the design.

For designing systems engineering educational courses, general topics are addressed: the general systematic top-down design process; analysis for design; and systems engineering project management. Specific topics are: establishment and analysis of the top-level need with attention to customer desires; functional decomposition; development of a hierarchical arranged function structure; determination of functional and performance requirements; identification of interfaces and design parameters; development of conceptual designs using brainstorming and parameter analysis; selection of criteria for the evaluation of designs; trade studies and down-selection of best concept; parametric analysis; and preliminary and detailed designs. Application of engineering analysis includes the depth and detail required at various phases during the design process. Systems engineering management procedures—such as failure modes and effects analysis, interface control documents, work breakdown structures, safety and risk analysis, cost analysis, and total quality management—are discussed and illustrated with reference to student projects.

In summary, due to NASA's efforts in systems engineering, these principles were transferred to undergraduate engineering courses.

Graduate Student Science Education

The Space Shuttle's impact on science and engineering is well documented. For scientists, the shuttle enabled the microgravity environment to be used as a tool to study fundamental processes and phenomena ranging from combustion science to biotechnology. The impact of the microgravity life and physical science research programs on graduate education should not be overlooked.

Many graduate students were involved in the thousands of experiments conducted in space and on the ground. A comparable number of undergraduates were exposed to the program. Perusal of task books for microgravity and life science programs reveals that, between 1995 and 2003, flight and microgravity research in the life and physical sciences involved an average of 744 graduate students per year. Thus, the shuttle provided thousands of young scientists with the opportunity to contribute to the design and implementation of experiments in the unique laboratory environment provided by a spacecraft in low-Earth orbit. Such experiments required not only an appreciation of a specific scientific discipline, but also an appreciation of the nature of the microgravity and how weightlessness influences phenomena or processes under investigation.

In addition to mainstream investigations, shuttle flight opportunities such as the self-contained payloads program—Get Away Specials—benefited students and proved to be an excellent mechanism for engineering colleges and private corporations to join together in programs oriented toward the development of spaceflight hardware.

All shuttle science programs significantly enhanced graduate education in the physical and life sciences and trained students to work in interdisciplinary teams, thus contributing to US leadership in space science, space engineering, and space health-related disciplines.

Industries and Spin-offs

Industries and Spin-offs

Industries

Aerospace Industry
Jennifer Ross-Nazzal
Helen Lane

Commercial Users
Charles Walker

Small Businesses
Glen Curtis

Spin-offs

NASA Helps Strengthen the "Bridge for Heart Transplants"
Jennifer Fogarty

Making Oxygen Systems Safe
Joel Stoltzfus
Steve Peralta
Sarah Smith

Preventing Land Mine Explosions—Saving Lives with Rocket Power
Brad Cragun

LifeShear Cutters to the Rescue—Powerful Jaws Move Life-threatening Concrete
Jim Butler

The Ultimate Test Cable Testing Device
Pedro Medelius

Keeping Stored Water Safe to Drink—Microbial Check Valve
Richard Sauer

"Green" Lubricant—An Environmentally Friendly Option for Shuttle Transport
Carole-Sue Feagan
Perry Becker
Daniel Drake

In the late 1960s, many of America's aerospace companies were on the brink of economic disaster. The problems stemmed from cutbacks in the space agency's budget and significant declines in military and commercial orders for aircraft. President Richard Nixon's approval of the Space Shuttle Program came along just in time for an industry whose future depended on securing lucrative NASA contracts.

The competition for a piece of the new program was fierce. For the Space Shuttle Main Engines, the agency selected North American Rockwell's Rocketdyne Division. The biggest financial contract of the program, estimated at $2.6 billion, also went to North American Rockwell Corporation to build the Orbiter. The announcement was one bright spot in a depressed economy, and California-based Rockwell allocated work to rivals in other parts of the country. Grumman of Long Island, New York, which had built the Lunar Module, constructed the Orbiter's wings. Fairchild Industries in Germantown, Maryland, manufactured the vertical tail fin. NASA chose Martin Marietta of Denver, Colorado, to build the External Tank, which would be manufactured at the Michoud Assembly Facility in Louisiana. Thiokol Chemical Corporation, based in Utah, won the Solid Rocket Motor contract. In addition to these giants, smaller aerospace companies played a role. Over the next 2 decades, NASA placed an increased emphasis on awarding contracts to small and minority-owned businesses, such as Cimarron Software Services Inc. (Houston, Texas), a woman-owned business.

Shuttle engineering and science sparked numerous innovations that have become commercial products called spin-offs. This section offers seven examples of such technological innovations that have been commercialized and that benefit many people. Shuttle-derived technologies, ranging from medical to industrial applications, are used by a variety of companies and institutions.

Industries

Aerospace Industry

Concurrent with the emphasis placed on reduced costs, policy makers began studying the issue of privatizing the shuttle and turning over routine operations to the private sector. Complete and total privatization of the shuttle failed to come to fruition, but economic studies suggested that contract consolidation would simplify oversight and save funds. In 1980, NASA decided to consolidate Kennedy Space Center (KSC) contracts, and 3 years later, KSC awarded the Shuttle Processing Contract. Johnson Space Center followed KSC's lead in 1985 by awarding the Space Transportation System Operations Contract, which consolidated mission operations work. Industry giants Lockheed and Rockwell won these plums.

Space Shuttle Program Active Flight Hardware Suppliers Distribution by State—12/30/00 to 12/30/04

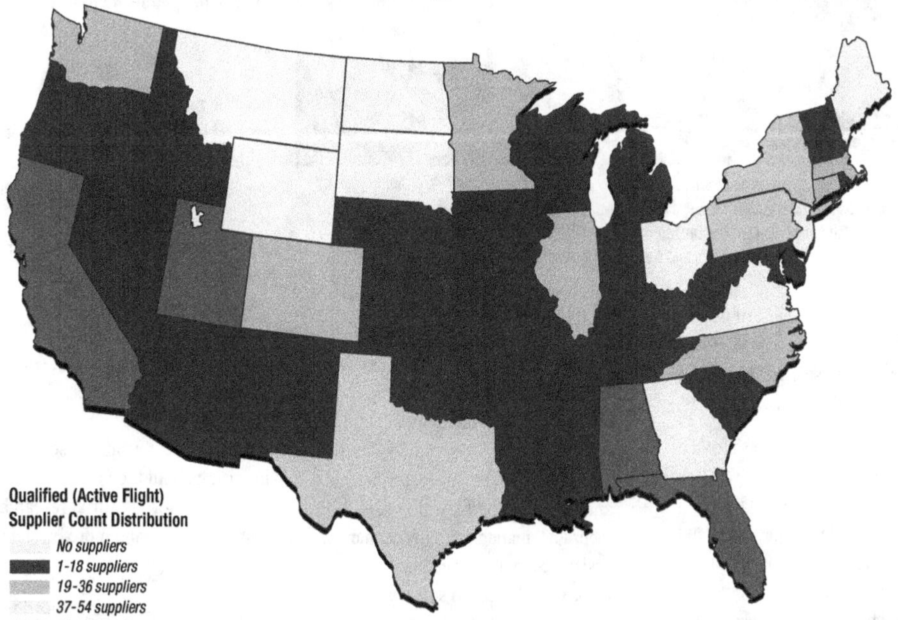

Qualified (Active Flight) Supplier Count Distribution

- *No suppliers*
- *1-18 suppliers*
- *19-36 suppliers*
- *37-54 suppliers*
- *55-72 suppliers*
- *72+ suppliers*

Number of Supplier Companies per Major Component

Orbiter	*817*
Main Engines	*147*
Solid Rocket Boosters	*119*
Motors	*147*
External Tank	*68*
Total	***1,541***

Space Shuttle Program by Contractor—Fiscal Year 2007 - $2.932 Billion

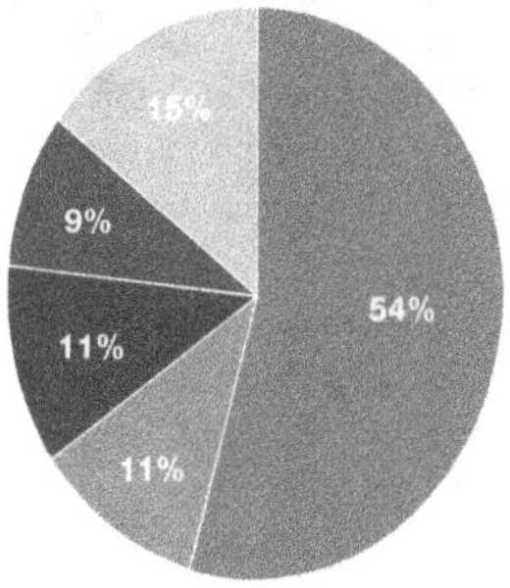

- *United Space Alliance—Space Program Operations Contract*
- *Lockheed Martin—External Tank and Missions Operations Contract*
- *Alliant Techsystems/Thiokol—Solid Rocket Booster Motor Contract*
- *Pratt & Whitney Rocketdyne—Shuttle Main Engine Contract*
- *Other Contracts—Jacobs Technology; InDyne, Inc.; Computer Sciences Corporation; and SGS*

Wyle Laboratories, Inc. works with scientists for the payloads on Neurolab (STS-90 [1998]). The experiment shown is the kinematic, eye tracking, vertical ground reaction force study in March 2002. In the foreground are test operators Chris Miller (left) and Ann Marshburn. The test subject in the harness is Jason Richards and the spotter is Jeremy House.

NASA introduced a host of new privatization contracts in the 1990s to further increase efficiency in operations and decrease costs.

Over the years, companies provided the day-to-day engineering for the shuttle and its science payloads. For instance, Hamilton Sundstrand and ILC Dover were instrumental companies for spacesuit design and maintenance. Lockheed Martin and Jacobs Engineering provided much of the engineering needed to routinely fly the shuttle. Both Lockheed Martin and Wyle Laboratories, Inc. are examples of companies that assured the science payloads operations were successful.

Commercial Users

US industry, aerospace, and others found ways to participate in the Space Shuttle project. Hundreds of large and small companies provided NASA with hardware, software, services, and supplies. Industry also provided technical, management, and financial assistance to academia pursuing government-granted science and technology research in Earth orbit. Yet, a basic drive of industry is to develop new, profitable business.

Beginning in the late 1970s, NASA encouraged American businesses to develop profitable uses of space. This meant conceiving of privately funded, perhaps unique, products for both government and commercial customers—termed "dual use"—as well as for purely commercial consumers. While several aerospace companies were inspired by earlier work in American space projects, a few had ideas for the use of space entirely founded in the unique characteristics of orbital spaceflight. These included launching commercial-use satellites, such as two communications satellites—Anik C-2 and Palapa Bl—launched from Space Transportation System (STS)-7 (1983). The shuttle phased out launching commercial satellites after the Challenger accident in 1986.

Non-aerospace firms, such as pharmaceutical manufacturers, also became interested in developing profitable uses for space. Compared to those of previous spacecraft, the capabilities of the shuttle provided new opportunities for innovation and entrepreneurship. Private capital was invested because of these prospects: regular transport to orbit; lengthy periods of flight; and, if needed, frequent human-tended research and development. Even before the first flight of the shuttle, US private sector businesses were inquiring about the vehicle's availability for industrial research, manufacturing, and more, in space.

During the 30-year Space Shuttle Program, companies interested in microgravity sciences provided commercial payloads, such as a latex reactor experiment performed on STS-3 (1982). These industry-funded payloads continued into the International Space Station Program.

Although the shuttle did not prove to be the best vehicle to enhance commercial research efforts, it was the stepping-stone for commercial use of spacecraft.

Small Businesses Provided Critical Services for the Space Shuttle

As of 2010, government statistics indicated that almost 85% of Americans were employed by businesses with 250 employees or fewer. Such "small businesses" are the backbone of the United States. They also play an important role in America's space program, and were instrumental during the shuttle era. For example, the manufacture and refurbishment of Solid Rocket Motors required the dedication and commitment of many commercial suppliers. Small business provided nearly a fourth of the total dollar value of those contracts. Two examples include: Kyzen Corporation, Nashville, Tennessee; and PT Technologies, Tucker, Georgia.

Kyzen Corporation enabled NASA's goal to eliminate ozone-depleting chemicals by providing a cleaning solvent. This solvent, designed for precision cleaning for the electronics industry, was ideal for dissolving solid rocket propellant from the manufacturing cleaning tooling. The company instituted the rigid controls necessary to ensure product integrity and eliminate contamination.

PT Technologies manufactured precision-cleaning solvent with non-ozone-depleting chemicals. This solvent was designed for use in the telephone and electrical supply industry to clean cables. It also proved to perform well in the production of Solid Rocket Motors.

Small business enterprises are adaptive, creative, and supportive, and their partnerships with NASA have helped our nation achieve its success in space.

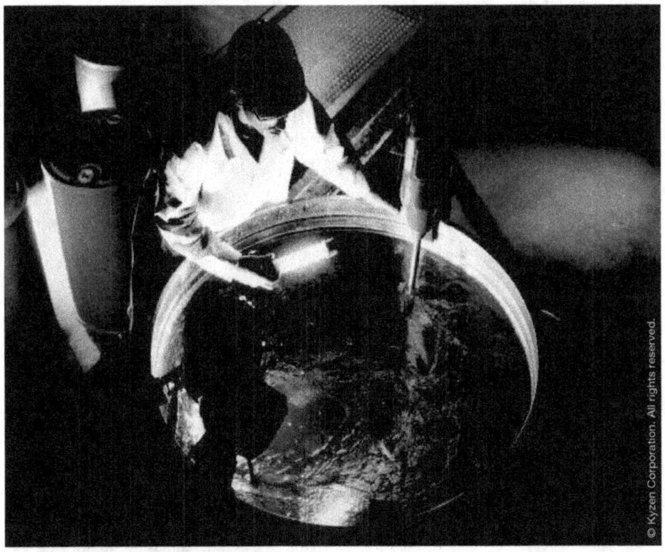

A mixing tank used to produce the cleaning solvent for dissolving solid rocket propellant at Kyzen Corporation. This solvent was free of ozone-depleting chemicals.

Spin-offs

NASA Helps Strengthen the "Bridge for Heart Transplants"

Innovation can occur for many reasons. It can arise from the most unlikely places at the most unlikely times, such as at the margins of disciplines, and it can occur because the right person was at the right place at the right time. The story of David Saucier illustrates all of these points.

Dave Saucier sought medical care for his failing heart and received a heart transplant in 1984 from Drs. DeBakey and Noon at the DeBakey Heart Center at Baylor College of Medicine, Houston, Texas. After his transplant, Dave felt compelled to use his engineering expertise and the expertise of other engineers at Johnson Space Center (JSC) to contribute to the development of a ventricular assist device (VAD)—a project of Dr. DeBakey, Dr. Noon, and colleagues. A VAD is a device that is implanted in the body and helps propel blood from the heart throughout the body. The device was intended to be a bridge to transplant. This successful collaboration also brought in computational expertise from NASA Advanced Supercomputing Division at Ames Research Center (Moffett Field, California).

This far-reaching collaboration of some unlikely partners resulted in an efficient, lightweight VAD. VAD had successful clinical testing and is implemented in Europe for children and adults. In the United States, VAD is used in children and is being tested for adults.

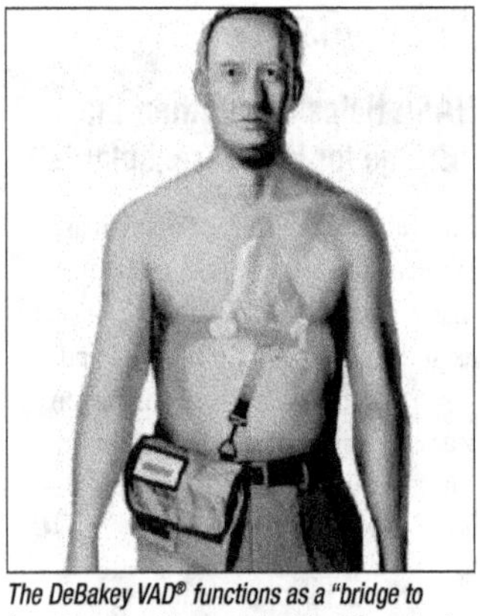

The DeBakey VAD® functions as a "bridge to heart transplant" by pumping blood throughout the body to keep critically ill patients alive until a donor heart is available.

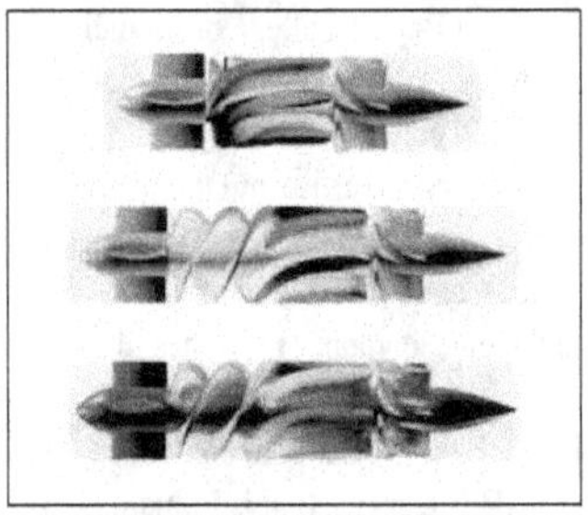

These illustrations show a visual comparison of the original ventricular assist device (top) and the unit after modifications by NASA researchers (center and bottom). Adding the NASA improvements to the MicroMed DeBakey VAD® eliminated the dangerous backflow of blood by increasing pressure and making flow more continuous. The highest pressure around the blade tips are shown in magenta. The blue/green colors illustrate lower pressures.

So, what was it that Dave Saucier and the other engineers at JSC thought they knew that could help make a VAD work better, be smaller, and help thousands of people seriously ill with heart failure and waiting for a transplant? Well, these folks had worked on and optimized the turbopumps for the shuttle main engines that happen to have requirements in common with VAD. The turbopumps needed to manage high flow rates, minimize turbulence, and eliminate air bubbles. These are also requirements demanded of a VAD by the blood and body.

In the beginning, VADs had problems such as damaging red blood cells and having stagnant areas leading to the increased likelihood of blood clot development. Red blood cells are essential for carrying oxygen to the tissues of the body. Clots can prevent blood from getting to a tissue, resulting in lack of oxygenation and buildup of toxic waste products that lead to tissue death. Once engineers resolved the VAD-induced damage to red blood cells and clot formation, the device could enter a new realm of clinical application. In 1996 and 1999, engineers from JSC and NASA Ames Research Center and medical colleagues from the Baylor College of Medicine were awarded US patents for a method to reduce pumping damage to red blood cells and for the design of a continuous flow heart pump, respectively. Both of these were exclusively licensed to MicroMed Cardiovascular, Inc. (Houston, Texas) for the further development of the small, implantable DeBakey VAD®.

MicroMed successfully implanted the first DeBakey VAD® in 1998 in Europe and, to date, has implanted 440 VADs. MicroMed's HeartAssist5® (the 2009 version of the DeBakey VAD®) weighs less than 100 grams (3.5 oz), is implanted in the chest cavity in the pericardial space, which reduces surgical complications such as infections, and can operate for as many as 9 hours on battery power, thereby resulting in greater patient freedom. This device not only acts as a bridge to transplant, allowing patients to live longer and better lives while waiting for a donor heart, it is now a destination therapy. People are living out their lives with the implanted device and some are even experiencing recovery, which means they can have the device explanted and not require a transplant.

Making Oxygen Systems Safe

Hospitals, ambulances, industrial complexes, and NASA all use 100% oxygen and all have experienced tragic fires in oxygen-enriched atmospheres. Such fires demonstrated the need for knowledge related to the use of materials in oxygen-enriched atmospheres. In fact, on April 18, 1980, an extravehicular mobility unit planned for use in the Space Shuttle Program was destroyed in a dramatic fire during acceptance testing. In response to these fire events, NASA developed a test method and procedures that significantly reduced the danger. The method and procedures are now national and international industrial standards. NASA White Sands Test Facility (WSTF) also offered courses on oxygen safety to industry and government agencies.

During the shuttle era, NASA made significant advances in testing and selecting materials for use in high-pressure, oxygen-enriched atmospheres. Early in the shuttle era, engineers became concerned that small metal particles could lead to ignition if the particles were entrained in the 277°C (530°F) oxygen that flowed through the shuttle's Main Propulsion System gaseous oxygen flow control valve. After developing a particle impact test, NASA determined that the stainless-steel valve was vulnerable to particle impact ignition. Later testing revealed that a second gaseous oxygen flow control valve, fabricated from an alloy with nickel chromium, Inconel® 718, was also vulnerable to particle impact ignition. Finally, engineers showed that an alloy with nickel-copper, Monel®, was invulnerable to ignition by particle impact and consequently was flown in the Main Propulsion System from the mid 1980s onward.

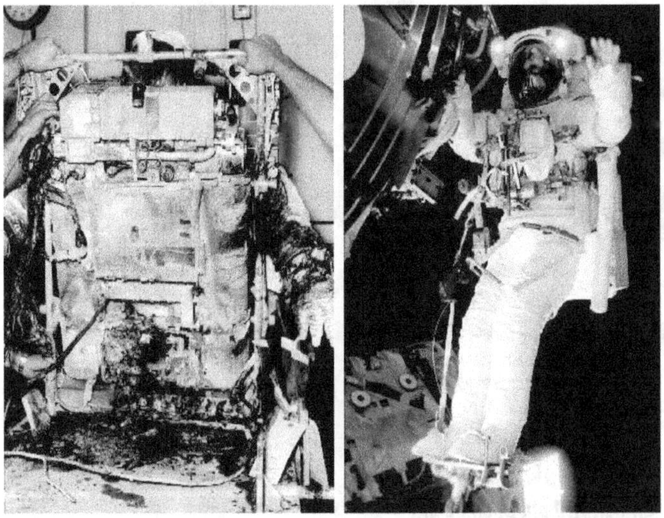

The original shuttle extravehicular mobility unit with an aluminum secondary oxygen pack isolation valve and first-stage regulator ignited and burned during acceptance ground testing on an unoccupied unit in 1980 (left). The redesigned unit with a nickel-copper alloy secondary oxygen pack isolation valve and first-stage regulator is being used with much success (right).

NASA's activities led to a combustion test patent (US Patent Number 4990312) that demonstrated the superior burn resistance of a nickel-copper alloy used in the redesigned, high-pressure oxygen supply system. Member companies of the American Society for Testing and Materials (ASTM) Committee G-4 pooled their resources and requested that NASA use the promoted combustion test method to determine the relative flammability of alloys being used in industry oxygen systems. Ultimately, this test method was standardized as ASTM G124.

NASA developed an oxygen compatibility assessment protocol to assist engineers in applying test data to the oxygen component and system designs. This protocol was codified in ASTM's Manual 36 and in the National Fire Protection Association *Fire Protection Handbook*, and has gained international acceptance.

Pretest.

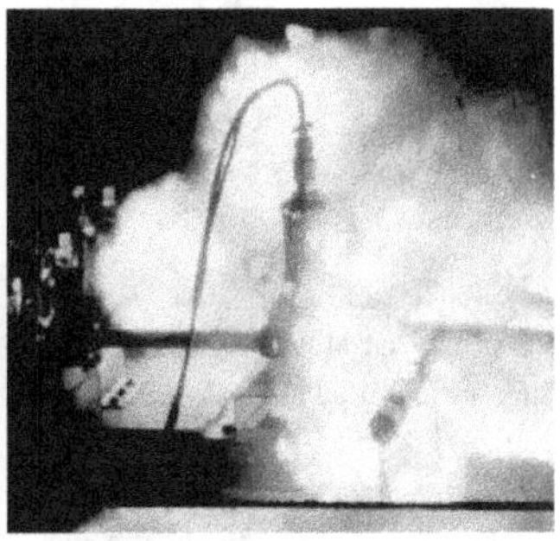

Ignition by particle impact.

This gaseous oxygen valve was found to be vulnerable to ignition when small metal particles were ingested into the valve. The test method developed for this is being used today by the aerospace and industrial oxygen communities.

Another significant technology transfer from the Space Shuttle Program to other industries is related to fires in medical oxygen systems. From 1995 through 2000, more than 70 fires occurred in pressure-regulating valves on oxygen cylinders used by firefighters, emergency medical responders, nurses, and therapeutic-oxygen patients. The Food and Drug Administration approached NASA and requested that a test be developed to ensure that only the most ignition- and burn-resistant, pressure-regulating valves be allowed for use in these medical systems. With the help of a forensic engineering firm in Las Cruces, New Mexico, the WSTF team developed ASTM G175, entitled *Standard Test Method for Evaluating the Ignition Sensitivity and Fault Tolerance of Oxygen Regulators Used for Medical and Emergency Applications*. Since the development and application of this test method, the occurrence of these fires has diminished dramatically.

This spin-off was a significant development of the technology and processes to control fire hazards in pressurized oxygen systems. Oxygen System Consultants, Inc., in Tulsa, Oklahoma, OXYCHECK™ Pty Ltd in Australia, and the Oxygen Safety Engineering division at Wendell Hull & Associates, Inc., in Las Cruces, New Mexico, are examples of companies that performed materials and component tests related to pressurized oxygen systems. These businesses are prime examples of successful technology transfer from the shuttle activities. Those involved in the oxygen production, distribution, and user community worldwide recognized that particle impact ignition of metal alloys in pressurized oxygen systems was a significant ignition threat.

Preventing Land Mine Explosions—Saving Lives with Rocket Power

Every month, approximately 500 people—including civilians and children—are killed or maimed by accidental contact with land mines. Estimates indicate as many as 60 to 120 million active land mines are scattered across more than 70 countries, including areas where hostilities have ceased. Worldwide, many of the more than 473,000 surviving victims require lifelong care.

In 1990, the US Army solicited existing or short-term solutions to in-field mine neutralization with the ideal solution identified as a device that was effective, versatile, inexpensive, easy to carry, and easy to use, but not easily converted to a military weapon.

Rocket Science—An Intelligent Solution

The idea of using leftover shuttle propellant to address this humanitarian crisis can be traced back to late 1998 when shuttle contractor Thiokol (Utah) suggested that a flare, loaded with propellant, could do the job. To validate the concept, engineers tested their idea on small motors. These miniature rocket motors, no larger than a D-size battery, were used in research and development efforts for ballistics characterization. With some refinements, by late 1999, the flare evolved into a de-mining device that measures 133 mm (5 in.) in length by 26 mm (1 in.) in diameter, weighs only 90 grams (3.2 oz), and burns for approximately 60 seconds. NASA and Thiokol defined an agreement to use the excess propellant.

The Thiokol de-mining flare used excess shuttle propellant resulting from Solid Rocket Motor casting operations to burn through land mine casings and safely ignite the explosives contained within. The flares were activated with an electric match or a pyrotechnic fuse.

Ignition Without Detonation—How It Works

The de-mining flare device is ignited by an electric match or a pyrotechnic fuse; it neutralizes mines by quickly burning through the casing and igniting the explosive fill without detonation. The benefit of this process includes minimizing the destructive effect of demolition, thereby preventing shrapnel from forming out of metallic and thick-cased targets. The flares are simple and safe to use, and require minimal training. The flare tube can be mounted on a three-legged stand for better positioning against the target case.

These de-mining flares were tested against a variety of mines at various installations. These trials went well and generated much interest. Thiokol funded further development to improve production methods and ease deployment.

All branches of the US armed services have purchased the flare. It has been successfully used in Kosovo, Lebanon, Jordan, Ethiopia, Eritrea, Djibouti, Nicaragua, Iraq, and Afghanistan, and has been shown to be highly effective.

LifeShear Cutters to the Rescue—Powerful Jaws Move Life-threatening Concrete

Hi-Shear Technology Corporation of Torrance, California, used NASA-derived technology to develop a pyrotechnic-driven cutting tool that neutralized a potentially life-threatening situation in the bombed Alfred P. Murrah Federal Building in Oklahoma City, Oklahoma, in April 1995. Using Jaws of Life™ heavy-duty rescue cutters, a firefighter from the Federal Emergency Management Agency Task Force team sliced through steel reinforcing cables that suspended an 1,814.4-kg (2-ton) slab of concrete, dropping the slab six stories. It took only 30 seconds to set up and use the cutters.

The shuttle used pyrotechnic charges to release the vehicle from its hold-down posts on the launch pad, the Solid Rocket Boosters from the External Tank after their solid fuel was spent, and the tank from the shuttle just prior to orbit. This type of pyrotechnical separation technology was applied in the early 1990s to the development of a new generation of lightweight portable emergency rescue cutters for freeing accident victims from wreckage. Known as LifeShear cutters, they were developed under a cooperative agreement that teamed NASA and Hi-Shear Technology Corporation. Hi-Shear incorporated this pyrotechnic feature into their Jaws of Life™ heavy-duty rescue cutters. The development project was undertaken to meet the need of some 40,000 US fire departments for modern, low-cost emergency cutting equipment.

Hi-Shear Technology Corporation developed, manufactured, and supplied pyrotechnically actuated thrusters, explosive bolts, pin pullers, and cutters, and supplied such equipment for a number of NASA deep-space missions plus the Apollo/Saturn, Skylab, and shuttle.

The key technology for the LifeShear cutter is a tailored power cartridge—a miniature version of the cartridges that actuated pyrotechnic separation devices aboard the shuttle. Standard cutting equipment employs expensive gasoline-powered hydraulic pumps, hoses, and cutters for use in accident extraction. The Jaws of Life™ rescue tool requires no pumps or hoses, and takes only about 30 seconds to ready for use. It can sever automotive clutch and brake pedals or cut quickly through roof posts and pillars to remove the roof of an automobile. Firefighters can clear an egress route through a building by cutting through reinforcement cable and bars in a collapsed structure situation.

NASA-developed tool, licensed under the name "LifeShear," used at the bombed Alfred P. Murrah Federal Building (1995), Oklahoma City, Oklahoma.

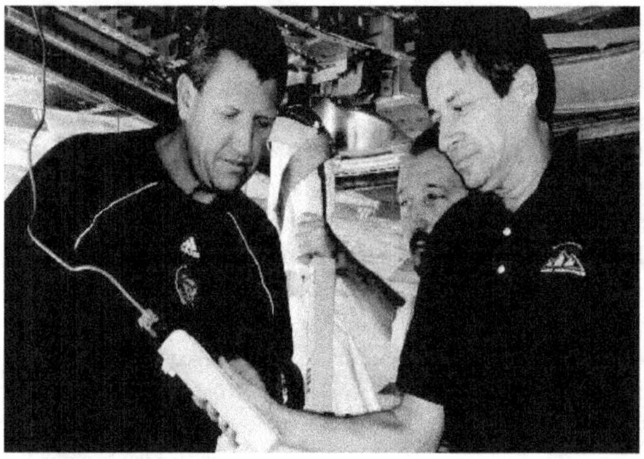

Kennedy Space Center engineers conduct wire fault testing using portable Standing Wave Reflectometer. From left to right: Ken Hosterman; John Jones; and Pedro Medelius (inventor).

The Ultimate Test Cable Testing Device

It's hard to imagine, when looking at a massive launch vehicle or aircraft, that a problem with one tiny wire could paralyze performance. Faults in wiring are a serious concern for the aerospace and aeronautic (commercial, military, and civil) industries. The shuttle had circuits go down because of faulty insulation on wiring. STS-93 (1999) experienced a loss of power when one engine experienced a primary power circuit failure and a second engine had a backup power circuit fault. A number of accidents occurred as a result of faulty wiring creating shorts or opens, causing the loss of control of the aircraft or arcing and leading to fires and explosions. Some of those accidents resulted in loss of lives, such as in the highly publicized TWA Flight 800 accident in 1996.

With the portable Standing Wave Reflectometer cable tester, it was possible to accurately pinpoint malfunctions within cables and wires to reliably verify conditions of electrical power and signal distribution. This included locating problems inside shuttle. One of its first applications at Kennedy Space Center (KSC) was to detect intermittent wire failures in a cable used in the Solid Rocket Boosters.

The Standing Wave Reflectometer cable tester checked a cable with minimal disruption to the system under test. Personnel frequently had to de-mate both ends of cables when troubleshooting a potential instrument problem to verify that the cable was not the source of the problem. Once a cable was de-mated, all systems that had a wire passing through the connector had to be retested when the cable was reconnected. This resulted in many labor-hours of revalidation testing on systems that were unrelated to the original problem. The cost was exorbitant for retesting procedures. The same is true for aeronautical systems, where airplanes have to be checked frequently for faulty cables and sensors. The most useful method and advantage of the Standing Wave Reflectometer technology over other existent types of technologies is the ability to measure from one end of a cable, and to do comparative-type testing with components and avionics still installed.

Eclypse International Corporation, Corona, California, licensed and marketed two commercial versions of the Standing Wave Reflectometers based on the prototype designed and patented by KSC. One called ESP provided technicians with a simple, plain-English response as to where the electrical fault was located from the point at which the technicians were testing. A second product, ESP+, provided added memory and software for looking at reflections from the aircraft, which was useful in determining some level of "soft fault"—faults that are not open or shorted wires.

The technology was evaluated by the US Navy, US Marines, and US Air Force to test for its ruggedness for deployment in Afghanistan. The country was known for a fine grade of sand and dusty conditions—a taxing combination rarely found in the United States. The model underwent operational evaluation by the US Navy, US Marines, and US Air Force, and the US Army put these instruments into the battle damage and repair kits that went to Afghanistan, Iraq, and other parts of the world where helicopter support is required. This innovation has proved to be versatile in saving time and lives.

The Ultimate Test

In Bagram, Afghanistan, October 2004, one particular Northrop Grumman EA-6B Prowler aircraft was exhibiting intermittent problems on a critical cockpit display panel. To make matters worse, these problems were seldom seen during troubleshooting but occurred multiple times on nearly every flight. It was a major safety problem, especially when flying at night in a war zone in mountainous terrain. Squadron maintainers had been troubleshooting for weeks, changing all associated removable components and performing wire checks with no discernable success. After approximately 60 hours of troubleshooting, which included phone consultation with engineering and the manufacturer of the electronic system that was providing intermittent symptoms, the Naval Air Technical Data & Engineering Service Command decided to try the Standing Wave Reflectometer and immediately observed a measured change of conductor length as compared with similar paths on the same aircraft. Technicians were able to isolate the problem and replace the faulty wire.

Keeping Stored Water Safe to Drink—Microbial Check Valve

The Space Shuttle system for purifying water has helped the world's need for safe water, especially for disaster situations, backpackers, and remote water systems where power and active monitoring were limited. This well-tested system, called the Microbial Check Valve, is also used on the International Space Station. This valve is ideal for such applications since it can be stored for a long period of time and is easily activated.

The licensee and co-inventor, with NASA, of the Microbial Check Valve was Umpqua Research Company (Myrtle Creek, Oregon). The system was used on all shuttle flights to prevent growth of pathogens in the crew drinking water supply. The valve is a flow-through cartridge containing an iodinated polymer, which provides a rapid contact microbial kill and also imparts a small quantity of dissolved iodine into the effluent stream. This prevents further microbial growth and maintains water safety.

The Microbial Check Valve—measuring 5.1 cm (2 in.) in diameter, 12.7 cm (5 in.) in length—is a stainless-steel cylinder with connections on its ends that facilitated its installation in the shuttle water system line. The cylinder is packed with iodinated ion exchange resin (the base resin is Dowex SBR®). A perforated plate backed by a spring presses against the resin and keeps it compacted to prevent short-circuiting of the water as it flows through the resin.

Treatment of uncontrolled microbial growth in stored water was essential in the shuttle because water was produced through the fuel cells of oxygen and hydrogen, and the resultant water was stored in large tanks. The shuttle was reused and, therefore, some residual water always remained in the tanks between launches. Iodine, like chlorine, prevents microbial growth, is easy to administer, and has long-life effectiveness as it is much less volatile than chlorine.

The innovation was a long-shelf-life iodinated resin. When water passed through the resin, iodine was released to produce acceptable drinking water. This system inactivated seven bacteria, yeasts and molds and three different viruses, including polio. The costs were also very reasonable.

The volume of the resin in the valve was selected to treat five 30-day shuttle equivalent missions (3,000 L [793 gal]: based on 2.8 L [0.7 gal]/day/person use rate for a seven-person crew) for the maximum shuttle fuel cell water production rate of 120 L (31.7 gal)/hr. All in-flight-produced water flowed through the microbial check valve to impart a small iodine residual to prevent microbial growth during storage and back contaminations, further contributing to the safety and purification of drinking water during shuttle missions.

"Green" Lubricant—An Environmentally Friendly Option for Shuttle Transport

In the mid 1990s, NASA uncovered an environmental problem with the material used to lubricate the system used to transport the shuttle. The agency initiated an effort to identify an environmentally friendly lubricant as a replacement.

The Mobile Launcher Platform at KSC provided a transportable launch base for the shuttle. NASA used a vehicle called a "crawler" with a massive track system to transport the platform and a shuttle. During transport, lubricants had to withstand pressures as high as 5,443 metric tons (6,000 tons). Lubrication reduced wear and noise, lengthened component life, and provided protection from corrosive sand and heat.

NASA personnel injected low-viscosity lubricant on the pins that structurally linked 57 individual track "shoes" together to form an individual tread belt. Periodic application during transport minimized crankshafting of individual pins inside the shoe lug holes, thus reducing the risk of structural damage and/or failure of the tread belt system. The performance parameters of the original lubricant resulted in a need for operators to spray the pins approximately every mile the transporter traveled.

Lockheed Martin Space Operations, NASA's contractor for launch operations at KSC, turned to Sun Coast Chemicals of Daytona, Inc. (Daytona Beach, Florida) for assistance with co-developing a biodegradable, nontoxic lubricant that would meet all Environmental Protection Agency and NASA requirements while providing superior lubricating qualities. Sun Coast Chemicals of Daytona, Inc. assembled a team of researchers, production personnel, and consultants who met with NASA personnel and contractors. This team produced a novel formulation that was tested and certified for trial, then tested directly on the crawler.

The new lubricant—Crawler Track Lube—had a longer service life than previous lubricants, and was injected at longer intervals as the transporter was being operated. Additionally, the product was not an attractive food source to wildlife. Success with its initial product and the Crawler Track Lube led to an industrial product line of 19 separate specialty lubricants.

The Mobile Launch Platform transported the shuttle to the launch pad. Inset photo shows the dispenser that injects the lubricant on the pins, which are necessary for the treadbelt.

The Shuttle Continuum, Role of Human Spaceflight

The Shuttle Continuum, Role of Human Spaceflight

President George H.W. Bush

Pam Leestma and Neme Alperstein
Elementary school teachers

Norman Augustine
Former president and CEO of Lockheed Martin Corporation

John Logsdon
Former director of Space Policy Institute The George Washington University

Canadian Space Agency

General John Dailey
Director of Smithsonian National Air and Space Museum

Leah Jamieson
Dean of the College of Engineering Purdue University

Michael Griffin
Former NASA administrator

The theme of this book is the scientific and engineering accomplishments of the Space Shuttle Program. The end of this longest-running human spaceflight program marks the end of an era for our nation. At this juncture, it is natural to ask: Why human spaceflight? What is the future of human spaceflight? What space exploration initiatives should we engage in, in the future?

The editor in chief of this publication invited some noted leaders from the government and industry, educators, students, and others to share their views and thoughts on these questions. Each contributor provided his or her own unique perspective. The editors are pleased and grateful for their contribution.

GEORGE BUSH

First and foremost, I am pleased to contribute a few words to this worthwhile project on the legacy of the Space Shuttle program because of my respect for the remarkable men and women who have shaped the program, and led it, and made it one of the most vital forces for scientific discovery and progress in our world.

To me, there are few public endeavors that best exemplify the American spirit of innovation and daring than does our Space Shuttle program. Like the manned space flight programs that preceded it — indeed, as was the case with each and every explorer throughout the ages — the Americans leading the Shuttle program yesterday, today, and tomorrow are drawn to challenge. They seek to push back the horizon of discovery. And, yes, maybe as creator Gene Roddenberry of *Star Trek* fame suggested, they also seek to "go boldly where no man has gone before."

Was it not the same spirit that inspired the pioneers of old to take to the ancient spice trails of Asia, or to alight from the ports of medieval Europe for the highly uncertain journey ahead?

Just as important to me is the way NASA and our government has opened our Shuttle and Space Station program to our partners and allies across the world, to ensure that the exploration of space benefits not just America — but mankind as a whole.

This, then, is the essence of American leadership, your leadership, that today is expanding our scientific awareness even as it brings our world closer together.

President Kennedy was right: we do these things not because they are easy, but because they are hard. But because they are hard, and because you continue to persevere and succeed, and because, furthermore, you succeed based on the values that have always made America a force for goodness and progress in our world, you also help to continue inspiring our world and capturing our imagination.

So to the heirs of our manned space program, keep up the wonderful work. Keep pushing back that horizon, and boldly seeking new places to go. And in the process, help us to keep setting higher standards for the kinds of scientific research and courageous exploration that make this a better world than we found it. Succeeding generations of Americans have blessed us in such a manner; now it is your time to answer the call.

May God bless you all, and our United States.

G. Bush

Inspiring Students Through Human Spaceflight

Pam Leestma and Neme Alperstein

Pam Leestma taught elementary school for 30 years at Valley Christian Elementary School, Bellflower, California. She won the 2008 National American Star Teaching Award.

Neme Alperstein taught for 22 years at the Harry Eichler School, a New York City public school. She was the New York City Teacher of the Year in 2000.

Neil Armstrong's "one small step for man, one giant leap for mankind" changed the course of history in our quest to explore space. "Failure is not an option" was the Apollo Program's vision to inspire the nation and is the space agency's legacy for the next generation.

Today we are a global community with international space partners exploring a new frontier filled with imagination and innovation. Scientific discoveries, human spaceflight, space tourism, moon colonies, and the exploration of Mars and beyond will be the vehicles that will continue to find common ground for transcending borders through understanding, respect, friendship, and peace.

NASA's education programs have provided the powerful resources to engage young minds. Their essential 21st century tools have brought our youth closer to those on the frontier of exploration through numerous multimedia interactive technologies. Some ways that we, as educators, have been able to get our students "up close and personal" with NASA include speaking with an astronaut aboard the International Space Station in real time (a downlink), using the facilities of a local California city hall and a New York City community center for a NASA first coast-to-coast downlink, videoconferencing with NASA's Digital Learning Network experts and astronauts living and training under water off the Florida coast (NASA's Extreme Environment Missions Operations), growing basil seeds flown in space with astronaut and educator Barbara Morgan, participating in NASA's live webcasts, watching NASA TV during coverage of Space Shuttle launches and landings, and organizing stargazing family nights for the school community. The impact of these extraordinary experiences has been life changing.

The unimaginable has become the world of infinite possibilities in science, technology, engineering, and mathematics. Human spaceflight missions reflect the diversity of our global community and the best that such collaboration offers mankind. This diversity reaches out to all students who see increased opportunities for participation. They see the potential to create the next generation of "spinoffs" that will improve daily life as a result of NASA research and development. They include medical breakthroughs, the development of robotics in exploration and in everyday life, materials science in the creation of materials with new properties (i.e., spacesuits), researching the effect of extreme environments, and the quest for cures and developing new medicines in microgravity.

NASA continues to support teachers through its professional development, conferences, workshops, content across the curriculum, and its willingness to provide access to its scientific community and experts. We never cease to be amazed by NASA's generosity of spirit ever present at the Space Exploration Educators Conference we always attend. Teachers return to their classrooms inspired. It's a ripple effect.

NASA's vision has provided the spark that ignites the excitement and wonder of exploration and discovery. Our students see themselves as the next explorers of this new frontier. It is an imperative that we continue human spaceflight if for no other reason than to improve life here on Earth and foster cooperation within the global community. Space exploration offers our children hope for the future.

What's Next for Human Spaceflight?

Norman Augustine
Former president and CEO of Lockheed Martin Corporation and recipient of many honors for his national defense, homeland security, and science policy accomplishments.

Parachuting an instrument package onto the summit of Mt. Everest would, without question, have been a significant and exciting scientific contribution. But would it have had the broad impact of Sir Edmund Hillary and Tenzing Norgay standing atop the 29,035 ft peak?

There are many important missions that can and should be accomplished with robotic spacecraft, but when it comes to inspiring a nation, motivating young would-be scientists and engineers and adaptively exploring new frontiers, there is nothing like a human presence. But humans best serve a nation's space goals when employed not as truck drivers but rather when they have the opportunity to exploit that marvelous human trait: flexibility. A prime example is the on-orbit repair of the Hubble Space Telescope using the shuttle. Without that capability for in situ human intervention, Hubble, itself a monumental accomplishment, would have been judged a failure. Indeed, there are important missions for both humans and robots in space—but each is at its best when it does not try to invade the other's territory.

So what is next for human spaceflight? There is a whole spectrum of interesting possibilities that range from exploring Mars, Demos, or Phoebus, to establishing a station on the moon or at a neutral gravity point. It would seem that the 1990 recommendations of the White House/NASA commission on the Future of the U.S. Space Program still make a lot of sense. These include designating Mars as the primary long-term objective of the human space program, most likely with the moon as a scientific base and stepping-off point, and getting on with developing a new heavy-lift launch capability (probably based on the shuttle's External Tank).

The cost of space transportation was, and is today, the most intransigent impediment to human space travel. The mission traffic models are sparse; the development costs large; the hazard of infant mortality of new vehicles daunting; and the arithmetic of discounted cost accounting and amortization intimidating. Thus, at least in my opinion, the true breakthrough in human spaceflight will occur only when space tourism becomes a reality. Yes, space tourism. There is a close parallel to the circumstance when World War II solved the chicken and egg problem of commercial air travel.

By space tourism I do not refer to a few wealthy individuals experiencing a few moments of exposure to high altitudes and zero g's. Rather, I mean a day or two on orbit for large numbers of people, peering through telescopes, taking photographs, eating, and exercising. There are, of course, those who would dismiss any such notion as fantasy—but what might the Wright Brothers have said if told that within the century the entire population of Houston would each day climb aboard an airplane somewhere in the US and complain that they had already seen the movie? Or Scott and Amundsen if informed that 14,000 people would visit

Antarctica each summer and 50 would live at the South Pole? Or James Wesley Powell if advised that 15,000 people would raft the Grand Canyon each year? Or Sir Edmund Hillary if told that 40 people would stand on top of Mount Everest one morning? In short, to be human is to be curious, and to be curious is to explore. And if there is any one thing we have learned about space pursuits, it is that they are a lot like heart surgery…if you are going to do any of it, it is wise to do a lot of it.

We have of course learned many other important things from the Space Shuttle Program. Those include how to integrate extraordinarily complex systems so as to operate in very unforgiving environments; that high traffic rates can and must be satisfied with reusability; that subsystems intended to be redundant are redundant only when they are independent; that long-term exposure to space can be tolerable for humans, at least in near-Earth orbit; and that the problems you expect (read tiles) can be overcome, while the problems you don't expect can overcome you (read seals and high-velocity, low-density fragment impacts). These and other lessons from the Space Shuttle human space programs have had a major effect on engineering discipline throughout the aerospace industry and much of the electronics industry as well.

There is a noteworthy parallel between the situation in which America found itself just after the Sputnik wake-up call and the circumstance that exists today just after the toxic mortgage wake-up call. In the former instance, much attention was turned to our nation's shortcomings in education, in producing future scientists and engineers, and in underinvestment in basic research. After Sputnik, the human space program became the centerpiece in an effort to reverse the above situation and helped underpin several decades of unparalleled prosperity. Today, the nation once again suffers these same ailments and once again is in need of "centerpieces" to focus our attention and efforts. And to this end nothing inspires young would-be scientists and engineers like space and dinosaurs—and we are noticeably short of the latter.

As for me, nothing other than the birth of my children and grandchildren has seemed more exciting than standing at the Cape and watching friends climb aboard those early shuttles, atop several hundred thousand gallons of liquid hydrogen and liquid oxygen, and then fly off into space.

My mother lived to be 105 and had friends who crossed the prairies in covered wagons. She also met friends of mine who had walked on the moon. Given those genes I may still have a shot at buying a round-trip ticket to take my grandchildren to Earth orbit instead of going to Disney World. And the Space Shuttle Program provided important parts of the groundwork for that adventure. All I need is enough "runway" remaining.

Global Community Through Space Exploration

John Logsdon, PhD

Former director of Space Policy Institute and professor, The George Washington University, and member of major space boards and advisory committees including the NASA Columbia Accident Investigation Board.

The Space Shuttle has been a remarkable machine. It has demonstrated the many benefits of operations in low-Earth orbit, most notably the ability to carry large pieces of equipment into space and assemble them into the International Space Station (ISS). Past research aboard the shuttle and especially future research on the ISS could have significant benefits for people on Earth. But research in low-Earth orbit is not exploration. In my view, it is past time for humans once again to leave low-Earth orbit and restart exploration of the moon, Mars, and beyond. President George W. Bush's January 2004 call for a return to the moon and then a journey to Mars and other deep space destinations is the policy that should guide US government human spaceflight activities in the years to come.

The 2004 exploration policy announced by President Bush also called for international participation in the US exploration initiative. The experience of the ISS shows the value of international partnerships in large-scale space undertakings. While the specifics of the ISS partnership are probably not appropriate for an open-ended exploration partnership, the spirit and experience of 16 countries working together for many years and through difficult challenges certainly is a positive harbinger of how future space exploration activities can be organized.

Since 2006, 14 national space agencies have been working together to chart that future. While the United States is so far the only country formally committed to human exploration, other space agencies are working hard to convince their governments to follow the US lead and join with the United States in a multinational exploration effort. One product of the cooperation to date is a "Global Exploration Strategy" document that was approved by all 14 agency heads and issued in May 2007. That document reflects on the current situation with words that I resonate with: "Opportunities like this come rarely. The human migration into space is still in its infancy. For the most part, we have remained just a few kilometers above the Earth's surface—not much more than camping out in the backyard."

It is indeed time to go beyond the "camping out" phase of human space activity, which has kept us in low-Earth orbit for 35 years. Certainly the United States should capitalize on its large investment in the ISS and carry out a broadly based program of research on this orbiting laboratory. But I agree with the conclusions of a recent White Paper prepared by the Space, Policy, and Society Research Group at MIT: "A primary objective of human spaceflight has been, and should be, exploration." The Group argues that "Exploration is an expansion of human experience, bringing people into new places, situations, and environments, expanding and redefining what it means to be human." It is exploration, so defined, that provides the compelling rationale for continuing a government-funded program of human spaceflight.

I believe that the new exploration phase of human spaceflight should begin with a return to the moon. I think the reasons to go back to the moon are both that it is the closest place to go and it is an interesting place in its own right. We are not technologically ready for human missions to Mars, and the moon is a more understandable destination than just flying to a libration point in space or to a near-Earth object. The moon is like an offshore island of the planet Earth, and it only takes 3 days to get there. During the Apollo Program, the United States went to the surface of the moon six times between 1969 and 1972; the lunar crews explored only the equatorial region of the moon on the side that always faces the Earth. So we have never visited 85 to 90 percent of the moon's surface, and there are lots of areas yet to explore. The far side of the moon may be the best place in the solar system for radio astronomy. Most people who are looking at the issue now think that one of the poles of the moon, probably the South Pole, is a very interesting place scientifically, and that there may be resources there that can be developed for use in further space exploration. So the moon is an interesting object to study, and to do science from, and perhaps as a place to carry out economically productive activity.

The Space Shuttle has left us a legacy of exciting and valuable exploits in low-Earth orbit. But it is now time to go explore.

The Legacy of the "Space Shuttle"

Views of the Canadian Space Agency

The Space Transportation System; a.k.a. the "Space Shuttle"; is the vehicle that arguably brought Canada to maturity as a global space power. Canada was an early advocate in recognizing the importance that space could play in building the country. Initially, this was achieved through the development of small indigenous scientific satellites to study the Earth's upper atmosphere, beginning with Alouette, launched by NASA in 1962, which positioned Canada as the third nation, after the Soviet Union and the United States of America, to have its own satellite successfully operate in the harsh and largely unknown environment of space. The follow-on Alouette-II and ISIS series of satellites (1965 to 1971) built national competence and expertise and set the foundation for Canada's major contributions to the rapidly developing field of satellite communications (Anik series and Hermes), to using Earth Observation data to meet national needs, as well as to the development of signature technologies that were the basis of Canada's space industry (e.g., STEM* deployable systems, antennas). By the mid-1970s, however, Canada's emerging space program was at a crossroads: space communications were becoming commercialized, Canada was not yet ready to commit to the development of an Earth Observation Satellite, and no new scientific satellites or payloads were approved. This situation changed dramatically in 1974 when the Government of Canada approved the development of a robotic arm as a contribution to the Space Shuttle Program initiated by NASA two years earlier. This Shuttle Remote Manipulator System was designed to deploy and retrieve satellites from and to the Shuttle orbiter's payload bay, as well as support and move extra-vehicular astronauts and payloads within the payload bay. The first "Canadarm" was paid for by Canada and first flew on the second Shuttle flight in November 1981. Originally planned by NASA to be flown only occasionally, Canadarm has become a semi-permanent fixture due to its versatility and reliability, especially in support of extra-vehicular activities; i.e., spacewalks; and, more recently, as an essential element in the construction and servicing of the International Space Station and the detailed remote inspection of the Shuttle after each launch that is now a mandatory feature of each mission. Canadarm has become an important and very visible global symbol of Canadian technical competence, a fact celebrated in a recent 2008 poll of Canadians that identified the Canadarm as the top defining accomplishment of the country over the last century.

Returning to scientific endeavours, the Shuttle's legacy with respect to the space sciences in Canada was more circuitous. Towards the end of the 1970s, following the successful Alouette/ISIS series, Canada turned its attention to defining its next indigenous scientific satellite mission. As the merits of a candidate satellite called Polaire were debated, Canadian

**STEM—storage tubular extendible member*

scientists were encouraged to propose experiments in response to an Announcement of Opportunity released by NASA in 1978 to fly future missions on the Shuttle. This was during the heady days when a Shuttle mission was proposed to fly every couple of weeks with rapid change-out of payloads—the "space truck" concept—and with the possibility to utilize the formidable advantage of the Shuttle to launch and return scientific payloads leading to multiple mission scenarios for the same experiment or facility. Three Canadian proposals to fly sophisticated, complex experiments in the Shuttle payload bay were accepted by NASA—an Energetic Ion Mass Spectrometer to measure the charged particle environment; an ambitious topside-sounder experiment called Waves In Space Plasmas, a follow-on to the Alouette/ISIS program, to measure the propagation of radio waves through and within the Earth's atmosphere; and an optical measurement of atmospheric winds from space called Wide Angle Michelson Doppler Imaging Interferometer. Ironically, none of these three experiments flew on the Shuttle, all falling to the reality of a technically challenging program where missions every few months became the norm rather than every couple of weeks. However, the impetus to the Canadian scientific community of this stimulus through the infusion of new funds and opportunities enabled the community to flourish that, in turn, led to the international success of the space science program that is recognized today. Since 1978, Canada has successfully flown well over 100 scientific experiments in space with practically a 100% success rate based on the metric of useful data returned to investigators. The other contribution to science that Canada's partnership in the Shuttle Program provided was the possibility to develop new fields related to the investigation of how living systems and materials and fluids behave in space, especially the understanding of the effects of gravity and exposure to increased radiation. The possibility to fly such experiments on the Shuttle was reinforced in 1983 when, during the welcoming ceremony for the Shuttle Enterprise in Ottawa, the Administrator of NASA formally and publically invited Canada to fly two Canadians as payload specialists on future missions and the Minister of Science and Technology accepted on behalf of the Government of Canada. Canada responded by launching a nation-wide search for six individuals to join a newly formed Canadian Astronaut Program. In October 1984, now 25 years ago, Marc Garneau successfully flew a suite of six Canadian investigations called CANEX* that was put together in approximately 9 months—a development schedule that, today, would be practically impossible. Since that time, Canadian scientists have flown approximately 35 more experiments on the Shuttle, all producing excellent results for the scientific teams and significantly advancing our understanding of the way that living and physical systems behave in space.

**CANEX—Canadian experiments in space science, space technology, and life sciences*

continued on next page

The Canadian astronaut program has been a remarkable success for Canada, not only in relation to the excellent support that the outstanding individuals who make up the corps have provided to the overall program but also by virtue of the visibility the individuals and missions have generated, especially within Canada. Canadian astronauts remain inspirational figures for Canadians, with every mission being widely covered in the media and appearances continuing to draw significant interest. It is a notable fact that after the Soviet Union/Russia and the USA, more Canadians have flown Shuttle missions than any other single country, fourteen such missions as of 2009.

In conclusion, it is fair to say that Canada's contribution to the Space Shuttle Program has dramatically changed the way that Canada participates in space activities. Over the past 35 years, since Canada initially decided to "throw its hat into the ring" in support of this new and revolutionary concept of a "space plane," Canada has become a leading player in global space endeavours. It can be argued credibly that Canada would not today be at the forefront of space science activities, space technology leadership, human spaceflight excellence and as a key partner in the International Space Station program if it had not been for the possibilities opened up by the Space Shuttle Program. A great debt of gratitude goes to those who saw and delivered on the promise of this program and to NASA for its generosity in believing in Canada's potential to contribute as a valuable and valued partner. Both gained enormously from this mutual trust and support and Canada continues to reap the benefits from this confidence in our program today. As we finish building and emphasize the scientific and technological use of the International Space Station, we look forward collectively to taking our first tentative steps as a species beyond our home planet. As we do so, the Space Shuttle will be looked upon as the vehicle that made all of this possible. Ad astra!

What is the Legacy of the Space Shuttle Program?

General John Dailey (USMC, Ret.)
Director
Smithsonian National Air and Space Museum

John Young, commander of the first space shuttle mission, pegged the shuttle perfectly as "a remarkable flying machine." Arising from the American traditions of ingenuity and innovation, the Space Shuttle expanded the range of human activity in near-Earth space. Serving as a cargo carrier, satellite deployment and servicing station, research laboratory, construction platform, and intermittent space station, the versatile shuttle gave scores of people an opportunity to live and do meaningful work in space. One of the most complex technology systems ever developed and the only reusable spacecraft ever operated, the shuttle was America's first attempt to make human spaceflight routine. For more than 30 years and more than 125 missions, the Space Shuttle kept the United States at the forefront of spaceflight and engaged people here and around the world with its achievements and its tragedies. The experience gained from the Space Shuttle Program will no doubt infuse future spacecraft design and spaceflight operations for years to come.

Inspiring Generations

Leah Jamieson, PhD
Dean of the College of Engineering
Purdue University

The space race, set in motion by the 1957 launch of Sputnik and reaching its pinnacle with the Apollo 11 landing on the moon, is credited with inspiring a generation of engineers. In the United States, Congress in 1958 provided funding for college students and improvements in science, mathematics, and foreign-language instruction at elementary and secondary schools. Math and science curricula flourished. University enrollment in science and engineering programs grew dramatically. For over a decade, not only engineers themselves, but policy makers and the public genuinely believed that the future depended on engineers and scientists and that education would have to inspire young people to pursue those careers.

Almost as if they were icing on the cake, innovation and technology directly or indirectly inspired by the space program began to shape the way we live and work: satellite communications, satellite navigation, photovoltaics, robotics, fault-tolerant computing, countless specialty materials, biomedical sensors, and consumer products all advanced through the space program.

Over the 30-year era of the Space Shuttle, it sometimes seems that we've come to take space flight for granted. Interest in technology has declined: bachelor's degrees awarded in engineering in the US peaked in 1985. Reports such as the Rising Above the Gathering Storm (National Academies Press, 2007) urge a massive improvement in K-12 math, science, and technology education in order to fuel innovation and ensure future prosperity. Engineering educators are looking to the National Academy of Engineering's "grand challenges" (NAE, 2008) not only to transform the world, but to inspire the next generation of students.

Has space exploration lost the ability to inspire? I don't think so. Over the past five years, I have talked about engineering careers with more than 6,000 first-year engineering students at Purdue University, asking them what engineers do and why they are studying engineering. Not a session has gone by without at least one student saying "I'm studying engineering because I want to be an astronaut." Purdue students come by this ambition honestly: 22 Purdue graduates have become astronauts, including Neil Armstrong, the first man to walk on the moon, and Eugene Cernan, the last—or as he prefers to say, "the most recent." A remarkable 18 of the 22 (all except Armstrong, Cernan, Grissom, and Chaffee) have flown Space Shuttle missions, for a total of 56 missions. Inspiration lives.

I've also talked with hundreds of IEEE* student leaders in Europe, Africa, Latin America, and Asia, asking them, as well as the Purdue undergrads, what their generation's technological legacy might be. In every session on every continent, without exception, students have talked about space exploration. Their aspirations range from settlements on the moon to human missions to Mars. These students, however, add a layer of intent that goes beyond the simple "we'll go because it's there." They talk about extraterrestrial settlements as part of the solution to Earth's grand challenges of population growth, dwindling resources, and growing poverty. More nuanced, perhaps, and more idealistic—but again, evidence of the power to inspire.

These students are telling us that space exploration is about dreaming, but it's also about doing. This isn't a new message, but it's one that is worth remembering. It's unlikely that the inspiration for the next generation of engineers will come from one galvanizing goal, as it did in the Sputnik and Apollo era. Yet, space exploration has the exquisite ability to stretch both our physical and spiritual horizons, combined with the proven ability to foster life-changing advances in our daily lives. This combination ensures that human exploration of space will continue to be a grand challenge that inspires. As the Space Shuttle era draws to a close, it's a fitting time to celebrate the Space Shuttle Program's achievements, at the same time that we ask today's students—tomorrow's engineers—"what's next?" I believe that we'll be inspired by their answers.

**The Institute of Electrical and Electronics Engineers* →

The Legacy of the Space Shuttle

Michael Griffin, PhD*
NASA administrator, 2005-2009

When I was asked by Wayne Hale to provide an essay on the topic of this paper, I was as nearly speechless as I ever become. Wayne is a former Space Shuttle Program Manager and Shuttle Flight Director. In the latter capacity, he holds the record—which cannot now be broken—for directing shuttle ascents and re-entries, generally the most dynamic portion of any shuttle mission. His knowledge of the Space Shuttle system and its history, capabilities, and limitations is encyclopedic.

In contrast, I didn't work on the shuttle until, on April 14, 2005, I became responsible for it. Forrest Gump's mother's observation that "life is like a box of chocolates; you never know what you're going to get," certainly comes to mind in this connection. But more to the point, what could I possibly say that would be of any value to Wayne? But, of course, I am determined to try.

The first thing I might note is that, whether I worked on it or not, the shuttle has dominated my professional life. Some connections are obvious. In my earlier and more productive years, I worked on systems that flew into space aboard shuttle. As I matured—meaning that I offered less and less value at higher and higher organizational levels—I acquired higher level responsibility for programs and missions flying on shuttle. I first met Mike Coats, director of the Johnson Space Center, through just such a connection. Mike commanded STS-39, a Strategic Defense Initiative mission for which I was responsible. Later, as NASA Chief Engineer in the early '90s, I led one of the Space Station Freedom redesign teams; the biggest factor influencing station design and operations was the constraint to fly on shuttle.

My professional connections with the Space Shuttle are hopelessly intertwined with more personal ones. Many of the engineers closest to me, friends and colleagues I value most highly, have worked with shuttle for decades. And, over the years, the roster of shuttle astronauts has included some of the closest friends I have. A hundred others have been classmates and professional colleagues, supervisors and subordinates, people I see every day, or people I see once a year. Speaking a bit tongue-in-cheek, I once told long-time friend Joe Engle that I loved hearing his stories about flying the X-15 because, I said, they were different; my other friends had all flown on shuttle.

From time to time, I make it a point to remember that two of them died on it.

Most of us have similar connections to the Space Shuttle, no matter what part of the space business in which we have worked. But the influence of the shuttle on the American

* Written in 2009 while serving as NASA administrator.

space program goes far beyond individual events, or even their sum, because the legacy of the Space Shuttle is a case where the whole truly is more than the sum of the parts.

Because of its duration at the center of human spaceflight plans and activities, because of the gap between promise and performance, because of the money that has been spent on it, because of what it can do and what it cannot do, because of its stunning successes and its tragic failures, the Space Shuttle has dominated the professional lives of most of us who are still young enough to be working in the space business. I'm 59 years old as I write this, and closer to retirement than I would like to be. Anyone my age or younger who worked on Apollo had to have done so in a very junior role. After Apollo, there were the all-too-brief years of Skylab, the single Apollo-Soyuz mission, and then—Space Shuttle. So, if you're still working today and spent any time in manned spaceflight over the course of your career, you worked with shuttle. And even if you never worked in human spaceflight, the shuttle has profoundly influenced your career.

So, as the shuttle approaches retirement, as we design for the future, what can we learn from having built and flown it, loved and feared it, exploited and been frustrated by it?

If the shuttle is retired by the end of 2010, as presently planned, we will have been designing, building, and flying it for more than 4 decades, four-fifths of NASA's existence. This is typical; aerospace systems normally have very long life cycles. It was Apollo that was an aberration. We must remember this as we design the new systems that will, one day, be commanded by the grandchildren of the astronauts who first fly them. We must resist making compromises now, just because budgets are tight. When a system is intended to be used for decades, it is more sensible to slip initial deployment schedules to accommodate budget cuts than to compromise technical performance or operational utility. "Late" is ugly until you launch; "wrong" is ugly forever.

The shuttle is far and away the most amazingly capable space vehicle the world has yet seen, more so than any of us around today will likely ever see again. Starting with a "clean sheet of paper" less than a decade after the first suborbital Mercury flight, its designers set—and achieved—technological goals as far beyond Apollo as Apollo was beyond Mercury. What it can do seems even now to be the stuff of science fiction.

But it is also operationally fragile and logistically undependable. Its demonstrated reliability is orders of magnitude worse than predicted, and certainly no better than the expendable vehicles it was designed to replace. It does not degrade gracefully. It can be flown safely and well, but

continued on next page

only with the greatest possible attention to every single detail, to the consequences both intended and unintended of every single decision made along the path to every single flight. The people who launch it and fly it are the best engineers, technicians, and pilots in the world, and most of the time they make it look easy. It isn't. They work knowing that they are always one misstep away from tragedy.

It was not intended to be this way; the shuttle was intended to be a robust, reliable vehicle, ready to fly dozens of times per year at a lower cost and a higher level of dependability than any expendable vehicle could ever hope to achieve. It simply didn't happen. What shuttle does is stunning, but it is stunningly less than what was predicted.

If it is true that "satisfaction equals results minus expectations," and if ultimately we have been unsatisfied, maybe where we went wrong was not with the performance achieved, but with the goals that were set. What if we had not tried for such an enormous technological leap all in one step? What if the goal had been to build an experimental prototype or two, fly them, and learn what would work and what was not likely to? Then, with that knowledge in hand, we could have proceeded to design and build a more operationally satisfactory system. What if we had kept the systems we had until we were certain we had something better, not letting go of one handhold until possessed of another?

That we did not, of course, was not NASA's fault alone. There was absolutely no money to follow the more prudent course outlined above. After the cancellation of Apollo by President Nixon, the NASA managers of the time were confronted with a cruel choice: try to achieve the goals that had been set for the shuttle, with far less money than was believed necessary, or cease US manned spaceflight. They chose the former, and we have been dealing with the consequences ever since. That they were forced to such a choice was a failure of national leadership, hardly the only one stemming from the Nixon era. But the lesson for the future is clear: in the face of hard choices, technical truth must hold sway, because it does so in the end, whether one accepts that or not.

I will end by commenting on the angst that seems to accompany our efforts to move in an orderly and disciplined manner to retire the shuttle. In my view we are missing the point, and maybe more than one point.

First, the shuttle has been an enormously productive step along the path to becoming a spacefaring civilization. But it does not lie at the end of that path, and never could have.

It was an enormous leap in human progress. The shuttle wasn't perfect, and we will make more such leaps, but none of them will be perfect, either.

Second, even if the shuttle had accomplished perfectly that which it was designed to do, we must move on because of what it cannot do and was never designed to do. The shuttle was designed to go to low orbit, and no more. NASA's funding is not such that we can afford to own and operate two human spaceflight systems at the same time. It never has been. There were gaps between Mercury and Gemini, Gemini and Apollo, Apollo and Space Shuttle. There will be a gap between Space Shuttle and Constellation*. So, if we can have only one space transportation system at a time—and I wish wholeheartedly that it were otherwise—then in my opinion it must be designed primarily to reach beyond low-Earth orbit.

If we are indeed to become a spacefaring civilization our future lies, figuratively, beyond the coastal shoals. It lies outward, beyond sight of land, where the water is deep and blue. The shuttle can't take us there. Our Constellation systems can.

So, yes, we are approaching the end of an era, an era comprising over 80% of NASA's history. We should recognize and celebrate what has been accomplished in that era. But we should not be sad, because by bringing this era to an end, we are creating the option for our children and grandchildren to live in a new and richer one. We are creating the future that we wanted to see.

**Constellation refers to the NASA program designed to build the capability to leave low-Earth orbit.*

Appendix

Flight Information

Orbiter Enterprise
Approach and Landing Test Flights

Captive-Active Flights—*High-speed taxi tests that proved the Shuttle Carrier Aircraft, mated to Enterprise, could steer and brake with the Orbiter perched on top of the airframe. These fights featured two-man crews.*

Captive-Active Flight No.	Crew Members	Test Date	Mission Length
1	Fred Haise (Cdr) Gordon Fullerton (Plt)	6/18/1977	55 min 46 s
2	Joseph Engle (Cdr) Richard Truly (Plt)	6/28/1977	62 min 0 s
3	Fred Haise (Cdr) Gordon Fullerton (Plt)	7/26/1977	59 min 53 s

Free Flights—*Flights during which Enterprise separated from the Shuttle Carrier Aircraft and landed at the hands of a two-man crew.*

Free Flight No.	Crew Members	Test Date	Mission Length
1	Fred Haise (Cdr) Gordon Fullerton (Plt)	8/12/1977	5 min 21 s
2	Joseph Engle (Cdr) Richard Truly (Plt)	9/13/1977	5 min 28 s
3	Fred Haise (Cdr) Gordon Fullerton (Plt)	9/23/1977	5 min 34 s
4	Joseph Engle (Cdr) Richard Truly (Plt)	10/12/1977	2 min 34 s
5	Fred Haise (Cdr) Gordon Fullerton (Plt)	10/26/1977	2 min 1 s

The Space Shuttle Numbering System

The first nine Space Shuttle flights were numbered in sequence from STS-1 to STS-9. Following STS-9, NASA changed the flight numbering system. The next flight became STS-41B instead of being designated STS-10. This new numbering system was designed to be more specific. The first numeral stood for the fiscal year in which the launch was to take place (i.e., "4" stood for "1984" in the STS-41B example). The second numeral represented the launch site—"1" for Kennedy Space Center, Florida, and "2" for Vandenberg Air Force Base, California. The letter represented the order of launch assignments. Following STS-51L, NASA reestablished the original numerical numbering system, therefore the next flight was designated STS-26 as it represented the 26th Space Shuttle mission.

Abbreviations, Acronyms, and Definitions

Cdr— *Commander*
Plt— *Pilot*
MS— *Mission Specialist (a career astronaut)*
PS— *Payload Specialist (an individual selected and trained for a specific mission)*
UP— *Crew member was taken up on the shuttle*
DN— *Crew member was brought down on the shuttle*

STS Flight No. and Crew Patch	Orbiter Name	Crew Members	Launch Date	Approx. Mission Days
1 	Columbia	John Young (Cdr) Robert Crippen (Plt)	4/12/1981	2
2 	Columbia	Joe Engle (Cdr) Richard Truly (Plt)	11/12/1981	2
3 	Columbia	Jack Lousma (Cdr) Gordon Fullerton (Plt)	3/22/1982	8
4 	Columbia	Thomas Mattingly (Cdr) Henry Hartsfield (Plt)	6/27/1982	7
5 	Columbia	Vance Brand (Cdr) Robert Overmyer (Plt) William Lenoir (MS) Joseph Allen (MS)	11/11/1982	5
6 	Challenger	Paul Weitz (Cdr) Karol Bobko (Plt) Story Musgrave (MS) Donald Peterson (MS)	4/4/1983	5
7 	Challenger	Robert Crippen (Cdr) Frederick Hauck (Plt) John Fabian (MS) Sally Ride (MS) Norman Thagard (MS)	6/18/1983	6
8 	Challenger	Richard Truly (Cdr) Daniel Brandenstein (Plt) Guion Bluford, Jr. (MS) Dale Gardner (MS) William Thornton (MS)	8/30/1983	6
9 	Columbia	John Young (Cdr) Brewster Shaw (Plt) Owen Garriott (MS) Robert Parker (MS) Byron Lichtenberg (PS) Ulf Merbold (PS)	11/28/1983	10

Flight Information

STS Flight No. and Crew Patch	Orbiter Name	Crew Members	Launch Date	Approx. Mission Days
41B 	Challenger	Vance Brand (Cdr) Robert Gibson (Plt) Bruce McCandless (MS) Ronald McNair (MS) Robert Stewart (MS)	2/3/1984	8
41C 	Challenger	Robert Crippen (Cdr) Francis Scobee (Plt) Terry Hart (MS) James van Hoften (MS) George Nelson (MS)	4/6/1984	7
41D 	Discovery	Henry Hartsfield (Cdr) Michael Coats (Plt) Judith Resnik (MS) Steven Hawley (MS) Richard Mullane (MS) Charles Walker (PS)	8/30/1984	6
41G 	Challenger	Robert Crippen (Cdr) Jon McBride (Plt) Kathryn Sullivan (MS) Sally Ride (MS) David Leestma (MS) Paul Scully-Power (PS) Marc Garneau (PS) Canada	10/5/1984	8
51A 	Discovery	Frederick Hauck, (Cdr) David Walker (Plt) Joseph Allen (MS) Anna Fisher (MS) Dale Gardner (MS)	11/8/1984	8
51C 	Discovery	Thomas Mattingly (Cdr) Loren Shriver (Plt) Ellison Onizuka (MS) James Buchli (MS) Gary Payton (PS)	1/24/1985	3
51D 	Discovery	Karol Bobko (Cdr) Donald Williams (Plt) Rhea Seddon (MS) David Griggs (MS) Jeffrey Hoffman (MS) Jake Garn (PS) Charles Walker (PS)	4/12/1985	7
51B 	Challenger	Robert Overmyer (Cdr) Frederick Gregory (Plt) Don Lind (MS) Norman Thagard (MS) William Thornton (MS) Lodewijk van den Berg (PS) Germany Taylor Wang (PS)	4/29/1985	7
51G 	Discovery	Daniel Brandenstein (Cdr) John Creighton (Plt) John Fabian (MS) Steven Nagel (MS) Shannon Lucid (MS) Patrick Baudry (PS) France Sultan Al-Saud (PS) Saudi Arabia	6/17/1985	7
51F 	Challenger	Gordon Fullerton (Cdr) Roy Bridges (Plt) Karl Henize (MS) Anthony England (MS) Story Musgrave (MS) Loren Acton (PS) John-David Bartoe (PS)	7/29/1985	8
51I 	Discovery	Joe Engle (Cdr) Richard Covey (Plt) James van Hoften (MS) John Lounge (MS) William Fisher (MS)	8/27/1985	7
51J 	Atlantis	Karol Bobko (Cdr) Ronald Grabe (Plt) Robert Stewart (MS) David Hilmers (MS) William Pailes (PS)	10/3/1985	4
61A 	Challenger	Henry Hartsfield (Cdr) Steven Nagel (Plt) Bonnie Dunbar (MS) James Buchli (MS) Guion Bluford (MS) Ernst Messerschmid (PS) Germany Reinhard Furrer (PS) Germany Wubbo Ockels (PS) Netherlands	10/30/1985	7
61B 	Atlantis	Brewster Shaw (Cdr) Bryan O'Connor (Plt) Sherwood Spring (MS) Mary Cleave (MS) Jerry Ross (MS) Rodolfo Neri Vela (PS) Charles Walker (PS)	11/26/1985	7
61C 	Columbia	Robert Gibson (Cdr) Charles Bolden (Plt) George Nelson (MS) Steven Hawley (MS) Franklin Chang-Diaz (MS) Robert Cenker (PS) C.William Nelson (PS)	1/12/1986	6
51L 	Challenger	Francis Scobee (Cdr) Michael Smith (Plt) Judith Resnik (MS) Ellison Onizuka (MS) Ronald McNair (MS) Gregory Jarvis (PS) Christa McAuliffe (PS)	1/28/1986	0

Flight Information

STS Flight No. and Crew Patch	Orbiter Name	Crew Members	Launch Date	Approx. Mission Days
26 	Discovery	Frederick Hauck (Cdr) Richard Covey (Plt) John Lounge (MS) George Nelson (MS) David Hilmers (MS)	9/29/1988	4
27 	Atlantis	Robert Gibson (Cdr) Guy Gardner (Plt) Richard Mullane (MS) Jerry Ross (MS) William Shepherd (MS)	12/2/1988	4
29 	Discovery	Michael Coats (Cdr) John Blaha (Plt) James Buchli (MS) Robert Springer (MS) James Bagian (MS)	3/13/1989	5
30 	Atlantis	David Walker (Cdr) Ronald Grabe (Plt) Norman Thagard (MS) Mary Cleave (MS) Mark Lee (MS)	5/4/1989	4
28 	Columbia	Brewster Shaw (Cdr) Richard Richards (Plt) James Adamson (MS) David Leestma (MS) Mark Brown (MS)	8/8/1989	5
34 	Atlantis	Donald Williams (Cdr) Michael McCulley (Plt) Shannon Lucid (MS) Franklin Chang-Diaz (MS) Ellen Baker (MS)	10/18/1989	5
33 	Discovery	Frederick Gregory (Cdr) John Blaha (Plt) Manley Carter (MS) Story Musgrave (MS) Kathryn Thornton (MS)	11/22/1989	5
32 	Columbia	Daniel Brandenstein (Cdr) James Wetherbee (Plt) Bonnie Dunbar (MS) Marsha Ivins (MS) David Low (MS)	1/9/1990	11
36 	Atlantis	John Creighton, (Cdr) John Casper (Plt) David Hilmers (MS) Richard Mullane (MS) Pierre Thuot (MS)	2/28/1990	4
31 	Discovery	Loren Shriver (Cdr) Charles Bolden (Plt) Bruce McCandless (MS) Steven Hawley (MS) Kathryn Sullivan (MS)	4/24/1990	5
41 	Discovery	Richard Richards (Cdr) Robert Cabana (Plt) Bruce Melnick (MS) William Shepherd (MS) Thomas Akers (MS)	10/6/1990	4
38 	Atlantis	Richard Covey (Cdr) Frank Culbertson (Plt) Carle Meade (MS) Robert Springer (MS) Charles Gemar (MS)	11/15/1990	5
35 	Columbia	Vance Brand (Cdr) Guy Gardner (Plt) Jeffrey Hoffman (MS) John Lounge (MS) Robert Parker (MS) Samuel Durrance (PS) Ronald Parise (PS)	12/2/1990	9
37 	Atlantis	Steven Nagel (Cdr) Kenneth Cameron (Plt) Linda Godwin (MS) Jerry Ross (MS) Jay Apt (MS)	4/5/1991	6
39 	Discovery	Michael Coats (Cdr) Blaine Hammond (Plt) Gregory Harbaugh (MS) Donald McMonagle (MS) Guion Bluford (MS) Charles Veach (MS) Richard Hieb (MS)	4/28/1991	8
40 	Columbia	Bryan O'Connor (Cdr) Sidney Gutierrez (Plt) James Bagian (MS) Tamara Jernigan (MS) Rhea Seddon (MS) Drew Gaffney (PS) Millie Hughes-Fulford (PS)	6/5/1991	9
43 	Atlantis	John Blaha (Cdr) Michael Baker (Plt) Shannon Lucid (MS) David Low (MS) James Adamson (MS)	8/2/1991	9

Flight Information

STS Flight No. and Crew Patch	Orbiter Name	Crew Members	Launch Date	Approx. Mission Days
48 	Discovery	John Creighton (Cdr) Kenneth Reightler (Plt) Charles Gemar (MS) James Buchli (MS) Mark Brown (MS)	9/12/1991	5
44 	Atlantis	Frederick Gregory (Cdr) Terence Henricks (Plt) James Voss (MS) Story Musgrave (MS) Mario Runco (MS) Thomas Hennen (PS)	11/24/1991	7
42 	Discovery	Ronald Grabe (Cdr) Stephen Oswald (Plt) Norman Thagard (MS) William Readdy (MS) David Hilmers (MS) Roberta Bondar (PS) Canada Ulf Merbold (PS) Germany	1/22/1992	8
45 	Atlantis	Charles Bolden (Cdr) Brian Duffy (Plt) Kathryn Sullivan (MS) David Leestma (MS) Michael Foale (MS) Dirk Frimout (PS) Belgium Bryon Lichtenberg (PS)	3/24/1992	9
49 	Endeavour	Daniel Brandenstein (Cdr) Kevin Chilton (Plt) Bruce Melnick (MS) Pierre Thuot (MS) Richard Hieb (MS) Kathryn Thornton (MS) Thomas Akers (MS)	5/7/1992	9
50 	Columbia	Richard Richards (Cdr) Kenneth Bowersox (Plt) Bonnie Dunbar (MS) Ellen Baker (MS) Carl Meade (MS) Lawrence DeLucas (PS) Eugene Trinh (PS)	6/25/1992	14
46 	Atlantis	Loren Shriver (Cdr) Andrew Allen (Plt) Claude Nicollier (MS) Switzerland Marsha Ivins (MS) Jeffrey Hoffman (MS) Franklin Chang-Diaz (MS) Franco Malerba (PS) Italy	7/31/1992	8
47 	Endeavour	Robert Gibson (Cdr) Curtis Brown (Plt) Mark Lee (MS) Jay Apt (MS) Jan Davis (MS) Mae Jemison (MS) Mamoru Mohri (PS) Japan	9/12/1992	8
52 	Columbia	James Wetherbee (Cdr) Michael Baker (Plt) Charles Veach (MS) William Shepherd (MS) Tamara Jernigan (MS) Steven MacLean (PS)	10/22/1992	10
53 	Discovery	David Walker (Cdr) Robert Cabana (Plt) Guion Bluford (MS) Michael Clifford (MS) James Voss (MS)	12/2/1992	7
54 	Endeavour	John Casper (Cdr) Donald McMonagle (Plt) Mario Runco (MS) Gregory Harbaugh (MS) Susan Helms (MS)	1/13/1993	6
56 	Discovery	Kenneth Cameron (Cdr) Stephen Oswald (Plt) Michael Foale (MS) Kenneth Cockrell (MS) Ellen Ochoa (MS)	4/8/1993	9
55 	Columbia	Steven Nagel (Cdr) Terence Henricks (Plt) Jerry Ross (MS) Charles Precourt (MS) Bernard Harris (MS) Ulrich Walter (PS) Germany Hans Schlegel (PS) Germany	4/26/1993	10
57 	Endeavour	Ronald Grabe (Cdr) Brian Duffy (Plt) David Low (MS) Nancy Sherlock (MS) Peter Wisoff (MS) Janice Voss (MS)	6/21/1993	10
51 	Discovery	Frank Culbertson (Cdr) William Readdy (Plt) James Newman (MS) Daniel Bursch (MS) Carl Walz (MS)	9/12/1993	10
58 	Columbia	John Blaha (Cdr) Richard Searfoss (Plt) Rhea Seddon (MS) William McArthur (MS) David Wolf (MS) Shannon Lucid (MS) Martin Fettman (PS)	10/18/1993	14

Flight Information

STS Flight No. and Crew Patch	Orbiter Name	Crew Members	Launch Date	Approx. Mission Days
61 	Endeavour	Richard Covey (Cdr) Kenneth Bowersox (Plt) Kathryn Thornton (MS) Claude Nicollier (MS) Switzerland Jeffrey Hoffman (MS) Story Musgrave (MS) Thomas Akers (MS)	12/2/1993	11
60 	Discovery	Charles Bolden (Cdr) Kenneth Reightler (Plt) Jan Davis (MS) Ronald Sega (MS) Franklin Chang-Diaz (MS) Sergei Krikalev (MS) Russia	2/3/1994	8
62 	Columbia	John Casper (Cdr) Andrew Allen (Plt) Pierre Thuot (MS) Charles Gemar (MS) Marsha Ivins (MS)	3/4/1994	14
59 	Endeavour	Sidney Gutierrez (Cdr) Kevin Chilton (Plt) Jay Apt (MS) Michael Clifford (MS) Linda Godwin (MS) Thomas Jones (MS)	4/9/1994	11
65 	Columbia	Robert Cabana (Cdr) James Halsell (Plt) Richard Hieb (MS) Carl Walz (MS) Leroy Chiao (MS) Donald Thomas (MS) Chiaki Mukai (PS) Japan	7/8/1994	15
64 	Discovery	Richard Richards (Cdr) Blaine Hammond (Plt) Jerry Linenger (MS) Susan Helms (MS) Carl Meade (MS) Mark Lee (MS)	9/9/1994	11
68 	Endeavour	Michael Baker (Cdr) Terrence Wilcutt (Plt) Steven Smith (MS) Daniel Bursch (MS) Peter Wisoff (MS) Thomas Jones (MS)	9/30/1994	11
66 	Atlantis	Donald McMonagle (Cdr) Curtis Brown (Plt) Ellen Ochoa (MS) Joseph Tanner (MS) Jean-Francois Clervoy (MS) France Scott Parazynski (MS)	11/3/1994	11
63 	Discovery	James Wetherbee (Cdr) Eileen Collins (Plt) Bernard Harris (MS) Michael Foale (MS) Janice Voss (MS) Vladimir Titov (MS) Russia	2/3/1995	8
67 	Endeavour	Stephen Oswald (Cdr) William Gregory (Plt) John Grunsfeld (MS) Wendy Lawrence (MS) Tamara Jernigan (MS) Samuel Durrance (PS) Ronald Parise (PS)	3/2/1995	17
71 	Atlantis	Robert Gibson (Cdr) Charles Precourt (Plt) Ellen Baker (MS) Gregory Harbaugh (MS) Bonnie Dunbar (MS) Anatoly Solovyev (UP) Russia Nikolai Budarin (UP) Russia Vladimir Dezhurov (DN) Russia Gennady Strekalov (DN) Russia Norman Thagard (MS, DN)	6/27/1995	10
70 	Discovery	Terence Henricks (Cdr) Kevin Kregel (Plt) Donald Thomas (MS) Nancy Currie (MS) Mary Ellen Weber (MS)	7/13/1995	9
69 	Endeavour	David Walker (Cdr) Kenneth Cockrell (Plt) James Voss (MS) James Newman (MS) Michael Gernhardt (MS)	9/7/1995	11
73 	Columbia	Kenneth Bowersox (Cdr) Kent Rominger (Plt) Catherine Coleman (MS) Michael Lopez-Alegria (MS) Kathryn Thornton (MS) Fred Leslie (PS) Albert Sacco (PS)	10/20/1995	16
74 	Atlantis	Kenneth Cameron (Cdr) James Halsell (Plt) Chris Hadfield (MS) Canada Jerry Ross (MS) William McArthur (MS)	11/12/1995	8
72 	Endeavour	Brian Duffy (Cdr) Brent Jett (Plt) Leroy Chiao (MS) Winston Scott (MS) Koichi Wakata (MS) Japan Daniel Barry (MS)	1/11/1996	9

Flight Information

STS Flight No. and Crew Patch	Orbiter Name	Crew Members	Launch Date	Approx. Mission Days
75 	Columbia	Andrew Allen (Cdr) Scott Horowitz (Plt) Jeffrey Hoffman (MS) Maurizio Cheli (MS) Italy Claude Nicollier (MS) Switzerland Franklin Chang-Diaz (MS) Umberto Guidoni (PS) Italy	2/22/1996	16
76 	Atlantis	Kevin Chilton (Cdr) Richard Searfoss (Plt) Ronald Sega (MS) Michael Clifford (MS) Linda Godwin (MS) Shannon Lucid (MS, UP)	3/22/1996	9
77 	Endeavour	John Casper (Cdr) Curtis Brown (Plt) Andrew Thomas (MS) Daniel Bursch (MS) Mario Runco (MS) Marc Garneau (MS) Canada	5/19/1996	10
78 	Columbia	Terence Henricks (Cdr) Kevin Kregel (Plt) Richard Linnehan (MS) Susan Helms (MS) Charles Brady (MS) Jean-Jacques Favier (PS) France Robert Thirsk (PS) Canada	6/20/1996	17
79 	Atlantis	William Readdy (Cdr) Terrence Wilcutt (Plt) Jay Apt (MS) Thomas Akers (MS) Carl Walz (MS) John Blaha (MS, UP) Shannon Lucid (MS, DN)	9/16/1996	10
80 	Columbia	Kenneth Cockrell (Cdr) Kent Rominger (Plt) Tamara Jernigan (MS) Thomas Jones (MS) Story Musgrave (MS)	11/19/1996	18
81 	Atlantis	Michael Baker (Cdr) Brent Jett (Plt) Peter Wisoff (MS) John Grunsfeld (MS) Marsha Ivins (MS) Jerry Linenger (MS, UP) John Blaha (MS, DN)	1/12/1997	10
82 	Discovery	Kenneth Bowersox (Cdr) Scott Horowitz (Plt) Joseph Tanner (MS) Steven Hawley (MS) Gregory Harbaugh (MS) Mark Lee (MS) Steven Smith (MS)	2/11/1997	10
83 	Columbia	James Halsell (Cdr) Susan Still (Plt) Janice Voss (MS) Michael Gernhardt (MS) Donald Thomas (MS) Roger Crouch (PS) Gregory Linteris (PS)	4/4/1997	4
84 	Atlantis	Charles Precourt (Cdr) Eileen Collins (Plt) Jean-Francois Clervoy (MS) France Carlos Noriega (MS) Edward Lu (MS) Elena Kondakova (MS) Russia Michael Foale (MS, UP) Jerry Linenger (MS, DN)	5/15/1997	10
94 	Columbia	James Halsell (Cdr) Susan Still (Plt) Janice Voss (MS) Michael Gernhardt (MS) Donald Thomas (MS) Roger Crouch (PS) Gregory Linteris (PS)	7/1/1997	16
85 	Discovery	Curtis Brown (Cdr) Kent Rominger (Plt) Jan Davis (MS) Robert Curbeam (MS) Stephen Robinson (MS) Bjarni Tryggvason (PS) Canada	8/7/1997	12
86 	Atlantis	James Wetherbee (Cdr) Michael Bloomfield (Plt) Vladimir Titov (MS) Russia Scott Parazynski (MS) Jean-Loup Chretien (MS) France Wendy Lawrence (MS) David Wolf (MS, UP) Michael Foale (MS, DN)	9/25/1997	11
87 	Columbia	Kevin Kregel (Cdr) Steven Lindsey (Plt) Kalpana Chawla (MS) Winston Scott (MS) Takao Doi (MS) Japan Leonid Kadenyuk (PS) Ukraine	11/19/1997	16
89 	Endeavour	Terrence Wilcutt (Cdr) Joe Edwards (Plt) James Reilly (MS) Michael Anderson (MS) Bonnie Dunbar (MS) Salizhan Sharipov (MS) Russia Andrew Thomas (MS, UP) David Wolf (MS, DN)	1/22/1998	9
90 	Columbia	Richard Searfoss (Cdr) Scott Altman (Plt) Richard Linnehan (MS) Kathryn Hire (MS) Dafydd Williams (MS) Canada Jay Buckey (PS) James Pawelczyk (PS)	4/17/1998	16

Flight Information

STS Flight No. and Crew Patch	Orbiter Name	Crew Members	Launch Date	Approx. Mission Days
91 	Discovery	Charles Precourt (Cdr) Dominic Gorie (Plt) Franklin Chang-Diaz (MS) Wendy Lawrence (MS) Janet Kavandi (MS) Valery Ryumin (MS) Russia Andrew Thomas (MS, DN)	6/2/1998	10
95 	Discovery	Curtis Brown (Cdr) Steven Lindsey (Plt) Stephen Robinson (MS) Scott Parazynski (MS) Pedro Duque (MS) Spain Chiaki Mukai (PS) Japan John Glenn (PS)	10/29/1998	10
88 	Endeavour	Robert Cabana (Cdr) Frederick Sturckow (Plt) Jerry Ross (MS) Nancy Currie (MS) James Newman (MS) Sergei Krikalev (MS) Russia	12/4/1998	12
96 	Discovery	Kent Rominger (Cdr) Rick Husband (Plt) Tamara Jernigan (MS) Ellen Ochoa (MS) Daniel Berry (MS) Julie Payette (MS) Canada Valery Tokarev (MS) Russia	5/27/1999	10
93 	Columbia	Eileen Collins (Cdr) Jeffrey Ashby (Plt) Catherine Coleman (MS) Steven Hawley (MS) Michel Tognini (MS) France	7/23/1999	5
103 	Discovery	Curtis Brown (Cdr) Scott Kelly (Plt) Steven Smith (MS) Jean-Francois Clervoy (MS) France John Grunsfeld (MS) Michael Foale (MS) Claude Nicollier (MS) Switzerland	12/19/1999	8
99 	Endeavour	Kevin Kregel (Cdr) Dominic Gorie (Plt) Gerhard Thiele (MS) Germany Janet Kavandi (MS) Janice Voss (MS) Mamoru Mohri (MS) Japan	2/11/2000	11
101 	Atlantis	James Halsell (Cdr) Scott Horowitz (Plt) Mary Ellen Weber (MS) Jeffrey Williams (MS) James Voss (MS) Susan Helms (MS) Yury Usachev (MS) Russia	5/19/2000	10
106 	Atlantis	Terrence Wilcutt (Cdr) Scott Altman (Plt) Edward Lu (MS) Richard Mastracchio (MS) Daniel Burbank (MS) Yuri Malenchenko (MS) Russia Boris Morukov (MS) Russia	9/8/2000	12
92 	Discovery	Bryan Duffy (Cdr) Pamela Melroy (Plt) Leroy Chiao (MS) William McArthur (MS) Peter Wisoff (MS) Michael Lopez-Alegria (MS) Koichi Wakata (MS) Japan	10/11/2000	12
97 	Endeavour	Brent Jett (Cdr) Michael Bloomfield (Plt) Joseph Tanner (MS) Marc Garneau (MS) Canada Carlos Noriega (MS)	11/30/2000	11
98 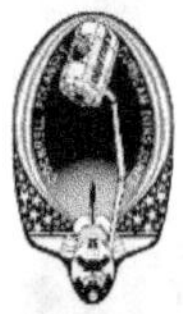	Atlantis	Kenneth Cockrell (Cdr) Mark Polansky (Plt) Robert Curbeam (MS) Marsha Ivins (MS) Thomas Jones (MS)	2/7/2001	13
102 	Discovery	James Wetherbee (Cdr) James Kelly (Plt) Andrew Thomas (MS) Paul Richards (MS) James Voss (MS, UP) Susan Helms (MS, UP) Yury Usachev (MS, UP) Russia Sergei Krikalev (MS, DN) Russia William Shepherd (MS, DN) Yuri Gidzenko (MS, DN) Russia	3/8/2001	13
100 	Endeavour	Kent Rominger (Cdr) Jeffrey Ashby (Plt) Chris Hadfield (MS) Canada John Phillips (MS) Canada Scott Parazynski (MS) Umberto Guidoni (MS) Italy Yuri Lonchakov (MS) Russia	4/19/2001	12
104 	Atlantis	Steven Lindsey (Cdr) Charles Hobaugh (Plt) Michael Gernhardt (MS) Janet Kavandi (MS) James Reilly (MS)	7/12/2001	13
105 	Discovery	Scott Horowitz (Cdr) Frederick Sturckow (Plt) Patrick Forrester (MS) Daniel Barry (MS) Frank Culbertson (MS, UP) Vladimir Dezhurov (MS, UP) Russia Mikhail Tyurin (MS, UP) Russia Yuri Usachev (MS, DN) Russia James Voss (MS, DN) Susan Helms (MS, DN)	8/10/2001	12

Flight Information

STS Flight No. and Crew Patch	Orbiter Name	Crew Members	Launch Date	Approx. Mission Days
108 	Endeavour	Dominic Gorie (Cdr) Mark Kelly (Plt) Linda Godwin (MS) Daniel Tani (MS) Yuri Onufrienko (MS, UP) Russia Daniel Bursch (MS, UP) Carl Walz (MS, UP) Frank Culbertson (MS, DN) Vladimir Dezhurov (MS, DN) Russia Mikhail Tyurin (MS, DN) Russia	12/5/2001	12
109 	Columbia	Scott Altman (Cdr) Duane Carey (Plt) John Grunsfeld (MS) Nancy Currie (MS) Richard Linnehan (MS) James Newman (MS) Michael Massimino (MS)	3/1/2002	11
110 	Atlantis	Michael Bloomfield (Cdr) Stephen Frick (Plt) Rex Walheim (MS) Ellen Ochoa (MS) Lee Morin (MS) Jerry Ross (MS) Steven Smith (MS)	4/8/2002	11
111 	Endeavour	Kenneth Cockrell (Cdr) Paul Lockhart (Plt) Franklin Chang-Diaz (MS) Philippe Perrin (MS) France Valery Korzun (MS, UP) Russia Peggy Whitson (MS, UP) Sergei Treschev (MS, UP) Russia Yuri Onufrienko (MS, DN) Russia Daniel Bursch (MS, DN) Carl Walz (MS, DN)	6/5/2002	14
112 	Atlantis	Jeffrey Ashby (Cdr) Pamela Melroy (Plt) David Wolf (MS) Sandra Magnus (MS) Piers Sellers (MS) Fyodor Yurchikhin (MS) Russia	10/7/2002	11
113 	Endeavour	James Wetherbee (Cdr) Paul Lockhart (Plt) Michael Lopez-Alegria (MS) John Herrington (MS) Kenneth Bowersox (MS, UP) Nikolai Budarin (MS, UP) Russia Donald Pettit (MS, UP) Valery Korzun (MS, DN) Russia Sergei Treschev (MS, DN) Russia Peggy Whitson (MS, DN)	11/23/2002	14
107 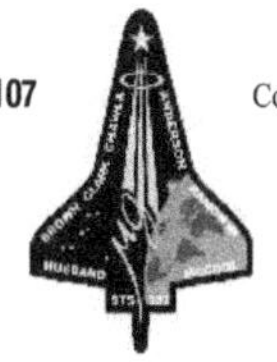	Columbia	Rick Husband (Cdr) William McCool (Plt) Michael Anderson (MS) David Brown (MS) Kalpana Chawla (MS) Laurel Clark (MS) Ilan Ramon (PS) Israel	1/16/2003	16
114 	Discovery	Eileen Collins (Cdr) James Kelly (Plt) Soichi Noguchi (MS) Japan Stephen Robinson (MS) Andrew Thomas (MS) Wendy Lawrence (MS) Charles Camarda (MS)	7/26/2005	14
121 	Discovery	Steven Lindsey (Cdr) Mark Kelly (Plt) Michael Fossum (MS) Lisa Nowak (MS) Stephanie Wilson (MS) Piers Sellers (MS) Thomas Reiter (MS, UP) Germany	7/4/2006	13
115 	Atlantis	Brent Jett (Cdr) Christopher Ferguson (Plt) Joseph Tanner (MS) Daniel Burbank (MS) Heidemarie Stefanyshyn-Piper (MS) Steven MacLean (MS) Canada	9/9/2006	12
116 	Discovery	Mark Polansky (Cdr) William Oefelein (Plt) Nicholas Patrick (MS) Robert Curbeam (MS) Christer Fuglesang (MS) Sweden Joan Higginbotham (MS) Sunita Williams (MS, UP) Thomas Reiter (MS, DN) Germany	12/9/2006	13
117 	Atlantis	Frederick Sturkow (Cdr) Lee Archambault (Plt) Patrick Forrester (MS) Steven Swanson (MS) John Olivas (MS) James Reilly (MS) Clayton Anderson (MS, UP) Sunita Williams (MS, DN)	6/8/2007	14
118 	Endeavour	Scott Kelly (Cdr) Charles Hobaugh (Plt) Tracy Caldwell (MS) Richard Mastracchio (MS) Dafydd Williams (MS) Canada Barbara Morgan (MS) Benjamin Drew (MS)	8/8/2007	14
120 	Discovery	Pamela Melroy (Cdr) George Zamka (Plt) Scott Parazynski (MS) Stephanie Wilson (MS) Douglas Wheelock (MS) Paolo Nespoli (MS) Italy Daniel Tani (MS, UP) Clayton Anderson (MS, DN)	10/23/2007	15

Flight Information

STS Flight No. and Crew Patch	Orbiter Name	Crew Members	Launch Date	Approx. Mission Days
122 	Atlantis	Stephen Frick (Cdr) Alan Poindexter (Plt) Leland Melvin (MS) Rex Walheim (MS) Hans Schlegel (MS) Germany Stanley Love (MS) Leopold Eyharts (MS, UP) France Daniel Tani (MS, DN)	2/7/2008	13
123 	Endeavour	Dominic Gorie (Cdr) Gregory H. Johnson (Plt) Robert Behnken (MS) Michael Foreman (MS) Takao Doi (MS) Japan Richard Linnehan (MS) Garrett Reisman (MS, UP) Leopold Eyharts (MS, DN) France	3/11/2008	16
124 	Discovery	Mark Kelly (Cdr) Kennneth Ham (Plt) Karen Nyberg (MS) Ronald Garan (MS) Michael Fossum (MS) Akihiko Hoshide (MS) Japan Gregory Chamitoff (MS, UP) Garrett Reisman (MS, DN)	5/31/2008	14
126 	Endeavour	Christopher Ferguson (Cdr) Eric Boe (Plt) Donald Pettit (MS) Stephen Bowen (MS) Heidemarie Stefanyshyn-Piper (MS) Shane Kimbrough (MS) Sandra Magnus (MS, UP) Gregory Chamitoff (MS, DN)	11/14/2008	16
119 	Discovery	Lee Archambault (Cdr) Dominic Antonelli (Plt) Joseph Acaba (MS) Steven Swanson (MS) Richard Arnold (MS) John Phillips (MS) Koichi Wakata (MS, UP) Japan Sandra Magnus (MS, DN)	3/15/2009	13
125 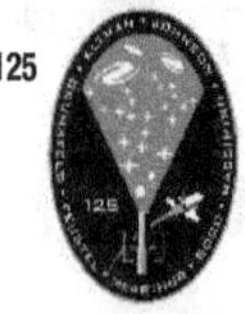	Atlantis	Scott Altman (Cdr) Gregory C. Johnson (Plt) Michael Good (MS) Megan McArthur (MS) John Grunsfeld (MS) Michael Massimino (MS) Andrew Feustel (MS)	5/11/09	13
127 	Endeavour	Mark Polansky (Cdr) Douglas Hurley (Plt) Christopher Cassidy (MS) Julie Payette (MS) Canada Thomas Marshburn (MS) David Wolf (MS) Timothy Kopra (MS, UP) Koichi Wakata (MS, DN) Japan	7/15/09	16

STS Flight No. and Crew Patch	Orbiter Name	Crew Members	Launch Date	Approx. Mission Days
128 	Discovery	Frederick Sturckow (Cdr) Kevin Ford (Plt) Patrick Forrester (MS) Jose Hernandez (MS) John Olivas (MS) Christer Fuglesang (MS) Sweden Nicole Stott (MS, UP) Timothy Kopra (MS, DN)	8/28/09	15
129 	Atlantis	Charles Hobaugh (Cdr) Barry Wilmore (Plt) Randolph Bresnik (MS) Michael Foreman (MS) Leland Melvin (MS) Robert Satcher (MS) Nicole Stott (MS, DN)	11/16/09	11
130 	Endeavour	George Zamka (Cdr) Terry Virts (Plt) Robert Behnken (MS) Nicholas Patrick (MS) Kathryn Hire (MS) Stephen Robinson (MS)	2/8/10	13
131 	Discovery	Alan Poindexter (Cdr) James Dutton (Plt) Richard Mastracchio (MS) Naoko Yamazaki (MS) Japan Clayton Anderson (MS) Dorothy Metcalf-Lindenburger (MS) Stephanie Wilson (MS)	4/5/10	15
132 	Atlantis	Kenneth Ham (Cdr) Dominic Antonelli (Plt) Stephen Bowen (MS) Michael Good (MS) Piers Sellers (MS) Garrett Reisman (MS)	5/14/10	12
133 	Discovery	Steven Lindsey (Cdr) Eric Boe (Plt) Benjamin Drew (MS) Michael Barratt (MS) Stephen Bowen (MS) Nicole Stott (MS)	2/24/11	12
134 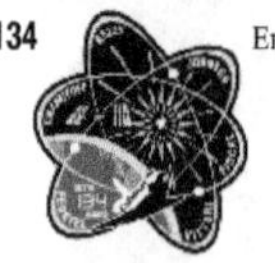	Endeavour	Mark Kelly (Cdr) Gregory H. Johnson (Plt) Andrew Feustel (MS) Michael Fincke (MS) Gregory Chamitoff (MS) Roberto Vittori (MS) Italy	5/16/11	16
135 	Atlantis	Christopher Ferguson (Cdr) Douglas Hurley (Plt) Sandra Magnus (MS) Rex Walheim (MS)	7/8/11	12

Payloads and Experiments per Space Shuttle Flight

STS Flight No.	Test Flights	US Department of Defense	International Payloads and Astronauts	Education Payloads and Student-Teacher Interactions	Earth Science	Space Science	Microgravity Science	Space Biology	Astronaut Health and Performance	Commercial Payloads and Satellites	Engineering Tests	Construction of International Space Station
1	●								●		●	
2	●				●				●		●	
3	●		●		●	●	●	●	●	●	●	
4	●	●		●	●		●		●	●	●	
5			●	●	●		●		●	●	●	
6			●	●	●		●		●	●	●	
7		●	●	●			●		●	●	●	
8		●	●	●	●		●	●	●	●	●	
9		●	●	●	●	●	●		●		●	
41B		●	●	●			●		●	●	●	
41C		●		●		●	●		●	●	●	
41D		●		●	●		●		●	●	●	●
41G		●	●	●	●	●	●		●	●	●	●
51A		●	●				●		●	●	●	
51C		●							●			
51D		●	●	●		●	●		●	●	●	
51B			●	●	●	●	●	●	●		●	
51G		●	●	●		●	●		●	●	●	
51F			●	●	●	●	●	●	●		●	
51I		●	●				●		●	●	●	
51J		●							●			
61A		●	●				●	●	●		●	
61B			●	●			●		●	●	●	●
61C		●		●			●		●	●	●	
51L												
26	●	●		●	●		●		●	●	●	
27		●							●		●	
29		●		●			●	●	●	●	●	
30		●			●	●	●		●	●	●	
28		●							●		●	
34		●		●	●	●	●	●	●	●	●	
33		●							●		●	
32		●			●		●	●	●	●	●	
36		●							●		●	
31		●		●		●	●		●	●	●	
41		●	●		●	●	●	●	●	●	●	
38		●			●				●		●	
35		●		●		●			●		●	
37		●		●		●	●		●	●	●	●
39		●	●						●		●	
40			●	●			●	●	●	●	●	
43		●			●		●		●	●	●	
48		●			●	●	●	●	●	●	●	
44		●				●	●		●		●	
42		●	●	●			●	●	●	●	●	

Payloads and Experiments per Space Shuttle Flight

STS Flight No.	Test Flights	US Department of Defense	International Payloads and Astronauts	Education Payloads and Student-Teacher Interactions	Earth Science	Space Science	Microgravity Science	Space Biology	Astronaut Health and Performance	Commercial Payloads and Satellites	Engineering Tests	Construction of International Space Station
45		●	●	●	●	●	●		●	●	●	
49							●		●	●	●	
50				●			●		●	●	●	
46			●		●	●	●	●	●	●	●	
47			●	●			●	●	●	●	●	
52			●	●	●		●	●	●	●	●	
53		●			●		●		●	●	●	
54				●		●	●	●	●	●	●	
56		●		●	●	●	●	●	●	●	●	
55			●	●	●	●	●	●	●		●	
57		●	●	●	●	●	●	●	●	●	●	
51		●	●			●	●	●	●	●	●	
58				●				●	●		●	
61		●	●			●			●	●	●	
60		●	●	●			●	●	●	●	●	
62		●	●		●		●	●	●	●	●	
59		●	●	●	●		●	●	●	●	●	
65		●	●	●		●	●	●	●	●	●	
64		●	●	●	●	●	●	●	●	●	●	
68		●	●	●	●	●	●	●	●	●	●	
66		●	●		●		●	●	●		●	
63		●	●		●	●	●	●	●	●	●	
67			●	●		●	●		●	●	●	
71			●	●				●	●	●	●	
70		●		●	●		●	●	●	●	●	
69		●	●	●	●	●	●	●	●	●	●	
73			●	●			●		●	●	●	
74		●	●	●	●				●	●	●	
72			●		●		●	●	●	●	●	
75			●	●			●		●	●	●	
76		●	●	●	●			●	●		●	
77		●	●	●			●	●	●	●	●	
78		●	●	●			●	●	●		●	
79		●	●	●	●	●	●	●	●	●	●	
80		●	●	●		●	●	●	●	●	●	
81		●	●	●	●			●	●		●	
82						●			●		●	
83		●	●				●		●	●	●	
84		●	●		●	●	●	●	●	●	●	
94		●	●	●			●		●	●	●	
85		●	●	●	●	●	●	●	●	●	●	
86		●	●	●	●	●	●		●	●	●	
87		●	●	●	●	●	●	●	●	●	●	
89		●	●	●	●		●	●	●		●	
90			●	●	●		●	●	●	●	●	

Payloads and Experiments per Space Shuttle Flight

STS Flight No.	Test Flights	US Department of Defense	International Payloads and Astronauts	Education Payloads and Student-Teacher Interactions	Earth Science	Space Science	Microgravity Science	Space Biology	Astronaut Health and Performance	Commercial Payloads and Satellites	Engineering Tests	Construction of International Space Station
91		●	●	●	●	●	●		●	●	●	
95		●	●	●	●	●	●	●	●	●	●	
88		●	●		●		●		●	●	●	●
96			●	●					●		●	●
93		●	●	●		●	●	●	●	●	●	
103			●			●			●		●	
99			●	●	●				●		●	
101			●	●			●		●	●	●	●
106			●	●			●	●	●	●	●	●
92			●						●	●	●	●
97			●	●					●	●	●	●
98				●	●				●		●	●
102			●	●					●		●	●
100			●						●	●	●	●
104									●	●	●	●
105			●	●			●		●		●	●
108		●	●	●			●	●	●	●	●	●
109				●		●			●		●	
110		●		●				●	●	●	●	●
111		●	●						●		●	●
112		●	●		●		●		●		●	●
113		●	●						●			●
107		●	●	●	●		●	●	●	●		
114	●	●	●						●		●	●
121	●		●					●	●		●	●
115		●	●					●	●		●	●
116		●	●	●					●			●
117		●							●			●
118		●	●	●				●	●			●
120		●	●						●			●
122		●	●		●			●	●			●
123		●	●					●	●		●	●
124			●						●			●
126		●	●	●				●	●		●	●
119		●	●	●					●		●	●
125						●			●	●	●	
127		●	●	●				●	●	●	●	●
128		●	●					●	●		●	●
129		●	●					●	●		●	●
130			●		●			●	●		●	●
131			●	●				●	●		●	●
132		●	●					●	●		●	●
133		●				●	●	●	●	●	●	●
134		●	●	●		●	●	●	●	●	●	●
135		●	●	●			●	●	●	●	●	●

Space Shuttle Program Managers

John Shannon
February 2008 – August 2011

Wayne Hale
September 2005 – February 2008

William Parsons
July 2003 – September 2005

Ronald Dittemore
April 1999 – July 2003

Thomas Holloway
November 1995 – April 1999

Brewster Shaw
March 1993 – November 1995

Leonard Nicholson
June 1989 – March 1993

Richard Kohrs
November 1986 – June 1989

Arnold Aldrich
June 1985 – November 1986

Glynn Lunney
June 1981 – June 1985

Robert Thompson
February 1970 – June 1981

Acknowledgments

We would like to extend a special "thank you" to the following individuals for their invaluable contributions to this book.

Research interns:

Jared Donnelly, Hannah Kohler, Tiffany Lewis, Jason Miller, and Jonathan Torres.

Technical, legal, budgetary, procurement, secretarial, photography, publication, and public affairs:

John Aaron, Randall Adams, Robin Allen, Carol Andrews, Lauren Artman, Robert Atkins, Joan Baker, Jonathan Baker, Timothy Bayline, Wayne Bingham, Gregory Blackburn, Jamie Bolton, Eric Bordelon, Jim Brazda, Jack Brazzel, Rebecca Bresnik, Frank Brody, Deborah Byerly, Vicki Cantrell, William Carr, John Casper, Norman Chaffee, Ruth Ann Chicoine, Randle Clay, Nicole Cloutier, John Coggeshall, Deborah Conder, Mark Craig, Maryann Cresap, Roger Crouch, Francis Cuccinota, Hunt Culver, Michael Curie, Benjamin Daniel, Dennis Davidson, Alexander Dawn, Alex De La Torre, William Dowdell, Cynthia Draughon, Roger Elliot, Stephen Elsner, Cliff Farmer, Edward Fein, Howard Flynn, Jerry Forney, Marcus Friske, Stephen Garber, Roberto Garcia, Joe Gensler, Cory George, Charles Ginnega, John Golden, Sharon Goza, Cathy Graham, Megan Grande, Laura Gross, John Grunsfeld, Michael Gunson, Mark Hammerschmidt, David Hanson, Mary Jo Harris, James Hartsfield, Daniel Hausman, Eileen Hawley, Sharon Hecht, Johnny Heflin, Mack Henderson, Edward Henderson, Fredrick Henn, Francisco Hernandez, Ben Higgins, Michael Hiltz, Jeff Hoffman, William Hoffman, Steve Holmes, Doris Hood, Christy Howard, Christopher Iannello, John Irving, Bob Jacobs, Brian Johnson, Janet Johnson, Katelyn Johnson, Nicholas Johnson, Perry Johnson-Green, Wesley Johnson, Kathleen Kaminski, David Kanipe, David Kendall, Gary Kitmacher, Peter Klonowski, Tommy Knight, Joseph Kosmo, Julie Kramer-White, John Kress, Michael Kuta, Keelee Kyles, Meghan LaCroix, Robert Lambdin, Barbara Langston, James Larocque, Kirby Lawless, Diane Laymon, Steven Lindsey, Steven Lloyd, Christopher Madden, Lynnette Madison, Raquel Madrigal, Lisa Malone, Charles Martin, Ryan Martin, Naoko Matsuo, Samantha McDonald, James Mceuen, Alexander McPherson, Marshall Mellard, Messia Miller, Jessica Miller, Katherine Mims, Danielle Mondoux, Owen Morris, Jeff Mosit, Paul Munafo, Margaret Nemerov, Peter Nickolenko, Lorna Onizuka, Michael Orr, James Owen, Kathy Padgett, Michael Pedley, Brian Peterson, Douglas Peterson, John Petty, Steve Poulos, Donald Prevett, Maureen Priddy, Alison Protz, Lisa Rasco, Dorothy Rasco, Brett Raulerson, Mark Richards, Timothy Riley, Thomas Roberts, Benjamin Robertson, Ned Robinson, Jennifer Rochlis, Patricia Ross, James Rostohar, Steven Roy, Gary Ruff, Robert Ryan, Ted Schaffner, Calvin Schomburg, Susan Scogin, Barbara Shannon, John Shannon, Jody Singer, Alice Slay, Jean Snowden, Eileen Stansbery, Mike Sterling, Victoria Stowe, Russ Stowe, David Sutherland, Macie Sutton, Robert Synder, Donald Tillian, Bert Timmerman, Robert Trevino, David Urban, Paula Vargas, Andy Warren, Kathy Weisskopf, Shayne Westover, Mary Wilkerson, Justin Wilkinson, Martin Wilson, Sean Wilson, Cynthia Wimberly, James Wise, Lybrease Woodard, Gary Woods, Dwight Woolhouse, Peggy Wooten, Roy Worthy, Rebecca Wright, and Martin Zell.

Selected Readings

General Information

Astronaut Biographies:
http://www.jsc.nasa.gov/Bios/

Johnson Space Center Oral History Project:
http://www.jsc.nasa.gov/history/oral_histories/oral_histories.htm

NASA History Program Office:
http://history.nasa.gov/

Space Shuttle Press Kits:
http://www.shuttlepresskit.com/

Mission Archives:
http://spaceflight.nasa.gov/shuttle/archives

NASA Scientific and Technical Information:
http://www.sti.nasa.gov/STI-public-homepage.html

Shuttle-Mir:
http://spaceflight.nasa.gov/history/shuttle-mir/

Spin-offs:
http://www.sti.nasa.gov/tto/

Small Business Innovative Research/Small Business Technology Transfer:
http://www.sba.gov/aboutsba/sbaprograms/sbir/sbirstir/index.html

NASA Centers:

Ames Research Center:
http://www.nasa.gov/centers/ames/home/index.html

Dryden Flight Research Center:
http://www.nasa.gov/centers/dryden/home/index.html

Glenn Research Center:
http://www.nasa.gov/centers/glenn/home/index.html

Goddard Space Flight Center:
http://www.nasa.gov/centers/goddard/home/index.html

Jet Propulsion Laboratory:
http://www.jpl.nasa.gov/

Johnson Space Center:
http://www.nasa.gov/centers/johnson/home/index.html

Kennedy Space Center:
http://www.nasa.gov/centers/kennedy/home/index.html

Langley Research Center:
http://www.nasa.gov/centers/langley/home/index.html

Marshall Space Flight Center:
http://www.nasa.gov/centers/marshall/home/index.html

Michoud Assembly Facility:
http://www.nasa.gov/centers/marshall/michoud/index.html

NASA Headquarters:
http://www.nasa.gov/centers/hq/home/index.html

Stennis Space Center:
http://www.nasa.gov/centers/stennis/home/index.html

Wallops Flight Facility:
http://www.nasa.gov/centers/wallops/home/index.html

White Sands Test Facility:
http://www.nasa.gov/centers/wstf/home/index.html

Magnificent Flying Machine—A Cathedral to Technology

Publications and Web links:

NASA's First 50 years – Historical Perspectives. Dick, S. NASA, Washington, DC. NASA/SP-2010-4704.

Remembering the Space Age. Proceedings of the 50th Anniversary Conference. Dick, S, editor. NASA, Washington DC. NASA/SP-2008-4703.
http://ntrs.nasa.gov/archive/nasa/casi.ntrs.nasa.gov/20090013341_2009005513.pdf

Leadership in Space. Selected Speeches of NASA Administrator Michael Griffin, May 2005-October 2008. Griffin, M. NASA/SP-2008-564.
http://ntrs.nasa.gov/archive/nasa/casi.ntrs.nasa.gov/20090009154_2009002630.pdf

Critical Issues in the History of Spaceflight. Dick, S and Launius, R, editors. NASA, Washington, DC. NASA/SP-2006-4702.
http://ntrs.nasa.gov/archive/nasa/casi.ntrs.nasa.gov/20060022843_2006166766.pdf

The Story of the Space Shuttle. Harland, DM. Springer, Praxis Publishing Ltd., 2004.

Additional Web links:

1903 Wright Flyer: *http://www.nasm.si.edu/exhibitions/gal100/wright1903.html*

LAGEOS: *http://msl.jpl.nasa.gov/QuickLooks/lageosQL.html*

The Historical Legacy

Milestones

Publications and Web links:

Remembering the Space Age. Proceedings of the 50th Anniversary Conference. Dick, S, editor. NASA, Washington DC. NASA/SP-2008-4703.
http://ntrs.nasa.gov/archive/nasa/casi.ntrs.nasa.gov/20090013341_2009005513.pdf

Critical Issues in the History of Spaceflight. Dick, S and Launius, R, editors. NASA, Washington, DC. NASA/SP-2006-4702.
http://ntrs.nasa.gov/archive/nasa/casi.ntrs.nasa.gov/20060022843_2006166766.pdf

Leadership in Space. Selected Speeches of NASA Administrator Michael Griffin, May 2005-October 2008. Griffin, M. NASA/SP-2008-564.
http://ntrs.nasa.gov/archive/nasa/casi.ntrs.nasa.gov/20090009154_2009002630.pdf

Space Shuttle Decision 1965-1972. Heppenheimer, TA. Smithsonian Institution Press, Washington, DC, 2002.

Development of the Space Shuttle 1972-1981. Heppenheimer, TA. Smithsonian Institution Press, Washington, DC, 2002.

Space Shuttle: The History of the National Space Transportation System, The First 100 Missions. Jenkins, DR, Cape Canaveral, Florida, 2001.

Toward a History of the Space Shuttle: An Annotated Bibliography. Compiled by Launius, RD and Gillette, AK, 1992.
http://www.hq.nasa.gov/office/pao/History/Shuttlebib/contents.html

The Accidents: A Nation's Tragedy, NASA's Challenge

Publications and Web links:

Report of the Presidential Commission on the Space Shuttle Challenger Accident:
http://history.nasa.gov/rogersrep/51lcover.htm

Columbia Crew Survival Investigation Report. NASA/SP-2008-566:
http://www.nasa.gov/pdf/298870main_SP-2008-565.pdf

Additional Web links:

Columbia Accident Investigation Board: *http://caib.nasa.gov/*

NASA sites—Challenger (STS-51L) Accident: *http://history.nasa.gov/sts51l.html*

Selected Readings

National Security

Publications and Web links:

Corona Between the Sun and the Earth: The First NRO Reconnaissance Eye in Space. McDonald, R, editor. American Society for Photogrammetry and Remote Sensing, 1997.

The Soviet Space Race with Apollo. Siddiqi, A. University of Florida Press, 2000.

Challenge to Apollo: The Soviet Union and the Space Race, 1945-1974. Siddiqi, A. NASA History Division, Washington, DC. NASA SP-2000-4408.

Space and National Security. Stares, P. Washington Brookings Institution Press, 1987.

The Politics of Space Security: Strategic Restraint and Pursuit of National Interests. Moltz, J. Stanford University Press, 2008.

Militarization of Space: US Policy, 1945-1984. Stares, P. Cornell University Press, Ithaca, NY, 1985.

"Secret Space Shuttles" Cassutt, M. *Air & Space Magazine*, August 1, 2009. *http://www.airspacemag.com/space-exploration/Secret-Space-Shuttles.html*

The Space Shuttle and Its Operations

The Space Shuttle

Publication and Web link:

Space Shuttle: The History of the National Space Transportation System, The First 100 Missions. Jenkins, DR. Cape Canaveral, Florida, 2001.

Additional Web link:

Typical Mission Profile: *http://history.nasa.gov/SP-407/part1.htm*

Processing the Shuttle for Flight

Web links:

Bill Parsons: *http://www.nasa.gov/centers/kennedy/about/biographies/parsons.html*

Lightning Delays Launch (STS-115): *http://www.nasa.gov/mission_pages/shuttle/behindscenes/115_mission_overview.html*

US National Lightning Detection Network Database: *http://gcmd.nasa.gov/records/GCMD_NLDN.html*

Flight Operations

Web links:

Shuttle Training Aircraft—Test Drive: *http://www.nasa.gov/vision/space/preparingtravel/rtf_week5_sta.html*

Payload Communication System: *http://spaceflight.nasa.gov/shuttle/reference/shutref/orbiter/comm/orbcomm/plcomm.html*

Extravehicular Activity Operations and Advancements

Web links:

Neutral Buoyancy Laboratory Training: *http://spaceflight.nasa.gov/shuttle/support/training/nbl/*

Suit Environment as Compared to Space Environment: *http://www.nsbri.org/HumanPhysSpace/introduction/intro-environment-atmosphere.html*

Hubble Servicing Missions: *http://hubblesite.org/the_telescope/team_hubble/servicing_missions.php*

Shuttle Builds the International Space Station

Publications and Web links:

Living and Working in Space: A History of Skylab. Compton, DW and Benson, CD. NASA, Washington, DC, SP-4208, 1983. *http://history.nasa.gov/SP-4208/sp4208.htm*

Reference Guide to the International Space Station. Kitmacher, GH. NASA-SP-2006-557. *http://www.nasa.gov/mission_pages/station/news/ISS_Reference_Guide.html*

Engineering Innovations

Propulsion

Publications and Web links:

Space Shuttle Main Engine: The First Twenty Years and Beyond. Biggs, RE. AAS History Series, Vol. 29. San Diego, CA, 2008. *http://www.univelt.com/htmlHS/htmlMisc/v29hiscon.pdf*

Facing the Heat Barrier: A History of Hypersonics. Heppenheimer, TA. NASA SP-2007-4232. *http://ntrs.nasa.gov/archive/nasa/casi.ntrs.nasa.gov/20070035924_2007036871.pdf*

Additional Web links:

Shuttle Thermal Protection System: *http://www.centennialofflight.gov/essay/Evolution_of_Technology/TPS/Tech41.htm*

Aerogel Beads as Cryogenic Thermal Insulation System: *http://rtreport.ksc.nasa.gov/techreports/2002report/600%20Fluid%20Systems/604.html*

Aerogels Insulate Missions and Consumer Products: *http://www.sti.nasa.gov/tto/Spinoff2008/ch_9.html*

Materials and Manufacturing

Publications and Web links:

"Oxygen Interaction with Materials III: Data Interpretation via Computer Simulation." Roussel, J and Bourdon, A. *Journal of Spacecraft and Rockets*, Vol. 37, No. 3, May–June 2000. *http://pdf.aiaa.org/jaPreview/JSR/2000/PVJAIMP3582.pdf*

Advances in Friction Stir Welding for Aerospace Applications: *http://pdf.aiaa.org/preview/CDReadyMATIO06_1322/PV2006_7730.pdf*

Aerodynamics and Flight Dynamics

Web links:

Boundary Layer Transition: *http://www.nas.nasa.gov/SC09/PDF/Datasheets/Tang_boundarylayer.pdf*

Early Conceptual Designs for the Orbiter: *http://history.nasa.gov/SP-432/ch4.htm*

The Space Shuttle's First Flight: STS-1: *http://history.nasa.gov/SP-4219/Chapter12.html*

Avionics, Navigation, and Instrumentation

Web link:

Computers in the Space Shuttle Avionics System: *http://history.nasa.gov/computers/Ch4-1.html*

Selected Readings

Structural Design

Web links:

Crack Models and Material Properties Required for Fracture Analyses: *http://www.swri.edu/4org/d18/mateng/matint/nasgro/New/NASGRO%20v6%20release%20notes.pdf*

Orbiter Structure and Thermal Protection System/Review of Design and Development: *http://ocw.mit.edu/courses/aeronautics-and-astronautics/16-885j-aircraft-systems-engineering-fall-2005/lecture-notes/mosr_strctrs_tps.pdf*

Orbiter Structure—Structural Arrangement: *http://history.nasa.gov/SP-4225/diagrams/shuttle/shuttle-diagram.htm*

Forward Fuselage/Crew Compartment:
http://history.nasa.gov/SP-4225/diagrams/shuttle/shuttle-diagram-5.htm
http://spaceflight.nasa.gov/shuttle/reference/shutref/structure/crew.html

Systems Engineering for Life Cycle of Complex Systems

Web links:

Calspan-University of Buffalo Research Center: *http://www.cubrc.org/*

Alliant Techsystems, Inc.: *http://www.atk.com/*

United Space Alliance: *http://www.unitedspacealliance.com/*

Pratt & Whitney Rocketdyne: *http://www.pw.utc.com/Products/Pratt+%26+Whitney+Rocketdyne*

Boeing: *http://www.boeing.com/defense-space/space_exploration/index.html*

Major Scientific Studies

The Space Shuttle and Great Observatories

Publication:

Hubble: A Journey Through Space and Time. Weiler, E. Abrams, NY, 2010.

Web links:

The Hubble Space Telescope: *http://hubble.nasa.gov/*

Space Telescope Science Institute/Hubble Space Telescope: *http://www.stsci.edu/hst/*

Atmospheric Observations and Earth Imaging

Publication:

Calibration and Radiometric Stability of the Shuttle Solar Backscatter Ultraviolet (SSBUV) Experiment. Hilsenrath, E; Williams, DE; Caffrey, RT; Cebula, RP; and Hynes, SJ. *Metrologia*, Issue 4, Vol. 30, 1993.

Web links:

Upper Atmosphere Research Satellite Project Science Office: *http://umpgal.gsfc.nasa.gov/www_root/homepage/uars-science.html*

Mediterranean Israeli Dust Experiment: *http://library01.gsfc.nasa.gov/host/hitchhiker/meidex.html*

Mapping the Earth: Radars and Topography

Publication:

"Shuttle Radar Topography Mission produces a wealth of data." Farr, TG and Kobrick, M. *American Geophysical Union Eos*, v. 81, p. 583-585, 2000.

Web links:

Jet Propulsion Laboratory—Shuttle Radar Topography Mission: *http://www2.jpl.nasa.gov/srtm/*

US Geological Survey—Shuttle Radar Topography Mission: *http://srtm.usgs.gov/*

Astronaut Health and Performance

Publications and Web links:

Neuroscience in Space. Clement, G and Reschke, MF. Springer Science+Business Media, LLC, 2008.

The Neurolab Spacelab Mission: Neuroscience Research in Space. Buckey, JC and Homick JL. NASA, Washington, DC, NASA SP-2003-535, 2003.

"Muscle, Genes and Athletic Performance." Andersen, J; Schjerling, P; and Saltin, B. *Scientific American.* September 2000.

Skeletal Muscle Structure, Function, & Plasticity: The Physiologic Basis of Rehabilitation, 2nd ed. Lieber, RL. Lippincott Williams & Wilkins, 2002.

Spacefaring: The Human Dimension. Harrison, A. University of California Press, Berkeley, CA, 2002.

Habitability in Living Aloft: Human Requirements for Extended Spaceflight. Connors, M; Harrison, A; and Akins, F. NASA SP-483, NASA Scientific and Technical Information Branch, Washington, DC, 1985. *http://history.nasa.gov/SP-483/contents.htm*

Principles of Clinical Medicine for Space Flight. Barratt, MR and Pool, SL. Springer, New York, NY, 2008.

Spacecraft Maximum Allowable Concentrations for Selected Airborne Contaminants: Volume 4. National Academy Press, Washington, DC, 2000. *http://www.nap.edu/catalog.php?record_id=9786#toc*

Additional Web links:

Effect of Prolonged Space Flight on Cardiac Function and Dimensions: *http://lsda.jsc.nasa.gov/books/skylab/Ch35.htm*

Life Sciences Data Base—Human Research Program Data: *http://lsda.jsc.nasa.gov/*

The Space Shuttle: A Platform That Expanded the Frontiers of Biology

Publications and Web links:

Animals In Space: From Research Rockets to the Space Shuttle. Burgess, C and Dubbs, C. Springer Praxis Books, 2007.

"Vertebrate Biology in Microgravity." Wassersug, R. *American Scientist*: 89:46-53, 2001. *https://www.americanscientist.org/issues/feature/vertebrate-biology-in-microgravity*

Life Into Space: Space Life Sciences Experiments, Ames Research Center, 1965-1990. Souza, K; Hogan, R; and Ballard R, editors. NASA RP-1372, 1995. *http://lis.arc.nasa.gov/*

Life Into Space: Space Life Sciences Experiments, Ames Research Center, Kennedy Space Center, 1991-1998. Souza, K; Etheridge G; and Callahan, P, editors. NASA SP-2000-534. *http://lis.arc.nasa.gov/*

Cell Biology and Biotechnology in Space. Cogoli, A, editor. Elsevier, 2002.

US and Russian Cooperation in Space Biology and Medicine. Volume V. Sawin, C; Hanson, S; House, N; and Pestov, I. editors. AIAA, 2009.

Advances in Space Biology and Medicine. Volume 1. Bonting, S, editor. Elsevier, 1991.

Selected Readings

Microgravity Research in the Space Shuttle Era

Publications and Web links:

Cell Growth in Microgravity. Sundaresan, A; Risin, D; and Pellis, NR. *Encyclopedia of Molecular Cell Biology and Molecular Medicine, Vol. 2,* pp 303-321, Edited by Meyers, RA; Sendtko, A; and Henheik, P. Wiley-VCH, Weinheim, Germany, 2004.

"Genes in Microgravity," Rayl, AJS. *DISCOVER*, Vol. 22, No. 9, September 2001. *http://discovermagazine.com/2001/sep/featgenes*

Spacelab Science Results Study. Naumann, RJ; Lundquist, CA; Tandberg-Hanssen, E; Horwitz, JL; Cruise, JF; Lewis, ML; and Murphy, KL. NASA/CR-2009-215740. *http://ntrs.nasa.gov/archive/nasa/casi.ntrs.nasa.gov/20090023425_2009021429.pdf*

Spacelab 3 Mission Science Review. NASA Conference Publication 2429. Fichtl, GH; Theon, JS; Hill, KC; and Vaughan, OH, editors. *http://ntrs.nasa.gov/archive/nasa/casi.ntrs.nasa.gov/19870012670_1987012670.pdf*

First International Microgravity Laboratory. McMahan, T; Shea, C; Wiginton, M; Neal, V; Gately, M; Hunt, L; Graben, J; and Tiderman, J; Accardi, D. NASA TM-108007, 1993. *http://ntrs.nasa.gov/archive/nasa/casi.ntrs.nasa.gov/19930003925_1993003925.pdf*

First International Microgravity Laboratory Experiment Descriptions. Miller, TY. TM-4353, 1992. *http://ntrs.nasa.gov/archive/nasa/casi.ntrs.nasa.gov/19920014357_1992014357.pdf*

Microgravity: A Teacher's Guide With Activities in Science, Mathematics, and Technology. Rogers, JB; Vogt, GL; and Wargo, MJ. EG-1997-08-1100-HQ. *http://teacherlink.ed.usu.edu/tlnasa/units/Microgravity/04.pdf*

Joint Launch + One Year Science Review of USML-1 and USMP-1 with the Microgravity Measurement Group. Volume I and II. Ramachandran, N; Frazier, DO; Lehoczky, SL; and Baugher, CR, editors. NASA-CP-3272-VOL-I and NASA-CP-3272-VOL-II.
Volume I:
http://www.ntrs.nasa.gov/archive/nasa/casi.ntrs.nasa.gov/19950007793_1995107793.pdf
Volume II:
http://ntrs.nasa.gov/archive/nasa/casi.ntrs.nasa.gov/20030075796_2003085850.pdf

The First United States Microgravity Laboratory. Shea, C; McMahan, T; Accardi, D; and Mikatarian, J. NASA-TM-107980, 1993. *http://ntrs.nasa.gov/archive/nasa/casi.ntrs.nasa.gov/19930003763_1993003763.pdf*

Second United States Microgravity Payload: One Year Report. Curreri, PA and McCauley, DE. NASA-TM-4737, 1996. *http://ntrs.nasa.gov/archive/nasa/casi.ntrs.nasa.gov/19960038726_1996063204.pdf*

Second International Microgravity Laboratory (IML-2) Final Report. Snyder, R, compiler. NASA/RP-1405, 1997. *http://ntrs.nasa.gov/archive/nasa/casi.ntrs.nasa.gov/19970035095_1997064524.pdf*

Second United States Microgravity Laboratory (USML-2) One Year Report, Volume I. Vlasse, M; McCauley, D; and Walker, C. NASA/TM-1998-208697, 1998. *http://ntrs.nasa.gov/archive/nasa/casi.ntrs.nasa.gov/19990018868_1998415108.pdf*

Second United States Microgravity Laboratory (USML-2) One Year Report, Volume 2. Vlasse, M; McCauley, D; and Walker, C. NASA/TM-1998-208697/VOL2. *http://ntrs.nasa.gov/archive/nasa/casi.ntrs.nasa.gov/19990009671_1998415144.pdf*

Get Away Special... the first ten years. NASA-TM-102921, 1989. *http://ntrs.nasa.gov/archive/nasa/casi.ntrs.nasa.gov/19900007459_1990007459.pdf*

Additional Web links:

European Experiments: Erasmus Experiment Archive—Erasmus Centre—ESA: *http://eea.spaceflight.esa.int/?pg=explore&cat=sh*

Get Away Special Web site: *http://library01.gsfc.nasa.gov/host/hitchhiker/gas.html*

Social, Cultural, and Educational Legacies

NASA Reflects America's Changing Opportunities; NASA Impacts US Culture

Publication:

Societal Impact of Spaceflight. Dick, SJ and Launius, RD. NASA, Washington, DC, NASA SP-2007-4801.

Education: Inspiring Students as Only NASA Can

Web links:

EarthKAM: *https://earthkam.ucsd.edu*
http://geoearthkam.tamu.edu/EarthKAM_AM.ppt
http://www.ncsu.edu/earthkam/simulation/

Toys in Space: *http://quest.nasa.gov/space/teachers/liftoff/toys.html*

Challenger Center: *http://www.challenger.org/*

Resources for Educators: *http://www.nasa.gov/audience/foreducators/*

Project Starshine: *http://spacekids.hq.nasa.gov/starshine/*

Get Away Special Program—Historical Information: *http://library01.gsfc.nasa.gov/host/hitchhiker/history.html*

Shuttle Amateur Radio Experiment: *http://www.qsl.net/w2vtm/shuttle.html*

Instrumentation Technology Associates, Inc. (ITA) Student Outreach Program: *http://www.itaspace.com/students.html*

Industries and Spin-offs

Web links:

MicroMed Cardiovascular, Inc.: *http://www.micromedcv.com/united_states/index.html*

NASA-developed Tool—LifeShear: *http://ipp.nasa.gov/innovation/Innovation34/Rescue.html*

Microbial Check Valve: *http://www.urc.cc/rmcv.htm*

Acronyms

AIDS	Acquired Immunodeficiency Syndrome
ANDE	Atmospheric Neutral Density Experiment
ASTM	American Society for Testing and Materials
ATLAS	Atmospheric Laboratory for Applications and Science
BIRD	Bird Investigation Review and Deterrent
Ca	calcium
CAT	computerized axial tomography
CFC	chlorofluorocarbon
CIRRIS	Cryogenic Infrared Radiance Instrumentation for Shuttle
CO_2	carbon dioxide
CPR	Chemical Products Research
DAC	digital to analog converter
DFRC	Dryden Flight Research Center
DNA	deoxyribonucleic acid
DoD	Department of Defense
DOUG	Dynamic Onboard Ubiquitous Graphics
DSMC	Direct Simulation Monte Carlo
EarthKAM	Earth Knowledge Acquired by Middle School Students
EDGE	Engineering DOUG Graphics for Exploration
EROS	Earth Resources Observation and Science
ESA	European Space Agency
ET	External Tank
EVA	extravehicular activity
FAA	Federal Aviation Administration
Fe	iron
FGB	Functional Cargo Block
g	gravitational force (eg, 3*g*)
g-suits	gravity suits
GLS	ground launch sequence
GPS	Global Positioning Satellite
GSFC	Goddard Space Flight Center
HAL/S	high-order software language
HCFC	hydrochlorofluorocarbon
HEPA	high-efficiency particulate air
hp	horsepower
IBM	International Business Machines
Intelsat	International Telecommunications Satellite Organization
ISO	International Standards Organization
ISS	International Space Station
JATO	jet-assisted takeoff
JAXA	Japan Aerospace Exploration Agency
JSC	Johnson Space Center
K	potassium
kph	kilometers per hour
KSC	Kennedy Space Center
LAURA	Langley Aerothermodynamic Upwind Relaxation Algorithm
LED	light-emitting diode
LiOH	lithium hydroxide
MOTEL	Microgravity Opportunity To Enhance Learning
MRI	magnetic resonance imaging
MSFC	Marshall Space Flight Center
NEXRAD	next-generation weather radar
Na	sodium
NASA	National Aeronautics and Space Administration
nm	nanometers
NOAA	National Oceanic and Atmospheric Administration
NSS	National Security Space
O_2	oxygen
PCGOAL	Personal Computer Ground Operations Aerospace Language
psi	pounds per square inch
psia	pounds per square inch, absolute
REM	Rapid Eye Movement
rem	roentgen-equivalent man
SAFER	Simplified Aid for EVA Rescue
SI	*Système International*
SLA	Super-Lightweight Ablator
SolarMax	Solar Maximum Satellite
SRB	Solid Rocket Booster
SSME	Space Shuttle Main Engine
STS	Space Transportation System
USA	United Space Alliance
USAF	US Air Force
USSR	Union of Soviet Socialist Republics
UV	ultraviolet
VAD	ventricular assist device
Vdc	volts, direct current
WSTF	White Sands Test Facility

Contributers' Biographies

Alexander, Iwan – Professor and Chair of Mechanical and Aerospace Engineering at Case Western Reserve University. Investigator for five space experiments, semiconductor crystal growth, liquid diffusion experiment, and an acceleration measurement. Director of the National Center for Space for 5 years.

Alfrey, Clarence – Professor at Baylor College of Medicine and former chief of hematology and medical director of the regional blood center. MD from Baylor College of Medicine with residency in internal medicine at State University of Iowa and fellow in hematology at the Mayo Clinic.

Armor, James – Major General, US Air Force (retired). Selected as a military spaceflight engineering program astronaut, but never flew as program discontinued.

Bacon, John – Systems engineer in the International Space Station (ISS) Program Office. For 20 years, he held assignments in the integration of all US international partner systems in the ISS Program at NASA. PhD, University of Rochester.

Bains, Elizabeth – PhD. Leads engineering analysis of Shuttle Robotic Arm operations. Co-chairs a panel overseeing Shuttle Robotic Arm model accuracy. Worked in many areas of Shuttle Robotic Arm software, from testing simulation dynamics models to requirements definition and verification testing for the arm control software.

Baldwin, Kenneth – PhD. Professor at University of California, Irvine. Principal investigator for four shuttle missions and numerous ground-based NASA research projects. Muscle team lead for the National Space Biomedical Research Institute for 8 years.

Barger, Laura – Instructor in medicine at Harvard Medical School. Associate physiologist at Brigham and Women's Hospital. Co-principal investigator of the sleep study conducted aboard shuttle flights from 2000-2011. Conducted sleep studies on the International Space Station.

Bauer, Paul – Thermal analyst at ATK. Led the Reusable Solid Rocket Motor Carbon Fiber Rope implementation team. Worked in design engineering for Electronic Specialty, producer of space-bound relays and switches. BS in Mechanical Engineering, Washington State University.

Becker, Perry – NASA, chief of the Engineering Directorate Ground Systems Structures Mechanisms. Twenty-five years of service. Served as crawler systems engineer, transporting over 100 shuttles to the launch pad. Master's degree in Mechanical Engineering, and an MBA.

Beek, Joachim – Manages the NASGRO project. Member of the Fracture Control Board at Johnson Space Center. MS in Aerospace Engineering, Texas A&M University.

Bell, Bradley – Responsibilities include development and maintenance of the visual simulation systems used in astronaut training, including the rendering software and the helmet-mounted display hardware at Johnson Space Center.

Blumberg, Baruch – Professor at Fox Chase Cancer Center, Pennsylvania. Former director NASA Astrobiology Institute. Received the 1976 Nobel Prize in Medicine for identification of hepatitis B virus. MD from Columbia, New York.

Bordano, Aldo – Retired from NASA in 2000 after 37 years of engineering service at Johnson Space Center. Chief of the Aeroscience and Flight Mechanics Division (1991-2000). Expertise in vehicle guidance and flight mechanics was critical to the design and development of shuttle spacecraft.

Brown, Steve – Started at Johnson Space Center in 1974 with the McDonnell Douglas Corporation. Supported the Space Shuttle Program in aerodynamics throughout career. Worked in the area of wind tunnel testing, and verification of the aerodynamic database for the simulators.

Brown, Robert – Lead electrical controls engineer. More than 11 years experience working electrical control upgrades for all mobile launcher platform and pad ground support equipment at Kennedy Space Center. BS in Electrical Engineering, University of Central Florida.

Bryant, Lee – Started as a NASA contractor in 1982 in Mission Planning and Analysis Division after graduating from the University of Texas. Flight Mechanics and Trajectory Design. Joined NASA in 1987 as an engineer in the guidance analysis section of Mission Planning and Analysis Division.

Buning, Pieter – PhD. Joined NASA in 1979 as a researcher in computational fluid dynamics. Developed computational tools for aerospace vehicles from helicopters and commercial airliners to hypersonic research vehicles and the shuttle, first at NASA Ames Research Center and then at NASA Langley Research Center.

Burkholder, Jonathan – Engineer in the Damage Tolerance Assessment Branch at Marshall Space Flight Center (MSFC). Technical secretary of the MSFC Fracture Control Board. BS in Mechanical Engineering, University of Alabama in Huntsville.

Burns, Bradley – More than 20 years experience at Kennedy Space Center developing ground support equipment and shop aids for the Space Shuttle Program. BS in Electrical Engineering, University of Central Florida.

Butler, Jim – Writer for United Space Alliance at Marshall Space Flight Center. Managed writing assignments for Computer Sciences Corporation, Intergraph, and the US Army prior to joining the NASA team. BA in English and History, University of Alabama in Huntsville.

Campbell, Charles – PhD. Began career with Johnson Space Center in 1987 as a cooperative education student, joining the Engineering Directorate in 1990 after graduating from the University of Minnesota with a bachelor's degree. Became the lead for Orbiter aerothermodynamics as the NASA subsystem engineer in 2003.

Captain, Janine – Works for NASA at Kennedy Space Center (since 2005), focusing on in-situ resource utilization technologies and sensors for field deployment. PhD in Chemistry, Georgia Institute of Technology.

Caron, Dan – Curriculum specialist for Engineering by Design. Teaches aerospace/technology education at Kingswood Regional High School in Wolfeboro, New Hampshire. Led the NASA Educational Workshops at Goddard Space Flight Center and Wallops Flight Facility (1997-1999).

Carpenter, Bradley – Works in the Space Operations Mission Directorate at NASA Headquarters. Lead scientist in the Microgravity Research Division of NASA from 1996-2005. PhD in Chemical Engineering, Stanford University.

Castner, Willard – Metallurgical engineer who, during his 30+ years at Johnson Space Center, specialized in nondestructive testing, materials testing, and failure analysis. Active member of the American Society for Nondestructive Testing during NASA career.

Chandler, Michael – Deputy branch chief of medical operations at Johnson Space Center. Member of the Department of Defense Space Transportation System contingency support office during the Challenger accident. Member of the NASA Mishap Investigation Team following the Columbia Accident.

Chapline, Gail – Worked primarily at Johnson Space Center as a materials engineer. Supervised the materials branch. Also worked in the Shuttle Program Office, NASA Headquarters, National Transportation and Safety Board, and NASA White Sands Test Facility. MS in Materials Engineering, Northwestern University.

Charles, John – Program scientist for NASA's Human Research Program at Johnson Space Center. Principal investigator for several investigations into the changes in the cardiovascular system. PhD in Physiology and Biophysics, University of Kentucky.

Christian, Carol – PhD. Deputy of the Community Missions Office and an astronomer at the Space Telescope Science Institute at Baltimore, Maryland. Served as head of the Office of Public Outreach for Hubble Space Telescope for many years, and has researched stellar populations in nearby galaxies.

Christiansen, Eric – PhD. NASA Micro-Meteoroid and Orbital Debris (MMOD) Protection lead at Johnson Space Center. Holds a patent for the Stuffed-Whipple shield used extensively on the International Space Station. Developed a number of design and operational methods to reduce MMOD risk to NASA spacecraft.

Coglitore, Sebastian – Brigadier General, retired from US Air Force. Program manager of the first Department of Defense spacecraft to fly on the Space Shuttle.

Cohen, Aaron – Worked for NASA from 1962-1993. Served as center director (1986-1993), then returned to Texas A&M University to a distinguished engineering chair. MS in Applied Math, the Stevens Institute of Technology.

Collins, David – Deputy associate director of Technology Development and chief of the Instrumentation Section for Development Engineering at Kennedy Space Center. MS in Electrical Engineering, Georgia Tech.

Connolly, Janis – Project manager for NASA's Human Research Program and its Space Human Factors Engineering Project at Johnson Space Center. MS in Architecture, University of Wisconsin-Milwaukee.

Cort, Robert – Associate manager-technical at NASA White Sands Test Facility. Began working on ground testing of Space Shuttle Orbiter Maneuvering System and reaction control subsystems in 1987, and managed repair and overhaul of flight hardware for those systems/subsystems at White Sands Test Facility.

Cragun, Brad – ATK scientist. Formulated propellants and pyrotechnics for ATK's Castor 120® rocket motor and Boeing's Sea Lance missile. Inducted into the Space Technology Hall of Fame for developing a demining flare based on shuttle propellant technology. Graduate of Weber State University.

Contributers' Biographies

Cross, Jeffrey – Aeronautical engineer involved in rotorcraft flight research for 16 years. Public outreach lead and visitor center curator for 10 years. Member of the NASA Ames Research Center's Office of Education for 3 years.

Crucian, Brian – Senior scientist with Wyle Laboratories at Johnson Space Center. Expertise in spaceflight-associated immune dysregulation, flow cytometry assay development, and immunology research in extreme environments. PhD, University of South Florida.

Curtis, Glen – ATK program manager over Reusable Solid Rocket Motor supply chain, process control, and program transition. Twenty-two-year career has included duties as a proposal manager, supervisor in industrial engineering, and manager of budgets, proposals, and training for operations. Space Shuttle Program Star Award.

Czeisler, Charles – PhD, MD, the Baldino Professor of Sleep Medicine, and director of the Division of Sleep Medicine at Harvard Medical School. Chief of the Division of Sleep Medicine at Brigham and Women's Hospital. Principal investigator of multiple sleep studies.

DeTroye, Jeff – Works for the CIA (2003-present). Worked for NASA (1985-1998). Commander of the National Reconnaissance Office Aerospace Defense Facility – East. Officer in US Air Force (1977-1985). MS, University of Houston-Clear Lake.

Ding, Robert – Welding engineer at NASA Marshall Space Flight Center (MSFC). Currently works in the Material and Processes Laboratory at MSFC in welding process development. Master's degree in Engineering Management.

Dolman, Everett – PhD. Professor of Comparative International Studies at the US Air Force's School of Advanced Air and Space Studies. Formerly an intelligence analyst, National Security Agency. Published works include Astropolitik, The Warrior State, and Pure Strategy.

Dorsey, Geminesse – Mechanical engineer at Johnson Space Center. Worked as a test director and technical area lead of the Battery Systems Test Facility in the Energy Systems Test Area. Worked on numerous test programs to certify and evaluate batteries used on-orbit.

Drake, Daniel – United Space Alliance, lead mechanical engineer. Twenty-six years of service at Kennedy Space Center. Primarily responsible for the hydraulic systems of the crawlers. Holds certifications as driver, jacking console operator, and local test conductor.

Ecord, Glenn – Materials Branch, Engineering Directorate at Johnson Space Center. Served as integration technical manager for Fracture Control and for Pressure Vessels and Pressurized Systems, Orbiter, and payloads.

Faile, Gwyn – Former chief of the Marshall Space Flight Center Structural Integrity Branch. Served as co-chair of the NASA Fracture Control Analytical Methodology Panel. Currently works for the Qualis Corporation on the Jacobs Engineering team supporting the Marshall Space Flight Center Damage Tolerant Assessment Branch.

Feagan, Carole-Sue – Twenty-five years management and human resource experience in private industry. Came to Kennedy Space Center in 2008 to support the director of vehicle operations, planning development with United Space Alliance. Joined a contractor in support of the NASA chief engineer of launch vehicle processing.

Feeback, Daniel – Head of the Muscle Research Laboratory, Johnson Space Center, until 2010. Adjunct associate professor, Department of Biochemistry, Institute of Biosciences Bioengineering at Rice University. PhD, University of Oklahoma.

Fiorucci, Tony – Aerospace engineer at Marshall Space Flight Center. Responsible for vibration analysis and redline methodology algorithm development and integration for the Space Shuttle Main Engine, Advanced Health Management System. BS in Engineering Science, University of Tennessee.

Fish, Ozzie – Works in the NASA Instrumentation Branch. Has served as a Hazardous Warning System engineer since 1988. BS in Electrical Engineering, University of Central Florida.

Fitts, David – Chief, Habitability and Human Factors Branch in Johnson Space Center's Space Life Sciences Directorate (2003-present). An architect by formal education, he focused on NASA becoming a product-based and design-solution organization.

Flores, Rose – Led the Shuttle Remote Manipulator System analysis, flight hardware and software activities for the Flight Robotic Systems Branch at Johnson Space Center. Co-chaired the Robotics Analysis Working Group and was the shuttle robotics chief engineer. MS in Systems Engineering.

Fogarty, Jennifer – Innovation and development lead for Johnson Space Center Space Life Sciences. PhD in Cardiovascular Research, Texas A&M University.

Folensbee, Al – Worked at Kennedy Space Center, performing and overseeing the development, automation, and testing of ground application software for the Space Shuttle Program. Master's degree in Computer Science, Florida Institute of Technology.

Forman, Royce – Served as the primary NASA technical expert at Johnson Space Center on fracture control and fracture mechanics technology, initiated formation and co-chaired the NASA Fracture Control Methodology Panel, and performed the majority of fracture mechanics experimental efforts at the center.

Forth, Scott – Chairs the Johnson Space Center Fracture Control Board and works with the pressure vessel for manned spaceflight. PhD in Mechanical Engineering, Clarkson University.

Fowler, Michael – Worked as a materials engineer at Johnson Space Center for 23 years. PhD in Chemical Engineering, University of Texas.

Fraley, John – Has worked at Kennedy Space Center for 32 years. Served as an Apollo Structural Systems engineer in spacecraft operations, then as chief, Orbiter Structures, Handling Access Systems Section. BS in Mechanical Engineering, University of Kentucky.

Frandsen, Jon – Engineer with Pratt & Whitney Rocketdyne, working with the Space Shuttle Main Engine (SSME). Specialized in fracture mechanics and hydrogen embrittlement materials testing as they relate to the SSME. MS, UCLA.

Galvez, Roberto – Started career at NASA as a shuttle flight controller in the Guidance, Navigation & Control Systems. Served as manager of the Space Shuttle Program Flight Management Office. BS in Electrical Engineering, Louisiana State University.

Gardze, Eric – Pratt & Whitney Rocketdyne Kennedy Space Center (KSC) senior engineering manager. Supported Space Shuttle Main Engine since 1973. Supported combustion devices development at Canoga Park, California, the first engine hot fire testing at Stennis Space Center, and launch operations at KSC since STS-1.

Gaylor, Stephen – Began career with Rockwell Shuttle Operations and joined NASA in 1990. Was responsible for shuttle flight definition and mission performance analysis. Served as a flight manager in the Space Shuttle Program. Degree in Mechanical Engineering, Texas A&M University.

Gibson, Cecil – Began career at the Army Ballistic Missile Agency. Transferred to Johnson Space Center Propulsion and Power Division and became Apollo Service Propulsion System manager and, later, Ascent Engine manager. Supervised propulsion development and mission activities for the Space Shuttle and station until he retired.

Gnoffo, Peter – Senior research engineer in the Aerothermodynamics Branch at Langley Research Center. Has worked in the area of computational aerothermodynamics since joining NASA in 1974.

Gomez, Reynaldo – Member of Johnson Space Center Engineering Directorate since May 1985, after graduating from Rice University. Space Shuttle Ascent Aerosciences Technical Panel chairman since 1993.

Greene, Ben – Engineering project manager for the Reinforced Carbon-Carbon Repair Team at Johnson Space Center (JSC). Has been developing extravehicular activity tools and equipment at JSC for spacewalking astronauts for 15 years. BS in Mechanical Engineering, University of Houston.

Grogan, James – Colonel, retired, US Air Force.

Hale, Wayne – Shuttle flight director for 41 missions at Johnson Space Center. Kennedy Space Center shuttle launch integration manager, shuttle deputy program manager, and Space Shuttle Program manager. MS in Engineering, Purdue University.

Hall, Jennifer – More than 20 years of technical and managerial experience at Kennedy Space Center. Deputy director of the Florida Program Office. BS in Industrial Engineering, University of Central Florida. MBA, Florida Tech.

Hallett, Charles – Worked for 20 years with manufacturing systems in New York and started at Kennedy Space Center in 1990. Introduced many standard manufacturing concepts to shuttle business processes and has been Collaborative Integrated Processing Solutions project manager since its inception. Graduated from University of Buffalo.

Hamel, Michael – Lieutenant General, retired, US Air Force.

Harris, Yolanda – Technical representative for the Marshall Space Flight Center Ares First Stage Office. Served as technical assistant to the Space Shuttle Program deputy manager for propulsion. Juris Doctor Degree, University of Alabama.

Hayes, Judith – Exercise physiologist at Johnson Space Center. Deputy division chief, Human Adaptation & Countermeasures. Master of Public Health. MS in Exercise Physiology, West Virginia University.

Helms, Bill – Retired NASA physicist, 35 years Kennedy Space Center (KSC) designing launch complex instrumentation for the Space Shuttle and the Hazardous Gas Detection System. Managed KSC Instrumentation Development Labs for 20 years.

Herron, Marissa – Began career at Johnson Space Center in 2000 as a flight controller in the Flight Design and Dynamics Division. MS in Aerospace Engineering, University of Colorado at Boulder.

Contributers' Biographies

Herst, Terri – More than 26 years of shuttle processing technical and managerial experiences at Kennedy Space Center. Serves as Shuttle Project Engineer and is responsible for leading integrated technical issues to resolution during the launch countdown.

Hess, David – Director, Department of Defense (DoD) Human Space Flight Payloads Office, Johnson Space Center. Responsible for all actions related to access to space aboard human-rated spacecraft on DoD's behalf.

Hill, Arthur – Member of the Pratt & Whitney Rocketdyne technical staff since 1975. Led the development and implementation of the Space Shuttle Main Engine instrumentation system for over 30 years. BS in Electrical Engineering, UCLA.

Hill, Paul – Director of Mission Operations for Space Shuttle and International Space Station at Johnson Space Center. MS in Aerospace Engineering, Texas A&M University.

Hilsenrath, Ernest – PhD. Retired from Goddard Space Flight Center (GSFC). Served as principal investigator for several remote sensing satellite and shuttle missions of the Earth's atmosphere and was director of GSFC's Radiometric Calibration and Development Laboratory.

Hirko, John – Worked on Kennedy Space Center's Operational Intercommunication System – Digital (OIS-D) development team starting in 1987. Contributed to design, build, integration, testing, installation, operation, and troubleshooting throughout OIS-D's 21-year history at that center. Graduated from University of Pittsburgh's School of Engineering.

Hoblit, Jeffrey – Has served as the contractor task lead of Johnson Space Center's Integrated Extravehicular Activity Radiation Monitoring Virtual Reality Laboratory since the mid 1990s. BS in Aerospace Engineering, University of Cincinnati.

Holland, Albert – PhD. Senior operations psychologist at Johnson Space Center. Worked with astronauts and their families for over 25 years, including during the Shuttle-Mir Program, International Space Station, and analog environments such as winter over in Antarctica. Credited with numerous publications.

Homan, David – Manager of the Integrated Extravehicular Activity Robotics Virtual Reality Simulation Facility at Johnson Space Center. BS in Mechanical Engineering, Iowa State University.

Horvath, Thomas – Senior research engineer in the Research Technology Directorate at Langley Research Center, where he has worked since 1989. Primary area of expertise includes experimental research to determine and optimize the aerodynamic characteristics and heating environments for aerospace vehicles.

Howell, Patricia – Aerospace engineer with 20 years of experience in nondestructive evaluation research at NASA Langley Research Center, specializing in thermal modeling and data analysis for defect detection methods. NASA's Silver Snoopy Award. NASA Exceptional Achievement Medal.

Huss, Terry – Senior materials and processes engineer for United Space Alliance. Responsibilities include automation and robotic process development for shuttle and Ares Solid Rocket Booster elements. Graduate of the University of Colorado at Boulder's Aerospace Engineering Program.

James, John – PhD in Pathology and a Diplomat of the American Board of Toxicology. NASA chief toxicologist at Johnson Space Center. NASA Exceptional Service Medal and Shuttle Star Award. Authored or co-authored more than 100 articles and numerous book chapters.

Johnson, Dexer – Began career with Rockwell Shuttle Operations and joined NASA Johnson Space Center in 1989 in the Cargo Integration Office. Served as technical monitor representative for the Shuttle Middeck Integration contract. BS in Physics, Michigan State University.

Johnson, Steve – PhD. Professional Engineer. Member of Space Radiation Analysis Group, which is responsible for radiation monitoring and operational support in mission control for shuttle and International Space Station (ISS) missions. Participated in radiation investigations conducted on shuttle, Mir, and ISS during his 20 years at Johnson Space Center.

Jones, Samuel – Division chief engineer for the Space Shuttle. Mechanical engineer at Johnson Space Center in the Energy Systems Division. During 35 years experience, has served as test manager in the Energy Systems Test Area for test programs involving pyrotechnic devices, fuel cell components, and cryogenics.

Jordan, Coy – ATK design engineer. Responsible for the nozzle flexible bearing and bearing Thermal Protection System for the Reusable Solid Rocket Motor and the Ares rocket motor. Employed with Raytech Corporation, prior to ATK. BS in Mechanical Engineering, Arizona State University.

Jorgensen, Glenn – Worked on the Shuttle Robotic Arm with Spar Aerospace as a systems engineer and then a project manager. Participated in design upgrades to the arm and has supported shuttle missions throughout the program. Assigned as subsystem manager for the Shuttle Robotic Arm with NASA in 2007.

Jue, Fred – Performs strategic analysis and business development for the Pratt & Whitney Rocketdyne Space Shuttle Main Engine (SSME) program. Began career with Rocketdyne as an SSME turbomachinery engineer. Served as resident manager for development of the alternate turbopumps at the Pratt & Whitney Florida facility.

Kahl, Bob – Director of Palmdale Shuttle Operations for Boeing Explorations, and part of the Space Shuttle Program since 1975. Operations director of Orbiter Assembly Test and Logistic Spares (1997-present).

Kauffman, Larry – Director of California Operations for Boeing Space Exploration. Part of the Space Shuttle Program since 1979. Associate program director of Orbiter production (1996-2000).

Kaupp, Henry – Part of the NASA team that evaluated Canadian ability to build the Shuttle Robotic Arm. Followed the shuttle arm development and supported early missions. Served as shuttle division chief engineer for the Robotics Division, and was prime point of contact for the Shuttle Robotic Arm until his retirement.

Kaye, Jack – PhD. Associate director for research, Earth Science Division, NASA Headquarters. Program scientist for Atmospheric Laboratory of Applications and Science missions, Cryogenic Infrared Spectrometers & Telescopes for the Atmosphere-Shuttle Palette Satellite, Mediterranean Israeli Dust Experiment, and Solar Shuttle Backscatter Ultraviolet Experiment.

Kelly, Mark – Captain, US Navy. NASA astronaut. Assigned to command crew of STS-134 (2011). Commander on STS-124 (2008). Pilot on STS-121 (2006) and STS-108 (2001). Has received several awards and honors. MS in Aeronautical Engineering, US Naval Postgraduate School.

Killpack, Michael – Manages the analytical chemistry department within the ATK Launch Systems research and development laboratory in Promontory, Utah, where he has been employed for more than 10 years. Prior to joining ATK, retired as a Lieutenant Colonel following a 20-year career with the US Air Force.

Kirazes, John – Chief of the Communications and Tracking Branch at Kennedy Space Center. Started working on shuttle navigation systems with NASA in 1985. MS in Electrical Engineering, Florida Institute of Technology.

Kirk, Benjamin – Joined the Aerosciences & Flight Mechanics Division at Johnson Space Center in 2003. Heavily supported Thermal Protection System repair technique development and implementation for the Orbiter. PhD in Aerospace Engineering.

Kloeris, Vickie – Food scientist with a concentration in food microbiology. Manager of the Space Food Systems Laboratory at Johnson Space Center. Manages the International Space Station food system. Additionally, managed the shuttle food system (1989-2005). MS, Texas A&M University.

Knight, Jack – Forty years hands-on and management experience in human spaceflight programs at Johnson Space Center. Includes spaceflight operations procedures and planning, real-time vehicle command and control, and facility development project management for simulators and mission control centers.

Kobrick, Michael – PhD. Senior scientist at NASA's Jet Propulsion Laboratory, Pasadena, California. Served as the director of the Shuttle Radar Topography Mission.

Koontz, Steven – PhD. Works in the Materials and Processes Branch at Johnson Space Center. System manager and expert for spaceflight environment effects on spacecraft performance.

Kosmo, Joseph – Senior project engineer in the Extravehicular Activity & Space Suit Systems Branch at Johnson Space Center. Started career at the NASA-Langley Space Task Group in 1961. Involved in design, development, and testing of all major spacesuit assemblies, from Mercury to the International Space Station Program.

Kuo, Y.M. – PhD. Modeler of dynamics of on-orbit systems, particularly manipulators, including certification of the Shuttle Robotic Arm model that added capabilities such as constrained motion and end effector dynamics. Leads analyses of manipulator on-orbit performance at Johnson Space Center.

Lamb, Holly – Manager of community relations for aerospace and defense manufacturer ATK. Oversees efforts to inspire the next generation of scientists and engineers through education outreach initiatives. Degree in Professional Writing, Carnegie Mellon University.

Lane, Helen – Registered Dietician. Served as lead for Johnson Space Center for nutritional biochemistry laboratory, clinical research laboratories, branch chief, engineering interface, and manager of University Research and Affairs. Research focus is nutrition and biochemistry. PhD in Nutrition, University of Florida.

Contributers' Biographies

LeBeau, Gerald – Joined Johnson Space Center as a cooperative education student in 1987. Focus of career was in the area of computational aerosciences, specializing in the development and application of rarefied gas dynamics tools. Served as the chief of the Applied Aeroscience and Computational Fluid Dynamics Branch since 2006.

Leckrone, David – Part of the Hubble Space Telescope Project since 1976, first as scientific instruments project scientist, then deputy senior project scientist, and later as chief engineer. Lead project scientist at Johnson Space Center "mission control" during the Hubble servicing missions (1993, 1997, 1999, 2008). PhD in Astronomy, UCLA.

Leger, Lubert – Served as chief of the Materials Branch, Engineering Directorate at Johnson Space Center.

Levin, Zev – PhD. The J. Goldemberg chair professor in Atmospheric Physics. Principal investigator of the Mediterranean Israeli Dust Experiment on board the Space Shuttle Columbia on its last flight. Served as dean of research and vice president of research at Tel Aviv University, Israel.

Lewis, Marilyn – EdD. Education Specialist with WILL Technology, Inc. working in support of the Marshall Space Flight Center Office of Human Capital contract. Coordinates Minority University Research and Education Projects for the Marshall Academic Affairs Office.

Limero, Thomas – Johnson Space Center Toxicology Laboratory supervisor (1990-present). Expert in measurement of trace volatile organics in closed environments. Served as lead scientist for development of several spacecraft air quality monitors. PhD in Analytical Chemistry, University of Houston.

Lingbloom, Mike – Served as lead ATK engineer for Reusable Solid Rocket Motor optically simulated electron emission technology. Holds Level III certifications in magnetic particle, liquid penetrant, and laser shearography via the American Society for Nondestructive Testing. Associate of Science degree in Electronic Technology.

Locke, James – Joined NASA in 1999 as a flight surgeon. Has worked in the NASA Flight Medicine Clinic at Johnson Space Center, and served as a crew surgeon on numerous shuttle and International Space Station missions. MD, University of Wisconsin Medical School. Completed medical residencies in Emergency Medicine and Aerospace Medicine.

Loveall, James – Has served as the division chief engineer for shuttle flight software in the Johnson Space Center Engineering Directorate since 2003. Serves as deputy branch chief for the Operational Space Systems Integration Branch in the Avionic Systems Division.

Lucid, Shannon – Flew on STS-51G, STS-34, STS-43, STS-57, STS-76, and STS-79, and spent 6 months on Russian space station Mir. Was one of seven women chosen for the first astronaut class that accepted women. PhD in Biochemistry, University of Oklahoma.

Lulla, Kamlesh – PhD. Served as chief scientist for Earth Observations and Astronaut Training in Earth Observations for the Space Shuttle and the International Space Station. Conducted experiments in human-directed remote sensing and technology development at Johnson Space Center for the past 23 years.

Lumpkin, Forrest – Began career at NASA Ames Research Center in 1990. Joined Johnson Space Center 1994. Career has focused on rarefied gas dynamics emphasizing on plumes. PhD, Stanford University.

Madura, John – Over 29 years of weather analysis and research experiences working both for NASA and the Air Force. Serves as manager for the Kennedy Space Center weather office. MS in Meteorology, University of Michigan.

Manning, Samantha – Assistant launch vehicle processing chief engineer. Worked at Johnson Space Center for 5 years before going to Kennedy Space Center. Worked Main Propulsion and Max Launch Abort System for 2 years each. Degree in Aeronautical and Astronautical Engineering, University of Illinois at Urbana-Champaign.

Martin, Fred – Orbiter NASA subsystem engineer for aerodynamics, and Aeroscience and Flight Mechanics Division chief engineer for aerosciences. Began career at Johnson Space Center in 1980. Led the development of the Space Shuttle Launch Vehicle computational fluid dynamics analysis (1989-1993).

McArthur, Cynthia – Lead for Teaching From Space, a NASA K-12 education office located in the Astronaut Office at Johnson Space Center. Teaching From Space facilitates on-orbit education opportunities that use the unique environment of spaceflight, including in-flight education downlinks and education payload operations.

McClellan, Wayne – Lead system engineer for ground instrumentation and controls at Kennedy Space Center. BS in Electrical Engineering, Florida Atlantic University.

McCormick, Patrick – PhD. Professor and co-director, Center for Atmospheric Sciences, Hampton University. Principal investigator for series of Earth science satellite experiments. Co-principal investigator for Apollo-Soyuz Stratospheric Aerosol Measurement and Cloud-Aerosol Lidar and Infrared Pathfinder Satellite Observation experiments.

McGill, Preston – Structural materials engineer in the Damage Tolerance Assessment Branch at Marshall Space Flight Center (MSFC). Serves on the MSFC Fracture Control Board. Doctorate in Civil Engineering, Auburn University.

McKelvey, Timothy – NASA lead computer engineer for the Launch Processing System. Has worked at Kennedy Space Center since 1987. BS in Electrical Engineering, University of South Florida. MS in Engineering Management, Florida Institute of Technology.

McPeters, Richard – PhD. Atmospheric physicist at Goddard Space Flight Center. Closely involved in the measurement of ozone from space from a series of Task Order Management System and Solar Backscatter Ultraviolet Instrument since the 1970s.

Medelius, Pedro – Has worked at Kennedy Space Center for 18 years—last 7 years with ASRC Aerospace Corporation. Responsible for research and development activities in various aerospace-related areas, applied physics, and real-time signal processing. PhD, University of Florida.

Mehta, Satish – Senior scientist at the Microbiology Department of Johnson Space Center. Since 1992, his research focused on reactivation and shedding of Herpes viruses in space and space analogs. PhD, Guru Nanak Dev University.

Meinhold, Anne – Principal senior engineer with International Trade Bridge, Inc. MS in Environmental Science, University of North Carolina at Chapel Hill.

Merceret, Francis – Director of research for the Kennedy Space Center Weather Office. Specializes in meteorological observation and data analysis with emphasis on winds and lightning. Authored over 100 professional papers (more than 40 peer-reviewed). PhD in Atmospheric Physics, Johns Hopkins University.

Miller, Glenn – Senior technical expert working structural design projects. Began career at Johnson Space Center in 1984 as structural engineer in the field of structural analysis and certification. BS in Civil Engineering, Texas A&M University.

Miralles, Evelyn – Principal software engineer of the Virtual Reality Laboratory, an astronaut training facility, at Johnson Space Center. BS in Computer Science.

Mizell, Richard – Associate director for Management Launch Vehicle Processing Directorate at Kennedy Space Center. Worked at NASA for more than 20 years as a systems engineer on various flight and ground systems, including 10 years on the Hazard Warning Systems beginning during the Main Propulsion System leaks in 1990.

Modlin, Tom – Worked at Johnson Space Center in structural analysis. Supported the Mercury, Gemini, Apollo, and Space Shuttle Programs as a structural analysis expert. Served as the chief of the Structural Mechanics Branch.

Moore, Gilbert – Retired Thiokol engineer, Utah State physics professor, and US Air Force Academy astronautics professor, where he helped develop the cadet satellite program. Director of Project Starshine. Served as lead for the first canister of Get Away Special experiments and first Space Shuttle student satellite.

Moore, Dennis – Chief engineer for Space Shuttle Reusable Solid Rocket Motor at Marshall Space Flight Center. MS, University of Alabama.

Morgan, Barbara – Mission specialist and teacher in space on STS-118. Worked as an elementary school teacher in Idaho and educator in residence at Boise State University.

Moser, Thomas – Held key positions at Johnson Space Center, including head of structural design, deputy manager Orbiter Project, director of engineering, deputy associate administrator for spaceflight and space station, and director of Space Station Program at Headquarters. MS, University of Pennsylvania.

Muratore, John – Teaches at University of Tennessee Space Institute. Supported the Space Shuttle for 28 years, both with the US Air Force and NASA. Worked at Vandenberg Air Force Base, Kennedy Space Center, and Johnson Space Center. Served as manager of Space Shuttle Systems Engineering and Integration following Columbia accident.

Nickerson, Cheryl – PhD. Associate professor at The Biodesign Institute, Arizona State University. An expert in mechanisms of microbial pathogenesis. Pioneered discovery of molecular genetic and virulence changes in Salmonella and other pathogens in response to spaceflight.

Contributers' Biographies

Nickolenko, Peter – Has worked at Kennedy Space Center for more than 20 years in shuttle processing operations. Launch director for STS-127 and STS-128. Served in both technical and managerial positions planning launch and landing operations. Degree in Engineering from Military Academy-West Point.

Norbraten, Lee – Joined NASA in 1967 as an Apollo mission designer at Johnson Space Center. Led project teams to improve ascent structural safety margins, payload capability, and launch probability for the International Space Station during the shuttle era. MS in Mathematics, University of Houston.

O'Neill, Patrick – Has worked in the design and analysis of Guidance, Navigation, & Control Systems at Johnson Space Center. Served as "Radiation Effects scientist," responsible for planning radiation testing, modeling natural space radiation environments, and predicting radiation effects on performance of systems.

Ott, Mark – PhD. Microbiologist. Supports spaceflight program operations at Johnson Space Center Microbiology Laboratory. Extensive experience in the assessment of infectious disease risk to the crew during spaceflight missions.

Paloski, William – Professor of Health and Human Performance at the University of Houston. Spent 23 years as a neurosciences researcher at Johnson Space Center, studying sensory-motor adaptation to spaceflight. PhD in Biomedical Engineering, Rensselaer Polytechnic Institute.

Patrick, Nancy – Started as a NASA shuttle contractor in 1983 in the Mission Operations Directorate after graduating from the University of Notre Dame. Joined NASA in 1990 as an assembly planner for the International Space Station. Worked in the Extravehicular Activity (EVA) office as EVA staff engineer (1996–2008).

Payne, Stephen – NASA Payload Operations, Discovery lead for Kennedy Space Center Vehicle Integration Test Team office, NASA test director, ground operations manager for transatlantic abort landing deployments, tanking test director, and shuttle test director for eight launches. MS in Engineering Management, University of Central Florida.

Payton, Gary – Lieutenant General, retired, US Air Force. Deputy, Under Secretary of Air Force for Special Program and military payload specialist on STS-51C.

Pellis, Neal – Senior scientist at Johnson Space Center. Led the Biotechnology Program and the Biological Systems Office, and was International Space Station Program scientist, following a 21-year career in academics. PhD in Microbiology, Miami University. Postdoctoral fellowship at Stanford University.

Peralta, Steven – Technical expert on identifying and controlling fire hazards in oxygen systems. Started career as an engineer and project manager at NASA's White Sands Test Facility in 1999. BS in Mechanical Engineering, New Mexico State University.

Perkins, Fred – ATK chief engineer for the Reusable Solid Rocket Motor. Held leadership positions in both design and reliability engineering. MS in Mechanical Engineering, University of Utah.

Pessin, Myron – Consultant with Jacobs on the ARES Program. Former NASA External Tank chief engineer. Served as a Space Shuttle Main Engine propulsion engineer. BS in Mechanical Engineering, Tulane University.

Pham, Chau – Johnson Space Center Crew and System Division chief engineer for Orbiter Environment Control and Life Support Systems. BS in Aerospace Engineering, University of Texas.

Pierson, Duane – NASA's senior microbiologist at Johnson Space Center. Agency's expert on the many microbiological aspects of spaceflight. PhD, Oklahoma State University.

Pilet, Jeffrey – Chief Engineer for Lockheed Martin Michoud Assembly Facility on the External Tank Project.

Platts, Steven – Head of the Cardiovascular Research Laboratory at Johnson Space Center. PhD in Cardiovascular Physiology, Texas A&M University. Postdoctoral Fellowship, University of Virginia.

Richmond, Dena – Employed by United Space Alliance on the Collaborative Integrated Processing Solutions team and is a Solumina subject matter expert.

Ride, Sally – PhD. NASA astronaut. First American woman to fly in space. Flew on STS-7 and STS-41G. President of Sally Ride Science – a company that promotes education in science, technology, engineering, and mathematics.

Ring, Richard – Employed with United Space Alliance. More than 25 years in the aerospace industry as a design engineer.

Rivera, Jorge – Deputy chief engineer for shuttle processing. More than 28 years of technical and managerial experiences at Kennedy Space Center. BS in Industrial Engineering, University of Puerto Rico – Mayaguez.

Roberson, Luke – His research at NASA deals with the development, application, and evaluation of conductive polymers, microelectronic devices, and nanocomposite polymeric materials. PhD, Georgia Institute of Technology.

Roberts, Katherine – Brigadier General, retired, US Air Force. An original military astronaut for manned spaceflight engineering program, MSE-2, before program was cancelled.

Rodriguez, Alvaro – Supported the Space Shuttle Program at Johnson Space Center as the NASA subsystem engineer for the Leading Edge Structural Subsystem using expertise in thermal analysis and testing of Thermal Protection System. Masters of Mechanical Engineering, Rice University.

Rohan, Richard – System analyst specialist for Jacobs Technology. Worked supporting NASA for the past 22 years. Provides both 2-D and 3-D graphics and technical drawings for the Johnson Space Center Flight Mechanics Laboratory, in addition to building and maintaining high-performance computer clusters.

Romere, Paul – Started career at the Manned Spacecraft Center (now Johnson Space Center). Part of the Shuttle Skunk Works. Served as shuttle aerodynamics subsystem manger for 10 years.

Ross-Nazzal, Jennifer – Johnson Space Center historian. Her biography of Emma Smith DeVoe – *Winning the West for Women: The Life of Emma Smith DeVoe* – was published by the University of Washington Press. Her essay, "From Farm to Fork," is included the *Societal Impact of Spaceflight*. PhD in History, Washington State University.

Ruiz, Jose – Guidance, Navigation, & Control engineer at Johnson Space Center. Supported rendezvous operations for four shuttle missions in 2007 and 2008 from Mission Control. MS in Aeronautics and Astronautics, Massachusetts Institute of Technology.

Russo, Dane – PhD. Scientist-manager at Johnson Space Center/Space Life Sciences Directorate. For more than 30 years, managed the Space Human Factors and Habitability Element and the Advanced Human Support Technology Program.

Sams, Clarence – PhD. Biochemist. Director of Johnson Space Center Immunology Laboratories. Scientific and technical lead (element scientist) for the International Space Station Medical Project.

Sauer, Richard – NASA inventor of the year for the microbial check valve that resulted in a patent and license. Major contributor to providing safe water for shuttle crews as the Johnson Space Center lead for the water laboratory and deputy branch chief. He has numerous publications.

Saunders, Melanie – Associate director, Johnson Space Center. Served as a member of the NASA negotiation teams for the International Space Station. Main author of the barters for shuttle launch of the European and Japanese labs, the Balance of Contributions with Russia, and the Code of Conduct for Space Station Crew. Juris Doctor, University of California, Davis.

Scarpa, Jack – Manager of the Productivity Enhancement Materials Development at Marshall Space Flight Center. Responsibilities included design, materials development, and testing of Thermal Protection System materials and non-metallic materials for the shuttle Solid Rocket Booster.

Schneider, William – Expertise in mathematical engineering mechanics, structural and mechanical design, spacecraft entry Thermal Protection Systems, and large space structures. PhD in Mechanical Engineering, Rice University.

Schuh, Joseph – Started career as part of the Orbiter Electrical Engineering group and moved to supporting the design of the Ares I and Ares V/Heavy Launch Vehicle at Kennedy Space Center.

Scobee Rogers, June – Founding chairman of Challenger Center for Space Sciences. Taught every grade level from kindergarten through college. Married Dick Scobee, who perished during the Challenger accident (1986). PhD, Texas A&M University.

Scott, Carl – Supported thermal protection material testing, aerothermodynamics, and flow diagnostics at Johnson Space Center Was the first to determine the temperature dependent catalytic atom recombination on shuttle tiles. PhD in Physics, University of Texas.

Scully, Robert – Lead engineer of the Johnson Space Center Electromagnetics Compatibility Group. Co-chair of the Shuttle Electromagnetic Environmental Effects (E3) Control Tech Panel, and co-lead of the Constellation Program E3 Working Group.

Smith, Sarah – Worked at Johnson Space Center White Sands Test Facility in oxygen hazard analysis as well as in the development of tests and test systems for evaluating ignition and combustion of materials in oxygen-enriched environments. BS in Mechanical Engineering, New Mexico State University.

Smith, Scott – Chief of Nutritional Biochemistry Laboratory at Johnson Space Center since 1992 with research in bone metabolism. PhD in Nutrition, Penn State University.

Contributers' Biographies

Snapp, Cooper – Supported the Space Shuttle Program at Kennedy Space Center as a thermal protection engineer prior to becoming the NASA subsystem engineer. Aided the development of tile inspection, analysis, and repair techniques used after the Columbia accident. MS in Engineering, University of Central Florida.

Sollock, Paul – Worked in human spaceflight for 42 years at Johnson Space Center. Worked with the hardware and software, which eventually became known as Avionics. Had first hand key roles in the design, development, and verification of critical Avionic Systems on Apollo and the Space Shuttle.

Souza, Kenneth – Retired as the deputy director of space research at NASA Ames Research Center. Was responsible for animal and plants payloads. Served as senior scientist for the SETI Institute and Logyx, LLC. PhD, University of California, Berkeley.

Sparks, J. Scott – NASA External Tank assistant chief engineer. Served in Marshall Space Flight Center's Materials and Processes Laboratory and specialized in non-metallic materials. MS, Georgia Institute of Technology.

Spiker, Ivan – Expert in polymer materials, composites, and bonding. Member of the Materials Branch, Johnson Space Center.

Steinetz, Bruce – Expert on seal technology and tribology for aeronautic and space applications. Widely published, and holds 10 patents for seal development work. Twenty-three years experience at NASA Glenn Research Center.

Stepaniak, Philip – NASA flight surgeon and lead for the Space Shuttle Program Medical Operations at Johnson Space Center. MD, Northeastern Ohio University, Rootstown. Residency in aerospace and emergency medicine, Wright State University, Dayton.

Stevenson, Charles – Worked for NASA for over 43 years. Wide range of experience in management and technical direction for all engineering aspects of integration, test, checkout, documentation, and launch preparation of space vehicles. Served as principal advisor-coordinator and program interface.

Stone, Randy – Served in mission operations during the Apollo, Skylab, Apollo Soyuz, Space Shuttle, and International Space Station Programs. Served as flight controller during the early programs, shuttle flight director, director of mission operations, and retired as the deputy center director at Johnson Space Center after 37 years of service.

Stoltzfus, Joel – Began his career at NASA's White Sands Test Facility in 1978, developing tests to ignite and burn metals in high-pressure oxygen. Serves as a senior technical expert on identifying and controlling fire hazards in oxygen systems. BS in Mechanical Engineering, New Mexico State University.

Stull, Edith – Writer and editor who has worked at Kennedy Space Center since 1973 in technical and public affairs writing. Works for United Space Alliance. Previously worked as a magazine and newspaper writer and editor.

Sullivan, Steven – Chief engineer for shuttle processing. More than 25 years of engineering experience in Kennedy Space Center shuttle ground operations preparing the Space Shuttle for flight. MS in Management, Florida Tech.

Sumners, Carolyn – EdD. Director for Astronomy at the Houston Museum of Natural Science. Served as the principal investigator for "Toys in Space" payload on two Space Shuttle missions in 1985 and 1993.

Swanson, Gregory – PhD. Engineer in the Damage Tolerance Assessment Branch at Marshall Space Flight Center (MSFC). More than 25 years experience in spaceflight systems structural and fracture mechanics. Chairs the MSFC Fracture Control Board. Co-chairs the NASA Fracture Control Methodology Panel.

Tigges, Michael – Entry guidance subsystem manager for the crew exploration vehicle at Johnson Space Center. Started as a NASA contractor in 1982 in the Mission Planning and Analysis Division (MPAD) after graduating with an MS from Georgia Tech. Joined NASA in 1985 as a guidance engineer for MPAD.

Trevino, Robert – Professional Engineer. Worked on Space Shuttle, International Space Station, and Constellation Programs' extravehicular activity programs at Johnson Space Center. MS in Space Studies, University of North Dakota.

Trevino, Luis – Thermal lead engineer in the Extravehicular Activity and Space Suit Systems Branch at Johnson Space Center. BS in Mechanical Engineering, University of Texas.

Ulrich, Richard – Engineer for Boeing Mission Planning and Analysis Division at Johnson Space Center. Developed ascent guidance software for Solid Rocket Booster dispersions, Day of Launch I-load Update, and First Stage Engine Out.

Upton, Avis – Software engineer at Kennedy Space Center since 1985. Oversees the development, testing, and deployment of advisory software for the Space Shuttle Program. Bachelor's degree in Mathematics, Norfolk State University.

Van Hooser, Katherine – For 14 years, worked at Marshall Space Flight Center on the Space Shuttle Main Engine (SSME) high-pressure turbopumps. Served as Turbomachinery branch chief and SSME deputy chief engineer before becoming SSME chief engineer in 2008. BS in Aerospace Engineering, University of Tennessee (1991).

Velez, Ivan – Worked for more than 31 years in the Mechanical Systems Division at Kennedy Space Center in various roles. Involved in testing, repairs, and flight preparations for Orbiter mechanical systems. Participated in the application of new technologies to improve the flight readiness of these systems.

Vellinger, John – Executive vice president and chief operating officer of Techshot, Inc. Principal investigator for the shuttle student involvement project that developed avian housing for shuttle.

Vicker, Darby – Started engineering career in the Applied Aeroscience and Computational Fluid Dynamics Branch at Johnson Space Center supporting various programs with Computational Fluid Dynamics analysis. Graduated from Iowa State University (2000).

Walker, Charles – First commercial payload specialist. Was employed by McDonnell Douglas Astronautics Company and a member of the space manufacturing team. Led the microgravity research on STS-41D, STS-51D, and STS-61D. BS in Engineering, Purdue University.

Walker, James – Member of the Nondestructive Evaluation Team at Marshall Space Flight Center since 1999, specializing in the field of nontraditional nondestructive evaluation methods and composite structures. Active member of the American Society for Nondestructive Testing.

Webb, Dennis – Served in Mission Operations at Johnson Space Center in the Skylab, Space Shuttle, International Space Station, and Constellation Programs. Electrical engineer from the University of Houston. Received NASA's Outstanding Leadership Medal.

Welzyn, Kenneth – Served as NASA External Tank chief engineer beginning with STS-121 through the end of the Space Shuttle Program at Marshall Space Flight Center. MS in Mechanical Engineering, University of Alabama.

Whipps, Patrick – Deputy project manager for the External Tank Project and resident manager at Michoud Assembly Facility. Served as senior engineer, design integration lead, and materials and processes engineer.

White, Harold – More than a decade of experience with flight hardware at Johnson Space Center. Served as Shuttle Remote Manipulator System subsystem manager during Return to Flight. NASA Exceptional Achievement Medal. PhD in Physics, Rice University.

White, Susan – Education director for Johnson Space Center's Office of External Relations. Math educator, having taught at Pearland High School in Pearland, Texas, for 10 years. MS in Math Education, University of Houston.

Whitten, Mary – Served as assistant professor of chemistry at University of the Virgin Islands prior to employment at Kennedy Space Center. PhD in Chemistry, Northern Illinois University.

Williams, Martha – Lead polymer scientist in the Polymer and Chemical Analysis Branch at Kennedy Space Center. Principal investigator for several wire repair and fault detection systems activities. PhD in Polymer Chemistry.

Wood, David – Chief engineer for the shuttle Reusable Solid Rocket Booster since 2003. Auburn University graduate whose 24-year career has been dedicated to supporting NASA programs, including 20 with the Reusable Solid Rocket Booster.

Young, Charles – Started career at NASA as a shuttle mission flight controller in the Shuttle Propulsion System. Managed the preliminary mission analysis process responsible for defining the mission parameters for each shuttle mission. Degree in Aerospace Engineering, Texas A&M University.

Young, Laurence – Apollo Program Professor of Astronautics and Professor of Health Sciences and Technology. Principal investigator on neurovestibular studies. Founding director of the National Space Biomedical Research Institute. PhD, Massachusetts Institute of Technology.

Youngquist, Robert – Lead of the Kennedy Space Center (KSC) Applied Physics Lab. Taught at University College London, then joined KSC in 1988. Multiple publications and patents resulting from his work on the Space Shuttle Program. PhD in Applied Physics, Stanford University.

Zapp, Neal – PhD. Manager of the Space Radiation Analysis Group at Johnson Space Center. International Space Station (ISS) Radiation System manager, managing the technical baseline for radiation protection aboard the ISS. Background in particle physics, space radiation dosimetry.

Index

T

The final Space Shuttle flight, STS-135, returning to Earth. Astronaut Michael Fossum, aboard the International Space Station, snapped this remarkable image of Atlantis as it descended into Earth's atmosphere on the way to landing at Kennedy Space Center. The blazing heat of re-entry leaves a trail of hot plasma behind the shuttle—a phenomenon visible in this photo.

www.ingramcontent.com/pod-product-compliance
Lightning Source LLC
Chambersburg PA
CBHW081232180526
45171CB00005B/403